Finite Eleme[illegible]
Analysis and Design

COMPUTATIONAL MATHEMATICS AND APPLICATIONS

Series Editors

J. R. WHITEMAN
Institute of Computational Mathematics, Brunel University, UK

and J. H. DAVENPORT
School of Mathematical Sciences, University of Bath, UK

E. HINTON and D. R. J. OWEN: Finite Element Programming
M. A. JASWON and G. T. SYMM: Integral Equation Methods in Potential Theory and Elastostatics
J. R. CASH: Stable Recursions: with applications to the numerical solution of stiff systems
H. ENGELS: Numerical Quadrature and Cubature
L. M. DELVES and T. L. FREEMAN: Analysis of Global Expansion Methods: weakly asymptotically diagonal systems
J. E. AKIN: Application and Implementation of Finite Element Methods
J. T. MARTI: Introduction to Sobolev Spaces and Finite Element Solution of Elliptic Boundary Value Problems
H. R. SCHWARZ: Finite Element Methods
J. DELLA DORA and J. FITCH: Computer Algebra and Parallelism
E. TOURNIER: Computer Algebra and Differential Equations
A. ŽENÍŠEK: Nonlinear Elliptic and Evolution Problems and Their Finite Element Approximations
M. SINGER: Differential Equations and Computer Algebra
G. CHEN and J. ZHOU: Boundary Element Methods
B. R. DONALD, D. KAPUR and J. L. MUNDY: Symbolic and Numerical Computation for Artificial Intelligence
J. E. AKIN: Finite Elements for Analysis and Design

Dedication

To my family.

Finite Elements for Analysis and Design

BY

J. E. AKIN

Department of Mechanical Engineering
and Materials Science
Rice University, Houston, Texas, USA

ACADEMIC PRESS

Harcourt Brace & Company, Publishers

London San Diego New York
Boston Sydney Tokyo Toronto

ACADEMIC PRESS LIMITED
24–28 Oval Road
London NW1 7DX

United States Edition published by
ACADEMIC PRESS INC
San Diego, CA 92101

This book is printed on acid-free paper

A catalogue record for this book is available from the British Library

ISBN 0-12-047653-3 (Hbk)
ISBN 0-12-047654-1 (Pbk)

Printed and bound in Great Britain by
Hartnolls Limited, Bodmin, Cornwall

Contents

Appendices:

---

* Denotes sections or chapters that can be omitted for a first reading or shorter course.

PREFACE

There are several good texts on finite element analysis techniques that each have some special attribute that makes them worthwhile. My personal experience was that I did not really understand the theory until I could successfully implement it in a computer program. Thus, this text includes computational procedures, as well as the basic theory and its mathematical foundations, and several example applications. It is based on courses that I have taught on design and finite element theory.

This book is primarily intended for advanced undergraduate engineering students and beginning graduate students. The text contains more material than could be covered in a single quarter or semester course. Therefore, a number of chapters or sections that could be omitted in a first course have been marked with an asterisk (*). Most of the subject matter deals with linear, static, heat transfer and elementary stress analysis. Other novel applications are presented in the later chapters that deal mainly with computational concepts.

The future of finite element analysis will probably heavily involve adaptive analysis and design methods. One should have a course in Functional Analysis to best understand those techniques. Most undergraduate curriculums do not contain such courses. Therefore, chapters on mathematical preliminaries, error estimators, and adaptive methods are included.

A disk is provided that includes all of the source code covered in the text. It also has the source, data and output files for most of the examples, and typical applications presented in the text. Most of the computational details are included in the last third of the text, but the basic computational needs are introduced early in the text. For simplicity, most of the theoretical examples are presented in terms of relatively simple linear elements or quadratic elements. The corresponding numerical examples are given on the disk. Example extensions to higher order elements are often included on the disk but not covered in the text. Thus, after completing a chapter, the reader should also review the corresponding chapter files in the example files directory. A complete UNIX version of MODEL is available via Email to `softlib@cs.rice.edu` with the message, "send catalogue".

I would like to thank many current and former students, colleagues, and friends at Rice University for their constructive criticisms and comments during the evolution of this book. My son, Jeffrey, prepared many of the drawings. The text was prepared in a camera-ready form by Mr. Don Schroeder employing the UNIX Troff systems.

Chapter 1

INTRODUCTION

1.1 Finite Element Methods

In modern engineering design it is rare to find a project that does not require some type of finite element analysis (FEA). When not actually required, FEA can usually be utilized to improve a design. The practical advantages of FEA in stress analysis and structural dynamics have made it the accepted design tool for the last two decades. It is also heavily employed in thermal analysis, especially in connection with thermal stress analysis. Its use in Computational Fluid Dynamics (CFD) is rapidly becoming commonplace.

Clearly, the greatest advantage of FEA is its ability to handle truly arbitrary geometry. Probably its next most important features are the ability to deal with general boundary conditions and to include nonhomogeneous materials. These features alone mean that we can treat systems of arbitrary shape that are made up of numerous different material regions. Each material could have constant properties or the properties could vary with spatial location. To these very desirable features we can add a large amount of freedom in prescribing the loading conditions and in the postprocessing of items such as the stresses and strains. For elliptical boundary value problems the FEA procedures offer significant computational and storage efficiencies that further enhance its use. These classes of problems include stress analysis, heat conduction, electrical fields, magnetic fields, ideal fluid flow, etc. FEA also gives us an important solution technique for other problem classes such as the nonlinear Navier-Stokes equations for fluid dynamics, and for plasticity in nonlinear solids.

Here we will show what FEA has to offer the designer and illustrate some of its theoretical formulations and practical applications. The modern designer should study finite element methods in more detail than we can consider here. It is still an active area of research. The current trends are toward the use of error estimators and automatic adaptive FEA procedures that give the maximum accuracy for the minimum computational cost. This is also closely tied to shape modification and optimization procedures.

1.2 Capabilities of FEA

There are many commercial and public-domain finite element systems that are available to the designer. To summarize the typical capabilities, several of the most widely used software systems have been compared to identify what they have in common. Often we find about 90% of the options are available in all the systems. Some offer very specialized capabilities such as aeroelastic flutter or hydroelastic lubrication. The mainstream capabilities to be listed here are found to be included in the majority of the commercial systems of ABACUS, ANSYS, MARC, and NASTRAN. The newer adaptive systems like ADAPT, PHLEX, PROBE, and RASNA may have fewer options installed but they are rapidly adding features common to those given above. Most of these systems are available on engineering workstations and personal computers as well as mainframes and supercomputers. The extent of the usefulness of a FEA system is directly related to the extent of its element library. The typical elements found within a single system usually include membrane, solid, and axisymmetric elements that offer linear, quadratic, and cubic approximations with a fixed number of unknowns per node. The new hierarchical elements have relatively few basic shapes but they do offer a potentially large number of unknowns per node (up to 81 for a solid). Thus, the actual effective element library size is extremely large.

In the finite element method, the boundary and interior of the region are subdivided by lines (or surfaces) into a finite number of discrete sized subregions or finite elements. A number of nodal points are established with the mesh. These nodal points can lie anywhere along, or inside, the subdividing mesh, but they are usually located at intersecting mesh lines (or surfaces). The elements may have straight boundaries and thus, some geometric approximations will be introduced in the geometric idealization if the actual region of interest has curvilinear boundaries.

The nodal points and elements are assigned identifying integer numbers beginning with unity and ranging to some maximum value. The assignment of the nodal numbers and element numbers will have a significant effect on the solution time and storage requirements. The analyst assigns a number of generalized degrees of freedom to each and every node. These are the unknown nodal parameters that have been chosen by the analyst to govern the formulation of the problem of interest. Common nodal parameters are displacement components, temperatures, and velocity components. The nodal parameters do not have to have a physical meaning, although they usually do. For example, the hierarchical elements typically use the derivatives up to order six as the midside nodal parameters. This idealization procedure defines the total number of degrees of freedom associated with a typical node, a typical element, and the total system. Data must be supplied to define the spatial coordinates of each nodal point. It is common to associate some code to each node to indicate which, if any, of the parameters at the node have boundary constraints specified. In the new adaptive systems the number of nodes, elements, and parameters per node usually all change with each new iteration.

Another important concept is that of *element connectivity*, i.e., the list of global node numbers that are attached to an element. The element connectivity data defines the topology of the (initial) mesh, which is used, in turn, to assemble the system algebraic equations. Thus, for each element it is necessary to input, in some consistent order, the node numbers that are associated with that particular element. The list of node numbers connected to a particular element is usually referred to as the element incident list for that element. We usually also associate a material code with each element.

Finite element analysis can require very large amounts of input data. Thus, most FEA systems, and some CAD systems, offer the user significant data generation or supplementation capabilities. The common data generation and validation options include the generation and/or replication of coordinate systems, node locations, element connectivity, loading sets, restraint conditions, etc. The verification of such extensive

Table 1.1 Typical variables in finite element analysis

Application	Primary	Associated	Secondary
Stress analysis	Displacement, Rotation	Force, Moment	Stress, Failure criterion Error estimates
Heat transfer	Temperature	Flux	Interior flux Error estimates
Potential flow	Potential function	Normal velocity	Interior velocity Error estimates
Navier-Stokes	Velocity	Pressure	Error estimates

amounts of input and generated data is greatly enhanced by the use of computer graphics.

In the hierarchical methods we must also compute the error indicators, error estimators, and various energy norms. All these quantities are output at 1 to 27 points in each of thousands of elements. Thus, stress file editors are usually provided to allow the designer to selectively extract such data. Most of the output options from an FEA system are available in graphical form. The most commonly needed information in the design process is the state of temperatures or stresses and displacements. Thus, almost every system offers linear static stress analysis capabilities, and linear thermal analysis capabilities for conduction and convection that are often needed to provide temperature distributions for thermal stress analysis. Usually the same mesh geometry is used for the temperature analysis and the thermal stress analysis. Of course, some designs require information on the natural frequencies of vibration or the response to dynamic forces or the effect of frequency driven excitations. Thus, dynamic analysis options are usually available.

Today efficient utilization of materials in the design processes often requires us to employ nonlinear material properties and/or nonlinear equations. Such resources require a more experienced and sophisticated user. The usual nonlinear stress analysis features in large commercial FEA systems include buckling, creep, large deflections, and plasticity.

Mechanical design may also require the use of *computational fluid dynamics*. There are a small number of FEA systems that offer such analysis and design aids. For example, the FIDAP product for CFD FEA offers numerous practical incompressible flow features, such as isothermal Newtonian and non-Newtonian flows, free, forced, or mixed convection, flows in saturated porous media, advection-diffusion problems, etc. Similar adaptive codes like ADAPT and PHLEX offer the incompressible Navier-Stokes equations, porous media flow, etc.

There are certain features of finite element systems which are so important from a practical point of view that, essentially, we cannot get along without them. Basically we have the ability to handle completely arbitrary geometries, which is essential to practical engineering design. Almost all the structural analysis, whether static, dynamic, linear or

Table 1.2 Typical given variables and corresponding reactions

Application	Given	Reaction
Stress analysis	Displacement	Force
	Rotation	Moment
	Force	Displacement
	Couple	Rotation
Heat transfer	Temperature	Heat flux
	Heat flux	Temperature
Potential flow	Potential	Normal velocity
	Normal velocity	Potential

nonlinear, is done by finite element techniques on large problems. The other abilities provide a lot of flexibility in specifying loading and restraints (support capabilities). Typically, we will have several different materials at different arbitrary locations within an object and we automatically have the capability to handle these nonhomogeneous materials. Just as importantly, the boundary conditions that attach one material to another are automatic, and we don't have to do anything to describe them unless it is possible for gaps to open between materials. Most important, or practical, engineering components are made up of more than one material, and we need an easy way to handle that. What takes place less often is the fact that we have *anisotropic materials* (one whose properties vary with direction, instead of being the same in all directions). There is a great wealth of materials that have this behavior, although at the undergraduate level, anisotropic materials are rarely mentioned. Many materials, such as reinforced concrete, plywood, any filament-wound material, and composite materials, are essentially anisotropic. Likewise, for heat-transfer problems, we will have thermal conductivities that are directionally dependent and, therefore, we would have to enter two or three thermal conductivities that indicate how this material is directionally dependent. Thus, these things mean that for practical use in design, finite element analysis is very important to us.

The biggest disadvantage of the finite element method is that it has so much power there is the potential that large amounts of data and computation will be required. On small problems with about two thousand unknowns, many personal computers are available that can run an FEA system. For moderate problems with 10,000 to 15,000 equations, we need an engineering workstation or superminicomputer. Above 25,000 unknowns, we usually need to use a mainframe. A supercomputer or mini-supercomputer is usually necessary when we have more than 100,000 unknowns. All these systems should provide access to good graphical displays.

All components employed in a design are three-dimensional but several common special cases have been defined that allow two-dimensional studies to provide useful design procedures. The most common examples in solid mechanics are the states of *plane stress* (covered in undergraduate mechanics of materials) and *plane strain*, the *axisymmetric solid* model, the *thin-plate* model, and the *thin-shell* model. The latter is defined in terms of two parametric surface coordinates even though the shell exists in three dimensions. The *thin beam* can be thought of as a degenerate case of the thin-plate model. Similar concepts are used in CFD to avoid three-dimensional models. We often encounter plane and axisymmetric flows in various design projects. Even though today's solid modellers can generate three-dimensional meshes relative easily one should learn to approach such problems carefully. A well planned series of two-dimensional approximations can provide important insight into planning a good three-dimensional model. They also provide good "ballpark" checks on the three-dimensional answers. Of course, use of basic handbook calculations in estimating the answer before approaching a FEA system is also recommended.

1.3 Outline of Finite Element Procedures

From the mathematical point of view the finite element method is an integral formulation. Modern finite element integral formulations are usually obtained by either of two different procedures: *variational formulations* or *weighted residual* formulations. The following sections briefly outline the common procedures for establishing finite element models. It is fortunate that all these techniques use the same bookkeeping operations to generate the final assembly of algebraic equations that must be solved for the unknowns.

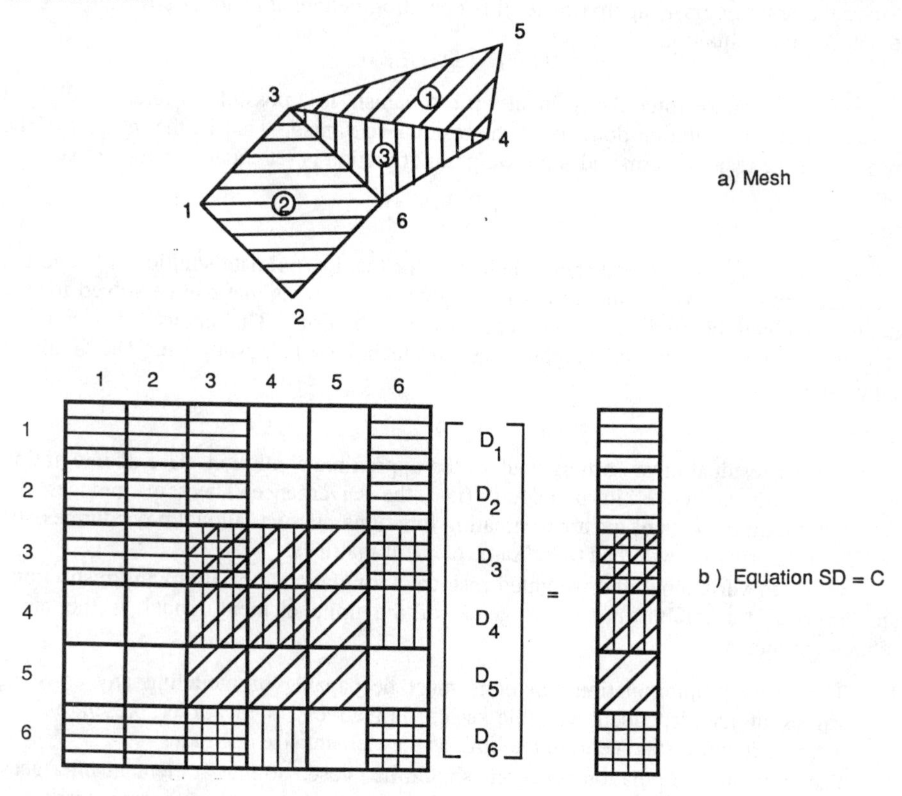

Figure 1.3.1 Graphical assembly

The earliest formulations for finite element models were based on variational techniques. This is especially true in the areas of structural mechanics and stress analysis. Modern analysis in these areas has come to rely on FEA almost exclusively. Variational models find the nodal parameters that yield a minimum (or stationary) value of an integral known as a functional. In most cases it is possible to assign a physical meaning to the integral. For example, in solid mechanics the integral represents the *total potential energy*, whereas in a fluid mechanics problem it may correspond to the rate of entropy production. Most physical problems with variational formulations result in quadratic forms that yield algebraic equations for the system which are symmetric and positive definite. The solution that yields a minimum value of the integral functional and satisfies the essential boundary conditions is equivalent to the solution of an associated differential equation, known as the Euler equation.

The generation of finite element models by the utilization of weighted residual techniques is increasingly important in the solution of differential equations for non-structural applications. The weighted residual method starts with the governing differential equation

$$L(\phi) = Q,$$

where L denotes a differential operator acting on ϕ. Generally we assume an approximate solution, say ϕ^*, and substitute this solution into the differential equation. Since the assumption is approximate, this operation defines a residual error term, R, in the differential equation

$$L(\phi^*) - Q = R \neq 0.$$

Although we cannot force the residual term to vanish, it is possible to force a weighted integral, over the solution domain, of the residual to vanish. That is, the integral of the product of the residual term and some weighting function is set equal to zero, so that

$$I = \int_V R W \, dV = 0.$$

Substituting an assumed spatial behavior for the approximate solution, ϕ^*, and the weighting function, W, results in a set of algebraic equations that can be solved for the unknown nodal coefficients in the approximate solution. The choice of weighting function defines the type of weighted residual technique being utilized. The Galerkin criterion selects

$$W = \phi^*,$$

to make the residual error "orthogonal" to the approximate solution. Use of integration by parts with the Galerkin procedure (i.e., the Divergence Theorem) reduces the continuity requirements of the approximating functions. If a variational procedure exists, the Galerkin criterion will lead to the same element matrices.

For both variational and weighted residual formulations, the following restrictions are accepted for establishing convergence of the finite element model as the mesh refinement increases:

1. The element interpolation functions must be capable of modeling any constant values of the dependent variable or its derivatives, to the order present in the defining integral statement, in the limit as the element size decreases.
2. The element interpolation functions should be chosen so that at element interfaces the dependent variable and its derivatives, of one order less than those occurring in the defining integral statement, are continuous.

```
      SUBROUTINE  STORCL (NDFREE, NELFRE, INDEX, C, CC)
C     * * * * * * * * * * * * * * * * * * * * * * * * * *
C     STORE ELEMENT COLUMN MATRIX IN SYSTEM COLUMN MATRIX
C     * * * * * * * * * * * * * * * * * * * * * * * * * *
CDP   IMPLICIT REAL*8(A-H,O-Z)
      DIMENSION  C(NELFRE), CC(NDFREE), INDEX(NELFRE)
C     INDEX  = SYSTEM DOF NOS OF THE ELEMENT DOF
C     C      = ELEMENT COLUMN MATRIX
C     CC     = SYSTEM COLUMN MATRIX
C     NDFREE = NO DEGREES OF FREEDOM IN THE SYSTEM
C     NELFRE = NUMBER OF DEGREES OF FREEDOM PER ELEMENT
      DO 10 I = 1,NELFRE
        J = INDEX(I)
        IF ( J .GT. 0 )  CC(J) = CC(J) + C(I)
 10   CONTINUE
      RETURN
      END

      SUBROUTINE  STRFUL (NDFREE, NELFRE, S, SS, INDEX)
C     * * * * * * * * * * * * * * * * * * * * * * * * * *
C     STORE ELEMENT SQ MATRIX IN FULL SYSTEM SQ MATRIX
C     * * * * * * * * * * * * * * * * * * * * * * * * * *
CDP   IMPLICIT REAL*8 (A-H,O-Z)
      DIMENSION  S(NELFRE,NELFRE), SS(NDFREE,NDFREE),
     1           INDEX(NELFRE)
C     NELFRE = NO DEGREES OF FREEDOM PER ELEMENT
C     NDFREE = TOTAL NO OF SYSTEM DEGREES OF FREEDOM
C     SS     = FULL SYSTEM SQUARE MATRIX
C     S      = FULL ELEMENT SQUARE MATRIX
C     INDEX  = SYSTEM DOF NOS OF ELEMENT PARAMETERS
      DO 20  I = 1,NELFRE
        II = INDEX(I)
        IF ( II .GT. 0 )  THEN
          DO 10  J = 1,NELFRE
            JJ = INDEX(J)
            IF ( JJ .GT. 0 ) THEN
              SS(II,JJ) = SS(II,JJ) + S(I,J)
            ENDIF
 10       CONTINUE
        ENDIF
 20   CONTINUE
      RETURN
      END
```

Figure 1.3.2 Assembly of element arrays into system arrays

The variables of interest and their derivatives are uniquely specified throughout the solution domain by the nodal parameters associated with the nodal points of the system. The parameters at a particular node directly influence only the elements connected to that particular node. A spatial interpolation, or blending, function is assumed for the purpose of relating the quantity of interest within the element in terms of the values of the nodal parameters at the nodes connected to that particular element.

After the element behavior has been described by spatial assumptions, then the derivatives of the space functions are used to approximate the spatial derivatives required in the integral form. The remaining fundamental problem is to establish the element matrices, $\mathbf{S}^e$ and $\mathbf{C}^e$. This involves substituting the space functions and their derivatives into the governing integral form. Historically, the resulting matrices have been called the element stiffness matrix and load vector, respectively.

Once the element equations have been established the contribution of each element is added, using its topology (or connectivity), to form the system equations. The system of algebraic equations resulting from FEA will be of the form $\mathbf{SD} = \mathbf{C}$. The vector **D** contains the unknown nodal parameters, and the matrices **S** and **C** are obtained by assembling the known element matrices, $\mathbf{S}^e$ and $\mathbf{C}^e$, respectively. In the majority of problems $\mathbf{S}^e$, and thus, **S**, will be symmetric. For CFD problems $\mathbf{S}^e$ and thus **S** are often nonsymmetric matrices. Also, the system square matrix, **S**, is usually banded about the diagonal or at least *sparse*. If **S** is unsymmetric its upper and lower triangles have the same sparsity. The assembly process is graphically illustrated in Fig. 1.3.1 for a three-element mesh consisting of a four-node quadrilateral and two three-node triangles, with one parameter per node. The top of the figure shows an assembly of the system **S** and **C** matrices that is coded to denote the sources of the contributing terms but not their values. A hatched area indicates a term that was added in from an element that has the same hash code. For example, the load vector term C(6) is seen to be the sum of contributions from elements 2 and 3, which are hatched with horizontal (–) and oblique (/) lines, respectively. By way of comparison, the term C(1) has a contribution only from element 2. Figure 1.3.2 shows how the assembly can be implemented for column matrices (subroutine STORCL) and full matrices (STRFUL) if one has an integer index that relates the local element degrees of freedom (dof). Figure 1.3.3 shows how that index could be computed for the common case where the number of generalized degrees of freedom per node, NG, is everywhere constant. A more complicated indexing is needed when the number of dof per node is variable (as with hierarchical elements).

After the system equations have been assembled, it is necessary to apply the *essential boundary constraints* before solving for the unknown nodal parameters. The most common types of essential boundary conditions are (1) defining explicit values of the parameter at a node and (2) defining constraint equations that are linear combinations of the unknown nodal quantities. An essential boundary condition should not be confused with a forcing condition of the type that involves a flux or traction on the boundary of one or more elements. These element boundary source, or forcing, terms contribute additional terms to the governing integral form and thus to the element square and/or column matrices for the elements on which the sources were applied. Thus, although these (*Neumann-type*) conditions do enter into the system equations, their presence may not be obvious at the system level.

The sparseness of the square matrix, **S**, is an important consideration. It depends on the numbering of the nodes (or the elements). If the FEA system being employed does not have an automatic renumbering system to increase sparseness, then the user must learn how to number nodes (or elements) efficiently. After the system algebraic equations have been solved for the unknown nodal parameters, it is usually necessary to output the parameters, **D**. For every essential boundary condition on **D**, there is a corresponding unknown *reaction* term in **C** that can be computed after **D** is known. These usually have physical meanings and should be output to help the designer check the results.

In some cases the problem would be considered completed at this point, but in most cases it is necessary to use the calculated values of the nodal parameters to calculate other quantities of interest. For example, in stress analysis we use the calculated nodal displacements to solve for the strains and stresses. All adaptive programs must do a very large amount of postprocessing to be sure that the solution, **D**, has been obtained to the

```
      SUBROUTINE  INDXPT (IPT, NG, INDEX)
C     * * * * * * * * * * * * * * * * * * * * * * * * *
C     DETERMINE DEGREES OF FREEDOM NUMBERS AT A NODE
C     * * * * * * * * * * * * * * * * * * * * * * * * *
      DIMENSION  INDEX(NG)
C     IPT   = SYSTEM NODE NUMBER
C     NG    = NUMBER OF PARAMETERS (DOF) PER NODE
C     INDEX = SYSTEM DOF NOS OF NODAL DOF
      NGIM1 = NG*(IPT - 1)
      DO 10  J = 1, NG
C       INDEX(J) = NG*(IPT - 1) + J
        INDEX(J) = NGIM1 + J
 10   CONTINUE
      RETURN
      END
      SUBROUTINE  INDXEL (N, LEMFRE, NG, LNODE, INDEX)
C     * * * * * * * * * * * * * * * * * * * * * * * * *
C     DETERMINE DEGREES OF FREEDOM NUMBERS OF ELEMENT
C     * * * * * * * * * * * * * * * * * * * * * * * * *
      DIMENSION  INDEX(LEMFRE), LNODE(N)
C     N      = NUMBER OF NODES PER ELEMENT
C     NG     = NUMBER OF PARAMETERS (DOF) PER NODE
C     LEMFRE = N*NG = NUMBER OF DOF PER ELEMENT
C     LNODE  = NODAL INCIDENCES OF THE ELEMENT
C     INDEX  = SYSTEM DOF NOS OF ELEMENT DOF
C      LOOP OVER NODES OF ELEMENT
      DO 20  K = 1, N
        IDOF = -NG
        IF ( LNODE(K) .GT. 0 ) IDOF = IDOF + NG*LNODE(K)
        NGKM1 = NG*(K-1)
C         LOOP OVER GENERALIZED DEGREES OF FREEDOM
        DO 10  IG = 1, NG
          IELM = NGKM1 + IG
C         INDEX(NG*(K-1)+IG) = NG*(LNODE(K)-1) + IG
 10     INDEX(IELM) = IDOF + IG
 20   CONTINUE
      RETURN
      END
```

Figure 1.3.3 Computing equation numbers for homogeneous nodal dof

level of accuracy specified by the designer. The preceding described data input flow and analysis steps are combined in an FEA flowchart in Fig. 1.3.4. Some of the preceding concepts are also given in Tables 1.1 and 1.2.

1.4 Optimization Concepts

The process of engineering design, or synthesis, may often be viewed as an optimization task. Computer assisted engineering typically gives the analyst the ability to investigate several "what if" questions during the synthesis process. Such questions can involve many variables and are often subject to various constraints or design rules. The total group of variables needed for the design are called the *analysis variables*. The quantities for which values are to be chosen to produce the design are a subset of the analysis variables typically called the *design variables, design parameters*, or the *trial vector*. We denote this list or vector of n quantities as

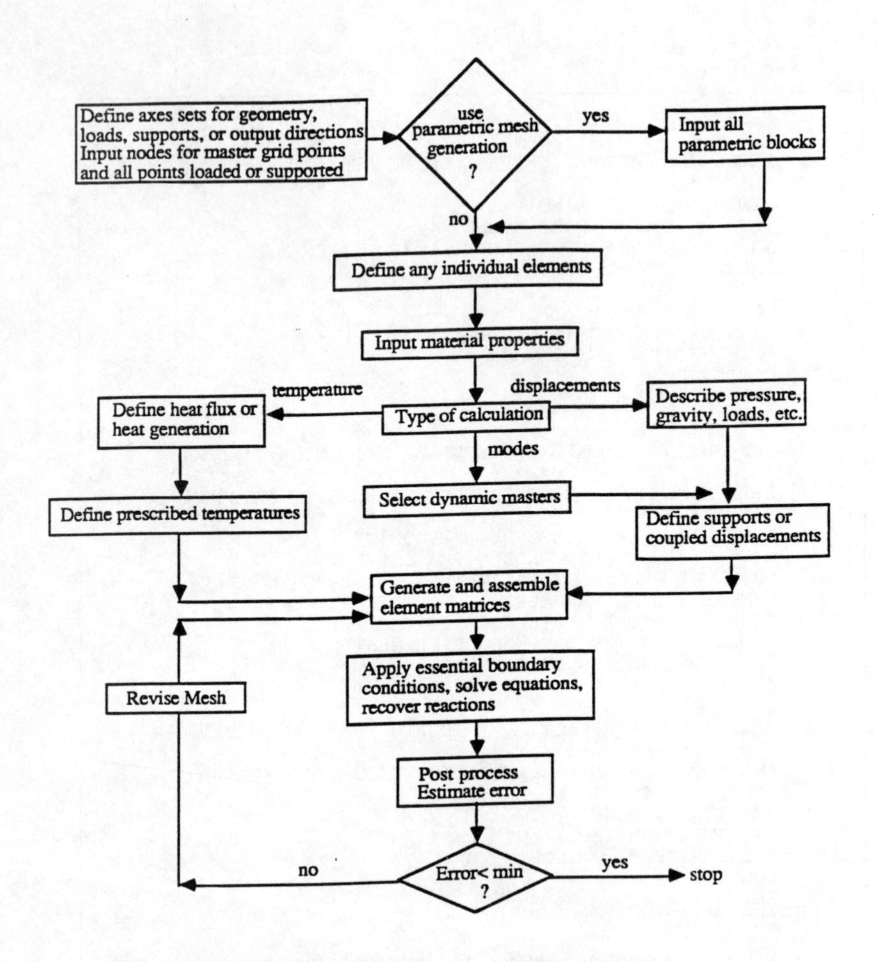

Figure 1.3.4 Analysis flow chart

$$\mathbf{x} = (x_1, x_2, \ldots, x_n)^T.$$

For example, if we were optimizing a helical spring we might consider its modulus of elasticity, E, the diameter of the wire, d, the number of coils, N, the mean diameter of the spring, D, its free length, L, and its cost, C. Then the trial vector would be

$$\mathbf{x}^T = (E, d, N, D, L, C) = (x_1, x_2, x_3, x_4, x_5, x_6).$$

Some of the analysis variables that affect a design will be known and are not subject to change. We will denote these quantities as *data* or *prescribed variables*. Fixed data items will not be included in the trial vector.

Almost any engineering design is going to require some analysis. A design is analyzed by evaluating a mathematical model of the system by calculating various functions that we call *analysis functions*, or *state variables*. The designer selects other items, conditions, or functions to be utilized in evaluating the objective or merit function.

These will be called the *design functions*. Here we use the word function in a mathematical sense and do not directly refer to the performance of the system being designed. The selection of the design variables and functions from the analysis space is often referred to as mapping between the analysis and design space.

The mathematical optimization of a design assumes that a single *measure of merit, objective function*, or *merit function* exists and can be expressed formally in terms of the design parameters. We denote this scalar merit function by $F(\mathbf{x})$. Thus, the optimization problem can be simply stated as: find $\mathbf{x}$ that satisfies the constraints and minimizes $F(\mathbf{x})$. It is assumed that there exist one or more trial vectors, $\mathbf{x}^*$, at which the function F has a local minimum.

Of course, the optimization process is not quite as simple as the preceding remarks suggest. We have already mentioned that certain *constraints* or design restrictions must be satisfied while selecting the trial vector, $\mathbf{x}$. A design that meets all requirements is called a *feasible design*, or an *acceptable design*. Otherwise, it is an infeasible or unacceptable design. A constraint that restricts the upper and lower limits on the design parameters will be called a *regional constraint*, or *side constraint*. A constraint derived from performance or functional requirements that are explicitly considered is called a *behavior constraint*, or *functional constraint*.

Thus, we will rephrase the previous statement and say that we search for a feasible design given by the design vector, $\mathbf{x}^*$, such that

$$F(\mathbf{x}^*) = \min F(\mathbf{x})$$

subject to the regional constraints that

$$G_i(\mathbf{x}^*) < 0 \quad \text{for } i = 1, 2, \ldots, p$$

and to the functional constraints

$$H_j(\mathbf{x}^*) = 0 \quad \text{for } j = 1, 2, \ldots, q .$$

Here $\mathbf{x}^* \in R$ where R denotes the *design space* made up by varying each of the n design parameters. Each local minimum corresponds to an optimal set of design variables. The best design must be selected from the feasible local minima. The global minimum is the lowest value of any of the local minima. In the synthesis of engineering systems, there is no assurance that the global minimum corresponds to the optimal feasible design. Each local feasible design must be considered on its own merits. In other words, a solution may not exist, and if one exists, it may not be unique.

There are many approaches to the solution of an optimization. Basically, there are search methods that use the function values, F, alone; gradient methods that use the function values and its first partial derivatives, $\partial F/\partial x_i$; and finally, methods that also require the second partial derivatives, $\partial^2 F/\partial x_i\,\partial x_j$. Search methods include: exhaustive search, bisection search, grid search, random search, golden search, simplex search. Popular gradient methods are: steepest descent, conjugate gradients, Newton's method.

It is important to distinguish between design and analysis. Generally a *design* will utilize *analysis variables* in one or more *analysis functions*. Design is a process of selecting the values or limits of the analysis variables. The design variables are a subset of the analysis variables which the designer selects for manipulation in order to attempt to improve the design merit function. During the design process the designer may find it necessary to change the allowable range of design variables, select a new set of design variables, change the design constraints, or interchange analysis variables with an independent analysis program.

More advanced software aids include interactive graphics and interactive input to improve the user interface. The *OPTDES.BYU* program is an example of an inexpensive public-domain software product that offers such features. It offers: Contour plots or color surfaces of the design space or its subsets, A history plot of design variables versus iteration number, Interactive adjustments of the design variables, Interactive selection of design variables from the list of analysis variables, Sensitivity studies of design variables, Option to save and recall design data, and "black box" interfaces to major analysis or graphics software. Quite often the "black box" referred to above is a finite element analysis system. Thus, it is becoming common for such codes to provide output on sensitivity studies.

As we have already stated, the design process frequently involves utilizing a finite element analysis to assist with the product development. The merit function is often defined as an output from an FEA system. Example quantities may include the maximum temperature, deflection, or stress, or related items such as the natural frequency, the strain-energy density, or the weight. Thus, we note that any FEA output data, or postprocessing combinations of those data can be employed in an arbitrary fashion by the designer to describe the merit function. Many optimization software systems provide for an interface to a general analysis program that will return the merit function. However, FEA programs are generally quite large, and each merit function evaluation may be very expensive to evaluate. The merit function in a FEA system can be difficult to define and recover, since FEA systems were not originally designed with such an option in mind. Thus, there are special considerations and trends in analysis for combining FEA and optimization procedures. A few commercial FEA systems are beginning to offer optimization options to their already numerous capabilities. These options usually offer the designer a *parametric language* by which the design variables, analysis functions, or state variables and the merit function may be selected. A parametric design language lets a designer access, manipulate, and recover almost any quantity in the data base. A parametric language also allows for regional or functional constraints to be defined. The availability of such a parametric design tool in an FEA system will become increasingly common and will significantly enhance the state of computer assisted mechanical design.

The expense of evaluating the merit function via an FEA system also causes the designer to utilize *approximate optimization* procedures. Since the simplex search algorithm often needs thousands of merit function evaluations, its heavy use with FEA systems may have to await the next generation of computers. Since the ANSYS system for FEA analysis is widely utilized, its approximate optimization capability will be described here as an example of what is commercially available today. The characteristics of the analysis functions are determined by calculating their actual values for a given set of design variables. Usually a complete FEA is employed for each set of design variables in order to produce the preceding values for the analysis functions. The resulting data are then fit with surfaces (or curves) in the design variable space. A similar fit for the merit function is obtained by either using the values returned by the FEA system or by postprocessing the returned analysis function (state variable) values. Each surface fit to the FEA results defines an approximation. An individual approximation surface, in terms of the design variables, is formed for the merit function and each analysis variable selected through the parametric design language. Thus, all these approximations are dependent quantities that act as functions of the design variables. To construct the initial approximations the range of the design variables is usually sampled at random. The approximating surfaces are fitted to those results by a least squares technique. During each optimization loop and new FEA run, the approximations are updated or refit to account for the new data. It is the *merit function approximation* that is actually minimized. Likewise, it is the approximations to the analysis functions that are checked against the functional constraints or are modified with penalties to include the constraint data.

Clearly using the approximation will be much faster (more economical) than dealing exclusively with the FEA system during the design optimization. Regardless of the optimization algorithm used, the location of the minimum merit function (e.g., the set of optimum design variables) indicated by these approximations will differ from the location of the true minimum. However, the approximation location of the minimum is usually close to the location of the true minimum, even though the function values at the two locations may be noticeably different. The ANSYS system employs the approximations to locate the minimum, but the actual value of the merit function is calculated by returning to the FEA system with the design variables that were found at the minimum location. If the optimization loop is to proceed, then this new and accurate merit function value is used to update the approximations for the next location search. Within ANSYS the designer is allowed to select approximation surfaces in the design space that are either linear, incomplete quadratic, or quadratic. Given a trial vector, $\mathbf{x}$, with N design variables the approximating surfaces have the form

$$A(\mathbf{x}) = a_o + \sum_{j=1}^{N} a_j x_j + \sum_{j=1}^{N} b_j x_j^2 + \sum_{j=1}^{N} \sum_{k=j+1}^{N} c_{jk} x_j x_k .$$

The coefficients a_j, b_j, and c_{jk} are found by minimizing the weighted least squares error for the variable A:

$$E^2 = \sum_{n=1}^{S} w_n \left[A_n - A(\mathbf{x}_n) \right]^2$$

where S denotes the total number of sets of design variables. The weights, w, are often selected to bias the fit in favor of sets that are closest to the previous estimate of the location of the minimum. The quadratic approximation of a cubic merit function often gives a good prediction for the location of the minimum. The actual minimum value would be calculated by using the true cubic at that point. True design optimization often involves the concept of *design sensitivity*. These data can be very expensive to compute, especially when a large FEA is used to define the merit function. The design sensitivities are defined as the partial derivatives, with respect to the design variables, of the merit function and the analysis functions (state variables). The preceding approximation surfaces can be used to calculate those derivatives at the local minimum location, and thus to estimate the design sensitivities. In that case we use the approximate merit function to define the slopes at the optimal point but not the value there. Chapter 15 will deal with theoretical and computational aspects of sensitivity calculations.

1.5 References

[1] Akin, J.E., *Computer Assisted Mechanical Design*, Englewood Cliffs: Prentice-Hall (1990).

[2] Bathe, K.J., *Finite Element Procedures in Engineering Analysis*, Englewood Cliffs: Prentice-Hall (1982).

[3] Becker, E.B., Carey, G.F., and Oden, J.T., *Finite Elements – An Introduction*, Englewood Cliffs: Prentice-Hall (1981).

[4] Cook, R.D., *Concepts and Applications of Finite Element Analysis*, New York: John Wiley (1974).

[5] Desai, C.S., *Elementary Finite Element Method*, Englewood Cliffs: Prentice-Hall (1979).

[6] Huebner, K.H. and Thornton, E.A., *Finite Element Method for Engineers*, New York: John Wiley (1982).

[7] Hughes, T.J.R., *The Finite Element Method*, Englewood Cliffs: Prentice-Hall (1987).

[8] Imgrund, M.C., "Using the Approximate Optimization Algorithm in ANSYS for the Solution of Optimum Convective Surfaces and Other Thermal Design Problems," Fifth Intern. Conf. on Thermal Problems, Montreal (1986).

[9] Noor, A.K. and Pilkey, W.D., "State of the Art Surveys on Finite Element Technology," ASME Publ. H000290, New York (1983).

[10] Norrie, D.H. and DeVries, G., *Finite Element Bibliography*, New York: Plenum Press (1976).

[11] Segerlind, L.J., *Applied Finite Element Analysis*, New York: John Wiley (1984).

[12] Silvester, P.P. and Ferrari, R.L., *Finite Elements for Electrical Engineers*, Cambridge: Cambridge University Press (1983).

[13] Tong, P. and Rossettos, J.N., *Finite Element Method: Basic Techniques and Implementation*, MIT Press (1977).

[14] Whiteman, J.R., *A Bibliography of Finite Elements*, London: Academic Press (1975).

[15] Zienkiewicz, O.C., *The Finite Element Method,* 3rd edition, New York: McGraw-Hill (1979).

Chapter 2

MATHEMATICAL PRELIMINARIES

2.1 Introduction

The early forms of finite element analysis were based on physical intuition with little recourse to higher mathematics. As the range of applications expanded, for example to the theory of plates and shells, some physical approaches failed and some succeeded. The use of higher mathematics such as variational calculus explained why the successful methods worked. At the same time the mathematicians were attracted by this new field of study. In the last few years the mathematical theory of finite element analysis has grown quite large. Since the state of the art now depends heavily on error estimators and error indicators it is necessary for an engineer to be aware of some basic mathematical topics of finite element analysis. We will consider load vectors and solution vectors, and residuals of various weak forms. All of these require us to define some method to measure these entities. For the above linear vectors (or vector spaces) with discrete coefficients, $\mathbf{V}^T = [\, V_1 \; V_2 \; \cdots \; V_n \,]$, we might want to use a measure like the root mean square, *RMS*:

$$RMS^2 = \frac{1}{n} \sum_{i=1}^{n} V_i^2$$

which we will come to call a norm of the linear vector space. Other quantities vary with spatial position and appear in integrals over the solution domain and/or its boundaries. Thus, we will have to introduce various other norms to "measure" these integral quantities.

The finite element method always involves integrals so it is useful to recall here some integral identities such as Gauss' Theorem (Divergence Theorem):

$$\int_{\Omega} \nabla \cdot \mathbf{u} \, d\Omega = \int_{\Gamma} \mathbf{u} \cdot \mathbf{n} \, d\Gamma = \int_{\Gamma} \frac{\partial u}{\partial n} \, d\Gamma$$

which is expressed in Cartesian tensor form as

$$\int_{\Omega} u_{i,i} \, d\Omega = \int_{\Gamma} u_i \, n_i \, d\Gamma$$

where there is an implied summation over subscripts that occurs an even number of times and a comma denotes partial differentiation with respect to the directions that follow it. That is, $(\,)_{,i} = \partial(\,)/\partial x_i$. The above theorem can be generalized to a tensor with any number of subscripts:

$$\int_{\Omega} A_{ijk...q,r}\, d\Omega = \int_{\Gamma} A_{ijk...q}\, n_r\, d\Gamma .$$

We will often have need for one of the Green's Theorems:

$$\int_{\Omega} (\nabla A \cdot \nabla B + A \nabla^2 B)\, d\Omega = \int_{\Gamma} A \frac{\partial B}{\partial n}\, d\Gamma$$

and

$$\int_{\Gamma} (A \nabla^2 B - B \nabla^2 A)\, d\Omega = \int_{\Gamma} (A \nabla B - B \nabla A) \cdot \mathbf{n}\, d\Gamma$$

which in Cartesian tensor form are

$$\int_{\Omega} (A_{,i}\, B_{,i} + AB_{,ii})\, d\Omega = \int_{\Gamma} AB_{,i}\, n_i\, d\Gamma$$

and

$$\int_{\Omega} (AB_{,ii} - BA_{,ii})\, d\Omega = \int_{\Gamma} (AB_{,i} - BA_{,i})\, n_i\, d\Gamma .$$

We need these relations to derive the Galerkin weak form statements and to manipulate the associated error estimators. Usually, we are interested in removing the highest derivative term in an integral and use the second from last equation in the form

$$\int_{\Omega} AB_{,ii}\, d\Omega = \int_{\Gamma} AB_{,i}\, n_i\, d\Gamma - \int_{\Omega} A_{,i}\, B_{,i}\, d\Omega . \tag{2.1}$$

Error estimators are often based on proofs utilizing inequalities like the Schwarz inequality

$$|\mathbf{a} \cdot \mathbf{b}| \le |\mathbf{a}|\ |\mathbf{b}| \tag{2.2}$$

and the triangle inequality

$$|\mathbf{a} + \mathbf{b}| \le |\mathbf{a}| + |\mathbf{b}| . \tag{2.3}$$

Similar inequalities exist for summations and integrals. Finite element error estimates often use the Minkowski inequality

$$\left[\sum_{i=1}^{n} |x_i \pm y_i|^p \right]^{1/p} \le \left[\sum_{i=1}^{n} |x_i|^p \right]^{1/p} + \left[\sum_{i=1}^{n} |y_i|^p \right]^{1/p} , \quad 1 < p < \infty, \tag{2.4}$$

and the corresponding integral inequality

$$\left[\int_{\Omega} |x \pm y|^p\, d\Omega \right]^{1/p} \le \left[\int_{\Omega} |x|^p\, d\Omega \right]^{1/p} + \left[\int_{\Omega} |y|^p\, d\Omega \right]^{1/p} , \quad 1 < p < \infty. \tag{2.5}$$

We will begin the preliminary concepts by introducing linear spaces. These are a collection of objects for which the operations of addition and scalar multiplication are defined in a simple and logical fashion.

2.2 Linear Spaces and Norms

The increased practical importance of error estimates and adaptive methods makes the use of *functional analysis* a necessary tool in finite element analysis. Today's student should consider taking a course in functional analysis, or studying texts such as those of Liusternik [5], Nowinski [7], or Oden [9]. This chapter will only cover certain basic topics. Other related advanced works, such as that of Hughes [4], should also be consulted.

We will almost always be seeking to approximate a more complicated solution by a finite element solution. To develop a feel for the "closeness" or "distance between" these solutions, we need to have some basic mathematical tools. Since the approximation and the true solution vary throughout the spatial domain of interest, we are not interested in examining their difference at an arbitrarily selected point. Instead, we will want to examine integrals of the solutions, or integrals of differences between the solutions. This leads us naturally into the concepts of linear spaces and norms. We will also be interested in integrals of the derivatives of the solution. That will lead us to the Sobolev norm which includes both the function and its derivatives.

Consider a set of functions

$$\phi_1(x)_1 \;\; \phi_2(x)_1 \;\; \cdots \;\; \phi_n(x).$$

If the functions can be linearly combined they are called elements of a *linear space*. Figure 2.2.1 illustrates such a space. The following properties hold:

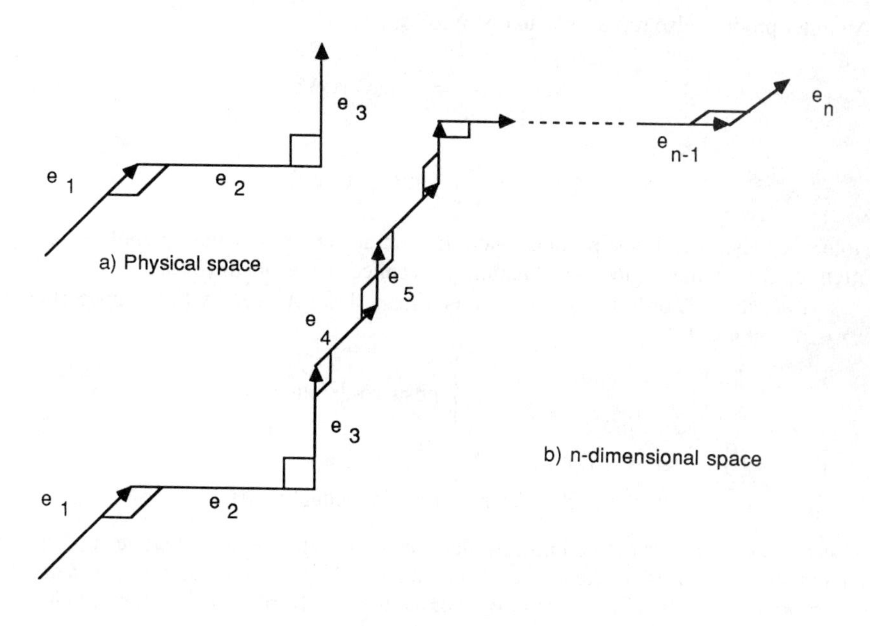

Figure 2.2.1 Linear vector space

$$
\begin{aligned}
&\alpha, \beta \in R \\
&\phi_1 + \phi_2 = \phi_2 + \phi_1 \\
&(\alpha + \beta)\,\phi = \alpha\,\phi + \beta\,\phi \\
&\alpha\,(\phi_1 + \phi_2) = \alpha\,\phi_1 + \alpha\,\phi_2 .
\end{aligned} \tag{2.6}
$$

An *inner product,* $< \bullet , \bullet >$, on a real linear space A is a map that assigns to an ordered pair $x, y \in A$ a real number R denoted by $< x, y >$:

$$< \bullet, \bullet > \; : A \times A \rightarrow R .$$

It has the following properties

$$
\begin{array}{lll}
\text{i.} & < x, y > \; = \; < y, x > & \text{symmetry} \\
\text{ii.} & < \alpha\, x, y > \; = \alpha < y, x > & \multirow{2}{*}{linearity} \\
\text{iii.} & < (x + y), z > \; = \; < x, z > + < y, z > & \\
\text{iv.} & < x, x > \; \geq 0 \text{ and} & \multirow{2}{*}{positive-definiteness,} \\
 & < x, x > \; = 0 \text{ iff } x = 0 &
\end{array}
$$

The pair $x, y \in A$ are said to be orthogonal if $< x, y > \; = 0$. A useful property is the Schwarz inequality:

$$< x, y >^2 \; \leq \; < x, x > < y, y > .$$

An inner product also represents an operation such as

$$< u, v > \; = \int_{x_1}^{x_2} u(x)\, v(x)\, dx \tag{2.7}$$

or

$$< u, v > \; = \int_{o}^{t} u(t - \tau)\, v(\tau)\, d\tau .$$

Note that when the inner product operations is an integration the symbol $< u, v >$ is often replace by the symbol (u, v) and may be called the *bi-linear form.*

A *norm,* $\| \bullet \|$, on a linear space A is a map, $\| \bullet \| : A \rightarrow R$, with the properties (for $x, y \in A$ and $\alpha \in R$)

$$
\begin{array}{lll}
\text{i.} & \| x \| \geq 0 \text{ and} & \multirow{2}{*}{positive-definiteness} \\
 & \| x \| = 0 \text{ iff } x = 0 & \\
\text{ii.} & \| \alpha\, x \| = | \alpha | \, \| x \| & \\
\text{iii.} & \| x + y \| \leq \| x \| + \| y \|, & \text{triangle inequality.}
\end{array} \tag{2.8}
$$

A *semi-norm,* $| x |$, is defined in a similar manner except that it is positive semi-definite. That is, condition i is weakened so we can have $|x| = 0$ for x not zero. A *measure* or *natural norm* of a function x can be taken as the square root of the inner product with itself. This is denoted as

$$\| x \| = \; < x, x >^{1/2} \tag{2.9}$$

2.3 Sobolev Norms *

The $L_2(\Omega)$ inner product norm involves only the inner product of the functions, and no derivatives:

$$(u, v) = \int_\Omega uv \, d\Omega$$

where $\Omega \subset R^n$, $n \geq 1$. Then the norm is

$$\| u \|_{L_2} = \| u \|_0 = (u, u)^{1/2} = \left[\int_\Omega u^2 \, d\Omega \right]^{1/2} . \tag{2.10}$$

The $H^1(\Omega)$ inner product and norm includes both the functions and their first derivatives

$$(u, v)_1 = \int_\Omega \left[uv + \sum_{i=1}^{n} u_{,i} v_{,i} \right] d\Omega$$

where

$$(\ )_{,i} = \partial (\ )/\partial x_i \qquad \text{and}$$

$$\| u \|_H^1 = \| u \|_1 = (u, u)_1^{1/2} = \left[\int_\Omega \left(u^2 + \sum_{i=1}^{n} u_{,i}^2 \right) d\Omega \right]^{1/2} . \tag{2.11}$$

Note $H^0 = L_2$. Likewise, we can extend $H^s(\Omega)$ to include the *S-th* order derivatives and create $\| u \|_s$ norms.

2.4 Self-Adjointness *

Consider the differential operator represented by the homogeneous equations

$$L(u) = 0 \in \Omega . \tag{2.12}$$

We form the inner product of $L(u)$ with another function, say v, to get $< L(u), v >$. If we integrate by parts (sometimes repeatedly) we obtain the alternate form

$$< L(u), v > = < u, L^*(v) > + \int_\Gamma \left[F(v) \, G(u) - F(u) \, G^*(v) \right] d\Gamma , \tag{2.13}$$

where F and G are differential operators whose forms follow naturally from integration by parts. The operator L^* is the *adjoint* of L. If $L^* = L$ then L is *self-adjoint* and $G^* = G$, also. The $F(u)$ are called the *essential boundary conditions* and $G(u)$ are the *natural boundary conditions*. When $L^* = L$, $F(u)$ is prescribed on Γ_1, and $G(u)$ is prescribed on Γ_2 where $\Gamma = \Gamma_1 \cup \Gamma_2$, $\Gamma_1 \cap \Gamma_2 = \emptyset$. We say $< L(u), u > \ > 0$ is positive definite *iff* $L^* = L$, $u \neq 0$. For example, the model differential equation

$$L(u) = \frac{d^2 u}{dx^2} \qquad x \in \,]0, 1[$$

has the inner product

$$(v, L(u)) = \int_0^1 v\, L(u)\, dx = \int_0^1 v \frac{d^2 u}{dx^2}\, dx\,.$$

Recall integration by parts: $\int_a^b p dq = pq \Big|_a^b - \int_a^b q\, dp$. Here we let $p = v$ so $dp = (dv/dx)\, dx$, and $dq = (d^2u/dx^2)\, dx$, so $q = du/dx$, such that

$$(v, L(u)) = v \frac{du}{dx}\Big|_0^1 - \int_0^1 \frac{du}{dx} \frac{dv}{dx}\, dx\,.$$

Integrate by parts again

$$(v, L(u)) = v \frac{du}{dx}\Big|_0^1 - \left[u \frac{dv}{dx}\Big|_0^1 - \int_0^1 u \frac{d^2 v}{dx^2}\, dx \right]$$

$$= (L^*(v),\, u) + \left[v \frac{du}{dx} - u \frac{dv}{dx} \right] \Big|_0^1\,.$$

The ODE requires two boundary conditions. Our options are: **a)** give u at $x=0$ and $x=1$ and recover du/dx at $x=0$ and $x=1$ from the solution, **b)** give u at $x=0$ and du/dx at $x=1$ (or vice versa). We compute u for all x and recover du/dx at $x=0$, **c)** give du/dx at $x=0$ and $x=1$. This determines u to within an arbitrary constant.

2.5 Weighted Residuals

Consider the following model equation:

$$L(u) = \frac{d^2 u}{dx^2} + u + x = 0, \quad x \in \,]0, 1[\tag{2.15}$$

with the essential boundary conditions $u=0$ at $x=0$ and $u=0$ at $x=1$ so that the exact solution is $u = Sin\, x/Sin\, 1 - x$. We want to find a global approximate solution involving constants Δ_i, $1 \le i \le n$ that will lead to a set of n simultaneous equations. For homogeneous essential boundary conditions we usually pick a global product approximation of the form

$$u^* = g(x)\, f(x, \Delta_i) \tag{2.16}$$

where $g(x) \equiv 0$ on Γ. Here the boundary is $x=0$ and $x-1=0$ so we select a form such as

$$g_1(x) = x(1-x)$$

or

$$g_2(x) = x - \frac{Sin\, x}{Sin\, 1}\,.$$

We could pick $f(x, \Delta_i)$ as a polynomial

$$f(x) = \Delta_1 + \Delta_2 x + \cdots \Delta_n x^{(n-1)}\,.$$

For simplicity, select $n = 2$ and use $g_1(x)$ so the approximate solution is

$$u^*(x) = x(1 - x)(\Delta_1 + \Delta_2 x). \tag{2.17}$$

Here we will employ the method of *weighted residuals* to find the Δ's. From Eq. (2.17) we see that the residual error at any point is $R(x) = u'' + u + x$, or in expanded form:

$$R(x) = x + (-2 + x - x^2)\,\Delta_1 + (2 - 6x + x^2 - x^3)\,\Delta_2 \neq 0. \tag{2.18}$$

Note for future reference that the partial derivatives of the residual with respect to the unknown degrees of freedom are:

$$\frac{\partial R}{\partial \Delta_1} = (-2 + x - x^2), \qquad \frac{\partial R}{\partial \Delta_2} = (2 - 6x + x^2 - x^3).$$

The residual error will vanish everywhere only if we guess the exact solution. The method of weighted residuals requires that a weighted integral of the residual vanish, that is,

$$\int_0^1 R(x)\, w(x)\, dx \equiv 0 \tag{2.19}$$

where $w(x)$ is a weighting function. For an approximate solution with n constants we can split R into parts including and independent of the Δ_j, say

$$R = R_0 + \sum_{j=1}^{n} h_j(x)\,\Delta_j. \tag{2.20}$$

We use n weights to get the necessary algebraic equations

$$\int_\Omega R\, w_i\, d\Omega = \int_\Omega \left[R_0 + \sum_{j=1}^{n} h_j(x)\,\Delta_j \right] w_i\, d\Omega = 0_i, \quad 1 \le i \le n$$

or

$$\sum_{j=1}^{n} \int_\Omega h_j(x)\, w_i(x)\, \Delta_j\, d\Omega = -\int_\Omega R_0(x)\, w_i(x)\, d\Omega, \quad 1 \le i \le n. \tag{2.21}$$

In matrix form this system of equations is written as:

$$\underset{n \times n}{\mathbf{S}} \quad \underset{n \times 1}{\boldsymbol{\Delta}} = \underset{n \times 1.}{\mathbf{C}} \tag{2.22}$$

Clearly, there are many ways to pick the weighting functions, w_i. Mathematical analysis and engineering experience have lead to the following five most common choices of the weights used in various weighted residual methods:

A) Collocation Method: For this method we force the residual error to vanish at n points. Thus, we select

$$w(x) = \delta(x - x_i), \quad 1 \le i \le n \tag{2.23}$$

where the Dirac Delta distribution $\delta(x - x_i)$ which has the properties

$$\delta(x - x_i) = \begin{cases} 0 & x \neq x_i \\ \infty & x = x_i \end{cases}$$

and

$$\int_{-\infty}^{\infty} \delta(x-x_i)\,dx = \int_{x_i-a}^{x_i+a} \delta(x-x_i)\,dx = 1\,.$$

and for any function $f(x)$ continuous at x_i

$$\int_{-\infty}^{\infty} \delta(x-x_i)\,f(x)\,dx = \int_{x_i-a}^{x_i+a} \delta(x-x_i)\,f(x)\,dx = f(x_i)\,. \tag{2.24}$$

Our problem is that we have an infinite number of choices for the collocation points, x_i. For $n=2$, we could pick two points where R is large, or the third point, or the Gauss points, etc. For this example pick the two collocation points as $x_1 = 1/4$ and $x_2 = 1/2$; then

$$\begin{bmatrix} \frac{29}{16} & -\frac{35}{64} \\ \frac{7}{4} & \frac{7}{8} \end{bmatrix} \begin{Bmatrix} \Delta_1 \\ \Delta_2 \end{Bmatrix} = \begin{Bmatrix} \frac{1}{4} \\ \frac{1}{2} \end{Bmatrix}$$

is our unsymmetric algebraic system. Since the essential boundary conditions have already been satisfied by the assumed solution we can solve these equations without additional modifications. Here we obtain $\Delta_1 = 6/31$ and $\Delta_2 = 40/217$ so that our first approximate solution is

$$u^* = x(1-x)\,(42+40\,x)/217\,.$$

Selected interior results compared to the exact solution are :

x	u	u^*
1/4	0.044	0.045
1/2	0.070	0.071
3/4	0.060	0.062

Note that $u(x_i) - u^*(x_i) \neq 0$ even though $R(x_i) = 0$. That is, the error in the differential equation is zero at these collocation points, but the error in the solution is not zero.

B) Least Squares Method : For the n equations pick

$$\int_0^1 R(x)\,w_i(x)\,dx = 0, \qquad 1 \le i \le n$$

with the weights defined as

$$w_i(x) = \frac{\partial R(x)}{\partial \Delta_i}\,. \tag{2.25}$$

This is equivalent to

$$\int_0^1 R^2(x)\,dx \;\rightarrow\; \text{stationary (minimum)}\,. \tag{2.26}$$

For this example

$$\int_0^1 R(x)\,\frac{\partial R}{\partial \Delta_1}\,dx = 0, \quad \int_0^1 R(x)\,\frac{\partial R}{\partial \Delta_2}\,dx = 0$$

and substitutions from Eq. (2.18) gives

$$\frac{202}{60}\Delta_1 + \frac{101}{60}\Delta_2 = \frac{55}{60}$$

$$\frac{101}{60}\Delta_1 + \frac{1532}{60}\Delta_2 = \frac{393}{60}.$$

It should be noted from Eq. (2.18) that this procedure yields a square matrix which is always symmetric. Solving gives $\Delta_1 = 0.192$, $\Delta_2 = 0.165$ and selected results at the three interior points of: 0.043, 0.068, and 0.059, respectively.

C) Galerkin Method: The concept here is to make the residual error orthogonal to the functions associated with the spatial influence of the constants. That is, let

$$u^*(x) = g(x)\, f(x, \Delta_i) = \sum_{i=1}^{n} h_i(x)\, \Delta_i .$$

Then for $n = 2$ and $h_1 = (x - x^2)$ and $h_2 = (x^2 - x^3)$, we set

$$w_i(x) \equiv h_i(x) \tag{2.27}$$

so that we require

$$\int_0^1 R(x)\, h_1(x)\, dx = 0, \quad \int_0^1 R(x)\, h_2(x)\, dx = 0 \tag{2.28}$$

so that Eq. (2.18) yields

$$\frac{3}{10}\Delta_1 + \frac{3}{20}\Delta_2 = \frac{1}{12}$$

$$\frac{3}{20}\Delta_1 + \frac{13}{105}\Delta_2 = \frac{1}{20}$$

which is again symmetric (for the self-adjoint equation). Solving gives degree of freedom values of $\Delta_1 = 71/369$, $\Delta_2 = 7/41$ and selected results at the three interior points of: 0.044, 0.070, and 0.060, respectively.

D) Method of Moments: Pick a spatial coordinate "lever arm" as a weight:

$$w_i(x) \equiv x^{(i-1)} \tag{2.29}$$

so that

$$\int_0^1 R(x)\, x^0\, dx = 0, \quad \int_0^1 R(x)\, x^1\, dx = 0 \tag{2.30}$$

gives the unsymmetric algebraic system

$$\frac{11}{6}\Delta_1 + \frac{11}{6}\Delta_2 = \frac{1}{2}$$

$$\frac{11}{12}\Delta_1 + \frac{19}{20}\Delta_2 = \frac{1}{3}$$

with the solution $\Delta_1 = 122/649$, $\Delta_2 = 110/649$ and selected results at the three interior points of: 0.043, 0.068, and 0.059, respectively.

E) Subdomain Method: For this method we split the solution domain, Ω, into n non-overlapping subdomains such that

$$\Omega = \bigcup_{i=1}^{n} \Omega_i \tag{2.31}$$

with $\Omega_i \cap \Omega_j = \phi$, if $i \neq j$. Then we define

$$w_i(x) \equiv 1 \quad \text{for } x \in \Omega_i . \tag{2.32}$$

This makes the residual error vanish on each of n different regions. Here $n = 2$, so we arbitrarily pick $\Omega_1 =]0, 1/2[$ and $\Omega_2 =]1/2, 1[$. Then

$$\int_{\Omega_1} R(x)\,dx = 0, \qquad \int_{\Omega_2} R(x)\,dx = 0 \tag{2.33}$$

yields the unsymmetric algebraic system

$$\begin{bmatrix} \frac{11}{6} & \frac{11}{6} \\ \frac{11}{22} & \frac{19}{20} \end{bmatrix} \begin{Bmatrix} \Delta_1 \\ \Delta_2 \end{Bmatrix} = \begin{Bmatrix} \frac{1}{2} \\ \frac{1}{3} \end{Bmatrix}.$$

This results in $\Delta_1 = 122/649$, $\Delta_2 = 110/649$ and selected results at the three interior points of: 0.043, 0.068, and 0.059, respectively.

The above examples illustrate how analytical approximations can be obtained for differential equations. These approximate methods offer some practical advantages. Instead of solving a differential equation we are now presented with the easier problem of solving an integral relation. The weighted residual procedure is valid of any number of spatial dimensions N. The procedure is valid for any shaped domain Ω. It allows non-homogeneous coefficients. That is, the coefficient multiplying the derivatives in L can vary with location. Note that so far we have not yet made any references to finite element methods. Later, you may look back on these examples as special cases of a single element solution. These simple hand examples could have been solved with the elementary inversion routines I2BY2 and I3BY3 included on the disk. For larger symmetric systems one might want to use subroutine SYMINV, while non-symmetric forms require a general inversion routine such as INVERT. Both of these are discussed in Appendix I. In practice, inversions are much too computationally expressive, and one must solve the equations by iterative methods or by a factorization process such as the Crout Factorization shown in Fig. 2.5.1, and represented graphically in Fig. 2.5.2.

Since numerical integration simply replaces an integral with a special summation this approach has the potential for automation. Then we can include thousands of unknown coefficients, Δ_i, in our test solution. Here we are dealing with polynomials. It is well known that in one-dimension Gaussian quadrature with N terms will exactly integrate a polynomial of order $(2N-1)$. Thus, we could replace the above integrals with a two-point Gauss rule. (This will be considered in full detail later.) For example, in the Galerkin approach the source term is

$$\int_0^1 x\, h_1(x)\, dx = \sum_{j=1}^{N} x_j\, h_1(x_j)\, w_j$$

where the x_j and w_j are tabulated. For $N = 2$ on $\Omega =]0, 1[$ we have $w_1 = w_2 = 1/2$ and $x_j = (1 \pm 1/\sqrt{3})/2$, or $x_1 = 0.21133248$ and $x_2 = 0.7886751$. Thus,

```
      SUBROUTINE  FULFAC (NDFREE, S)
C     * * * * * * * * * * * * * * * * * * * * * * * * *
C     CROUT FACTORIZATION OF FULL EQS, S = L*D*LT
C     * * * * * * * * * * * * * * * * * * * * * * * * *
      DIMENSION  S(NDFREE,NDFREE)
C     D      = DIAGONAL MATRIX STORED ON S
C     L      = LOWER TRIANGULAR MATRIX STORED ON S
C     NDFREE = TOTAL NUMBER OF DEGREES OF FREEDOM
C     S      = ORIGINAL FULL SYMMETRIC MATRIX
C     NOTE: INEFFICIENT STORAGE, ONLY UPPER TRI USED
C      D1 = S11 BY DEFAULT
      DO 40  I = 2, NDFREE
        DO 20  J = 1, (I - 1)
          SUM = 0.D0
          IF ( (J - 1) .EQ. 0 )  GO TO 20
C            FACTOR COLUMN
            DO 10  K = 1,(J -1)
  10        SUM = SUM + S(K,K)*S(I,K)*S(J,K)
  20    S(I,J) = ( S(I,J) - SUM )/S(J,J)
C        FACTOR DIAGONAL
        SUM = 0.D0
        DO 30  K = 1,(I - 1)
  30    SUM = SUM + S(K,K)*S(I,K)**2
  40  S(I,I) = S(I,I) - SUM
      RETURN
      END

      SUBROUTINE  FULSOL (NDFREE, S, C, D)
C     * * * * * * * * * * * * * * * * * * * * * * * * *
C     FORWARD, BACK CROUT SUBSTITUTION FOR D, S*D = C
C     * * * * * * * * * * * * * * * * * * * * * * * * *
      DIMENSION  S(NDFREE,NDFREE), C(NDFREE), D(NDFREE)
C     C      = SOURCE OR FORCE VECTOR
C     D      = SOLUTION VECTOR, RETURNED
C     DIA    = DIAGONAL MATRIX STORED ON S
C     L      = LOWER TRIANGULAR MATRIX STORED ON S
C     NDFREE = TOTAL NUMBER OF DEGREES OF FREEDOM
C     S      = FULL FACTORED MATRIX OF L*DIA*L^T
C      FORWARD SUBSTITUTION
      DO 20  I = 1, NDFREE
        SUM = 0.D0
        IF ( (I - 1) .EQ. 0 )  GO TO 20
          DO 10  K = 1,(I - 1)
  10      SUM = SUM + D(K)*S(I,K)
  20  D(I) = C(I) - SUM
C      BACK SUBSTITUTION
      DO 40  I = NDFREE,1,-1
        SUM = 0.D0
        IF ( (NDFREE - I) .EQ. 0 )  GO TO 40
          DO 30  K = 1,(NDFREE-I)
  30      SUM = SUM + D(I+K)*S(I+K,I)
  40  D(I) = D(I)/S(I,I) - SUM
      RETURN
      END
```

Figure 2.5.1 Factorization solution of full equations

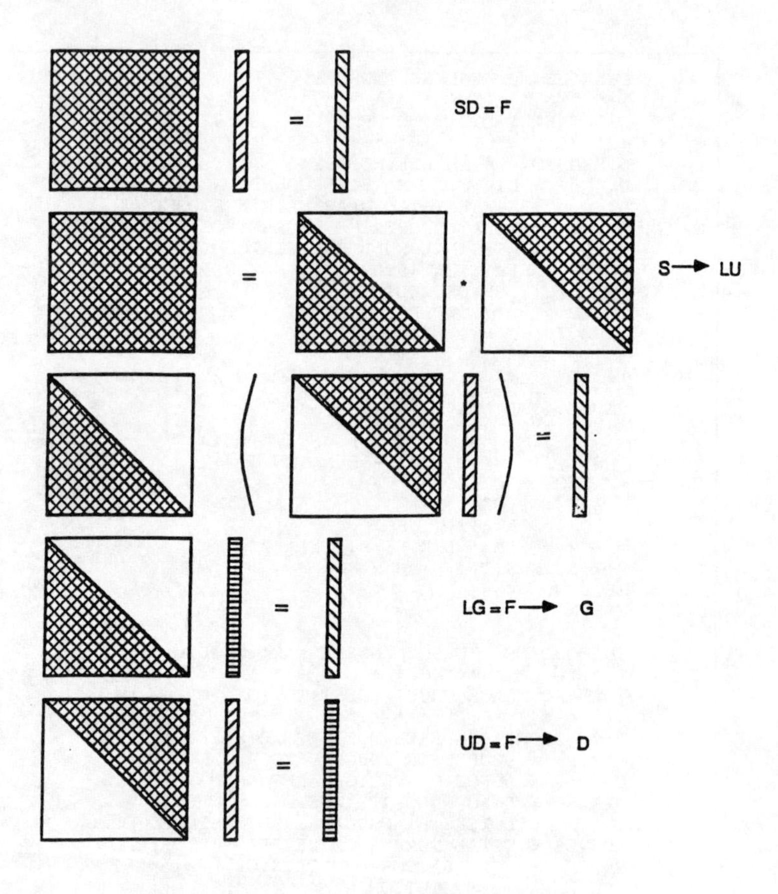

Figure 2.5.2 Steps in the factorization process

$$\int_0^1 x(x - x^2)\, dx = \sum_{j=1}^{N} (x_j^2 - x_j^3)\, w_j = \sum_{j=1}^{2} x_j^2 (1 - x_j)\, w_j$$

$$= \left[(0.2113248)^2 (0.7886751)\ 1/2 + (0.7886751)^2 (0.2113248)\ 1/2 \right]$$

$$= (0.16666667)\ 1/2 = 0.083333 .$$

If we had an infinite word length machine this process would yield the exact value of 1/12 which was previously found in Eq. (2.28).

2.6 Essential Boundary Conditions

If the assumed solution does not satisfy the essential boundary conditions then we must extend the assumed approximate solution to include additional constants to be used to satisfy the essential boundary conditions. Usually these conditions are invoked prior to or during the solution of the algebraic equations. Sometimes they are satisfied only in a weak sense. To illustrate the algebraic procedure that is usually used we will solve the ODE in Eq. (2.15) with an approximation that does not initially satisfy the boundary condition at $x = 0$. In Chap. 20 we will present other options for use with iterative equation solvers.

To illustrate the new concept of applying the boundary condition after we have selected an approximate solution we will use Eq. (2.16) and pick $g(x) = (1 - x)$ so that only the boundary condition at $x = 1$ is satisfied in advance. We add another constant to $f(x)$ to allow any boundary condition at $x = 0$ to be applied:

$$f(x) = \Delta_1 + \Delta_2 x + \Delta_3 x^2.$$

The new residual error for any x is

$$R = x + (1 - x)\,\Delta_1 + (x + 2 - x^2)\,\Delta_2 + (x^2 + 2 - 6x - x^3)\,\Delta_3.$$

Since we now have three unknown degrees of freedom, $\mathbf{\Delta}$, we must have three weighted residual equations. For simplicity we will choose the collocation method and pick three equally spaced collocation points. Evaluating the residual at the quarter points and multiplying by the common denominator gives the three equations

$$\begin{bmatrix} -48 & 116 & -35 \\ -32 & 112 & 56 \\ -16 & 116 & 151 \end{bmatrix} \begin{Bmatrix} \Delta_1 \\ \Delta_2 \\ \Delta_3 \end{Bmatrix} = \begin{Bmatrix} 16 \\ 32 \\ 48 \end{Bmatrix}.$$

Note that since we know u at $x = 0$ these unknowns are not independent. Substituting $x = 0$ into our approximate solution and equating it to the assign boundary value there gives $u(0) = \Delta_1$. We call this an *essential boundary condition* on Δ_1. There are an infinite number of possible boundary conditions and we gain flexibility by allowing extra constants to satisfy them. Since Δ_1 will be a known number only the last two rows are independent for determining the remaining terms in $\mathbf{\Delta}$. Note that the first column of numbers is now multiplied by a known value and thus they can be carried to the right hand side to give the reduced algebraic system:

$$\begin{bmatrix} 112 & 56 \\ 116 & 151 \end{bmatrix} \begin{Bmatrix} \Delta_2 \\ \Delta_3 \end{Bmatrix} = \begin{Bmatrix} 32 \\ 48 \end{Bmatrix} + \begin{Bmatrix} 32 \\ 16 \end{Bmatrix} \Delta_1.$$

An equivalent matrix modification routine, MODFUL, (on the disk) deletes the redundant coefficients, but keeps the matrix the same size to avoid re-ordering all the coefficients as done above. For the common original boundary condition of $u(0) = 0$, we have $\Delta_1 = 0$ and the changes to the RHS are not necessary. But the above form also allows us the option of specifying any non-zero boundary condition we need. Using the zero value gives a solution of $\Delta_2 = 0.2058$ and $\Delta_3 = 0.1598$. The resulting values at the interior quarter points are 0.046, 0.071, and 0.061, respectively. These compare well with the previous results.

If the second boundary condition had been applied at any other point then we would have a more complicated relation between the $\mathbf{\Delta}$. For example, assume we move the boundary condition to $x = 0.5$. Then evaluating the approximate solution there yields

$$u(0.5) = 0.5\Delta_1 + 0.25\Delta_2 = 0.125\Delta_3$$

which is called a *linear constraint equation* on $\mathbf{\Delta}$. In other words we would have to solve the weighted residual algebraic system subject to a linear constraint. This is a fairly common situation in practical design problems and adaptive analysis procedures. The computational details for enforcing the above essential boundary conditions are discussed in detail in Chap. 20.

2.7 Elliptic Boundary Value Problems

2.7.1 One-Dimensional Equations

In order to review the previous introductory concepts consider the following one-dimensional model problem which will serve as an analytical example. The differential equation of interest is

$$k\,\frac{d^2 t}{dx^2} + Q = 0$$

on the closed domain, $x \in \,]0, L[$, and is subjected to the boundary conditions $t(L) = t_L$, and $dt/dx\,(0) = q_0$. The corresponding governing integral statement to be used for the finite element model is obtained from the *Galerkin weighted residual method.* In this case, it states that the function, t, which satisfies the essential condition, $t(L) = t_L$, and satisfies

$$I = \int_0^L \left[(dt/dx)^2 - tQ \right] dx + q_0 t(0) - q_L t(L) = 0, \tag{2.34}$$

also satisfies Eq. (2.15). In a finite element model, I is assumed to be the sum of the NE element and NB boundary segment contributions so that

$$I = \sum_{e=1}^{NE} I^e + \sum_{b=1}^{NB} I^b, \tag{2.35}$$

where here $NB = 2$ and consists of the last two terms in I. A typical element contribution is

$$I^e = \int_{L^e} \left[k^e (dt^e / dx)^2 - Q^e t^e \right] dx, \tag{2.36}$$

where L^e is the length of the element. To evaluate such a typical element contribution, it is necessary to introduce a set of interpolation functions, $\mathbf{H}$, such that $t^e(x) = \mathbf{H}^e(x)\,\mathbf{T}^e$, and

$$\frac{dt^e}{dx} = \frac{d\mathbf{H}^e}{dx}\,\mathbf{T}^e = \mathbf{T}^{e^T}\,\frac{d\mathbf{H}^{e^T}}{dx},$$

where $\mathbf{T}^e$ denotes the nodal values of t for element e. One of the few standard notations in finite element analysis is to denote the result of the differential operator acting on the interpolation functions, $\mathbf{H}$, by the symbol $\mathbf{B}$. That is, $\mathbf{B}^e \equiv d\mathbf{H}^e/dx$. Thus, a typical element contribution is

$$I^e = \mathbf{T}^{e^T} \mathbf{S}^e \mathbf{T}^e - \mathbf{T}^{e^T} \mathbf{C}^e, \tag{2.37}$$

where

$$\mathbf{S}^e \equiv \int_{L^e} k^e \frac{d\mathbf{H}^{e^T}}{dx} \frac{d\mathbf{H}^e}{dx} dx = \int_{L^e} \mathbf{B}^{e^T} k^e \mathbf{B}^e dx, \quad \mathbf{C}^e \equiv \int_{L^e} Q^e \mathbf{H}^{e^T} dx.$$

Clearly, both the element degrees of freedom, $\mathbf{T}^e$, and the boundary degrees of freedom, $\mathbf{T}^b$, are subsets of the total vector of unknown parameters, $\mathbf{T}$. That is, $\mathbf{T}^e \subseteq \mathbf{T}$ and $\mathbf{T}^b \subset \mathbf{T}$. Of course, the $\mathbf{T}^b$ are usually a subset of the $\mathbf{T}^e$ (i.e., $\mathbf{T}^b \subset \mathbf{T}^e$ and in higher dimensional problems $\mathbf{H}^b \subset \mathbf{H}^e$). The main point here is that $I = I(\mathbf{T})$, and that fact must be considered in the summation. The consideration of the subset relations is merely a bookkeeping problem. This allows Eq. (2.37) to be written as

$$I = \mathbf{T}^T \mathbf{S} \mathbf{T} - \mathbf{T}^T \mathbf{C} = \mathbf{T}^T (\mathbf{S} \mathbf{T} - \mathbf{C}) = 0 \tag{2.38}$$

where

$$\mathbf{S} = \sum_{e=1}^{NE} \beta^{e^T} \mathbf{S}^e \beta^e, \quad \mathbf{C} = \sum_{e=1}^{NE} \beta^{e^T} \mathbf{C}^e + \beta^{b^T} \mathbf{C}^b,$$

and where β denotes a set of symbolic bookkeeping operations. The combination of the summations and bookkeeping is commonly referred to as the *assembly process.*

It is easily shown that for a non-trivial solution, $\mathbf{T} \neq \mathbf{O}$, we must have $\mathbf{O} = \mathbf{S}\mathbf{T} - \mathbf{C}$, as the governing algebraic equations to be solved for the unknown nodal parameters, $\mathbf{T}$. To be specific, consider a linear interpolation element with $N = 2$ nodes per element. If the element length is $L^e = x_2 - x_1$, then the element interpolation is

$$\mathbf{H}^e(x) = \begin{bmatrix} \dfrac{(x_2 - x)}{L^e} & \dfrac{(x - x_1)}{L^e} \end{bmatrix},$$

so that

$$\mathbf{B}^e = \frac{d\mathbf{H}^e}{dx} = \begin{bmatrix} -\dfrac{1}{L^e} & \dfrac{1}{L^e} \end{bmatrix}.$$

Therefore, the element square matrix is simply

$$\mathbf{S}^e = \frac{k^e}{L^e} \begin{bmatrix} 1 & -1 \\ -1 & 1 \end{bmatrix},$$

while the element column matrix is

$$\mathbf{C}^e = \int_{L^e} \frac{Q^e}{L^e} \begin{Bmatrix} (x_2 - x) \\ (x - x_1) \end{Bmatrix} dx.$$

Assuming that $Q = Q_0$, a constant, this simplifies to

$$\mathbf{C}^e = \frac{Q_0 \, L^e}{2} \begin{Bmatrix} 1 \\ 1 \end{Bmatrix}.$$

That is, the finite element model replaces the continuously distributed source by lumping half its resultant, $Q_0 L^e$, at each of the two nodes of the element.

The exact solution of the original problem for $Q = Q_0$ is

$$kt(x) = kt_L + q(x - L) + \frac{Q_0(L^2 - x^2)}{2}. \tag{2.39}$$

Since for $Q_0 \neq 0$ the exact value is quadratic and the selected element is linear, our finite element model can give only an approximate solution. However, for the homogeneous problem $Q_0 = 0$, the model can (and does) give an exact solution. To compare a finite element solution with the exact one, select a two element model. Let the elements be of equal length, $L^e = L/2$. Then the element matrices are the same for both elements.

The assembly process yields, $\mathbf{S}\,\mathbf{T} = \mathbf{C}$, as

$$\frac{2k}{L}\begin{bmatrix} 1 & -1 & 0 \\ -1 & (1+1) & -1 \\ 0 & -1 & 1 \end{bmatrix}\begin{Bmatrix} T_1 \\ T_2 \\ T_3 \end{Bmatrix} = \frac{Q_0 L}{4}\begin{Bmatrix} 1 \\ (1+1) \\ 1 \end{Bmatrix} - \begin{Bmatrix} q_0 \\ 0 \\ -q_L \end{Bmatrix}.$$

However, these equations do not yet satisfy the essential boundary condition of $t(L) = T_3 = t_L$. That is, $\mathbf{S}$ is singular. After applying this condition, the reduced equations are

$$\frac{2k}{L}\begin{bmatrix} 1 & -1 \\ -1 & 2 \end{bmatrix}\begin{Bmatrix} T_1 \\ T_2 \end{Bmatrix} = \frac{Q_2 L}{4}\begin{Bmatrix} 1 \\ 2 \end{Bmatrix} - \begin{Bmatrix} q \\ 0 \end{Bmatrix} + \frac{2t_L}{L}\begin{Bmatrix} 0 \\ 1 \end{Bmatrix},$$

or $\mathbf{S}_r\,\mathbf{T}_r = \mathbf{C}_r$. Solving for $\mathbf{T}_r = \mathbf{S}_r^{-1}\,\mathbf{C}_r$ yields

$$\mathbf{S}_r^{-1} = \frac{L}{2k}\begin{bmatrix} 2 & 1 \\ 1 & 1 \end{bmatrix}, \quad \mathbf{T}_r = \begin{Bmatrix} T_1 \\ T_2 \end{Bmatrix} = \frac{Q_0 L^2}{8k}\begin{Bmatrix} 4 \\ 3 \end{Bmatrix} - \frac{qL}{2k}\begin{Bmatrix} 2 \\ 1 \end{Bmatrix} + t_L\begin{Bmatrix} 1 \\ 1 \end{Bmatrix}.$$

These are the exact nodal values as can be verified by evaluating the solution at $x = 0$ and $x L/2$, respectively. Thus, our finite element solution is giving an *interpolate* solution. That is, it interpolates the solution exactly at the node points, and is approximate at all other points on the interior of the element. For the homogeneous problem, $Q_0 = 0$, the finite element solution is exact at all points. These properties are common to other finite element problems. The exact and finite element solutions are illustrated in Fig. 2.7.1. Note that the derivatives are also exact at least at one point in each element. The optimal derivative sampling points will be considered in a later section. For this differential operation, it can be shown that the center point gives a derivative estimate accurate to $0(L^{e2})$, while all other points are only $0(L^e)$ accurate. For $Q = Q_0$, the center point derivatives are exact in the example.

Next, we want to utilize the last equation from the weighted residual algebraic system to recover the flux "reaction" that is necessary to enforce the essential boundary condition at $x = L$:

$$\frac{2k}{L}\left[0 - 1\,T_2 + 1\,T_3\right] = \frac{QL}{4} - (-q_L)$$

or

$$-\,QL + q_0 = q_L,$$

which states the flux equilibrium: that which was generated internally, QL, minus that which exited at $x = 0$, *must* equal that which exits at $x = L$.

If one is going to save the reaction data for post-solution recovery, then one could use programs like SAVFUL and REACTS to store and later recover the associated rows

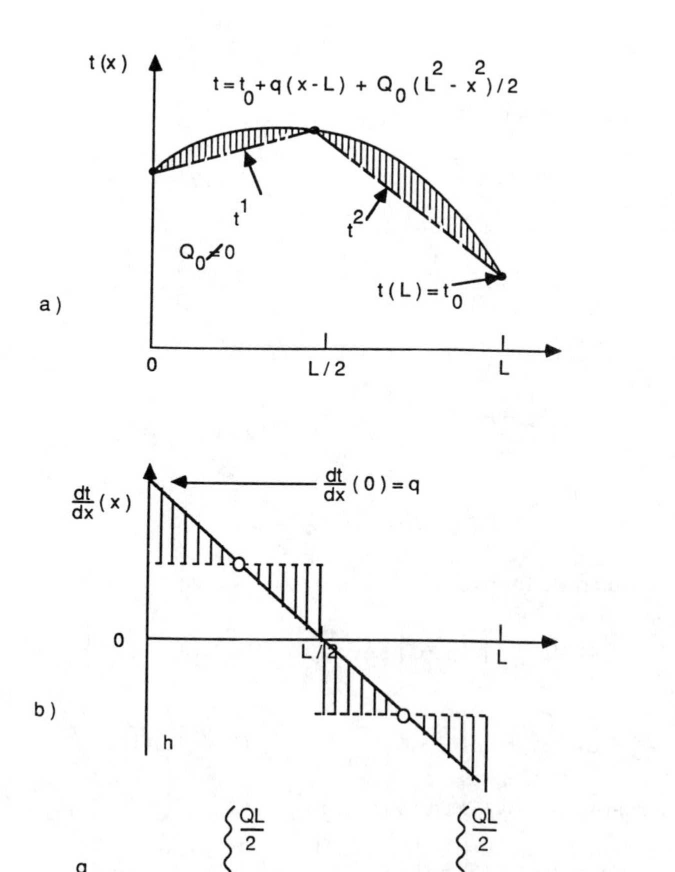

Figure 2.7.1 Example interpolate solution

of the square matrix. SAVFUL would have to be used for each boundary condition before using MODFUL mentioned earlier.

2.7.2 Two-Dimensional Equations

As an example of the utilization of Galerkin's method in higher dimensions, consider the following model linear operator:

$$L(u) \;=\; \frac{\partial}{\partial x}\left[k_x\,\frac{\partial u}{\partial x}\right] + \frac{\partial}{\partial y}\left[k_y\,\frac{\partial u}{\partial y}\right] - Q - \zeta\,\frac{\partial u}{\partial t} \;=\; 0\,. \tag{2.40}$$

Multiplying by a test function, $v(x)$, and integrating gives

$$I = \int_\Omega L(u)\, v\, d\Omega = 0 .$$

The last two terms are simply

$$I_Q = \int_\Omega Q\, v\, d\Omega, \qquad I_\zeta = \int_\Omega \zeta\, v \frac{\partial u}{\partial t}\, d\Omega .$$

Our main interest is with the first two terms

$$I_k = \int_\Omega \left[v \frac{\partial}{\partial x} \left(k_x \frac{\partial u}{\partial x} \right) + v \frac{\partial}{\partial y} \left(k_y \frac{\partial u}{\partial y} \right) \right] d\Omega$$

which involve the second order partial derivatives. We can remove them by invoking Green's theorem

$$\int_\Omega \left[\frac{\partial \Psi}{\partial x} - \frac{\partial \phi}{\partial y} \right] d\Omega = \int_\Gamma \left[\phi\, dx + \Psi\, dy \right]$$

where we define

$$\Psi = v\, k_x \frac{\partial u}{\partial x}, \qquad \phi = -\, v\, k_y \frac{\partial u}{\partial y} .$$

We note that an alternate form of I_k is

$$I_k = \int_\Omega \left[\frac{\partial}{\partial x} \left(v\, k_x \frac{\partial u}{\partial x} \right) + \frac{\partial}{\partial y} \left(v\, k_y \frac{\partial u}{\partial y} \right) \right] d\Omega$$

$$- \int_\Omega \left[\frac{\partial v}{\partial x} k_x \frac{\partial u}{\partial x} + \frac{\partial v}{\partial y} k_y \frac{\partial u}{\partial y} \right] d\Omega .$$

The first term is re-written by Green's theorem

$$I_k = \int_\Gamma \left[-\left(v\, k_y \frac{\partial u}{\partial y} \right) dx + \left(v\, k_x \frac{\partial u}{\partial x} \right) dy \right]$$

$$- \int_\Omega \left[k_x \frac{\partial v}{\partial x} \frac{\partial u}{\partial x} + k_y \frac{\partial v}{\partial y} \frac{\partial u}{\partial y} \right] d\Omega .$$

For a contour integral with a unit normal vector, **n**, with components $[\, n_x \quad n_y \,]$, we note the geometric relations that

$$-\, dx = ds \operatorname{Cos} \Theta_y = ds\, n_y, \qquad dy = ds \operatorname{Cos} \Theta_x = ds\, n_x$$

which reduces the boundary integral to

$$\int_\Gamma v \left[k_x \frac{\partial u}{\partial x} n_x + k_y \frac{\partial u}{\partial y} n_y \right] ds = \int_\Gamma v\, k_n \frac{\partial u}{\partial n}\, ds$$

where $\partial u / \partial n$ is the normal gradient of u, that is $\nabla u \cdot \mathbf{n}$, and k_n is the value of the orthotropic k in the direction of the normal. The resulting Galerkin form is

$$-I = \int_{\Omega} \left[k_x \frac{\partial v}{\partial x} \frac{\partial u}{\partial x} + k_y \frac{\partial v}{\partial y} \frac{\partial u}{\partial x} \right] d\Omega - \int_{\Omega} Q v \, d\Omega$$
$$+ \int_{\Omega} \zeta v \frac{\partial u}{\partial t} d\Omega - \int_{\Gamma} v k_n \frac{\partial u}{\partial n} ds = 0 . \quad (2.41)$$

Note that Green's theorem brought in the information on the conditions of the surface. If we split this into element domains so that $\Omega = \underset{e}{\cup} \Omega^e$ and $\Gamma = \underset{b}{\cup} \Gamma^b$, we generate four typical matrices from these terms. We interpolate, as before, with

$$v(x, y) = \mathbf{H}^e(x, y) \, \mathbf{V}^e, \quad u(x, y, t) = \mathbf{H}^e(x, y) \, \mathbf{U}^e(t) \ .$$

That is, the unknown coefficients, $\mathbf{U}^e(t)$, are time dependent, and we describe their spatial form in the usual way. The only new concern is the time derivative

$$\frac{\partial u}{\partial t} = \mathbf{H}^e(x, y) \frac{\partial}{\partial t} \mathbf{U}^e(t) = \mathbf{H}^e(x, y) \, \dot{\mathbf{U}}^e .$$

These results in two square matrices

$$\mathbf{S}^e = \int_{\Omega^e} \left[\mathbf{H}^{e^T}_{,x} k_x \mathbf{H}^{e^T}_{,x} + \mathbf{H}^{e^T}_{,y} k_y \mathbf{H}^{e}_{,y} \right] d\Omega, \quad \mathbf{M}^e = \int_{\Omega^e} \mathbf{H}^{e^T} \mathbf{H}^e \zeta^e \, d\Omega$$

and the two source vectors

$$\mathbf{C}^e_Q = \int_{\Omega^e} \mathbf{H}^{e^T} Q^e \, d\Omega, \quad \mathbf{C}^b_q = \int_{\Gamma^b} \mathbf{H}^{b^T} k^b_n \frac{\partial u^b}{\partial n} d\Gamma = \int_{\Gamma^b} \mathbf{H}^{b^T} q^b_n \, d\Gamma,$$

which when assembled yield the system of ordinary differential equations such that

$$\mathbf{V}^T \left[\mathbf{M}\dot{\mathbf{U}} + \mathbf{S}\mathbf{U} - \mathbf{C} \right] = 0,$$

so that for arbitrary $\mathbf{V} \neq \mathbf{0}$, we have

$$\mathbf{M}\dot{\mathbf{U}} + \mathbf{S}\mathbf{U} = \mathbf{C}(t), \quad (2.42)$$

which is an initial value problem to be solved from the state at $\mathbf{U}(t=0)$. We will not consider time dependent problems until later. Here we will note that $\mathbf{M}^e$ and $\mathbf{M}$ are usually called the *element* and *system mass* (or *capacity*) *matrices*, respectively. For the steady state case, $\partial/\partial t = 0$, this reduces to the system of algebraic equations $\mathbf{S}\mathbf{U} = \mathbf{C}$ considered earlier. The same procedure is easily carried out in three dimensions. Matrix $\mathbf{S}^e$ changes by having a third term involving z and for $\mathbf{C}^b_q$ we think of $d\Gamma$ as a surface area instead of a line segment.

If we had also included a second time derivative term, we could then see by inspection that it would simply introduce another mass matrix (with a different coefficient) times the second time derivative of the nodal degrees of freedom. This would then represent the *wave equation* whose solution in time could be accomplished by techniques to be presented later. If the time derivative terms are not present, and if the term $c\,u$ is included in the PDE and if the coefficient c is an unknown global constant, then this reduces to an *eigenproblem.* That is, we wish to determine a set of *eigenvalues*, c_i, and a corresponding set of *eigenvectors* , or *mode shapes.*

2.7.3 Equivalent Forms*

Analysis problems can be stated in different formats or forms. In finite element methods, we do not deal with the original differential equation (which is called the *strong form*). Instead, we convert to some equivalent integral form. In this optional section, we go through some mathematical manipulations in the one-dimensional case to assure the reader that they are, indeed, mathematically equivalent approaches to the solution. Here we will consider equivalent forms of certain model differential equations. For most elliptic Boundary Value Problems (*BVP*) we will find that there are three equivalent forms; the original strong form (*S*), the variational form (*V*), and the weak form, (*W*). The latter two integral forms will be reduced to a matrix form, (*M*). As an example, consider the one-dimensional two-point boundary value problem where, $(\,)' = d(\,)/dx$:

$$(S)\quad \left[\begin{array}{ll} & x \in\,]0,1[\\ & -u''(x) = f(x) \\ \text{with} & u(0) = 0 \\ & u(1) = g\,. \end{array}\right.$$

That is, given $f: \Omega \to R$ and $g \in R$ find $u: \overline{\Omega} \to R$ such that $u'' + f = 0$ on Ω and $u(0) = 0$, $u(1) = g$. Recall the Hilbert space properties that the function, u, and its first derivative must be square integrable

$$H^1 = \left[u;\ \int_0^1 [\, u^2 + u_x^2 \,]\, dx < \infty \right],$$

and introduce a linear trial solution space with the properties that it satisfies certain boundary conditions, is a Hilbert space, and produces a real number when evaluated on the closure of the solution domain:

$$S = \left[u;\ u: \overline{\Omega} \to R,\ \ u \in H^1,\ \ u(1) = g,\ \ u(0) = 0 \right]$$

and a similar linear space of weighting functions with a different set of boundary conditions:

$$V = \left[\omega;\ \omega: \overline{\Omega} \to R,\ \ \omega \in H^1,\ \ \omega(0) = 0,\ \ \omega(1) = 0 \right].$$

The Variational Problem is: given a linear functional, F, that maps S into a real number, $F: S \to R$, given by

$$(V)\quad \left[\begin{array}{l} F(\omega) = \tfrac{1}{2} < \omega', \omega' > - < f, \omega > \\ \\ \text{Find } u \in S \text{ such that } F(u) \le F(\omega) \text{ for all } \omega \in S \end{array}\right.$$

$$(W)\quad \left[\begin{array}{l} \text{The Weak Problem is find } u \in S \text{ such that for all } \omega \in V \\ \\ < u', \omega' > \; = \; < f, \omega > . \end{array}\right.$$

For this problem, we can show that the strong, variational, and variational form imply the existence of each other: (*S*) $\Leftrightarrow$ (*V*) $\Leftrightarrow$ (*W*). In solid mechanics applications (*S*) would represent the differential equations of equilibrium, (*V*) would denote the Principal of Minimum Total Potential Energy, and (*W*) would be the Principal of Virtual Work.

Other physical applications have similar interpretations but the three equivalent forms often have no physical meaning. Here, we will introduce the common symbols for the *bi-linear form*

$$a(u, v) \equiv \; < u', v' > \; = \int_0^1 u' v' \, dx$$

and the *linear form*

$$(f, v) \; = \; < f, v > \; = \int_0^1 f v \, dx .$$

Here we want to prove the relation that the strong and weak forms imply the existence of each other, $(S) \Leftrightarrow (W)$. Let u be a solution of (S). Note that $u \in S$ since $u(1) = g$. Assume that $\omega \in V$. Now set $0 = (u'' + f)$; then we can say

$$0 \; = \; - \int_0^1 (u'' + f) \, \omega \, dx \; = \; - \int_0^1 u'' \, \omega \, dx - \int_0^1 f \omega \, dx$$

integrating by parts

$$0 \; = \; - u' \omega \Big|_0^1 + \int_0^1 u' \omega' \, dx - \int_0^1 f w \, dx .$$

Note that since $\omega \in V$ we have $\omega(1) = 0 = \omega(0)$ so that

$$u' \omega \Big|_0^1 \; = \; u'(1) \, \omega(1) - u'(0) \, \omega(0) \; \equiv \; 0 ,$$

and finally

$$0 \; = \; \int_0^1 u' \omega' \, dx - \int_0^1 f \omega \, dx .$$

Thus, we conclude that $u(x)$ is a solution of the weak problem, (W). That is, $a(u, \omega) = (f, \omega)$ for all $\omega \in V$. Next, we want to verify that the weak solution also implies the existence of the strong form, $(W) \Leftrightarrow (S)$. Assume that $u(x) \in S$ so that $u(0) = 0$, $u(1) = g$, and assume that $\omega \in V$:

$$\int_0^1 u' \omega' \, dx \; = \; \int_0^1 f \omega \, dx \qquad \text{for all} \quad \omega \in V .$$

Integrating by parts

$$- \int_0^1 (u'' + f) \, \omega \, dx + u' \omega \Big|_0^1 \; = \; 0$$

but,

$$u' \omega \Big|_0^1 \; = \; u'(1) \, \omega(1) - u'(0) \, \omega(0) \; \equiv \; 0 ,$$

since $\omega \in V$, and thus

$$0 \; = \; \int_0^1 (u'' + f) \, \omega(x) \, dx , \qquad \text{for all} \quad \omega \in V .$$

Now we pick $\omega(x) \equiv \phi(x) \, (u'' + f)$ such that ϕ produces a real positive number when evaluated on the domain, $\phi : \overline{\Omega} \to R$, and $\phi(x) > 0$, when $x \in \,]0, 1[$ and $\phi(0) = \phi(1) = 0$.

Is $\omega \in V$? Since $\omega(0) = 0$ and $\omega(1) = 0$, we see that it is and proceed with that substitution,

$$0 = \int_0^1 (u'' + f)^2 \, \phi(x)\, dx = \int (\geq 0)\,(>0)\, dx$$

which implies $(u'' + f) \equiv 0$. Thus, the weak form does imply the strong form, $(W) \Rightarrow (S)$ and combining the above two results we find $(S) \Leftrightarrow (W)$.

Next, we will consider the proposition that the variational form implies the existence of the weak form, $(V) \Leftrightarrow (W)$. Assume $u \in S$ is a solution to the weak form (W) and note that $v(1) = \omega(1) + u(1) = 0 + g$. Therefore, $v \in S$. Set $\omega = v - u$ such that $v = \omega + u$, and $\omega \in V$. Given

$$F(v) = \tfrac{1}{2} < v', v' > - < f, v >$$

then

$$F(v) = F(u + \omega) = \tfrac{1}{2} < (u' + \omega'),\ (u' + \omega') > - < f,\ u + \omega >$$

$$= \tfrac{1}{2} < u', u' > + < u', \omega' > + \tfrac{1}{2} < \omega', \omega' > .$$

But, since u is a solution to (W)

$$< u', \omega' > - < f, \omega > \equiv 0$$

and the above simplifies to

$$F(v) = F(u) + \tfrac{1}{2} < \omega', \omega' >$$

so that $F(v) \geq F(u)$ for all $v \in S$ since $< \omega', \omega' > \geq 0$. This shows that u is also a solution of the variational form, (V). That is, $(W) \Leftrightarrow (V)$, which is what we wished to show. Finally, we verify the uniqueness of the weak form, (W). Assume that there are two solutions u_1 and u_2 that are both in the space S. Then

$$a(u_1, v) = (f, v), \quad a(u_2, v) = (f, v) \qquad \forall\ v \in V$$

and subtracting the second from the first

$$a(u_1 - u_2, v) = 0, \qquad < (u_1{}' - u_2{}'), v' > = 0 \qquad \forall\ v \in V.$$

Consider the choice $v(x) \equiv u_1(x) - u_2(x)$. Is it in V? We note that $v(0) = 0 - 0 = 0$ and $v(1) = g - g = 0$, so we proceed with this choice. Thus, $v' = u_1' - u_2'$ and the inner product is

$$< (u_1' - u_2'),\ (u_1' - u_2') > = 0 \int_0^1 \left[u_1'(x) - u_2'(x) \right]^2 dx .$$

This means $u_1(x) - u_2(x) = c$. The constant, c, is evaluated from the boundary condition at $x = 0$ so $u_1(0) - u_2(0) = 0 - 0 = c$ which means that $u_1(x) = u_2(x)$, and the weak form solution, (W), is unique.

2.8 References

[1] Adams, R.A., *Sobolev Spaces*, New York: Academic Press (1975).

[2] Aziz, A.K., *The Mathematical Foundations of the Finite Element Method with Applications to Partial Differential Equations*, New York: Academic Press (1972).

[3] DeBoor, C., ed., *Mathematical Aspects of Finite Elements in Partial Differential Equations*, London: Academic Press (1974).

[4] Hughes, T.J.R., *The Finite Element Method*, Englewood Cliffs: Prentice-Hall (1987).

[5] Liusternik, L.A. and Sobolev, V.J., *Elements of Functional Analysis*, New York: Frederick Ungar (1961).

[6] Mitchell, A.R. and Wait, R., *The Finite Element Method in Partial Differential Equations*, London: John Wiley (1977).

[7] Nowinski, J.L., *Applications of Functional Analysis in Engineering*, New York: Plenum Press (1981).

[8] Oden, J.T. and Reddy, J.N., *An Introduction to the Mathematical Theory of Finite Elements*, New York: John Wiley (1976).

[9] Oden, J.T., *Applied Functional Analysis*, Englewood Cliffs: Prentice-Hall (1979).

[10] Strang, W.G. and Fix, G.J., *An Analysis of the Finite Element Method*, Englewood Cliffs: Prentice-Hall (1973).

[11] Szabo, B. and Babuska, I., *Finite Element Analysis*, New York: John Wiley (1991).

[12] Whiteman, J.R., "Some Aspects of the Mathematics of Finite Elements," pp. 25–42 in *The Mathematics of Finite Elements and Applications, Vol. II*, ed. J.R. Whiteman, London: Academic Press (1976).

[13] Zienkiewicz, O.C. and Morgan, K., *Finite Elements and Approximation*, Chichester: John Wiley (1983).

Chapter 3

VARIATIONAL METHODS

3.1 Introduction

The Galerkin method given in the previous chapter can be shown to produce element matrix integral definitions that would be identical to those obtained from a variational form, if one exists. Most non-linear problems do not have a variational form, yet the Galerkin method and other weighted residual methods can still be used. Thus, one might ask, "Why consider variational methods?" There are several reasons for using them. One is that if the variational integral form is known, one does not have to derive the corresponding differential equation. Also, most of the important variational statements for problems in engineering and physics have been known for over 200 years. Another important feature of variational methods is that often dual principles exist that allow one to establish both an upper bound estimate and a lower bound estimate for an approximate solution. These can be very helpful in establishing accurate error estimates for adaptive solutions. Thus, the variational methods still deserve serious study — especially the energy methods of solid mechanics.

We have seen that the weighted residual methods provide several approaches for generating approximate (or exact) solutions based on equivalent integral formulations of the original partial differential equations. *Variational Methods*, or the *Calculus of Variations* have given us another widely used set of tools for equivalent integral formulations. The were developed by the famous mathematician Euler in the mid-1700's. Since that time the variational forms of most elliptic partial differential equations have been known. It has been proved that a variational form and the Galerkin method yield the same integral formulations when a governing variational principal exists. Variational methods have thus been used to solve problems in elasticity, heat transfer, electricity, magnetism, ideal fluids, etc. Thus, it is logical to expect that numerical approximations based on these methods should be very fruitful. They have been very widely employed in elasticity and structural mechanics so we begin with that topic. Then we will introduce the finite element techniques as logical extensions of the various classical integral formulations.

3.2 Structural Mechanics

Modern structural analysis relies extensively on the finite element method. Its integral formulation, based on the variational calculus of Euler, is the *Principal of Minimum Total Potential Energy*. (This is also known as the principal of virtual work.)

Basically, it states that the displacement field that satisfies the essential displacement boundary conditions and minimizes the total potential energy is the one that corresponds to the state of static equilibrium. This implies that displacements are our primary unknowns. They will be interpolated in space as will their derivatives, the strains. The total potential energy, Π, is the strain energy, U, of the structure minus the mechanical work, W, done by the applied forces. Recall from introductory mechanics that the mechanical work, W, done by a force is the scalar dot product of the force vector, $\mathbf{F}$, and the displacement vector, $\mathbf{u}$, at its point of application.

To illustrate the concept of energy formulations we will review the equilibrium of the well-known linear spring. Figure 3.2.1 shows a typical spring of stiffness k that has an applied force, F, at the free end. That end undergoes a displacement of Δ. The work done by the single force is

$$W = \vec{\Delta} \cdot \vec{F} = \Delta F. \tag{3.1}$$

Recall that the spring stores potential energy due to its deformation. Here we call that strain energy. That energy is given by

$$U = \frac{1}{2} k \Delta^2. \tag{3.2}$$

Therefore, the total potential energy for the loaded spring is

$$\Pi(\Delta) = U - W = \frac{1}{2} k \Delta^2 - \Delta F. \tag{3.3}$$

The equation of equilibrium is obtained by minimizing Π with respect to the displacement; that is, $\partial \Pi / \partial \Delta = 0$. This simplifies to $k\Delta = F$, which is the well-known equilibrium equation for a linear spring. This example was slightly simplified, since we started with the condition that the left end of the spring had no displacement (an essential boundary condition). Next we will consider a spring where either end can be fixed or

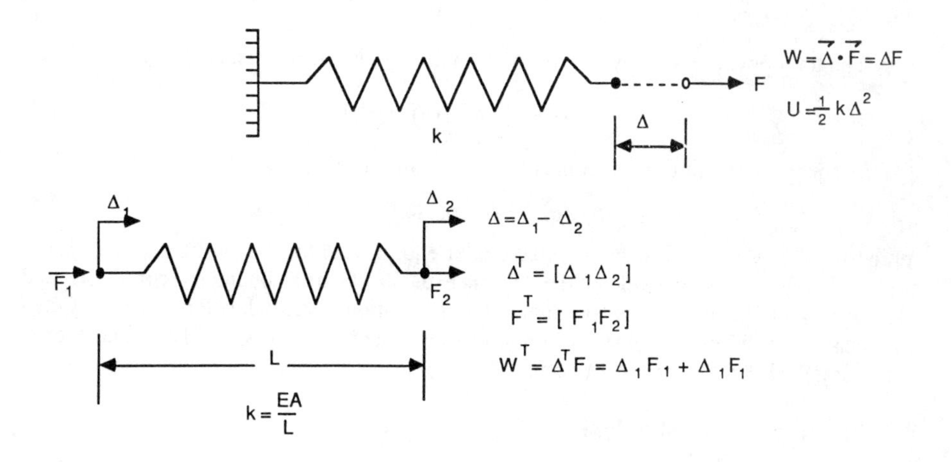

Figure 3.2.1 A simple linear spring

free to move.

The elastic bar is often modeled as a linear spring. In introductory mechanics of materials the axial stiffness of a bar is defined as $k = EA/L$ where it has a length of L, an area A, and an elastic modulus of E. Our spring model of the bar (see Fig. 3.2.1) has two end displacements, Δ_1 and Δ_2, and two associated axial forces, F_1 and F_2. The net deformation of the bar is

$$\Delta = \Delta_2 - \Delta_1 .$$

We denote the total vector of displacements as

$$\mathbf{\Delta}^T = [\, \Delta_1 \;\; \Delta_2 \,]$$

and the associated vector of forces as

$$\mathbf{F}^T = [\, F_1 \;\; F_2 \,] .$$

Then the work done on the bar is

$$W = \mathbf{\Delta}^T \mathbf{F} = \Delta_1 F_1 + \Delta_2 F_2 .$$

The net displacement will be expressed in matrix form to compare with later mathematical formulations. It is

$$\Delta = [\, -1 \quad 1 \,] \, \mathbf{\Delta} .$$

Then the strain energy can be written as

$$U = \frac{1}{2} k \Delta^2 = \frac{EA}{2L} \mathbf{\Delta}^T \begin{Bmatrix} -1 \\ 1 \end{Bmatrix} [\, -1 \quad 1 \,] \, \mathbf{\Delta}^T = \frac{1}{2} \mathbf{\Delta}^T \mathbf{K} \, \mathbf{\Delta}$$

where the bar stiffness is

$$\mathbf{K} = \frac{EA}{L} \begin{bmatrix} 1 & -1 \\ -1 & 1 \end{bmatrix} .$$

The total potential energy, Π, depends on all the displacements, $\mathbf{\Delta}$:

$$\Pi(\mathbf{\Delta}) = \frac{1}{2} \mathbf{\Delta}^T \mathbf{K} \, \mathbf{\Delta} - \mathbf{\Delta}^T \mathbf{F} \tag{3.4}$$

and the equation of equilibrium comes from the minimization

$$\partial \Pi / \partial \mathbf{\Delta} = \mathbf{0}, \text{ or } \mathbf{K} \, \mathbf{\Delta} = \mathbf{F} \tag{3.5}$$

represents the system of algebraic equations of equilibrium for the elastic system. These two equations do not yet reflect the presence of an essential boundary condition, and $|\mathbf{K}| \equiv 0$ and the system is singular. These relations were developed on physical arguments and did not involve any finite element theory. Next we will see that a one-dimensional FEA yields the same forms.

3.3 Finite Element Analysis

Up to this point we have considered equivalent integral forms in the classical sense and not invoked their enhancement by finite element methods. We have seen that the resulting algebraic equation systems based on a global approximate solution are fully coupled. That is, the coefficient matrix is not *sparse* (not highly populated with zeros) so that the solution or inversion cost would be high. Also, some of the methods will always yield an unsymmetric system of algebraic equations, even when the differential equation is self-adjoint. This means that the computer storage requirements will be high.

While we have looked so far mainly at one-dimensional problems in the general case we should be able to see that the residual error will involve volume integrals as well as surface integrals over part of the surface of the volume. For complicated shapes encountered in solving practical problems it is almost impossible to assume a global solution that would satisfy the boundary conditions. Even if we could do that the computational expense would probably prevent us from obtaining a solution. Both of these important practical limitations can be overcome if we utilize a *piecewise approximation* that has only local support in space. That is part of what *finite element analysis* offers us.

The basic concept is that we split the actual solution domain into a series of sub-domains, or *finite elements*, that are interconnected in such a way that we can split the required integrals into a summation of integrals over all the element domains. If we restrict the approximation to a function that exists only within the element domain then the algebraic system becomes sparse because an element only directly interacts with those elements that are connected to it. By restricting the element to a single shape, or to a small library of shapes, we can do the required integrals over that shape and use the results repeatedly to build up the integral contributions over the entire solution domain. The main additional piece of work that results is the requirement that we do some bookkeeping to keep up with the contribution of each element. We refer to this as the equation *assembly*. That topic was illustrated in Fig. 1.3.1, and will be discussed in more detail in Chap. 20. In today's terminology of computer science the assembly procedure and the postprocessing procedures are a series of *gather* and *scatter* operations.

Many finite element problems can be expressed symbolically in terms of a scalar quantity, I, such as

$$I(\mathbf{\Delta}) = \frac{1}{2}\,\mathbf{\Delta}^T\mathbf{K}\,\mathbf{\Delta} + \mathbf{\Delta}^T\mathbf{C} \rightarrow \min \tag{3.6}$$

where $\mathbf{\Delta}$ is a vector containing the unknown nodal parameters associated with the problem, and $\mathbf{K}$ and $\mathbf{C}$ are matrices defined in terms of the element properties and geometry. The above quantity is known as a *quadratic form.* If one uses a variational formulation then the solution of the finite element problem is usually required to satisfy the following system equations: $\partial I / \partial \mathbf{\Delta} = \mathbf{0}$.

In the finite element analysis one assumes that the (scalar) value of I is given by the sum of the element contributions. That is, one assumes

$$I(\mathbf{\Delta}) = \sum_{e=1}^{\mathrm{NE}} I^e(\mathbf{\Delta})$$

where I^e is the contribution of element number 'e', and NE represents the total number of elements in the system. Usually one can (but does not in practice) define I^e in terms of $\mathbf{\Delta}$ such that

$$I^e(\mathbf{\Delta}) = \frac{1}{2}\,\mathbf{\Delta}^T\mathbf{K}^e\mathbf{\Delta} + \mathbf{\Delta}^T\mathbf{C}^e \tag{3.7}$$

where the $\mathbf{K}^e$ are the same size as $\mathbf{K}$, but very sparse. Therefore, Eq. (3.7) can be expressed as

$$I(\Delta) = \frac{1}{2}\Delta^T\left[\sum_{e=1}^{NE}\mathbf{K}^e\right]\Delta + \Delta^T\left[\sum_{e=1}^{NE}\mathbf{C}^e\right] \tag{3.8}$$

and comparing this with Eq. (3.8) one can identify the relations

$$\mathbf{K} = \sum_{e=1}^{NE}\mathbf{K}^e, \quad \mathbf{C} = \sum_{e=1}^{NE}\mathbf{C}^e. \tag{3.9}$$

If NDFREE represents the total number of unknowns in the system, then the size of these matrices are $\text{NDFREE} \times \text{NDFREE}$ and $\text{NDFREE} \times 1$, respectively.

As a result of Eq. (3.9) one often sees the statement, "the system matrices are simply the sum of the corresponding element matrices." This is true, and indeed the symbolic operations depicted in the last equation are simple but one should ask (while preparing for the ensuing handwaving), "in practice, how are the element matrices obtained and how does one carry out the summations?" Before attempting to answer this question, it will be useful to backtrack a little. First, recall that it has been assumed that an element's behavior, and thus its contribution to the problem, depends *only* on those nodal parameters that are associated with the element. In practice, the number of parameters associated with a single element usually lies between a minimum of 2 and a maximum of 96; with the most common range in the past being from three to eight. (However, for hierarchical elements in three-dimensions it is possible for it to be 27!) By way of comparison, NDFREE can easily reach a value of several thousand, or several hundred thousand. Consider an example of a system where NDFREE = 5000. Let this system consist of one-dimensional elements with two parameters per element. A typical matrix $\mathbf{C}_e$ will contain 5000 terms and all but two of these terms will be identically zero since only those two terms of Δ, 5000×1, associated with element 'e' are of any significance to element 'e'. In a similar manner one concludes that, for the present example, only four of the 25,000 terms of $\mathbf{K}^e$ would not be identically zero. Therefore, it becomes obvious that the symbolic procedure introduced here is not numerically efficient and would not be used in practice.

There are some educational uses of the symbolic procedure that justify pursuing it a little further. Recalling that it is assumed that the element behavior depends only on those parameters, says $\boldsymbol{\delta}^e$, that are associated with element 'e', it is logical to assume that

$$I^e = \frac{1}{2}\boldsymbol{\delta}^{e^T}\mathbf{k}^e\boldsymbol{\delta}^e + \boldsymbol{\delta}^{e^T}\mathbf{c}^e. \tag{3.10}$$

If NELFRE represents the number of degrees of freedom associated with the element then $\boldsymbol{\delta}^e$ and $\mathbf{c}^e$ are $\text{NELFRE} \times 1$ in size and the size of $\mathbf{k}^e$ is $\text{NELFRE} \times \text{NELFRE}$. Note that in practice NELFRE is much less than NDFREE. The matrices $\mathbf{k}^e$ and $\mathbf{c}^e$ are commonly known as *the* element matrices. For the one-dimensional element discussed in the previous example, $\mathbf{k}^e$ and $\mathbf{c}^e$ would be 2×2 and 2×1 in size, respectively, and would represent the only coefficients in $\mathbf{K}^e$ and $\mathbf{C}^e$ that are not identically zero.

All that remains is to relate $\mathbf{k}^e$ to $\mathbf{K}^e$ and $\mathbf{c}^e$ to $\mathbf{C}^e$. Obviously Eqs. (3.7) and (3.10) are equal and are the key to the desired relations. In order to utilize these equations, it is necessary to relate the degrees of freedom of the element $\boldsymbol{\delta}^e$, to the degrees of freedom of the total system Δ. This is done symbolically by introducing a $\text{NELFRE} \times \text{NDFREE}$ bookkeeping matrix, $\boldsymbol{\beta}^e$ (to be discussed later), such that the following identity is satisfied:

$$\boldsymbol{\delta}^e \equiv \boldsymbol{\beta}^e\Delta. \tag{3.11}$$

Substituting this identity, Eq. (3.10) becomes

$$I^e(\Delta) = \frac{1}{2}\,\Delta^T \beta^{e^T} \mathbf{k}^e \beta^e \Delta + \Delta^T \beta^{e^T} \mathbf{c}^e .$$

Comparing this relation with Eq. (3.7), one can establish the symbolic relationships

$$\mathbf{K}^e = \beta^{e^T} \mathbf{k}^e \beta^e, \qquad \mathbf{C}^e = \beta^{e^T} \mathbf{c}^e$$

and

$$\mathbf{K} = \sum_{e=1}^{NE} \beta^{e^T} \mathbf{k}^e \beta^e = \underset{e}{\mathbf{A}}\, \mathbf{k}^e, \qquad \mathbf{C} = \sum_{e=1}^{NE} \beta^{e^T} \mathbf{c}^e = \underset{e}{\mathbf{A}}\, \mathbf{c}^e . \tag{3.12}$$

Equation (3.12) can be considered as the symbolic definitions of the *assembly operator* and its procedures relating *the* element matrices, $\mathbf{k}^e$ and $\mathbf{c}^e$, to the total system matrices, $\mathbf{K}$ and $\mathbf{C}$. Note that these relations involve the element connectivity (topology), β^e, as well as the element behavior, $\mathbf{k}^e$ and $\mathbf{c}^e$. Although some programs do use this procedure, it is very inefficient and thus very expensive.

For the sake of completeness, the β^e matrix will be briefly considered. To simplify the discussion, it will be assumed that each nodal point has only a single unknown scalar nodal parameter (degree of freedom). Define a mesh consisting of four triangular elements. Figure 3.3.1 shows both the system and element degree of freedom numbers. The system degrees of freedom are defined as

$$\Delta^T = [\, \Delta_1 \;\; \Delta_2 \;\; \Delta_3 \;\; \Delta_4 \;\; \Delta_5 \;\; \Delta_6 \,]$$

and the degrees of freedom of element '*e*' are

$$\delta^{e^T} = [\, \delta_1 \;\; \delta_2 \;\; \delta_3 \,]^e .$$

The connectivity or topology data supplied for these elements are

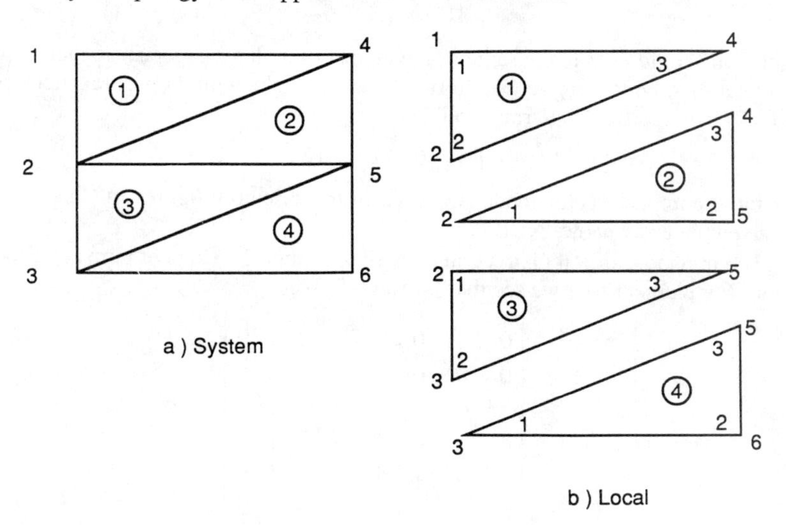

Figure 3.3.1 Relationship between system and element degrees of freedom

	e		
1	1	2	4
2	2	5	4
3	2	3	5
4	3	6	5

Thus, for element number four ($e = 4$), these quantities are related by

$$\boldsymbol{\delta}_4 = \begin{Bmatrix} \delta_1 \\ \delta_2 \\ \delta_3 \end{Bmatrix}_4 = \begin{Bmatrix} \Delta_3 \\ \Delta_6 \\ \Delta_5 \end{Bmatrix}$$

which can be expressed as the gather operation $\boldsymbol{\delta}_4 \equiv \boldsymbol{\beta}_4 \boldsymbol{\Delta}$ where

$$\boldsymbol{\beta}_4 \equiv \begin{bmatrix} 0 & 0 & 1 & 0 & 0 & 0 \\ 0 & 0 & 0 & 0 & 0 & 1 \\ 0 & 0 & 0 & 0 & 1 & 0 \end{bmatrix}.$$

The matrices $\boldsymbol{\beta}_1$, $\boldsymbol{\beta}_2$, and $\boldsymbol{\beta}_3$ can be defined in a similar manner. Since the matrix $\boldsymbol{\beta}^e$ contains only ones and zeros, it is called the element Boolean or binary matrix. Note that there is a single unity term in each row and all other coefficients are zero. Therefore, this NELFRE $\times$ NDFREE array will contain only NELFRE non-zero (unity) terms, and since NELFRE $\ll$ NDFREE, the matrix multiplications of any Boolean gather or scatter operation are numerically very inefficient. There is a common shorthand method for writing any Boolean matrix to save space. It simply is written as a vector with the column number that contains the unity term on any row. In that form we would write $\boldsymbol{\beta}_4$ as

$$\boldsymbol{\beta}_4^T \leftrightarrow [\,3 \quad 6 \quad 5\,]$$

which you should note is the same as the element topology list because there is only one parameter per node. If we had formed the example with two parameters per node (NG = 2) then the Boolean array would be

$$\boldsymbol{\beta}_4^T \leftrightarrow [\,5 \quad 6 \quad 11 \quad 12 \quad 9 \quad 10\,]\,.$$

This more compact vector mode was used in the assembly figures in Chap. 1. There it was given the array name INDEX.

The transpose of a $\boldsymbol{\beta}$ matrix can be used to scatter the element terms into the system vector. For element four we see that $\boldsymbol{\beta}_4^T \mathbf{c}_4 = \mathbf{C}_4^s$ gives

$$\begin{bmatrix} 0 & 0 & 0 \\ 0 & 0 & 0 \\ 1 & 0 & 0 \\ 0 & 0 & 0 \\ 0 & 0 & 1 \\ 0 & 1 & 0 \end{bmatrix} \begin{Bmatrix} c_1 \\ c_2 \\ c_3 \end{Bmatrix}_4 = \begin{Bmatrix} 0 \\ 0 \\ C_1 \\ 0 \\ C_2 \\ C_3 \end{Bmatrix}_4^s.$$

This helps us to see how the scatter and sum operation for $\mathbf{C}$ in Eq. (3.12) actually works. The Boolean arrays, $\boldsymbol{\beta}$, have other properties that are useful in understanding certain other element level operations. For future reference note that $\boldsymbol{\beta}_e \boldsymbol{\beta}_e^T = \mathbf{I}$, and that if elements i and j are not connected $\beta_i \beta_j^T = \mathbf{0} = \beta_j \beta_i^T$, and if they are connected then $\beta_i \beta_j^T = \mathbf{X}_{ij}$ where $\mathbf{X}_{ij}$ indicates the *dof* that are common to both. That is, the Boolean array $\mathbf{X}$ is zero except for those *dof* common to both element i and j.

Although these symbolic relations have certain educational uses their gross inefficiency for practical computer calculations led to the development of the equivalent programming procedure of the "direct method" of assembly that was discussed earlier, and illustrated in Figs. 1.3.1 to 1.3.3. It is useful at times to note that the identity (3.11) leads to the relation

$$\frac{\partial(\ )^e}{\partial \Delta} = \beta^{e^T} \frac{\partial(\ )^e}{\partial \delta^e}, \tag{3.13}$$

where $(\ )^e$ is some quantity associated with element 'e'. At this point we will begin to illustrate finite element domains and their piecewise local polynomial approximations to variational approximations by applying them to an elastic rod.

Before leaving the assembly relations for a while one should consider their extension to the case where there are more than one unknown per node. This is illustrated in Fig. 3.3.2 for three line elements with two nodes each and two dof per node.

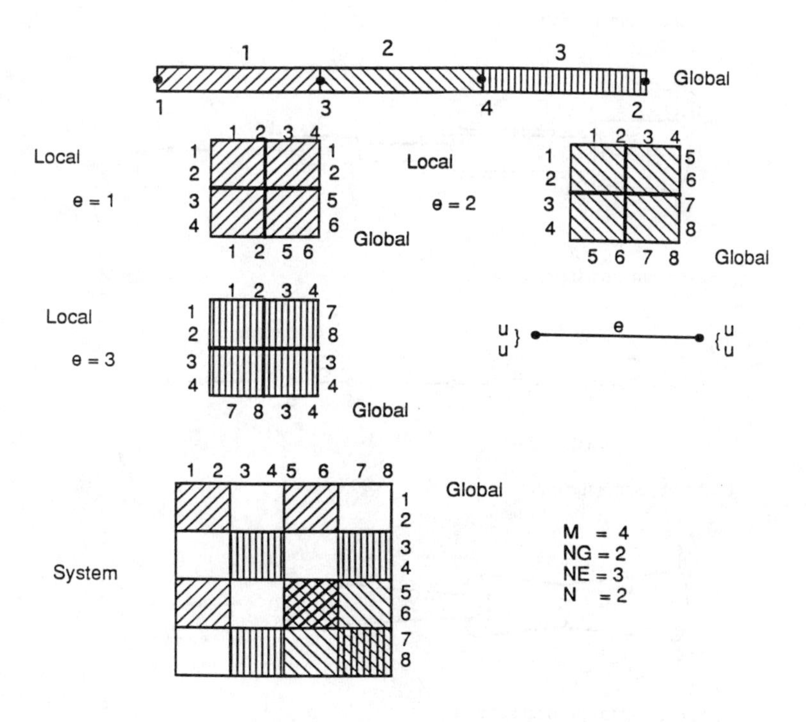

Figure 3.3.2 Graphical assembly for elements with two parameters per node

3.4 Continuous Elastic Bar

Consider an axisymmetric rod shown in Fig. 3.4.1. The cross-sectional area, $A(x)$, the perimeter, $p(x)$, the material modulus of elasticity, $E(x)$, and axial loading conditions would, in general, depend on the axial coordinate, x. The loading conditions could include surface tractions (shear) per unit area, $T(x)$, body forces per unit volume, $X(x)$, and concentrated point loads, P_i at point i. The axial displacement at a point will be denoted by $u(x)$, and its value at point i is u_i. Recall that the work done by a force is the product of the force, the displacement at the point of application of the force, and the cosine of the angle between the force and the displacement. Here the forces are all parallel so the cosine is either plus or minus one. Evaluating the work

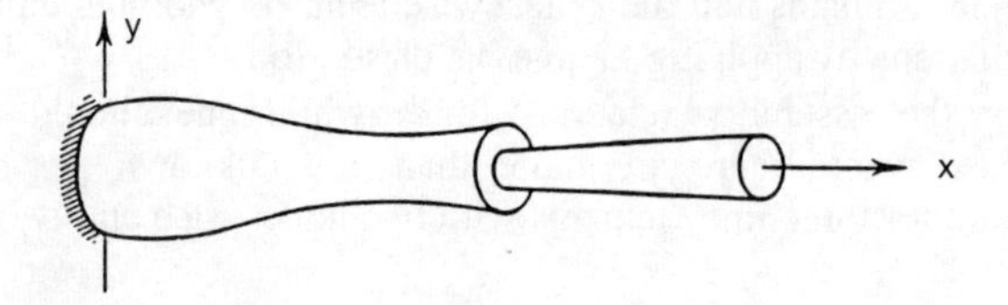

(a) An axisymmetric rod

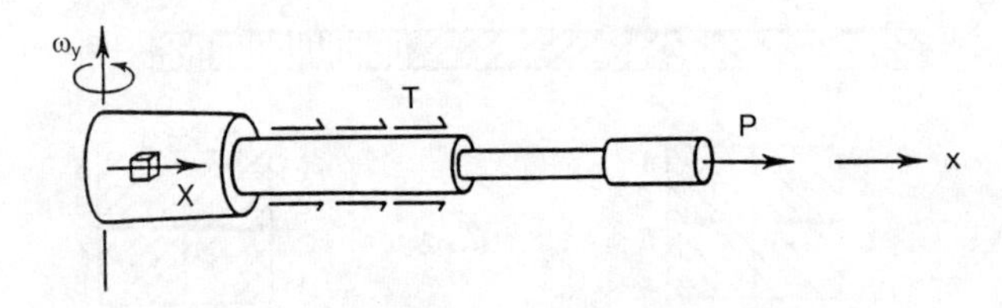

(b) Constant area approximation

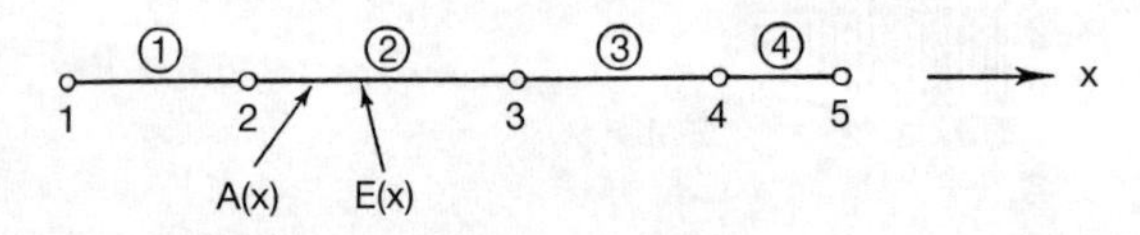

(c) A finite element mesh

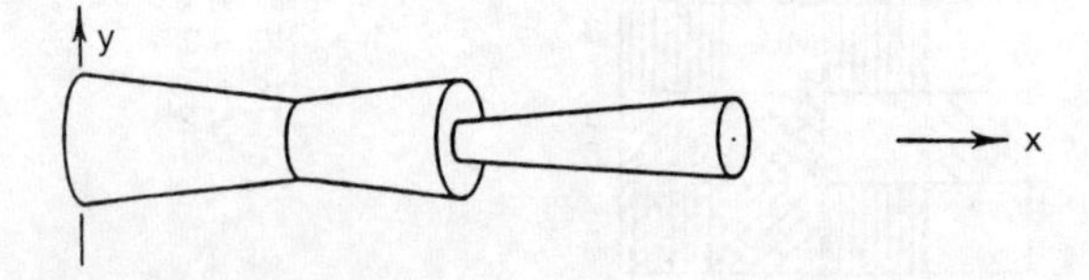

(d) Variable area approximation

Figure 3.4.1 A typical axially loaded bar

$$W = \int_0^L u(x)\, X(x)\, A(x)\, dx + \int_0^L u(x)\, T(x)\, p(x)\, dx + \sum_i u_i\, P_i . \tag{3.14}$$

As mentioned earlier the total potential energy, Π, includes the *strain energy*, U, and work of the externally applied forces, W. That is, $\Pi = U - W$. In a mechanics of materials course it is shown that the strain energy per unit volume is half the product of the stress and strain. The axial strain and stress will be denoted by $\sigma(x)$ and $\varepsilon(x)$, respectively. Thus, the strain energy is

$$U = \frac{1}{2} \int_0^L \sigma(x)\, \varepsilon(x)\, A(x)\, dx . \tag{3.15}$$

The latter two equations have used $dV = A\, dx$ and $dS = p\, dx$ where dS is an exterior surface area. The work is clearly defined in terms of the displacement, u, since the loads would be given quantities. For example, the body force could be gravity, $X = \rho(x)\, g$, or a centrifugal load due to rotation about the y-axis, $X = \rho(x)\, x\omega^2$. Surface tractions are less common in 1-D but it could be due to a very viscous fluid flowing over the outer surface and in the x-direction.

Our goal is to develop a displacement formulation. Thus, we also need to relate both the stress and strain to the displacement, u. To begin with we recall the *strain-displacement relation*

$$\varepsilon(x) = \frac{du(x)}{dx}$$

which relates the strain to the derivative of the displacement. The stress at a point is directly proportional to the strain at the point. Thus, it is also dependent on the *displacement gradient*. The relation between stress and strain is a *constitutive relation* known as *Hooke's law* $\sigma(x) = E(x)\, \varepsilon(x)$. Therefore, we now see that the total potential energy depends on the unknown displacements and displacement gradients. We are searching for the displacement configuration that minimizes the total potential energy since that configuration corresponds to the state of stable equilibrium. As we have suggested above, a finite element model can be introduced to approximate the displacements and their derivatives. Here, we begin by selecting the simplest model possible. That is, the two node line element with assumed linear displacement variation with x. As suggested, we now assume that all the integrals can be represented as the sum of the corresponding element integrals (and the intersection of those elements with the boundary of the solution domain) :

$$\Pi = \sum_{e=1}^{\mathrm{NE}} \Pi^e + \sum_{b=1}^{\mathrm{NB}} \Pi^b + \sum_i P_i u_i \tag{3.16}$$

where Π^e is the typical element domain contribution, and Π^b is the contribution from a typical boundary segment domain. This one-dimensional example is somewhat misleading since in general a surface traction, T, acts only on a small portion of the exterior boundary. Thus, the number of boundary domains, NB, is usually much less than the number of elements, NE. Here, however, NB = NE and the distinction between the two may not become clear until two-dimensional problems are considered.

Substituting Eqs. (3.14) and (3.15) into Eq. (3.3) and equating to Eq. (3.16) yields

$$\Pi^e = \tfrac{1}{2} \int_{L^e} \sigma^e \varepsilon^e A^e\, dx - \int_{L^e} u^e X^e A^e\, dx, \qquad \Pi^b = -\int_{L^b} u^b T^b p^b\, dx$$

where $u^e(x)$ and $u^b(x)$ denote the approximated displacements in the element and on the boundary surface, respectively. In this special example, $u^b = u^e$ but that is not usually true. Symbolically we interpolate such that

$$u^e(x) = \mathbf{H}^e(x)\,\mathbf{u}^e = \mathbf{u}^{e^T}\,\mathbf{H}^{e^T}(x)$$

and likewise if we degenerate this to a portion (sub-set) of the boundary of the element

$$u^b(x) = \mathbf{H}^b(x)\,\mathbf{u}^b = \mathbf{u}^{b^T}\,\mathbf{H}^{b^T}(x).$$

In the example we have the unusual case that $\mathbf{u}^b = \mathbf{u}^e$ and $\mathbf{H}^b = \mathbf{H}^e$. Generally $\mathbf{u}^b$ is a subset of $\mathbf{u}^e$ (i.e., $\mathbf{u}^b \subset \mathbf{u}^e$) and $\mathbf{H}^b$ is a subset of $\mathbf{H}^e$. Substituting these interpolation relationships gives the strain approximation in an element as:

$$\varepsilon^e(x) = \frac{du^e}{dx} = \frac{d\mathbf{H}^e(x)}{dx}\,\mathbf{u}^e = \mathbf{u}^{e^T}\,\frac{d\mathbf{H}^{e^T}(x)}{dx}$$

or in more common notation

$$\varepsilon^e = \mathbf{B}^e(x)\mathbf{u}^e \tag{3.17}$$

$$\sigma^e(x) = E^e(x)\,\varepsilon^e(x) \tag{3.18}$$

so that

$$\Pi^e = \frac{1}{2}\,\mathbf{u}^{e^T}\int_{L^e}\mathbf{B}^{e^T}E^e\,\mathbf{B}^e A^e\,\mathbf{u}^e\,dx - \mathbf{u}^{e^T}\int_{L^e}\mathbf{H}^{e^T}X^e A^e\,dx$$

$$\Pi^b = -\,\mathbf{u}^{b^T}\int_{L^b}\mathbf{H}^{b^T}T^b p^b\,dx.$$

The latter two relations can be written symbolically as

$$\Pi^e = \frac{1}{2}\,\mathbf{u}^{e^T}\,\mathbf{S}^e\,\mathbf{u}^e - \mathbf{u}^{e^T}\,\mathbf{C}_x^e, \qquad \Pi^b = -\,\mathbf{u}^{b^T}\,\mathbf{C}_T^b \tag{3.19}$$

where the element stiffness matrix is

$$\mathbf{S}^e = \int_{L^e}\mathbf{B}^{e^T}E^e\,\mathbf{B}^e A^e\,dx, \tag{3.20}$$

the element body force vector is

$$\mathbf{C}_x^e = \int_{L^e}\mathbf{H}^{e^T}X^e A^e\,dx, \tag{3.21}$$

and the boundary segment traction vector is

$$\mathbf{C}_T^b = \int_{L^e}\mathbf{H}^{b^T}T^b p^b\,dx. \tag{3.22}$$

The total potential energy of the system is

$$\Pi = \frac{1}{2}\sum_e \mathbf{u}^{e^T}\,\mathbf{S}^e\,\mathbf{u}^e - \sum_e \mathbf{u}^{e^T}\,\mathbf{C}_x^e - \sum_b \mathbf{u}^{b^T}\mathbf{C}_T^b - \mathbf{u}^T P \tag{3.23}$$

where $\mathbf{u}$ is the vector of all of the unknown nodal displacements. Here we have assumed that the external point loads are applied at node points only. The last term represents the scalar, or dot, product of the nodal displacement and nodal forces. That is,

$$\mathbf{u}^T\mathbf{P} = \mathbf{P}^T\mathbf{u} = \sum_i u_i P_i.$$

Of course, in practice most of the P_i are zero. By again applying the direct assembly procedure, or from the Boolean assembly operations of Sec. 20.3 the total potential energy is

$$\Pi(\mathbf{u}) = \frac{1}{2}\mathbf{u}^T\mathbf{S}\mathbf{u} - \mathbf{u}^T\mathbf{C}$$

and minimizing with respect to all the unknown displacements, **u**, gives the algebraic equilibrium equations for the entire structure $\mathbf{S}\mathbf{u} = \mathbf{C}$. Therefore, we see that our variational principle has lead to a very general and powerful formulation for this class of structures. It automatically includes features such as variable material properties, variable loads, etc. These were difficult to treat when relying solely on physical intuition. Although we will utilize the simple linear element none of our equations are restricted to that definition of **H** and the above symbolic formulation is valid for any linear elastic solid of any shape.

If we substitute $\mathbf{H}^e$ for the linear element (see Sec. 4.2) and assume constant properties, then the element and boundary matrices are simple to integrate. The results are

$$\mathbf{S}^e = \frac{E^e A^E}{L^e}\begin{bmatrix} 1 & -1 \\ -1 & 1 \end{bmatrix}$$

$$\mathbf{C}_x^e = \frac{X^e A^e L^e}{2}\begin{Bmatrix} 1 \\ 1 \end{Bmatrix}, \qquad \mathbf{C}_T^b = \frac{T^e P^e L^e}{2}\begin{Bmatrix} 1 \\ 1 \end{Bmatrix}.$$

Also in this case one obtains the strain-displacement relation

$$\varepsilon^e = \frac{1}{L^e}\,[-1 \quad 1]\begin{Bmatrix} u_1^e \\ u_2^e \end{Bmatrix} \tag{3.24}$$

which means that the strain is constant in the element but the displacement approximation is linear. It is common to refer to this element as the constant strain line element, CSL. The above stiffness matrix is the same as that obtained in Sec. 3.1. The load vectors take the resultant element, or boundary, force and place half at each node. That logical result does not carry over to more complicated load conditions, or interpolation functions and it then becomes necessary to rely on the mathematics of Eqs. (3.21) and (3.22).

As an example of a slightly more difficult loading condition consider a case where the body force varies linearly with x. This could include the case of centrifugal loading mentioned earlier. For simplicity assume a constant area A and let us define the value of the body force at each node of the element. To define the body force at any point in the element we again utilize the interpolation function and set

$$X^e(x) = \mathbf{H}^e(x)\,\mathbf{X}^e \tag{3.25}$$

where $\mathbf{X}^e$ are the defined nodal values of the body force. For these assumptions the body force vector becomes

$$\mathbf{C}_x^e = A^e \int_{L^e} \mathbf{H}^{e^T}\mathbf{H}^e\,dx\,\mathbf{X}^e .$$

For the linear element the integration reduces to (see Sec. 5.3)

$$\mathbf{C}_x^e = \frac{A^e L^e}{6}\begin{bmatrix} 2 & 1 \\ 1 & 2 \end{bmatrix}\begin{Bmatrix} X_1^e \\ X_2^e \end{Bmatrix}.$$

This agrees with our previous result for constant loads since if $X_1^e = X_2^e = X^e$, then

$$\mathbf{C}_x^e = \frac{A^e L^e X^e}{6}\begin{Bmatrix} 2+1 \\ 1+2 \end{Bmatrix} = \frac{A^e L^e X^e}{2}\begin{Bmatrix} 1 \\ 1 \end{Bmatrix}.$$

A more common problem is that illustrated in Fig. 3.4.1 where the area of the member varies along the length. To approximate that case, with constant properties, one could interpolate for the area at any point as $A^e(x) = \mathbf{H}^e(x)\,\mathbf{A}^e$, then the stiffness in Eq. (3.20) becomes

$$\mathbf{S}^e = \frac{E^e}{(L^e)^2}\begin{bmatrix} 1 & -1 \\ -1 & 1 \end{bmatrix} V^e$$

where

$$V^e = \int_{L^e} A^e\,dx = \int_{L^e} \mathbf{H}^e(x)\,dx\,\mathbf{A}^e = \frac{L^e}{2}[1 \quad 1]\begin{Bmatrix} A_1^e \\ A_2^e \end{Bmatrix} = \frac{L^e(A_1^e + A_2^e)}{2}$$

is the average volume of the element. In a similar manner the body force vector would be

$$\mathbf{C}_x^e = \frac{X^e}{2}\begin{Bmatrix} 1 \\ 1 \end{Bmatrix} V^e = \frac{X^e L^e(A_1^e + A_2^e)}{4}\begin{Bmatrix} 1 \\ 1 \end{Bmatrix}.$$

The above approximations should be reasonably accurate. However, we recall that the area is related to the radius by $A = \pi r^2$. Thus, it would be slightly more accurate to describe the radius at each end and interpolate $r^e(x) = \mathbf{H}^e(x)\,\mathbf{r}^e$ so that

$$V^e = \int_{L^e} A^e\,dx = \pi\int_{L^e} r^e(x)^2\,dx = \mathbf{r}^{e^T}\pi\int_{L^e} \mathbf{H}^{e^T}\mathbf{H}^e\,\mathbf{dx}\,\mathbf{r}^e$$

$$= \mathbf{r}^{e^T}\frac{\pi L^e}{6}\begin{bmatrix} 2 & 1 \\ 1 & 2 \end{bmatrix}\mathbf{r}^e = \pi L^e(r_1^2 + r_1 r_2 + r_2^2)/3\,.$$

Clearly, as one utilizes more advanced interpolation functions, the integrals involved in Eqs. (3.20) to (3.22) become more difficult to evaluate. An example to illustrate the use of these element matrices and to introduce the benefits of post-solution calculations follows. Consider a prismatic bar of steel rigidly fixed to a bar of brass and subjected to a vertical load of $P = 10{,}000$ lbs., as shown in Fig. 3.4.2. The structure is supported at the top point and is also subjected to a gravity (body force) load. We wish to determine the deflections, reactions, and stresses for the properties :

Element	L^e	A^e	E^e	X^e
1	420"	10 sq. in.	30×10^6 psi	0.283 lb/in^3
2	240"	8 sq. in.	13×10^6 psi	0.300 lb/in^3

The first element has a stiffness constant of $EA/L = 0.7143 \times 10^6$ lb/in. and the body

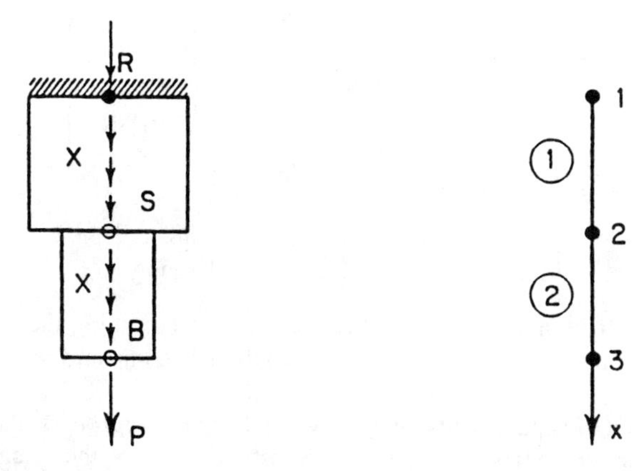

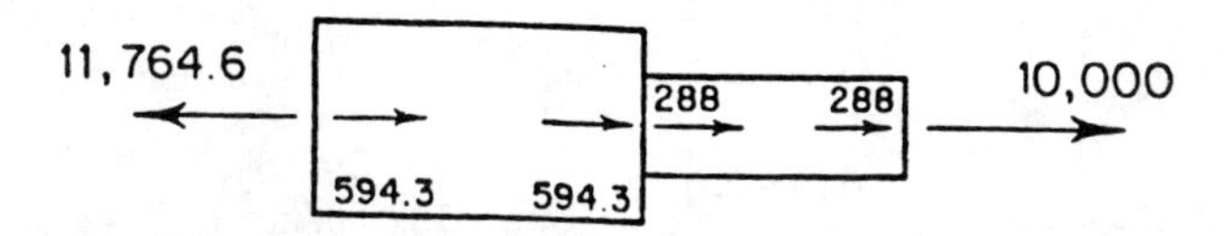

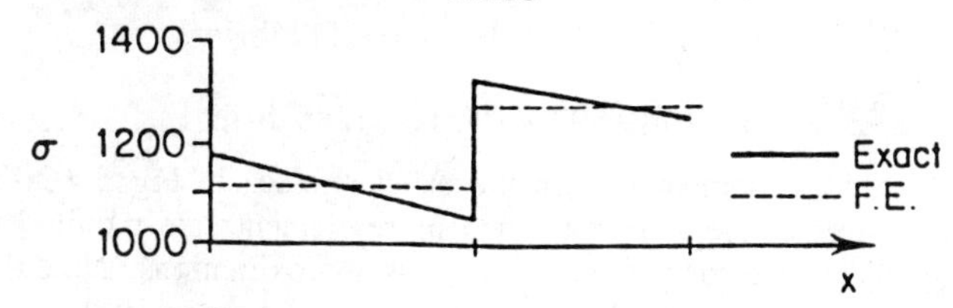

Figure 3.4.2 An axially loaded system

force is $XAL = 1188.6$ lbs. while for the second the corresponding terms are 0.4333×10^6 lb/in. and 576 lbs., respectively. The system nodal force vector is $P^T = [R \;\; 0 \;\; 10{,}000\,]^{\text{lb}}$. Where R is the unknown reaction at node 1. Assemblying the equations gives

$$10^5 \begin{bmatrix} 7.143 & -7.143 & 0 \\ -7.143 & 7.143 + 4.333 & -4.333 \\ 0 & -4.333 & 4.333 \end{bmatrix} \begin{Bmatrix} u_1 \\ u_2 \\ u_3 \end{Bmatrix}$$

$$= \begin{Bmatrix} R \\ 0 \\ 10{,}000 \end{Bmatrix} + \frac{1}{2} \begin{Bmatrix} 1188.6 \\ 1188.6 + 576. \\ 576. \end{Bmatrix} = \begin{Bmatrix} R + 594.3 \\ 882.3 \\ 10288 \end{Bmatrix}.$$

Applying the essential condition that $u_1 = 0$

$$10^5 \begin{bmatrix} 11.476 & -4.333 \\ -4.333 & 4.333 \end{bmatrix} \begin{Bmatrix} u_2 \\ u_3 \end{Bmatrix} = \begin{Bmatrix} 882.3 \\ 10288 \end{Bmatrix}$$

so that $u_2 = 1.5638 \times 10^{-2}$ in., $u_3 = 3.9381 \times 10^{-2}$ in., and determining the reaction from the first system equation: $R = -11764.6$ lb. This reaction is compared with the applied loads in Fig. 3.4.2c.

Now that all the displacements are known we can *postprocess* the results to determine the other quantities of interest. Substituting into the element strain-displacement relation, Eq. (3.24) gives

$$\varepsilon^1 = \frac{1}{420} [-1 \quad 1] \begin{Bmatrix} 0.0 \\ 0.01564 \end{Bmatrix} = 3.724 \times 10^{-5} \text{ in/in}$$

$$\varepsilon^2 = \frac{1}{240} [-1 \quad 1] \begin{Bmatrix} 0.01564 \\ 0.03938 \end{Bmatrix} = 9.892 \times 10^{-5} \text{ in/in}$$

and from Eq. (3.18) the element stresses are

$$\sigma^1 = E^1 \varepsilon^1 = 30 \times 10^6 (3.724 \times 10^{-5}) = 1117 \text{ lb/in}^2$$

$$\sigma^2 = E^2 \varepsilon^2 = 13 \times 10^6 (9.892 \times 10^{-5}) = 1286 \text{ lb/in}^2 .$$

These approximate stresses are compared with the exact stresses in Fig. 3.4.2d. This suggests that if accurate stresses are important then more elements are required to get good estimates from the piecewise constant element stress approximations. Note that the element stresses are exact if they are considered to act only at the element center.

3.5 Flux Recovery for an Element

Regardless of whether we use variational methods or weighted residual methods we are often interested in postprocessing to get the *flux recovery* data for some or all of the elements in the system. Once the assembled system has been solved for the primary nodal unknowns $\boldsymbol{\delta}$, we are often interested in also computing the nodal forces (or fluxes) that act on each individual element. Recall that for structural equilibrium, or thermal equilibrium, or a general Galerkin statement the algebraic equations are of the form

$$\mathbf{S}\boldsymbol{\Delta} = \mathbf{C} + \mathbf{P} \tag{3.26}$$

where the square matrix $\mathbf{S}$ is the assembly of the element square matrices $\mathbf{S}^e$ and the column matrix $\mathbf{C}$ is the sum of the consistent element force (or flux) matrices, $\mathbf{C}^e$, due to spatially distributed forces (or fluxes). Finally, $\mathbf{P}$ is the vector of externally applied concentrated point forces (or fluxes). The vector $\mathbf{P}$ can also be thought of as an assembly of point sources on the elements, $\mathbf{P}^e$. This is always done if one is employing an element wavefront equation solving system. Most of the $\mathbf{P}^e$ are identically zero. When P_j is applied to the j-th node of the system we simply find the element, e, where that node makes its first appearance in the data. Then, P_j is inserted in $\mathbf{P}^e$ for that element and no entries are made in any other elements. If degree of freedom δ_j is given then recall that P_j is an unknown reaction. To recover the concentrated "external" nodal forces or fluxes associated with a specific element we make the *assumption* that a similar expression wholes for the element. That is,

$$\mathbf{S}^e \mathbf{\Delta}^e = \mathbf{C}^e + \mathbf{P}^e \tag{3.27}$$

This is clearly exact if the system has only one element. Otherwise, it is a reasonable approximation. When we use an energy method to require equilibrium of an assembled system, we do not exactly enforce equilibrium in every element that makes up that system. Solving the reasonable approximation gives

$$\mathbf{P}^e = \mathbf{S}^e \mathbf{\Delta}^e - \mathbf{C}^e \tag{3.28}$$

where everything on the right hand side is known, since $\mathbf{\Delta}^e \subset \mathbf{\Delta}$ can be recovered as a gather operation. To illustrate these calculations consider the one-dimensional stepped bar system given above in Fig. 3.4.2 where

$$\mathbf{S} = \frac{E^e A^e}{L^e}\begin{bmatrix} 1 & -1 \\ -1 & 1 \end{bmatrix}, \qquad \mathbf{C}^e = \frac{\mathbf{X}^e A^e L^e}{2}\begin{Bmatrix} 1 \\ 1 \end{Bmatrix}.$$

Now that all of the $\mathbf{\Delta}$ are known the $\mathbf{\Delta}^e$ can be extracted and substituted into Eq. (3.28) for each of the elements. For the first element Eq. (3.28) gives

$$\begin{aligned} \mathbf{P}^e &= 7.143 \times 10^5 \begin{bmatrix} 1 & -1 \\ -1 & 1 \end{bmatrix} \begin{Bmatrix} 0 \\ 1.5638 \times 10^{-2} \end{Bmatrix} - \begin{Bmatrix} 594.3 \\ 594.3 \end{Bmatrix} \\ &= \begin{Bmatrix} -11{,}170.3 \\ 11{,}170.3 \end{Bmatrix} - \begin{Bmatrix} 594.3 \\ 594.3 \end{Bmatrix} = \begin{Bmatrix} -11{,}764.6 \\ 10{,}576.0 \end{Bmatrix} \text{lb.} \end{aligned} \tag{3.29}$$

Likewise for the second element

$$\begin{aligned} \mathbf{P}^e &= 4.333 \times 10^5 \begin{bmatrix} 1 & -1 \\ -1 & 1 \end{bmatrix} \begin{Bmatrix} 1.5638 \times 10^{-2} \\ 3.9381 \times 10^{-2} \end{Bmatrix} - \begin{Bmatrix} 288.0 \\ 288.0 \end{Bmatrix} \\ &= \begin{Bmatrix} -10{,}576.0 \\ 10{,}000.0 \end{Bmatrix} \text{lb.} \end{aligned} \tag{3.30}$$

Note that if we choose to assemble these element $\mathbf{P}^e$ we obtain the system $\mathbf{P}$. That is because element contributions at all unloaded nodes are equal and opposite (Newton's Third Law) and cancel when assembled. Figure 3.5.1. shows these "external" element forces when viewed on each element, as well as their assembly which matches the original system.

The reader is warned to remember that these calculations in Eqs. (3.28) have been carried out in the global coordinate systems. In more advanced structural applications it is often desirable to transform the $\mathbf{P}^e$ back to element local coordinate system. For example, with a general truss member we are more interested in the force along the line of action of the bar rather than its x and y components. Sometimes we list both results and the user selects which is most useful. Recall that the necessary transformation is of the form

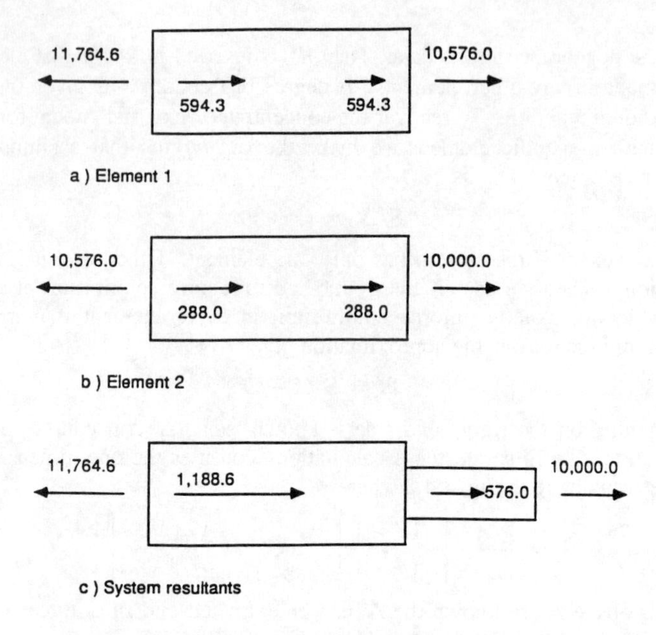

Figure 3.5.1 Element equilibrium reaction recovery

$$\mathbf{P}_L^e = \mathbf{R}^e \, \mathbf{P}_g^e \tag{3.31}$$

where the square rotation matrix, **R**, contains the direction cosines between the global axis, g, and the element local axis, L (see Sec. 7.3).

3.6 Thermal Loads on a Bar *

Before leaving the bar element it may be useful to note that another common loading condition can be included, that is the loading due to an initial thermal strain. Recall that the thermal strain, ε_t, due to a temperature rise of Δt is $\varepsilon_t = \alpha \, \Delta t$ where α is the coefficient of thermal expansion. The work term in Eq. (3.14) is extended to include this effect by adding an *initial strain* contribution

$$W_t = \int_0^L \sigma \, \varepsilon_t \, A(x) \, dx \, .$$

This defines an element thermal force vector

$$\mathbf{C}_t^e = \int_{L^e} \mathbf{B}^{e^T}(x) \, E^e(x) \, \alpha^e(x) \, \Delta t(x) \, A^e(x) \, dx$$

or for constant properties and uniform temperature rise

$$\mathbf{C}_t^e = E^e \, \alpha^e \, \Delta t^e \, A^e \begin{Bmatrix} -1 \\ +1 \end{Bmatrix} . \tag{3.32}$$

There is a corresponding change in the constitutive law such that $\boldsymbol{\sigma} = \mathbf{E}\,(\varepsilon - \varepsilon_t)$. As a numerical example of this loading consider the previous model with the statically

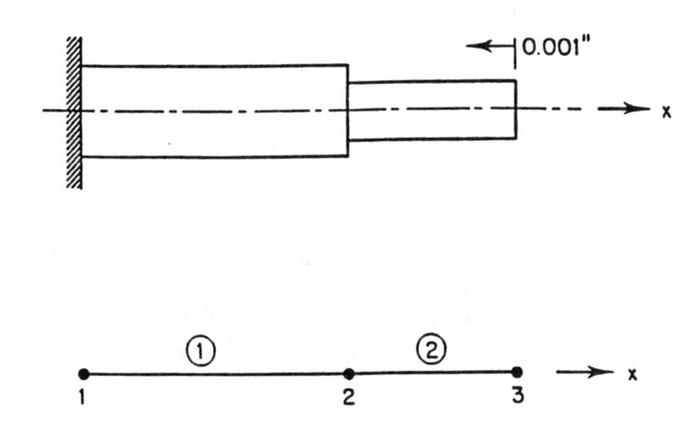

Figure 3.6.1 A thermally loaded structure

indeterminate supports in Fig. 3.6.1. The left support is fixed but the right support is displaced to the left 0.001 in. and the system is cooled by 35° F. Find the stress in each member if $\alpha_1 = 6.7 \times 10^{-6}$ and $\alpha_2 = 12.5 \times 10^{-6}$ in/in F. The assembled equations are

$$10^5 \begin{bmatrix} 7.143 & -7.143 & 0 \\ -7.143 & 11.476 & -4.333 \\ 0 & -4.333 & 4.333 \end{bmatrix} \begin{Bmatrix} u_1 \\ u_2 \\ u_3 \end{Bmatrix} = 10^4 \begin{Bmatrix} -7.035 \\ 7.035 - 4.550 \\ +4.550 \end{Bmatrix} + \begin{Bmatrix} R_1 \\ 0 \\ R_2 \end{Bmatrix}$$

applying the boundary conditions that $u_1 = 0$, and $u_3 = -0.001$ in. and solving for u_2 yields $u_2 = -0.02203$ in. The reactions at points 1 and 3 are $R_1 = -54613$ lb., and $R_3 = +54613$ lb. Substituting into the element strain-displacement matrices yields element mechanical strains of -5.245×10^{-5} in/in., and $+8.763 \times 10^{-5}$ in/in., respectively. But, the initial thermal strains were -2.345×10^{-4}, and -4.375×10^{-4}, respectively so together they result in net tensile stresses in the elements of 5.46×10^3 and 6.83×10^3 psi, respectively. That is, the tension due to the temperature reduction exceeds the compression due to the support movement. The coding, data, and output for the last two examples are on the text disk.

3.7 Heat Transfer in a Rod

A problem closely related to the previous problem is that of steady state heat transfer. Consider the heat transfer in a slender rod that has a specified temperature, t_0, at $x = 0$ and is insulated at the other end, $x = L$. The rod has cross-sectional area, A, with a thermal conductivity of K. Thus, the rod conducts heat along its length. The rod is also surrounded by a convecting medium with a uniform temperature of t_∞. Thus, the rod also convects heat on its outer surface area. Let the convective transfer coefficient be h and the outer perimeter of the rod be P. To simplify the model we will assume that the external reference temperature is $t_\infty = 0$. The governing differential equation for the temperature, $t(x)$, is given by Myers [12] as

$$KA \frac{d^2t}{dx^2} - hPt = 0, \quad 0 < x < L \tag{3.33}$$

with the essential condition $t(0) = t_0$ and the natural boundary condition $dt/dx\,(L) = 0$. The exact solution can be shown to $t(x) = \text{Cos } h[\,m(L-x)\,]/\text{Cos } h(mL)$ where $m^2 = hP/KA$. This problem can be identified as the Euler condition of a variational principle. This principle will lead to system equations that are structured differently from our previous example with the bar. In that case, the boundary integral contributions (tractions) defined a column matrix and thus went on the right hand side of the system equations. Here we will see that the boundary contributions (convection) will also define a square matrix. Thus, they will go into the system coefficients on the left hand side of the system equations.

Generally a variational formulation of steady state head transfer involves volume integrals containing conduction terms and surface integrals with boundary heat flux, e.g., convection, terms. In our one-dimensional example both the volume and surface definitions involve an integral along the length of the rod. Thus, the distinction between volume and surface terms is less clear and the governing functional given by Myers is simply stated as a line integral. Specifically, one must render stationary the functional

$$I(t) = \tfrac{1}{2} \int_0^L [KA\,(dt/dx)^2 + hPt^2]\,dx \tag{3.34}$$

subject to the essential condition at $x = 0$. Divide the rod into a number of nodes and elements and introduce a finite element model where we assume

$$I = \sum_{e=1}^{NE} I^e + \sum_{b=1}^{NB} I^b$$

where we have defined a typical element volume contribution of

$$I^e = \tfrac{1}{2} \int_{L^e} K^e A^e\,(dt^e/dx)^2\,dx \tag{3.35}$$

and typical boundary contribution is

$$I^b = \tfrac{1}{2} \int_{L^b} h^b\,P^b\,t^{b^2}\,dx\,. \tag{3.36}$$

Using our interpolation relations as before $t^e(x) = \mathbf{H}^e(x)\,\mathbf{t}^e = \mathbf{t}^{e^T}\,\mathbf{H}^{e^T}$, and again in this special case $t^b(x) = t^e(x)$. Thus, these can be written symbolically as

$$I^e = \tfrac{1}{2}\,\mathbf{t}^{e^T}\,\mathbf{S}^e\,\mathbf{t}^e, \qquad I^b = \tfrac{1}{2}\,\mathbf{t}^{b^T}\,\mathbf{S}^b\,\mathbf{t}^b\,.$$

Assuming constant properties and the linear interpolation in Eq. (3.24) the square matrices reduce to

$$\mathbf{S}^e = \frac{K^e A^e}{L^e}\begin{bmatrix} 1 & -1 \\ -1 & 1 \end{bmatrix}, \qquad \mathbf{S}^b = \frac{h^b P^b L^b}{6}\begin{bmatrix} 2 & 1 \\ 1 & 2 \end{bmatrix}.$$

Note that there is no column matrix defined so the equations will be homogeneous, that is, the assembled system equations are $\mathbf{ST} = \mathbf{0}$ where $\mathbf{S}$ is the direct assembly of $\mathbf{S}^e$ and $\mathbf{S}^b$. Another aspect of interest here is how to *postprocess* the results so as to determine the convective heat loss. Recall from heat transfer that at any point the convection heat loss is $dq = hP(t - t_\infty)\,dx$ where $(t - t_\infty)$ is the surface temperature difference. If we again assume that the heat loss on a typical boundary segment is to a surrounding fluid at a temperature of $t_\infty = 0$ then this simplifies to

$$Q^b = \int_{L^b} dq = h^b P^b \int_{L^b} t^b(x)\,dx = h^b P^b \int_{L^b} \mathbf{H}^b(x)\,dx\,\mathbf{t}^b \tag{3.37}$$

or simply

$$Q^b = \tfrac{1}{2}\, h^b P^b L^b [\,1 \quad 1\,]\,\mathbf{t}^b = \mathbf{G}^b\,\mathbf{t}^b. \tag{3.38}$$

Thus, if the constant array $\mathbf{G}^b$ is computed and stored for each segment then once the temperatures are computed $\mathbf{t}^b$ can be recovered along with $\mathbf{G}^b$ to compute the loss Q^b. Summing on the total boundary gives the total heat loss. That value would, of course, equal the heat entering at the end $x = 0$. As a numerical example let $L = 4$ ft., $A = 0.01389$ ft^2, $h = 2$ BTU/hr–ft^{2}F, $K = 120$ BTU/hr–ft F, $P = 0.5$ ft., and $t_0 = 10$ F. The mesh selected for this analysis are shown in Fig. 3.7.1 along with the results of the finite element analysis given with full programming details in Sec. 21.2.

3.8 Gradient Estimates

In our finite element calculations we often have a need for accurate estimates of the derivatives of the primary variable. For example, in plane stress or plane strain analysis, the primary unknowns which we compute are the displacement components of the nodes. However, we often are equally concerned about the strains and stresses which are computed from the derivatives of the displacements. Likewise, when we model an ideal fluid with a velocity potential, we actually have little or no interest in the computed potential; but we are very interested in the velocity components which are the derivatives of the potential. A logical question at this point is: what location in the element will give me the most accurate estimate of derivatives? Such points are called *optimal points* or *Barlow points* [3]. A heuristic argument for determining their location can be easily presented. Let us begin by recalling some of our previous observations. In Sec.s 2.6.2 and 3.4, we found that our finite element solution example was an *interpolate* solution, that is, it was exact at the node points and approximate elsewhere. Such accuracy is rare but, in general, one finds that the computed values of the primary variable are most accurate at the node points. Thus, for the sake of simplicity we will assume that the element's nodal values are exact.

Recall that we have taken our finite element approximation to be a polynomial of some specific order, say m. If the exact solution is also a polynomial of order m, then our finite element solution will be exact everywhere in the element. In addition, the finite element derivative estimates will also be exact. It is rare to have such good luck. In general, we must expect our results to only be approximate. However, we can hope for the next best thing to an exact solution. That would be where the exact solution is a polynomial that is one order higher, say $n = m + 1$, than our finite element polynomial. Let the subscripts E and F denote the exact and finite element solutions, respectively. Consider a one-dimensional formulation in natural coordinates, $-1 < a < +1$. Then the exact solution could be written as

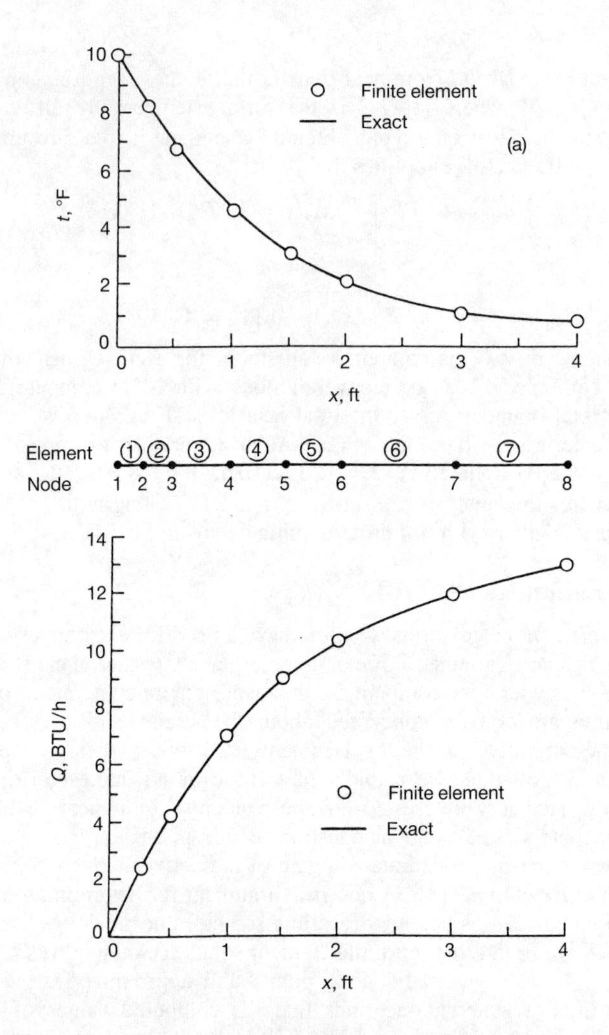

Figure 3.7.1 Temperatures and convection losses in a slender rod

$$U_E(a) = \mathbf{P}_E(a)\,\mathbf{V}_E = \begin{bmatrix} 1 & a & a^2 \cdots a^n \end{bmatrix} \begin{Bmatrix} V_1 \\ V_2 \\ \vdots \\ V_n \end{Bmatrix}_E$$

and our approximate finite element polynominal solution would be

$$U_F(a) = \mathbf{P}_F(a)\,\mathbf{V}_F = \begin{bmatrix} 1 & a & a^2 \cdots a^m \end{bmatrix} \begin{Bmatrix} V_1 \\ V_2 \\ \vdots \\ V_m \end{Bmatrix}_F$$

where $n = (m+1)$, as assumed above. In the above $\mathbf{V}_E$ and $\mathbf{V}_F$ represent different vectors of unknown constants. In the domain of a typical element, these two forms should be almost equal. If we assume that they are equal at the nodes, then we can equate $u_E(a_j) = u_F(a_j)$ where a_j is the local coordinate of node j. Then the following identities are obtained: $\mathbf{P}_F(a_j)\,\mathbf{V}_F = \mathbf{P}_E(a_j)\,\mathbf{V}_E$ or in expanded form

$$\begin{bmatrix} \mathbf{P}_F(a_1) \\ \mathbf{P}_F(a_2) \\ \vdots \\ \mathbf{P}_F(a_m) \end{bmatrix} \mathbf{V}_F = \begin{bmatrix} \mathbf{P}_E(a_1) \\ \mathbf{P}_E(a_2) \\ \vdots \\ \mathbf{P}_E(a_m) \end{bmatrix} \mathbf{V}_E,$$

or symbolically

$$\mathbf{A}_F\,\mathbf{V}_F = \mathbf{A}_E\,\mathbf{V}_E \tag{3.39}$$

where $\mathbf{A}_E$ has one more column than the square matrix $\mathbf{A}_F$. Indeed, upon closer inspection we should observe that $\mathbf{A}_E$ can be partitioned into a square matrix and column so that

$$\mathbf{A}_E = \begin{bmatrix} \mathbf{A}_F \mid \mathbf{C}_E \end{bmatrix}$$

where the column is $\mathbf{C}_E^T = [a_1^n \ a_2^n \ a_3^n \cdots a_m^n]$. If we solve Eq. (3.39) we can relate the finite element constants, $\mathbf{V}_F$, to the exact constants, $\mathbf{V}_E$, at the nodes of the element. Then inverting $\mathbf{A}_F$, Eq. (3.39) gives

$$\mathbf{V}_F = \mathbf{A}_F^{-1}\,\mathbf{A}_E\,\mathbf{V}_E = \begin{bmatrix} \mathbf{I} \mid \mathbf{A}_F^{-1}\,\mathbf{C}_E \end{bmatrix} \mathbf{V}_E = \begin{bmatrix} \mathbf{I} \mid \mathbf{E} \end{bmatrix} \mathbf{V}_E$$

or simply

$$\mathbf{V}_F = \mathbf{K}\,\mathbf{V}_E \tag{3.40}$$

where $\mathbf{K} = \mathbf{A}_F^{-1}\,\mathbf{A}_E$ is a rectangular matrix with constant coefficients. Therefore, we can return to Eq. (3.39) and relate everything to $\mathbf{V}_E$. This gives

$$u_F(a) = \mathbf{P}_F(a)\,\mathbf{K}\mathbf{V}_E = \mathbf{P}_E(a)\,\mathbf{V}_E = u_E(a)$$

so that for arbitrary $\mathbf{V}_E$, one probably has the finite element polynomial and the exact polynomial related by $\mathbf{P}_F(a)\,\mathbf{K} = \mathbf{P}_E(a)$. Likewise, the derivatives of this relation should be approximately equal.

As an example, assume a quadratic finite element in one-dimensional natural coordinates, $-1 < a < +1$. The exact solution is assumed to be cubic. Therefore,

$$\mathbf{P}_F = \begin{bmatrix} 1 & a & a^2 \end{bmatrix}, \qquad \mathbf{V}_F^T = \begin{bmatrix} V_1 & V_2 & V_3 \end{bmatrix}_F,$$

$$\mathbf{P}_E = \begin{bmatrix} 1 & a & a^2 \mid a^3 \end{bmatrix}, \qquad \mathbf{V}_E^T = \begin{bmatrix} V_1 & V_2 & V_3 \mid V_4 \end{bmatrix}_E.$$

Selecting the nodes at the standard positions of $a_1 = -1$, $a_2 = 0$, and $a_3 = 1$ gives:

$$\mathbf{A}_F = \begin{bmatrix} 1 & -1 & 1 \\ 1 & 0 & 0 \\ 1 & 1 & 1 \end{bmatrix}, \qquad \mathbf{A}_F^{-1} = \frac{1}{2}\begin{bmatrix} 0 & 2 & 0 \\ -1 & 0 & 1 \\ 1 & -2 & 1 \end{bmatrix},$$

$$\mathbf{A}_E = \left[\begin{array}{ccc|c} 1 & -1 & 1 & -1 \\ 1 & 0 & 0 & 0 \\ 1 & 1 & 1 & 1 \end{array}\right], \qquad \mathbf{C}_E = \begin{Bmatrix} -1 \\ 0 \\ 1 \end{Bmatrix},$$

$$\mathbf{A}_F^{-1}\,\mathbf{C}_E = \begin{Bmatrix} 0 \\ 1 \\ 0 \end{Bmatrix} \equiv \mathbf{E}, \quad \mathbf{K} = \left[\begin{array}{ccc|c} 1 & 0 & 0 & 0 \\ 0 & 1 & 0 & 1 \\ 0 & 0 & 1 & 0 \end{array}\right].$$

For an interpolate solution, the two equivalent forms are exact at the three nodes ($a = \pm 1, a = 0$) and inside the element

$$u(a) = \begin{bmatrix} 1 & a & a^2 \end{bmatrix} \begin{Bmatrix} V_1 \\ V_2 + V_4 \\ V_3 \end{Bmatrix}_E = \begin{bmatrix} 1 & a & a^2 & a^3 \end{bmatrix} \begin{Bmatrix} V_1 \\ V_2 \\ V_3 \\ V_4 \end{Bmatrix}_E$$

or

$$\begin{bmatrix} 1 & a & a^2 & a \end{bmatrix} \begin{Bmatrix} V_1 \\ V_2 \\ V_3 \\ V_4 \end{Bmatrix}_E = \begin{bmatrix} 1 & a & a^2 & a^3 \end{bmatrix} \begin{Bmatrix} V_1 \\ V_2 \\ V_3 \\ V_4 \end{Bmatrix}_E.$$

Only the last polynomial term differs. By inspection we see that $a\,V_4 = a^3\,V_4$ when a is evaluated at any of the three nodes. Equating the first derivatives at the optimum point a_0,

$$\begin{bmatrix} 0 & 1 & 2a_0 & 1 \end{bmatrix} = \begin{bmatrix} 0 & 1 & 2a_0 & 3a_0^2 \end{bmatrix},$$

or simply $1 = 3a_0^2$ so that $a_0 = \pm 1/\sqrt{3}$. These are the usual Gauss points used in the two point integration rule. Similarly, the *optimal location*, a_s, for the second derivative is found from

$$\begin{bmatrix} 0 & 0 & 2 & 0 \end{bmatrix} = \begin{bmatrix} 0 & 0 & 2 & 6a_s \end{bmatrix},$$

so that $a_s = 0$, the center of the element. The same sort of procedure can be applied to 2-D and 3-D elements. Generally, we find that derivative estimates are least accurate at the nodes. The derivative estimates are usually most accurate at the tabulated integration points. That is indeed fortunate, since it means we get a good approximation of the element square matrix. The typical sampling positions for the quadratic elements are shown in Fig. 3.8.1. It is easy to show that the center of the linear element is the optimum position for sampling the first derivative. Since the front of $\mathbf{K}$ is an identity matrix, $\mathbf{I}$, we are really saying that an exact nodal interpolate solution implies that

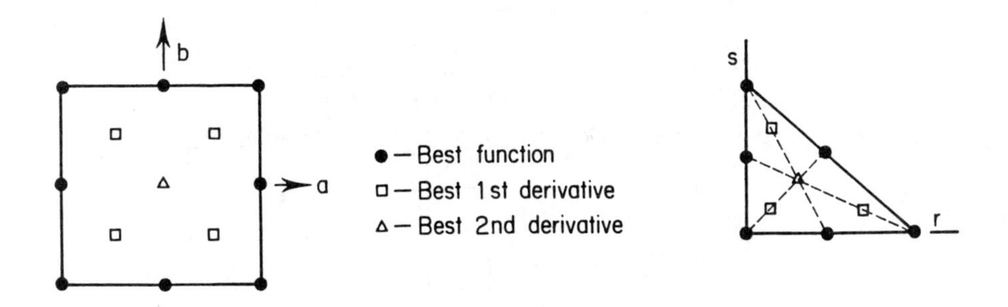

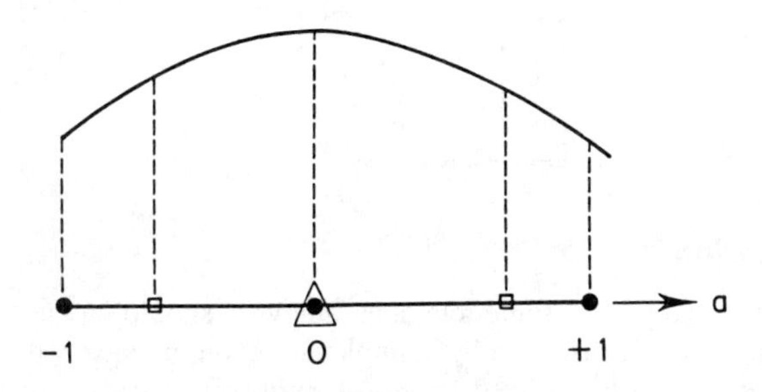

Figure 3.8.1 Barlow points for quadratic elements

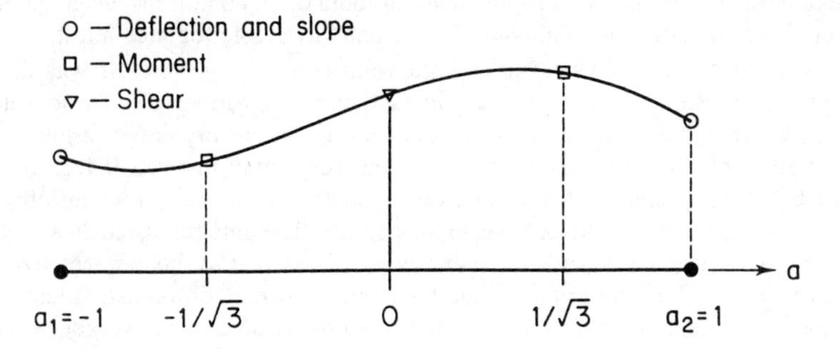

Figure 3.8.2 Barlow points for C^1 Hermite element

$$\mathbf{P}_F(a)\,\mathbf{A}_F^{-1}\,\mathbf{C}_E = a^n .$$

Let the vector $\mathbf{A}_F^{-1}\,\mathbf{C}_E$ denote an extrapolation vector, say $\mathbf{E}$. Then, the derivatives would be the same in the two systems at points where

$$\left[\frac{d^k}{da^k}\,\mathbf{P}_F(a)\right]\mathbf{E} = \left[\frac{d^k}{da^k}\,a^n\right], \qquad 0 \le k \le n-1 . \tag{3.41}$$

For example, the above quadratic element interpolate of a cubic solution gave

$$\begin{matrix} k=0, & [1 & a & a^2] \\ k=1, & [0 & 1 & 2a] \\ k=2, & [0 & 0 & 2\ ] \end{matrix} \begin{Bmatrix} 0 \\ 1 \\ 0 \end{Bmatrix} = \begin{matrix} a^3 \\ 3a^2 \\ 6a \end{matrix}$$

which are only satisfied for

k	a_k
0	$-1,\ 0,\ +1$
1	$\pm 1/\sqrt{3}$
2	0

3.9 Element Validation *

The successful application of finite element analysis should always include a validation of the element to be used and its implementation in a specific computer program. Usually, the elements utilized in most problem classes are very well understood and tested. However, some applications can be difficult to model, and the elements used for their analysis may be more prone to numerical difficulties. Therefore, one should subject elements to be used to a series of element validation tests. Two of the most common and important tests are the *patch test* introduced by Irons [8, 9] and the *single-element tests* proposed by Robinson [14]. The single-element tests generally show the effects of element geometrical parameters such as convexity, aspect ratio, skewness, taper, and out-of-plane warpage. It is most commonly utilized to test for a sensitivity to element aspect ratio. The single-element test usually consists of taking a single element in rectangular, triangular, or line form, considering it as a complete domain, and then investigating its behavior for various load or boundary conditions as a geometrical parameter is varied. An analytical solution is usually available for such a test.

The patch test has been proven to be a valid convergence test. It was developed from physical intuition and later written in mathematical forms. The basic concept is fairly simple. Imagine what happens as one introduces a very large, almost infinite, number of elements. Clearly, they would become very small in size. If we think of the quantities being integrated to form the element matrices, we can make an observation about how the solution would behave in this limit. The integrand, such as the strain energy, contains derivative terms that would become constant as the element size shrinks toward zero. Thus, to be valid in the limit, the element formulation must, at least, be able to yield the correct results in that state. That is, to be assured of convergence one must be able to exactly satisfy the state where the derivatives, in the governing integral statement, take on constant or zero values. This condition can be stated as a mathematical test or as a simple numerical test. The latter option is what we want here. The patch test provides a simple numerical way for a user to test an element, or complete computer program, to verify that it behaves as it should.

We define a patch of elements to be a mesh where at least one node is completely surrounded by elements. Any node of this type is referred to as an *interior node*. The other nodes are referred to as *exterior* or *perimeter nodes*. We will compute the dependent variable at all interior nodes. The derivatives of the dependent variable will be computed in each element. The perimeter nodes are utilized to introduce the essential boundary conditions and/or loads required by the test. Assume that the governing integral statement has derivatives of order n. We would like to find boundary conditions that would make those derivatives constant. This can be done by selecting an arbitrary n-th order polynomial function of the global coordinates to describe the dependent variable in the global space that is covered by the patch mesh. Clearly, the n-th order derivatives of such a function would be constant as desired. The assumed polynomial is used to define the essential boundary conditions on the perimeter nodes of the patch mesh.

This is done by substituting the input coordinates at the perimeter nodes into the assumed function and computing the required value of the dependent variable at each such node. Once all of the perimeter boundary conditions are known, the solution can be numerically executed. The resulting values of the dependent variable are computed at each interior node. To pass the patch test, these computed internal values must agree with the value found when their internal nodal coordinates are substituted into the assumed global polynomial. However, the real test is that when each element is checked, the calculated n-th order derivatives must agree with the arbitrary assumed values used to generate the global function. If an element does not satisfy this test, it should not be used. The patch test can also be used for other purposes. For example, the analyst may wish to distort the element shape and/or change the numerical integration rule to see what effect that has on the numerical accuracy of the patch test.

As a simple elementary example of an analytic solution of the patch test, consider the bar element. The smallest possible patch is one with two line elements. Such a patch has two exterior nodes and one interior node. For simplicity, let the lengths of the two elements be equal and have a value of L. The governing integral statement contains only the first derivative of u. Thus, an arbitrary linear function can be selected for the patch test, since it would have a constant first derivative. Therefore, select $u(x) = a + bx$ for $0 \le x \le 2L$, where a and b are arbitrary constants. Assembling the two-element patch gives

$$\frac{AE}{L}\begin{bmatrix} 1 & -1 & 0 \\ -1 & (1+1) & -1 \\ 0 & -1 & 1 \end{bmatrix} \begin{Bmatrix} u_1 \\ u_2 \\ u_3 \end{Bmatrix} = \begin{Bmatrix} F_1 \\ 0 \\ F_3 \end{Bmatrix}$$

where F_1 and F_3 are the unknown reactions associated with the prescribed external displacements. These two exterior patch boundary conditions are obtained by substituting their nodal coordinates into the assumed patch solution:

$$u_1 = u(x_1) = a + b(0) = a, \quad u_3 = u(x_2) = a + b(2L) = a + 2bL.$$

Modify the assembled equations to include the patch essential boundary conditions gives

$$\frac{AE}{L}\begin{bmatrix}0 & -1 & 0\\ 0 & 2 & 0\\ 0 & -1 & 0\end{bmatrix}\begin{Bmatrix}a\\ u_2\\ a+2bL\end{Bmatrix}=\begin{Bmatrix}F_1\\ 0\\ F_3\end{Bmatrix}-\frac{aAE}{L}\begin{Bmatrix}1\\ -1\\ 0\end{Bmatrix}-\frac{(a+bL)\,AE}{L}\begin{Bmatrix}0\\ -1\\ 1\end{Bmatrix}.$$

Retaining the independent second equation gives the displacement relation

$$\frac{2AE}{L}u_2 = 0 + \frac{aAE}{L} + \frac{(a+2bL)\,AE}{L}.$$

Thus, the internal patch displacement is

$$u_2 = \frac{(2a+2bL)}{2} = (a+bL).$$

The value required by the patch test is

$$u(x_2) = (a+bx_2) = (a+bL).$$

This agrees with the computed solution, as required by a valid element. Recall that the element strains are defined as follows :

$$e = 1: \qquad \varepsilon = \frac{(u_2-u_1)}{L} = \frac{[\,(a+bL)-a\,]}{L} = b$$

$$e = 2: \qquad \varepsilon = \frac{(u_3-u_2)}{L} = \frac{[\,(a+2bL)-(a+bL)\,]}{L} = b.$$

Thus, all element derivatives are constant. However, these constants must agree with the constant assumed in the patch. That value is

$$\varepsilon = \frac{du}{dx} = \frac{d(a+bx)}{dx} = b.$$

Therefore, the patch test is completely satisfied. At times one also wishes to compute the reactions, i.e., F_1 and F_3. To check for possible rank deficiency in the element formulation, one should repeat the test with only enough displacements prescribed to prevent rigid body motion. (That is, to render the square matrix non-singular.) Then, the other outer perimeter nodes are loaded with the reactions found in the precious patch test. In the above example, substituting u_1 and u_2 into the previously discarded first equation yields the reaction $F_1 = -bAE$. Likewise, the third equation gives $F_3 = -F_1$, as expected. Thus, the above test could be repeated by prescribing u_1 and F_3, or F_1 and u_3. The same results should be obtained in each case. A major advantage of the patch test is that it can be carried out numerically. In the above case, the constants a and b could have been assigned arbitrary numerical values. Inputting the required numerical values of A, E and L would give a complete numerical description that could be tested in a standard program. Such a procedure also verifies that the computer program satisfies certain minimum requirements. A problem with some elements is that they can pass the patch test for a uniform mesh, but fail when an arbitrary irregular mesh is employed. Thus, as a general rule, one should try to avoid conducting the test with a regular mesh, such as that given in the above example. It would have been wiser to use unequal element lengths such as L and αL, where α is an arbitrary constant. The linear bar element should pass the test for any scaling ratio, α. However, for α near zero, numerical ill-conditioning begins to affect the answers.

3.10 Element Distortion *

The effects of distorting various types of elements can be serious, and most codes do not adequately validate data in this respect. As an example, consider a quadratic isoparametric line element. As shown in Fig. 3.10.1, let the three nodes be located in physical (x) space at points 0, ah, and h, where h is the element length, and $0 \leq a \leq 1$ is a location constant. The element is defined in a local unit space where $0 \leq s \leq 1$. The relation between x and s is easily shown to be

$$x(s) = h(4a - 1)\, s + h(2 - 4a)\, s^2$$

and the two coordinates have derivatives related by

$$\frac{\partial x}{\partial s} = h(4a - 1) + 4h(1 - 2a)\, s\,.$$

The Jacobian of the transformation, J, is the inverse relation; that is, $J = \partial s / \partial x$. The integrals required to evaluate the element matrices utilize this Jacobian. The mathematical principles require that J be positive definite. Distortion of the elements can cause J to go to zero or become negative. This possibility is easily seen in the present 1-D example. If one locates the interior ($s = 1/2$) node at the standard midpoint position, then $a = 1/2$ so that $\partial x / \partial s = h$ and J is constant throughout the element. Such an element is generally well formulated. However, if the interior node is distorted to any other position, the Jacobian will not be constant and the accuracy of the element may suffer. Generally, there will be points where $\partial x / \partial s$ goes to zero, so that the stiffness becomes singular due to division by zero. For slightly distorted elements, say $0.4 < a < 0.6$, the singular points lie outside the element domain. As the distortion increases, the singularities move to the element boundary, e.g., $a = 1/4$ or $a = 3/4$. Eventually, the distortions cause singularities of J inside the element. Such situations can cause poor stiffness matrices and very bad stress estimates, unless the true solution has the same singularity.

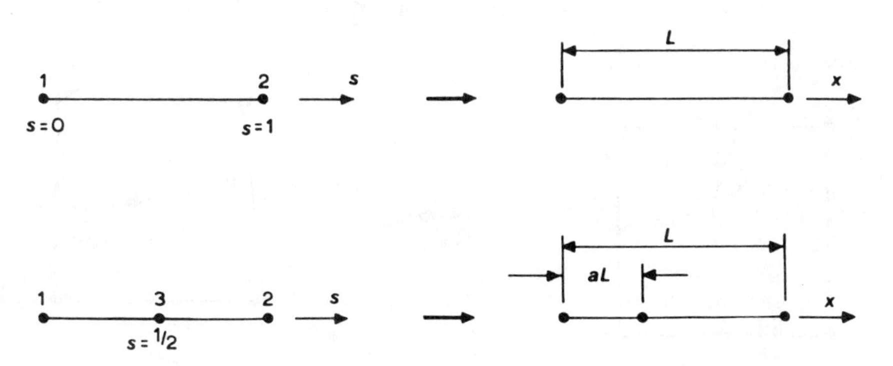

Figure 3.10.1 Constant Jacobian and distorted elements

The effects of distortions of two- or three-dimensional elements are similar. For example, the edge of a quadratic element may have the non-corner node displaced in a similar way, or it may be moved normal to the line between the corners. Similar analytic singularities can be developed for such elements. However, the presence of singularities due to element distortions can easily be checked by numerical experiments. Several such analytic and numerical studies have led to useful criteria for checking the element geometry for undesirable effects. For example, Fig. 3.10.2 shows a typical edge of a two- or three-dimensional quadratic element. Let L be the cord length, D the normal displacement of the mid-side node, and α the angle between the corner tangent and the cord line. The criteria show:

$$\text{warning range:} \quad 1/7 < D/L < 1/3, \quad \alpha \le 30°$$

$$\text{error range:} \quad 1/3 < D/L, \quad \alpha \ge 53°.$$

These values are obtained when only one edge is considered. If more than one edge of a single element causes a warning state, then the warnings should be considered more serious. Other parameters influence the seriousness of element distortion. Let R be a measure of the aspect ratio, that is, R is the ratio of the longest side to the shortest side. Let the minimum and maximum angles between corner cord lines be denoted by θ and γ, respectively. Define H, the lack of flatness, to be the perpendicular distance of a fourth node from the plane of the first three divided by the maximum side length. Then the following guidelines in Table 3.1 should be considered when validating geometric data for membrane or solid elements.

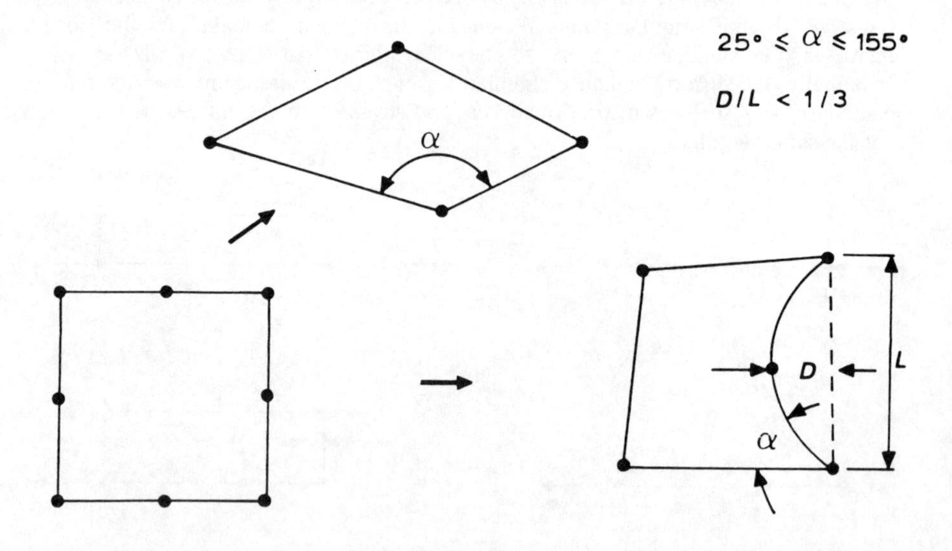

Figure 3.10.2 Distorted quadrilaterals

Table 3.1 Geometric criteria for two- and three-dimensional elements

Shape	*Warning state*	*Error state*
Triangle	$5 < R < 15$	$R > 15$
	$15° < \theta < 30°$	$\theta \leq 15°$
	$150° < \gamma < 165°$	$\gamma \geq 165°$
Quadrilateral	$5 < R < 15$	$R \geq 15$
	$25° < \theta < 45°$	$\theta \leq 25°$
	$135° < \gamma < 155°$	$\gamma \geq 155°$
	$10^{-5} < H < 10^{-2}$	$H \geq 10^{-2}$

Solids: The above limits on R, θ and γ are checked on each face.

3.11 Euler's Equations of Variational Calculus *

Euler's Theorem of Variational Calculus states that the solution, u, that satisfies the essential boundary conditions and renders stationary the functional

$$I = \int_{\Omega} f\left(x, y, z, u, \frac{\partial u}{\partial x}, \frac{\partial u}{\partial y}, \frac{\partial u}{\partial z}\right) d\Omega + \int_{\Gamma} \left(qu + au^2/2\right) d\Gamma \tag{3.42}$$

also satisfies the partial differential equation

$$\frac{\partial f}{\partial u} - \frac{\partial}{\partial x}\frac{\partial f}{\partial(\partial u/\partial x)} - \frac{\partial}{\partial y}\frac{\partial f}{\partial(\partial u/\partial y)} - \frac{\partial}{\partial z}\frac{\partial f}{\partial(\partial u/\partial z)} = 0 \tag{3.43}$$

in Ω, and satisfies the natural boundary condition that

$$n_x \frac{\partial f}{\partial(\partial u/\partial x)} + n_y \frac{\partial f}{\partial(\partial u/\partial y)} + n_z \frac{\partial f}{\partial(\partial u/\partial z)} + q + au = 0 \tag{3.44}$$

on Γ that is not subject to an essential boundary. Here n_x, n_y, n_z are the components of the normal vector on u. For example, consider a two-dimensional functional

$$I = \tfrac{1}{2} \int_{\Omega} \left[K_x \left(\frac{\partial u}{\partial x}\right)^2 + K_y \left(\frac{\partial u}{\partial y}\right)^2 - 2Qu\right] d\Omega + \int_{\Gamma} \left(qu + au^2/2\right) d\Gamma .$$

Then, $\partial f/\partial u = -Q$ and

$$\frac{\partial f}{\partial(\partial u/\partial x)} = 2K_x \frac{\partial u}{\partial x}, \qquad \frac{\partial f}{\partial(\partial u/\partial y)} = 2K_y \frac{\partial u}{\partial y},$$

so that the partial differential equation is

$$-Q - \frac{\partial}{\partial x}\left(K_x \frac{\partial u}{\partial x}\right) - \frac{\partial}{\partial y}\left(K_y \frac{\partial u}{\partial y}\right) = 0 \qquad \varepsilon\ \Omega$$

which is the classic Poisson equation. Iff the K_x and K_y are isotropic $(K_x = K_y = K)$ and K is the same constant everywhere in Ω, then it becomes

$$\frac{\partial^2 u}{\partial x^2} + \frac{\partial^2 u}{\partial y^2} + \frac{Q}{K} = 0 \qquad \varepsilon\ \Omega,$$

which for no source, $Q = 0$, reduces to the Laplace equation

$$\frac{\partial^2 u}{\partial x^2} + \frac{\partial^2 u}{\partial y^2} = 0, \qquad \varepsilon\ \Omega.$$

The corresponding natural boundary condition for this example is

$$n_x K_x \frac{\partial u}{\partial x} + n_y K_y \frac{\partial u}{\partial y} + q + a u = 0 \qquad \varepsilon\ \Gamma$$

or simply

$$K_n \frac{\partial u}{\partial n} + q + a u = 0 \qquad \varepsilon\ \Gamma,$$

where K_n is the coefficient in the direction of the unit normal, **n**. Likewise, one could reduce this to a one-dimensional form

$$I = \int_a^b f\left(x, u, \frac{du}{dx}\right) dx + (q + au)\Bigg|_{x=b} - (q + au)\Bigg|_{x=a} \tag{3.45}$$

so the ODE is

$$\frac{\partial f}{\partial u} - \frac{d}{dx}\frac{\partial f}{\partial (du/dx)} = 0 \tag{3.46}$$

and the natural boundary condition if u is not prescribed is

$$n_x \frac{\partial f}{\partial (du/dx)} + q + a u = 0. \tag{3.47}$$

Many fourth order equations are formulated in the same way. For example, the functional

$$I(u) = \int_a^b f\left(x, u, \frac{du}{dx}, \frac{d^2u}{dx^2}\right) dx \tag{3.48}$$

involving the second derivative of u has the Euler equation

$$\frac{\partial f}{\partial u} - \frac{d}{dx}\frac{\partial f}{\partial (du/dx)} + \frac{d^2}{dx^2}\frac{\partial f}{\partial (d^2u/dx^2)} = 0, \tag{3.49}$$

and the natural boundary condition of

$$\frac{\partial f}{\partial (du/dx)} - \frac{d}{dx}\frac{\partial f}{\partial (d^2u/dx^2)} = 0 \tag{3.50}$$

if du/dx is specified (and u unknown), and the natural condition of

$$\frac{\partial f}{\partial (d^2u/dx^2)} = 0 \tag{3.51}$$

if u is specified (and du/dx is unknown). A common example is

$$I(u) = \int_0^L \left[EI\left(\frac{d^2u}{dx^2}\right)^2 - 2uQ \right] dx$$

that has the Euler relations

$$\frac{\partial f}{\partial u} = -2Q, \qquad \frac{\partial f}{\partial (du/dx)} = 0, \qquad \frac{\partial f}{\partial (d^2u/dx^2)} = 2EI\frac{d^2u}{dx^2}$$

which yield the ODE

$$\frac{d^2}{dx^2}\left[EI\frac{d^2u}{dx^2}\right] = Q,$$

which is the fourth order equation for the deflection, u, of a beam subject to a transverse load per unit length of Q. Here EI denotes the bending stiffness of the member. This example requires four boundary conditions. At any boundary condition point, one may prescribe u and/or du/dx. If u is not specified, the natural boundary condition is that

$$-\frac{d}{dx}\left[EI\frac{d^2u}{dx^2}\right] = 0,$$

while, if u is not specified, the natural condition is that $d^2u/dx^2 = 0$. They correspond to conditions on the shear and moment, respectively.

3.12 System Validation*

There are a number of analysis or input options that can lead to numerical ill-conditioning of the algebraic system to be solved. A number of procedures can be used to assess the existence of such difficulties in the analysis (see [1] and [2]). One useful procedure is to invoke a *rigid body check*. Let **K** denote the system stiffness matrix, **F** the generalized nodal forces, and **R** a set of displacement components that represent a rigid body motion (zero gradient solution). For example, **R** may correspond to a rigid body translation in the x direction. Then **R** would be zero except for constants, say c, associated with each x displacement degree of freedom. We know that such a rigid body displacement should cause no external nodal forces and no internal stresses. Thus, before solving the assembled equations, we execute a rigid body check:

$$\mathbf{K}\mathbf{R} = \mathbf{F} \overset{?}{=} \mathbf{0}. \tag{3.52}$$

This should identify local errors associated with poor elements or incorrect constraints. Since we generally have three rigid body translations and three rigid body rotations, the array **R** usually contains six columns. Thus, the above result should be a null rectangular array. If we applied this test to **K** previously assembled for our patch test, we would find $\mathbf{R}^T = [\, c\ c\ c\,]$, and

$$\frac{EA}{L}\begin{bmatrix} 1 & -1 & 0 \\ -1 & 2 & -1 \\ 0 & -1 & 1 \end{bmatrix}\begin{Bmatrix} c \\ c \\ c \end{Bmatrix} = \begin{Bmatrix} 0 \\ 0 \\ 0 \end{Bmatrix}$$

as expected. If we had run the distorted patch test, then the left-hand side would be

$$\frac{EA}{L}\begin{bmatrix} 1 & -1 & 0 \\ -1 & (1+1/\alpha) & -1/\alpha \\ 0 & -1/\alpha & 1/\alpha \end{bmatrix}\begin{Bmatrix} c \\ c \\ c \end{Bmatrix}.$$

Depending on the number of significant digits used in a numerical calculation, extreme values of α may lead to non-zero residual forces. In addition to the errors in the residual forces, we may want to compute a check on the potential energy for each of the rigid body motion sets :

$$\mathbf{E} = \mathbf{R}^T \mathbf{K} \mathbf{R} \stackrel{?}{=} \mathbf{0}, \tag{3.53}$$

which should result in a square null matrix. Of course, these rigid body checks can be applied at the element level, if desired. At the element level, we usually also have access to the strain-displacement matrix, **B**, so that

$$\varepsilon = \mathbf{B}\,\delta,$$

where ε denotes the strain vector and δ the element nodal displacements. Usually, we compute the element results

$$\mathbf{K}^e = \int \mathbf{B}^T \mathbf{D} \mathbf{B}\, d\Omega V.$$

If we move the element as a rigid body, $\delta = \mathbf{R}^e$, then the strain, ε, should be zero at all points (including the integration points) in the element. Thus, we might use

$$\varepsilon^e = \mathbf{B}^e \mathbf{R}^e \stackrel{?}{=} \mathbf{0} \tag{3.54}$$

to try to assess a bad element before assembly begins. For the bar element used in the patch test, we know that $\mathbf{B}^e = [-1 \quad 1]/L$ so that a rigid body x-translation, $\mathbf{R}^e = [\,c\; c\,]^T$, gives

$$\varepsilon = \frac{1}{L}[-1 \quad 1]\begin{Bmatrix} c \\ c \end{Bmatrix} = \mathbf{0}$$

as expected. For numerically integrated elements, this could be checked at each quadrature point.

As mentioned above, another item of concern is the possibility of ill-conditioning of the system matrix **K** and its condition number (see [5]). If we return to the distorted patch assembly and restrain the third node, then **K** has the form

$$\mathbf{K} = \begin{bmatrix} K_A & -K_A \\ -K_A & (K_A + K_B) \end{bmatrix}. \tag{3.55}$$

If the second element were very long, $\alpha \gg 1$, compared to the first, then $K_A \gg K_B$, so that K_A dominates the structural stiffness, K. However, the inverse, K^{-1}, is dominated by K_B, since

$$\mathbf{K}^{-1} = \frac{1}{K_B}\begin{bmatrix} \left(1 + \dfrac{K_B}{K_A}\right) & 1 \\ 1 & 1 \end{bmatrix}. \tag{3.56}$$

This means that the displacements, and thus the stresses, are governed by K_B. This illustrates that a common cause of ill-conditioned problems in thin-walled structures is a large ratio of element stiffness to supporting structure stiffness. Similar problems often occur when large elements adjoin small ones. There is a trend to use 'pre-conditioned' iterative solvers to help overcome the accuracy problems of such formulations where the condition number of the system is large.

Each system stiffness matrix, **K**, will have a condition number, denoted here as $C(\mathbf{K})$. It has been shown that if p decimal digits per computer word are used to represent the stiffness terms, then the computed displacements will be correct to q decimal digits, where $q \geq p - \log_{10} C(\mathbf{K})$.

3.13 References

[1] Akin, J.E., "Verification Checks of Finite Element Models," pp. 43–50 in *Computing in Civil Engineering*, New York: A.S.C.E. (1981).

[2] Baier, H., "Checking Techniques for Complex Finite Element Models," pp. 145–156 in *Accuracy, Reliability and Training in FEM Technology*, ed. J. Robinson, Dorset, U.K.: Robinson and Associates (1984).

[3] Barlow, J., "Optimal Stress Locations in Finite Element Models," *Int. J. Num. Meth. Eng.*, **10**, pp. 243–251 (1976).

[4] Bathe, K.J., *Finite Element Procedures in Engineering Analysis*, Englewood Cliffs: Prentice-Hall (1982).

[5] Cook, R.D., *Concepts and Applications of Finite Element Analysis*, New York: John Wiley (1974).

[6] Gallagher, R.H., *Finite Element Analysis Fundamentals*, Englewood Cliffs: Prentice-Hall (1975).

[7] Hughes, O.F., *Ship Structural Design*, New York: John Wiley (1983).

[8] Irons, B.M. and Razzaque, A., "Experience with the Patch Test for Convergence of the Finite Element Method," pp. 557–587 in *Mathematical Foundation of the Finite Element Method*, ed. A.R. Aziz, New York: Academic Press (1972).

[9] Irons, B.M. and Ahmad, S., *Techniques of Finite Elements*, New York: John Wiley (1980).

[10] Kawai, T., "The Application of Finite Element Methods to Ship Structures," *Computers and Structures*, **3**, pp. 1175–1194 (1973).

[11] Mitchell, A.R. and Wait, R., *The Finite Element Method in Partial Differential Equations*, London: John Wiley (1977).

[12] Myers, G.E., *Analytical Methods in Conduction Heat Transfer*, New York: McGraw-Hill (1971).

[13] Noor, A.K. and Pilkey, W.D., "State of the Art Surveys on Finite Element Technology," ASME Publ. H000290, New York (1983).

[14] Robinson, J., "A Single Element Test," *Comp. Math. Appl. Mech. Engng.*, **7**, pp. 191–200 (1976).

[15] Weaver, W.F., Jr. and Johnston, P.R., *Finite Elements for Structural Analysis*, Englewood Cliffs: Prentice-Hall (1984).

[16] Zienkiewicz, O.C., *The Finite Element Method,* 3rd edition, New York: McGraw-Hill (1979).

Chapter 4

ELEMENT INTERPOLATION AND LOCAL COORDINATES

4.1 Introduction

Up to this point we have relied on the use of a linear interpolation relation that was expressed in *global coordinates* and given by inspection. In the previous chapter we saw numerous uses of these interpolation functions. By introducing more advanced interpolation functions, $\mathbf{H}$, we can obtain more accurate solutions. Here we will show how the common interpolation functions are derived. Then a number of expansions will be given without proof. Also, we will introduce the concept of non-dimensional *local* or element *coordinate* systems. These will help simplify the algebra and make it practical to automate some of the integration procedures.

4.2 Linear Interpolation

Assume that we desire to define a quantity, u, by interpolating in space, from certain given values, $\mathbf{u}$. The simplest interpolation would be linear and the simplest space is the line, e.g. x-axis. Thus to define $u(x)$ in terms of its values, $\mathbf{u}^e$, at selected points on an element we could choose a linear polynomial in x. That is:

$$u^e(x) = c_1^e + c_2^e x = \mathbf{P}(x)\,\mathbf{c}^e \tag{4.1}$$

where $\mathbf{P} = [\,1 \quad x\,]$ denotes the linear polynomial behavior in space and $\mathbf{c}^{e^T} = [\,c_1^e \quad c_2^e\,]$ are undetermined constants that relate to the given values, $\mathbf{u}^e$. Referring to Fig. 4.2.1, we note that the element has a physical length of L^e and we have defined the nodal values such that $u^e(x_1) = u_1^e$ and $u^e(x_2) = u_2^e$. To be useful, Eq. (4.1) will be required to be valid at all points on the element, including the nodes. Evaluating Eq. (4.1) at each node of the element gives the set of identities:

$$u^e(x_1^e) = u_1^e = c_1^e + c_2^e x_1^e, \quad u^e(x_2^e) = u_2^e = c_1^e + c_2^e x_2^e, \quad \text{or } \mathbf{u}^e = \mathbf{g}^e\,\mathbf{c}^e \tag{4.2}$$

where

$$\mathbf{g}^e = \begin{bmatrix} 1 & x_1^e \\ 1 & x_2^e \end{bmatrix}. \tag{4.3}$$

This shows that the physical constants, $\mathbf{u}^e$, are related to the polynomial constants, $\mathbf{c}^e$ by

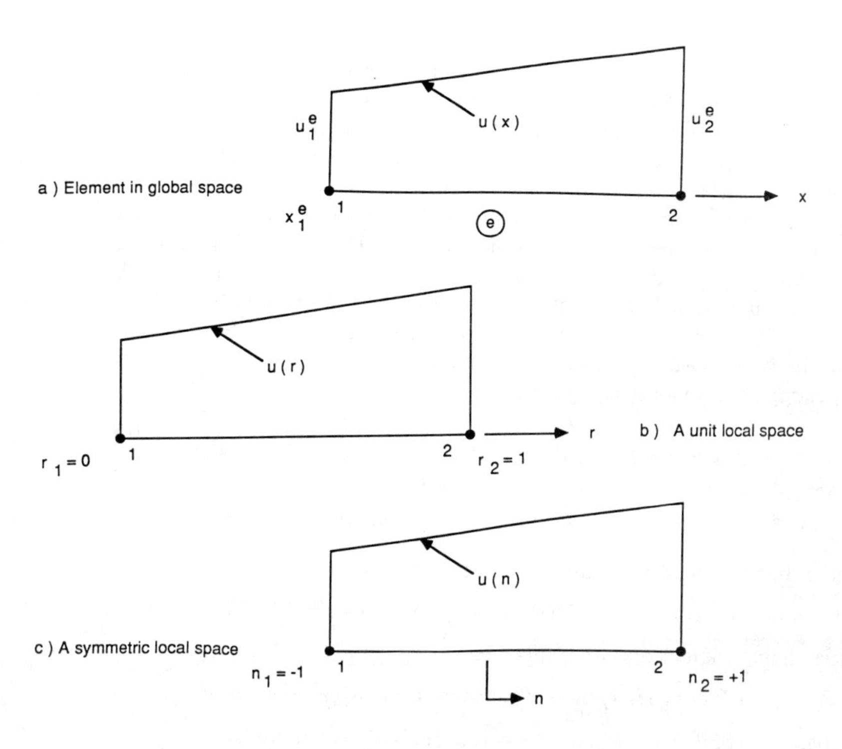

Figure 4.2.1 One-dimensional linear interpolation

information on the geometry of the element, $\mathbf{g}^e$. Since $\mathbf{g}^e$ is a square matrix we can (usually) solve Eq. (4.2) to get the polynomial constants:

$$\mathbf{c}^e = \mathbf{g}^{e^{-1}} \mathbf{u}^e . \tag{4.4}$$

In this case the element geometry matrix can be easily inverted to give

$$\mathbf{g}^{e^{-1}} = \frac{1}{x_2^e - x_1^e} \begin{bmatrix} x_2^e & -x_1^e \\ -1 & 1 \end{bmatrix} . \tag{4.5}$$

By putting these results into our original assumption, Eq. (4.1), it is possible to write $u^e(x)$ directly in terms of $\mathbf{u}^e$. That is,

$$u^e(x) = \mathbf{P}(x)\, \mathbf{g}^{e^{-1}} \mathbf{u}^e = \mathbf{H}^e(x)\, \mathbf{u}^e \tag{4.6}$$

or

$$u^e(x) = [\,1 \quad x\,] \frac{1}{L^e} \begin{bmatrix} x_2^e & -x_1^e \\ -1 & 1 \end{bmatrix} \begin{Bmatrix} u_1^e \\ u_2^e \end{Bmatrix} = \begin{bmatrix} \dfrac{x_2^e - x}{L^e} & \dfrac{x - x_1^e}{L^e} \end{bmatrix} \{u^e\} \tag{4.7}$$

where H^e is called the element interplation array. Clearly

$$\mathbf{H}^e(x) = \mathbf{P}(x)\, \mathbf{g}^{e^{-1}} . \tag{4.8}$$

From Eq. (4.6) we can see that the approximate value, $u^e(x)$ depends on the assumed behavior in space, $\mathbf{P}$, the element geometry, $\mathbf{g}^e$, and the element nodal parameters, $\mathbf{u}^e$. This is also true in two- and three-dimensional problems.

Since this element interpolation has been defined in a global or physical space the geometry matrix, $\mathbf{g}^e$, and thus $\mathbf{H}^e$ will be different for every element. Of course, the algebraic form is common but the numerical terms differ from element to element. For a given type of element it is possible to make $\mathbf{H}$ unique if a local non-dimensional coordinate is utilized. This will also help reduce the amount of calculus that must be done by hand. Local coordinates are usually selected to range from 0 to 1, or from -1 to $+1$. These two options are also illustrated in Fig. 4.2.1. For example, consider the *unit coordinates* shown in Fig. 4.2.1 where the linear polynomial is now $\mathbf{P} = [1 \quad r]$. Repeating the previous steps yields

$$u^e(r) = \mathbf{P}(r)\,\mathbf{g}^{-1}\,\mathbf{u}^e, \qquad \mathbf{g} = \begin{bmatrix} 1 & 0 \\ 1 & 1 \end{bmatrix}, \qquad \mathbf{g}^{-1} = \begin{bmatrix} 1 & 0 \\ -1 & 1 \end{bmatrix} \tag{4.9}$$

so that

$$u^e(r) = \mathbf{H}(r)\,\mathbf{u}^e \tag{4.10}$$

where the unit coordinate interpolation function is

$$\mathbf{H}(r) = [(1-r) \quad r] = \mathbf{P}\,\mathbf{g}^{-1}. \tag{4.11}$$

Expanding back to scalar form this means

$$u^e(r) = H_1(r)\,u_1^e + H_2(r)\,u_2^e = (1-r)\,u_1^e + r u_2^e = u_1^e + r(u_2^e - u_1^e)$$

so that at $r = 0$, $u^e(0) = u_1^e$ and at $r = 1$, $u^e(1) = u_2^e$ as required.

A possible problem here is that while this simplifies $\mathbf{H}$ one may not know "where" a given r point is located in global or physical space. In other words, what is x when r is given? One simple way to solve this problem is to note that the nodal values of the global coordinates of the nodes, x^e, are given data. Therefore, we can use the concepts in Eq. (4.10) and define $x^e(r) = \mathbf{H}(r)\,\mathbf{x}^e$, or

$$x^e(r) = (1-r)\,x_1^e + r x_2^e = x_1^e + L^e r \tag{4.12}$$

for any r in a given element, e. If we make this popular choice for relating the local and global coordinates, we call this an *isoparametric* element. The name implies that a single (iso) set of parametric relations, $\mathbf{H}(r)$, is to be used to define the geometry, $x(r)$, as well as the primary unknowns, $u(r)$.

If we select the symmetric, or Gaussian, local coordinates such that $-1 \le n \le +1$, then a similar set of interpolation functions are obtained. Specifically, $u^e(n) = \mathbf{H}(n)\,\mathbf{u}^e$ with $H_1(n) = (1-n)/2$, $H_2(n) = (1+n)/2$, or simply

$$H_i(n) = (1 + n_i\,n)/2 \tag{4.13}$$

where n_i is the local coordinate of node i. This coordinate system is often called a *natural* coordinate system. Of course, the relation to the global system is

$$x^e(n) = \mathbf{H}(n)\,\mathbf{x}^e. \tag{4.14}$$

The relationship between the unit and natural coordinates is $r = (1+n)/2$. This will sometimes by useful in converting tabulated data in one system to the other. The above local coordinates can be used to define how an approximation changes in space. They also allow one to calculate derivatives. For example, from Eq. (4.10)

$$\frac{du^e}{dr} = \frac{d\mathbf{H}(r)}{dr}\mathbf{u}^e \tag{4.15}$$

and similarly for other quantities of interest. Another quantity that we will find very important is dx/dr. In a typical linear element, Eq. (4.12) gives

$$\frac{dx^e(r)}{dr} = \frac{dH_1}{dr}x_1^e + \frac{dH_2}{dr}x_2^e = -x_1^e + x_2^e$$

or simply $dx^e/dr = L^e$. By way of comparison, if the natural coordinate is utilized

$$\frac{dx^e(n)}{dn} = \frac{L^e}{2}. \tag{4.16}$$

This illustrates that the choice of the local coordinates has more effect on the derivatives than it does on the interpolation itself. The use of unit coordinates is more popular with *simplex elements*. These are elements where the number of nodes is one higher than the dimension of the space. The generalization of unit coordinates for common simplex elements is illustrated in Fig. 4.2.2. For simplex elements the natural coordinates becomes *area coordinates* and *volume coordinates*, which the author finds rather unnatural. Both unit and natural coordinates are effective for use on squares or cubes in the local space. In global space these shapes become quadrilaterals or hexahedra as illustrated in Fig. 4.2.3. The natural coordinates are more popular for these shapes.

4.3 Quadratic Interpolation

The next logical spatial form to pick is that of a quadratic polynomial. Select three nodes on the line element, two at the ends and the third inside the element. In local space the third node is at the element center. Thus, the local unit coordinates are $r_1 = 0$, $r_3 = ½$, and $r_2 = 1$. It is usually desirable to have x_3 also at the center of the element in global space. If we repeat the previous procedure using $u(r) = c_1 + c_2 r + c_3 r^2$, then the element interpolation functions are found to be

$$\begin{aligned} H_1(r) &= 1 - 3r + 2r^2 \\ H_2(r) &= -r + 2r^2 \\ H_3(r) &= 4r - 4r^2 \\ \hline \Sigma H_i(r) &= 1 \end{aligned} \tag{4.17}$$

These quadratic functions are completely different from the linear functions. Note that these functions have a sum that is unity at any point, r, in the element. These three functions illustrate another common feature of C^0 interpolation functions. They are unity at one node and zero at all others:

$$H_i(r_j) = \delta_{ij} = \begin{cases} 1 & \text{if } i = j \\ 0 & \text{if } i \neq j \end{cases}.$$

These functions expressed in natural coordinates are

$$H_1(n) = n(n-1)/2, \quad H_2(n) = n(n+1)/2, \quad H_3(n) = 1 - n^2. \tag{4.18}$$

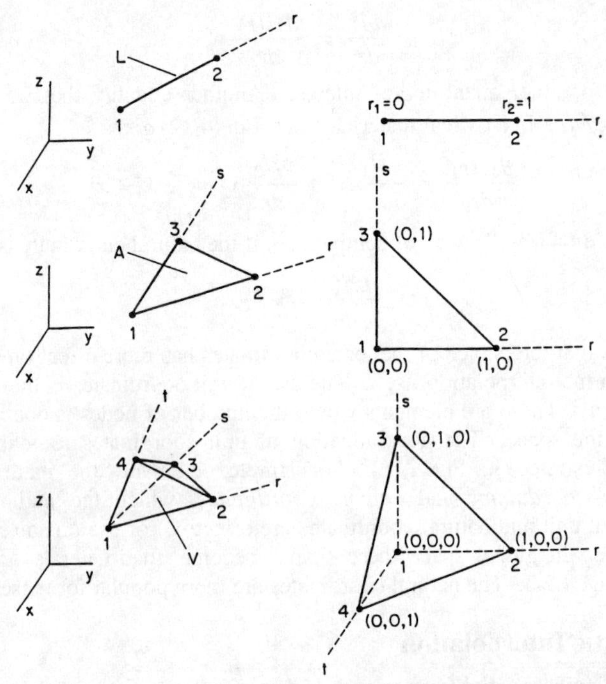

Figure 4.2.2 The simplex element family in unit coordinates

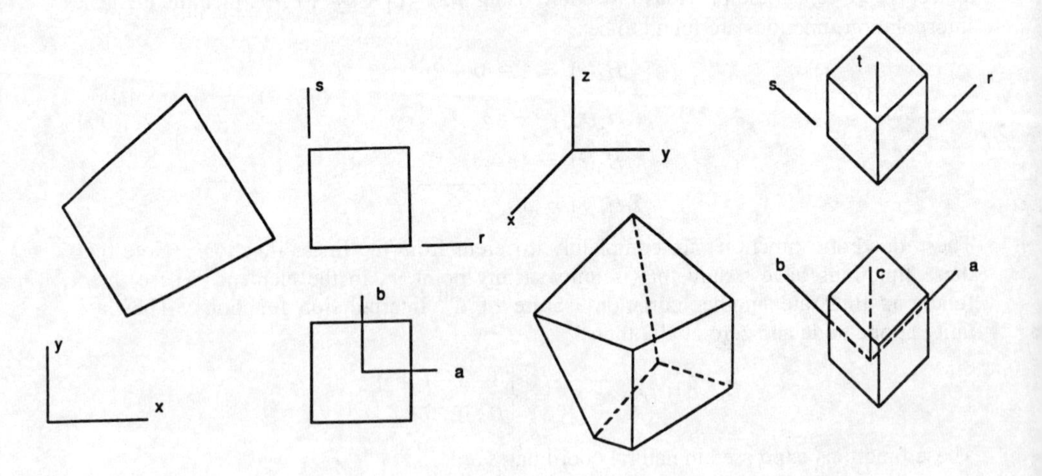

Figure 4.2.3 Global and local spaces for quadrilaterals and hexahedra

$-1 < n < 1$	$0 < r < 1$
a) Linear 1 - - - - - - - - 2	
$H_1 = (1-n)/2$	$H_1 = (1-r)$
$H_2 = (1+n)/2$	$H_2 = r$
b) Quadratic 1 - - - 3 - - - 2	
$H_1 = n(n-1)/2$	$H_1 = (r-1)(2r-1)$
$H_2 = n(n+1)/2$	$H_2 = r(2r-1)$
$H_3 = (1+n)(1-n)$	$H_3 = 4r(1-r)$
c) Cubic 1 - - 3 - - 4 - - 2	
$H_1 = (1-n)(3n+1)(3n-1)/16$	$H_1 = (1-r)(r-3r)(1-3r)/2$
$H_2 = (1+n)(3n+1)(3n-1)/16$	$H_2 = r(2-3r)(1-3r)/2$
$H_3 = 9(1+n)(n-1)(3n-1)/16$	$H_3 = 9r(1-r)(2-3r)/2$
$H_4 = 9(1+n)(1-n)(3n+1)/16$	$H_4 = 9r(1-r)(3r-1)/2$

Figure 4.4.1 Typical Lagrange interpolations

4.4 Lagrange Interpolation

Clearly this one dimensional procedure can be readily extended by adding more nodes to the interior of the element. Usually the additional nodes are equally spaced along the element. However, they can be placed in arbitrary locations. The interpolation function for such an element is known as the Lagrange interpolation polynomial. The one-dimensional m-th order *Lagrange interpolation* polynomial is the ratio of two products. For an element with $(m+1)$ nodes, r_i, $i = 1, 2, \ldots, (m+1)$, the interpolation function for the k-th node is

$$H_k^m(n) = \prod_{\substack{i=1 \\ i \neq k}}^{m+1} (n - n_i) \bigg/ \prod_{\substack{i=1 \\ i \neq k}}^{m+1} (n_k - n_i). \tag{4.19}$$

This is a complete m-th order polynomial in one dimension. It has the property that $H_k(n_i) = \delta_{ik}$. For local coordinates given on the domain $[-1, 1]$, a typical term for three equally spaced nodes is

$$H_3(n) = \frac{(n-(-1))\,(n-1)}{(0-(-1))\,(0-1)} = (1-n^2).$$

Similarly, $H_1(n) = n(n-1)/2$ and $H_2 = n(n+1)/2$ and their sum is unity. Figure 4.4.1 shows typical node locations and interpolation functions for members of this family of complete polynomial functions on simplex elements. Figure 4.4.2 shows the typical coding of a quadratic line element (subroutines SHP3L and DER3L).

```
      SUBROUTINE  SHP3L (X, H)
C     * * * * * * * * * * * * * * * * * * * * * * * * * *
C     CALCULATE SHAPE FUNCTIONS OF A 3 NODE LINE ELEMENT
C                 IN NATURAL COORDINATES
C     * * * * * * * * * * * * * * * * * * * * * * * * * *
CDP   IMPLICIT REAL*8 (A-H,O-Z)
      DIMENSION  H(3)
C     H = ELEMENT SHAPE FUNCTIONS
C     X = LOCAL COORDINATE OF POINT,     -1 TO +1
C     LOCAL NODE COORD. ARE -1,0,+1.    1-----2-----3
      H(1) = 0.5*(X*X - X)
      H(2) = 1. - X*X
      H(3) = 0.5*(X*X + X)
      RETURN
      END
      SUBROUTINE  DER3L (X, DH)
C     * * * * * * * * * * * * * * * * * * * * * * * * * *
C     FIND LOCAL DERIVATIVES FOR A 3 NODE LINE ELEMENT
C                 IN NATURAL COORDINATES
C     * * * * * * * * * * * * * * * * * * * * * * * * * *
CDP   IMPLICIT REAL*8 (A-H,O-Z)
      DIMENSION  DH(3)
C     DH = LOCAL DERIVATIVES OF SHAPE FUNCTIONS (SHP3L)
C     X  = LOCAL COORDINATE OF POINT,     -1 TO +1
C     LOCAL NODE COORD. ARE -1,0,+1.    1----2----3
      DH(1) = X - 0.5
      DH(2) = -2.*X
      DH(3) = X + 0.5
      RETURN
      END
```

Figure 4.4.2 Coding a Lagrangian quadratic

4.5 Hermitian Interpolation

All of the interpolation functions considered so far have C^0 continuity between elements. That is, the function being approximated is continuous between elements but its derivative is discontinuous. However, know that some applications, such as a beam analysis, also require that their derivative be continuous. These C^1 functions are most easily generated by using derivatives, or slopes, as nodal degrees of freedom.

The simplest element in this family is the two node line element where both y and dy/dx are taken as nodal degrees of freedom. Note that a global derivative has been selected as a degree of freedom. Since there are two nodes with two dof each, the interpolation function has four constants, thus, it is a cubic polynomial. The form of this *Hermite polynomial* is well known. The element is shown in Fig. 4.5.1 along with the interpolation functions and their global derivatives. The latter quantities are obtained from the relation between local and global coordinates, e.g., Eq. (4.12). On rare occasions one may also need to have the second derivatives continuous between elements. Typical C^2 equations of this type are also given in Fig. 4.5.1 and elsewhere. Since derivatives have also been introduced as nodal parameters, the previous statement that $\Sigma H_i = 1$ is no longer true.

$$x = Lr \qquad (\ )' = d(\ )/dx$$

a) $C^1: U = H_1 U_1 + H_2 U_1' + H_3 U_2 + H_4 U_2'$

$$\begin{aligned} H_1(r) &= (2r^3 - 3r^2 + 1) \\ H_2(r) &= (r^3 - 2r^2 + r) L \\ H_3(r) &= (3r^2 - 2r^3) \\ H_4(r) &= (r^3 - r^2) L \end{aligned}$$

b) $C^2: U = H_1 U_1 + H_2 U_1' + H_3 U_1'' + H_4 U_2 + H_5 U_2' + H_6 U_2''$

$$\begin{aligned} H_1 &= (1 - 10r^3 + 15r^4 - 6r^5) \\ H_2 &= (r - 6r^3 + 8r^4 - 3r^5) L \\ H_3 &= (r^2 - 3r^3 + 3r^4 - r^5) L^2/2 \\ H_4 &= (10r^3 - 15r^4 + 6r^5) \\ H_5 &= (7r^4 - 3r^5 - 4r^3) L \\ H_6 &= (r^3 - 2r^4 + r^5) L^2/2 \end{aligned}$$

c) $C^3: U = H_1 U_1 + H_2 U_1' + H_3 U_1'' + H_4 U_1''' + H_5 U_2 + H_6 U_2' + H_7 U_2'' + H_8 U_2'''$

$$\begin{aligned} H_1 &= (1 - 35r^4 + 84r^5 - 70r^6 + 20r^7) \\ H_2 &= (r - 20r^4 + 45r^5 - 36r^6 + 10r^7)/L \\ H_3 &= (r^2 - 10r^4 + 20r^5 - 15r^6 + 4r^7) L^2/2 \\ H_4 &= (r^3 - 4r^4 + 6r^5 - 4r^6 + r^7) L^3/6 \\ H_5 &= (35r^4 - 84r^5 + 70r^6 - 20r^7) \\ H_6 &= (10r^7 - 34r^6 + 39r^5 - 15r^4) L \\ H_7 &= (5r^4 - 14r^5 + 13r^6 - 4r^7) L^2/2 \\ H_8 &= (r^7 - 3r^6 + 3r^5 - r^4) L^3/6 \end{aligned}$$

Figure 4.5.1 Hermitian interpolation in unit coordinates

4.6 Hierarchical Interpolation

Recently some alternate types of interpolation have become popular. They are called *hierarchical functions*. The unique feature of these polynomials is that the higher order polynomials contain the lower order ones. This concept is shown in Fig. 4.6.1. Thus, to get new functions you simply add some terms to the old functions. To illustrate this concept let us return to the linear element in local natural coordinates. In that element

$$u^e(n) = H_1(n)\, u_1^e + H_2(n)\, u_2^e \tag{4.20}$$

where the two H_i are given in Eq. (4.10). We want to generate a quadratic interpolation form that will not destroy these H_i as Eq. (4.17) did. The key to accomplishing this goal is to note that the second derivative of (4.10) is everywhere zero. Thus, if we introduce an additional degree of freedom related to the second derivative of u it will not affect the linear terms. Figure 4.6.2 shows the linear element where we have added a third midpoint ($n = 0$) control node to be associated with the quadratic additions. At the third node let the degree of freedom be the second local derivative, d^2u/dr^2. Upgrade the approximation by setting

$$u(n) = H_1(n)\, u_1^e + H_2(n)\, u_2^e + Q_3(n)\, \frac{d^2 u^e}{dn^2} \qquad (4.21)$$

where the hierarchical quadratic addition is: $H_3(n) = c_1 + c_2 n + c_3 n^2$. The three constants are found from the conditions that it vanishes at the two original nodes, so as not to change H_1 and H_2, and the second derivative is unity at the new midpoint node. The result is

$$H_3(n) = (n^2 - 1)/2. \qquad (4.22)$$

Figure 4.6.3 illustrates how the concept is extended to a cubic hierarchical element.

The higher order hierarchical functions are becoming increasingly popular. They utilize the higher derivatives at the center node. We introduce the notation $m \rightarrow n$ to denote the presence of consecutive tangential derivatives from order m to order n. The value of the function is implied by $m = 0$. These functions must vanish at the end nodes. Finally, we usually want the function $H_{p+1}(n)$, $p \geq 2$ to have its p-th derivative take on a value of unity at the center node. The resulting functions are not unique. A common set of hierarchical functions in natural coordinates $-1 \leq n \leq 1$ are

$$H_p(n) = (n^p - b)/p!\ , \qquad p \geq 2 \qquad (4.23)$$

where $b = 1$ if p is even, and $b = n$ if p is odd. The first eight members of this family are shown in Fig. 4.6.4. Note that the even functions approach a rectangular shape as $p \rightarrow \infty$, but there is not much change in their form beyond the fourth order polynomial. Likewise, the odd functions approach a sawtooth as $p \rightarrow \infty$, but they change relatively little after the cubic order polynomial. These observations suggest that for the above hierarchical choice it may be better to stop at the fourth order polynomial and refine the mesh rather than adding more hierarchical degrees of freedom. However, this form might capture shape boundary layers or shocks better than other choices. These relations are zero at the ends, $n = \pm 1$. The first derivatives of these functions are

$$H'_{p+1} = [pn^{(p-1)} - b']/p!$$

and since b'' is always zero, the second derivatives are

$$H''_{p+1} = p(p-1)\, n^{(p-2)}/p! = n^{(p-2)}/(p-2)!\,.$$

Proceeding in this manner it is easy to show by induction that the m-th derivative for $m \geq 2$ is

$$H_{p+1}^{(m)}(n) = n^{(p-m)}/(p-m)!\,. \qquad (4.24)$$

At the center point, $n = 0$, the derivative has a value of

$$H_{p+1}^{(m)}(0) = \begin{cases} 0 & \text{if } m \neq p \\ 1 & \text{if } m = p. \end{cases}$$

We will see later that when hierarchical functions are utilized, the element matrices for a p-th order polynomial are partitions of the element matrices for a $(p+1)$ order polynomial. A typical cubic element, shown in Fig. 4.6.3, would be built by using the first two hierarchical functions shown in the previous figure.

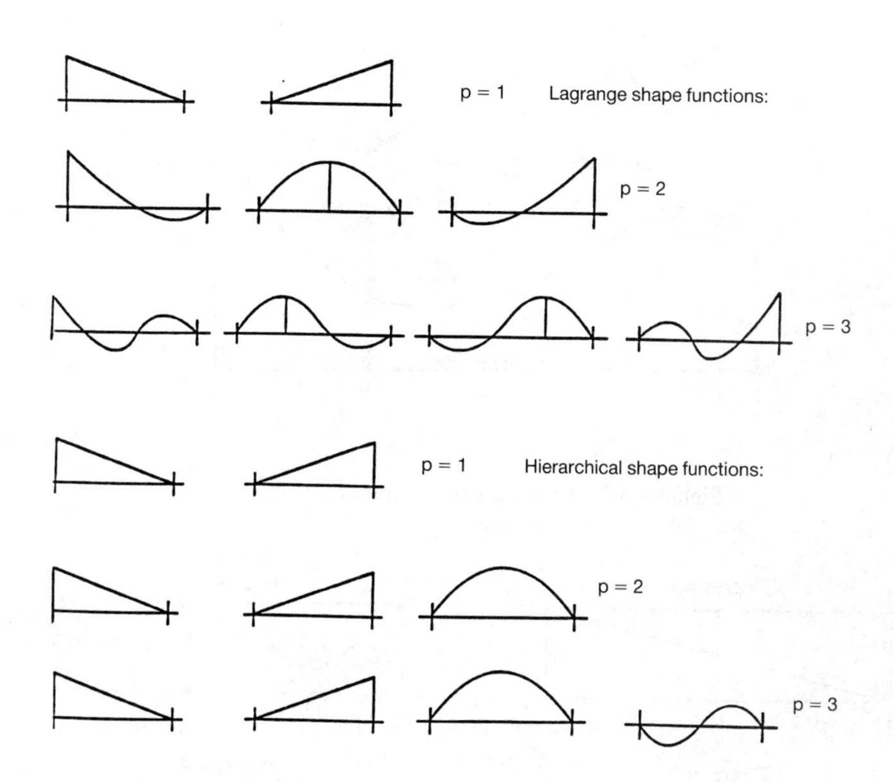

Figure 4.6.1 Concept of hierarchical shape functions

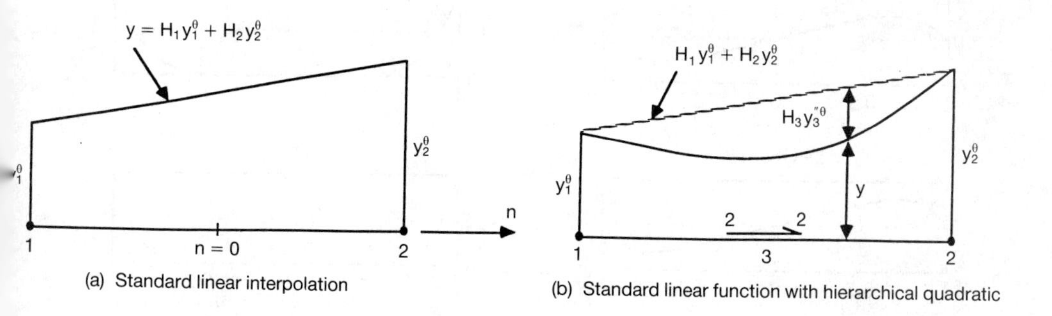

Figure 4.6.2 A quadratic hierarchical element

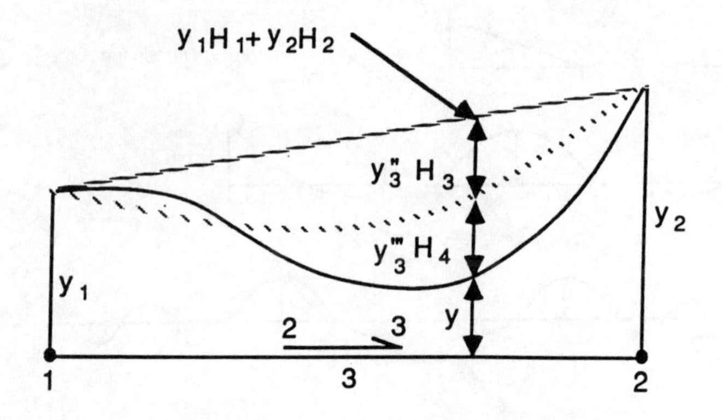

Figure 4.6.3 A typical cubic hierarchical element

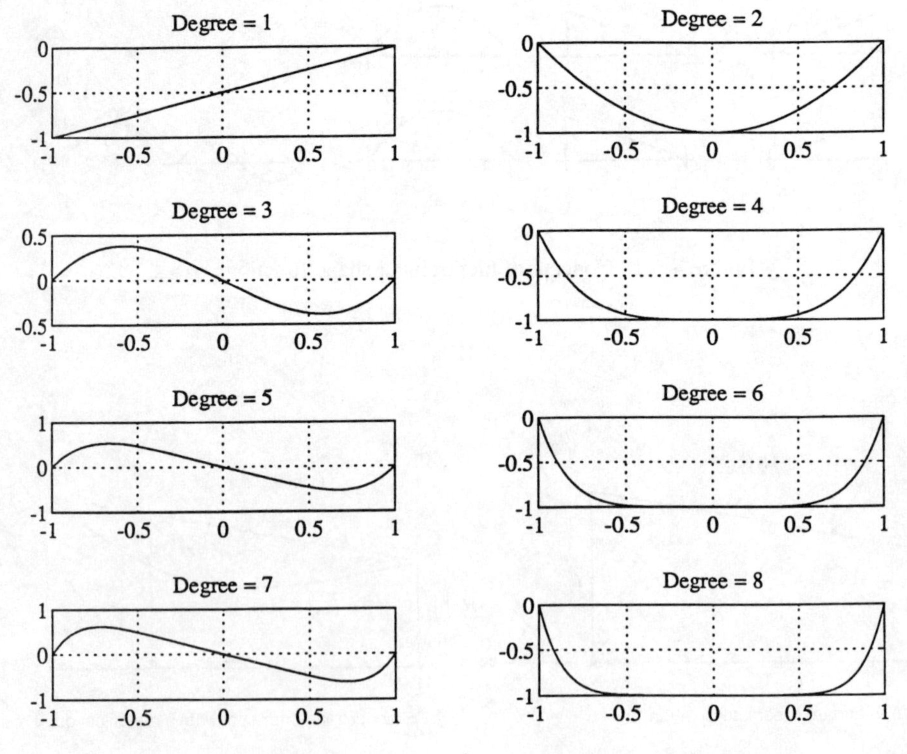

Figure 4.6.4 A C^0 hierarchical family

The element square matrix will always involve an integral of the product of the derivatives of the interpolation functions. If those derivatives were orthogonal then they would result in a diagonal square matrix. That would be very desirable. Thus, it is becoming popular to search for interpolation functions whose derivatives are close to being orthogonal. It is well known that integrals of products of *Legendre polynomials* are orthogonal. This suggests that a useful trick would be to pick interpolation functions that are integrals of Legendre polynomials so that their derivatives are Legendre polynomials. Such a trick is very useful in the so-called *p-method* and *hp-method* of adaptive finite element analysis. For future reference we will observe that the first ten Legendre polynomials on the domain of $-1 \le x \le 1$ are [1, 4]:

$$\begin{aligned}
P_0(x) &= 1 \\
P_1(x) &= x \\
P_2(x) &= (3x^2 - 1)/2 \\
P_3(x) &= (5x^3 - 3x)/2 \\
P_4(x) &= (35x^4 - 30x^2 + 3)/8 \\
P_5(x) &= (63x^5 - 70x^3 + 15x)/8 \\
P_6(x) &= (231x^6 - 315x^4 + 105x^2 - 5)/16 \\
P_7(x) &= (429x^7 - 693x^5 + 315x^3 - 35x)/16 \\
P_8(x) &= (6435x^8 - 12012x^6 + 6930x^4 - 1260x^2 + 35)/128 \\
P_9(x) &= (1260x - 18480x^3 + 72072x^5 - 102960x^7 + 48620x^9)/512 \\
P_{10}(x) &= (184756x^{10} - 437580x^8 + 360360x^6 - 120120x^4 + 13860x^2 - 252)/1024 .
\end{aligned} \tag{4.25}$$

Legendre polynomials can be generated from the *recursion formula*:

$$(n+1)\,P_{n+1}(x) = (2n+1)\,x\,P_n(x) - nP_{n-1}(x), \qquad n \ge 1$$

and

$$n\,P'_{n+1}(x) = (2n+1)\,x\,P'_n(x) - (n+1)\,P'_{n-1}(x). \tag{4.26}$$

To avoid roundoff error and unnecessary calculations, these recursion relations should be used instead of Eq. (4.25) when computing these polynomials. They have the orthogonality property:

$$\int_{-1}^{+1} P_i(x)\,P_j(x)\,dx = \begin{cases} \dfrac{2}{2i+1} & \text{for } i = j \\ 0 & \text{for } i \ne j. \end{cases} \tag{4.27}$$

To create a family of functions for potential use as hierarchical interpolation functions we next consider the integral of the above polynomials. Define a new function

$$\gamma_j(x) = \int_{-1}^{x} P_{j-1}(t)\,dt. \tag{4.28}$$

A handbook of mathematical functions shows the useful relation for Legendre polynomials that

$$(2j-1)\,P_{j-1}(t) = P'_j(t) - P'_{j-2}(t) \tag{4.29}$$

where ( )′ denotes dP/dt. The integral of the derivative is evaluated by inspection and yields

$$\gamma_j(x) = [\,P_j(x) - P_{j-2}(x)\,]\,/(2j-1) \tag{4.30}$$

since the lower limit terms cancel each other because

$$P_j(-1) = \begin{cases} 1 & j \text{ even} \\ -1 & j \text{ odd}. \end{cases}$$

We may want to multiply by a constant to scale such a function in a special way. For example, to make its second derivative unity at $x = 0$. Thus, for use as interpolation functions we will consider the family of functions defined as

$$\phi_j(x) = [\,P_j(x) - P_{j-2}(x)\,]\,/\lambda_j \equiv \psi_j(x)/\lambda_j \tag{4.31}$$

where λ_j is a constant to be selected later. From the definition of the Legendre polynomials, we see that the first few values of $\psi_j(x)$ that are of interest are:

$$\begin{aligned}
\psi_2(x) &= 3(x^2-1)/2 \\
\psi_3(x) &= 5(x^3-x)/2 \\
\psi_4(x) &= 7(5x^4-6x^2+1)/8 \\
\psi_5(x) &= 9(7x^5-10x^3+3x)/8 \\
\psi_6(x) &= 11(21x^6-35x^4+15x^2-1)/16 \\
\psi_7(x) &= 13(33x^7-63x^5+35x^3-5x)/16 \\
\psi_8(x) &= 15(429x^8-924x^6+630x^4-140x^2+5)/128\,.
\end{aligned} \tag{4.32}$$

These functions are shown in Fig. 4.6.5 along with a linear polynomial. Note that each function has its number of roots (zero values) equal to the order of the polynomial. The previous set had only two roots for the even order polynomials and three roots for the odd order polynomials (excluding the linear one). Thus, this is clearly a different type of function for hierarchical use. These would be more expensive to integrate numerically since there are more terms in each function. Note that the $\psi_j(x)$ have the property that they vanish at the ends of the domain: $\psi_j(\pm 1) \equiv 0,\ j \geq 2$. A popular choice for the midpoint hierarchical interpolation functions is to pick

$$H_j(x) = \phi_{j-1}(x), \qquad j \geq 3 \tag{4.33}$$

where the scaling is chosen to be

$$\lambda_j \equiv \sqrt{4j-2}\,. \tag{4.34}$$

The reader should note for future reference that if the above domain of $-1 \leq x \leq 1$ was the edge of a two-dimensional element then the above derivatives would be viewed as tangential derivatives on that edge. The same is true for edges of solid elements. If one wanted to use extra nodes to allow for curved line segments and use them also as degrees of freedom, then one can envision a family of hierarchical elements as shown in Fig. 4.6.6. Hierarchical enrichment is not just restricted to C^0 functions, but have also been used with Hermite functions as well. Earlier we saw the C^1 cubic Hermite and the C^2 fifth order Hermite polynomials. The cubic has nodal dof that are the value and

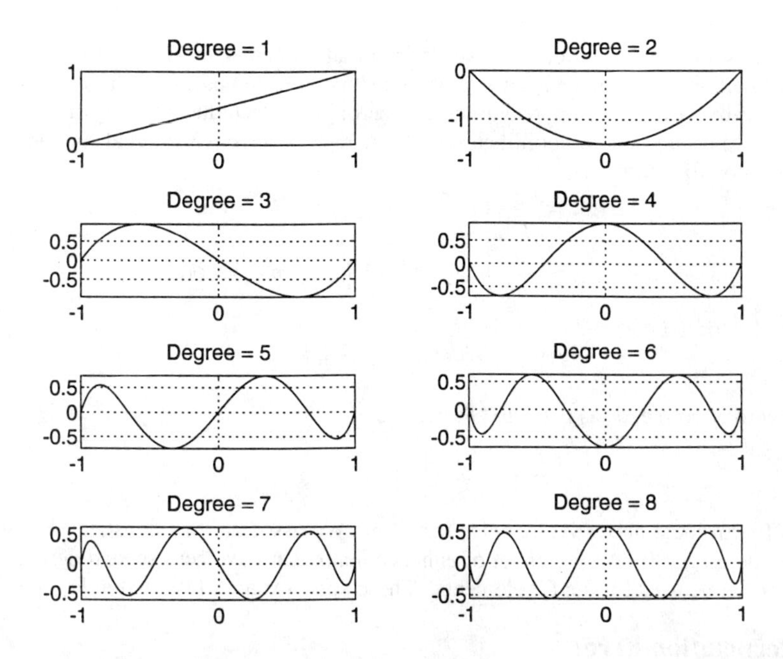

Figure 4.6.5 Integrals of Legendre polynomials

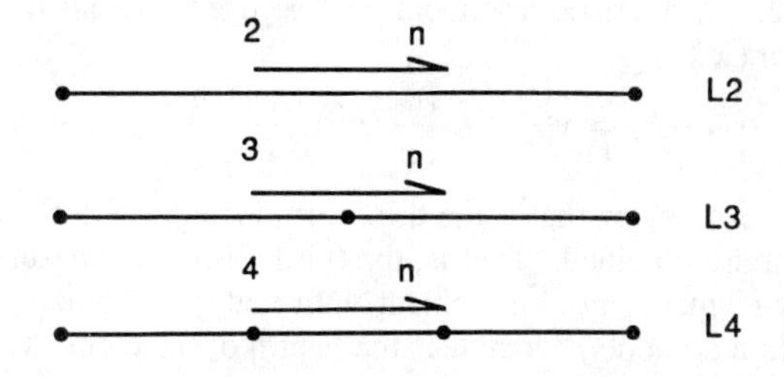

Figure 4.6.6 Other Lagrangian hierarchical families

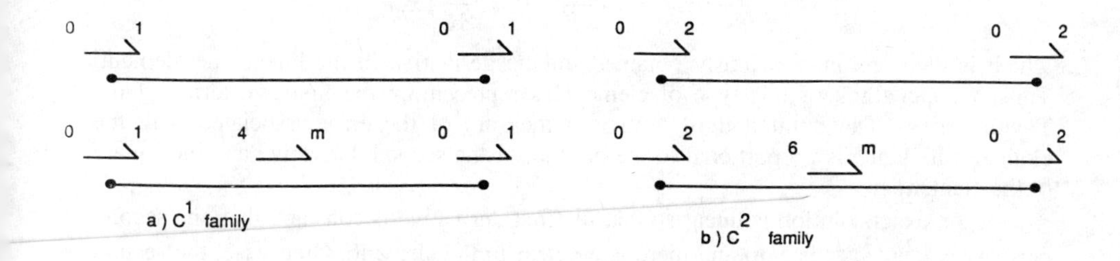

Figure 4.6.7 Hermite hierarchical elements

slope of the solution at each end. If we desire to add a center hierarchical enrichment, then that function should have a zero value and slope at each end. In addition, since the fourth derivative of the cubic polynomial is zero, we select that quantity as the first hierarchical dof. In natural coordinates $-1 \le a \le 1$, we have $p-3$ internal functions for $p \ge 3$. One possible choice is

$$H_p^{(0)} = \frac{1}{p!}\left[a^{p/2} - 1\right]^2, \quad p \ge 4, \text{ even}$$
$$\frac{1}{p!}\left[a^{(p-1)/2} - 1\right]^2, \quad p \ge 5, \text{ odd}. \tag{4.35}$$

For example, for $p = 4$

$$H_p^{(0)} = \frac{1}{24}\left[a^4 - 2a^2 + 1\right],$$

which is zero at both ends as is its first derivative

$$\frac{dH_p^{(0)}}{da} = \frac{1}{24}\left[4a^3 - 4a\right]$$

while its fourth local derivative is unity for all a. We associate that constant dof with the center point, $a = 0$. A similar set of enhancements that have zero second derivatives at the ends can be used for the C^2 elements. These concepts are illustrated in Fig. 4.6.7.

4.7 Interpolation Error *

To obtain a physical feel for the typical errors involved, we consider a one-dimensional model. A hueristic argument will be used. Recall the Taylor's series of a function v at a point x :

$$v(x+h) = v(x) + h\frac{\partial v}{\partial x}(x) + \frac{h^2}{2}\frac{\partial^2 v}{\partial x^2}(x) + \cdots \tag{4.36}$$

The objective here is to show that if the third term is neglected, then the relations for the linear line element are obtained. That is, the third term is a measure of the interpolation error in the linear element. For an element with nodes at i and j, we use Eq. (4.36) to estimate the function at node j when h is the length of the element:

$$v_j = v_i + h\frac{\partial v}{\partial x}(x_i).$$

Solving for the gradient at node i yields

$$\frac{\partial v}{\partial x}(x_i) = \frac{(v_j - v_i)}{h} = \frac{\partial v}{\partial x}(x_j)$$

which is the constant previously obtained for the derivative in the linear line element. Thus, we can think of this type of element as representing the first two terms of the Taylor series. The omitted third term is a measure of the error associated with the element. Its value is proportional to the product of the second derivative and the square of the element size.

If the exact solution is linear so that the first derivative is constant, then the second derivative, $\partial^2 v/\partial x^2$, is zero and there is no error in the element. Otherwise, the second derivative and element error do not vanish. If the user wishes to exercise control over this relative error, then the element size, h, must be varied, or we must use a higher degree interpolation for the element. If we think in terms of the bar element, then v and

$\partial v / \partial x$ represent the displacement and strain, respectively. The contribution to the error represents the strain gradient (and stress gradient). Therefore, we must use our engineering judgment to make the element size, h, small in regions of large strain gradients (stress concentrations). Conversely, where the strain gradients are small, we can increase the element size, h, to reduce the computational cost. A similar argument can be stated for the heat conduction problem. Then, v is the temperature, $\partial v / \partial x$ describes the temperature gradient (heat flux), and the error is proportional to the flux gradient. If one does not wish to vary the element sizes, h, then to reduce the error, one must add higher order polynomial terms of the element interpolation functions so that the second derivative is present in the element. These two approaches to error control are known as the *h-method* and the *p-method,* respectively.

The previous comments have assumed the use of a uniform mesh, that is, h was the same for all elements in the mesh. Thus, the above error discussions have not considered the interaction of adjacent elements. The effects of adjacent element sizes have been evaluated for the case of a continuous bar subject to an axial load. An error term, in the governing differential equation, due to the finite element approximation at node j has been shown to be

$$E = -\frac{h}{3}(1-a)\frac{\partial^3 v}{\partial x^3}(x_j) + \frac{h^2}{12}\left[\frac{1+a^3}{1+a}\right]\frac{\partial^4 v}{\partial x^4}(x_j) + \cdots$$

where h is the size of one element and ah is the size of the adjacent element. Here it is seen that for a smooth variation ($a \doteq 1$) or a uniform mesh ($a = 1$), the error in the approximated ODE is of order h squared. However, if the adjacent element sizes differ greatly ($a \neq 1$), then a larger error of order h is present. This suggests that it is desirable to have a gradual change in element sizes when possible. One should avoid placing a small element adjacent to one that is many times larger.

4.8 References

[1] Abramowitz, M. and Stegun, I.A., *Handbook of Mathematical Functions*, National Bureau of Standards (1964).

[2] Babuska, I., Griebel, M., and Pitkaranta, J., "The Problem of Selecting Shape Functions for a p-Type Finite Element," *Int. J. Num. Meth. Eng.*, **28**, pp. 1891–1908 (1989).

[3] Desai, C.S., *Elementary Finite Element Method*, Englewood Cliffs: Prentice-Hall (1979).

[4] Szabo, B. and Babuska, I., *Finite Element Analysis*, New York: John Wiley (1991).

[5] Zienkiewicz, O.C., *The Finite Element Method,* 3rd edition, New York: McGraw-Hill (1979).

Chapter 5

ONE-DIMENSIONAL INTEGRATION

5.1 Introduction

Since the finite element method is based on integral relations it is logical to expect that one should strive to carry out the integrations as efficiently as possible. In some cases we will employ exact integration. In others we may find that the integrals can become too complicated to integrate exactly. In such cases the use of numerical integration will prove useful or essential. The important topics of local coordinate integration and Gaussian quadratures will be introduced here. They will prove useful when dealing with higher order interpolation functions in complicated element integrals.

5.2 Local Coordinate Jacobian

We have previously seen that the utilization of local element coordinates can greatly reduce the algebra required to establish a set of interpolation functions. Later we will see that some 2-D elements must be formulated in local coordinates in order to meet the interelement continuity requirements. Thus, we should expect to often encounter local coordinate interpolation. However, the governing integral expressions must be evaluated with respect to a unique global or physical coordinate system. Clearly, these two coordinate systems must be related. The relationship for integration with a change of variable (change of coordinate) was defined in elementary concepts from calculus. At this point it would be useful to review these concepts from calculus. Consider a definite integral

$$I = \int_a^b f(x)\, dx\,, \quad a < x < b \tag{5.1}$$

where a new variable of integration, r, is to be introduced such that $x = x(r)$. Here it is required that the function $x(r)$ be continuous and have a continuous derivative in the interval $\alpha \le r \le \beta$. The region of r directly corresponds to the region of x such that when r varies between α and β, then x varies between $a = x(\alpha)$ and $b = x(\beta)$. In that case

$$I = \int_a^b f(x)\, dx = \int_\alpha^\beta f(\,x(r)\,)\, \frac{dx}{dr}\, dr \tag{5.2}$$

or

$$I = \int_{\alpha}^{\beta} f(r)\, J\, dr \tag{5.3}$$

where $J = dx/dr$ is called the *Jacobian* of the coordinate transformation.

5.3 Exact Polynomial Integration *

If we utilize the unit coordinates, then $\alpha = 0$ and $\beta = 1$. Then from Sec. (4.2), the Jacobian is $J = L^e$ in an element domain defined by linear interpolation. By way of comparison, if one employs natural coordinates, then $\alpha = -1$, $\beta = +1$, and from Eq. (4.16) $J = L^e/2$. Generally, we will use interpolation functions that are polynomials. Thus, the element integrals of them and/or their derivatives will also contain polynomial terms. Therefore, it will be useful to consider expressions related to typical polynomial terms. A typical polynomial term is r^m where m is an integer. Thus, from the above

$$I = \int_{x_1^e}^{x_2^e} r^m\, dx = \int_0^1 r^m L^e\, dr = L^e \left. \frac{r^{(1+m)}}{1+m} \right|_0^1 = \frac{L^e}{(1+m)}\,. \tag{5.4}$$

A similar expression can be developed for the natural coordinates. It gives

$$I = \int_{L^e} n^m\, dx = \frac{L^e}{m+1} \begin{cases} 0 & \text{if } n \text{ is odd} \\ 1 & \text{if } n \text{ is even}\,. \end{cases} \tag{5.5}$$

Later we will tabulate the extension of these concepts to two- and three-dimensional integrals. As an example of the use of Eq. (5.5), consider the integration

$$I = \int_{L^e} \mathbf{H}^{e^T} \mathbf{H}^e\, dx\,.$$

Recall that the integral of a matrix is the matrix resulting from the integration of each of the elements of the original matrix. If linear interpolation is selected for $\mathbf{H}^e$ on a line element then typical terms will include H_1^2, $H_1 H_2$, etc. Thus, one obtains:

$$I_{11} = \int_{L^e} H_1^2(r)\, dx = \int_{L^e} (1-r)^2\, dx = L^e(1 - 2/2 + 1/3) = \frac{L^e}{3}$$

$$I_{12} = \int_{L^e} H_1 H_2\, dx = \int_{L^e} (1-r)\, r dx = L^e(1/2 - 1/3) = \frac{L^e}{6}$$

$$I_{22} = \int_{L^e} H_2^2\, dx = \int_{L^e} r^2\, dx = \frac{L^e}{3}$$

so that

$$I = \frac{L^e}{6} \begin{bmatrix} 2 & 1 \\ 1 & 2 \end{bmatrix}. \tag{5.6}$$

Similarly, if one employs the Lagrangian quadratic $\mathbf{H}$ in Eq. (4.17) one obtains:

$$I = \frac{L^e}{30} \begin{bmatrix} 4 & -1 & 2 \\ -1 & 4 & 2 \\ 2 & 2 & 16 \end{bmatrix}. \tag{5.7}$$

By way of comparison, if one selects the hierarchical quadratic polynomial in Eq. (4.22) the above integral becomes

$$I = \frac{L^e}{6} \begin{bmatrix} 2 & 1 & | & -1/4 \\ 1 & 2 & | & -1/4 \\ - & - & | & - \\ -1/4 & -1/4 & | & 1/10 \end{bmatrix} .$$

Note that the top left portion of this equation is the same as Eq. (5.7) which was obtained from the linear polynomial. This desirable feature of hierarchical elements was mentioned in Sec. 4.6.

Before leaving the subject of simplex integrations one should give consideration to the common special case of axisymmetric geometries, with coordinates (ρ, z). Recall from calculus that the Theorem of Pappus relates a differential volume and surface area to a differential area and length in the (ρ, z) plane of symmetry, respectively. That is, $dv = 2\pi\rho\, dA$ and $dS = 2\pi\rho dl$, where ρ denotes the radial distance to the differential element. Thus, typical axisymmetric surface integrals reduce to

$$\mathbf{I}_S = \int_S \mathbf{H}^T dS = 2\pi \int_{L^e} \mathbf{H}^T \rho\, dl = 2\pi \left(\int_{L^e} \mathbf{H}^T\mathbf{H}\, dl \right) \rho^e = \frac{2\pi L^e}{6} \begin{bmatrix} 2 & 1 \\ 1 & 2 \end{bmatrix} \rho^e ,$$

since $\rho = \mathbf{H}\rho^e$. Many workers like to omit the 2π term and work on a per-unit-radian basis so that they can more easily do both two-dimensional and axisymmetric calculations with a single program.

5.4 Numerical Integration

Numerical integration is simply a procedure that approximates (usually) an integral by a summation. To review this subject we refer to Fig. 5.4.1. Recall that the integral

$$I = \int_a^b f(x)\, dx \tag{5.8}$$

can be viewed graphically as the area between the x-axis and the curve $y = f(x)$ in the region of the limits of integration. Thus, we can interpret numerical integration as an approximation of that area. The *trapezoidal rule* of numerical integration simply approximates the area by the sum of several equally spaced trapezoids under the curve between the limits of a and b. The height of a trapezoid is found from the integrand, $y_j = y(x_j)$, evaluated at equally spaced points, x_j and x_{j+1}. Thus, a typical contribution is $A = h(y_j + y_{j+1})/2$, where $h = x_{j+1} - x_j$ is the spacing. Thus, for n points (and $n-1$ spaces), the well-known approximation is

$$I \approx h \left(\frac{1}{2} y_1 + y_2 + y_3 + \cdots + y_{n-1} + \frac{1}{2} y_n \right), \quad I \approx \sum_{j=1}^{n} w_j\, f(x_j) \tag{5.9}$$

where $w_j = h$, except $w_1 = w_n = h/2$. A geometrical interpretation of this is that the area under curve, I, is the sum of the products of certain heights, $f(x_j)$ times some corresponding widths, w_j. In the terminology of numerical integration, the locations of the points, x_j, where the heights are computed are called *abscissae* and the widths, w_j, are called *weights*. Another well-known approximation is the *Simpson rule*, which uses parabolic segments in the area approximation. For most functions the above rules may require 20 to 40 terms in the summation to yield acceptable accuracy. We want to carry

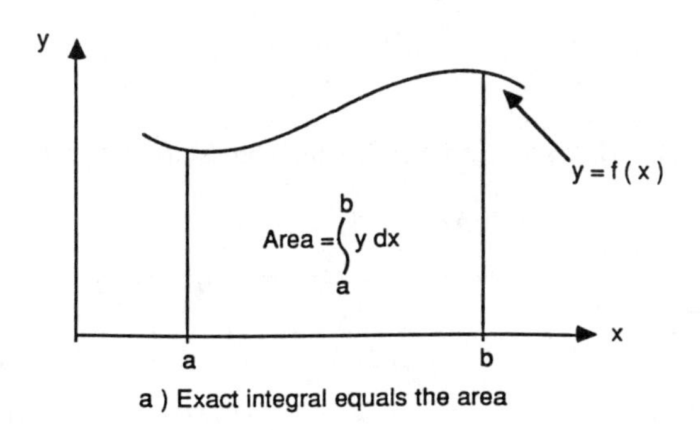

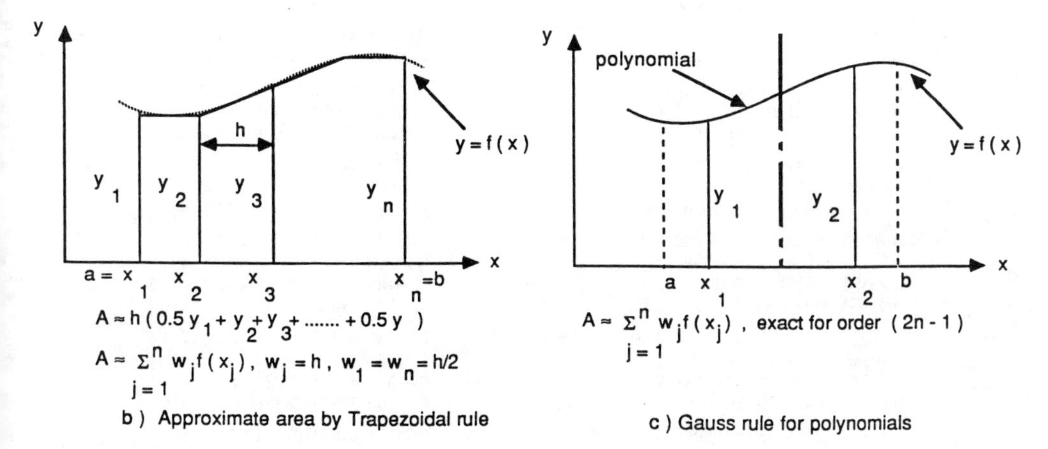

Figure 5.4.1 One-dimensional numerical integration

out the summation with the minimum number of terms, n, in order to reduce the computational cost. What is the minimum number of terms? The answer depends on the form of the integrand $f(x)$. Since the parametric geometry usually involves polynomials we will consider that common special case for $f(x)$.

The famous mathematician Gauss posed this question: What is the minimum number of points, n, required to exactly integrate a polynomial, and what are the corresponding abscissae and weights? If we require the summation to be exact when $f(x)$ is any one of the power functions $1, x, x^2, \ldots, x^{2n-1}$, we obtain a set of $2n$ conditions that allow us to determine the n abscissae, x_i, and their corresponding n weights, w_j. The n *Gaussian quadrature* points are symmetrically placed with respect to the center of the interval, and will exactly integrate a polynomial of order $(2n + 1)$. The center point location is included in the abscissae list when n is odd, but the end points are never utilized in a Gauss rule. The Gauss rule data are usually tabulated for a non-dimensional *unit coordinate* range of $0 \leq t \leq 1$, or for a *natural coordinate* range of $-1 \leq t \leq +1$. Table 5.1 presents the low-order Gauss rule data. A two-point Gauss rule can often exceed the accuracy of a 20-point trapezoidal rule.

Table 5.1 Abscissas and weights for Gaussian quadrature

$$\int_{-1}^{+1} f(x)\,dx = \sum_{i=1}^{n} w_i f(x_i)$$

$\pm x_i$		w_i
0.00000 00000 00000 00000 0000	$n = 1$	2.00000 00000 00000 00000 000
0.57735 02691 89625 76450 9149	$n = 2$	1.00000 00000 00000 00000 000
0.77459 66692 41483 37703 5835	$n = 3$	0.55555 55555 55555 55555 556
0.00000 00000 00000 00000 0000		0.88888 88888 88888 88888 889
0.86113 63115 94052 57522 3946	$n = 4$	0.34785 48451 37453 85737 306
0.33998 10435 84856 26480 2666		0.65214 51548 62546 14262 694
0.90617 98459 38663 99279 7627	$n = 5$	0.23692 68850 56189 08751 426
0.53846 93101 05683 09103 6314		0.47862 86704 99366 46804 129
0.00000 00000 00000 00000 0000		0.56888 88888 88888 88888 889
0.93246 95142 03152 02781 2302	$n = 6$	0.17132 44923 79170 34504 030
0.66120 93864 66264 51366 1400		0.36076 15730 48138 60756 983
0.23861 91860 83196 90863 0502		0.46791 39345 72691 04738 987
0.94910 79123 42758 52452 6190	$n = 7$	0.12948 49661 68869 69327 061
0.74153 11855 99394 43986 3865		0.27970 53914 89276 66790 147
0.40584 51513 77397 16690 6607		0.38183 00505 05118 94495 037
0.00000 00000 00000 00000 0000		0.41795 91836 73469 38775 510
0.96028 98564 97536 23168 3561	$n = 8$	0.10122 85362 90376 25915 253
0.79666 64774 13626 73959 1554		0.22238 10344 53374 47054 436
0.52553 24099 16328 98581 7739		0.31370 66458 77887 28733 796
0.18343 46424 95649 80493 9476		0.36268 37833 78361 98296 515
0.96816 02395 07626 08983 5576	$n = 9$	0.08127 43883 61574 41197 189
0.83603 11073 26635 79429 9430		0.18064 81606 94857 40405 847
0.61337 14327 00590 39730 8702		0.26061 06964 02935 46231 874
0.32425 34234 03808 92903 8538		0.31234 70770 40002 84006 863
0.00000 00000 00000 00000 0000		0.33023 93550 01259 76316 453
0.97390 65285 17171 72007 7964	$n = 10$	0.06667 13443 08688 13759 357
0.86506 33666 88984 51073 2097		0.14945 13491 50580 59314 578
0.67940 95682 99024 40623 4327		0.21908 63625 15982 04399 554
0.43339 53941 29247 19079 9266		0.26926 67193 09996 35509 123
0.14887 43389 81631 21088 4826		0.29552 42247 14752 87017 389
0.98722 86581 46056 99280 3938	$n = 11$	0.05566 85671 16173 66648 275
0.88706 25997 68095 29907 5158		0.12558 03694 64904 62463 469
0.73015 20055 74049 32409 3416		0.18629 02109 27734 25142 610
0.51909 61292 06811 81592 5726		0.23319 37645 91990 47991 852
0.26954 31559 52344 97233 1532		0.26280 45445 10246 66218 069
0.00000 00000 00000 00000 0000		0.27292 50867 77900 63071 448
0.98156 06342 46719 25069 0549	$n = 12$	0.04717 53363 86511 82719 462
0.90411 72563 70474 85667 8466		0.10693 93259 95318 43096 025
0.76900 26741 94304 68703 6894		0.16007 83285 43346 22633 465
0.58731 79542 86617 44729 6702		0.20316 74267 23065 92174 906
0.36783 14989 98180 19375 2692		0.23349 25365 38354 80876 085
0.12523 34085 11468 91547 2441		0.24914 70458 13402 78500 056

Sometimes it is desirable to have a numerical integration rule that specifically includes the two end points in the abscissae list when $(n \geq 2)$. The *Lobatto rule* is such an alternate choice. Its n points will exactly integrate a polynomial of order $(2n - 3)$ for $n > 2$. Its data are included in Table 5.2. It is usually less accurate than the Gauss rule but it can be useful. Mathematical handbooks give tables of Gauss or Lobatto data for much higher values of n. Some results of Gauss's work are outlined below. Let y denote $f(x)$ in the integral to be computed. Define a change of variable

$$x(r) = 1/2\,(b - a)\,r + 1/2\,(b + a) \tag{5.10}$$

so that the non-dimensional limits of integration of r become -1 and $+1$. The new value of $y(r)$ is

$$y = f(x) = f[\,1/2\,(b - a)\,r + 1/2\,(b + a)\,] = \Phi(r)\,. \tag{5.11}$$

Table 5.2. Abscissas and weight factors for Lobatto integration

$$\int_{-1}^{+1} f(x)\,dx \approx \sum_{i=1}^{n} w_i\, f(x_i)$$

$\pm x_i$		w_i
0.00000 00000 00000	$n = 1$	2.00000 00000 00000
1.00000 00000 00000	$n = 2$	1.00000 00000 00000
1.00000 00000 00000	$n = 3$	0.33333 33333 33333
0.00000 00000 00000		1.33333 33333 33333
1.00000 00000 00000	$n = 4$	0.16666 66666 66667
0.44721 35954 99958		0.83333 33333 33333
1.00000 00000 00000	$n = 5$	0.10000 00000 00000
0.65465 36707 07977		0.54444 44444 44444
0.00000 00000 00000		0.71111 11111 11111
1.00000 00000 00000	$n = 6$	0.06666 66666 66667
0.76505 53239 29465		0.37847 49562 97847
0.28523 15164 80645		0.55485 83770 35486
1.00000 00000 00000	$n = 7$	0.04761 90476 19048
0.83022 38962 78567		0.27682 60473 61566
0.46884 87934 70714		0.43174 53812 09863
0.00000 00000 00000		0.48761 90476 19048
1.00000 00000 00000	$n = 8$	0.03571 42857 14286
0.87174 01485 09607		0.21070 42271 43506
0.59170 01814 33142		0.34112 26924 83504
0.20929 92179 02479		0.41245 87946 58704
1.00000 00000 00000	$n = 9$	0.02777 77777 77778
0.89975 79954 11460		0.16549 53615 60805
0.67718 62795 10738		0.27453 87125 00162
0.36311 74638 26178		0.34642 85109 73406
0.00000 00000 00000		0.37151 92743 76417
1.00000 00000 00000	$n = 10$	0.02222 22222 22222
0.91953 39081 66459		0.13330 59908 51070
0.73877 38651 05505		0.22488 93420 63126
0.47792 49498 10444		0.29204 26836 79684
0.16527 89576 66387		0.32753 97611 83897

Noting from Eq. (5.10) that $dx = 1/2\,(b-a)\,dr$, the original integral becomes

$$I = \frac{1}{2}(b-a)\int_{-1}^{1} \Phi(r)\,dr. \tag{5.12}$$

Gauss showed that the integral in Eq. (5.12) is given by

$$\int_{-1}^{1} \Phi(r)\,dr = \sum_{i=1}^{n} W_i\,\Phi(r_i),$$

where W_i and r_i represent tabulated values of the *weight functions* and *abscissae* associated with the n points in the non-dimensional interval (–1, 1). Thus, the final result is

$$I = \frac{1}{2}(b-a)\sum_{i=1}^{n} W_i\,\Phi(r_i) = \sum_{i=1}^{n} f\left[x(r_i)\right] W_i. \tag{5.13}$$

Gauss also showed that this equation will exactly integrate a polynomial of degree $(2n-1)$. For a higher number of space dimensions (which range from -1 to $+1$), one obtains a multiple summation. Since Gaussian quadrature data are often tabulated in references for the range $-1 \le r \le +1$, it is popular to use the natural coordinates in defining element integrals. However, one can convert the tabulated data to any convenient system such as the unit coordinate system where $0 \le r \le 1$. The latter may be more useful on triangular regions. As an example of Gaussian quadratures, consider the following one-dimensional integral:

$$I = \int_{1}^{2} \begin{bmatrix} 2 & 2x \\ 2x & (1+2x^2) \end{bmatrix} dx = \int_{1}^{2} \mathbf{F}(x)\,dx.$$

If two Gauss points are selected, then the tabulated values from Table 5.1 give $W_1 = W_2 = 1$ and $r_1 = 0.57735 = -r_2$. The change of variable gives $x(r) = (r+3)/2$, so that $x(r_1) = 1.788675$ and $x(r_2) = 1.211325$. Therefore, from Eq. (5.13)

$$I = 1/2\,(2-1)\left[W_1\mathbf{F}\left[x(r_1)\right] + W_2\mathbf{F}\left[x(r_2)\right]\right]$$

$$= 1/2\,(1)\left\{(1)\begin{bmatrix} 2 & 2(1.788675) \\ sym. & 1+2(1.788675)^2 \end{bmatrix} + (1)\begin{bmatrix} 2 & 2(1.211325) \\ sym. & 1+2(1.211325)^2 \end{bmatrix}\right\}$$

$$I = \begin{bmatrix} 2.00000 & 3.00000 \\ 3.00000 & 5.66667 \end{bmatrix},$$

which is easily shown to be in good agreement with the exact solution. As another example consider a typical term in Eq. (5.8). Specifically, from Eqs. (4.16) and (4.18)

$$I_{33} = \int_{L^e} H_3^2\,dx = \frac{L^e}{2}\int_{-1}^{+1} (1-n^2)^2\,dn.$$

Since the polynomial terms to be integrated are fourth order, we should select $(2m-1) = 4$, or $m = 3$ Gaussian points. Then,

$$I_{33} = \int_{L^e} H_3^2\, dx = \frac{L^e}{2} \int_{-1}^{+1} H_3^2(n)\, dn$$

$$I_{33} = \frac{L^e}{2} \left[0.55556 \left[1.00000 - (0.77459)^2 \right]^2 \right.$$

$$\left. + \; 0.88889 \left[1.00000 - (0.0)^2 \right]^2 + 0.55556 \left[1.00000 - (+0.77459)^2 \right]^2 \right]$$

$$I_{33} = \frac{L^e}{2} (0.08889 + 0.88889 + 0.08889) = 0.5333\, L^e,$$

which agrees well with the exact value of $16\, L^e / 30$ given in Eq. (5.7).

5.5 Variable Jacobians

When the parametric space and physical space have the same number of dimensions then the Jacobian is a square matrix. Otherwise, we need to use more calculus to evaluate the integrals. For example, we often find the need to execute integrations along a two-dimensional curve defined by our one-dimensional parametric representation. Consider a planar curve in the xy-plane. We may need to know its length and first moments (centroid), which are defined as

$$L = \int_L ds\,, \qquad \bar{x} L = \int_L x(t)\, ds\,, \qquad \bar{y} L = \int_L y(t)\, ds\,,$$

respectively, where ds denotes the physical length of a segment, dt, of the parametric length. To evaluate these quantities we need to convert to an integral in the parametric space. For example,

$$L = \int_L ds = \int_0^1 \frac{ds}{dt}\, dt\,.$$

To relate the physical and parametric length scales, we must first recall the planar relation that

$$ds^2 = dx^2 + dy^2\,.$$

Since both x and y are defined in terms of t, we can extend this identity to the needed quantity

$$\left(\frac{ds}{dt} \right)^2 = \left(\frac{dx}{dt} \right)^2 + \left(\frac{dy}{dt} \right)^2$$

where dx/dt can be found from the spatial interpolation functions, etc. for dy/dt. Thus, our physical length is defined in terms of the parametric coordinate, t, and the spatial data for the nodes defining the curve location in the xy-plane (i.e., x^e and y^e) :

$$L = \int_0^1 \sqrt{\left(\frac{dx}{dt} \right)^2 + \left(\frac{dy}{dt} \right)^2}\; dt$$

where

$$\frac{dx(t)}{dt} = \sum_{i=1}^{N} \frac{dH_i(t)}{dt} x_i^e, \qquad dy\frac{(t)}{dt} = \sum_{i=1}^{N} \frac{dH_i(t)}{dt} y_i^e .$$

Note that this does preserve the proper units for L, and it has what we could refer to as a variable Jacobian. The preceding integral is trivial only when the planar curve is a straight line ($N=2$). Then, from linear geometric interpolation $dx/dt = x_2^e - x_1^e$ and $dy/dt = y_2^e - y_1^e$ are both constant, and the result simplifies to

$$L^e = \sqrt{\left[x_2^e - x_1^e\right]^2 + \left[y_2^e - y_1^e\right]^2} \int_0^1 dt = L$$

which, by inspection, is exact. For any other curve shape, these integrals become unpleasant to evaluate, and we would consider their automation by means of numerical integration.

To complete this section we outline an algorithm to automate the calculation of the y-centroid:

1. Recover the N points describing the curve, $\mathbf{x}^e$, $\mathbf{y}^e$.
2. Recover the M-point quadrature data, t_j and w_j.
3. Zero the integrals: $L = 0, \bar{y}L = 0$.
4. Loop over all the quadrature points: $1 \le j \le M$. At each local quadrature point, t_j:

A. Find the length scales (i.e., Jacobian):

(i) Compute local derivatives of the N interpolation functions

$$\mathbf{DH}_j \equiv \frac{\partial \mathbf{H}}{\partial t}\bigg|_{t=t_j}$$

(ii) Get x- and y-derivatives from curve data

$$\frac{dx}{dt_j} \equiv \mathbf{DH}_j\, \mathbf{x}^e = \sum_{i=1}^{N} \frac{\partial H_i(t_j)}{\partial t} x_i^e, \qquad \frac{dy}{dt_j} \equiv \mathbf{DH}_j\, \mathbf{y}^e$$

(iii) Find length scale at point t_j

$$\frac{ds}{dt_j} = \sqrt{\left[\frac{dx}{dt}\right]_j^2 + \left[\frac{dy}{dt}\right]_j^2}$$

B. Evaluate the integrand at point t_j:

(i) Evaluate the N interpolation functions:

$$\mathbf{H}_j \equiv \mathbf{H}(t_j)$$

(ii) Evaluate y from curve data

$$y_j = \mathbf{H}_j\, \mathbf{y}^e = \sum_{i=1}^{N} H_i(t_j)\, y_i^e$$

C. Form products and add to previous values

$$L = L + \frac{ds}{dt_j} w_j, \qquad \bar{y}L = \bar{y}L + y_j \frac{ds}{dt_j} w_j .$$

5. Evaluate items computed from the completed integrals : $\bar{y} = \bar{y}L/L$, etc. for $\bar{x}$.

Note that to automate the integration we simply need: (1) storage of the quadrature data, t_j and w_j, (2) access to the curve data, $\mathbf{x}^e$, $\mathbf{y}^e$, (3) a subroutine to find the parametric derivative of the interpolation functions at any point, t, and (4) a function program to evaluate the integrand(s) at any point, t. This usually requires (see Step **4B**) a subroutine to evaluate the interpolation functions at any point, t. The evaluations of the interpolation products, at a point t_j, can be thought of as a dot product of the data array $\mathbf{x}^e$ and the evaluated interpolation quantities $\mathbf{H}_j$ and $\mathbf{DH}_j$.

5.6 References

[1] Abramowitz, M. and Stegun, I.A., *Handbook of Mathematical Functions*, National Bureau of Standards (1964).

[2] Carey, G.F. and Oden, J.T., *Finite Elements – Computational Aspects*, Englewood Cliffs: Prentice-Hall (1984).

[3] Hinton, E. and Owen, D.R.J., *Finite Element Programming*, London: Academic Press (1977).

[4] Hughes, T.J.R., *The Finite Element Method*, Englewood Cliffs: Prentice-Hall (1987).

[5] Stroud, A.H. and Secrest, D., *Gaussian Quadrature Formulas*, Englewood Cliffs: Prentice-Hall (1966).

[6] Zienkiewicz, O.C., *The Finite Element Method*, London: McGraw-Hill (1977).

Chapter 6

BEAM ANALYSIS

6.1 Introduction

A common structural system considered in engineering is that of the elastic beam. Such a beam is shown in Fig. 6.1.1. In mechanics of materials a number of common assumptions are made in order to reduce the analysis to a one-dimensional formulation. The most common assumption is that planes in the beam, normal to a fiber along the x-axis, remain normal to that fiber in its deformed state. This assumption makes that axial displacement, u, and the axial strain, ε, vary linearly with the transverse coordinate, y. Let v denote the transverse displacement, and $\theta = v'$ the slope of the beam. Then the axial displacement relation, for small slopes, and the axial strain are

$$u(x, y) = -yv' = -y\frac{dv}{dx}, \tag{6.1}$$

$$\varepsilon(x, y) = \frac{du}{dx} = -y\frac{d^2v}{dx^2} = -yv'', \tag{6.2}$$

respectively. For an elastic material the stress, σ, is defined by Hooke's law as

$$\sigma(x, y) = E(x)\,\varepsilon(x, y) \tag{6.3}$$

where E is the elastic modulus of the material. These quantities could vary with both of the spatial coordinates. We desire to formulate a one-dimensional model. We will define a *generalized strain* and a *generalized stress* to accomplish this goal.

From calculus we should recall that the quantity $v''(x) = 1/\rho$ is known as the *curvature* of the deflected beam and ρ is the radius of curvature. The signs in Eqs. (6.1) and (6.2) have been chosen so that a positive sign denotes tension. From statics one can show that the resultant axial load from the distributed stress is zero. However, there is a non-zero resultant moment. Its value is given by

$$M(x) = \int dm = \int -y\,dF = \int -y\,\sigma\,da = \int -y\,E\,\varepsilon\,da = \int_A E\,y^2\,v''\,da = EI\,v''(x)$$

where A is the cross-sectional area of the member and

$$I(x) = \int_A y^2\,da$$

is the moment of inertia of the cross sectional area. We will call this moment our

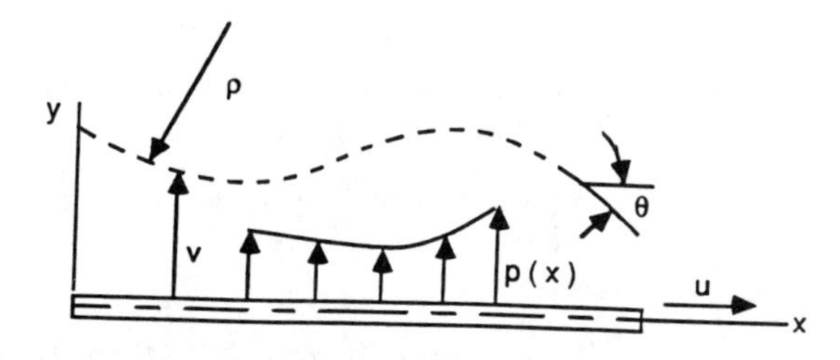

a) Coordinates and notation

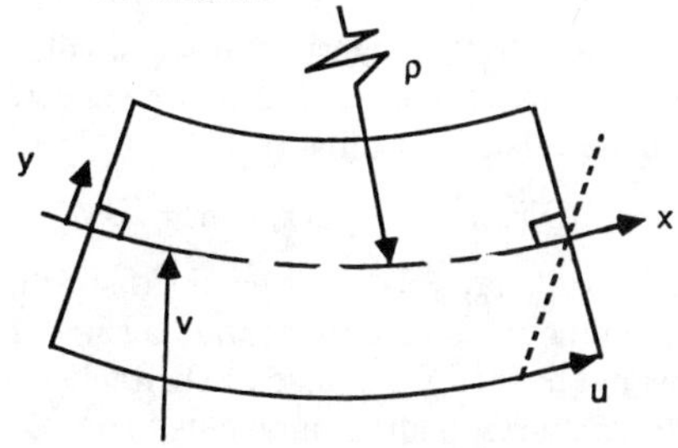

b) Displacements when plane sections remain plane

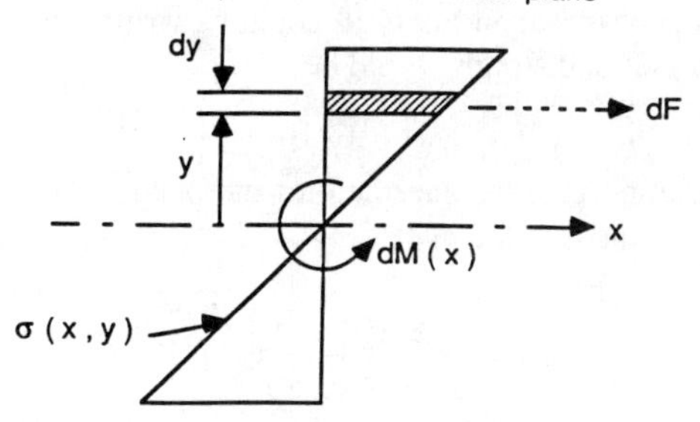

c) Moment as a generalized measure of stress

Figure 6.1.1 An elastic beam

generalized stress measure since it only depends on x. In mechanics of materials the deflections of the beam are determined by solving the differential equation of equilibrium :

$$\frac{d^2}{dx^2}\left[EI\,\frac{d^2v}{dx^2}\right] = p(x)$$

where $p(x)$ is the distributed transverse load per unit length. The four constants of integration are determined by satisfying the boundary conditions on the deflections, v, and slopes, θ. However, in our present study we need an integral formulation for our finite element model, that is equivalent to the solution of the differential equation.

6.2 Variational Procedure

A variational formulation for the elastic beam is related to minimizing the total energy and work in the system. One important term required for the analysis is the strain energy. That quantity is defined as half the volume integral of the product of the stress and strain. Here we wish to reduce this quantity to a function of x alone. The scalar strain energy is

$$U = \tfrac{1}{2} \int_V \sigma\, \varepsilon\, dV = \tfrac{1}{2} \int_L \int_{A(x)} (E\, y\, v'')(y\, v'')\, da\, dx\,. \qquad (6.4)$$

Since only v'' depends on x this reduces to

$$U = \tfrac{1}{2} \int_L E(x)\, I(x)\, (v'')^2\, dx = \tfrac{1}{2} \int_L M(x)\, v''(x)\, dx\,. \qquad (6.5)$$

Comparing Eqs. (6.4) and (6.5) suggests that we should select the curvature, v'', as our generalized strain measure. Having made this choice we can use Eq. (6.4) to define a generalized constitutive relation. Define

$$\sigma = M(x)\,, \quad \varepsilon = v''(x)\,, \quad \sigma = \mathbf{D}\, \varepsilon\,, \quad \mathbf{D} = E(x)\, I(x) \qquad (6.7)$$

as the generalized Hooke's law. The left hand side of the last three equations have been defined as arrays even though they only contain a single term. This is done to give some insight into what would happen for a plate or shell where there would be three curvatures of the surface, three corresponding moments, and $\mathbf{D}$ would become a 3×3 array involving the material properties and thickness of the section. If we consider a beam of width b and thickness h then Eq. (6.9) could be written as $D = Eh^3 b / 12$. For a plate the generalized stresses and strains would be

$$\sigma^T = [M_{xx} \;\; M_{yy} \;\; M_{xy}], \qquad \varepsilon^T = [w_{,\,xx} \;\; w_{,\,yy} \;\; w_{,\,xy}]$$

where w is the transverse displacement of the plate. The moments are written on a unit length basis ($b = 1$), so for the plate

$$\mathbf{D} = \frac{Eh^3}{12(1-\nu^2)} \begin{bmatrix} 1 & 0 & 0 \\ 0 & 1 & 0 \\ 0 & 0 & (1-\nu)/2 \end{bmatrix}.$$

While we do not plan to consider plates here we will use the generalized symbolism to give insight to such problems.

Equation (6.5) shows that our selections for generalized stress-strain measures will correctly define the strain energy in the system. Next, we need to define the work done by the applied loads, P_i, or couples, C_i. The work done by a transverse force is the product of the force and the transverse displacement. Likewise, the work done by a couple is the product of the couple and rotation (slope) at its point of application. These contributions define a work term, W, given by

$$W = \int_L v(x)\, p(x)\, dx + \sum_i v(x_i)\, P_i + \sum_j v'(x_j)\, C_j\,.$$

The last two terms represent work done by concentrated point loads or couples. Thus, the total potential energy, $\Pi = U - W$ is

$$\Pi = \tfrac{1}{2} \int_L EI\, (v''(x))^2\, dx - \int_L v(x)\, p(x)\, dx - \sum_i v_i\, P_i - \sum_j v'_j\, C_j\,. \qquad (6.7)$$

To determine the displacement field, $v(x)$, that corresponds to the equilibrium state we must minimize Π and satisfy the boundary conditions on v and $v' = \theta$.

6.3 Hermite Element Matrices

To introduce our finite elements we select a series of line segments to make up the region L. There are numerous elements that could be selected. First we will select an element with two nodes. Next, it is necessary to assume a displacement approximation so we can evaluate the potential energy in Eq. (6.5). That equation contains second derivatives and thus we need to assume a solution for v that will at least have both the deflection, v, and the slope, v', continuous between elements. The most common assumption is to select the cubic Hermite polynomial presented in Fig. 4.5.1. The unknowns at each of the two element nodes are v and $v' = \theta$. These quantities will be called our *generalized displacements* or the generalized degrees of freedom. Thus, our element interpolation functions are the Hermite form in Fig. 4.5.1:

$$v(x) = \begin{bmatrix} H_1^e(x) \, H_2^e(x) \, H_3^e(x) \, H_4^e(x) \end{bmatrix} \begin{Bmatrix} v_1 \\ v'_1 \\ v_2 \\ v'_2 \end{Bmatrix}^e$$

or $v(x) = \mathbf{H}^e(x)\,\mathbf{a}^e$, where $\mathbf{a}^e$ denotes the generalized displacements of the element. The derivatives of the displacements are

$$v'(x) = \theta(x) = \mathbf{H}^{e'}(x)\,\mathbf{a}^e, \qquad v''(x) = \mathbf{H}^{e''}(x)\,\mathbf{a}^e. \tag{6.8}$$

Since v'' and $\mathbf{a}^e$ have been selected as our generalized strains and generalized displacements we will use the notation of Eq. (6.8) and write Eq. (6.7) as

$$\varepsilon^e = \mathbf{B}^e\,\mathbf{a}^e$$

where $\varepsilon = v''$ in our present study. In the study of plates and shells additional curvature terms would be present in ε. Employing our generalized notation the stiffness matrix and distributed load vector can be written by inspection as

$$\mathbf{K}^e = \int_{L^e} \mathbf{B}^e(x)^T\,\mathbf{D}^e(x)\,\mathbf{B}^e(x)\,dx, \quad \mathbf{F}_p^e = \int_{L^e} \mathbf{H}^e(x)^T\,p^e(x)\,dx.$$

Here we will again use unit coordinates on the element and set $r = x/L$ so that $d(\ )/dx = d(\ )/dr \times 1/L$. Thus,

$$\mathbf{B}^e = \mathbf{H}^{e''} = \frac{1}{L^2}\frac{d^2\mathbf{H}}{dr^2}$$

so for the cubic Hermite in Fig. 4.5.1 this becomes

$$\mathbf{B}^e = \frac{1}{L^2}\begin{bmatrix} (12r-6) & L(6r-4) & (6-12r) & L(6r-2) \end{bmatrix}.$$

Recalling that

$$\int_L r^m\,dx = \frac{L}{(m+1)}$$

and assuming that $\mathbf{D}^e$ is a constant then the stiffness is

$$\mathbf{K}^e = \frac{EI}{L^e} \begin{bmatrix} 12 & & & \text{sym.} \\ 6L & L^2 & & \\ -12 & -6L & 12 & \\ 6L & 2L^2 & -6L & L^2 \end{bmatrix}. \tag{6.9}$$

If the lateral load, p^e, is constant then

$$\mathbf{F}_p^e = p^e \int_{L^e} \mathbf{H}^{e^T} dx = p^e L^e \int_0^1 \mathbf{H}^{e^T}(r)\, dr = p^e L^e \begin{Bmatrix} 1/2 \\ L^{e^T}/12 \\ 1/2 \\ -L^{e^T}/12 \end{Bmatrix}. \tag{6.10}$$

This result for the consistent nodal loads is illustrated in Fig. 6.3.1. Note that the distributed load puts half the resultant load at each end. It also causes equal and opposite nodal couples at each of the two nodes.

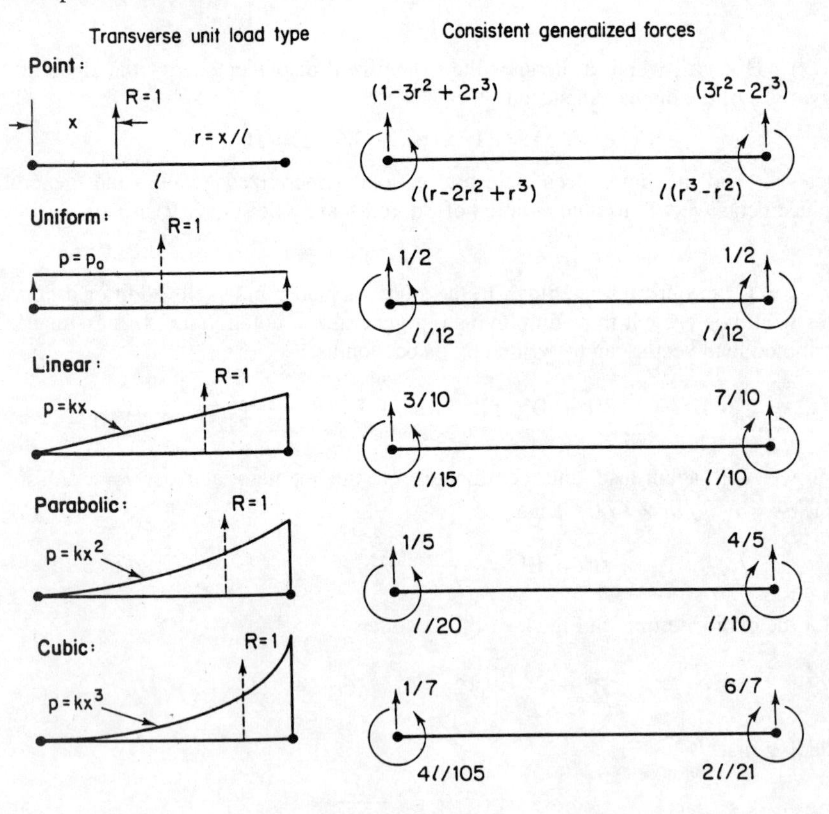

Figure 6.3.1 Consistent resultant nodal loads and couples

When we wrote Eq. (6.7) we assumed that point loads would only be applied at the node points. This may not always be true and we should consider such a load condition as a special case of a distributed load. In that case the length of the distributed load approaches zero and the magnitude of the force per unit length approaches infinity, but the resultant load P is constant. That is, we define the load to be $p(x) = P\,delta(x - x_0)$ where δ is the Dirac Delta distribution. Then the generalized load vector is

$$\mathbf{F}_p^e = \int_{L^e} \mathbf{H}^{e^T}(x) P\,\delta(x - x_0)\,dx$$

which is integrated by inspection by using the integral property of δ in Eq. (2.23) to yield $\mathbf{F}_p^e = P\,\mathbf{H}^e(r_0)$ where $r_0 = x_0/L$ is the point of application of the load. This result is also shown in Fig. 6.3.1. To check this concept, assume that the load is at Node 1. Then, $r_0 = 0$ so that

$$\mathbf{F}_p^{e^T} = P[1 \quad 0 \quad 0 \quad 0]$$

as expected. That is, all the force goes into Node 1 and no element nodal moments are generated. Other common loading conditions can be treated in the same way. For example, if $p(x)$ varies linearly from p_1^e to p_2^e at the nodes of element e then

$$p(x) = (1 - x/L)\,p_1^e + x/L\,p_2^e = \begin{bmatrix}(1-r) & r\end{bmatrix}\begin{Bmatrix}p_1\\ p_2\end{Bmatrix}^e$$

and

$$\mathbf{F}_p^e = \int_{L^e}\begin{Bmatrix}1 - 3r^2 + 2r^3\\ L(r - 2r^2 + r^3)\\ 3r^2 - 2r^3\\ L(r^3 - r^2)\end{Bmatrix} p(x)\,dx = \frac{L^e}{20}\begin{bmatrix}7 & 3\\ L & 2L/3\\ 3 & 7\\ -2L/3 & -L\end{bmatrix}\begin{Bmatrix}p_1\\ p_2\end{Bmatrix}^e. \tag{6.11}$$

If the load is constant so that $p_1^e = p_2^e = p_e$, then this reduces to Eq. (6.10), as expected. Likewise, if $p_1^e = 0$ and $p_2^e = p$, this defines a triangular load and

$$\mathbf{F}_p^{e^T} = \frac{PL}{20}\begin{bmatrix}3 & 2L/3 & 7 & -L\end{bmatrix}. \tag{6.12}$$

It is common to tabulate such results in terms of an applied unit resultant load. That resultant is

$$R^e = \int_{L^e} p^e(x)\,dx\,.$$

For common load variations, such as constant, linear, parabolic, and cubic forms where p varies in proportion to r^n, the resultant loads are $R^e = pL/(n+1)$. The location, $\bar{x}$, of the resultant applied load is found from

$$\bar{x}R^e = \int_{L^e} px\,dx$$

and the corresponding results are $\bar{x} = L(n+1)/(n+2)$. Thus, if we normalize Eq. (6.12) by dividing the resultant load, $pL/2$, the result is

$$\mathbf{f}_p^{e^T} = [\,3/10 \quad L/15 \quad 7/10 \quad -L/10\,]$$

which is the result in Fig. 6.3.1. We can also check the unit load results by applying statics to the data in that figure. To check the triangular load summary, we first take the sum of the moments about Node 1. This gives

$$+(2L/3)1 \;=\; 0 + (7/10)L + L/15 - L/10 \;= L(21+2-3)/30, \quad \text{OK}.$$

Similarly, the sum of the moments about Node 2 is verified.

6.4 Sample Application

To present an analytic example of this element consider a single element solution of the cantilever beam shown in Fig. 6.4.1 to determine the deflection and slope at the free end. Usually the deflected shape of a beam is defined by a fourth or fifth order polynomial in x. Thus, our cubic element solution will usually give only an approximate solution. For a single element the system equations are

$$\frac{EI}{L^3}\left[\begin{array}{cc|cc} 12 & 6L & -12 & 6L \\ 6L & 4L^2 & -6L & 2L^2 \\ \hline -12 & -6L & 12 & -6L \\ 6L & 2L^2 & -6L & 4L^2 \end{array}\right]\left\{\begin{array}{c} v_1 \\ \theta_1 \\ \hline v_2 \\ \theta_2 \end{array}\right\} = \frac{WL}{2}\left\{\begin{array}{c} 3/10 \\ L/15 \\ \hline 7/10 \\ -L/10 \end{array}\right\} + \left\{\begin{array}{c} 0 \\ 0 \\ F_2 \\ M_2 \end{array}\right\}.$$

The right side support requires that $v_2 = 0 = \theta_2$. The reduced equations become

$$\frac{EI}{L^3}\begin{bmatrix} 12 & 6L \\ 6L & L^2 \end{bmatrix}\begin{Bmatrix} v_1 \\ \theta_1 \end{Bmatrix} = \frac{WL}{2}\begin{Bmatrix} 3/10 \\ L/15 \end{Bmatrix} + \begin{Bmatrix} 0 \\ 0 \end{Bmatrix}$$

so that

$$\begin{Bmatrix} v_1 \\ \theta_1 \end{Bmatrix} = \frac{L^3}{12\,EIL^2}\begin{bmatrix} 4L^2 & -6L \\ -6L & 12 \end{bmatrix}\begin{Bmatrix} 3/10 \\ L/15 \end{Bmatrix}\frac{WL}{2} = \frac{WL^3}{EI}\begin{Bmatrix} L/30 \\ -1/24 \end{Bmatrix}.$$

The exact solution is $120\,EI\,v = wL^4[\,4 - 5x/L + (x/L)^5\,]$ so that the exact values of the maximum deflection and slope are $v = WL/(30\,EI)$ and $\theta = -WL^3/(24\,EI)$, respectively. Thus, our single element solution gives the exact values of both v and θ at the nodes, but is only approximate in the interior of the element. The last two equations give the exact reactions.

In practical analysis one can often utilize a partial model to reduce the data preparation and more importantly the analysis cost. One must be alert for planes where the geometry, material property, supports and loads are symmetric mirror images. Two such cases are illustrated in Fig. 6.4.2 to show how a half-symmetry model can be employed.

Often the loading conditions occur in an anti-symmetric, or negative mirror image, fashion so that one can still use a half portion model and simply recognize that the deflections and stresses on the omitted half have signs opposite from those in the partial model. These concepts are shown in Fig. 6.4.3. These concepts are easy to illustrate for a beam, but are usually employed in much larger and more complicated structural systems. Even with today's large memory computers one eventually finds a structure that is too big to execute. Then one searches for alternate procedures such as solving two have size problems and combining their results (with the appropriate sign changes). This

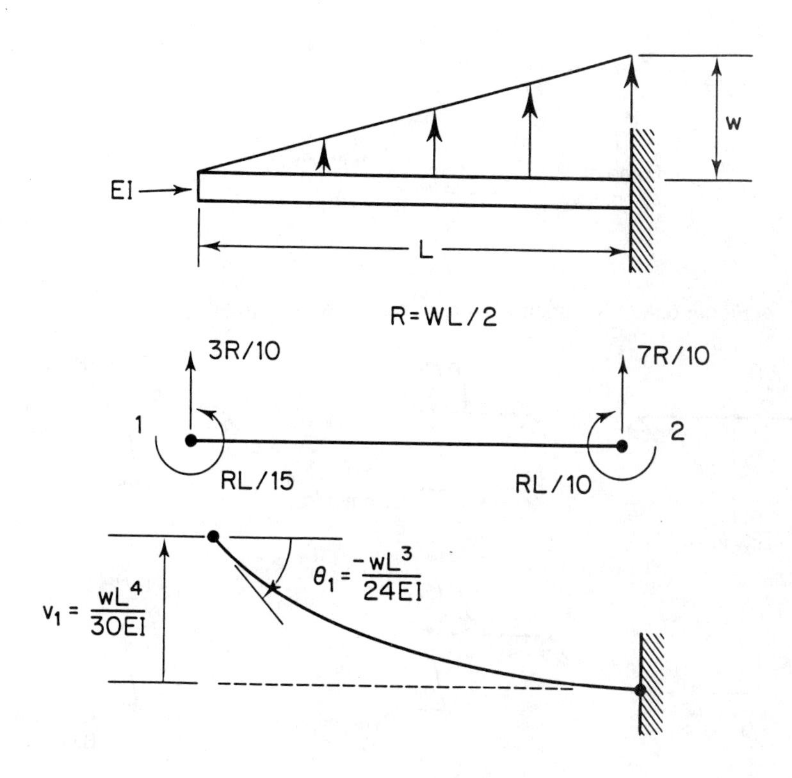

Figure 6.4.1 A single element approximate solution

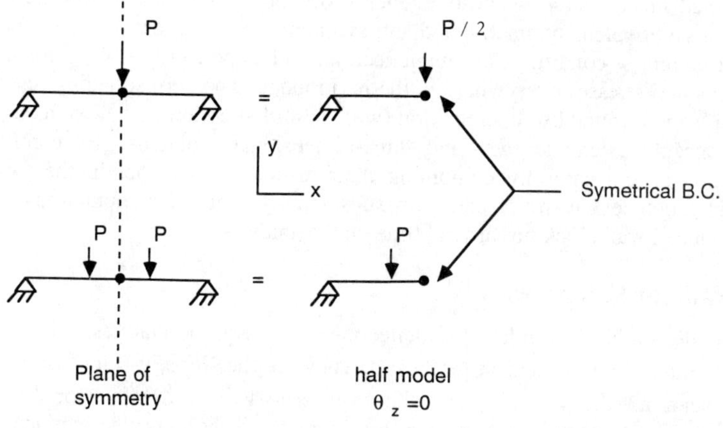

Figure 6.4.2 Symmetrical structure, symmetrical loading model

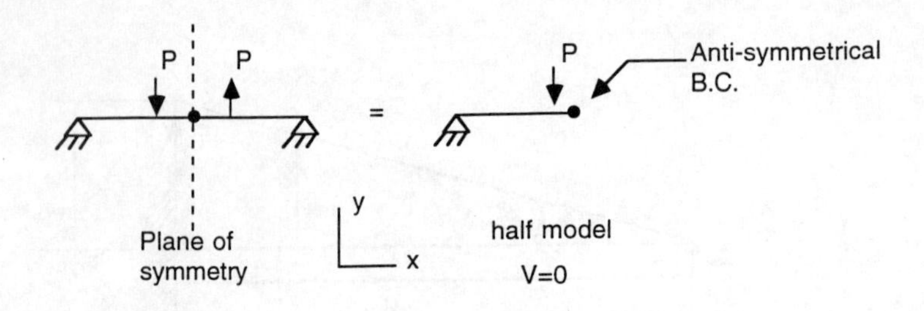

Figure 6.4.3 Symmetrical structure, anti-symmetrical loading model

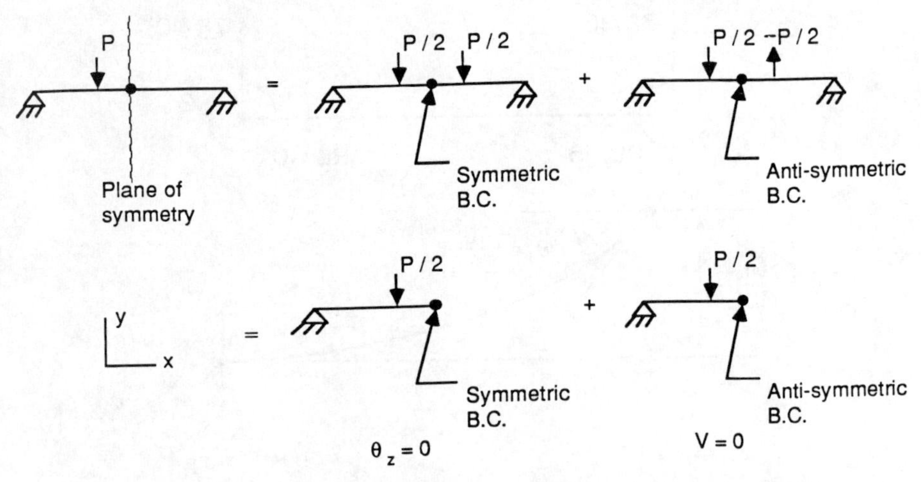

Figure 6.4.4 Symmetric structure, non-symmetric loading model

is illustrated in Fig. 6.4.4, where the general load on a otherwise symmetric problem is split into its equivalent symmetric and anti-symmetric parts so that two have models with different boundary conditions on the deflection, and slope can be solved for the desired deflections and stresses everywhere in the half model. The corresponding results for the other half can be found by superposition (with careful attention to their signs) and output. These concepts extend to two- and three-dimensional problems. By carefully using partial models and carefully combining their results one can obtain the full analysis results in much less memory and with substantially reduced computations. This also means reduced wall clock turnaround time on the analysis.

6.5 Gradient Estimates *

It is desirable to correlate the deflection, v, and its derivatives with the physical quantities that they represent in the beam. They are the slope, $v' = dv/dx = \theta$, moment, $EI\, v''$, shear, $dM/dx = EI\, v'''$, and the load, $q = dV/dx = EI\, v''''$ (for EI constant). Thus, quantities like moment and shear are represented by higher derivatives. They are often used to design a member. An important question is, where are the moments and shears most accurate? Or, where are v''' and v'''' most accurate? A similar procedure to that given in Sec. 3.8 can be utilized for estimating derivatives in beam elements. A common example is the use of the cubic *Hermitian elements*. Recall that this element has two *dof* at each of the two nodes. Those *dof* are the function, u, and its first derivative, du/da. Since the integral form requires a C^1 function, the required modification in the procedure is to set both the nodal parameter and its slope equal to the corresponding exact values at the nodes. That is,

$$u_E(a_j) = u_F(a_j), \qquad \frac{du_E(a_j)}{da} = \frac{du_F(a_j)}{da}$$

at each of the two end node locations, $a_j = \pm 1$. Since the element is cubic, we assume that the exact solution is of fourth order. Then

$$\mathbf{P}_F = \begin{bmatrix} 1 & a & a^2 & a^3 \end{bmatrix}, \qquad \mathbf{P}_E = \begin{bmatrix} 1 & a & a^2 & a^3 & a \end{bmatrix}$$

and

$$\frac{d\mathbf{P}_F}{da} = \begin{bmatrix} 0 & 1 & 2a & 3a^2 \end{bmatrix}, \qquad \frac{d\mathbf{P}_E}{da} = \begin{bmatrix} 0 & 1 & 2a & 3a^2 & a^3 \end{bmatrix}.$$

Setting these quantities equal at the nodes gives the identities

$$\begin{bmatrix} \mathbf{P}_F(a_1) \\ \dfrac{d\mathbf{P}_F}{da}(a_1) \\ \mathbf{P}_F(a_2) \\ \dfrac{d\mathbf{P}_F(a_2)}{da} \end{bmatrix} \mathbf{V}_F = \begin{bmatrix} \mathbf{P}_E(a_1) \\ \dfrac{d\mathbf{P}_E}{da}(a_1) \\ \mathbf{P}_E(a_2) \\ \dfrac{d\mathbf{P}_E(a_2)}{da} \end{bmatrix} \mathbf{V}_E$$

so that for $a_1 = -1$ and $a_2 = +1$:

$$\mathbf{A}_F = \begin{bmatrix} 1 & -1 & 1 & -1 \\ 0 & 1 & -2 & 3 \\ 1 & 1 & 1 & 1 \\ 0 & 1 & 2 & 3 \end{bmatrix}, \qquad \mathbf{B}_E = \begin{bmatrix} 1 & -1 & 1 & -1 & 1 \\ 0 & 1 & -2 & 3 & -4 \\ 1 & 1 & 1 & 1 & 1 \\ 0 & 1 & 2 & 3 & 4 \end{bmatrix}.$$

The corresponding generalized forms of **K** and **E** (defined in Sec. 3.8) are

$$\mathbf{K} = \left[\begin{array}{cccc|c} 1 & 0 & 0 & 0 & -1 \\ 0 & 1 & 0 & 0 & 0 \\ 0 & 0 & 1 & 0 & 2 \\ 0 & 0 & 0 & 1 & 0 \end{array}\right], \qquad \mathbf{E} = \begin{Bmatrix} -1 \\ 0 \\ 2 \\ 0 \end{Bmatrix}.$$

Therefore, equating the last term in each solution $(-1 + 2a^2)\, V_E = a^4\, V_E$. Thus, the exact and approximate functions are equal only when $a = \pm 1$ and at the center, $a = 0$; that is, only at the end nodes. Equating the first derivatives at a_f gives

$$(0 + 4a_f) = 4a_f^3,$$

which means that they are also best at the ends, $a_f = \pm 1$ and at the center, $a_f = 0$. Let the second derivatives agree at a_s. Then, $4 = 12a_s^2$ so that $a_s = \pm 1/\sqrt{3}$, which are the *Gauss points*. Finally, for the third derivatives to agree at a_t requires $0 = 24\,a_t$, or simply $a_t = 0$, which is the center of the element. The above analysis suggests that the deflection and slope (u and u') should be output at the two ends. The moment or curvature (u'') of an element should be output at the two Gauss points, and the shear (u''') should be output only at the center point of the element. Similar observations carry over to plate bending elements and shells. The conclusions for the beam element are illustrated in Fig. 3.8.2.

6.6 Element Equations via Galerkin's Method *

Here we will illustrate the development of the element matrices by applying Galerkin's Method to the governing differential equation. Recall that for a beam subject to a load $p(x)$, the differential equation describing the elastic curve is given by Eq. (6.3). If we substitute an approximate solution, $u(x)$, this gives

$$\frac{d^2}{dx^2}\left[EI\,\frac{d^2u}{dx^2} \right] - p(x) = R(x)$$

where $R(x)$ is the residual error term. Interpolating the approximate solution with $u(x) = \mathbf{H}^e(x)\,\mathbf{a}^e$ and applying Galerkin's criterion to the error term gives

$$0 = \int u(x)\,R(x)\,dx = \sum_e \int_{L^e} u(x)\,R(x)\,dx = \sum_e \mathbf{a}^{e^T} \int_{L^e} \mathbf{H}^{e^T}(x)\,R(x)\,dx\,.$$

But the array $\mathbf{a}^e$ is a vector of arbitrary constants. This implies that we require

$$\mathbf{0} = \int_{L^e} \mathbf{H}^{e^T}(x)\,R(x)\,dx = \int_{L^e} \mathbf{H}^{e^T} \left[\frac{d^2}{dx^2}\,EI\frac{d^2u}{dx^2} - p(x) \right] dx\,.$$

Let a prime denote a derivative, then twice integrating the first term by parts gives

$$\int_L EI\,\mathbf{H}''^T u''dx + EI\,\mathbf{H}^T u'''\Big|_0^L - EI\,\mathbf{H}'^T u''\Big|_0^L - \int_L \mathbf{H}^T p\,dx = \mathbf{0}\,.$$

Substituting the interpolation for u the first integral gives

$$\int_{L^e} EI\,\mathbf{H}''^T(x)\,\mathbf{H}''(x)\,dx\,\mathbf{a}^e = \mathbf{S}^e\,\mathbf{a}^e\,, \quad \text{where} \quad \mathbf{S}^e = \int_{L^e} EI\,\mathbf{H}''^T\,\mathbf{H}''\,dx$$

is the element stiffness matrix given earlier. The last integral is the consistent force vector given earlier in Eq. (6.11). The remaining two terms define the *natural boundary conditions* on the beam.

6.7 Beams on an Elastic Foundation *

The stresses in a statically indeterminate structure on an elastic foundation are influenced by the deformation of the foundation while the pressure distribution on the foundation is affected by the relative stiffness of the structure and the foundation medium. To allow for this structure-foundation interaction the finite element method is ideally suited. The problem has been studied by many authors. Most of those works use Winkler hypothesis, and assume that the soil adheres to the beam, i.e., the separation between beam and soil is not allowed. This is not true for many physical cases. For instance, when a beam or a beam-column rests on the soil foundation with some type of load on it, some parts of the beam might be lifted up. Because of soils lacking both adhesive and cohesive properties, gaps occur in those regions. The method presented in the study represents the foundation by a one-dimensional line finite element. The foundation is assumed to be of the two-parameter type. Also, the separation between structure and soil foundation is allowed when tension develops. The location of those regions is solved by iterating the solutions. The differential equation of equilibrium of a beam-column on an elastic foundation is

$$EI\, d^4w/dx^4 + N d^2w/dx^2 - k_s d^2w/dx^2 + k_w w = q(x)$$

in which k_w and k_s are Winkler modulus and second-parameter foundation coefficients, respectively, N is the axial tension force, q is the transverse load, and EI is the bending stiffness. This is the differential equation of an ordinary beam on a Winkler elastic foundation which is well known.

The analysis of beams, beam-columns and plates on elastic foundations is widespread in engineering. Hetenyi [3] extensively developed the classical differential equation solutions for this problem. Finite element approaches have been used extensively in the analysis of beams on an elastic foundation. Most of these works use the Winkler hypothesis. Thus, the foundation acts as if it consisted of infinitely many closely spaced linear springs. Bowles [1] formulates a stiffness matrix by combining a conventional beam element with discrete soil springs at the end of the beam. The degree of accuracy using this element is highly dependent on the number of elements modeled. Ting [8] derives the stiffness and flexibility matrices from the exact solution of the differential equation.

6.7.1 Foundation Models

As a result of a line load, $q(x)$, on the upper surface the beam deflects, causing the foundation to resist with a line load $p(x)$, whose units may be taken as N/mm. Various foundation models define $p(x)$ in various ways. The *Winkler Foundation* model has been used for a century. It assumes that the foundation applies only a reaction $p(x)$ normal to the beam, and the $p(x)$ is directly proportional to the beam deflection $w(x)$: $p(x) = k\,w(x)$. The Winkler foundation modulus k has units $N/mm/mm$. Effectively, this foundation is a row of closely-packed linear springs. Note that interactions between the springs are not considered, so it does not accurately represent the characteristics of many practical foundations. To improve the Winkler model some authors assumed interactions between the springs and added a second parameter. The *Filonenko-Borodich Foundation* assumes that the top ends of the springs are connected to an elastic membrane that is stretched by a constant tension. The *Pasternak Foundation* introduces shear interactions between the springs. He assumed that the top ends of the springs are connected to an incompressible layer that resists only transverse shear deformation. The *Generalized Foundation* model assumes that at each point of contact there is not only a pressure but also a moment applied to the beam by the foundation. The moment is assumed to be proportional to the angle of rotation (or slope) of the beam. Mathematically, all these models are equivalent. The only difference is the definition of the parameters, so we rewrite these equations in the form

$$p(x) = k\,w(x) - k_s d^2w(x)/dx^2$$

in which k is the first parameter (Winkler's modulus), and k_s is the second parameter.

6.7.2 Element Stiffness Matrices

The integral form mathematically equivalent to this problem is to find the extremum of the functional

$$\Pi = \frac{1}{2}\int_L \left[EI\,(w'')^2 + N\,(w')^2 - k_s(w')^2 + k_2 w^2 - 2\,q(x)\,w \right] dx$$

for all smooth functions w satisfying the boundary conditions. In a finite element formulation the assumed interpolation functions should be such that w and w' are continuous between elements. This assures that Π is defined and that we can write Π as the sum of contributions from the elements into which the region is divided. Substituting the usual interpolation matrices :

$$\Pi = \frac{1}{2}\sum_e \left(\mathbf{u}^T\mathbf{K}_1\mathbf{u} + \mathbf{u}^T\mathbf{K}_2\mathbf{u} + \mathbf{u}^T\mathbf{K}_3\mathbf{u} + \mathbf{u}^T\mathbf{K}_4\mathbf{u} - 2\,\mathbf{u}^T\mathbf{C} \right)^e$$

allows us to define the total element stiffness matrix as $\mathbf{K} = \mathbf{K}_1 + \mathbf{K}_2 + \mathbf{K}_3 + \mathbf{K}_4$ where $\mathbf{K}_1$ is the classical beam stiffness matrix, given earlier, which involves the beam flexural stiffness EI, $\mathbf{K}_2$ is the stiffness matrix which involves the axial load N, called *geometric stiffness matrix*, $\mathbf{K}_3$ is the second-parameter foundation stiffness matrix, and $\mathbf{K}_4$ is the Winkler foundation stiffness matrix. They are defined by the relations

$$\mathbf{u}^T\mathbf{K}_1\mathbf{u} = \int EI\,(w'')^2\,dx\,, \quad \mathbf{u}^T\mathbf{K}_2\mathbf{u} = \int -N(w')^2\,dx\,, \tag{6.13}$$

$$\mathbf{u}^T\mathbf{K}_3\mathbf{u} = \int -k_s(w')^2\,dx\,, \quad \mathbf{u}^T\mathbf{K}_4\mathbf{u} = \int k_w w^2\,dx\,, \quad \mathbf{u}^T\mathbf{C} = \int q(x)\,w\,dx$$

where the integrations are carried out over the element and where $\mathbf{u}$ denotes the element degrees of freedom. Assembling the elements stiffness matrices according to the connectivity of elements, the system equations can be obtained as $\mathbf{K}\,\mathbf{U} = \mathbf{F}$.

6.7.3 Cubic Beam Elements

The practical application of the finite element method depends on the use of various interpolation functions and their derivatives. Most of the interpolation functions for C^0 and C^1 continuity elements are well known. The C^0 functions are continuous across an inter-element boundary while the C^1 functions also have their first derivatives continuous across the boundary. Here we begin with a C^1 continuity element for the beam. This element is obtained by using Hermite interpolation polynomial and nodal variables that include derivatives as well. The geometrical nodes, function $\mathbf{H}$ and the Jacobian matrix $\mathbf{J}$ remain identical to those of the linear element (C^0 bar element). The resulting matrices are

$$\mathbf{K}_1 = E\frac{I}{L^3}\begin{bmatrix} 12 & & & \text{Sym.} \\ 6L & 4L^2 & & \\ -12 & -6L & 12 & \\ 6L & 2L^2 & -6L & 4L^2 \end{bmatrix}$$

$$\mathbf{K}_2 = \frac{-N}{30L}\begin{bmatrix} 36 & & & \text{Sym.} \\ 3L & 4L^2 & & \\ -36 & -3L & 36 & \\ 3L & -L^2 & -3L & 4L^2 \end{bmatrix}, \qquad \mathbf{K}_3 = \frac{k_s}{N}\mathbf{K}_2$$

$$\mathbf{K}_4 = \frac{k_w}{420}\begin{bmatrix} 156 & & & \text{Sym.} \\ 22L & 4L^2 & & \\ 54 & 13L & 156 & \\ -13L & -3L^2 & -22L & 4L^2 \end{bmatrix}.$$

6.7.4 Fifth Order Beam Element

The Hermite family includes members with additional derivatives at the two ends. To seek a more accurate solution we will next consider a fifth order Hermite polynomial. This has three variables per node: w_i, w'_i, w''_i. The corresponding generalized parameters are u_1, u_2, and u_3, respectively. For a two nodes line element, the equation of deflection is:

$$w(x) = H_1u_1 + H_2u_2 + H_3u_3 + H_4u_4 + H_5u_5 + H_6u_6 .$$

The shape function in local coordinates, from Fig. 4.5.1, are

$$\begin{aligned} H_1 &= 1 - 10r^3 + 15r - 6r^5 & H_4 &= 10r^3 - 15r + 6r^5 \\ H_2 &= (r - 6r^3 + 8r - 3r^5)\,L_e & H_5 &= (7r - 3r^5 - 4r^3)\,L \\ H_3 &= (r^2 - 3r^3 + 3r - r^5)\,L_e^2/2 & H_6 &= (r^3 - 2r + r^5)\,L_e^2/2 , \end{aligned} \tag{6.14}$$

where $L_e = x_2 - x_1$ is the length of the element, and $x = x_1 + rL_e$. By using numerical integration and cubic interpolation one obtains the numerical form equivalent to the element stiffness matrices given in Eq. (6.14). Likewise, if we employ the fifth order interpolation functions, the corresponding numerically integrated (6×6) matrices are obtained. Both forms are used in later examples.

6.7.5 Member and Foundation Reaction Recovery

In Sec. 3.4 the topic of flux or member reaction recovery was discussed. Here that topic is somewhat more complicated because both the beam and the foundation have reactions that we may want to compute. Recall that the member algebraic equilibrium equations in the form $\mathbf{Ku} = \mathbf{C} + \mathbf{p}$. Thus, we approximate the member joint forces as $\mathbf{p} = \mathbf{Ku} - \mathbf{C}$ where the matrices $\mathbf{K}$ and $\mathbf{C}$ must be stored for later use in computing the final member actions. Assume $\mathbf{K}$ is a sum of beam and foundation effects shown earlier. We want to know the reactions from the foundation, and the beam internal effects. We have

$$\mathbf{p} = \left(\mathbf{K}_1 + \mathbf{K}_2 + \mathbf{K}_3 + \mathbf{K}_4\right)\mathbf{u} - \mathbf{C}$$

and set

$$\mathbf{C}_f = - \left[\mathbf{K}_3 + \mathbf{K}_4 \right] \mathbf{u}$$

Here $\mathbf{C}_f$ are the reactions from the foundation and we obtain

$$\mathbf{p} = \left[\mathbf{K}_1 + \mathbf{K}_2 \right] \mathbf{u} - \mathbf{C} - \mathbf{C}_f.$$

After $\mathbf{u}$ has been computed the foundation reactions and member forces are recovered for each element from the last two equations, respectively.

Suppose a beam on elastic foundation is loaded by a concentrated force in midspan. The numerical values of the parameters are: $K_2 = 6.12\ N/mm$; $E = 5200\ N/mm^2$; $I = 3.413\,(10^7)\ mm^4$; $L = 5500\ mm$; $P = N$. The deflection curve in this case is symmetric. Member force recovery are computed for each element of the four-element half span. Both system and element equilibriums are satisfied as seen in Fig. 6.7.1.

6.7.6 The Treatment for Gaps

There are several real situations where separations between structures and soil foundations are allowed to develop. The structure-foundation interaction no longer exists in those sections, so the Winkler model cannot be used throughout the whole span

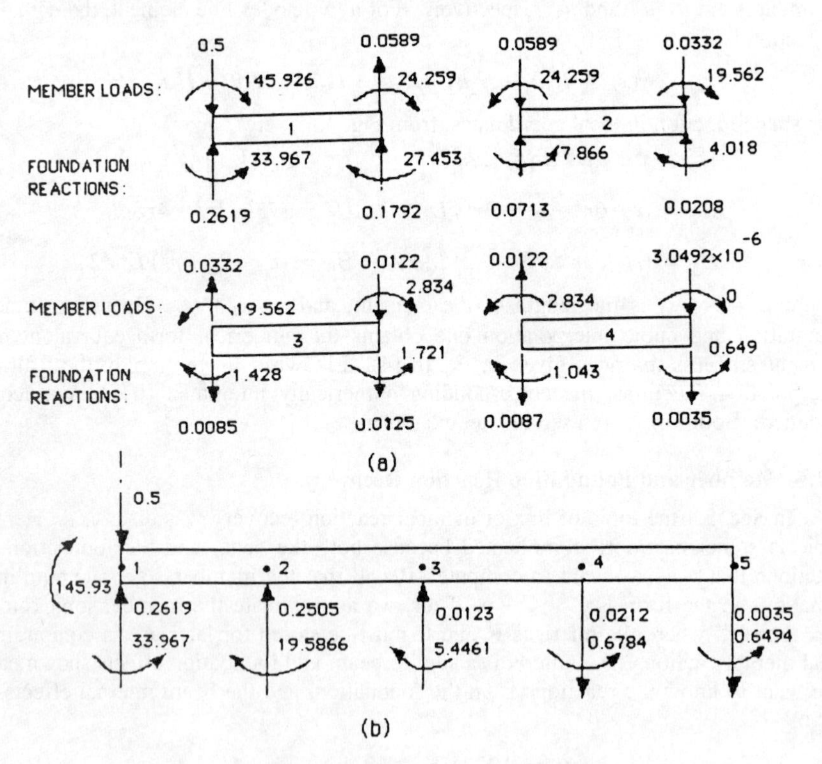

Figure 6.7.1 Beam and Foundation Reaction Recovery

of the beam. What we are interested in approximating is the locations of zero-deflection points from which we can find the regions of gaps. The differential equation for the gap section, with constant EI, is

$$EI\, d^4w/dx^4 = q(x)$$

and for the sections where the beam contacts the two-parameter foundation model. Now we have two kinds of differential equations. To solve the problem the finite element procedure is employed. In all cases, the six-point Gaussian quadrature is used for numerical integration. Testing each Gauss point, we can find if that point has lifted off the foundation. If so, we set the two foundation parameters to zero, that is, we treat the foundation coefficients as a function of x, even if they are constant. By increasing the number of Gauss points (and thereby decreasing the space between them) we can accurately bound the location of the lift-off points and form the correct foundation stiffness, even for those elements that only partially contact the foundation. By using the iterative procedure, the solution is repeated until nodal coordinates of zero-deflection points remain unchanged. Then, the deflection of the beam along the whole length will be accurate, as will the moments and shears. During this procedure the foundation moduli, k_w and k_s, are not constant but depend on location. Thus, the usual forms in Eq. (6.13) is no longer valid. The most practical way to treat partial contact over the element is to numerically integrate the contributions to $\mathbf{K}_3$ and $\mathbf{K}_4$. Using Gaussian quadratures also simplifies the programming required to implement both the C^2 and C^1 beam elements and their associated load resultants. To distinguish from the classical Winkler foundation model, this procedure is simply called a no-tension foundation model.

6.7.7 Numerical Example

Consider a long beam resting on the elastic foundation loaded by a concentrated force at the middle point of the beam. The numerical values for the parameters are: Young's modulus $E = 9100\, N/mm^2$; Winkler foundation modulus $K_w = .0 N/mm^2$; the second-parameter of foundation $K_s = 6 \times 10^5 N$; $P = 20000\, N$; $L = 18050\, mm$. One-parameter solutions using both C^1 and C^2 continuity element models are compared with the analytic infinite beam solution in Table 6.1. The C^2 continuity element model gives more accurate results than those obtained by using the C^1 continuity element model. Even the five-element solution of the C^2-model is better than that of a 20-element C^1-model. From the numerical results of Table 6.1, we can see that the deflection and the rotation at the far end point are very close to zero.

From the above examples we see that elements based on a cubic displacement function can give reasonable results for deflections, rotations, and bending moments by a moderately fine element mesh. A very fine mesh is needed to obtain good predictions for transverse shear force. Elements based on a fifth order displacement function can give exact solutions at most nodal points for deflections, rotations, bending moments, as well as transverse shear forces. When the second-parameter k_s is not very large the beam can be analyzed as if it rests on Winkler foundation. When k_s is large, especially when k_s is close to $(4KEI)^{1/2}$, the error caused by ignoring k_s may be appreciable. The iteration procedure, discussed above, has been found very efficient for solving the beam on elastic foundation problems with the consideration of gaps.

Table 6.1 Winkler model solution

NODE	INFINITE SOLUTION	C^1 CONTINUITY 20 ELEMENTS	C^2 CONTINUITY 20 ELEMENTS	C^2 CONTINUITY 5 ELEMENTS
		Deflections		
1	2.83271	2.83191	2.83271	2.83322
5	0.15856	0.15839	0.15856	0.15855
9	–0.06614	–0.06609	–0.06614	–0.06614
13	0.00516	0.00516	0.00516	0.00517
17	0.00050	0.00050	0.00050	0.00050
21	–0.00015	–0.00029	–0.00029	–0.00029
		Rotations		
1	0.0000E+00	0.0000E+00	0.0000E+00	0.0000E+00
5	–7.3866E–04	–7.3830E–04	–7.3866E–04	–7.3961E–04
9	8.7296E–05	8.7270E–05	8.7296E–05	8.6601E–05
13	2.0423E–06	2.0363E–06	2.0460E–06	2.2501E–06
17	–1.7020E–06	–1.7093E–06	–1.7114E–06	–1.7380E–06
21	1.6697E–07	1.1096E–08	1.0464E–08	1.1007E–08
		Moments		
1	4412730.00	4412340.00	4412730.00	4412540.00
5	–768508.00	–768336.00	–768508.00	–768600.00
9	16990.50	17022.30	16990.50	17000.00
13	10850.60	10835.80	10847.70	10853.30
17	–1566.64	–1540.58	–1541.25	–1543.01
21	3.60	0.00	0.00	0.00
		Shears		
1	1.0000E+04	1.0000E+04	1.0000E+04	1.0000E+04
5	–59091E+02	–59103E+02	–5.9091E+02	–5.9088E+02
9	–9.7484E+01	–9.7392E+01	–9.7483E+01	–9.7510E+01
13	2.1408E+01	2.1398E+01	2.1409E+01	2.1420E+01
17	–8.9895E–01	–9.4316E–01	–9.4189E–01	–9.4341E–01
21	–2.5196E–01	0.0000E+00	0.0000E+00	0.0000E+00

6.8 Exact Analytic Elements *

Recall that the analytic solution to a differential equation is generally viewed as the sum of a homogeneous solution and a particular solution. It has been proved by Tong [9] that if the finite element interpolation functions are the exact solution to the homogeneous differential equation, then the finite element solution of a non-homogeneous (no zero) source term will *always* be exact at the nodes. Clearly, this also means that if the source is zero, then this type of solution would be exact everywhere. It is well known that the exact solution of the non-homogeneous equations for the beam on an elastic foundation will generally involve the products and sums of trigeometric and hyperbolic functions. Therefore, if we replaced the previous polynomial interpolations with a more complicated set, we can assure ourselves of results that are at least exact at

Table 6.2 Homogeneous solution interpolation for semi-infinite axial bar on a foundation

a) PDE : $\frac{d}{dx}\left[EA\,\frac{du}{dx}\right] - ku = q, \quad m = k/EA, \quad k > 0$

b) Homogeneous Interpolation : $H_1 = e^{-mx}$

c) Stiffness Matrix : $K_{11} = \frac{m\,EA}{2} + \frac{k}{2\,m}$

d) Force Vector : $F_1 = q/m, \qquad q = \text{constant}$

e) Mass Matrix : $M_{11} = \rho/2\,m$

Table 6.3 Homogeneous solution interpolation for semi-infinite axial bar on foundation

a) PDE : $\frac{d}{dx}\left[EA\,\frac{du}{dx}\right] - ku = q, \quad m = k/EA, \quad k > 0$

b) Homogeneous Interpolation : $S = \sinh(mL), \qquad C = \cosh(mL)$

$$\mathbf{H} = \frac{1}{S}\left[\sinh[\,m(L-x)\,] \quad \sinh(mx)\right]$$

c) Stiffness Matrix : $a = k + EA\,m^2, \quad b = k - EA\,m^2$

$$\mathbf{K} = \frac{1}{2m\,S^2}\begin{bmatrix} (a\sinh(2\,ml) - b\,mL) & (b\sinh(2\,mL) - aS) \\ \text{symmetric} & (a\sinh(2\,mL) - b\,mL) \end{bmatrix}$$

d) Force Vector : $q = q_1(1 - x/2) + q_2\,x/L$

$$\mathbf{F} = \frac{1}{m}\begin{Bmatrix} \frac{q_1(C-1)}{S} + \frac{(q_2 - q_1)(1 - mL/S)}{mL} \\ \frac{q_1(C-1)}{S} + \frac{(q_2 - q_1)(mL\coth(mL) - 1)}{mL} \end{Bmatrix}$$

e) Mass Matrix : $\mathbf{M} = \frac{\zeta}{2m\,S^2}\begin{bmatrix} (\sinh(2\,mL) - mL) & (S + \sinh(2\,mL)) \\ \text{symmetric} & (\sinh(2\,mL) - mL) \end{bmatrix}$

Table 6.4 Homogeneous solution interpolation for semi-infinite beam on foundation

a) PDE : $\frac{d^2}{dx^2}\left[EI\,\frac{d^2v}{dx^2}\right] - N\,\frac{d^2v}{dx^2} + kv = q, \quad |N| \neq 2\sqrt{kEI}$

b) Homogeneous Interpolation : $\mathbf{H} = \frac{1}{a-b}\left[\,(ae^{bx} - be^{ax})\ \ (e^{ax} - e^{bx})\,\right]$

c) Stiffness Matrix : $c = a^2 + 3\,ab + b^2$

$$\mathbf{K} = \frac{1}{2\,ab}\begin{bmatrix} \dfrac{-EI\,a^3 b^3 - N a^2 b^2 - c\,k}{a+b} & EI\,a^2 b^2 - k \\ \text{symmetric} & \dfrac{-c\,EI\,a^3 b^3 - N a^2 b^2 - k}{a+b} \end{bmatrix}$$

d) Force Vector : $\mathbf{F} = \frac{q}{ab}\begin{Bmatrix} -a-b \\ 1 \end{Bmatrix}$

e) Mass Matrix : $\mathbf{M} = \frac{\rho}{2\,ab}\begin{bmatrix} \dfrac{-c}{a+b} & 1 \\ \text{sym.} & \dfrac{-c}{a+b} \end{bmatrix}$

$$d = \sqrt{\sqrt{\frac{k}{4\,EI}} + \frac{N}{4\,EI}}\,, \qquad e = \sqrt{\sqrt{\frac{k}{4\,EI}} - \frac{N}{4\,EI}}$$

$$a = -d + e, \qquad b = -d + 2e\,.$$

the nodes. In practice, using hyperbolic functions with large arguments can break down due to the way their values are computed in the operating system mathematics library. For the problems considered here, it can be shown that the typical element matrices obtained from interpolating with the exact homogeneous solutions are summarized in Tables 6.2 to 6.4.

6.9 References

[1] Bowles, J.E., *Foundation Analysis and Design*, New York: McGraw-Hill (1988).

[2] Gallagher, R.H., *Finite Element Analysis Fundamentals*, Englewood Cliffs: Prentice-Hall (1975).

[3] Hetenyi, M., *Beam on Elastic Foundation*, University of Michigan Press (1946).

[4] Miranda, C. and Nair, K., "Finite Beams on Elastic Foundation," *ASCE J. Structural Division*, **92**(ST2), pp. 131–142 (1966).

[5] Miyahara, F. and Ergatoudis, J.G., "Matrix Analysis of Structure-Foundation Interaction," *ASCE J. Structural Division*, **102**(ST1), pp. 251–266 (1976).

[6] Nogami, T. and O'Neill, M.W., "Beam on Generalized Two-parameter Foundation," *ASCE J. Engineering Mechanics*, **111**(5), pp. 664–679 (1985).

[7] Selvadurai, A.P.S., *Elastic Analysis of Soil-Foundation Interaction*, Amsterdam: Elsevier (1979).

[8] Ting, B.Y. and Mochry, E.F., "Beam on Elastic Foundation Finite Element," *ASCE J. Structural Engineering*, **110**(10), pp. 2324–2339 (1984).

[9] Tong, P. and Rossettos, J.N., *Finite Element Method: Basic Techniques and Implementation*, MIT Press (1977).

[10] Zhaohua, F. and Cook, R.D., "Beam Elements on Two-parameter Elastic Foundations," *ASCE J. Engineering Mechanics*, **109**(6), pp. 1390–1402 (1983).

Chapter 7

TRUSS ELEMENTS AND AXIS TRANSFORMATIONS *

7.1 Introduction

The truss element is a very common structural member. Recall that a truss element is a two force member. That is, it is loaded by two equal and opposite collinear forces. These two forces act along the line through the two connection points of the member. In elementary statics we compute the forces in truss elements as if they were rigid bodies. However, there was a class of problems, called statically indeterminant, that could not be solved by treating the members as rigid bodies. With the finite element approach we will be able to solve both classes of problems. In Sec. 3.2 the equilibrium equation for an elastic bar was developed. Clearly, the elastic bar is a special form of a truss member. To extend the previous work to include trusses in two- or three-dimensions basically requires some review of analytic geometry. Thus, we begin by reviewing that subject.

7.2 Direction Cosines

Consider a directed line segment in global space going from point 1 at (x_1, y_1, z_1) to point 2 at (x_2, y_2, z_2). Then the length of the line between the two points has components parallel to the axes of

$$L_x = x_2 - x_1, \qquad L_y = y_2 - y_1, \qquad L_z = z_2 - z_1$$

and the total length is $L^2 = (L_x^2 + L_y^2 + L_z^2)$. Specifying the end points of a line is a common way of locating its direction in space. Another common way to describe the direction is to give the *direction angles* or the corresponding *direction cosines*. Let the direction angles from the x-, y-, and z-axes to the line segment be denoted by ϕ_x, ϕ_y, and ϕ_z, respectively. Recall the relation between the total magnitude of a vector and its components, i.e., $L_x = L \operatorname{Cos} \phi_x$, etc. We generally will find the inverse geometric relation more useful. Specifically

$$\operatorname{Cos} \phi_x = L_x / L, \qquad \operatorname{Cos} \phi_y = L_y / L, \qquad \operatorname{Cos} \phi_z = L_z / L.$$

For two-dimensional problems we will assume that the structure lies in the global x-y plane so that $L_z = 0$, $\operatorname{Cos} \phi_z = 0$, and $\phi_z = 90$. In that special case only one angle is required to describe the direction rather than the usual three. It is common then to select

ϕ_x as the required angle and to omit reference to $\phi_y = 90 - \phi_x$ and to replace the second direction cosine with the relation $\text{Cos}\,\phi_y = \text{Sin}\,\phi_x$ (for $\phi_z = 90$). This is illustrated in Fig. 7.2.1. For two-dimensional problems one can utilize the simplicity of referring to a single angle. However, if one wants to automate the analysis for two- and three-dimensional problems then it is best in the long run to refer to the direction cosines.

7.3 Transformation of Displacement Components

To extend the bar element to a general truss element we need to consider the relations between a local coordinate system that is parallel and perpendicular to the element and the fixed global coordinate directions. Let the local x-axis lie along the member, that is, it passes through the two end points of the member. This means that the direction cosines of the local x-axis are the same as those for the line segment. The bar element had a single displacement, u, at any point. That local displacement vector will have components in the global space. Let the global displacements of a point be denoted by u_x, u_y, and u_z. To be consistent with this, one could also define three local components of the displacement. For a bar element the local y- and z-components are identically zero. Later we will consider members that have no zero local components. Thus, we will consider the general case of transformation of local displacement components. Referring to Fig. 7.2.1, one finds from geometry that the local x

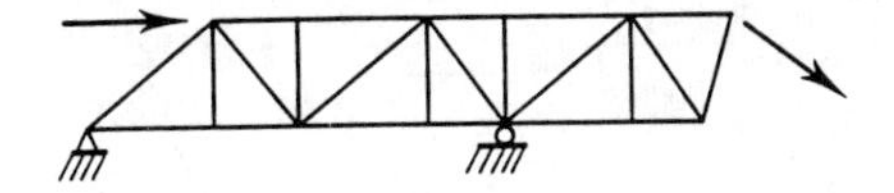

(a) A typical truss system

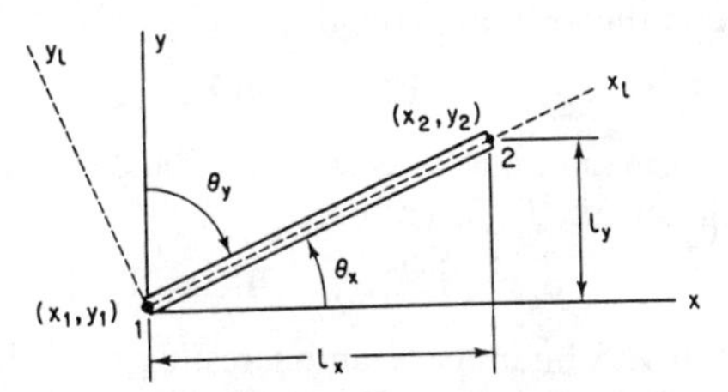

(b) Location of a typical element in local and global coordinates

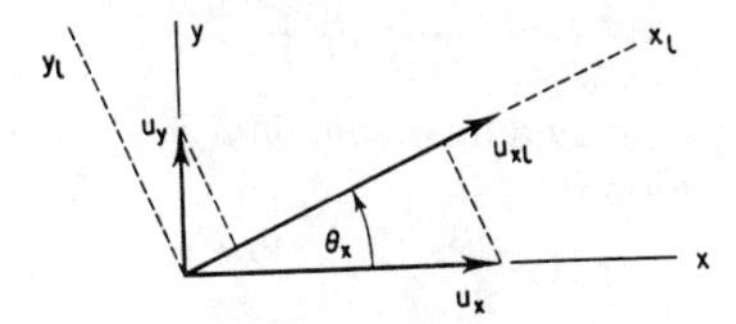

(c) Relation between local and global displacement components

Figure 7.2.1 A truss structure

displacement is related to the two-dimensional global displacements by

$$u_{xL} = u_{xg} \text{ Cos } \theta_x + u_{yg} \text{ Cos } \theta_y .$$

Similarly, if there was a local y-component of displacement it would be related to the global components by

$$u_{yL} = -u_{xg} \text{ Cos } \theta_y + u_{yg} \text{ Cos } \theta_x .$$

Writing these identities in a matrix form in terms of $\theta_x = \theta$

$$\begin{Bmatrix} u_x \\ u_y \end{Bmatrix}_L = \begin{bmatrix} \text{Cos } \theta & \text{Sin } \theta \\ -\text{ Sin } \theta & \text{Cos } \theta \end{bmatrix} \begin{Bmatrix} u_x \\ u_y \end{Bmatrix}_g \tag{7.1}$$

or symbolically this *transformation* is $\mathbf{u}_L = \mathbf{t}(\theta)\, \mathbf{u}_g$ where $\mathbf{t}$ is a nodal *transformation matrix* and $\mathbf{u}_g$ and $\mathbf{u}_L$ denote the global and local displacement components, respectively, at a point. If this relation is written at each node of the element it defines the element dof transformation matrix, $\mathbf{T}$. Specifically,

$$\begin{Bmatrix} u_{1x} \\ u_{1y} \\ \hline u_{2x} \\ u_{2y} \end{Bmatrix}_L^e = \left[\begin{array}{cc|cc} \text{Cos } \theta & \text{Sin } \theta & 0 & 0 \\ -\text{ Sin } \theta & \text{Cos } \theta & 0 & 0 \\ \hline 0 & 0 & \text{Cos } \theta & \text{Sin } \theta \\ 0 & 0 & -\text{ Sin } \theta & \text{Cos } \theta \end{array}\right] \begin{Bmatrix} u_{1x} \\ u_{1y} \\ \hline u_{2x} \\ u_{2y} \end{Bmatrix}_g^e$$

or

$$\mathbf{u}_L^e = \mathbf{T}(\theta)\, \mathbf{u}_g^e . \tag{7.2}$$

The same type of coordinate transformation will apply to components of the element force vector, $\mathbf{P}^e$, namely :

$$\mathbf{P}_L^e = \mathbf{T}(\theta)\, \mathbf{P}_g^e . \tag{7.3}$$

Notice that the transformation matrix is square. Thus, the inverse transformation can be found by inverting the matrix T. Therefore,

$$\mathbf{u}_g = \mathbf{T}^{-1}\mathbf{u}_L, \qquad \mathbf{P}_g = \mathbf{T}^{-1}\mathbf{P}_L . \tag{7.4}$$

If one carries out the inversion process, an interesting result is obtained. Specifically, we find that the inverse of the transformation is the same as its transpose. This is always true, and it makes our calculations much easier since we can write

$$\mathbf{T}^{-1} = \mathbf{T}^T . \tag{7.5}$$

A matrix with this property is called an *orthogonal* matrix. Therefore, the simple way to write the inverse transformation is

$$\mathbf{u}_g = \mathbf{T}^T \mathbf{u}_L . \tag{7.6}$$

7.4 Transformation of Element Matrices

Our ultimate goal is to solve the global equilibrium equations. This requires that all elements be referred to a single global coordinate system, and that the assembly of element contributions be relative to that system. Therefore, before we can assemble the element stiffness and load matrices they must be written relative to the global axes. This means that we need to define global versions of the element matrices, say $\mathbf{S}_g^e$ and $\mathbf{C}_g^e$.

Clearly, they are somehow related to the corresponding local element matrices, $\mathbf{S}_L^e$ and $\mathbf{C}_L^e$. To gain some insight into the relation between the two systems recall that the element behavior was defined in terms of the total potential energy, Π^e, of the element. Since that quantity is a scalar, its value must be the same regardless of whether it is computed in element coordinates or global coordinates. If we compute Π^e using Eq. (3.7) in local coordinates the result is

$$\Pi^e = \tfrac{1}{2}\,\mathbf{u}_L^{e^T}\,\mathbf{S}_L^e\,\mathbf{u}_L^e - \mathbf{u}_L^{e^T}\,\mathbf{C}_L^e . \tag{7.7}$$

By way of comparison, if it is calculated in global coordinates

$$\Pi^e = \tfrac{1}{2}\,\mathbf{u}_g^{e^T}\,\mathbf{S}_g^e\,\mathbf{u}_g^e - \mathbf{u}_g^{e^T}\,\mathbf{C}_g^e . \tag{7.8}$$

The two forms can be more easily compared if Eq. (7.7) is also written in terms of the global components of the displacements of the element. Before doing that, let us recall the form of the element stiffness and load matrices for a bar parallel to the x-axis:

$$\mathbf{S}_L^e = \frac{E^e A^e}{L^e}\begin{bmatrix} 1 & -1 \\ -1 & 1 \end{bmatrix}, \qquad \mathbf{C}_L^e = \begin{Bmatrix} C_{1x} \\ C_{2x} \end{Bmatrix}$$

where C_1 and C_2 represent the resultant loads along the local x-axis. Since the global structure will have two displacements per node it will be useful to rewrite the element equations in terms of two local displacements per node. Specifically, the expanded element equations for the equilibrium of a single element are

$$\frac{E^e A^e}{L^2}\begin{bmatrix} 1 & 0 & -1 & 0 \\ 0 & 0 & 0 & 0 \\ -1 & 0 & 1 & 0 \\ 0 & 0 & 0 & 0 \end{bmatrix}\begin{Bmatrix} u_{1x} \\ u_{1y} \\ u_{2x} \\ u_{2y} \end{Bmatrix}_L^e = \begin{Bmatrix} C_{1x}^e \\ 0 \\ C_{2x}^e \\ 0 \end{Bmatrix}_L .$$

Note that the stiffness matrix has been expanded by adding rows and columns of zeros to correspond to the local y displacement. That was done because the element cannot resist loads in the local y direction. The above expanded element matrices would be substituted into Eq. (7.7). Next, substituting the transformation identity of Eq. (7.2) into Eq. (7.7) yields

$$\Pi^e = \tfrac{1}{2}\,\mathbf{u}_g^{e^T}(\mathbf{T}^{e^T}\,\mathbf{S}_L^e\,\mathbf{T}^e)\,\mathbf{u}_g^e - \mathbf{u}_g^{e^T}(\mathbf{T}^{e^T}\,\mathbf{C}_g^e) .$$

Comparing this scalar with the same quantity in Eq. (7.8) gives the desired identities

$$\mathbf{S}_g^e = \mathbf{T}^{e^T}\,\mathbf{S}_L^e\,\mathbf{T}^e \tag{7.9}$$

and

$$\mathbf{C}_g^e = \mathbf{T}^{e^T}\,\mathbf{C}_L^e . \tag{7.10}$$

Of major importance here is that Eqs. (7.9) and (7.10) are not restricted to truss elements. For certain types of elements it would be simpler to form the global element matrices numerically by matrix multiplication.

For the truss element in two dimensions the products in these transformations are easily written out. The results are

$$\mathbf{S}^e = \frac{E^e A^e}{L^e} \begin{bmatrix} \lambda\lambda & \lambda\mu & -\lambda\lambda & -\lambda\mu \\ \lambda\mu & \mu\mu & -\lambda\mu & -\mu\mu \\ -\lambda\mu & -\lambda\mu & \lambda\lambda & \lambda\mu \\ -\lambda\mu & -\mu\mu & \lambda\mu & \mu\mu \end{bmatrix}^e \tag{7.11}$$

and

$$\mathbf{C}^e = \begin{Bmatrix} \lambda C_{1x} \\ -\mu C_{1x} \\ \lambda C_{2x} \\ -\mu C_{2x} \end{Bmatrix}^e \tag{7.12}$$

where $\lambda = \text{Cos}\ \phi_x = L_x/L$, $\mu = \text{Cos}\ \phi_y = L_y/L = \text{Sin}\ \phi_x$. A similar set of transformed global stiffness and force vectors can be obtained for a truss element located in three-dimensional space.

7.5 Example Structures

Consider the example shown in Fig. 7.5.1. Assume that all three members have the

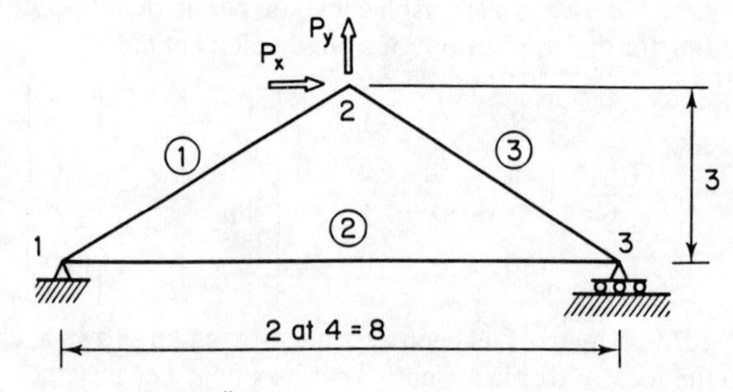

(a) Unsymmetric loading

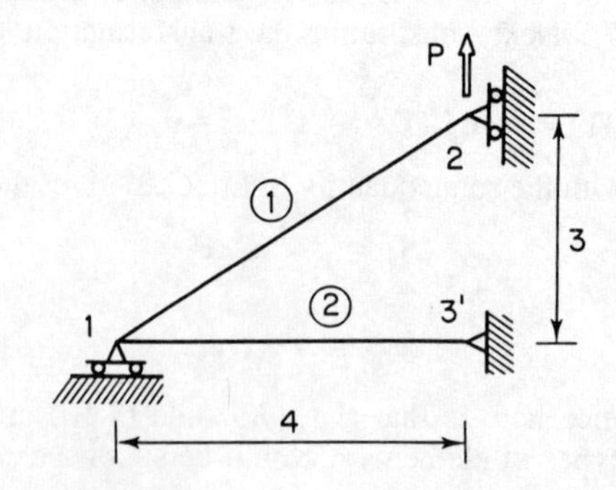

(b) Symmetric loading

Figure 7.5.1 A three-bar truss structure

same area and modulus of elasticity. The structure is described by

Element	L_x	L_y	L	Topology		EA
1	4	3	5	1	2	1000
2	8	0	8	1	3	1000
3	4	–3	5	2	3	1000

The structure is pinned at node 1 and on a horizontal roller at node 3. No distributed loads or thermal loads are considered on the bars. Thus, for each element $\mathbf{C}^e = \mathbf{0}$. Only nodal loads are externally applied. Their values, at node 2 are $P_x = 10$, and $P_y = -20$. From Eq. (7.13) the element stiffness matrices, when transformed to the global axes, have the values of

$$e = 1: \qquad S^e = \frac{1000}{125}\begin{bmatrix} 16 & 12 & -16 & -12 \\ 12 & 9 & -12 & -9 \\ -16 & -12 & 16 & 12 \\ -12 & -9 & 12 & 9 \end{bmatrix} \begin{matrix} \text{Global} \\ 1 \\ 2 \\ 3 \\ 4 \end{matrix}$$

$$e = 2: \qquad S^e = \frac{1000}{512}\begin{bmatrix} 64 & 0 & -64 & 0 \\ 0 & 0 & 0 & 0 \\ -64 & 0 & 64 & 0 \\ 0 & 0 & 0 & 0 \end{bmatrix} \begin{matrix} \text{Global} \\ 1 \\ 2 \\ 5 \\ 6 \end{matrix}$$

$$e = 3: \qquad S^e = \frac{1000}{125}\begin{bmatrix} 16 & -12 & -16 & 12 \\ -12 & 9 & 12 & -9 \\ -16 & 12 & 16 & -12 \\ 12 & -9 & -12 & 9 \end{bmatrix} \begin{matrix} \text{Global} \\ 3 \\ 4 \\ 5 \\ 6 \end{matrix}.$$

The assembled system equilibrium equations are

$$\begin{bmatrix} (128+125) & 96 & -128 & -96 & 0 \\ & 72 & -96 & -72 & 0 \\ & & (128+128) & (96-96) & 0 \\ & & & (72+72) & -72 \\ & & & & -96 \\ \text{symmetric} & & & & 72 \end{bmatrix} \begin{Bmatrix} u_1 \\ v_1 \\ u_2 \\ v_2 \\ u_3 \\ v_3 \end{Bmatrix} = \begin{Bmatrix} 0 \\ 0 \\ P_x \\ P_y \\ 0 \\ 0 \end{Bmatrix} + \begin{Bmatrix} R_1 \\ R_2 \\ 0 \\ 0 \\ 0 \\ R_3 \end{Bmatrix}.$$

However, three displacements are prescribed to be zero. Modifying the above equations to include the boundary conditions reduces them to

$$\begin{bmatrix} 256 & 0 & -128 \\ & 144 & 96 \\ \text{Sym.} & & 253 \end{bmatrix} \begin{Bmatrix} u_2 \\ v_2 \\ u_3 \end{Bmatrix} = \begin{Bmatrix} P_x \\ P_y \\ 0 \end{Bmatrix}.$$

These equations can be inverted by hand or by using subroutine I3BY3. The result is

$$\begin{Bmatrix} u_2 \\ v_2 \\ u_3 \end{Bmatrix} = \frac{1}{4.608 \times 10^6} \begin{bmatrix} 27216 & & \text{Sym.} \\ -12288 & 48384 & \\ 18432 & -24576 & 36864 \end{bmatrix} \begin{Bmatrix} P_x \\ P_y \\ 0 \end{Bmatrix}.$$

Substituting the given load values yields

$$[u_2 \;\; v_2 \;\; u_3] = [0.1124 \;\; -0.2367 \;\; 0.1467].$$

Note that if P_x had been zero, then $u_2 = u_3/2 = 0.0533$, and $v_2 = -0.21$. That is, the deformation of the structure would have been symmetric with respect to the center of the truss.

The concepts of symmetry and anti-symmetry are often useful in finite element analysis. It is common to find half, quarter, or one-eighth order symmetry conditions that can reduce the analysis cost to the square of the corresponding fractional part of a total analysis cost. Half symmetry was employed in the previous chapter in Sec. 6.4. For a truss we have no rotational degrees of freedom so we only have to consider the displacement components tangent or normal to the symmetry plane. Here we will apply symmetry to the above truss. First, we view the loads, members, and supports as viewed relative to a mirror placed at the symmetry section. The resulting partial model is shown in Fig. 7.5.1b. The applied loads and the stiffness of members lying in the symmetry plane are reduced by half. The nodes or member midpoints that lie in the symmetry plane are allowed to move only in that plane. Any supports that are not on the symmetry plane can be modified to support the structure in a consistent manner when viewed from the symmetry plane. This means that our simplified structure can be described as

Element	L_x	L_y	L	Topology		EA
1	4	3	5	1	2	1000
2	4	0	4	1	3	1000

The stiffness for the third element is no longer needed. The first member is unchanged. The length of the second member is cut in half so its stiffness doubles. The assembled elements give

$$\begin{bmatrix} (128{+}250) & 96 & -128 & -96 & -250 & 0 \\ & 72 & -96 & -72 & 0 & 0 \\ & & 128 & 96 & 0 & 0 \\ & & & 72 & 0 & 0 \\ & & & & 250 & 0 \\ \text{symmetric} & & & & & 0 \end{bmatrix} \begin{Bmatrix} u_1 \\ v_1 \\ u_2 \\ v_2 \\ u_{3'} \\ v_{3'} \end{Bmatrix} = \begin{Bmatrix} 0 \\ R_1 \\ R_2 \\ P_y/2 \\ R_{3'} \\ 0 \end{Bmatrix}.$$

Points in the plane of symmetry must always move in that plane. Thus, $u_{3'} = 0$. Conversely, node 1 must be able to move normal to the plane of symmetry. Thus, $u_1 \neq 0$. Node 2 has an external load, P, applied tangent to the plane of symmetry. Thus, it must be allowed to move tangent to the plane ($v_2 \neq 0$) so that the force can do work on the structure. That is, in a given direction one can specify either the force or the displacement at a point, but not both. Clearly, node 1 has an unknown reaction that is parallel to the symmetry plane. Thus, it must be restrained in that direction, $v_1 = 0$. The restrained structural stiffness is

$$\begin{bmatrix} 378 & -96 & 0 \\ & 72 & 0 \\ \text{sym.} & & 0 \end{bmatrix} \begin{Bmatrix} u_1 \\ v_2 \\ v_{3'} \end{Bmatrix} = \begin{Bmatrix} 0 \\ -20/2 \\ 0 \end{Bmatrix}.$$

However, these equations are still singular after the application of the usually symmetric conditions. Note that the third row and column are zero. This means that there is no stiffness associated with the displacement $v_{3'}$. From the original structure in Fig. 7.5.1, we note that the center of member 2 must have a zero vertical deflection. Employing this additional physical insight, we can now also state that $v_{3'} = 0$. Therefore, for a symmetric structure with symmetric loads the equilibrium equations, relative to the plane of symmetry are

$$\begin{bmatrix} 378 & -96 \\ -96 & 72 \end{bmatrix} \begin{Bmatrix} u_1 \\ v_2 \end{Bmatrix} = \begin{Bmatrix} 0 \\ -10 \end{Bmatrix}, \quad \begin{Bmatrix} u_1 \\ v_2 \end{Bmatrix} = \begin{Bmatrix} -0.05333 \\ -0.21 \end{Bmatrix}$$

as before, except for the sign change on u_1. This solution shows that for the above example there are only two degrees of freedom required when symmetry is available versus the three that were used before. For this simple example there was not much difference in the computational effort required in the symmetric and non-symmetric solutions. However, if there are hundreds or thousands of symmetric elements then the cost saving is very significant when a symmetric analysis can be utilized.

7.6 References

[1] Cook, R.D., *Concepts and Applications of Finite Element Analysis*, New York: John Wiley (1974).

[2] Fenner, R.T., *Finite Element Methods for Engineers*, London: Macmillan (1975).

[3] Martin, H.C. and Carey, G.F., *Introduction to Finite Element Analysis*, New York: McGraw-Hill (1974).

[4] Ural, O., *Matrix Operations and Use of Computers in Structural Engineering*, International Textbook (1971).

[5] Weaver, W.F., Jr. and Johnston, P.R., *Finite Elements for Structural Analysis*, Englewood Cliffs: Prentice-Hall (1984).

Chapter 8

CYLINDRICAL ANALYSIS PROBLEMS

8.1 Introduction

There are many problems that can accurately be modeled as being revolved about an axis. Many of these can be analyzed by employing a radial coordinate, R, and an axial coordinate, Z. Solids of revolution can be formulated in terms of the two-dimensional area that is revolved about the axis. This concept is illustrated in Fig. 8.1.1. Numerous other objects are very long in the axial direction and can be treated as segments of a cylinder. This reduces the analysis to a one-dimensional study in the radial direction. We will begin with that common special case. We will find that changing to these *cylindrical coordinates* will make small changes in the governing differential equations and the corresponding integral theorems that govern the finite element formulation. Also the volume and surface integrals take on special forms. These use the *Theorems of Pappus*. The first states that the surface area of a revolved arc is the product of the arc length and the distance traveled by the centroid of the arc. The second states that the volume of revolution of the generating area is the product of the area and the distance traveled by its centroid. In both cases the distance traveled by the centroid, in full revolution, is $2\pi\bar{R}$ where $\bar{R}$ is the centroidal radial coordinate of the arc or area. If we consider differential arcs or areas, then the corresponding differential surface or volume of revolution are $da = 2\pi R\, dL$ and $dv = 2\pi R\, dr\, dZ$.

8.2 Heat Conduction in a Cylinder

The previous one-dimensional heat transfer model becomes slightly more complicated here. When we consider a point on a radial line we must remember that it is a cross-section of a ring of material around the hoop of the cylinder. Thus as heat is conducted outward in the radial direction it passes through an ever increasing amount of material. The resulting differential equation for thermal equilibrium is well known:

$$\frac{1}{R}\frac{d}{dR}\left[Rk\frac{dT}{dR}\right] + Q = 0 \tag{8.1}$$

where R is the radial distance from the axis of revolution, k is the thermal conductivity, T is the temperature, and Q is the internal heat generation per unit volume. One can

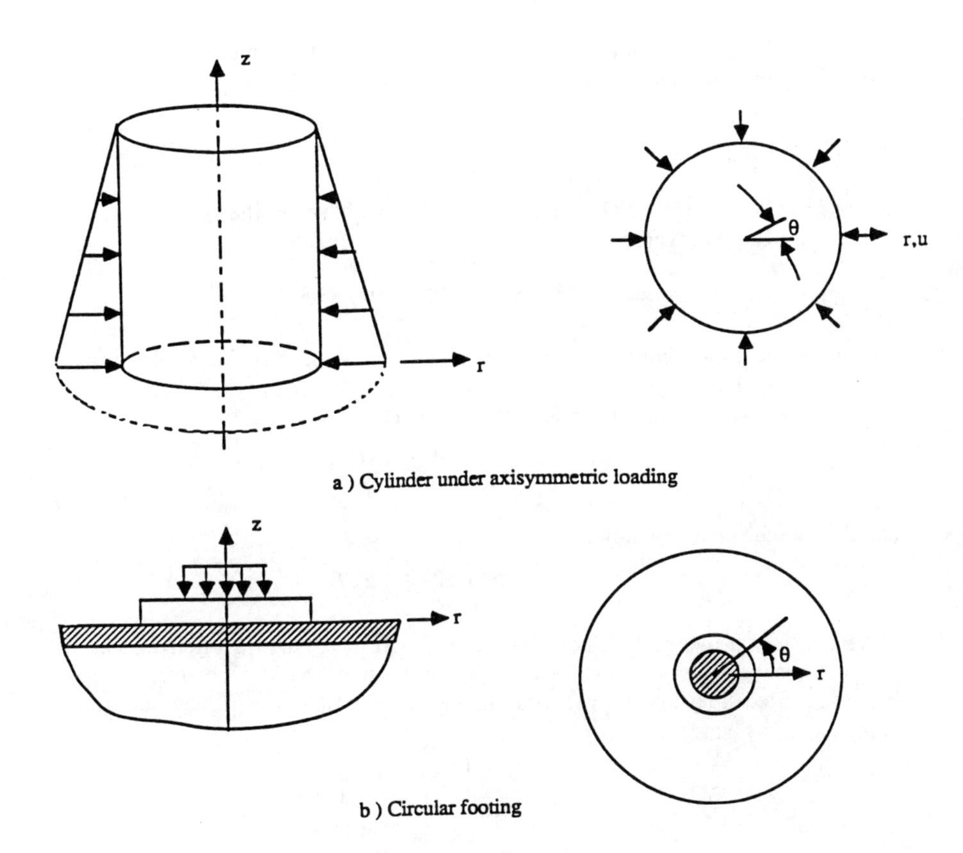

Figure 8.1.1 Axisymmetric problems

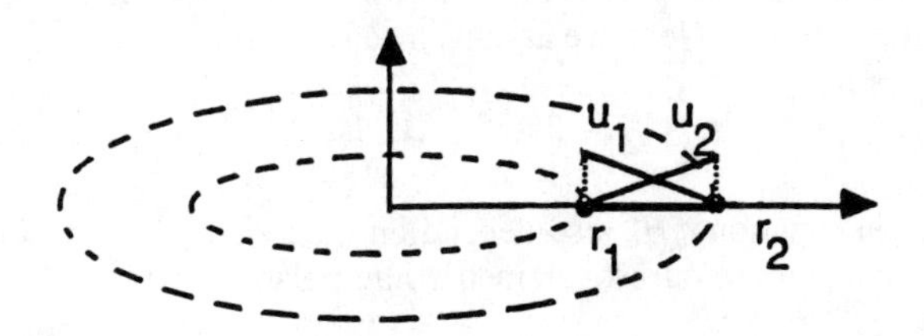

Figure 8.2.1 Linear element for cylindrical analysis

have essential boundary conditions where T is given or as a surface flux condition

$$- Rk \frac{dT}{dR} = q \tag{8.2}$$

where q is the flux normal to the surface (i.e., radially). If we multiply Eq. (8.1) by R, it would look like our previous one-dimensional form:

$$\frac{d}{dR}\left[k^* \frac{dT}{dR} \right] + Q^* = 0$$

where $k^* = Rk$ and $Q^* = RQ$ could be viewed as variable coefficients. This lets us find the required integral (variational) form by inspection. It is

$$I = 2\pi \Delta Z \int_L \tfrac{1}{2} \left[k(dT/dR)^2 - QT \right] R\, dR \quad \rightarrow \min \tag{8.3}$$

where the integration limits are the inner and outer radii of the cylindrical segment under study. The typical length in the axial direction, ΔZ, is usually defaulted to unity. The corresponding element square conduction matrix is

$$\mathbf{S}^e = 2\pi \int_{L^e} k^e \frac{d\mathbf{H}^{e^T}}{dR} \frac{d\mathbf{H}^e}{dR} R\, dR \tag{8.4}$$

and the source vector (if any) is

$$\mathbf{C}_Q^e = 2\pi \int_{L^e} \mathbf{H}^{e^T} Q^e R\, dR . \tag{8.5}$$

If we consider a two node (linear) line element in the radial direction we can use our previous results to write these matrices by inspection. Such an element is shown in Fig. 8.2.1. Recalling that $L^e = (R_2 - R_1)^e$ and assuming a constant material property, k, in the element gives

$$\mathbf{S}^e = 2\pi \frac{k^e}{(L^e)^2} \begin{bmatrix} 1 & -1 \\ -1 & 1 \end{bmatrix} \int_{R_1}^{R_2} R\, dR$$

$$\mathbf{S}^e = 2\pi \frac{k^e (R_2^2 - R_1^2)^e}{2(L^e)^2} \begin{bmatrix} 1 & -1 \\ -1 & 1 \end{bmatrix}, \qquad \mathbf{S}^e = \pi \frac{k^e (R_2 + R_1)^e}{(R_2 - R_1)^e} \begin{bmatrix} 1 & -1 \\ -1 & 1 \end{bmatrix}. \tag{8.6}$$

Thus, unlike the original one-dimensional case the conduction matrix depends on where the element is located, that is, it depends on how much material it includes (per unit length in the axial direction). Next, we assume a constant source term so

$$\mathbf{C}_Q^e = 2\pi Q^e \int_{L^e} \mathbf{H}^{e^T} R\, dR .$$

But the interpolation functions, H, also depend on the radial position. One approach to this integration is to again use our isoparametric interpolation and let $R = \mathbf{H}^e \mathbf{R}^e$. Then

$$\mathbf{C}_Q^e = 2\pi Q^e \int_{L^e} \mathbf{H}^{e^T} \mathbf{H}^e \, dR\, \mathbf{R}^e . \tag{8.7}$$

Therefore,

$$\mathbf{C}_Q^e = \frac{2\pi Q^e L^e}{6}\begin{bmatrix} 2 & 1 \\ 1 & 2 \end{bmatrix}\begin{Bmatrix} R_1 \\ R_2 \end{Bmatrix}^e = \frac{2\pi Q^e L^e}{6}\begin{bmatrix} 2R_1 + R_2 \\ R_1 + 2R_2 \end{bmatrix}^e .$$

As a simple numerical example consider a cylinder with constant properties, no internal heat generation, an inner radius temperature of $T = 100$ at $R = 1$, and an outer radius temperature of $T = 10$ at $R = 2$. Select a model with four equal length elements and five nodes. Numbering the nodes radially we have essential boundary conditions of $T_1 = 100$ and $T_5 = 10$. Considering the form in Eq. (8.6), we note that the element values of $(R_2 + R_1)^e / L^e$ are 9, 11, 13, and 15, respectively. Therefore, we can write the assembled system equations as

$$\pi k^e \begin{bmatrix} 9 & -9 & 0 & 0 & 0 \\ -9 & (9+11) & -11 & 0 & 0 \\ 0 & -11 & (11+13) & -13 & 0 \\ 0 & 0 & -13 & (13+15) & -15 \\ 0 & 0 & 0 & -15 & 15 \end{bmatrix}\begin{Bmatrix} T_1 \\ T_2 \\ T_3 \\ T_4 \\ T_5 \end{Bmatrix} = \begin{Bmatrix} q_1 \\ 0 \\ 0 \\ 0 \\ -q_5 \end{Bmatrix}.$$

Applying the essential boundary conditions, and dividing both sides by the leading constant gives the reduced set

$$\begin{bmatrix} 20 & -11 & 0 \\ -11 & 24 & -13 \\ 0 & -13 & 28 \end{bmatrix}\begin{Bmatrix} T_2 \\ T_3 \\ T_4 \end{Bmatrix} = 100\begin{Bmatrix} 9 \\ 0 \\ 0 \end{Bmatrix} + 10\begin{Bmatrix} 0 \\ 0 \\ 15 \end{Bmatrix}.$$

Solving yields the internal temperature distribution of $T_2 = 71.06$, $T_3 = 47.39$, and $T_4 = 27.34$. Comparing with the exact solution $T = [T_1 \ln(R_5/R) + T_5 \ln(R/R_1)] / \ln(R_5/R_1)$ shows that our approximation is accurate to at least three significant figures. Also note that both the exact and approximate temperature distributions are independent of the thermal conductivity k. This is true only because the internal heat generation Q was zero. Of course, k does have some effect on the two external heat fluxes (thermal reactions), q_1 and q_5, necessary to maintain the two prescribed surface temperatures. Substituting back into the first equation to recover the thermal reaction we obtain

$$\pi [\, 9(100) - 9(71.06) + 0 \,] = 818.3 = q_1 / k^e$$

entering at the inner radius. The fifth equation gives q_5 an equal amount exiting at the outer radius. Therefore, in this problem the heat flux is always in the positive radial direction. It should be noted that if we had used a higher order element then the integrals would have been much more complicated than the one-dimensional case. This is typical of most axisymmetric problems. Of course, in practice we use numerical integration to automate the evaluation of the element matrices.

8.3 Cylindrical Stress Analysis

Another common problem is the analysis of an axisymmetric solid with axisymmetric loads and supports. This becomes a two-dimensional analysis that is very

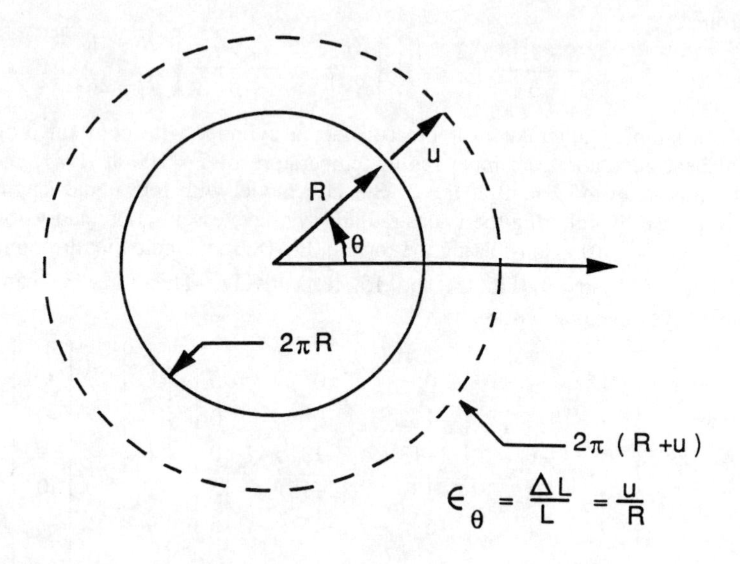

Figure 8.3.1 Hoop strain due to radial displacement

similar to plane strain analysis. The radial and axial displacement components will be denoted by u and v. These are the same unknowns used in the plane strain study. In addition to the previous strains there is another strain known as the *hoop strain*, ε_θ, and a corresponding hoop stress, σ_θ. The hoop strain results from the change in length of a fiber of material around the circumference of the solid. Recall the definition of strain as a change in length divided by the original length. The circumference at a typical radial position is $L = 2\pi R$. When such a point undergoes a radial displacement of u it occupies a new radial position of $(R + u)$, as shown in Fig. 8.3.1. It has a corresponding increase in circumference. The hoop strain becomes

$$\varepsilon_\theta = \frac{\Delta L}{L} = \frac{2\pi(R+u) - 2\pi R}{2\pi R} = \frac{u}{R} . \tag{8.8}$$

Then, our strains to be computed for the cylindrical analysis are denoted as

$$\varepsilon^T = [\varepsilon_R \ \ \varepsilon_\theta] \tag{8.9}$$

and the corresponding stress components are

$$\sigma^T = [\sigma_R \ \ \sigma_\theta] . \tag{8.10}$$

For an isotropic material the stress-strain law is like that for the plane strain case. In the axial, or Z, direction one can either set the stress or strain to zero, but not both. They are related by :

$$\varepsilon_Z = \frac{1}{E}(\sigma_Z - \nu\sigma_R - \nu\sigma_\theta) . \tag{8.11}$$

There is no shear stress or strain in the cylindrical case. In the case of an infinite cylinder the above formulation simplifies since $v = 0$ and $\partial/\partial z = 0$. Thus, we consider only the radial displacement, u, and the strains and stresses in the radial and hoop directions. This gives the two strain-displacement relations as

$$\boldsymbol{\varepsilon} = \mathbf{B}^e \mathbf{u}^e \tag{8.12}$$

where

$$\mathbf{B}^e = \begin{bmatrix} \partial \mathbf{H}/\partial R \\ \mathbf{H}/R \end{bmatrix}. \tag{8.13}$$

The radial and hoop stresses are $\boldsymbol{\sigma} = \mathbf{D}\,\boldsymbol{\varepsilon}$ where

$$\mathbf{D} = \begin{bmatrix} D_{11} & D_{12} \\ D_{21} & D_{22} \end{bmatrix} \tag{8.14}$$

and for an isotropic material

$$D_{11} = D_{22} = E(1-\nu)/(1+\nu)(1-2\nu)$$

$$D_{12} = D_{21} = E\nu/(1+\nu)(1-2\nu).$$

The stiffness matrix,

$$\mathbf{K}^e = 2\pi \int_{A^e} \mathbf{B}^{e^T} \mathbf{D}^e \mathbf{B}^e \, R \, dR \, dZ,$$

can be expanded to the form

$$\mathbf{K}^e = 2\pi\,\Delta z \int_{L^e} \left[D_{11} \frac{\partial \mathbf{H}^T}{\partial R} \frac{\partial \mathbf{H}}{\partial R} \right.$$
$$\left. + D_{12} \left(\frac{\partial \mathbf{H}^T}{\partial R} \mathbf{H} + \mathbf{H}^T \frac{\partial \mathbf{H}}{\partial R} \right) / R + D_{22} \mathbf{H}^T \mathbf{H} / R^2 \right] R \, dR. \tag{8.15}$$

The first integral we just evaluated and is given in Eqs. (8.4) and (8.6) if we let $\Delta z = 1$ and replace k with D_{11}. The second term we integrate by inspection since the R terms cancel. The result is

$$\mathbf{K}_{12}^e = 2\pi\,\Delta z\, D_{12}^e \begin{bmatrix} -1 & 0 \\ 0 & 1 \end{bmatrix}. \tag{8.16}$$

The remaining contribution is more difficult since it involves division by R. Assuming constant D_{22} we have

$$\mathbf{K}_{22} = 2\pi\,\Delta z\, D_{22} \int_{L^e} \frac{1}{R} \mathbf{H}^T \mathbf{H} \, dR \tag{8.17}$$

which requires analytic integration involving logarithms, or numerical integration. Using a one point (centroidal) quadrature rule gives the approximation

$$\mathbf{K}_{22}^e = \frac{2\pi\,\Delta z\, D_{22}^e L^e}{2(R_1 + R_2)^e} \begin{bmatrix} 1 & 1 \\ 1 & 1 \end{bmatrix}. \tag{8.18}$$

For a cylinder the loading would usually be a pressure acting on an outer surface or an internal centrifugal load due to a rotation about the z-axis. For a pressure load the resultant force at a nodal ring is the pressure times the surface area. Thus, $F_{p_i} = 2\pi\,\Delta z\, R_i p_i$. As a numerical example consider a single element solution for a cylinder with an internal pressure of $p = 1$ ksi on the inner radius $R_1 = 10$ in. Assume

$E = 10^4$ ksi and $\nu = 0.3$, and let the thickness of the cylinder by 1 in. Note that there is no essential boundary condition on the radial displacement. This is because the hoop effects prevent a rigid body radial motion. The numerical values of the above stiffness contributions are

$$\mathbf{K}_{11} = 2\pi\,\Delta z\,(1.413 \times 10^5)\begin{bmatrix} 1 & -1 \\ -1 & 1 \end{bmatrix}$$

$$\mathbf{K}_{12} = 2\pi\,\Delta z\,(5.769 \times 10^3)\begin{bmatrix} -1 & 0 \\ 0 & 1 \end{bmatrix}, \quad \mathbf{K}_{22} = 2\pi\,\Delta z\,(3.205 \times 10^2)\begin{bmatrix} 1 & 1 \\ 1 & 1 \end{bmatrix}$$

while the resultant force at the inner radius is $F_p = 2\pi\,\Delta z\,10$. Assembling and canceling the common constant gives

$$10^5\begin{bmatrix} 1.35897 & -1.41026 \\ -1.41026 & 1.47436 \end{bmatrix}\begin{Bmatrix} u_1 \\ u_2 \end{Bmatrix} = \begin{Bmatrix} 10 \\ 0 \end{Bmatrix}.$$

Solving gives $\mathbf{u} = [9.9642 \quad 9.5309] \times 10^{-3}$ in. This represents a displacement error of about 8% and 9%, respectively, at the two nodes. The maximum radial stress equals the applied pressure. The stresses can be found from Eq. (8.14). The hoop strain at node 1 is

$$\varepsilon_\theta = u_1 / R_1 = 9.964 \times 10^{-4}\ \text{in}/\text{in}.$$

The radial strain is

$$\varepsilon_R = \frac{\partial H_1}{\partial R}u_1 + \frac{\partial H_2}{\partial R}u_2.$$

The constant radial strain approximation is

$$\varepsilon_R = \frac{-u_1 + u_2}{R_2 - R_1} = -4.333 \times 10^{-4}\ \text{in}/\text{in}.$$

Therefore, the hoop stress at the first node is

$$\sigma_\theta = D_{12}\varepsilon_R + D_{22}\varepsilon_\theta = -2.500 + 13.413 = 10.91\ \text{ksi}.$$

This compares well with the exact value of 10.52 ksi. Note that the inner hoop stress is more than ten times the applied internal pressure. Since we have set ε_z to zero we should use Eq. (8.11) to determine the axial stress that results from the effect of Poisson's ratio. All three stresses would be used in evaluating a failure criterion like the Von Mises stress.

8.4 References

[1] Rockey, K.C., et al., *Finite Element Method – A Basic Introduction*, New York: Halsted Press (1975).

[2] Ross, C.T.F., *Finite Element Programs for Axisymmetric Problems in Engineering*, New York: Halsted Press (1984).

[3] Weaver, W.F., Jr. and Johnston, P.R., *Finite Elements for Structural Analysis*, Englewood Cliffs: Prentice-Hall (1984).

[4] Zienkiewicz, O.C., *The Finite Element Method in Structural and Continuum Mechanics*, New York: McGraw-Hill (1967).

Chapter 9

GENERAL INTERPOLATION

9.1 Introduction

The previous sections have illustrated the heavy dependence of finite element methods on both spatial interpolation and efficient integrations. In a one-dimensional problem it does not make a great deal of difference if one selects a local or global coordinate system for the interpolation equations, because the inter-element continuity requirements are relatively easy to satisfy. That is not true in higher dimensions. To obtain practical formulations it is almost essential to utilize local coordinate interpolations. Doing this does require a small amount of additional work in relating the derivatives in the two coordinate systems.

9.2 Unit Coordinate Interpolation

The use of unit coordinates have been previously mentioned in Chap. 4. Here some of the procedures for deriving the interpolation functions in unit coordinates will be presented. Consider the three-node triangular element shown in Fig. 4.2.2. The local coordinates of its three nodes are (0, 0), (1, 0), and (0, 1), respectively. Once again we wish to utilize polynomial functions for our interpolations. In two dimensions the simplest complete polynomial has three constants. Thus, this linear function can be related to the three nodal quantities of the element. Assume the polynomial for some quantity, u, is defined as:

$$u^e(r, s) = d_1^e + d_2^e r + d_3^e s = \mathbf{P}(r, s)\, \mathbf{d}^e . \tag{9.1}$$

If it is valid everywhere in the element then it is valid at its nodes. Substituting the local coordinates of a node into Eq. (9.1) gives an identity between the $\mathbf{d}^e$ and a nodal value of $\mathbf{u}$. Establishing these identities at all three nodes gives

$$\begin{Bmatrix} u_1^e \\ u_2^e \\ u_3^e \end{Bmatrix} = \begin{bmatrix} 1 & 0 & 0 \\ 1 & 1 & 0 \\ 1 & 0 & 1 \end{bmatrix} \begin{Bmatrix} d_1^e \\ d_2^e \\ d_3^e \end{Bmatrix}$$

or

$$\mathbf{u}^e = \mathbf{g}\,\mathbf{d}^e . \tag{9.2}$$

Iff the inverse exists, and it does here, this equation can be solved to yield

$$\mathbf{d}^e = \mathbf{g}^{-1}\mathbf{u}^e \tag{9.3}$$

and

$$u^e(r, s) = \mathbf{P}(r, s)\,\mathbf{g}^{-1}\,\mathbf{u}^e = \mathbf{H}(r, s)\mathbf{u}^e . \tag{9.4}$$

Here

$$\mathbf{g}^{-1} = \begin{bmatrix} 1 & 0 & 0 \\ -1 & 1 & 0 \\ -1 & 0 & 1 \end{bmatrix} \tag{9.5}$$

and

$$H_1(r, s) = 1 - r - s\ , \qquad H_2(r, s) = r\ , \qquad H_3(r, s) = s\ . \tag{9.6}$$

Typical coding for these relations and their local derivatives are shown as subroutines SHP3T and DER3T in Fig. 9.2.1.

Similarly, for the unit coordinate bilinear quadrilateral in Fig. 4.2.3 one could assume that

```
      SUBROUTINE  SHP3T (S, T, H)
C     * * * * * * * * * * * * * * * * * * * * * * * * *
C     SHAPE FUNCTIONS FOR A THREE NODE UNIT TRIANGLE
C     * * * * * * * * * * * * * * * * * * * * * * * * *
      DIMENSION  H(3)
C     S,T = LOCAL COORDINATES OF THE POINT     3     T
C     H   = SHAPE FUNCTIONS                    . .   .
C     NODAL COORDS 1-(0,0)  2-(1,0)  3-(0,1)  1..2  0..S
      H(1) = 1. - S - T
      H(2) = S
      H(3) = T
      RETURN
      END

      SUBROUTINE  DER3T (S, T, DH)
C     * * * * * * * * * * * * * * * * * * * * * * * * *
C     LOCAL DERIVATIVES OF A THREE NODE UNIT TRIANGLE
C                SEE SUBROUTINE SHP3T
C     * * * * * * * * * * * * * * * * * * * * * * * * *
      DIMENSION  DH(2,3)
C     S,T      = LOCAL COORDINATES OF THE POINT
C     DH(1,K)  = DH(K)/DS
C     DH(2,K)  = DH(K)/DT
C     NODAL COORDS ARE :  1-(0,0) 2-(1,0) 3-(0,1)
      DH(1,1)  = -1.
      DH(1,2)  = 1.
      DH(1,3)  = 0.0
      DH(2,1)  = -1.
      DH(2,2)  = 0.0
      DH(2,3)  = 1.
      RETURN
      END
```

Figure 9.2.1 Coding a linear unit coordinate triangle

$$u^e(r, s) = d_1^e + d_2^e r + d_3^e s + d_4^e rs \tag{9.7}$$

so that

$$\mathbf{g} = \begin{bmatrix} 1 & 0 & 0 & 0 \\ 1 & 1 & 0 & 0 \\ 1 & 1 & 1 & 1 \\ 1 & 0 & 1 & 0 \end{bmatrix} \tag{9.8}$$

and

$$\begin{aligned} H_1(r, s) &= 1 \quad -r \quad -s \quad +rs \\ H_2 &= \quad r \quad -rs \\ H_3 &= \quad rs \\ H_4 &= \quad s \quad -rs . \end{aligned} \tag{9.9}$$

However, for the quadrilateral it is more common to utilize the natural coordinates, as shown in Fig. 9.2.2. In that coordinate system $-1 \le a, b \le +1$ so that

$$\mathbf{g} = \begin{bmatrix} 1 & -1 & -1 & 1 \\ 1 & 1 & -1 & -1 \\ 1 & 1 & 1 & 1 \\ 1 & -1 & 1 & -1 \end{bmatrix}$$

and the alternate interpolation functions are

$$H_i(a, b) = (1 + aa_i)(1 + bb_i)/4, \qquad 1 \le i \le 4 \tag{9.10}$$

where (a_i, b_i) are the local coordinates of node i. These four functions and their local derivatives can be coded as shown in Fig. 9.2.3.

Note that up to this point we have utilized the local element coordinates for interpolation. Doing so makes the geometry matrix, $\mathbf{g}$, depend only on element type instead of element number. If we use global coordinates then the geometric matrix, $\mathbf{g}^e$ is

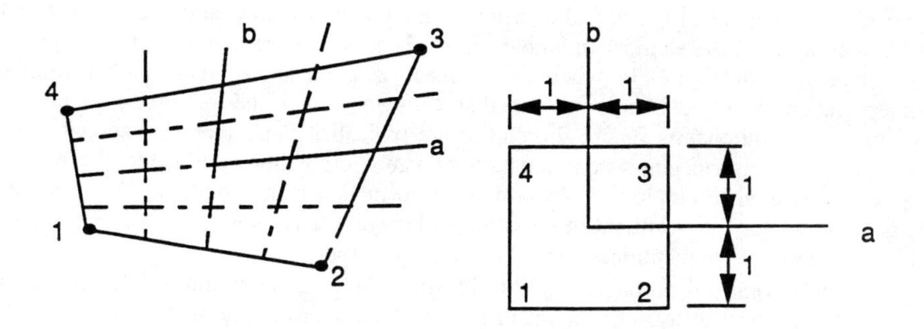

Figure 9.2.2 Natural coordinate quadrilateral element

always dependent on the element number, e. For example, if Eq. (9.1) is written in physical coordinates then

$$u^e(x, y) = d_1^e + d_2^e x + d_3^e y \tag{9.11}$$

so when the identities are evaluated at each node the result is

$$\mathbf{g}^e = \begin{bmatrix} 1 & x_1^e & y_1^e \\ 1 & x_2^e & y_2^e \\ 1 & x_3^e & y_3^e \end{bmatrix}. \tag{9.12}$$

Inverting and simplifying the algebra gives the global coordinate equivalent of Eq. (9.6) for a specific element :

$$H_i^e(x, y) = (a_i^e + b_i^e x + c_i^e y)/2A^e, \quad 1 \le i \le 3 \tag{9.13}$$

where the algebraic constants are

$$\begin{aligned} a_1^e &= x_2^e y_3^e - x_3^e y_2^e & b_1^e &= y_2^e - y_3^e & c_1^e &= x_3^e - x_2^e \\ a_2^e &= x_3^e y_1^e - x_1^e y_3^e & b_2^e &= y_3^e - y_1^e & c_2^e &= x_1^e - x_3^e \\ a_3^e &= x_1^e y_2^e - x_2^e y_1^e & b_3^e &= y_1^e - y_2^e & c_3^e &= x_2^e - x_1^e \end{aligned}$$

and A^e is the area of the element, that is, $A^e = (a_1^e + a_2^e + a_3^e)/2$, or

$$A^e = \left[x_1^e(y_2^e - y_3^e) + x_2^e(y_3^e - y_1^e) + x_3^e(y_1^e - y_2^e) \right] / 2 .$$

These algebraic forms assume that the three local nodes are numbered counter-clockwise from an arbitrarily selected corner. If the topology is defined in a clockwise order then the area, A^e, becomes negative.

It would be natural at this point to attempt to utilize a similar procedure to define the four node quadrilateral in the same manner. For example, if Eq. (9.7) is written as

$$u^e(x, y) = d_1^e + d_2^e x + d_3^e y + d_4^e xy . \tag{9.15}$$

However, we now find that for a general quadrilateral the inverse of matrix $\mathbf{g}^e$ may not exist. This means that the global coordinate interpolation is in general very sensitive to the orientation of the element in global space. That is very undesirable. This important disadvantage vanishes only when the element is a rectangle. This global form of interpolation also yields an element that fails to satisfy the required interelement continuity requirements. These difficulties are typical of those that are encountered in two- and three-dimensions when global coordinate interpolation is utilized. Therefore, it is most common to employ the local coordinate mode of interpolation. Doing so also easily allows for the treatment of curvilinear elements. That is done with *isoparametric elements* that will be mentioned later.

At this point it is probably useful to illustrate the lack of continuity that develops in the global coordinate form of the quadrilateral. First, consider the three-node triangular element and examine the interface or boundary where two elements connect. Along the interface between the two elements one has the geometric restriction that the edge is a straight line given by $y = m^b x + n^b$. Recall that the general form of the global coordinate interpolation functions for the triangle is

```
      SUBROUTINE  SHP4Q (R, S, H)
C     * * * * * * * * * * * * * * * * * * * * * * * *
C       SHAPE FUNCTIONS OF A 4 NODED QUADRILATERAL
C     * * * * * * * * * * * * * * * * * * * * * * * *
      DIMENSION H(4)
C     (R,S) = A POINT IN THE NATURAL COORDS       4--3
C     H     = LOCAL INTERPOLATION FUNCTIONS       I  I
C     H(I)  = 0.25*(1+R*R(I))*(1+S*S(I))          I  I
C     R(I)  = LOCAL R-COORDINATE OF NODE I        1--2
C     LOCAL COORDS, 1=(-1,-1)   3=(+1,+1)
      RP = 1. + R
      RM = 1. - R
      SP = 1. + S
      SM = 1. - S
      H(1) = 0.25*RM*SM
      H(2) = 0.25*RP*SM
      H(3) = 0.25*RP*SP
      H(4) = 0.25*RM*SP
      RETURN
      END
      SUBROUTINE  DER4Q (R, S, DELTA)
C     * * * * * * * * * * * * * * * * * * * * * * * *
C     LOCAL DERIVATIVES OF QUADRILATERAL WITH 4 NODES
C     * * * * * * * * * * * * * * * * * * * * * * * *
      DIMENSION  DELTA(2,4)
C     DELTA(1,I) = DH/DR, DELTA(2,I) = DH/DS, SEE SHP4Q
C     (R,S)      = A POINT IN THE LOCAL COORDINATES
C     HERE D(H(I))/DR = 0.25*R(I)*(1+S*S(I)), ETC.
      RP = 1. + R
      RM = 1. - R
      SP = 1. + S
      SM = 1. - S
      DELTA(1,1) = -0.25*SM
      DELTA(1,2) =  0.25*SM
      DELTA(1,3) =  0.25*SP
      DELTA(1,4) = -0.25*SP
      DELTA(2,1) = -0.25*RM
      DELTA(2,2) = -0.25*RP
      DELTA(2,3) =  0.25*RP
      DELTA(2,4) =  0.25*RM
      RETURN
      END
```

Figure 9.2.3 Coding a bi-linear quadrilateral

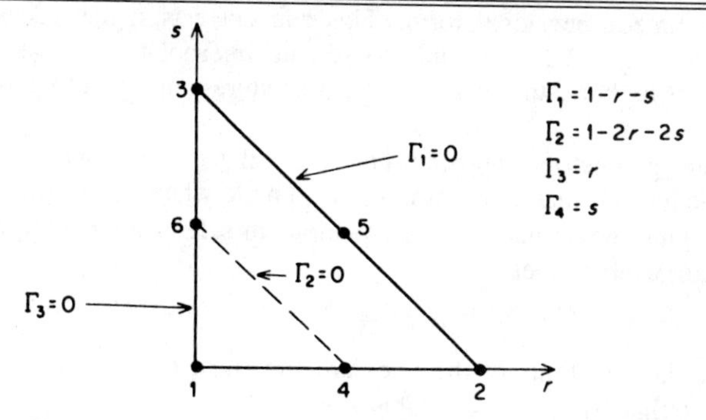

Figure 9.2.4 Boundary curves through element nodes

$$u(x, y) = d_1^e + d_2^e x + d_3^e y$$

where the g_i are element constants. Along the typical interface this reduces to

$$u = d_1^e + d_2^e x + d_3^e (m^b x + n^b)$$

$$u = (d_1^e + n^b d_3^e) + (g_2^e + m^b d_3^e)\, x,$$

or simply $u = f_1 + f_2 x$. Clearly, this shows that the boundary displacement is a linear function of x. The two constants, f_i, could be uniquely determined by noting that $u(x_1) = u_1$ and $u(x_2) = u_2$. Since those two quantities are common to both, elements the displacement, $u(x)$, will be continuous between the two elements.

By way of comparison when the same substitution is made in Eq. (9.15) the result for the quadrilateral element is

$$u = d_1^e + d_2^e x + d_3^e (m^b x + n^b) + d_4^e x (m^b x + n^b)$$

$$u = \left[d_1^e + (d_3^e + d_4^e)\, n^b \right] + x \left[d_2^e + d_3^e m^b \right] + x^2 d_4^e\, m^b = f_1 + f_2 x + f_3 x^2 .$$

This quadratic function cannot be uniquely defined by the two constants u_1 and u_2. Therefore, it is not possible to prove that the displacements will be continuous between elements. This is an undesirable feature of quadrilateral elements when formulated in global coordinates. If the quadrilateral interpolation is given in local coordinates such as Eq. (9.9) or Eq. (9.10), this problem does not occur. On the edge $s = 0$, Eq. (9.9) reduces to $u = f_1 + f_2 r$. A similar result occurs on the edge $s = 1$. Likewise, for the other two edges $u = f_1 + f_2 s$. Thus, in local coordinates the element degenerates to a linear function on any edge, and therefore will be uniquely defined by the two shared nodal displacements. In other words, the local coordinate four node quadrilateral will be compatible with elements of the same type and with the three-node triangle. The above observations suggest that global coordinates could be utilized for the four-node element only so long as it is a rectangle parallel to the global axes.

The extension of the unit coordinates to the three-dimensional tetrahedra illustrated in Fig. 4.2.2 is straightforward. In the result given below

$$\begin{aligned} H_1(r, s, t) &= 1 - r - s - t \qquad & H_2(r, s, t) &= r \\ H_3(r, s, t) &= s & H_4(r, s, t) &= t, \end{aligned} \tag{9.16}$$

and comparing this to Eqs. (9.6) and (4.11), we note that the 2-D and 1-D forms are contained in the three-dimensional form. This concept was suggested by the topology relations shown in Fig. 4.2.2. The unit coordinate interpolation is easily extended to quadratic, cubic, or higher interpolation. The procedure employed to generate Eq. (9.6) can be employed.

An alternate geometric approach can be utilized. We want to generate an interpolation function, H_i, that vanishes at the j-th node when $i \neq j$. Such a function can be obtained by taking the products of the equations of selected curves through the nodes on the element. For example, let

$$H_1(r, s) = C_1\, \Gamma_1\, \Gamma_2$$

where the Γ_i are the equations of the lines are shown in Fig. 9.2.4, and where C_1 is a constant chosen so that $H_1(r_1, s_1) = 1$. This yields

$$H_1 = (1 - 3r - 3s + 2r^2 + 4rs + 2s^2)\,.$$

Similarly, letting $H_4 = C_4 \Gamma_1 \Gamma_3$ gives $C_4 = 4$ and $H_4 = 4r(1 - r - s)$. This type of procedure is usually quite straightforward. However, there are times when there is not a unique choice of products, and then care must be employed to select the proper products. The resulting two-dimensional interpolation functions for the quadratic triangle are

$$\begin{aligned}
H_1(r,s) &= 1 - 3r + 2r^2 - 3s + 4rs + 2s^2 \\
H_2(r,s) &= -r + 2r^2 \\
H_3(r,s) &= -s + 2s^2 \\
H_4(r,s) &= 4r - 4r^2 - 4rs \\
H_5(r,s) &= 4rs \\
H_6(r,s) &= 4s - 4rs - 4s^2\,.
\end{aligned} \tag{9.17}$$

Once again, it is possible to obtain the one-dimensional quadratic interpolation on a typical edge by setting $s = 0$. Figure 9.2.5 shows the shape of the typical interpolation functions for a quadratic triangular element.

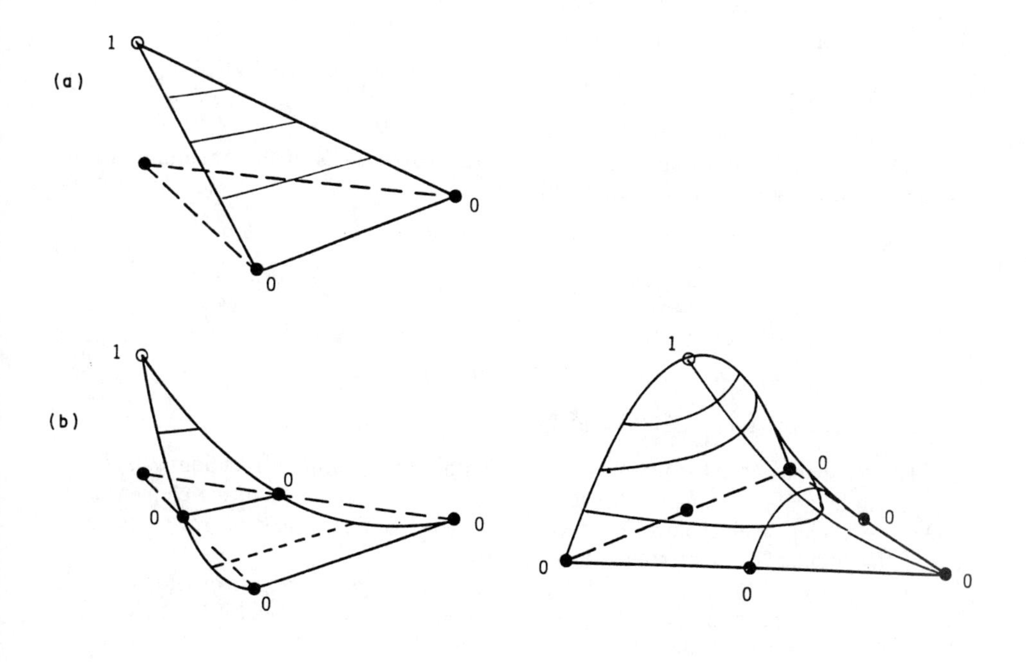

Figure 9.2.5 Linear and quadratic triangle interpolation

9.3 Natural Coordinates

The natural coordinate formulations for the interpolation functions can be generated in a similar manner to that illustrated in Eq. (9.10). However, the inverse geometric matrix, $\mathbf{G}^{-1}$, may not be unique. In such cases the commonly used functions are obtained when the inverse is obtained in a least squares sense. However, the most common functions have been known for several years and will be presented here in two groups. They are generally denoted as Lagrangian elements and as the Serendipity elements (see Tables 9.1 and 9.2). For the four-node quadrilateral element both forms yield Eq. (9.10). This is known as the bi-linear quadrilateral since it has linear interpolation on its edges and a bi-linear (incomplete quadratic) interpolation on its interior. This element is easily extended to the tri-linear hexahedra of Table 9.2 and Fig. 4.2.3. Its resulting interpolation functions are

$$H_i(a, b, c) = (1 + aa_i)(1 + bb_i)(1 + cc_i)/8, \tag{9.18}$$

for $1 \le i \le 8$ where (a_i, b_i, c_i) are the local coordinates of node i. On a given face, e.g., $c = \pm 1$, these degenerate to the four functions in Eq. (9.10) and four zero terms.

For quadratic (or higher) edge interpolation, the Lagrangian and Serendipity elements are different. The Serendipity interpolation functions for the corner quadratic nodes are

$$H_i(a, b) = (1 + aa_i)(1 + bb_i)(aa_i + bb_i - 1)/4, \tag{9.19}$$

where $1 \le i \le 4$ and for the mid-side nodes

$$H_i(a, b) = a_i^2(1 - b^2)(1 + a_i a)/2 + b_i^2(1 - a^2)(1 + b_i b)/2, \quad 5 \le i \le 8. \tag{9.20}$$

Other members of this family are listed in Tables 9.1 and 9.2.

The Lagrangian functions are obtained from the products of the one-dimensional equations. The resulting quadratic functions are

$$
\begin{aligned}
H_1(a,b) &= (a^2 - a)(b^2 - b)/4 & H_6(a,b) &= (a^2 + a)(1 - b^2)/2 \\
H_2(a,b) &= (a^2 + a)(b^2 - b)/4 & H_7(a,b) &= (1 - a^2)(b^2 + b)/2 \\
H_3(a,b) &= (a^2 + a)(b^2 + b)/4 & H_8(a,b) &= (a^2 - a)(1 - b^2)/2 \\
H_4(a,b) &= (a^2 - a)(b^2 + b)/4 & H_9(a,b) &= (1 - a^2)(1 - b^2) \\
H_5(a,b) &= (1 - a^2)(b^2 - b)/2.
\end{aligned}
$$

The typical shapes of these functions are shown in Fig. 9.2.6. The function $H_9(a, b)$ is referred to as a *bubble function* because it is zero on the boundary of the element and looks like a soap bubble blown up over the element. Similar functions are commonly used in hierarchical elements to be considered later.

It is possible to mix the order of interpolation on the edges of an element. Figure 9.2.7 illustrates the Serendipity interpolation functions for quadrilateral elements that can be either linear, quadratic, or cubic on any of its four sides. Such an element is often referred to as a *transition element*.

9.4 Isoparametric and Subparametric Elements

By introducing local coordinates to formulate the element interpolation functions we were able to satisfy certain continuity requirements that could not be satisfied by

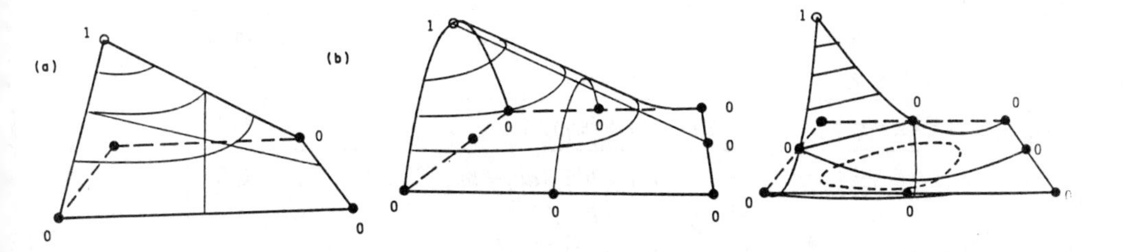

Figure 9.2.6 Quadratic Lagrangian quadrilateral interpolation

Topology:

```
4 – 11 – 7 – 3
|            |
8      S     10
|     *R     |
12           6
|            |
1 – 5 – 9 – 2
```

If Cubic Side : $i = 5, 9$, or $6, 10$ or $7, 11$ or $8, 12$

$$H_i(r, s) = (1 - s^2)(1 + 9ss_i)(1 + rr_i)\, 9/32$$

$$H_i(r, s) = (1 - r^2)(1 + 9rr_i)(1 + ss_i)\, 9/32$$

If Quadratic Side : $i = 5, 6, 7$, or 8

$$H_i(r, s) = (1 + rr_i)(1 - s^2)/2$$

$$H_i(r, s) = (1 + ss_i)(1 - r^2)/2$$

$$H_j = 0, \quad j = i + 4$$

If Linear Side :

$$H_j = H_k = 0, \quad j = i + 4, \quad k = i + 8, \quad i = 1, 2, 3, \text{ or } 4$$

If Corners : $i = 1,2,3,4$ $\quad H_i(r, s) = (P_r + P_s)(1 + ss_i)/4$

Order of Side	$P_r,\ s_i = \pm 1$	$P_s,\ r_i = \pm 1$
Linear	1/2	1/2
Quadratic	$rr_i - 1/2$	$ss_i - 1/2$
Cubic	$(9r^2 - 5)/8$	$(9s^2 - 5)/8$

Figure 9.2.7 Linear to cubic transition quadrilateral

Table 9.1. Serendipity quadrilaterals in natural coordinates

Node Location a_i	b_i	Interpolation Functions $H_i(a, b)$	Name
± 1	± 1	$(1 + aa_i)(1 + bb_i)/4$	Q4
± 1	± 1	$(1 + aa_i)(1 + bb_i)(aa_i + bb_i - 1)/4$	Q8
± 1	0	$(1 + aa_i)(1 - b^2)/2$	
0	± 1	$(1 + bb_i)(1 - a^2)/2$	
± 1	± 1	$(1 + aa_i)(1 + bb_i)[9(a^2 + b^2) - 10]/32$	Q12
± 1	$\pm 1/3$	$9(1 + aa_i)(1 - b^2)(1 + 9bb_i)/32$	
$\pm 1/3$	± 1	$9(1 + bb_i)(1 - a^2)(1 + 9aa_i)/32$	
± 1	± 1	$(1 + aa_i)(1 + bb_i)[4(a^2 - 1)aa_i$ $+ 4(b^2 - 1)bb_i + 3aba_ib_i]/12$	Q16
± 1	0	$2(1 + aa_i)(b^2 - 1)(b^2 - aa_i)/4$	
0	± 1	$2(1 + bb_i)(a^2 - 1)(a^2 - bb_i)/4$	
± 1	$\pm 1/2$	$4(1 + aa_i)(1 - b^2)(b^2 + bb_i)/3$	
$\pm 1/2$	± 1	$4(1 + bb_i)(1 - a^2)(a^2 + aa_i)/3$	
0	0	$(a^2 - 1)(b^2 - 1)$	

Table 9.2. Serendipity hexahedra in natural coordinates

Node Location a_i	b_i	c_i	Interpolation Functions $H_i(a, b, c)$	Name
± 1	± 1	± 1	$(1 + aa_i)(1 + bb_i)(1 + cc_i)/8$	H8
± 1	± 1	± 1	$(1 + aa_i)(1 + bb_i)(1 + cc_i)(aa_i + bb_i + cc_i - 2)/8$	H20
0	± 1	± 1	$(1 - a^2)(1 + bb_i)(1 + cc_i)/4$	
± 1	0	± 1	$(1 - b^2)(1 + aa_i)(1 + cc_i)/4$	
± 1	± 1	0	$(1 - c^2)(1 + aa_i)(1 + bb_i)/4$	
± 1	± 1	± 1	$(1 + aa_i)(1 + bb_i)(1 + cc_i)$ $[9(a^2 + b^2 + c^2) - 19]/64$	H32
$\pm 1/3$	± 1	± 1	$9(1 - a^2)(1 + 9aa_i)(1 + bb_i)(1 + cc_i)/64$	
± 1	$\pm 1/3$	± 1	$9(1 - b^2)(1 + 9bb_i)(1 + aa_i)(1 + cc_i)/64$	
± 1	± 1	$\pm 1/3$	$9(1 - c^2)(1 + 9cc_i)(1 + bb_i)(1 + aa_i)/64$	

global coordinate interpolation. We will soon see that a useful by-product of this approach is the ability to treat elements with curved edges. At this point there may be some concern about how one relates the local coordinates to the global coordinates. This must be done since the governing integral is presented in global (physical) coordinates and it involves derivatives with respect to the global coordinates. This can be accomplished with the popular *isoparametric elements*, and *subparametric* elements.

Isoparametric elements utilize a local coordinate system to formulate the element matrices. The local coordinates, say r, s, and t, are usually dimensionless and range from 0 to 1, or from -1 to 1. The latter range is usually preferred since it is directly compatible with the usual definition of abscissa utilized in numerical integration by Gaussian quadratures. The elements are called isoparametric since the same (iso) local coordinate parametric equations (interpolation functions) used to define any quantity of interest within the elements are also utilized to define the global coordinates of any point within the element in terms of the global spatial coordinates of the nodal points. If a lower order polynomial is used to describe the geometry then it is called a *subparametric element*. These are quite common when used with the newer hierarchical elements. Let the global spatial coordinates again be denoted by x, y, and z,. let the number of nodes per element be N. For simplicity, consider a single scalar quantity of interest, say $V(r, s, t)$. The value of this variable at any local point (r, s, t) within the element is assumed to be defined by the values at the N nodal points of the element (V_i^e, $1 \le i \le N$), and a set of interpolation functions ($H_i(r, s, t)$, $1 \le i \le N$). That is,

$$V(r, s, t) = \sum_{i=1}^{N} H_i(r, s, t)\, V_i^e = \mathbf{H}\, \mathbf{V}^e, \tag{9.21}$$

where $\mathbf{H}$ is a row vector. Generalizing this concept, the global coordinates are defined with a geometric interpolation, or blending, function, $\mathbf{G}$. If it interpolates between n points then it is subparametric if $n < N$, isoparametric if $n = N$ so $\mathbf{G} = \mathbf{H}$, and superparametric if $n > N$. Blending functions typically use geometric data everywhere on the edge of the geometric element. The geometric interpolation, or blending, is denoted as :

$$x(r, s, t) = \mathbf{G}\,\mathbf{x}^e, \qquad y = \mathbf{G}\,\mathbf{y}^e, \qquad z = \mathbf{G}\,\mathbf{z}^e .$$

Programming considerations make it desirable to write the last three relations as a position row matrix, $\mathbf{R}$, written in a partitioned form

$$\mathbf{R}(r, s, t) = \mathbf{G}(r, s, t)\, \mathbf{R}^e = \mathbf{G}\,[\,\mathbf{x}^e\; \mathbf{y}^e\; \mathbf{z}^e\,] \tag{9.22}$$

where the last matrix simply contains the spatial coordinates of the N nodal points incident with the element. If $\mathbf{G} = \mathbf{H}$, it is an isoparametric element. To illustrate a typical two-dimensional isoparametric element, consider a quadrilateral element with nodes at the four corners, as shown in Fig. 9.2.3. The global coordinates and local coordinates of a typical corner, i, are (x_i, y_i), and (r_i, s_i), respectively. The following local coordinate interpolation functions have been developed earlier for this element :

$$H_i(r, s) = \frac{1}{4}(1 + r r_i)(1 + s s_i), \qquad 1 \le i \le 4 .$$

Recall that

$$V(r, s) = \mathbf{H}(r, s)\,\mathbf{V}^e = \begin{bmatrix} H_1 & H_2 & H_3 & H_4 \end{bmatrix} \begin{Bmatrix} V_1 \\ V_2 \\ V_3 \\ V_4 \end{Bmatrix}^e .$$

Note that along an edge of the element ($r = \pm 1$ or $s = \pm 1$), these interpolation functions become linear and thus any of these three quantities can be uniquely defined by the two corresponding nodal values on that edge. If the adjacent element is of the same type (linear on the boundary), then these quantities will be continuous between elements since their values are uniquely defined by the shared nodal values on that edge. Since the variable of interest, V, varies linearly on the edge of the element, it is called the linear isoparametric quadrilateral although the interpolation functions are bilinear inside the element. If the (x, y) coordinates are also varying linearly with r or s on a side it means this element has straight sides.

For future reference, note that if one can define the interpolation functions in terms of the local coordinates then one can also define their partial derivatives with respect to the local coordinate system. For example, the local derivatives of the interpolation functions of the above element are

$$\frac{\partial H_i(r, s)}{\partial r} = \frac{1}{4} r_i (1 + s s_i), \qquad \frac{\partial H_i(r, s)}{\partial s} = \frac{1}{4} s_i (1 + r r_i).$$

In three dimensions, let the array containing the local derivatives of the interpolation functions be denoted by Δ, a $3 \times N$ matrix, where

$$\mathbf{\Delta}(r, s, t) = \begin{bmatrix} \dfrac{\partial}{\partial r}\mathbf{H} \\ \dfrac{\partial}{\partial s}\mathbf{H} \\ \dfrac{\partial}{\partial t}\mathbf{H} \end{bmatrix} = \partial_L \mathbf{H}. \tag{9.23}$$

Although x, y, and z can be defined in an isoparametric element in terms of the local coordinates, r, s, and t, a unique inverse transformation is not needed. Thus, one usually does not define r, s, and t in terms of x, y, and z. What one must have, however, are the relations between derivatives in the two coordinate systems. From calculus, it is known that the derivatives are related by the *Jacobian*. Recall that from the chain rule of calculus one can write, in general,

$$\frac{\partial}{\partial r} = \frac{\partial}{\partial x}\frac{\partial x}{\partial r} + \frac{\partial}{\partial y}\frac{\partial y}{\partial r} + \frac{\partial}{\partial z}\frac{\partial z}{\partial r}$$

with similar expressions for $\partial/\partial s$ and $\partial/\partial t$. In matrix form these identities become

$$\begin{Bmatrix} \frac{\partial}{\partial r} \\ \frac{\partial}{\partial s} \\ \frac{\partial}{\partial t} \end{Bmatrix} = \begin{bmatrix} \frac{\partial x}{\partial r} & \frac{\partial y}{\partial r} & \frac{\partial z}{\partial r} \\ \frac{\partial x}{\partial s} & \frac{\partial y}{\partial s} & \frac{\partial z}{\partial s} \\ \frac{\partial x}{\partial t} & \frac{\partial y}{\partial t} & \frac{\partial z}{\partial t} \end{bmatrix} \begin{Bmatrix} \frac{\partial}{\partial x} \\ \frac{\partial}{\partial y} \\ \frac{\partial}{\partial z} \end{Bmatrix} \tag{9.24}$$

where the square matrix is called the *Jacobian.* Symbolically, one can write the derivatives of a quantity, such as $V(r, s, t)$, which for convenience is written as $V(x, y, z)$ in the global coordinate system, in the following manner

$$\partial_L V = \mathbf{J}(r, s, t)\, \partial_g V,$$

where $\mathbf{J}$ is the Jacobian matrix, and where the subscripts L and g have been introduced to denote local and global derivatives, respectively. Similarly, the inverse relation is

$$\partial_g V = \mathbf{J}^{-1}\, \partial_L V. \tag{9.25}$$

Thus, to evaluate global and local derivatives, one must be able to establish the Jacobian, $\mathbf{J}$, of the geometric mapping and its inverse, $\mathbf{J}^{-1}$. In practical application, these two quantities usually are evaluated numerically. Consider the first term in $\mathbf{J}$ that relates the geometric mapping:

$$\frac{\partial x}{\partial r} = \frac{\partial}{\partial r}(\mathbf{G}\,\mathbf{x}^e) = \frac{\partial \mathbf{G}}{\partial r}\,\mathbf{x}^e.$$

Similarly, for any component in Eq. (9.22)

$$\frac{\partial \mathbf{R}}{\partial r} = \frac{\partial}{\partial r}\,\mathbf{G}\,\mathbf{R}^e.$$

Repeating for all local directions we find the identity that

$$\begin{bmatrix} \frac{\partial x}{\partial r} & \frac{\partial y}{\partial r} & \frac{\partial z}{\partial r} \\ \frac{\partial x}{\partial s} & \frac{\partial y}{\partial s} & \frac{\partial z}{\partial s} \\ \frac{\partial x}{\partial t} & \frac{\partial y}{\partial t} & \frac{\partial z}{\partial t} \end{bmatrix} = \begin{bmatrix} \frac{\partial}{\partial r}\mathbf{G} \\ \frac{\partial}{\partial s}\mathbf{G} \\ \frac{\partial}{\partial t}\mathbf{G} \end{bmatrix} \mathbf{R}^e$$

or, in symbolic form, the evaluation of the definition of the Jacobian within a specific element takes the form

$$\mathbf{J}^e(r, s, t) = \mathbf{\Delta}(r, s, t)\, \mathbf{R}^e. \tag{9.26}$$

This numerically defines the Jacobian matrix, $\mathbf{J}$, at a local point inside a typical element in terms of the spatial coordinates of the element's nodes, $\mathbf{R}^e$, which is referenced by the name *COORD* in the subroutines, and the local derivatives, $\mathbf{\Delta}$, of the geometric interpolation functions, $\mathbf{G}$. Thus, at any point (r, s, t) of interest, such as a numerical integration point, it is possible to define the values of $\mathbf{J}$, $\mathbf{J}^{-1}$, and the determinant of the Jacobian, $|\mathbf{J}|$. These operations are carried out by subroutine JACOB in Fig. 9.4.1.

We usually will consider two-dimensional problems. Then the Jacobian matrix is

```
      SUBROUTINE  JACOB (N, NSPACE, DELTA, COORD, AJ)
C     * * * * * * * * * * * * * * * * * * * * * * * * *
C      CALCULATE THE JACOBIAN MATRIX AT A LOCAL POINT
C     * * * * * * * * * * * * * * * * * * * * * * * * *
CDP   IMPLICIT REAL*8 (A-H,O-Z)
      DIMENSION  DELTA(NSPACE,N), COORD(N,NSPACE),
     1           AJ(NSPACE,NSPACE)
C     N      = NUMBER OF NODES PER ELEMENT
C     NSPACE = DIMENSION OF SPACE
C     DELTA  = LOCAL  DERIVATIVES OF N INTERPOLATION
C              FUNCTIONS AT POINT OF INTEREST.
C     COORD  = SPATIAL COORDINATES OF ELEMENT'S NODES
C     AJ     = JACOBIAN MATRIX = DELTA*COORD
      DO 30  I = 1, NSPACE
        DO 20  J = 1, NSPACE
          SUM = 0.0
          DO 10  K = 1, N
            SUM = SUM + DELTA(I,K)*COORD(K,J)
 10       CONTINUE
          AJ(I,J) = SUM
 20     CONTINUE
 30   CONTINUE
      RETURN
      END
```

Figure 9.4.1 Numerical evaluation of the Jacobian

$$\mathbf{J} = \begin{bmatrix} \frac{\partial x}{\partial r} & \frac{\partial y}{\partial r} \\ \frac{\partial x}{\partial s} & \frac{\partial y}{\partial s} \end{bmatrix}.$$

In general, the inverse Jacobian in two dimensions is

$$\mathbf{J}^{-1} = \frac{1}{|J|} \begin{bmatrix} \frac{\partial y}{\partial s} & -\frac{\partial y}{\partial r} \\ -\frac{\partial x}{\partial s} & \frac{\partial x}{\partial r} \end{bmatrix}, \quad \text{where} \quad |J| = x_{,r}\, y_{,s} - y_{,r}\, x_{,s}.$$

For future reference, note that the determinant and inverse of the three-dimensional Jacobian are

$$|\mathbf{J}| = x_{,r}(y_{,s}\, z_{,t} - y_{,t}\, z_{,s}) + x_{,s}(y_{,t}\, z_{,r} - y_{,r}\, z_{,t}) + x_{,t}(y_{,r}\, z_{,s} - y_{,s}\, z_{,r})$$

and

$$|\mathbf{J}| \times \mathbf{J}^{-1} = \begin{bmatrix} (y_{,s}\, z_{,t} - y_{,t}\, z_{,s}) & (y_{,t}\, z_{,r} - y_{,r}\, z_{,t}) & (y_{,r}\, z_{,s} - y_{,s}\, z_{,r}) \\ (x_{,t}\, z_{,s} - x_{,s}\, z_{,t}) & (x_{,r}\, z_{,t} - x_{,t}\, z_{,r}) & (x_{,s}\, z_{,r} - x_{,r}\, z_{,s}) \\ (x_{,s}\, y_{,t} - x_{,t}\, y_{,s}) & (x_{,t}\, y_{,r} - x_{,r}\, y_{,t}) & (x_{,r}\, y_{,s} - x_{,s}\, y_{,r}) \end{bmatrix},$$

where $(\;)_{,r} = \partial(\;)/\partial r$, etc.

Of course, one can in theory also establish the algebraic form of **J**. For simplicity consider the three-node isoparametric triangle in two dimensions. From Eq. (9.6) we note that the local derivatives of **G** are

$$\Delta = \begin{bmatrix} \partial \mathbf{G} / \partial r \\ \partial \mathbf{G} / \partial s \end{bmatrix} = \begin{bmatrix} -1 & 1 & 0 \\ -1 & 0 & 1 \end{bmatrix}. \tag{9.27}$$

Thus, the element has constant local derivatives since no functions of the local coordinates remain. Usually the local derivatives are also polynomial functions of the local coordinates. Employing Eq. (9.26) for a specific T3 element :

$$\mathbf{J}^e = \Delta \mathbf{R}^e = \begin{bmatrix} -1 & 1 & 0 \\ -1 & 0 & 1 \end{bmatrix} \begin{bmatrix} x_1 & y_1 \\ x_2 & y_2 \\ x_3 & y_3 \end{bmatrix}^e$$

or simply

$$\mathbf{J}^e = \begin{bmatrix} (x_2 - x_1) & (y_2 - y_1) \\ (x_3 - x_1) & (y_3 - y_1) \end{bmatrix}^e \tag{9.28}$$

which is also constant. The determinant of this 2×2 matrix is

$$|\mathbf{J}^e| = (x_2 - x_1)^e (y_3 - y_1)^e - (x_3 - x_1)^e (y_2 - y_1)^e = 2A^e,$$

which is twice the physical area of the element physical domain, Ω^e. For the above three-node triangle, the inverse relation is simply

$$\mathbf{J}^{e^{-1}} = \frac{1}{2A^e} \begin{bmatrix} (y_3 - y_1) & -(y_2 - y_1) \\ -(x_3 - x_1) & (x_2 - x_1) \end{bmatrix}^e = \frac{1}{2A^e} \begin{bmatrix} b_2 & b_3 \\ c_2 & c_3 \end{bmatrix}. \tag{9.29}$$

For most other elements it is common to form these quantities numerically by utilizing the numerical values of $\mathbf{R}^e$ given in the data.

The use of the local coordinates in effect represents a change of variables. In this sense the Jacobian has another important function. The determinant of the Jacobian, $|\mathbf{J}|$, relates differential changes in the two coordinate systems, that is,

$$dL = dx = |\mathbf{J}|\, dr$$

$$da = dx\, dy = |\mathbf{J}|\, dr\, ds$$

$$dv = dx\, dy\, dz = |\mathbf{J}|\, dr\, ds\, dt$$

in one-, two-, and three-dimensional problems. When the local and physical spaces have the same number of dimensions we can write this symbolically as

$$d\,\square^e = |\mathbf{J}|\, d\Omega^e.$$

The integral definitions of the element matrices usually involve the global derivatives of the quantity of interest. From Eq. (9.21) it is seen that the local derivatives of V are related to the nodal parameters by

$$\begin{Bmatrix} \dfrac{\partial V}{\partial r} \\ \dfrac{\partial V}{\partial s} \\ \dfrac{\partial V}{\partial t} \end{Bmatrix} = \begin{bmatrix} \dfrac{\partial}{\partial r}\mathbf{H} \\ \dfrac{\partial}{\partial s}\mathbf{H} \\ \dfrac{\partial}{\partial t}\mathbf{H} \end{bmatrix} \mathbf{V}^e,$$

or symbolically,

$$\partial_L V(r, s, t) = \mathbf{\Delta}(r, s, t)\,\mathbf{V}^e . \tag{9.30}$$

To relate the global derivatives of V to the nodal parameters, $\mathbf{V}^e$, one substitutes the above expression, and the geometry mapping Jacobian into Eq. (9.25) to obtain

$$\partial_g V = \mathbf{J}^{-1}\,\mathbf{\Delta}\,\mathbf{V}^e \equiv \mathbf{d}(r, s, t)\,\mathbf{V}^e ,$$

where

$$\mathbf{d}(r, s, t) = \mathbf{J}(r, s, t)^{-1}\mathbf{\Delta}(r, s, t) . \tag{9.31}$$

The matrix **d** is very important since it relates the global derivatives of the quantity of interest to the quantity's nodal values. Note that it depends on both the Jacobian of the geometric mapping and the local derivatives of the solution interpolation functions. For the sake of completeness, note that **d** can be partitioned as

$$\mathbf{d}(r, s, t) = \begin{bmatrix} \mathbf{d}_x \\ \hline \mathbf{d}_y \\ \hline \mathbf{d}_z \end{bmatrix} = \begin{bmatrix} \dfrac{\partial}{\partial x}\mathbf{H} \\ \hline \dfrac{\partial}{\partial y}\mathbf{H} \\ \hline \dfrac{\partial}{\partial z}\mathbf{H} \end{bmatrix} = \partial_g \mathbf{H} \tag{9.32}$$

so that each row represents a derivative of the solution interpolation functions with respect to a global coordinate direction. Sometimes it is desirable to compute and store the rows of **d** independently. In practice the **d** matrix usually exists only in numerical form at selected points. While the above operations are simply a matrix product subroutine, GDERIV is provided in Fig. 9.4.2 to serve as a useful reminder of this important step.

The three-node triangle is an exception since **J**, **Δ**, and the **d** are all constant. Substituting the results from Eqs. (9.27) and (9.29) into (9.31) yields

$$\mathbf{d}^e = \frac{1}{2A^e}\begin{bmatrix} (y_2 - y_3) & (y_3 - y_1) & (y_1 - y_2) \\ (x_3 - x_2) & (x_1 - x_3) & (x_2 - x_1) \end{bmatrix}^e = \frac{1}{2A^e}\begin{bmatrix} b_1 & b_2 & b_3 \\ c_1 & c_2 & c_3 \end{bmatrix}^e . \tag{9.33}$$

As expected for a linear triangle, all the terms are constant. This element is usually referred to as the Constant Strain Triangle (CST).

Any finite element analysis ultimately leads to the evaluation of the integrals that define the element and/or boundary segment matrices. The element matrices, $\mathbf{S}^e$ or $\mathbf{C}^e$, are usually defined by integrals of the symbolic form

```
      SUBROUTINE  GDERIV (NSPACE, N, AJINV, DELTA, GLOBAL)
C     * * * * * * * * * * * * * * * * * * * * * * * * *
C        NSPACE GLOBAL DERIVATIVES OF N INTERPOLATION
C               FUNCTIONS AT A LOCAL POINT.
C     * * * * * * * * * * * * * * * * * * * * * * * * *
CDP   IMPLICIT REAL*8 (A-H,O-Z)
      DIMENSION  AJINV(NSPACE,NSPACE), DELTA(NSPACE,N),
     1           GLOBAL(NSPACE,N)
C     NSPACE = DIMENSION OF SPACE
C     N      = NUMBER OF NODES PER ELEMENT
C     AJINV  = INVERSE JACOBIAN MATRIX AT LOCAL POINT
C     DELTA  = LOCAL COORD DERIV AT POINT OF INTEREST
C     GLOBAL = GLOBAL DERIVATIVES MATRIX AT LOCAL POINT
C                 GLOBAL = AJINV*DELTA
      DO 30  I = 1, NSPACE
        DO 20  J = 1, N
          SUM = 0.0
          DO 10  K = 1, NSPACE
            SUM = SUM + AJINV(I,K)*DELTA(K,J)
 10       CONTINUE
        GLOBAL(I,J) = SUM
 20     CONTINUE
 30   CONTINUE
      RETURN
      END
```

Figure 9.4.2 Numerical global derivative calculation

$$\mathbf{A}^e = \int\int\int_{\Omega^e} \mathbf{B}^e(x, y, z)\, dx\, dy\, dz \;=\; \int_{-1}^{1}\int_{-1}^{1}\int_{-1}^{1} \tilde{\mathbf{B}}^e(r, s, t)\, \left|\mathbf{J}^e(r, s, t)\right|\, dr\, ds\, dt, \tag{9.34}$$

where $\mathbf{B}^e$ is usually the sum of products of other matrices involving the element interpolation functions, **H**, their derivatives, **d**, and problem properties. In practice, on would often use numerical integration (see Chap.s 5 and 11) to obtain

$$\mathbf{A}^e = \sum_{i=1}^{NIP} W_i\, \tilde{\mathbf{B}}^e(r_i, s_i, t_i)\, |\mathbf{J}^e(r_i, s_i, t_i)| \tag{9.35}$$

where $\tilde{\mathbf{B}}^e$ and $|\mathbf{J}|$ are evaluated at each of the NIP integration points, and where (r_i, s_i, t_i) and W_i denote the tabulated abscissae and weights, respectively. These concepts will be considered in more detail later in Chap. 20. They are presented in subroutine ISOPAR in Fig. 9.4.3. It calls the "external" program NGRAND to supply the application-dependent integrand to be evaluated.

It should be noted that this type of coding makes repeated calls to the interpolation functions to evaluate them at the quadrature points. If the element type is constant, then the quadrature locations would not change. Thus, these computations are repetitious. Since machines have larger memories today, it would be more efficient to evaluate the interpolation functions and their local derivatives once at each quadrature point and store those data for later use. This is, done by adding an additional subscript to those arrays that correspond to the quadrature point number. This is illustrated in subroutine FILL in Fig. 9.4.4.

```
      SUBROUTINE  ISOPAR (N, NSPACE, NELFRE, NIP, SQ, COL,
     1                    QPT, QWT, H, DLH, DGH, COORD,
     2                    XPT, AJ, AJINV, NTAPE1, NGRAND)
C     * * * * * * * * * * * * * * * * * * * * * * * * * * *
C       NUMERICAL INTEGRATION IN AN ISOPARAMETRIC ELEMENT
C     * * * * * * * * * * * * * * * * * * * * * * * * * * *
      EXTERNAL  NGRAND
      DIMENSION COL(NELFRE), SQ(NELFRE,NELFRE), QWT(NIP),
     1          QPT(NSPACE,NIP), H(N), DLH(NSPACE,N),
     2          DGH(NSPACE,N), COORD(N,NSPACE), XPT(NSPACE),
     3          AJ(NSPACE,NSPACE), AJINV(NSPACE,NSPACE)
C     N      = NO NODES PER ELEMENT
C     NSPACE = NO OF SPATIAL DIMENSIONS
C     NELFRE = NO OF ELEMENT DEGREES OF FREEDOM
C     NIP    = NUMBER OF INTEGRATION POINTS
C     QPT    = QUADRATURE PT COORDS
C     QWT    = QUADRATURE PT WEIGHT
C     SQ     = PROB DEPENDENT SQ MATRIX
C     COL    = PROB DEPENDENT COLUMN MATRIX
C     H      = ELEMENT INTERPOLATION FUNCTIONS
C     DLH    = LOCAL DERIVATIVES OF H
C     DGH    = GLOBAL DERIVATIVES OF H
C     COORD  = GLOBAL COORD OF NODES OF ELEMENT
C     XPT    = GLOBAL COORD OF QUADRATURE POINT
C     AJ     = JACOBIAN MATRIX
C     AJINV  = JACOBIAN INVERSE
C     DET    = JACOBIAN DETERMINANT
C     NTAPE1 = STORAGE UNIT FOR POST SOLUTION DATA
C     NGRAND = 'EXTERNAL' PROB DEP INTEGRAND ROUTINE
C-->  ZERO INTEGRANDS
      CALL  ZEROA (NELFRE,COL)
      LSQ = NELFRE*NELFRE
      CALL  ZEROA (LSQ,SQ)
C-->  BEGIN INTEGRATION
      DO 100  IP = 1, NIP
C        EVALUATE INTERPOLATION FUNCTIONS
        CALL  SHAPE (QPT(1,IP),H,N,NSPACE)
C        FIND GLOBAL COORD, XPT = H*COORD
        CALL  MMULT (H,COORD,XPT,1,N,NSPACE)
C        FIND LOCAL DERIVATIVES
        CALL  DERIV (QPT(1,IP),DLH,N,NSPACE)
C        FIND JACOBIAN AT THE PT
        CALL  JACOB (N,NSPACE,DLH,COORD,AJ)
C        FORM INVERSE AND DETERMINATE OF JACOBIAN
        CALL  INVDET (AJ,AJINV,DET,NSPACE)
C        EVALUATE GLOBAL DERIVATIVES
        CALL  GDERIV (NSPACE,N,AJINV,DLH,DGH)
C       *** FORM PROBLEM DEPENDENT INTEGRANDS ***
        CALL  NGRAND (QWT(IP),DET,H,DGH,XPT,N,
     1                NSPACE,NELFRE,COL,SQ,NTAPE1)
  100 CONTINUE
      RETURN
      END
```

Figure 9.4.3 Forming isoparametric matrices

```
      SUBROUTINE  FILL (NIP, NL, NS, PT, WT, H, DLH)
C     * * * * * * * * * * * * * * * * * * * * * * * * * * * *
C     FILL INTERPOLATION AND LOCAL DERIVATIVE ARRAYS AT THE
C                   INTEGRATION POINTS.
C     * * * * * * * * * * * * * * * * * * * * * * * * * * * *
      DIMENSION  PT(NS,NIP), WT(NIP), H(NL,NIP), DLH(NS,NL,NIP)
C     DLH    = LOCAL DERIVATIVES OF INTERPOLATION FUNCTIONS
C     H      = ELEMENT INTERPOLATION ARRAY
C     NIP    = NUMBER OF INTEGRATION POINTS
C     NL     = NUMBER OF NODES PER ELEMENT
C     NS     = NUMBER OF LOCAL SPACE DIMENSIONS
C     PT     = QUADRATURE POINT LOCAL COORDINATES
C     WT     = QUADRATURE POINT WEIGHTS
C-->   LOOP OVER QUADRATURE POINTS
      DO 10  IQ = 1, NIP
        CALL  SHAPE (PT(1,IQ),H(1,IQ),NL,NS)
        CALL  DERIV (PT(1,IQ),DLH(1,1,IQ),NL,NS)
   10 CONTINUE
C      AVERAGES AND/OR SUMS COULD BE DONE HERE WITH WT
      RETURN
      END
```

Figure 9.4.4 Saving interpolation data at quadrature points

9.5 Hierarchical Interpolation

In Sec. 4.6 we introduced the typical hierarchical functions on line elements and let the mid-point tangential derivatives from order m to order n be denoted by $m \rightarrow n$. The exact same functions can be utilized on each edge of a two-dimensional or three-dimensional hierarchical element. We will begin by considering quadrilateral elements, or the quadrilateral faces of a solid element. To apply the previous one-dimensional element to each edge of the element requires an arbitrary choice of which way(s) we consider to the positive tangential direction. Our choice is to use the "right hand rule" so that the tangential derivatives are taken counterclockwise around the element. In other words, if we circle the fingers of our right hand in the direction of the tangential circuit, our thumb points in the direction of the outward normal vector perpendicular to that face. This rule also works for any solid element face where we consider the normal to act away from the solid.

Usually a (sub-parametric) four node element will be used to describe the geometry of the element. The element starts with the standard isoparametric form of four nodal values to begin the hierarchical approximation of the function. As needed, tangential derivatives of the unknown solution are added as additional degrees of freedom. It is well known that it is desirable to have complete polynomials included in the interpolation polynomials. Thus, at some point it becomes necessary to add internal (bubble) functions at the centroid of the element. There is more than one way to go about doing this. The main question is does one want to use the function value at the centroid as a dof or just its higher derivatives? The latter choice is simpler to automate if we begin with the Q4 element.

Since the hierarchical derivative interpolation functions are all zero at both ends of their edge they will also be zero on their two adjoining edges of the quadrilateral. Thus, to use these functions on the interior of the Q4 element we must multiply them by a function that is unity on the edge where the hierarchical functions are defined and zero on the opposite parallel edge. This process is sketched in Fig. 9.5.1. From the discussion of isoparametric elements it should be clear that on each of the four sides (see Table 9.1) the necessary functions (in natural coordinates a, b) are

$$N^{(1)}(b) = (1-b)/2, \quad N^{(3)}(b) = (1+b)/2$$
$$N^{(2)}(a) = (1+a)/2, \quad N^{(4)}(a) = (1-a)/2 \tag{9.36}$$

respectively, where $N^{(i)}$ denotes the interpolation normal to side i. If T_{ij} denotes the hierarchical tangential interpolations on side i and node j, then their net interior contributions are

$$H_{ij}(a, b) = N^{(i)} T_{ij}.$$

That is, the p-th degree edge interpolation enrichments of the Q4 element are

$$\begin{aligned} &\text{Side 1} \quad (b=-1) \quad H_p^{(1)}(a, b) = ½(1-b)\,\Psi_p(a) \\ &\text{Side 2} \quad (a=1) \quad H_p^{(2)}(a, b) = ½(1+a)\,\Psi_p(b) \\ &\text{Side 3} \quad (b=1) \quad H_p^{(3)}(a, b) = ½(1+b)\,\Psi_p(-a) \\ &\text{Side 4} \quad (a=-1) \quad H_p^{(4)}(a, b) = ½(1-a)\,\Psi_p(-b) \end{aligned} \tag{9.37}$$

where the $\Psi_p(a) = [P_p(a) - P_{p-2}(a)]\, 2p-1, \quad p \geq 2$. They are normalized such that their p-th tangential derivative is unity. Note that there are $4(p-1)$ such enrichments. Likewise, there are $(p-2)(p-3)/2$ internal enrichments for $p \geq 4$. They occur at the center (0, 0) of the element. Their degrees of freedom are the cross-partial derivatives

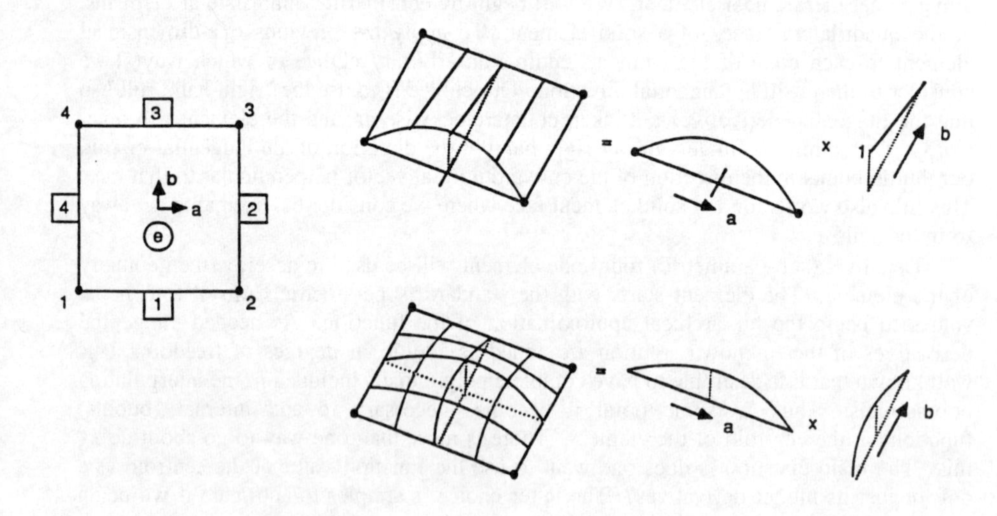

Figure 9.5.1 Extending hierarchical edge functions to the interior

$$\frac{\partial^{p-2}}{\partial a^j \, \partial b^k}, \qquad j+k = p-2 \,, \qquad 1 \le j,k \le p-3 \,.$$

The general form of the internal (centroid) enrichments are a product of "bubble functions" and other functions

$$H_p^{(0)}(a,b) = (1-a^2)(1-b^2) \; P_{p-4-j}(a) \; P_j(b) \,, \qquad j = 0, 1, \ldots, p\text{–}4 \,, \quad (9.38)$$

where $P_j(a)$ is the Legendre polynomial of degree j given in Eq. 4.25. The number of internal degrees of freedom, n, are

p	4	5	6	7	8	9	10
n	1	2	3	4	5	6	7
Total	1	3	6	10	15	21	28

so that we see the number of internal terms corresponds to the number of coefficients in a complete polynomial of degree $(p-3)$. The n terms for degree 4 to 10 are given in Table 9.3. It can be shown that the above combinations are equivalent to a complete polynomial of degree p, plus the two monomial terms $a^p b, \; a b^p$ for $p \ge 2$. This boundary and interior enrichment of the Q4 element is shown in Fig. 9.5.2. There p denotes the order of the edge polynomial, n is the total number of degrees of freedom (interpolation functions), and c is the number of dof needed for a complete polynomial form. For a quadrilateral we note that the total number of shape functions on any side is $n = p + 1$ for $p \ge 1$, and the number of interior nodes is $n_i = (p-2)(p-3)/2$ for $p \ge 4$, and the total for the element is

$$n_t = (p-2)(p-3)/2 + 4p = (p^2 + 3p + 6)/2 \quad \text{for } p \ge 4 \,.$$

Note that the number of dof grows rapidly and by the time $p = 9$ is reached the element has almost 15 times as many dof as it did originally.

At this point the reader should see that there is a very large number of alternate forms of this same element. Consider the case where an error estimator has predicted the need for a different polynomial order on each edge. This is called *anisotropic hierarchical p-enrichment.* For maximum value of $p = 8$ there are a total of 32 possible interpolation combinations, including the eight uniform ones shown in Fig. 9.5.2. It is likely that future codes will take advantage of anisotropic enrichment, although very few do so today.

If one is going to use a nine node quadrilateral (Q9) to describe the geometry then the same types of enrichments can be added to it as shown in Fig. 9.5.3. However, the Q4 form would have better orthogonality behavior, that is, it would produce square matrices that are more diagonally dominant. For triangular and tetrahedral elements one could generate different interpolation orders on each edge, and in the interior, by utilizing the enhancement procedures for Lagrangian elements that was described in Sec. 16.2. This is probably easier to do in baracentric coordinates.

Since these elements have so much potential power they tend to be relatively large in size, and/or distorted in shape, and small in number. That trend might begin to conflict with the major appeal of finite elements: the ability to match complicated shapes. Thus, the choice of describing the geometry (and it's Jacobian) by isoparametric, or sub-

Table 9.3. Quadrilateral hierarchical internal functions

$$\Psi_p(a,b) = (1 - a^2)(1 - b^2)\, P_m(a)\, P_n(b), \quad p \geq 3$$

p	m	n	j	k
4	0	0	1	1
5	1	0	1	2
	0	1	2	1
6	2	0	1	3
	1	1	2	2
	0	2	3	1
7	3	0	1	4
	2	1	2	3
	1	2	3	2
	0	3	4	1
8	4	0	1	5
	3	1	2	4
	2	2	3	3
	1	3	4	2
	0	4	5	1
9	5	0	1	6
	4	1	2	5
	3	2	3	4
	2	3	4	3
	1	4	5	2
	0	5	6	1
10	6	0	1	7
	5	1	2	6
	4	2	3	5
	3	3	4	4
	2	4	5	3
	1	5	6	2
	0	6	7	1

P_i = Legendre polynomial of degree i; $dof = \dfrac{\partial^{j+k}}{\partial a^j\, \partial b^k}$

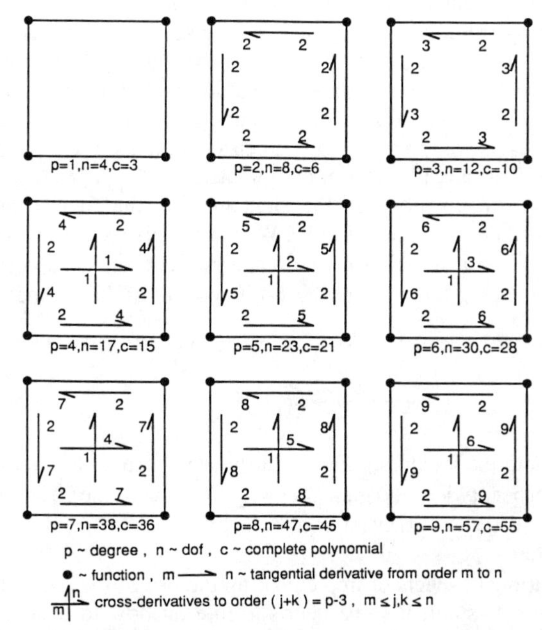

Figure 9.5.2 Hierarchical enrichments of the Q4 element

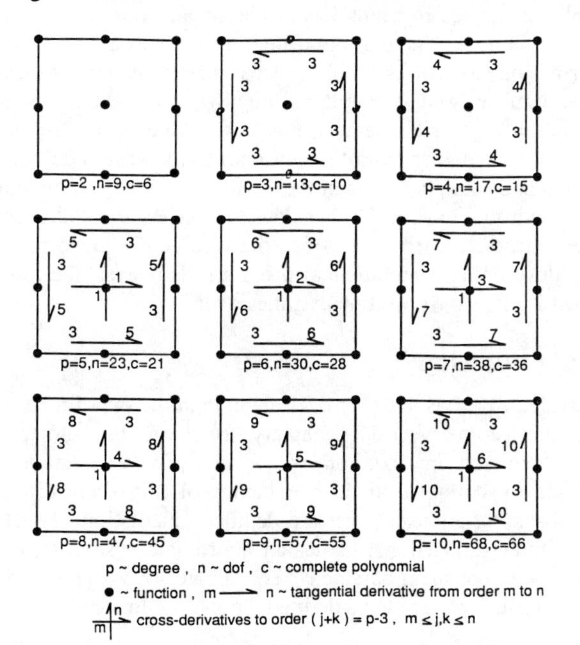

Figure 9.5.3 Hierarchical enrichments of the Q9 element

parametric methods might be dropped in favor of other geometric modeling methods. That is, the user may want to exactly match an ellipse or circle rather than approximate it with a parametric curve. One way to do that is to employ *blending functions* such as Coon's functions to describe the geometry. To do this we use local analytical functions to describe each physical coordinate on the edge of the element rather than 2, 3, or 4 discrete point values as we did with isoparametric elements in the previous sections. Let (a, b) denote the quadrilateral's natural coordinates, $-1 \le (a, b) \le 1$. Consider only the x physical coordinate of any point in the element. Let the four corner values of x be denoted by X_i. Number the sides in a CCW manner also starting from the first (LLH) corner node. Let x_j be a function of the tangential coordinate describing x on side j. Then the *Coon's blending function* for the x-component of the geometry is:

$$x(a, b) = \Big[x_1(a)(1-b) + x_2(b)(1+a) + x_3(a)(1+b) + x_4(a)(1-a) \Big] / 2 - \sum_{i=1}^{4} X_i (1 + a a_i)(1 + b b_i)/4 \tag{9.39}$$

where (a_i, b_i) denote the local coordinates of the i-th corner. Since the term in brackets includes each corner twice (e.g., $x_1(1) = x_2(-1) = X_2$), the last summation simply subtracts off one full set of corner contributions by using the standard Q4 isoparametric relation given earlier.

The computational aspects of implementing the use of the tangential derivatives are not trivial. One must establish some heuristic rule on how to handle the sign conflicts that can develop among different elements, or faces, on a common edge. As shown in Fig. 9.5.4, the above suggested right hand rule means that edges share degrees of freedom, but view them as having opposite signs. These sign conflicts must be accounted for during the element assembly process, or by invoking a different rule when assigning equation numbers so that shared dof are always viewed as having the same sign when viewed from any face or element on that edge. One could, for example, take the tangential derivative to be acting from the end with the lowest node number toward the end with the higher node number. Figure 9.5.5 shows that the same edge conflict can exist with the hexahedron element. Higher degree hexahedra develop the same type of conflict when they introduce derivative dof at the face center. One must plan for these difficulties before developing a hierarchical program. However, the returns on such an investment of effort is clearly worth it many times over.

9.6 Differential Geometry *

When the physical space is a higher dimension than the two-dimensional parametric surfaces defining the geometry it is necessary to utilize the subject of *differential geometry*. This is covered in texts on vector analysis or calculus. It is also an introductory topic in most books on the mechanics of thin shell structures. Here we cover most of the basic topics except for the detailed calculation of surface curvatures. Every surface in a three-dimensional Cartesian coordinate system (x, y, z) may also be expressed by a pair of independent parametric coordinates (r, s) that lie on the surface. In our geometric parametric form, we have defined the x-coordinate as

$$x(r, s) = \mathbf{G}(r, s)\, \mathbf{x}^e . \tag{9.40}$$

The y- and z-coordinates are defined similarly. The components of the *position vector* to a point on the surface

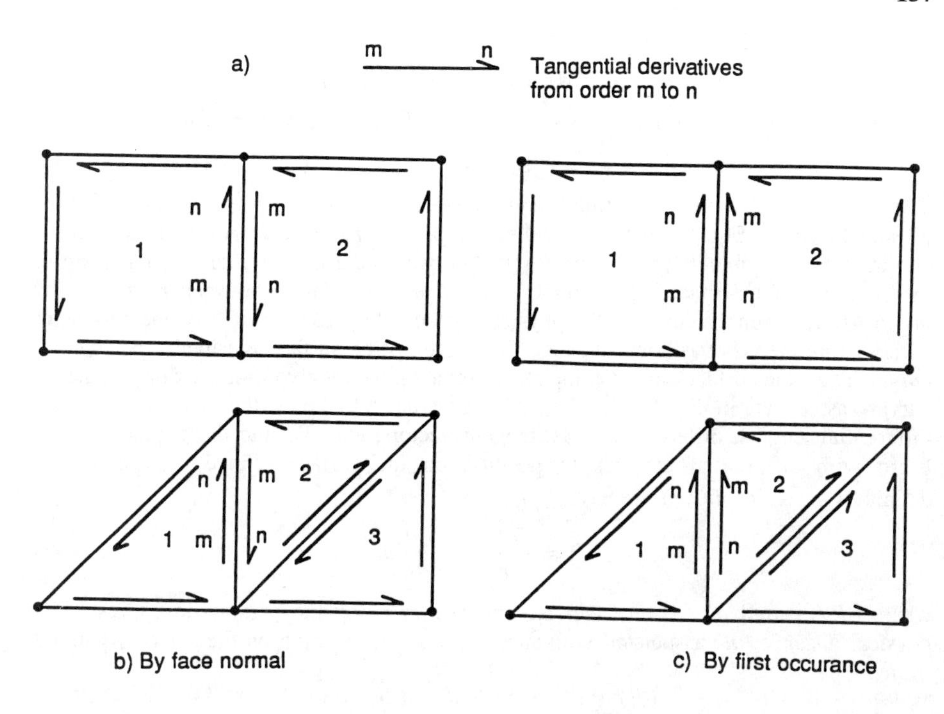

Figure 9.5.4 Sign conflicts for tangential derivatives

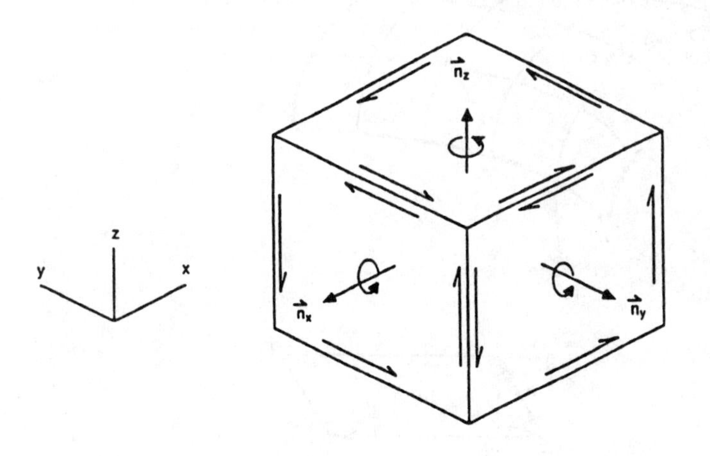

Figure 9.5.5 Typical faces on an H8 element

$$\vec{R}(r, s) = x(r, s)\,\hat{\imath} + y(r, s)\,\hat{\jmath} + z(r, s)\,\hat{k}, \tag{9.41}$$

where $\hat{\imath},\ \hat{\jmath},\ \hat{k}$ are the constant unit base vectors, could be written in array form as

$$\mathbf{R}^T = [\,x \quad y \quad z\,] = \mathbf{G}(r, s)\,[\,\mathbf{x}^e\ \mathbf{y}^e\ \mathbf{z}^e\,]\,. \tag{9.42}$$

The local parameters (r, s) constitute a system of curvilinear coordinates for points on the physical surface. Equation (9.41) is called the *parametric equation* of a surface. If we eliminate the parameters (r, s) from Eq. (9.41), we obtain the familiar implicit form of the equation of a surface, $f(x, y, z) = 0$. Likewise, any relation between r and s, say $g(r, s) = 0$, represents a curve on the physical surface. In particular, if only one parameter varies while the other is constant, then the curve on the surface is called a *parametric curve*. Thus, the surface can be completely defined by a doubly infinite set of parametric curves, as shown in Fig. 9.6.1. We will often need to know the relations between differential lengths, differential areas, tangent vectors, etc. To find these quantities we begin with differential changes in position on the surface. Since our parametric definition gives $\vec{R} = \vec{R}(r, s)$, we have

$$d\vec{R} = \frac{\partial \vec{R}}{\partial r}\,dr + \frac{\partial \vec{R}}{\partial s}\,ds \tag{9.43}$$

where $\partial \vec{R}/\partial r$ and $\partial \vec{R}/\partial s$ are the *tangent vectors* along the parametric curves. The physical distance, dl, associated with such a change in position on the surface is found from

$$(dl)^2 = dx^2 + dy^2 + dz^2 = d\vec{R} \cdot d\vec{R}\,. \tag{9.44}$$

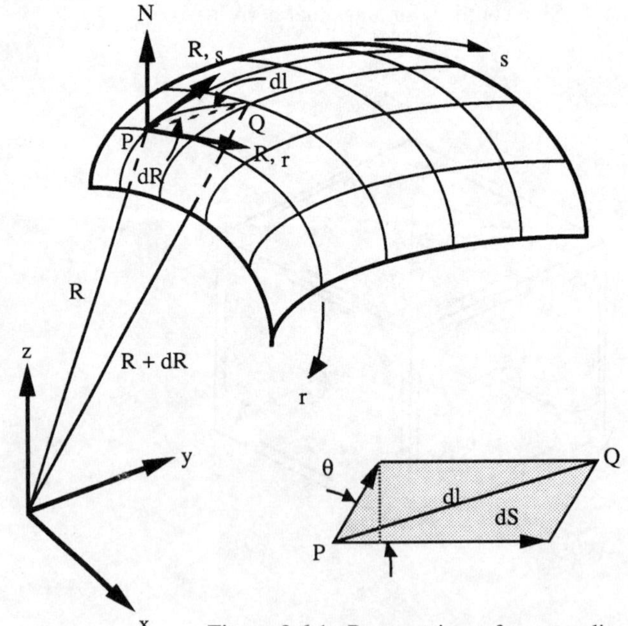

Figure 9.6.1 Parametric surface coordinates

This gives three contributions :

$$(dl)^2 = \left[\frac{\partial \vec{R}}{\partial r} \cdot \frac{\partial \vec{R}}{\partial r}\right] dr^2 + 2\left[\frac{\partial \vec{R}}{\partial r} \cdot \frac{\partial \vec{R}}{\partial s}\right] dr\, ds + \left[\frac{\partial \vec{R}}{\partial s} \cdot \frac{\partial \vec{R}}{\partial s}\right] ds^2 .$$

In the common notation of differential geometry this is called the *first fundamental form* of a surface, and is usually written as

$$(dl)^2 = E\,dr^2 + 2F\,dr\,ds + G\,ds^2 \tag{9.45}$$

where

$$E = \frac{\partial \vec{R}}{\partial r} \cdot \frac{\partial \vec{R}}{\partial r}, \qquad F = \frac{\partial \vec{R}}{\partial r} \cdot \frac{\partial \vec{R}}{\partial s}, \qquad G = \frac{\partial \vec{R}}{\partial s} \cdot \frac{\partial \vec{R}}{\partial s} \tag{9.46}$$

are called the first *fundamental magnitudes* (or metric tensor) of the surface. For future reference we will use this notation to note that the magnitudes of the surface tangent vectors are

$$\left| \frac{\partial \vec{R}}{\partial r} \right| = \sqrt{E}, \qquad \left| \frac{\partial \vec{R}}{\partial s} \right| = \sqrt{G} .$$

Of course, these magnitudes can be expressed in terms of the parametric derivatives of the surface coordinates, (x, y, z). For example, from Eq. (9.46),

$$F = \frac{\partial x}{\partial r}\frac{\partial x}{\partial s} + \frac{\partial y}{\partial r}\frac{\partial y}{\partial s} + \frac{\partial z}{\partial r}\frac{\partial z}{\partial s} \tag{9.47}$$

can be evaluated for an isoparametric surface by utilizing Eq. (9.42). Define a surface gradient array given by

$$\mathbf{g} = \begin{bmatrix} \frac{\partial x}{\partial r} & \frac{\partial y}{\partial r} & \frac{\partial z}{\partial r} \\ \frac{\partial x}{\partial s} & \frac{\partial y}{\partial s} & \frac{\partial z}{\partial s} \end{bmatrix}. \tag{9.48}$$

The rows contain the components of the tangent vectors along the parametric r and s curves, respectively. In the notation of Eq. (9.26), this becomes

$$\mathbf{g}(r, s) = [\boldsymbol{\partial}_l R] = \boldsymbol{\Delta}\mathbf{R}^e = \left[\boldsymbol{\partial}_l\, \mathbf{G}(r, s)\right] \left[\mathbf{x}^e\ \mathbf{y}^e\ \mathbf{z}^e\right]. \tag{9.49}$$

In other words, the surface gradient array at any point is the product of the parametric function derivatives evaluated at that point and the array of nodal data for the element of interest. We define the *metric array*, $\mathbf{m}$, as the product of the surface gradient with its transpose

$$\mathbf{m} \equiv \mathbf{g}\,\mathbf{g}^T = \begin{bmatrix} (x_{,r}^2 + y_{,r}^2 + z_{,r}^2) & (x_{,r}x_{,s} + y_{,r}y_{,s} + z_{,r}z_{,s}) \\ (x_{,r}x_{,s} + y_{,r}y_{,s} + z_{,r}z_{,s}) & (x_{,s}^2 + y_{,s}^2 + z_{,s}^2) \end{bmatrix} \tag{9.50}$$

where the subscripts denote partial derivatives with respect to the parametric coordinates. Comparing this relation with Eq. (9.46) we note that

$$\mathbf{m} = \begin{bmatrix} E & F \\ F & G \end{bmatrix} \tag{9.51}$$

contains the fundamental magnitudes of the surface. This surface metric has a determinant that is always positive. It is useful in later calculations and is usually denoted in differential geometry as

$$|\mathbf{m}| \equiv H^2 = EG - F^2 > 0 . \tag{9.52}$$

We can degenerate the differential length measure in Eq. (9.44) to the common special case where we are moving along a parametric curve, that is, $dr = 0$ or $ds = 0$. In the first case of $r = constant$, we have

$$(dl)^2 = G\,ds^2$$

where dl is a physical differential length on the surface and ds is a differential change in the parametric surface. Then

$$dl = \sqrt{G}\,ds \tag{9.53}$$

and for the parametric curve $s = constant$,

$$dl = \sqrt{E}\,dr . \tag{9.54}$$

The quantities $\sqrt{G}$ and $\sqrt{E}$ are known as the *Lame parameters*. They convert differential changes in the parametric coordinates to differential lengths on the surface when moving on a parametric curve. From Fig. 9.6.1 we note that the vector tangent to the parametric curves r and s are $\partial\vec{R}/\partial r$ and $\partial\vec{R}/\partial s$, respectively. While the isoparametric coordinates may be orthogonal, they generally will be nonorthogonal when displayed as parametric curves on the physical surface. The angle θ between the parametric curves on the surface can be found by using these tangent vectors and the definition of the dot product. Thus,

$$F \equiv \frac{\partial\vec{R}}{\partial r} \cdot \frac{\partial\vec{R}}{\partial s} = \sqrt{E}\,\sqrt{G}\,\text{Cos}\,\theta$$

and the angle at any point comes from

$$\text{Cos}\,\theta = \frac{F}{\sqrt{E}\,\sqrt{G}} . \tag{9.55}$$

Therefore, we see that the parametric curves form an orthogonal curvilinear coordinate system on the physical surface only when $F = 0$. Only in that case does Eq. (9.44) reduce to the orthogonal form

$$(dl)^2 = E\,dr^2 + G\,ds^2 .$$

Denote the parametric curve tangent vectors as $\vec{t}_r = \partial\vec{R}/\partial r$ and $\vec{t}_s = \partial\vec{R}/\partial s$. We have seen that the differential lengths in these two directions on the surface are $\sqrt{E}\,dr$ and $\sqrt{G}\,ds$. In a vector form, those lengths are $\vec{t}_r\,dr$ and $\vec{t}_s\,ds$, and they are separated by the angle θ. The corresponding differential surface area of the surface parallelogram is

$$dS = (\sqrt{E}\,dr)\,(\sqrt{G}\,ds\,\text{Sin}\,\theta) = \sqrt{E}\,\sqrt{G}\,\text{Sin}\,\theta\,dr\,ds . \tag{9.56}$$

By substituting the relation between Cos θ and the surface metric, this simplifies to

$$\begin{aligned} dS^2 &= EG\,\text{Sin}^2\theta\,dr^2\,ds^2 = EG\,(1 - \text{Cos}^2\,\theta)\,dr^2\,ds^2 \\ dS^2 &= (EG - F^2)\,dr^2\,ds^2 = \sqrt{H}\,dr\,ds . \end{aligned} \tag{9.57}$$

We also note that this calculation can be expressed as a vector cross product of the tangent vectors:

$$dS\,\vec{N} = \vec{t}_r \times \vec{t}_s\,dr\,ds \tag{9.58}$$

where $\vec{N}$ is a vector normal to the surface. We also note that the *normal vector* has a magnitude of

$$|\vec{N}| = |\vec{t}_r \times \vec{t}_s| = H. \tag{9.59}$$

Sometimes it is useful to note that the components of $\vec{N}$ are

$$\vec{N} = (y_{,r}\, z_{,s} - y_{,s}\, z_{,r})\,\hat{i} + (x_{,r}\, z_{,s} - x_{,s}\, z_{,r})\,\hat{j} + (x_{,r}\, y_{,s} - x_{,s}\, y_{,r})\,\hat{k}.$$

We often want the unit vector, $\vec{n}$, normal to the surface. It is

$$\vec{n} = \frac{\vec{N}}{H} = \frac{\vec{t}_r \times \vec{t}_s}{|\vec{t}_r \times \vec{t}_s|}. \tag{9.60}$$

By combining derivatives of the zero dot products of the normal and tangent vectors, we can derive relations for the *principal radii of curvature*, ρ_1 and ρ_2. They utilize the scalars

$$L = \frac{1}{H}\begin{vmatrix} x_{,rr} & y_{,rr} & z_{,rr} \\ x_{,r} & y_{,r} & z_{,r} \\ x_{,s} & y_{,s} & z_{,s} \end{vmatrix}, \quad M = \frac{1}{H}\begin{vmatrix} x_{,rs} & y_{,rs} & z_{,rs} \\ x_{,r} & y_{,r} & z_{,r} \\ x_{,s} & y_{,s} & z_{,s} \end{vmatrix}$$

$$N = \frac{1}{H}\begin{vmatrix} x_{,ss} & y_{,ss} & z_{,ss} \\ x_{,r} & y_{,r} & z_{,r} \\ x_{,s} & y_{,s} & z_{,s} \end{vmatrix} \tag{9.61}$$

to define the *total curvature*, K,

$$K \equiv \frac{1}{\rho_1 \rho_2} = \frac{LN - M^2}{H^2}. \tag{9.62}$$

We define the surface at the point to the elliptic if the radii are of the same sign, $K > 0$, and hyperbolic if their signs are different, $K < 0$. It is parabolic when one or both of the principal curvatures is zero, $K = 0$. These second fundamental magnitudes also define the *mean* curvature of the surface

$$\frac{1}{\rho_1} + \frac{1}{\rho_2} = \frac{EN + GL - 2FM}{H^2}. \tag{9.63}$$

9.7 Parametric Extrapolation *

Isoparametric interpolation as well as subparametric and superparametric methods are used very extensively in finite element analysis and design problems. It is less often recognized that isoparametric extrapolation can be very useful as well. For example, we often need to extrapolate derivative information from the interior integration points to the nodes of the element. These extrapolated data are then used for user output or sent to averaging routines to be used in smoothing the values on the total mesh in order to create element error estimates for an adaptive solution. Subroutines developed for parametric interpolation can also be used for parametric extrapolation.

To illustrate this point consider the bilinear quadrilateral, Q4, in natural coordinates, (r, s). Recall that a typical corner node, j, has local coordinate values of r_j and $s_j = \pm\, 1$. When that element, or the Q8, is integrated numerically we employ a 2×2 Gauss rule to assure sufficient rank of the element square matrix. The four Gauss points are placed symmetrically with respect to the origin. They will be referenced by Greek subscripts.

They have local coordinate values of r_α and $s_\alpha = \pm 1/\sqrt{3}$. Hinton and Owen [5] have presented a method for extrapolating from these four Gauss points to the four corner nodes by using a least square fit in the parametric space. We can get the same result by simply doing an isoparametric extrapolation. To do this we must simply change our point of view and define the nodal locations relative to Gauss point locations. This is easily done by using another set of natural coordinates (a, b) such that Gauss point α defines a_α and $b_\alpha = \pm 1$. The origin of both systems is the same physical point, but now the corner nodal coordinates have an absolute value that is greater than unity. In the current example node j has a local coordinate of a_j and $b_j = \pm\sqrt{3}$.

Let σ_α denote a derivative quantity at the Gauss point α that we want to extrapolate to a corresponding nodal value σ_j. We employ the old Q4 isoparametric interpolation functions $\mathbf{H}(r, s)$ to define

$$\sigma(a, b) = \sum_{\alpha=1}^{m=4} H_\alpha(a, b)\,\sigma_\alpha . \tag{9.64}$$

To find the value at the origin we simply substitute $a = 0, b = 0$ to obtain

$$\sigma(0, 0) = \frac{1}{4}\sum_{\alpha=1}^{4} \sigma_\alpha . \tag{9.65}$$

If the σ_α were all the same constant value, V, then at any point we obtain

$$\sigma(a, b) = V \sum_{\alpha=1}^{m} H_\alpha(a, b) = V \tag{9.66}$$

since the isoparametric H_α always sum to unity. To find the value of σ at an element node we simply substitute its extrapolated local nodal coordinates into Eq. (1) :

$$\sigma_j = \sigma(a_j, b_j) = \sum_{\alpha=1}^{m=4} H_\alpha(a_j, b_j)\,\sigma_\alpha \tag{9.67}$$

or in matrix form

$$\sigma_j = \mathbf{H}(a_j, b_j)\,\boldsymbol{\sigma}_\alpha = \mathbf{H}_j\,\boldsymbol{\sigma}_\alpha$$

where $\boldsymbol{\sigma}_\alpha$ denotes the vector of values sampled at the Gauss points. For example, let node $j = 3$ be at the top right corner so that $a_j = b_j = +\sqrt{3}$. Then

$$\sigma_j = \begin{bmatrix} d & e & c & e \end{bmatrix} \boldsymbol{\sigma}_\alpha \tag{9.68}$$

where $c = 1 + \sqrt{3}/2$, $d = 1 - \sqrt{3}/2$, and $e = -1/2$. If we pick any other corner node the same three numbers appear in the row matrix but the four column locations change in a cyclic permutation. Let $\boldsymbol{\sigma}_j$ be the four nodal values obtained from the four Gauss values $\boldsymbol{\sigma}_\alpha$. Then we find that

$$\boldsymbol{\sigma}_j = \mathbf{E}\,\boldsymbol{\sigma}_\alpha \tag{9.69}$$

where $\mathbf{E}$ is, in general, a rectangular array that extrapolates m Gauss values to n nodal values where $n \geq m$. Here $m = n = 4$ and $\mathbf{E}$ is the square array

$$\mathbf{E} = \begin{bmatrix} c & e & d & e \\ e & c & e & d \\ d & e & c & e \\ e & d & e & c \end{bmatrix} . \tag{9.70}$$

This extrapolation matrix is exactly the same result as to that obtained by Hinton and Owen's least squares fit method.

It is easy to extend this procedure to higher order elements. For example, the Q8 element has four mid-side nodes as well as the four corner nodes. Relative to the Gauss point coordinates, (a, b), each mid-side node $5 \le \alpha \le 8$ has one coordinate zero and the other is $\pm\sqrt{3}$. Then, $m = 4$ and $n = 8$ and

$$\sigma_j = \begin{bmatrix} \mathbf{E}_c \\ \mathbf{E}_s \end{bmatrix} \sigma_\alpha \tag{9.71}$$

where $\mathbf{E}$ has been partitioned into rows associated with corner nodes and side nodes. Here $\mathbf{E}_c$ is the square $\mathbf{E}$ given above, and if we order the mid-side nodes counter-clockwise from the bottom $(b = -1)$ then

$$\mathbf{E}_s = \begin{bmatrix} f & f & g & g \\ g & f & f & g \\ g & g & f & f \\ f & f & g & g \end{bmatrix} \tag{9.72}$$

where $f = (1 - \sqrt{3})/4$ and $g = (1 + \sqrt{3})/4$. In this case the extrapolation is linear on lines of $a = constant$ or $b = constant$ so the product of $\mathbf{E}_s$ and σ_α simply give mid-side values that are half the sum of the two corners that bound the mid-side. That is, for the Q8 element $f = (c + e)/2$ and $g = (d + e)/2$.

Clearly, there is no need to list all the entries in the $\mathbf{E}$ matrix as was done in Eqs. (9.70) and (9.71). A very simple procedure will allow the user to numerically compute and store $\mathbf{E}$ for any element type that uses a $m \times m$ symmetric integration rule. The procedure for each element type is:

1. Find the m-*th* order Lagrangian interpolation array $\mathbf{G}(a, b)$ that matches the $m \times m$ sampling points.
2. Scale all of the element nodes to generate the n coordinate pairs (a_j, b_j).
3. Set $\mathbf{E}$ to zero.
4. For each node j set the j-*th* row of $\mathbf{E}$ equal to $\mathbf{G}(a_j, b_j)$. That is, for each node j call the $\mathbf{G}$ subroutine with the scaled coordinate arguments (a_j, b_j) and place the n returned values in the j-*th* row of $\mathbf{E}$.

9.8 Mass Properties, Equivalent Geometry *

Mass properties and geometric properties are often needed in a design process. These computations provide a useful check on the model, and may also lead to reducing more complicated calculations by identifying geometrically equivalent elements. Whenever possible, we attempt to avoid computing element matrices that are identical to any that have already been computed. There are several geometrical properties we could compute to help in identifying elements of the same, identical geometric type. We may consider the area to define the relative size and the geometric centroidal inertia tensor or its principal axes to check the spatial orientation. For axisymmetric problems we would need to locate the radial centroid. These quantities are easily computed for an element. We can provide the program with a user selected set of "training elements" that are expected to appear frequently in the mesh or we can let the program automatically locate a maximum number of repeating shape types.

To illustrate the concept consider the following area, centroid, and inertia terms for a two- dimensional general curvilinear isoparametric element:

$$A = \int_A 1^2 \, da, \quad A\bar{x} = \int_A x1 \, da, \quad A\bar{y} = \int_A y1 \, da \tag{9.73}$$

$$I_{xx} = \int_A y^2 \, da, \quad -I_{xy} = \int_A xy \, da, \quad I_{yy} = \int_A x^2 \, da, \quad I_{zz} = I_{xx} + I_{yy}.$$

From the parallel axis theorem we know that

$$\bar{I}_{xx} = I_{xx} - \bar{y}^2 A, \quad \bar{I}_{xy} = I_{xy} + \bar{x}\,\bar{y} A, \quad \bar{I}_{yy} = I_{yy} - \bar{x}^2 A, \quad \bar{I}_{zz} = \bar{I}_{xx} + \bar{I}_{yy}.$$

Recall that the corresponding two general tensor definitions are

$$I_{ij} = \int_V (x_k x_k \delta_{ij} - x_i x_j) \, dV, \qquad \bar{I}_{ij} = I_{ij} - (\bar{x}_k \bar{x}_k \delta_{ij} - \bar{x}_i \bar{x}_j) \, V \tag{9.74}$$

where x_i are the components of the position vector of a point in volume, V and δ_{ij} is the Kronecker delta. Typically, elements that have the same area, and inertia tensor, relative to the element centroid will have the same square matrix integral if the properties do not depend of physical coordinates (x, y).

We want to illustrate these calculations in a finite element context for a two-dimensional geometry. For the parametric form in local coordinates (r, s) recall that

$$x(r, s) = \mathbf{G}(r, s)\, \mathbf{x}^e, \qquad y(r, s) = \mathbf{G}(r, s)\, \mathbf{y}^e$$

$$1 = \mathbf{G}(r, s)\, \mathbf{1} = \sum_i H_i(r, s)$$

where $\mathbf{1}$ is a vector of unity terms. Then the above measures become

$$A^e = \mathbf{1}^T \int_{A^e} \mathbf{G}^T \mathbf{G} \, dA \, \mathbf{1} = \mathbf{1}^T \mathbf{M}^e \, \mathbf{1}$$

where $\mathbf{M}^e$ is thought of as the element measure (or mass) matrix

$$A^e \bar{x}^e = \mathbf{1}^T \mathbf{M}^e \mathbf{x}^e, \qquad A^e \bar{y}^e = \mathbf{1}^T \mathbf{M}^e \mathbf{y}^e \tag{9.75}$$

$$I^e_{xx} = \mathbf{x}^{e^T} \mathbf{M}^e \mathbf{x}^e, \quad -I^e_{xy} = \mathbf{x}^{e^T} \mathbf{M}^e \mathbf{y}^e, \quad I^e_{yy} = \mathbf{y}^{e^T} \mathbf{M}^e \mathbf{y}^e.$$

The measure matrix is defined as :

$$\mathbf{M}^e = \int_{A^e} \mathbf{G}^T \mathbf{G} \, da = \int_{\square} \mathbf{G}^T \mathbf{G} \, |\mathbf{J}^e| \, d\square \tag{9.76}$$

where $\square$ denotes any non-dimensional parent domain (triangular or square) and $|\mathbf{J}^e|$ is the Jacobian of the transformation from $\square$ to A^e. For any straight sided triangular element it has a constant value of $|\mathbf{J}^e| = 2A^e$. Likewise, for a straight rectangular element or parallelogram element $|\mathbf{J}^e|$ is again constant. For a one-to-one geometric mapping, we always have the relation that

$$\mathbf{A}^e = \int_{A^e} da = \int_{\Box} |\mathbf{J}^e| \, d\Box$$

so that when $\mathbf{J}^e$ is constant $A^e = |\mathbf{J}^e| \, \Box_m$, and where here $\Box_m$ is the measure (volume) of the non-dimensional parent domain. For example, for the unit coordinate triangle we have $\Box_m = ½$ so that we get $A^e = (2A^e)(½)$, as expected. In the case of a constant or approximately constant Jacobian we may wish to consider a relative measure matrix

$$\mathbf{m}^e = \frac{\mathbf{M}^e}{|\mathbf{J}^e|} = \int_{\Box} \mathbf{G}^T \mathbf{G} \, d\Box. \tag{9.77}$$

Then $\mathbf{m}^e$ contains only known constants.

Let $\mathbf{R}^T = [x \;\; y]$ be the position vector to a point. Then the position vectors for all the element nodes are $\mathbf{R}^e = [\, \mathbf{R}_i \mid \mathbf{R}_j \mid \cdots \mid \mathbf{R}_n \,]^e$, or in other words,

$$\mathbf{R}^{e^T} = [\mathbf{x}^e \;\; \mathbf{y}^e].$$

A secondary product of interest is the rectangular array $\mathbf{s}^e = \mathbf{m}^e \mathbf{R}^{e^T}$ since the inertia products are defined with it as

$$\mathbf{I}^e = \mathbf{R}^e \mathbf{s}^e = \begin{bmatrix} I_{xx} & -I_{xy} \\ -I_{xy} & I_{yy} \end{bmatrix} \tag{9.78}$$

and the centroid vector is

$$\mathbf{c}^e = \mathbf{1}^T \mathbf{s}^e = [\bar{x} \;\; \bar{y}] \tag{9.79}$$

and

$$\mathbf{c}^T \mathbf{c} = \begin{bmatrix} \bar{x}^2 & \bar{x}\,\bar{y} \\ \bar{x}\,\bar{y} & \bar{y}^2 \end{bmatrix} \tag{9.80}$$

gives the terms necessary to determine the transfer terms for the centroidal inertia tensor. These computations are relatively cheap. This is especially true if we only use the corner nodes of the element.

9.9 Interpolation Error*

Here we will briefly outline how the concepts in Sec. 4.7 carry over to two-dimensions. Recall the Taylor expansion of a function, u, at a point (x, y) in two-dimensions:

$$\begin{aligned} u(x+h,\, y+k) &= u(x, y) + \left[h \frac{\partial u}{\partial x}(x, y) + k \frac{\partial u}{\partial y}(x, y) \right] \\ &+ \frac{1}{2!} \left[h^2 \frac{\partial^2 u}{\partial x^2} + 2hk \frac{\partial^2 u}{\partial x \, \partial y} + k^2 \frac{\partial^2 u}{\partial y^2} \right] + \cdots \end{aligned} \tag{9.81}$$

The objective here is to show that if the third term is neglected, then the relations for a linear interpolation triangle are obtained. That is, we will find that the third term is proportional to the error between the true solution and the interpolated solutions. Consider the triangle shown in Fig. 9.9.1. Employ Eq. (9.81) to estimate the nodal values u_j and u_m in terms of u_i :

$$u_j = u_i + \left[x_j \frac{\partial u}{\partial x}(x_i,\, y_i) + y_m \frac{\partial u}{\partial y}(x_i,\, y_i) \right]. \tag{9.82}$$

The value of $\partial u(x_i,\, y_i)/\partial x$ can be found by multiplying the first relation by y_m, and

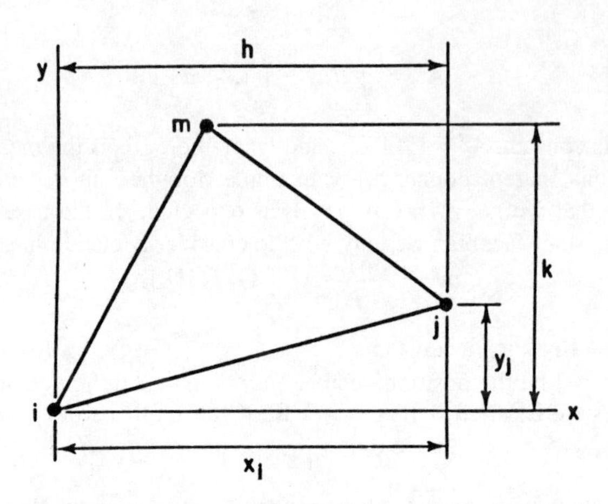

Figure 9.9.1 Interpolation error on an T3 element

subtracting the product of y_i and the second relation. The result is

$$\frac{\partial u}{\partial x}(x_i, y_i) = \frac{1}{2A}\left[u_i(y_j - y_m) + u_m(y_i - y_j) + u_j(y_m - y_i)\right] \tag{9.83}$$

where A is the area of the triangle. In a similar manner, if we compute this derivative at the other two nodes, we obtain

$$\frac{\partial u}{\partial x}(x_j, y_j) = \frac{\partial u}{\partial x}(x_m, y_m) = \frac{\partial u}{\partial x}(x_i, y_i).$$

That is, $\partial u/\partial x$ is a constant in the triangle. Likewise, $\partial u/\partial y$ is a constant.

We will see later that a linear interpolation triangle has constant derivatives that include Eq. (9.84). Thus, these common elements will represent the first two terms in Eq. (9.81). Thus, the element error is related to the third term:

$$E\,\alpha\left(h^2\frac{\partial^2 u}{\partial x^2} + 2hk\frac{\partial^2 u}{\partial x\,\partial y} + k^2\frac{\partial^2 u}{\partial y^2}\right) \tag{9.84}$$

where u is the exact solution, and h and k measure the element size in the x and y directions. Once again, we would find that these second derivatives are related to the strain and stress gradients. If the strains (e.g., $\varepsilon_x = \partial u/\partial x$) are constant, then the error is small or zero.

Before leaving these error comments, note that Eq. (9.84) could also be expressed in terms of the ratio (k/h). This is a measure of the relative shape of the element, and it is often called the *aspect ratio.* For an equilateral element, this ratio would be near unity. However, for a long narrow triangle, it could be quite large. Generally, it is best to keep the aspect ratio near unity (say < 5).

9.10 References

[1] Babuska, I., Griebel, M., and Pitkaranta, J., "The Problem of Selecting Shape Functions for a p-Type Finite Element," *Int. J. Num. Meth. Eng.*, **28**, pp. 1891–1908 (1989).

[2] Becker, E.B., Carey, G.F., and Oden, J.T., *Finite Elements – An Introduction*, Englewood Cliffs: Prentice-Hall (1981).

[3] Connor, J.C. and Brebbia, C.A., *Finite Element Techniques for Fluid Flow*, London: Butterworth (1976).

[4] El-Zafrany, A. and Cookson, R.A., "Derivation of Lagrangian and Hermite Shape Functions for Quadrilateral Elements," *Int. J. Num. Meth. Eng.*, **23**, pp. 1939–1958 (1986).

[5] Hinton, E. and Owen, D.R.J., *Finite Element Programming*, London: Academic Press (1977).

[6] Hu, K-K., Swartz, S.E., and Kirmser, P.G., "One Formula Generates Nth Order Shape Functions," *J. Eng. Mech.*, **110**(4), pp. 640–647 (1984).

[7] Hughes, T.J.R., *The Finite Element Method*, Englewood Cliffs: Prentice-Hall (1987).

[8] Segerlind, L.J., *Applied Finite Element Analysis*, New York: John Wiley (1984).

[9] Silvester, P.P. and Ferrari, R.L., *Finite Elements for Electrical Engineers*, Cambridge: Cambridge University Press (1983).

[10] Szabo, B. and Babuska, I., *Finite Element Analysis*, New York: John Wiley (1991).

[11] Zienkiewicz, O.C. and Morgan, K., *Finite Elements and Approximation*, Chichester: John Wiley (1983).

Chapter 10

ADAPTIVE ANALYSIS

10.1 Introduction

The use of uniform mesh refinement, shown in Fig. 10.1.1, has proven not to be cost effective. Today it appears that there will be significant future improvements and efficiencies based on the use of adaptive finite element methods for analysis and design. One price that we will pay for these new approaches is a more heavy reliance on mathematics such as functional analysis, norms, and interpolation theory. That is because of all of the adaptive methods are built on some concepts for estimating and automatically reducing the error in a solution. Another price is an increase in the complexity of the computational methods and the supporting data base. This text is not intended to be a complete introduction to adaptive analysis. Yet we will outline some of their basic theoretical and computational aspects that are needed for two- and three-dimensional analysis and design tools. The major developments in this area in the last decade follow from the works of Oden, et al. [10], Babuska, et al. [13], and Zienkiewicz, et al. [14, 16].

We generally think of the adaptive methods as falling into the categories of the:

1) h-method, where h is a measure of the size of the elements that use a fixed polynomial order, p. The mesh size, h, is the quantity to be locally adapted as shown in Fig. 10.1.2. It can involve refinement, or de-refinement (see Fig. 10.1.3).
2) p-method, where p is the polynomial order of some element of size h in a fixed mesh. The order, p, will be locally adapted. Figure 10.1.4 illustrates this idea where the number in each element represents the degree of its polynomial interpolation.
3) r-method, where r denotes a *relocation* of the nodes of the mesh, that is, the location of all nodes are adaptively changed to reduce the error estimates. This is shown in Fig. 10.1.5.
4) hp-method, where hp is a combination of methods 1) and 2). It locally adapts h and/or p so as to reduce the error estimates. This is shown in Fig. 10.1.6.

The r-method is not a commonly used method while the h-method is probably the most widely used adaptive method. The hp-method is currently the least commonly used method. However, it appears to be the optimal choice despite the additional computational complexity and data base complexity that it involves. It will probably be a predominate design tool in the future.

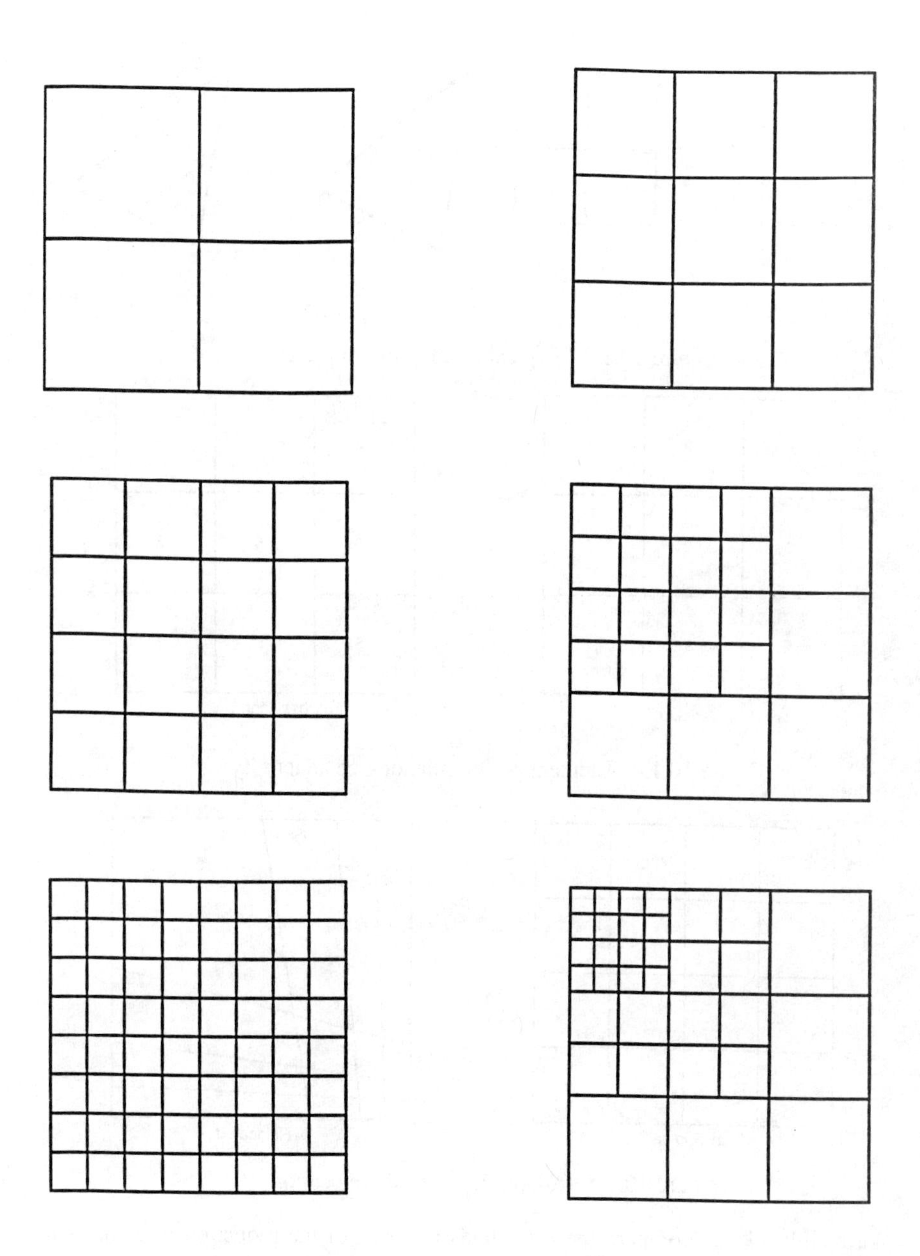

Figure 10.1.1 Uniform refinement

Figure 10.1.2 Selective refinement

Figure 10.1.3 Typical h-adaptivity in 2-D

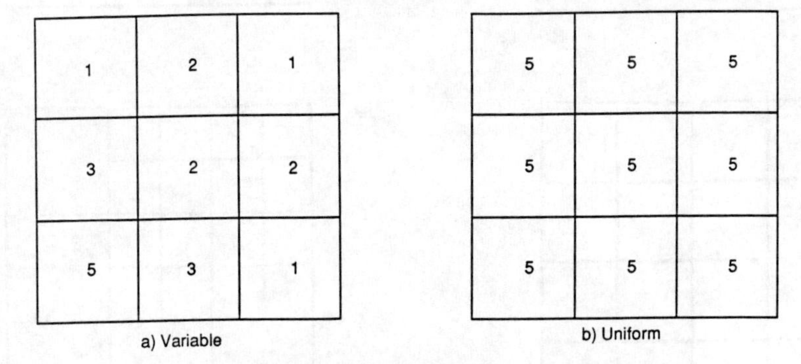

Figure 10.1.4 Element polynomial degree adaptivity

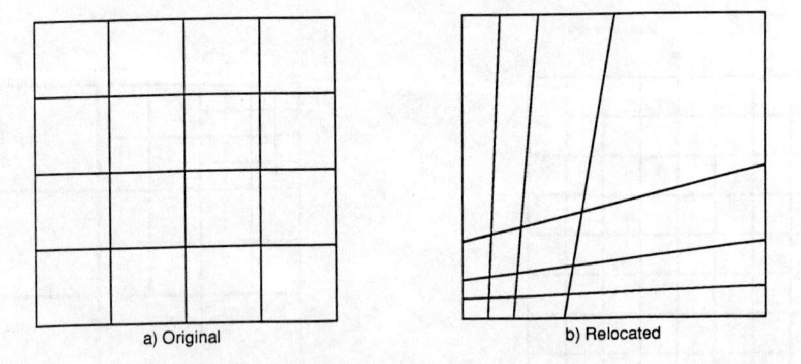

Figure 10.1.5 Adaptivity by node relocation

We mainly like to compare these methods in terms of the reduction in error as a function of the increase in the total number of degrees of freedom to be computed. Figure 10.1.7 shows a sketch of the logarithm of the error, E, (measured in some norm) versus the logarithm of the number of unknowns to be computed, N. You should note that the steeper the slope of a curve on this plot the more cost effective the calculation will be. Most of the methods reach a constant slope after a reasonable value of N has

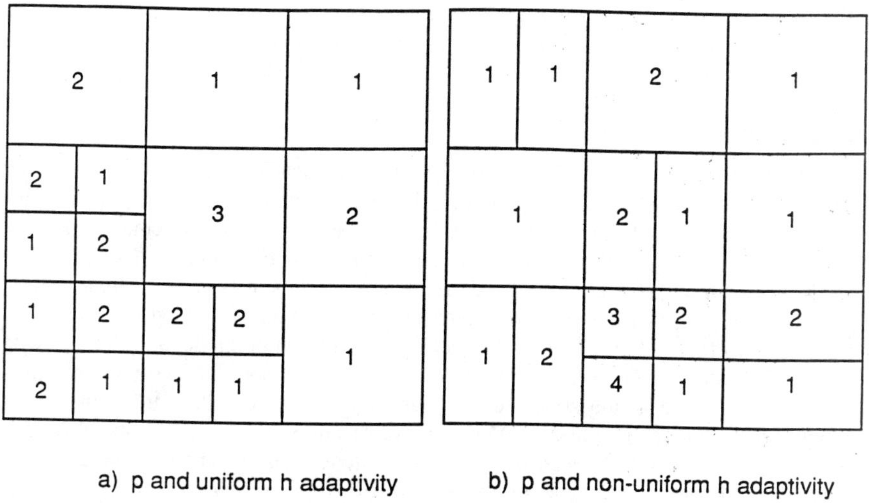

a) p and uniform h adaptivity

b) p and non-uniform h adaptivity

Figure 10.1.6 Adaptivity with h and p

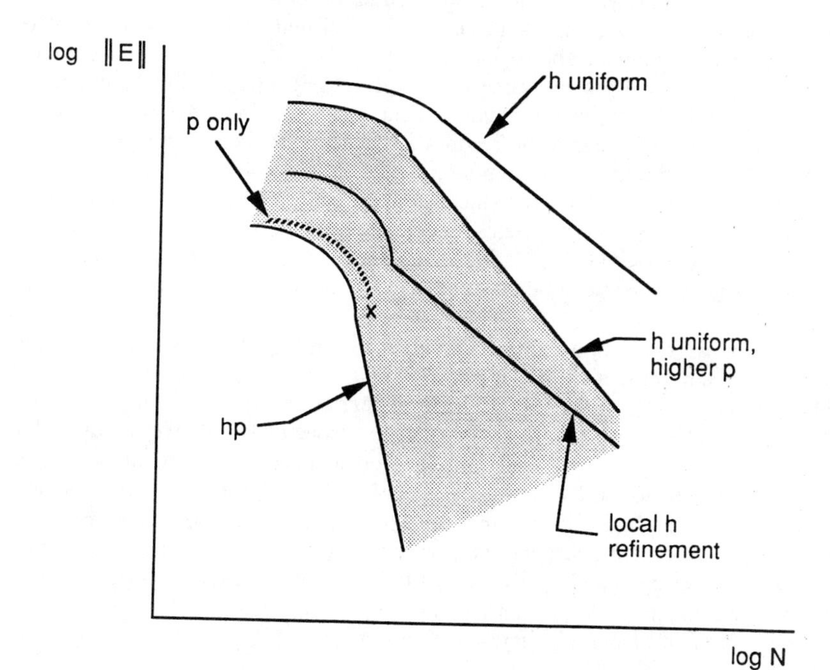

Figure 10.1.7 Behaviour of typical adaptive methods

been used. However, the *hp* solutions are often optimal because the error curve continues to curve down as N increases (as more degrees of freedom are added,. that is, there is an exponential decrease in the error as the number of unknowns is increased. The shaded area indicates a region where the adaptive methods pays off.

10.2 H-Adaptivity

We begin our introduction with the idea of locally splitting an element of relatively high error into a group of smaller elements. Usually we change the length of the side of the element by a factor of two in a refinement or de-refinement (fission or fusion) operation. This is illustrated in two-dimension in Fig. 10.1.3. There we have assumed that a parent produces four children that are one-fourth as large. Already we see that the bookkeeping difficulties have increased because we have added new nodes and new elements to the system. This will effect the storage requirements and the efficiency of our equation solver. To keep up with these children, that are also likely to be refined, we may need to build a tree structure into our data base or maintain a list of element neighbors and/or nodal neighbors. Most programs allow only a level one difference (factor of 2 length scale) between the child and parent. This means that the childrens' new node occurs at the midside of the parent element. We will soon see that this restriction can be lifted by further complicating the computational procedures.

After refinement we must be concerned about maintaining the continuity of the solution across the edge between a parent and a child. There are several methods that can be used but we need one that is general and consistent with our future goal of utilizing high order polynomial approximations. Thus, we begin with the assumption that both parent and child have the same order polynomial interpolation functions on their common edge (but not on their independent interiors). All that remains is to constrain the child to use the degrees of freedom associated with the parent's polynomial. Then there will be continuity of the solution, and of its tangential derivatives, on the common edge. This requires a well known constraint transformation matrix. It has been illustrated for common elements (beams, isoparametric, etc.) by Cook [6] and for hierarchical polynomials by Oden, et al. [10]. Here we will outline the development of such a matrix in a general sense and then simplify it for the usual level one restriction on child placement.

For simplicity we consider a bilinear Q4 element. In Fig. 10.2.1 we have shown a parent element, A, adjoining a child element, B, along a common side. Here side 2 of the parent connects to side 4 of the child. The local degrees of freedom of the parent and child are $[U_1 \;\; U_2 \;\; U_3 \;\; U_4]$ and $[u_1 \;\; u_2 \;\; u_3 \;\; u_4]$, respectively, if we assume for simplicity that there is only a single scalar unknown at each node. On the common edge, values u_1 and u_4 are tied to the parent values U_2 and U_3 by a constraint matrix, denoted by C in the figure. Thus, they are not independent degrees of freedom. Let the local edge coordinate of the parent be s, and the corresponding location of the attachment of the child's global nodes by s_5 and s_8. Recall that the value of any quantity on the edge of the parent (or any element) is defined only by the degrees of freedom it has on that edge and by the corresponding local edge interpolation functions, $\mathbf{H}^b(s)$. Thus, we can write the identity that

$$U(s) \;=\; H_2^b(s)U_2 + H_3^b(s)U_3$$

so at global node 5 (child local node 1), we have

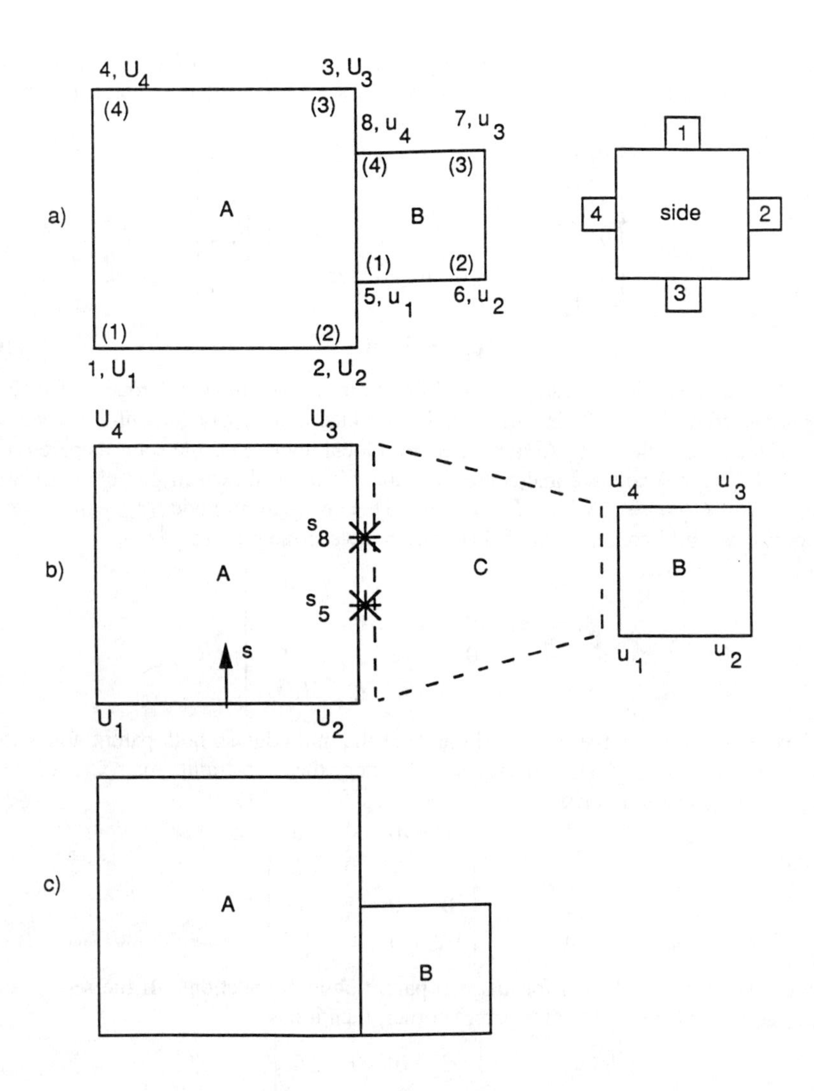

Figure 10.2.1 Child-parent side connections

$$U(s_5) \;=\; H_2^b(s_5)U_2 + H_3^b(s_5)U_3 \;=\; u_1\,,$$

and a similar expression for global node 8 (child local node 4). We could express these two equations in matrix form as

$$\begin{Bmatrix} U_5 \\ U_8 \end{Bmatrix} = \begin{Bmatrix} u_1^B \\ u_4^B \end{Bmatrix} = \begin{bmatrix} \mathbf{H}^b(s_5) \\ \mathbf{H}^b(s_8) \end{bmatrix} \begin{Bmatrix} U_2^A \\ U_3^A \end{Bmatrix}, \quad \mathbf{u}^b \;=\; \mathbf{T}^b\,\mathbf{U}^b\,, \tag{10.1}$$

where $\mathbf{T}^b$ is the constraint matrix on that boundary that ties the edge child dof, $\mathbf{u}^b$, to the parent edge dof $\mathbf{U}^b$. We may prefer to describe the child's dof in terms of its two independent values, U_6 and U_7, and the two values governed by the parent, U_2 and U_3, as

$$\begin{Bmatrix} u_1 \\ u_2 \\ u_3 \\ u_4 \end{Bmatrix} = \begin{bmatrix} H_2(s_5) & 0 & 0 & H_3(s_5) \\ 0 & 1 & 0 & 0 \\ 0 & 0 & 1 & 0 \\ H_2(s_8) & 0 & 0 & H_3(s_8) \end{bmatrix} \begin{Bmatrix} U_2 \\ U_6 \\ U_7 \\ U_3 \end{Bmatrix}$$

or

$$\mathbf{u}_c = \mathbf{R}^T \mathbf{u}_I, \tag{10.2}$$

where $\mathbf{R}^T$ is a constraint or restraint transformation that ties the child edge dof $\mathbf{u}_c$ to the independent edge dof $\mathbf{u}_I$. This simplifies if we place a corner of the child element at a corner of the parent element. Assume that child local node (1) is the same as parent local node (2), i.e. global nodes 2 and 5 are the same. This is shown in part c of the above figure. Then $H_2(s_5) \equiv 1$ while $H_3(s_5) \equiv 0$. That is, parent node (3) will make no contribution to child node (1), and $\mathbf{R}$ becomes a lower triangular matrix

$$\mathbf{R}^T = \begin{bmatrix} 1 & 0 & 0 & 0 \\ 0 & 1 & 0 & 0 \\ 0 & 0 & 1 & 0 \\ H_2(s_8) & 0 & 0 & H_3(s_8) \end{bmatrix},$$

and if we go on to place the other child node at the mid-edge so both parent nodes make equal contributions $H_2(s_8) = H_3(s_8) = 1/2$, then the constraint matrix becomes a constant lower triangle matrix

$$\mathbf{R}^T = \begin{bmatrix} 1 & 0 & 0 & 0 \\ 0 & 1 & 0 & 0 \\ 0 & 0 & 1 & 0 \\ 1/2 & 0 & 0 & 1/2 \end{bmatrix}$$

that could be stored and used for all such parent-child connections. If the second child attaches at the midside and the far parent corner, then it has

$$\mathbf{R}^T = \begin{bmatrix} 1/2 & 0 & 0 & 1/2 \\ 0 & 1 & 0 & 0 \\ 0 & 0 & 1 & 0 \\ 0 & 0 & 0 & 1 \end{bmatrix}.$$

To preserve the energy in the child formulation

$$\Pi = 1/2\, \mathbf{u}_c^T \mathbf{K}_c \mathbf{u}_c - \mathbf{u}_c^T \mathbf{F}_c$$

we substitute the transformation to independent dof

$$\Pi = 1/2\, (\mathbf{u}_I^T \mathbf{R}) \mathbf{K}_c (\mathbf{R}^T \mathbf{u}_I) - (\mathbf{u}_I^T \mathbf{R}) \mathbf{F}_c, \qquad \Pi = 1/2\, \mathbf{u}_I^T \mathbf{K}_I \mathbf{u}_I - \mathbf{u}_I^T \mathbf{F}_I$$

where

$$\mathbf{F}_I = \mathbf{R}\mathbf{F}_c, \qquad \mathbf{K}_I = \mathbf{R}\mathbf{K}_c\mathbf{R}^T \tag{10.3}$$

represent the operations of scattering the child's square and column matrix, $\mathbf{K}_c$ and $\mathbf{F}_c$, out to the actual independent square and column matrices. In practice, one would like to avoid the operations with the 0 and 1 coefficients, and consider only the other terms in **T** since they are the only ones that effect the change to independent degrees of freedom.

The above discussion was simplified because we considered only nodal values as unknown dof. Recall that the hierarchical functions used in most p-methods and hp-methods involve tangential derivatives on their edges. When one joins two such two- or three-dimensional elements together you find that you must account for a sign change in (some of) these local tangential dof, or in their interpolation functions. This is illustrated in Fig. 9.5.5, and implies yet more bookkeeping work to implement a p- or hp-formulation.

10.3 P - Adaptivity

The p-method keeps the mesh unchanged, but does vary the element interpolation order as indicated by the error indicators. Some codes upgrade to the maximum required degree and use it in all elements, as shown in Fig. 10.1.4b. That is relatively expensive and wasteful, unless you can automatically find *geometrically similar* elements and actually calculate the matrices only once. Otherwise, we want a different polynomial in each element. When two elements on a common interface have different polynomial orders, one must either enrich the edge interpolation functions of the lower order element, or degrade the higher order one. This can be done with Lagrangian or Hierarchical functions, although the latter are preferred by most analysts. Of course, the functions used to define the edge order of an element propagate into its interior as shown earlier in Fig. 9.5.1.

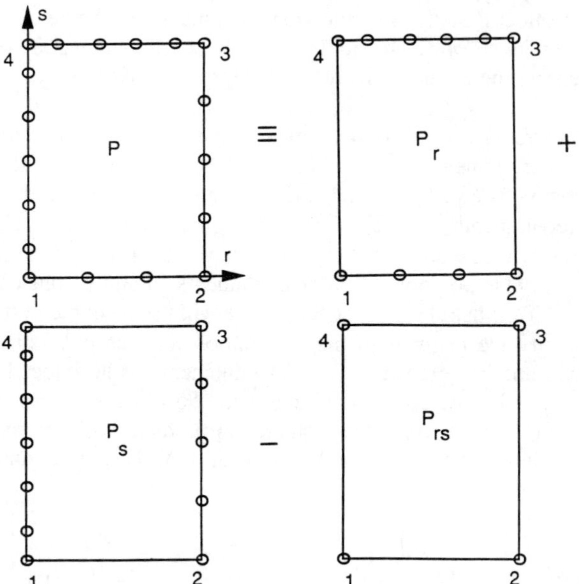

Figure 10.3.1 Superposition principle for quadrilateral elements

For Lagrangian interpolation functions there are various ways to implement different interpolation orders on the edges of an element. We refer to this as anisotropic interpolation. Detailed procedures for two-dimensional elements have been given by El-Zafrany and Cookson [7], while a large family of two- and three-dimensional elements has been given by Attenbach and Scholz [1]. The concept for combining different Lagrangian edge functions is illustrated in Fig. 10.3.1. A similar procedure for a linear-to-cubic quadrilateral was illustrated earlier. The disadvantage of this approach is that if we change the degree of the polynomial in the next adaptivity step, then all of the coefficients of the previously computed element matrices must be re-computed. The hierarchical methods require only the computation of the coefficients in rows and columns associated with the new enriched degrees of freedoms. This advantage might be offset somewhat due to the fact that matrix condition numbers can become very large in hierarchical formulations, especially for transient problems since they include the use of the "mass" matrix.

10.4 HP - Adaptivity

When the error estimator indicates a need to change both the orders of the polynomials in adjacent elements and their size, we are faced with two problems: 1) maintaining continuity between elements, and 2) enriching the interface interpolations. First, consider adjacent two-dimensional elements that share a common edge. We could have two small elements next to the large element.

Thus, we would expect to have three different polynomial orders along the edge. To maintain continuity, we must select a single polynomial order. At this time, we do not have a theoretical basis for the selection. We make the choice of order by a heuristic known as the *maximum rule*. On any interface between elements involving polynomials of different degree, the elements with lower degree on that interface are enriched to the higher degree.

This is illustrated in Fig. 10.4.1 for an edge between two small elements, say B and C, and a large element, say A. We will denote the polynomial orders of the large and small elements as P_A, P_B, and P_C, respectively. We assume that initially the error estimator has recommended values of $P_A = 3$, $P_B = 2$, and $P_C = 1$. The figure shows the initial discontinuity that would exist if these values were used. Since the highest order is 3 (cubic), the interface polynomial order in elements B and C must be enriched to that maximum level. To illustrate these concepts, we will use a simple numerical example.

For simplicity we begin with the assumption that the elements in Fig. 10.4.2 are Lagrangian quadratic line elements. The h-refinement has introduced additional degrees of freedom that need to be constrained together. To illustrate the process involved, we will consider a one-dimensional heat conduction problem with internal heat generation, f. For a single element model ($L^e = L$), we recall that the Lagrangian element matrices are

$$\mathbf{K}^e = \frac{K^e}{3L^e}\begin{bmatrix} 7 & 1 & -8 \\ 1 & 7 & -8 \\ -8 & -8 & 16 \end{bmatrix}, \quad \mathbf{F}^e = \frac{f^e L^e}{6}\begin{Bmatrix} 1 \\ 1 \\ 4 \end{Bmatrix}.$$

From Fig. 10.4.2 we see that

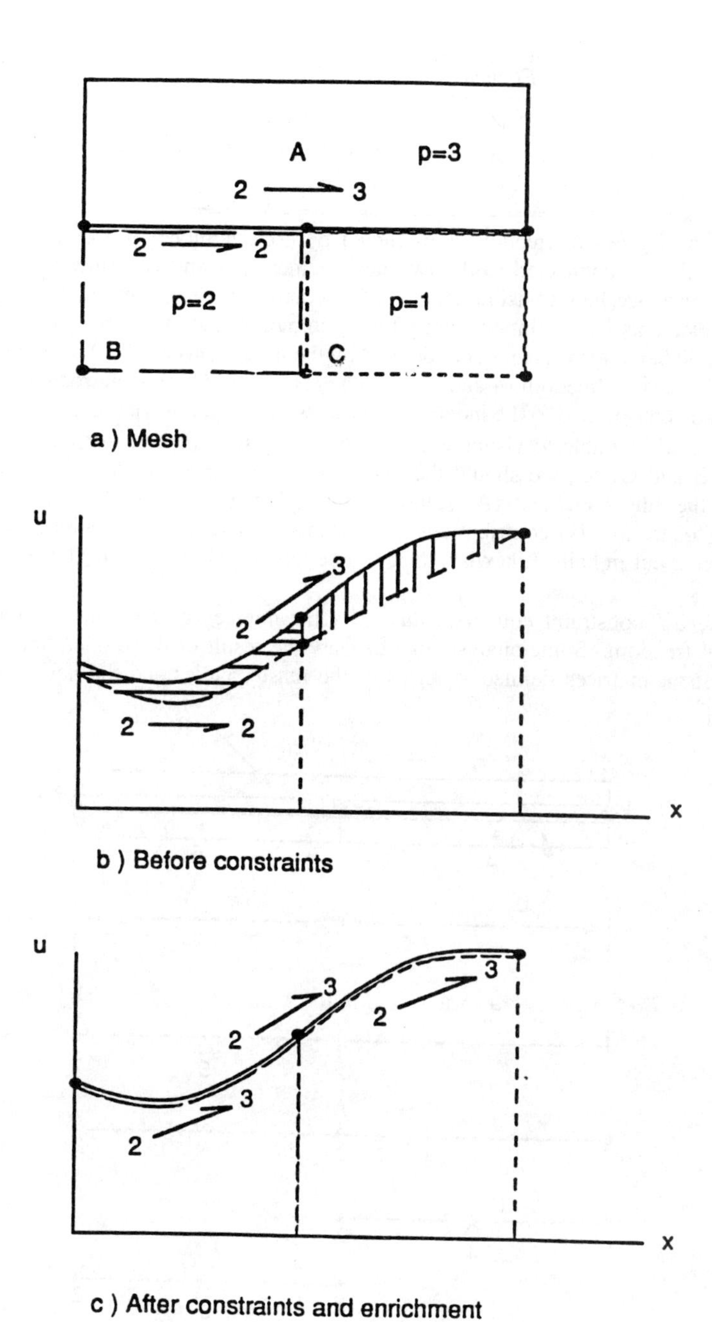

Figure 10.4.1 Invoking the maximum order rule

Element	*Connectivity*	*Length*
A	1, 2, 3	L
B	1, 3, 4	$L/2$
C	3, 2, 5	$L/2$

The original degrees of freedom are denoted by node numbers 1, 2, 3 on element A. They will also be connected to the two new elements, B and C. However, these new elements introduce new midside degrees of freedom at nodes 4 and 5. We require that the three elements be continuous along their common one-dimensional interface. Since the quadratic behavior was uniquely defined by the three constants (dof's) of element A, the new degrees of freedom (4 and 5) of elements B and C must be redundant, that is, they must be constrained to the independent values on the parent element A.

We should be able to visualize that, if we apply the proper constraints to the two-element (B and C) set, we should then generate the same assembled system equations found in the single element (A) solution. From Fig. 10.4.2, we should note that the stiffness (square matrix) contributions of elements B and C have doubled due to their lengths being cut in half. Likewise, their source terms (column matrix) have been cut in half.

There are constraint equations that tie some degrees of freedom to other master degrees of freedom. Some analysts like to view the result of these operations as a new set of element matrices defined in terms of the master nodes and the new independent

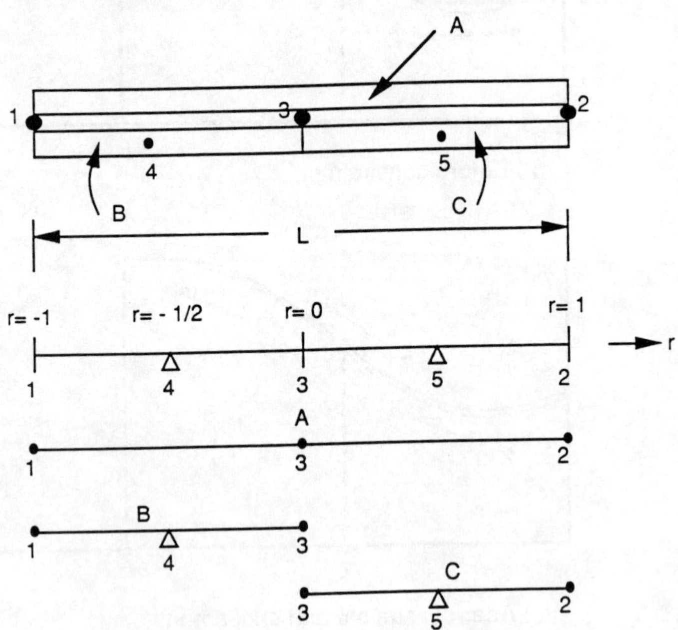

Figure 10.4.2 Constraining quadratic solutions

nodes. In other words, they change (and usually increase) the element topology by deleting the slave nodes and inserting the master nodes. For Lagrangian quadratic interpolation on A, we note (from Eq. 4.18) that the value at any point in natural coordinates is

$$u(a) = H_1(a)\,U_1 + H_2(a)\,U_2 + H_3(a)\,U_3, \qquad -1 \le a \le 1$$

$$u(a) = U_1 a(a-1)/2 + U_2 a(a+1)/2 + U_3(1+a)(1-a)\,.$$

The new nodes, 4 and 5, occur at $a_4 = -1/4$ and $a_5 = +1/4$, respectively. Thus, substitution gives

$$u_4 = u(a_4) = \frac{3}{8}u_1 - \frac{1}{8}u_2 + \frac{6}{8}u_3$$

and

$$u_5 = u(a_5) = -\frac{1}{8}u_1 + \frac{3}{8}u_2 + \frac{6}{8}u_3\,.$$

If we recast element B in terms of the master degrees of freedom of element A, the above identity for u_4 gives the following relations

$$\mathbf{u}_B = \begin{Bmatrix} u_1 \\ u_3 \\ u_4 \end{Bmatrix} = \begin{bmatrix} 1 & 0 & 0 \\ 0 & 0 & 1 \\ 0 & 0 & 0 \end{bmatrix} \begin{Bmatrix} u_1 \\ u_2 \\ u_3 \end{Bmatrix} + \frac{1}{8} \begin{bmatrix} 0 & 0 & 0 \\ 0 & 0 & 0 \\ 3 & -1 & 6 \end{bmatrix} \begin{Bmatrix} u_1 \\ u_2 \\ u_3 \end{Bmatrix}$$

or

$$\mathbf{u}_B = \mathbf{R}_u^B\,\mathbf{u}_A + \mathbf{R}_c^B\,\mathbf{u}_A\,, \tag{10.4}$$

where $\mathbf{R}_c$ denotes the constraints on element B in terms of the edge master degrees of freedom on element A. Matrix $\mathbf{R}_u$ indicates (by a Boolean operation) which degrees of freedom in element B are identically the same as those in A, or which are totally new independent degrees of freedom.

If all of the dof of B had been independent, we would have written its linear and quadratic contributions to the integrals (work and energy) as

$$I_1 = \mathbf{u}_B^T\,\mathbf{f} \qquad \text{and} \qquad I_2 = \mathbf{u}_B^T\,\mathbf{K}_B\,\mathbf{u}_B\,.$$

Since the $\mathbf{u}_B$ are not all independent, we must use Eq. (10.4) in these expressions. Therefore, we have

$$I_1 = \left[\mathbf{u}_A^T\,\mathbf{R}_u^T + \mathbf{u}_A^T\,\mathbf{R}_c^T\right]\mathbf{f}_B = \mathbf{u}_A^T\,\mathbf{f}_B^*\,, \tag{10.5}$$

where the constrained source terms are

$$\mathbf{f}_B^* = \left[\mathbf{R}_u^T + \mathbf{R}_c^T\right]\mathbf{f}_B\,. \tag{10.6}$$

For element B, with $L^e = L/2$, we obtain

$$\mathbf{f}_B^* = \frac{fL}{16}\begin{Bmatrix} 5 \\ -1 \\ 8 \end{Bmatrix}$$

and likewise for element C, we have

$$\mathbf{u}_c = \begin{Bmatrix} u_3 \\ u_2 \\ u_5 \end{Bmatrix} = \begin{bmatrix} 0 & 0 & 1 \\ 0 & 1 & 0 \\ 0 & 0 & 0 \end{bmatrix} \begin{Bmatrix} u_1 \\ u_2 \\ u_3 \end{Bmatrix} + \frac{1}{8} \begin{bmatrix} 0 & 0 & 0 \\ 0 & 0 & 0 \\ -1 & 3 & 6 \end{bmatrix} \begin{Bmatrix} u_1 \\ u_2 \\ u_3 \end{Bmatrix}$$

so that

$$\mathbf{f}_c^* = \frac{fL}{24} \begin{Bmatrix} -1 \\ 5 \\ 8 \end{Bmatrix}.$$

Their assembly gives $\mathbf{f} = \mathbf{f}_B^* + \mathbf{f}_c^*$;

$$\mathbf{f} = \frac{fL}{24} \begin{Bmatrix} 5 & -1 \\ -1 & +5 \\ 8 & +8 \end{Bmatrix} = \frac{fL}{6} \begin{Bmatrix} 1 \\ 1 \\ 4 \end{Bmatrix},$$

which is the same as the single parent element. We should have expected this result. Likewise, for the quadratic (energy) contributions, we obtain a constrained square matrix given by (see Oden [11]) :

$$\mathbf{K}^* = \mathbf{R}_u^T \mathbf{K} \mathbf{R}_u + \mathbf{R}_c^T \mathbf{K} \mathbf{R}_u + \mathbf{R}_u^T \mathbf{K} \mathbf{R}_c + \mathbf{R}_c^T \mathbf{K} \mathbf{R}_c. \tag{10.7}$$

In this example using $\mathbf{R}_u$ and $\mathbf{R}_c$ for element B yields

$$\mathbf{K}_B^* = \frac{k}{12L} \left(\begin{bmatrix} 56 & 0 & 8 \\ 0 & 0 & 0 \\ 8 & 0 & 56 \end{bmatrix} + \begin{bmatrix} -24 & 8 & -48 \\ 0 & 0 & 0 \\ -24 & 8 & -48 \end{bmatrix} \right.$$

$$\left. + \begin{bmatrix} -24 & 0 & -24 \\ 8 & 0 & 8 \\ -48 & 0 & -48 \end{bmatrix} + \begin{bmatrix} 18 & -6 & 36 \\ -6 & 2 & -12 \\ 36 & -12 & 72 \end{bmatrix} \right) = \frac{k}{6L} \begin{bmatrix} 13 & 1 & -14 \\ 1 & 1 & -2 \\ -14 & -2 & 16 \end{bmatrix}.$$

Likewise, for element C

$$\mathbf{K}_c^* = \frac{k}{12L} \left(\begin{bmatrix} 0 & 0 & 0 \\ 0 & 56 & 8 \\ 0 & 8 & 56 \end{bmatrix} + \begin{bmatrix} 0 & 0 & 0 \\ 8 & -24 & -48 \\ 8 & -24 & -48 \end{bmatrix} \right.$$

$$\left. + \begin{bmatrix} 0 & 8 & 8 \\ 0 & -24 & -24 \\ 0 & -48 & -48 \end{bmatrix} + \begin{bmatrix} 2 & -6 & -12 \\ -6 & 18 & 36 \\ -12 & 36 & 72 \end{bmatrix} \right) = \frac{k}{6L} \begin{bmatrix} 1 & 1 & -2 \\ 1 & 13 & -14 \\ -2 & -14 & 16 \end{bmatrix}.$$

The direct assembly of these two array gives

$$\mathbf{K}_{B+C}^* = \frac{k}{3L} \begin{bmatrix} 7 & 1 & -8 \\ 1 & 7 & -8 \\ -8 & -8 & 16 \end{bmatrix}$$

which should have been expected. That is, the two constrained quadratic elements are the same as a single unconstrained one with regard to both the source and stiffness contributions. In this case, we chose to use Lagrangian interpolation simply because more readers are familiar with those forms of **K** and **f**, and should recognize the validity of the result quicker. Also, this form gives checks by inspection that are not clear in hierarchical or Hermitian interpolation. For example, the assembled **f** vector clearly includes all of the 6/6 parts of the applied resultant sources acting on the element.

By way of comparison, if we used the quadratic hierarchical form, the single element matrices are

$$\mathbf{K}^e = \frac{A^e k^e}{3L^e} \begin{bmatrix} 3 & -3 & 0 \\ -3 & 3 & 0 \\ 0 & 0 & 4 \end{bmatrix}, \quad \mathbf{f}^e = \frac{f^e L^e A^e}{6} \begin{Bmatrix} 3 \\ 3 \\ -2 \end{Bmatrix}.$$

When checking the total source contribution in the last matrix, we must remember that the local degrees of freedom are

$$\mathbf{u}^{e^T} = [\, u_1 \quad u_2 \quad u_3'' \,]^e .$$

We do not include the derivative degrees of freedom in our source check (for a Poisson problem), and we then see that we do have all the necessary $(3+3)/6$ parts of the total source.

In closing, the reader is reminded that for an example with more nodes, these constraint matrices would assemble (scatter) the nine-element terms into a larger system **K**. In other words, the **R** matrices are generally not square.

10.5 Adaptive Example

To illustrate the advantages of adaptive analysis, we will consider the example of the Laplace equation on an L-shaped domain. This problem was solved with the commercial system PHLEX [5], which is an hp-adaptive system for general analysis. It also offers h-adaptivity and p-adaptivity. All of the options can be exercised under manual or automatic control to reach a specified level of accuracy.

The default-general weak form in PHLEX for second-order elliptical PDE's is

$$\begin{aligned}
&\int_\Omega \sum_{i=1}^{S} \sum_{j=1}^{S} Aij(k,l) \frac{\partial V_k}{\partial x_i} \frac{\partial U_l}{\partial x_j} \, d\Omega \\
&+ \int_\Omega \sum_{i=1}^{S} \left[Bi(k,l)\, V_k \frac{\partial U_l}{\partial x_i} + Di(k,l) \frac{\partial V_k}{\partial x_i} U_l \right] d\Omega + \int_\Omega Ci(k,l)\, V_k\, V_l \, d\Omega \\
&+ \int_\Gamma Att(k,l) \frac{\partial V_K}{\partial t} \frac{\partial V_l}{\partial t} \, d\Gamma + \int_\Gamma \left[Bt(k,l) \frac{\partial V_k}{\partial t} U_l + Pt(k,l)\, V_k \frac{\partial U_l}{\partial t} \right] d\Gamma \\
&+ \int_\Gamma \left[Pn(k,l)\, V_k \frac{\partial U_l}{\partial n} + C^*(k,l)\, V_k\, U_l \right] d\Gamma \\
&= \int_\Omega \left[F(k)\, V_k + \sum_{i=1}^{S} Fi(k) \frac{\partial V_k}{\partial x_i} \right] d\Omega + \int_\Gamma \left[G(k)\, V_k + Gt(k) \frac{\partial V_k}{\partial t} \right] d\Gamma
\end{aligned} \tag{10.8}$$

where S denotes the number of spatial dimensions, n is the direction normal to the boundary, t is the direction tangent to the boundary, k and l range over the number of components in the unknown **U**, and **V** denotes the components of the Galerkin weight function. The variables Aij, Bi, Di, C, Att, Bt, Pt, Pn, C^*, F, Fi, G, and Gt can be input as user-defined constants for each different material. They can also be defined by optional user-supplied subroutines, and thus spatially or time-dependent coefficients are easily treated. This also means that time-dependent parabolic and hyperbolic systems can be solved (as discussed in Chap. 17).

The constant coefficient Laplace equation

$$\frac{\partial^2 u}{\partial x^2} + \frac{\partial^2 u}{\partial y^2} = 0$$

is well known to have a weak formulation (see Chap. 12) of

$$\int_\Omega \left[\frac{\partial V}{\partial x}\frac{\partial U}{\partial x} + \frac{\partial V}{\partial y}\frac{\partial U}{\partial y} \right] dx - \int_\Gamma q_n V \frac{\partial U}{\partial n}\, d\Gamma = 0$$

where q_n is the normal flux on the boundary where U is not given as an essential boundary condition. The problem to be modeled has insulated boundaries so that $q_n \equiv 0$. Thus, our problem is obtained by inserting data that set $A_{11} = 1$, $A_{22} = 1$, and let all other coefficients in Eq. (10.8) default to zero.

The analysis domain is shown in Fig. 10.6.1 as modeled by three linear quadrilaterals. The left edge is held at $U = 1$ while the right edge is held at $U = 5$, and the other three edges have zero normal flux. Considering the essential boundary conditions, this first mesh has only two unknown degrees of freedom. Therefore, we expect the initial solution to have a large error. However, we compute a solution since it helps us to see that the adaptive algorithms are actually improving the solution. Note that since the normal gradient is zero on three of the edges, we know that the contours should be perpendicular to those boundaries where they intersect. From the initial contours in Fig. 10.6.1, we see that this does not happen — especially near corners at the centerline. We should expect the larger error to be predicted in that area. When PHLEX computes and displays the estimated error, it shows the element in that corner to have the highest predicted error (about 20%). While such predictions may be inaccurate on such a very crude mesh, one is reassured that the locations agree with our intuitive analysis. To improve our initial crude solution, we refine some elements and increase the polynomial orders in selected regions. The new mesh and degrees are shown in Fig. 10.6.2, along with the improved estimates of the error in the energy norm. The new contours are shown in Fig. 10.6.3. Clearly, they are becoming more nearly orthogonal to the insulated boundaries, as desired. Still, we want more accuracy, so we ask the program to automatically continue the hp-adaptivity until the maximum error estimate, in the energy norm, is less than 2%. The resulting contours are also shown in Fig. 10.6.4.

While it has been shown that the hp-method is optimal, here one could still have chosen pure h-adaptivity or pure p-adaptivity. Those two options result in solutions shown in Figs. 10.6.5 and 10.6.6, respectively. These adaptive procedures are also available and equally valid for three-dimensional problems.

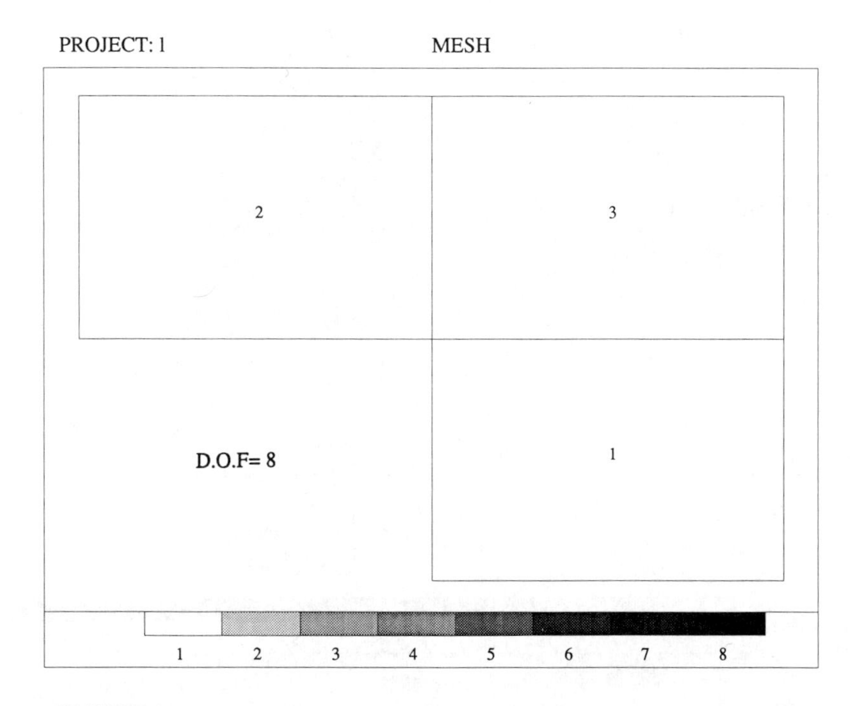

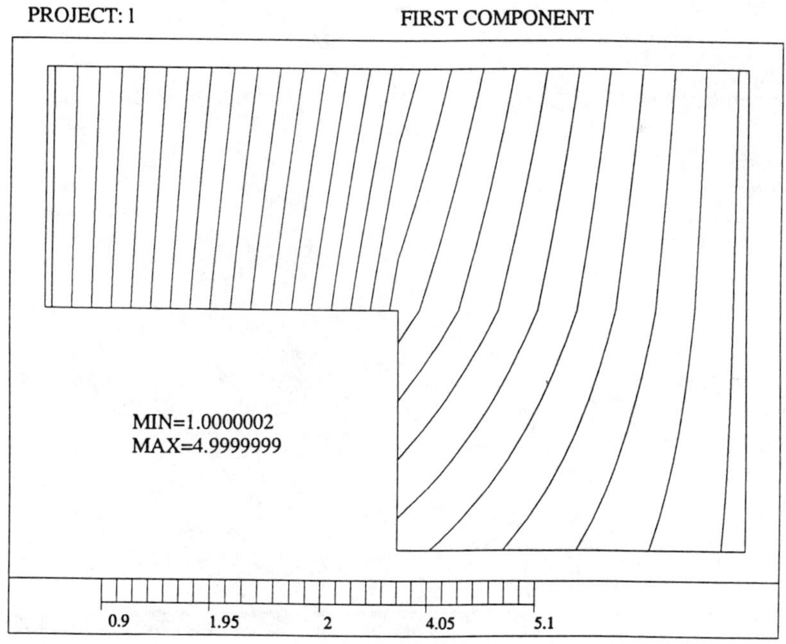

Figure 10.6.1 Initial mesh and crude countours

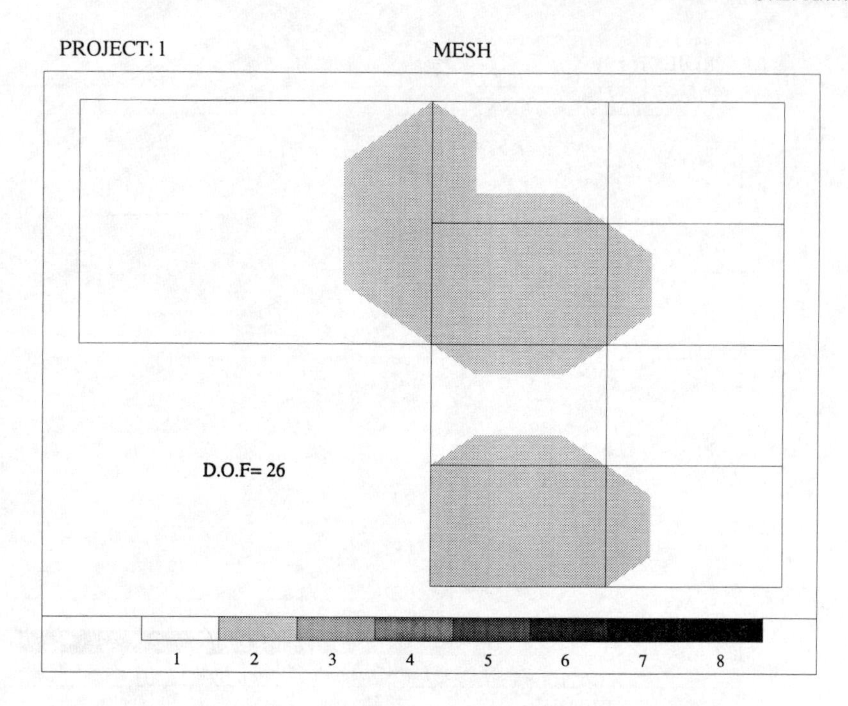

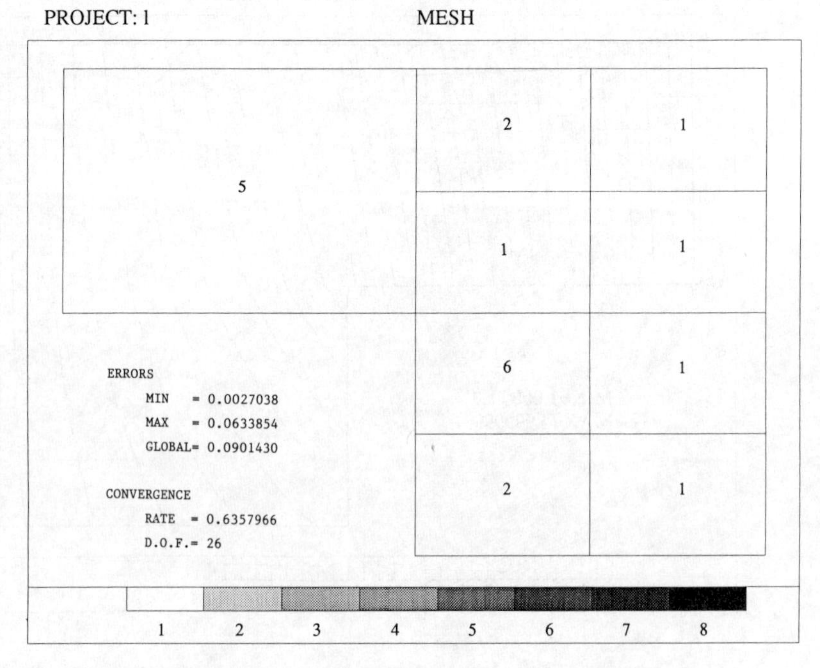

Figure 10.6.2 Manual enrichment and resulting error estimates

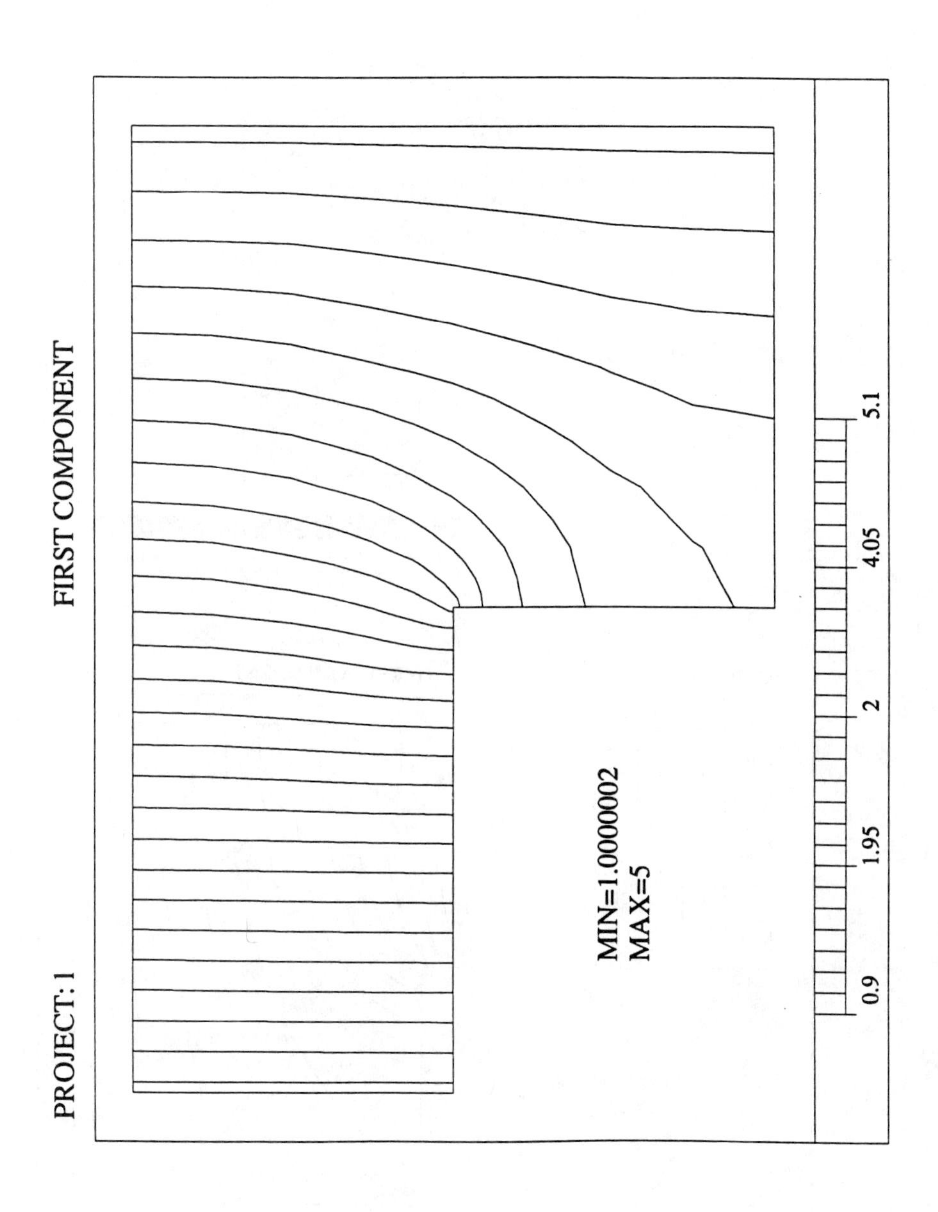

Figure 10.6.3 Improved countours via manual adaptivity

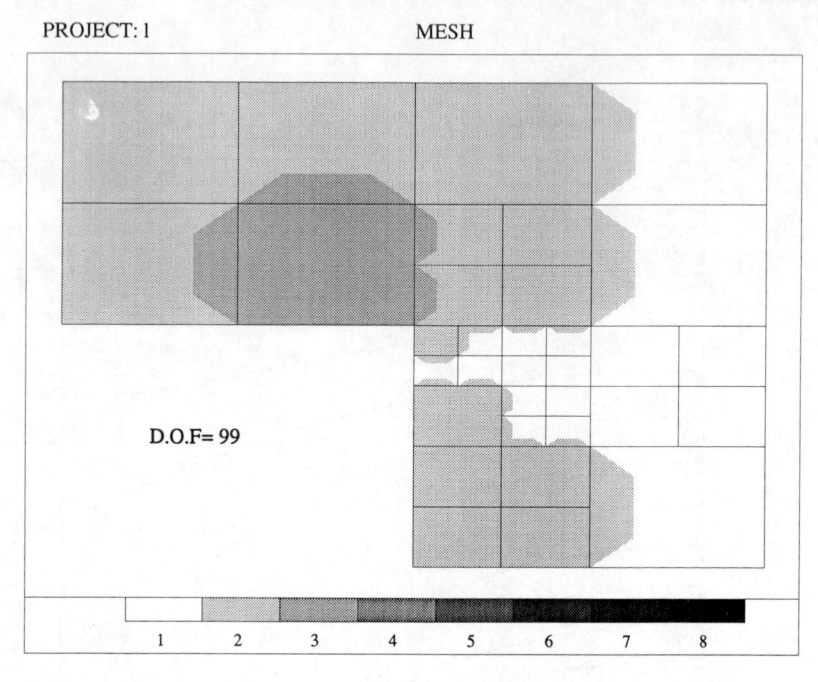

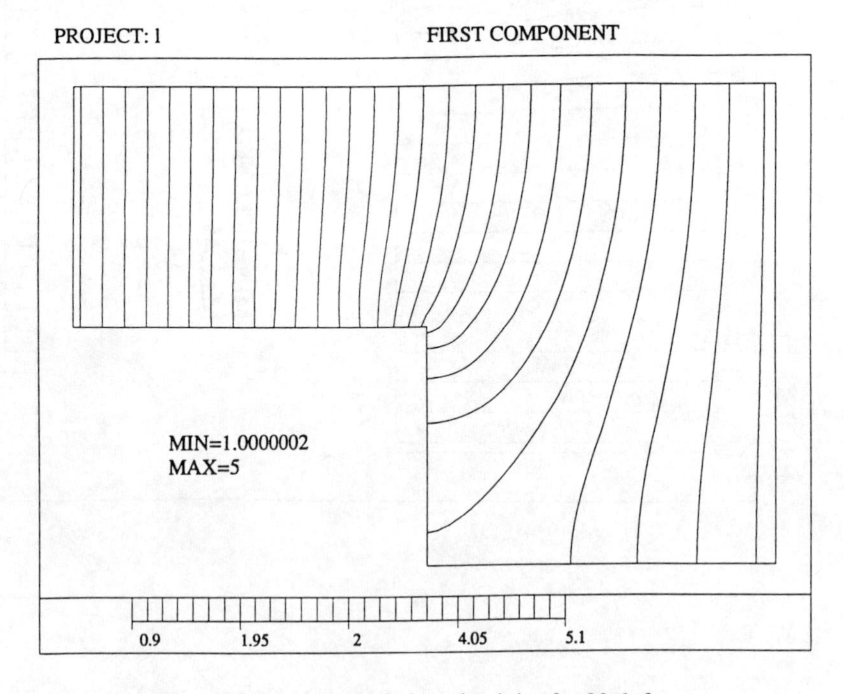

Figure 10.6.4 Automatic hp-adaptivity for 99 dof

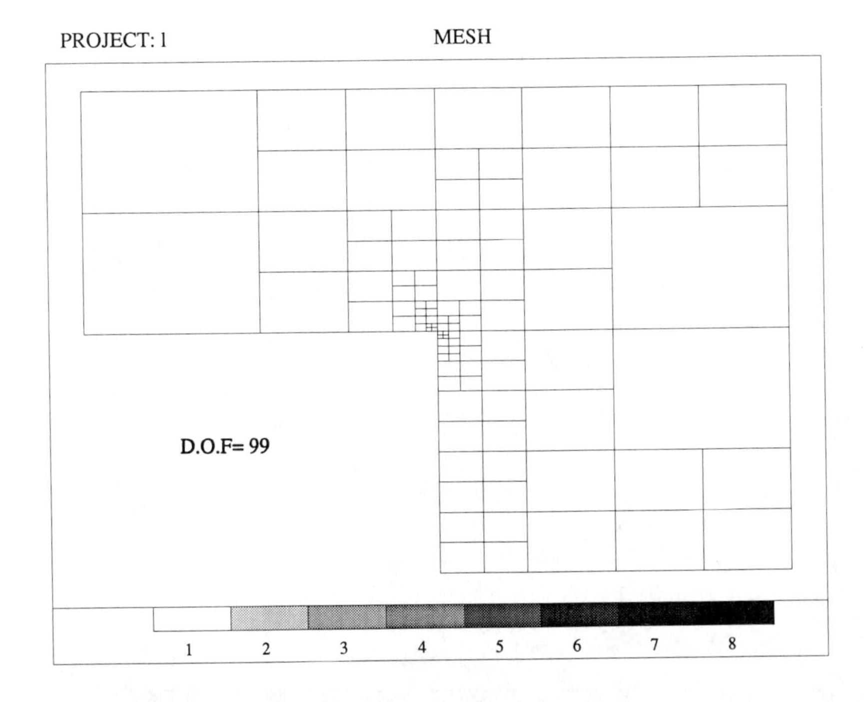

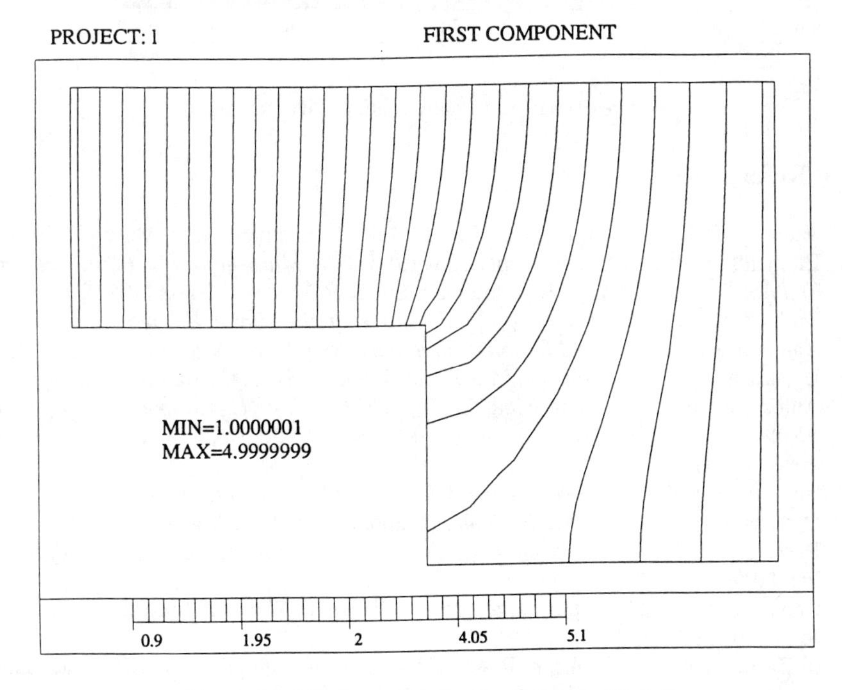

Figure 10.6.5 Automatic h-adaptivity for 99 dof

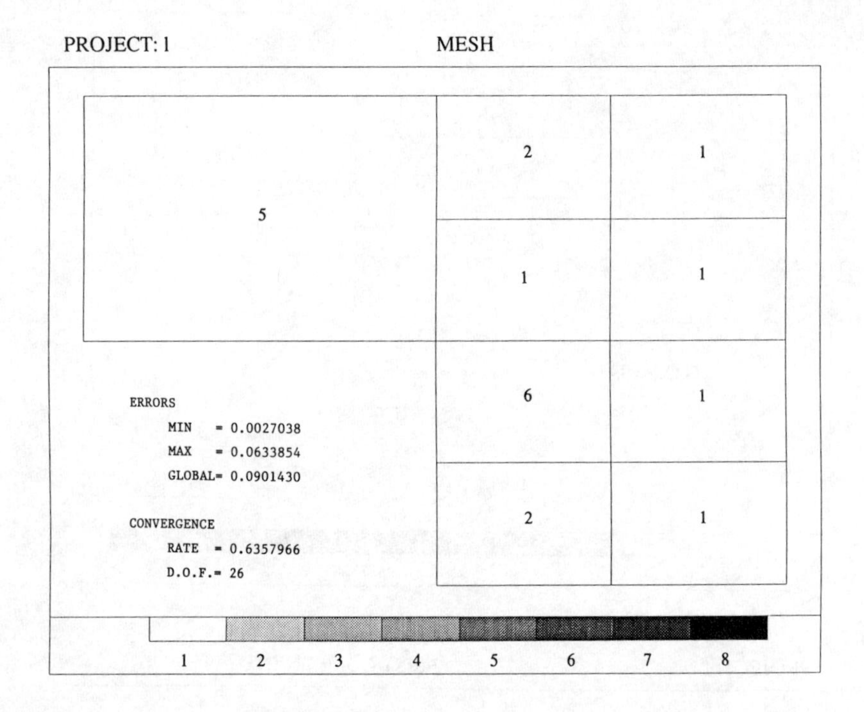

Figure 10.6.6 Automatic p-adaptivity for 96 dof

10.6 References

[1] Attenbach, J. and Scholz, E., "Local Mesh Refinements and Macro Transition Elements in Stress Analysis," pp. 123–135 in *The Mathematics of Finite Elements and Applications VI*, ed. J.R. Whiteman, London: Academic Press (1988).

[2] Aziz, A.K., *The Mathematical Foundations of the Finite Element Method with Applications to Partial Differential Equations*, New York: Academic Press (1972).

[3] Baehmann, P.L., Shephard, M.S., and Flaherty, J.E., "Adaptive Analysis for Automated Finite Element Modeling," pp. 521–532 in *The Mathematics of Finite Elements and Applications, VI*, ed. J.R. Whiteman, London: Academic Press (1988).

[4] Bank, R.E. and Weiser, A., "Some *A Posteriori* Error Estimates for Elliptic Partial Differential Equations," *Math. Comp.*, **44**, pp. 283–301 (1985).

[5] Bass, J., *PHLEX User's Manual*, Austin, TX: Computational Mechanics Co. Inc. (1990).

[6] Cook, R.D., *Concepts and Applications of Finite Element Analysis*, New York: John Wiley (1974).

[7] El-Zafrany, A. and Cookson, R.A., "Derivation of Lagrangian and Hermite Shape Functions for Quadrilateral Elements," *Int. J. Num. Meth. Eng.*, **23**, pp. 1939–1958 (1986).

[8] Oden, J.T. and Reddy, J.N., *An Introduction to the Mathematical Theory of Finite Elements*, New York: John Wiley (1976).

[9] Oden, J.T. and Demkowicz, L., "Adaptive Finite Element Methods for Complex Problems in Solid and Fluid Mechanics," pp. 1, 3–13 in *Finite Elements in Computational Mechanics*, ed. T. Kant, Oxford: Pergamon Press (1985).

[10] Oden, J.T., Demkowicz, L., Rochowicz, W., Westermann, T.A., and Hardy, O., "Toward a Universal *h-p* Adaptive Finite Element Strategy – Parts I, II, III," *Comp. Meth. Appl. Mech. Eng.*, **77**, pp. 79–212 (1989).

[11] Oden, J.T., "The Best FEM," *Finite Elements in Analysis and Design*, **7**, pp. 103–114 (1990).

[12] Shephard, M.S., "Approaches to the Automatic Generation and Control of Finite Element Meshes," *Appl. Mech. Rev.*, **41**(4), pp. 169–190 (Apr. 1988).

[13] Szabo, B. and Babuska, I., *Finite Element Analysis*, New York: John Wiley (1991).

[14] Zienkiewicz, O.C., *The Finite Element Method,* 3rd edition, New York: McGraw-Hill (1979).

[15] Zienkiewicz, O.C. and Morgan, K., *Finite Elements and Approximation*, Chichester: John Wiley (1983).

[16] Zienkiewicz, O.C., Kelley, D.W., Gago, J., and Babuska, I., "Hierarchal Finite Element Approaches Error Estimates and Adaptive Refinement," pp. 313–346 in *The Mathematics of Finite Elements and Applications, VI*, ed. J.R. Whiteman, London: Academic Press (1988).

Chapter 11

INTEGRATION METHODS

11.1 Introduction

Recall that the finite element analysis techniques are always based on an integral formulation. At the very minimum it will always be necessary to integrate at least an element square matrix. This means that every coefficient function in the matrix must be integrated. In the following sections various methods will be considered for evaluating the typical integrals that arise. Most simple finite element matrices for two-dimensional problems are based on the use of linear triangular or quadrilateral elements. Since a quadrilateral can be divided into two or more triangles, only exact integrals over arbitrary triangles will be considered here. Integrals over triangular elements commonly involve integrands of the form

$$I = \int_A x^m y^n \, dx\, dy \tag{11.1}$$

where A is the area of a typical triangle. When $0 \le (m+n) \le 2$, the above integral can easily be expressed in closed form in terms of the spatial coordinates of the three corner points. For a right-handed coordinate system, the corners must be numbered in counter-clockwise order. In this case, the above integrals are given in Table 11.1. These integrals should be recognized as the area, and first and second moments of the area. If one had a volume of revolution that had a triangular cross-section in the ρ–z plane, then one should recall that

$$I = \int_V \rho f(\rho, z)\, d\rho\, dz\, d\phi = 2\pi \int_A \rho f(\rho, z)\, d\rho\, dz$$

so that similar expressions could be used to evaluate the volume integrals. The above closed form integrals are included in subroutine TRGEOM. One enters this routine with the values of the integers m and n, the corner coordinates, and returns with the desired area (or volume) integral. Similar operations for quadrilaterals could be performed by splitting the quadrilateral into two triangles and making two calls to such a subroutine to evaluate the integral.

Table 11.1 Exact integrals for a triangle

m	n	I
0	0	$\int dA = A\left[x_1(y_2 - y_3) + x_2(y_3 - y_1) + x_3(y_1 - y_2)\right]/2$
0	1	$\int y\,dA = A\bar{y} = A(y_1 + y_2 + y_3)/3$
1	0	$\int x\,dA = A\bar{x} = A(x_1 + x_2 + x_3)/3$
0	2	$\int y^2\,dA = A(y_1^2 + y_2^2 + y_3^2 + 9\bar{y}^2)/12$
1	1	$\int xy\,dA = A(x_1 y_1 + x_2 y_2 + x_3 y_3 + 9\bar{x}\bar{y})/12$
2	0	$\int x^2\,dA = A(x_1^2 + x_2^2 + x_3^2 + 9\bar{x}^2)/12$

11.2 Unit Coordinate Integration

The utilization of global coordinate interpolation is becoming increasingly rare. However, as we have seen, the use of non-dimensional local coordinates is common. Thus we often see local coordinate polynomials integrated over the physical domain of an element. Sect. 5.3 presented some typical unit coordinate integrals in 1-D, written in exact closed form. These concepts can be extended to two- and three-dimensional elements. For example, consider an integration over a triangular element. It is known that for an element with a constant Jacobian

$$I = \int_A r^m s^n\,da = \frac{2A\,\Gamma(m+1)\,\Gamma(n+1)}{\Gamma(3+m+n)} \tag{11.2}$$

where Γ denote the Gamma function. Restricting consideration to positive integer values of the exponents, m and n, yields

$$I = 2A^e \frac{m!\,n!}{(2+m+n)!} = \frac{A^e}{K_{mn}}, \tag{11.3}$$

where ! denotes the factorial and K_{mn} is an integer constant given in Table 11.2 for common values of m and n. Similarly for the tetrahedron element

$$I^e = \int_{V^e} r^m s^n t^p\,dv = 6V^e \frac{m!\,n!\,p!}{(3+m+n+p)!}. \tag{11.4}$$

Thus, one notes that common integrals of this type can be evaluated by simply multiplying the element characteristic (i.e., global length, area, or volume) by known constants which could be stored in a data statement.

To illustrate the application of these equations in evaluating element matrices, we consider the following example for the three node triangle in unit coordinates:

$$I = \int_{A^e} \mathbf{H}^T da = \int_{A^e} \begin{Bmatrix} (1-r-s) \\ r \\ s \end{Bmatrix} da = \begin{Bmatrix} A^e - A^e/3 - A^e/3 \\ A^e/3 \\ A^e/3 \end{Bmatrix} = \frac{A^e}{3} \begin{Bmatrix} 1 \\ 1 \\ 1 \end{Bmatrix}.$$

$$\mathbf{I}_V = 2\pi \int_{A^e} \mathbf{H}^T \rho \, da = 2\pi \left[\int_{A^e} \mathbf{H}^T \mathbf{H} \, da \right] \rho^e = \frac{2\pi A^e}{12} \begin{bmatrix} 2 & 1 & 1 \\ 1 & 2 & 1 \\ 1 & 1 & 2 \end{bmatrix} \rho^e .$$

11.3 Simplex Coordinate Integration

A simplex region is one where the minimum number of vertices is one more than the dimension of the space. These were illustrated in Fig. 4.2.2. Some analysts like to define a set of *simplex coordinates* or *baracentric coordinates*. If there are N vertices then N non-dimensional coordinates, L_i, $1 \le i \le N$, are defined and constrained so that

$$1 = \sum_{i=1}^{N} L_i$$

at any point in space. Thus, they are not independent. However, they can be used to simplify certain recursion relations. In physical spaces these coordinates are sometimes called *line coordinates, area coordinates*, and *volume coordinates*. At a given point in the region we can define the simplex coordinate for node j, L_j, in a generalized manner. It is the ratio of the generalized volume from the point to all other vertices (other than j) and the total generalized volume of the simplex. This is illustrated in Fig. 11.3.1. If the simplex has a constant Jacobian (e.g., straight sides and flat faces), then the exact form of the integrals of the simplex coordinates are simple. They are

$$\begin{aligned} \int_L L_1^a L_2^b \, dL &= \frac{a!b!}{(a+b+1)!} (L) \\ \int_A L_1^a L_2^b L_3^c \, da &= \frac{a!b!c!}{(a+b+c+2)!} (2A) \\ \int_V L_1^a L_2^b L_3^c L_4^d \, dv &= \frac{a!b!c!d!}{(a+b+c+d+3)!} (6V) . \end{aligned} \tag{11.5}$$

The evaluation of partial derivatives in baracentric coordinates is not obvious since one coordinate is always dependent on the others. The independent coordinates are those we have generally referred to as the unit coordinates of an element. Since a lot of references make use of baracentric coordinates it is useful to learn how to manipulate them correctly. The baracentric coordinates, say L_j, essentially measure the percent of total volume contained in the region from the face (lower dimensional simplex) opposite to node j to any point in the simplex. Therefore, $L_j \equiv 0$ when the point lies on the opposite face and $L_j \equiv 1$ when the point is located at node j. Clearly, the sum of all these volumes is the total volume of the simplex.

We have referred to the independent coordinates in the set as the unit coordinates. For simplex elements, the use of baracentric coordinates simplifies the algebra needed to define the interpolation functions; however, it complicates the calculation of their derivatives. Baracentric coordinates are often used to tabulate numerical integration rules for simplex domains.

Table 11.2 Denominator for unit triangle $I = \int_A r^m s^n \, da = A/K$

M	N: 0	1	2	3	4	5	6	7	8
0	1	3	6	10	15	21	28	36	45
1	3	12	30	60	105	168	252	360	495
2	6	30	90	210	420	756	1260	1980	2970
3	10	60	210	560	1260	2520	4620	7920	12870
4	15	105	420	1260	3150	6930	13860	25740	45045
5	21	168	756	2520	6930	16632	36036	72072	135135
6	28	252	1260	4620	13860	36036	84084	180180	360360
7	36	360	1980	7920	25740	72072	180180	411840	875160
8	45	495	2970	12870	45045	135135	360360	875160	1969110

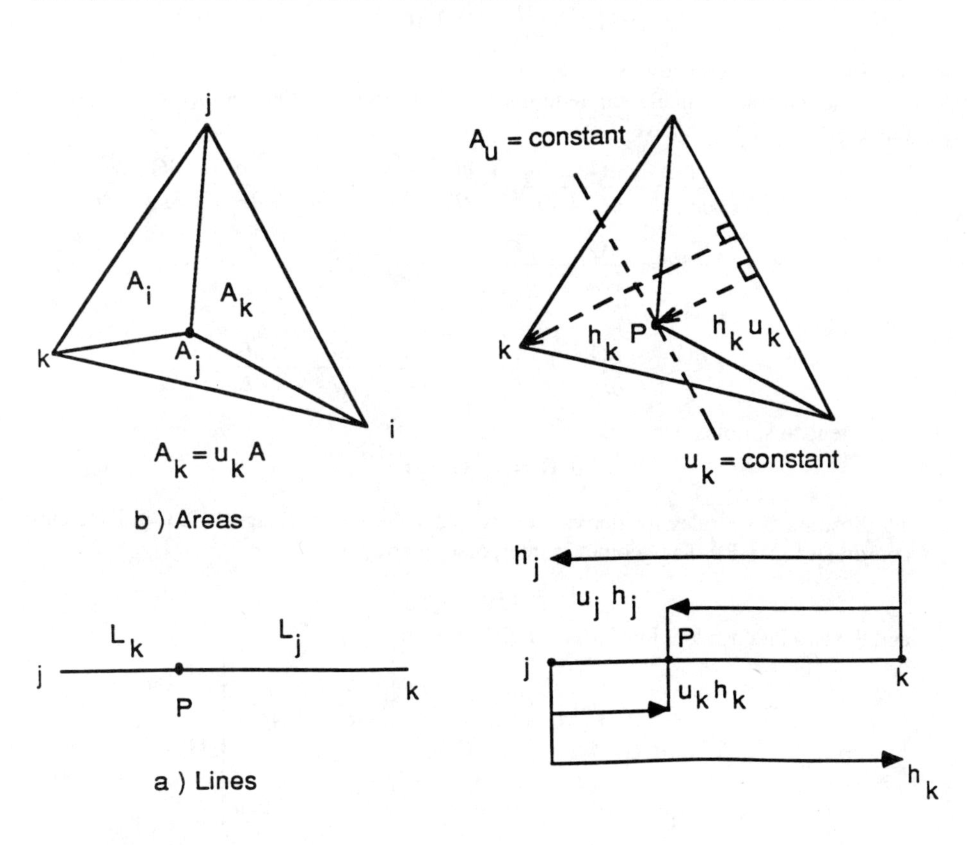

Figure 11.3.1 Area coordinates

For example, consider the three-dimensional case where $L_1 = r$, $L_2 = s$, $L_3 = t$, and $L_1 + L_2 + L_3 + L_4 = 1$. The interpolation functions for the linear tetrahedron (P4) are simply $G_j = L_j$. The expressions for the Lagrangian quadratic tetrahedron (P10) vertices are

$$G_1 = L_1(2L_1 - 1) \qquad G_2 = L_2(2L_2 - 1)$$
$$G_3 = L_3(2L_3 - 1) \qquad G_4 = L_4(2L_4 - 1)$$

and the six mid-edge values are

$$G_5 = 4L_1L_2 \qquad G_6 = 4L_1L_3$$
$$G_7 = 4L_1L_4 \qquad G_8 = 4L_2L_3$$
$$G_9 = 4L_3L_4 \qquad G_{10} = 4L_2L_4\,.$$

All the tetrahedra have the condition that

$$L_4 = 1 - L_1 - L_2 - L_3 = 1 - r - s - t$$

so that we can write the unit coordinate partial derivatives as

$$\frac{\partial L_j}{\partial r} = 1, 0, 0, -1, \qquad \frac{\partial L_j}{\partial s} = 0, 1, 0, -1, \qquad \frac{\partial L_j}{\partial t} = 0, 0, 1, -1$$

for $j = 1, 2, 3, 4$, respectively.

The Jacobian calculation requires the derivatives of the geometric interpolation functions, $\mathbf{G}$. Here we have

$$\frac{\partial \mathbf{G}}{\partial r} = \frac{\partial \mathbf{G}}{\partial L_1}\frac{\partial L_1}{\partial r} + \frac{\partial \mathbf{G}}{\partial L_2}\frac{\partial L_2}{\partial r} + \frac{\partial \mathbf{G}}{\partial L_3}\frac{\partial L_3}{\partial r} + \frac{\partial \mathbf{G}}{\partial L_4}\frac{\partial L_4}{\partial r}$$
$$= \frac{\partial \mathbf{G}}{\partial L_1} - \frac{\partial \mathbf{G}}{\partial L_4}\,.$$

Likewise,

$$\frac{\partial \mathbf{G}}{\partial s} = \frac{\partial \mathbf{G}}{\partial L_2} - \frac{\partial \mathbf{G}}{\partial L_4}, \qquad \frac{\partial \mathbf{G}}{\partial t} = \frac{\partial \mathbf{G}}{\partial L_3} - \frac{\partial \mathbf{G}}{\partial L_4}\,.$$

For a general simplex, we have

$$\partial_l \mathbf{G} = \partial_L \mathbf{G} - \mathbf{I}\frac{\partial \mathbf{G}}{\partial L}\,.$$

To illustrate these rules for derivatives, consider the linear triangle (T3) in baracentric coordinates ($N = 3$). The geometric interpolation array is

$$\mathbf{G} = [\,L_3 \quad L_1 \quad L_2\,]$$

and the two independent local space derivatives are

$$\boldsymbol{\Delta} = \partial_l \mathbf{G} = \begin{Bmatrix} \dfrac{\partial}{\partial r} \\ \dfrac{\partial}{\partial s} \end{Bmatrix} \mathbf{G} = \begin{Bmatrix} \dfrac{\partial}{\partial L_1} - \dfrac{\partial}{\partial L_3} \\ \dfrac{\partial}{\partial L_2} - \dfrac{\partial}{\partial L_3} \end{Bmatrix} \mathbf{G}$$

$$\Delta = \begin{bmatrix} (0-1) & (1-0) & (0-0) \\ (0-1) & (0-0) & (1-0) \end{bmatrix} = \begin{bmatrix} -1 & 1 & 0 \\ -1 & 0 & 1 \end{bmatrix},$$

which is the same as the previous result in Sect. 9.2 .

If one is willing to restrict the elements to having a constant Jacobian (straight edges and flat faces), then the inverse global to baracentric mapping is simple to develop. Then the global derivatives that we desire are easy to write

$$\frac{\partial}{\partial x} = \sum_{j=1}^{n+1} \frac{\partial}{\partial L_j} \frac{\partial L_j}{\partial x},$$

where $\partial L_j / \partial x$ is a known value, say V_j. For example, in 1-D we have

$$\begin{Bmatrix} L_1 \\ L_2 \end{Bmatrix} = \frac{1}{L^e} \begin{bmatrix} x_2 & -1 \\ -x_1 & 1 \end{bmatrix}^e \begin{Bmatrix} 1 \\ x \end{Bmatrix},$$

and in 2-D

$$\begin{Bmatrix} L_1 \\ L_2 \\ L_3 \end{Bmatrix} = \frac{1}{2A^e} \begin{bmatrix} 2A_{23} & (y_2 - y_3) & (x_3 - x_2) \\ 2A_{13} & (y_3 - y_1) & (x_1 - x_3) \\ 2A_{12} & (y_1 - y_2) & (x_2 - x_1) \end{bmatrix} \begin{Bmatrix} 1 \\ x \\ y \end{Bmatrix}$$

where A_{ij} is the triangular area enclosed by the origin $(0, 0)$ and nodes i and j.

11.4 Numerical Integration

In many cases it is impossible or impractical to integrate the expression in closed form and numerical integration must therefore be utilized. If one is using sophisticated elements, it is almost always necessary to use numerical integration. Similarly, if the application is complicated, e.g., the solution of a non-linear ordinary differential equation, then even simple one-dimensional elements can require numerical integration. Many analysts have found that the use of numerical integration simplifies the programming of the element matrices. This results from the fact that lengthy algebraic expressions are avoided and thus the chance of algebraic and/or programming errors is reduced. There are many numerical integration methods available. Only those methods commonly used in finite element applications will be considered here.

11.4.1 Unit Coordinate Quadrature

Numerical quadrature in one-dimension was introduced in Sec. 5.4. There we saw that an integral is replaced with a summation of functions evaluated at tabulated points and then multiplied by tabulated weights. The same procedure applies to all numerical integration rules. The main difficulty is to obtain the tabulated data. For triangular unit coordinate regions the weights, W_i, and abscissae (r_i, s_i) are less well known. Typical points for rules on the unit triangle are shown in Fig. 11.4.1. It presents rules that yield points that are symmetric with respect to all corners of the triangle. These low order data are placed in subroutine SYMRUL.

As before, one approximates an integral of $f(x, y) = F(r, s)$ over a triangle by

$$I = \int f(x, y)\, dx\, dy = \sum_{i=1}^{n} W_i F(r_i, s_i) \, |J_i| .$$

As a simple example of integration over a triangle, let $f = y$ and consider the integral over a triangle with its three vertices at (0, 0), (3, 0), and (0, 6), respectively, in (x, y) coordinates. Then the area $A = 9$ and the Jacobian is a constant $|J| = 18$. For a three point quadrature rule the integral is thus given by

$$I = \sum_{i=1}^{3} W_i y_i \, |J_i| .$$

Since our interpolation defines $y(r, s) = y_1 + (y_2 - y_1)r + (y_3 - y_1)s = 0 + 0 + 6s$, the transformed integrand is $F(r, s) = 6s$. Thus, at integration point, i, $F(r_i, s_i) = 6s_i$. Substituting a three-point quadrature rule and factoring out the constant Jacobian gives

$$I = 18 \left[(\, 6(1/6)\,)\, (1/6) + (\, 6(1/6)(1/6) + (6(2/3)\,)\, (1/6) \right] = 18$$

which is the exact solution.

Table 11.3 gives a tabulation of symmetric quadrature rules over the unit triangle. Decimal versions are given in subroutine SYMRUL of values of n up to 13. A similar set of rules for extension to the three-dimensional tetrahedra in unit coordinates are given in Tables 11.4a and 11.4b for polynomials up to degree five [5].

Quadrature rules for high degree polynomials on triangles have been published by Dunavant [4]. They are suitable for use with hierarchial elements. Those rules are given in Table 11.5 in area coordinates, since that form requires the smallest table size. Most of the lines are used multiple times by cycling through the area coordinates. The number N in the table indicates if the line is for the centroid, three symmetric points, or six symmetric locations. These data are expanded to their full form (up to 61 points for a polynomial of degree 17) in subroutine DQ. The corresponding unit triangle coordinate data are also given in subroutine DQRULE.

11.4.2 Natural Coordinate Quadrature

Here we assume that the coordinates are in the range of −1 to +1. In this space it is common to employ Gaussian quadratures. The one-dimensional rules were discussed in Sect. 5.4. For a higher number of space dimensions one obtains a multiple summation (tensor product) for evaluating the integral. For example, a typical integration in two dimensions

$$I = \int_{-1}^{1} \int_{-1}^{1} f(r, s)\, dr\, ds \approx \sum_{j=1}^{n} \sum_{k=1}^{n} f(r_j, s_k)\, W_j W_k$$

for n integration points in each dimension. This can be written as a single summation as

$$I \approx \sum_{i=1}^{m} f(r_i, s_i)\, W_i$$

where $m = n^2$, $i = j + (k - 1)n$, and where $r_i = \alpha_j$, $s_i = \alpha_k$, and $W_i = W_j W_k$. Here α_j and W_j denote the tabulated one-dimensional abscissae and weights given in Sect. 5.4. A similar rule can be given for a three-dimensional region. The result of the above summation is given in Table 11.6. The extension of the 1-D data to the quadrilateral and hexahedra are done by subroutines GAUS2D and GAUS3D (see Fig. 11.4.2).

Table 11.3. Symmetric quadrature for the unit triangle:

$$\int_0^1 \int_0^{1-r} f(r,s)\,dr\,ds = \sum_{i=1}^{n} f(r_i, s_i)\,W_i$$

n	p†	i	r_i	s_i	W_i
1	1	1	1/3	1/3	1/2
3	2	1	1/6	1/6	1/6
		2	2/3	1/6	1/6
		3	1/6	2/3	1/6
4	3	1	1/3	1/3	–9/32
		2	3/5	1/5	25/96
		3	1/5	3/5	25/96
		4	1/5	1/5	25/96
7	4	1	0	0	1/40
		2	1/2	0	1/15
		3	1	0	1/40
		4	1/2	1/2	1/15
		5	0	1	1/40
		6	0	1/2	1/15
		7	1/3	1/3	9/40

† P = Degree of Polynomial for exact integration.

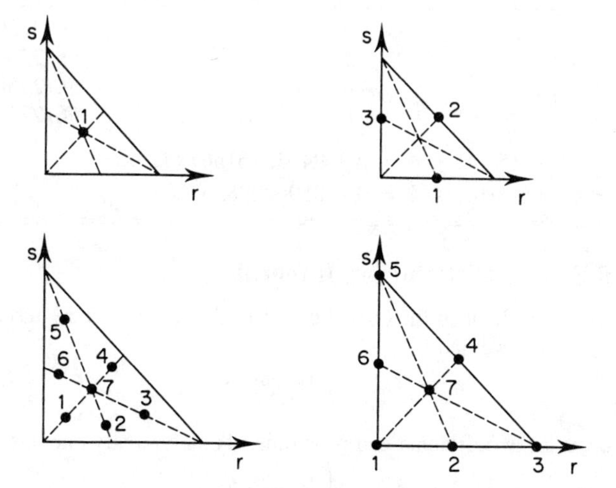

Figure 11.4.1 Symmetric quadrature locations for unit triangle

Table 11.4a. Quadrature for unit tetrahedra

Number of points N	Degree of precision	Unit coordinates r_i	s_i	t_i	Weights W_i
1	1	1/4	1/4	1/4	1/6
4	2	a	b	b	1/24
		b	a	b	1/24
		b	b	a	1/24
		b	b	b	1/24
$a = (5+3\sqrt{5})/20 = 0.5854101966249685$					
$b = (5-\sqrt{5})/20 = 0.1381966011250105$					
5	3	1/4	1/4	1/4	– 4/30
		1/2	1/6	1/6	9/120
		1/6	1/2	1/6	9/120
		1/6	1/6	1/2	9/120
		1/6	1/6	1/6	9/120
11	4	1/4	1/4	1/4	– 74/5625
		11/14	1/14	1/14	343/45000
		1/14	11/14	1/14	343/45000
		1/14	1/14	11/14	343/45000
		1/14	1/14	1/14	343/45000
		a	a	b	56/2250
		a	b	a	56/2250
		a	b	b	56/2250
		b	a	a	56/2250
		b	a	b	56/2250
		b	b	a	56/2250
$a = (1+\sqrt{(5/14)})/4 = 0.3994035761667992$					
$b = (1-\sqrt{(5/14)})/4 = 0.1005964238332008$					

11.5 Typical Source Distribution Integrals *

Previously we introduced the contributions of distributed source terms. For the C° continuity line elements we had

$$\mathbf{C}_Q^e = \int_{L^e} \mathbf{H}^{e^T} Q^e \, dx \, .$$

Similar forms occur in two-dimensional problems. Then typically one has

$$\mathbf{C}_Q^e = \int_{A^e} \mathbf{H}^{e^T} Q^e \, da \, .$$

If the typical source or forcing term, Q^e, varies with position we usually use the interpolation functions to define it in terms of the nodal values, Q^e, as

Table 11.4b. Quadrature for unit tetrahedra

Number of points N	Degree of precision	Unit coordinates r_i	s_i	t_i	Weights W_i
14	5	a	b	b	p
		b	a	b	p
		b	b	a	p
		b	b	b	p
		c	d	d	q
		d	c	d	q
		d	d	c	q
		d	d	d	q
		e	e	f	r
		e	f	e	r
		e	f	f	r
		f	e	e	r
		f	e	f	r
		f	f	e	r

$$g = 1/(46\sqrt{46}\,), \quad h = \mathrm{Cos}^{-1}(g) + (2/3)\,\mathrm{Sin}^{-1}(g)$$
$$k = (104 + 8\sqrt{46}\ \mathrm{Cos}\,(h)\,)/3, \quad s = \sqrt{(49-k)}$$
$$b = (7+s)/k = 0.3108859192633005$$
$$a = 1 - 3b = 0.0673422422100983$$
$$d = (7-s)/k = 0.0927352503108912$$
$$c = 1 - 3d = 0.7217942490673264$$
$$6p = (98 - k - 14s)/(1680s\,(b-a)^3\,) = 0.1126879257180162$$
$$6q = (98 - k + 14s)/(1680s\,(c-d)^3\,) = 0.0734930431163619$$
$$6r = (\,1 - 4\,(p+q)\,)/6 = 0.0425460207770812$$
$$e = (\,l + \sqrt[4]{(2/105r)}\,)/4 = 0.4544962958743506$$
$$f = (1-2e)/2 = 0.0455037041256494$$

$$Q^e = \mathbf{H}^{e^T}\mathbf{Q}^e. \tag{11.6}$$

Thus, a common element integral for the consistent nodal sources is

$$\mathbf{C}_Q^e = \int_{\Omega^e} \mathbf{H}^{e^T}\,\mathbf{H}^e\,d\Omega\,\mathbf{Q}^e. \tag{11.7}$$

The previous sections present analytic and numerical methods for evaluating these integrals. Figure 11.5.1 shows the typical analytic results for the two and three node line integrals. For linear or constant source distributions the normalized nodal resultants are summarized in Fig. 11.5.2. Once one goes beyond the linear (two-node) element the consistent results usually differ from physical intuition estimates. Thus, you must rely on the mathematics or the summaries in the above figures. Many programs will numerically integrate the source distributions for any element shape. If the source acts on an area

Table 11.5. Dunavant quadrature for area coordinate triangle

P	N	Wt	L_1	L_2	L_3
1	1	1.000000000000000	0.333333333333333	0.333333333333333	0.333333333333333
2	3	0.333333333333333	0.666666666666667	0.166666666666667	0.166666666666667
3	1	–0.562500000000000	0.333333333333333	0.333333333333333	0.333333333333333
	3	0.520833333333333	0.600000000000000	0.200000000000000	0.200000000000000
4	3	0.223381589678011	0.108103018168070	0.445948490915965	0.445948490915965
	3	0.109951743655322	0.816847572980459	0.091576213509771	0.091576213509771
5	1	0.225000000000000	0.333333333333333	0.333333333333333	0.333333333333333
	3	0.132394152788506	0.059715871789770	0.470142064105115	0.470142064105115
	3	0.125939180544827	0.797426985353087	0.101286507323456	0.101286507323456
6	3	0.116786275726379	0.501426509658179	0.249286745170910	0.249286745170910
	3	0.050844906370207	0.873821971016996	0.063089014491502	0.063089014491502
	6	0.082851075618374	0.053145049844817	0.310352451033784	0.636502499121399
7	1	–0.149570044467682	0.333333333333333	0.333333333333333	0.333333333333333
	3	0.175615257433208	0.479308067841920	0.260345966079040	0.260345966079040
	3	0.053347235608838	0.869739794195568	0.065130102902216	0.065130102902216
	6	0.077113760890257	0.048690315425316	0.312865496004874	0.638444188569810
8	1	0.144315607677787	0.333333333333333	0.333333333333333	0.333333333333333
	3	0.095091634267285	0.081414823414554	0.459292588292723	0.459292588292723
	3	0.103217370534718	0.658861384496480	0.170569307751760	0.170569307751760
	3	0.032458497623198	0.898905543365938	0.050547228317031	0.050547228317031
	3	0.027230314174435	0.008394777409958	0.263112829634638	0.728492392955404
9	1	0.097135796282799	0.333333333333333	0.333333333333333	0.333333333333333
	3	0.031334700227139	0.020634961602525	0.489682519198738	0.489682519198738
	3	0.077827541004774	0.125820817014127	0.437089591492937	0.437089591492937
	3	0.079647738927210	0.623592928761935	0.188203535619033	0.188203535619033
	3	0.025577675658698	0.910540973211095	0.044729513394453	0.044729513394453
	6	0.043283539377289	0.036838412054736	0.221962989160766	0.741198598784498
10	1	0.090817990382754	0.333333333333333	0.333333333333333	0.333333333333333
	3	0.036725957756467	0.028844733232685	0.485577633383657	0.485577633383657
	3	0.045321059435528	0.781036849029926	0.109481575485037	0.109481575485037
	6	0.072757916845420	0.141707219414880	0.307939838764121	0.550352941820999
	6	0.028327242531057	0.025003534762686	0.246672560639903	0.728323904597411
	6	0.009421666963733	0.009540815400299	0.066803251012200	0.923655933587500

P = Degree of complete polynomial exactly integrated
N = Number of cyclic uses
Wt = Weight at point
L_j = Area coordinates at the point
(See subroutine DQRULE for $P \leq 17$)

shaped like the parent element (constant Jacobian) then we can again easily evaluate the integrals analytically. For a uniform source over an area the consistent nodal contributions for quadrilaterals and triangles are shown in Figs. 11.5.3 and 11.5.4, respectively. Note that the Serendipity families can actually develop negative contributions. Triangular and Lagrangian elements do not have that behavior for uniform sources.

If the source varies linearly from zero then a more complicated set of nodal resultants are obtained as shown in Figs. 11.5.5 and 11.5.6 for triangles and rectangles, respectively. Of course, a general linear loading can be formed by combining the resultants for the constant and triangular sources.

11.6 Minimal, Optimal, Reduced and Selected Integration *

Since the numerical integration of the element square matrix can represent a large part of the total cost it is desirable to use low order integration rules. Care must be taken when selecting the *minimal order* of integration. Usually the integrand will contain global derivatives so that in the limit, as the element size h approaches zero, the integrand can be assumed to be constant, and then only the integral $I = \int dv = \int |J|\, dr\, ds\, dt$ remains to be integrated exactly. Such a rule could be considered the minimal order. However, the order is often too low to be practical since it may lead to a rank deficient element (and system) square matrix, if the rule does not exactly integrate the equations. Typical integrands involve terms such as the strain energy density per unit volume: $\mathbf{B}^T\mathbf{D}\mathbf{B}/2$.

Table 11.6. Gaussian quadrature on a quadrilateral

$$\int_{-1}^{1}\int_{-1}^{1} f(r,s)\, dr\, ds = \sum_{i=1}^{n} f(r_i, s_i)\, W_i$$

n	i	r_i	s_i	W_i
1	1	0	0	4
4	1	$-1/\sqrt{3}$	$-1/\sqrt{3}$	1
	2	$+1/\sqrt{3}$	$-1/\sqrt{3}$	1
	3	$-1/\sqrt{3}$	$+1/\sqrt{3}$	1
	4	$+1/\sqrt{3}$	$+1/\sqrt{3}$	1
9	1	$-\sqrt{3/5}$	$-\sqrt{3/5}$	25/81
	2	0	$-\sqrt{3/5}$	40/81
	3	$+\sqrt{3/5}$	$-\sqrt{3/5}$	25/81
	4	$-\sqrt{3/5}$	0	40/81
	5	0	0	64/81
	6	$+\sqrt{3/5}$	0	40/81
	7	$-\sqrt{3/5}$	$+\sqrt{3/5}$	25/81
	8	0	$+\sqrt{3/5}$	40/81
	9	$+\sqrt{3/5}$	$+\sqrt{3/5}$	25/81

```
      SUBROUTINE  GAUS3D (NQP, GPT, GWT, NIP, PT, WT)
C     * * * * * * * * * * * * * * * * * * * * * * * * *
C         USE 1-D GAUSSIAN DATA TO GENERATE
C           QUADRATURE DATA FOR A CUBE
C     * * * * * * * * * * * * * * * * * * * * * * * * *
CDP   IMPLICIT REAL*8 (A-H,O-Z)
      DIMENSION GPT(0:NQP), GWT(0:NQP), PT(3,0:NQP**3),
     1          WT(0:NQP**3)
C     NQP = NUMBER OF TABULATED 1-D POINTS
C     NIP = NQP**3 = NUMBER OF 3-D POINTS
C     GPT = TABULATED 1-D QUADRATURE POINTS
C     GWT = TABULATED 1-D QUADRATURE WEIGHTS
C     PT  = CALCULATED COORDS IN A CUBE
C     WT  = CALCULATED WEIGHTS IN A CUBE
C      GET TABLE DATA
      CALL  GAUSCO (NQP,GPT,GWT)
      NIP = NQP*NQP*NQP
      K = 0
C      LOOP OVER GENERATED POINTS
      DO 30  L = 1,NQP
        DO 20  I = 1,NQP
          DO 10  J = 1,NQP
            K = K + 1
            WT(K)  = GWT(I)*GWT(J)*GWT(L)
            PT(1,K)  = GPT(J)
            PT(2,K)  = GPT(I)
   10     PT(3,K)  = GPT(L)
   20   CONTINUE
   30 CONTINUE
      RETURN
      END
```

Figure 11.4.2 Gaussian rules for a cube

Let NQP denote the number of element integration points while NI represents the number of independent relations at each integration point; then the rank of the element is NQP*NI. Generally, NI corresponds to the number of rows in **B** in the usual symbolic integrand $\mathbf{B}^T\mathbf{D}\mathbf{B}$. For a typical element, we want NQP*(NI − NC) ≥ NELFRE, where NC represents the number of element constraints, if any. For a non-singular system matrix a similar expression is NE*(NQP*NI-NC) ≥ NDFREE-NR, where NR denotes the number of nodal parameter restraints (NR ≥ 1). these relations can be used as guides in selecting a minimal value of NQP. Consider a problem involving a governing integral statement with m-th order derivatives. If the interpolation (trial) functions are complete polynomials of order p then to maintain the theoretical convergence rate NQP should be selected [12] to give accuracy of order $0(h^{2(p-m)+1})$. That is, to integrate polynomial terms of order $(2p - m)$ exactly.

It has long been known that a finite element model gives a stiffness which is too high. Using reduced integration so as to underestimate the element stiffness has been accepted as one way to improve the results. These procedures have been investigated by several authors including Zienkiewicz [13], Zienkiewicz and Hinton [12], Hughes, Cohen and Haroun [7] and Malkus and Hughes [10]. Reduced integration has been especially useful in problems with constraints, such as material incompressibility. A danger of low order integration rules is that *zero energy modes* may arise in an element. That is,

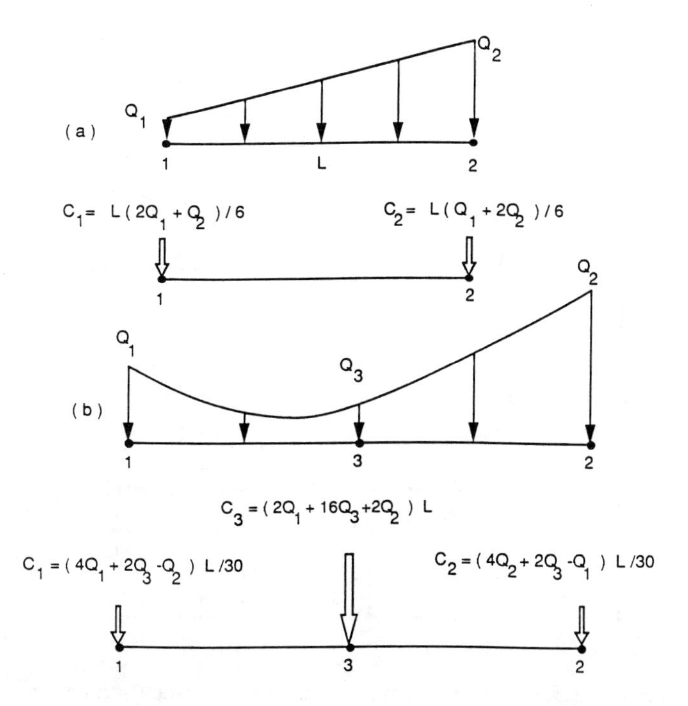

Figure 11.5.1 General consistent line sources

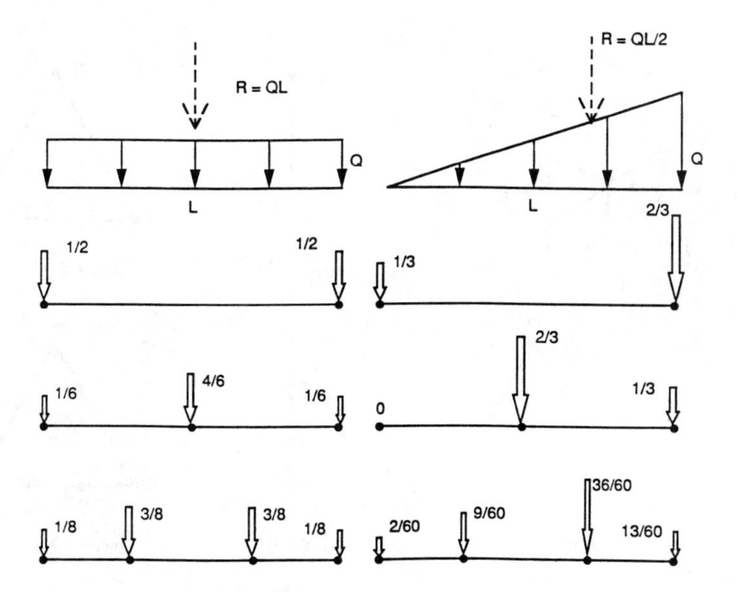

Figure 11.5.2 Consistent resultants for a unit source

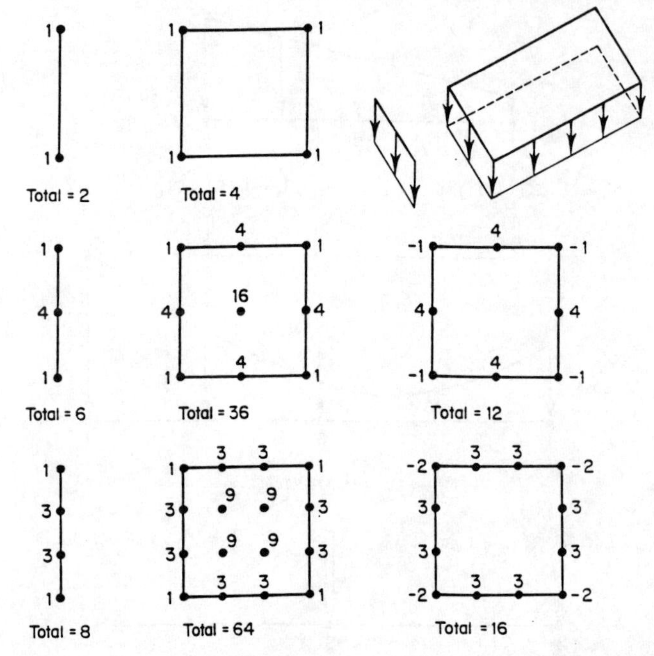

Figure 11.5.3 Quadrilateral resultants for a constant source

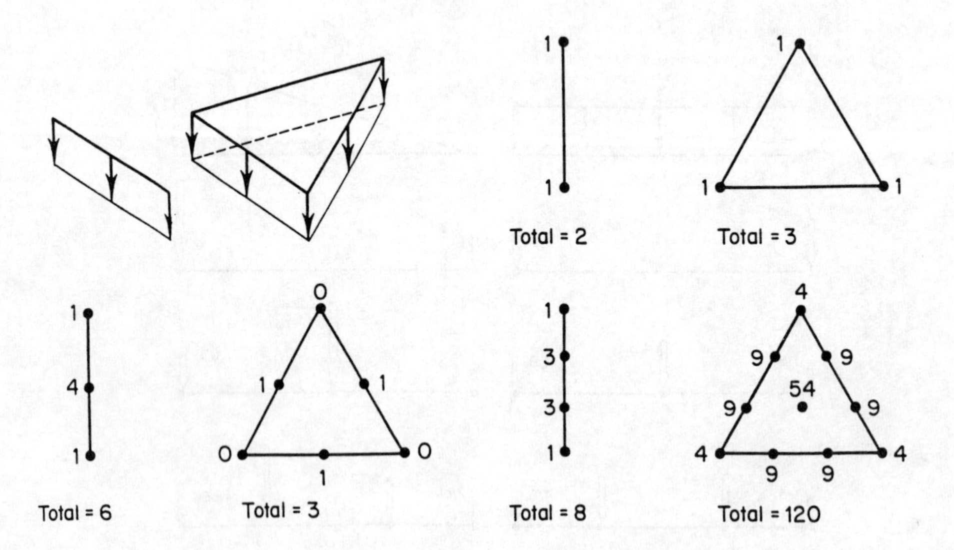

Figure 11.5.4 Nodal resultants for a uniform source on a triangle

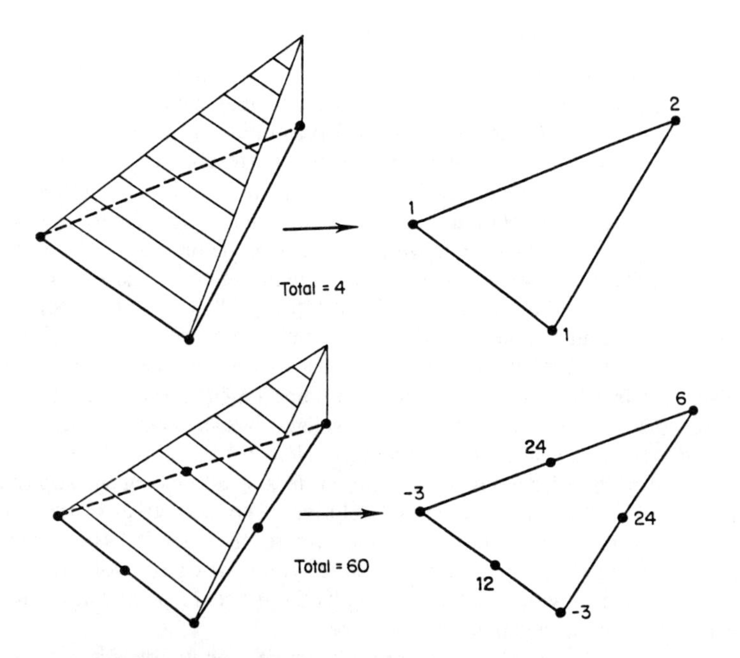

Figure 11.5.5 Nodal resultants for a linear source triangle

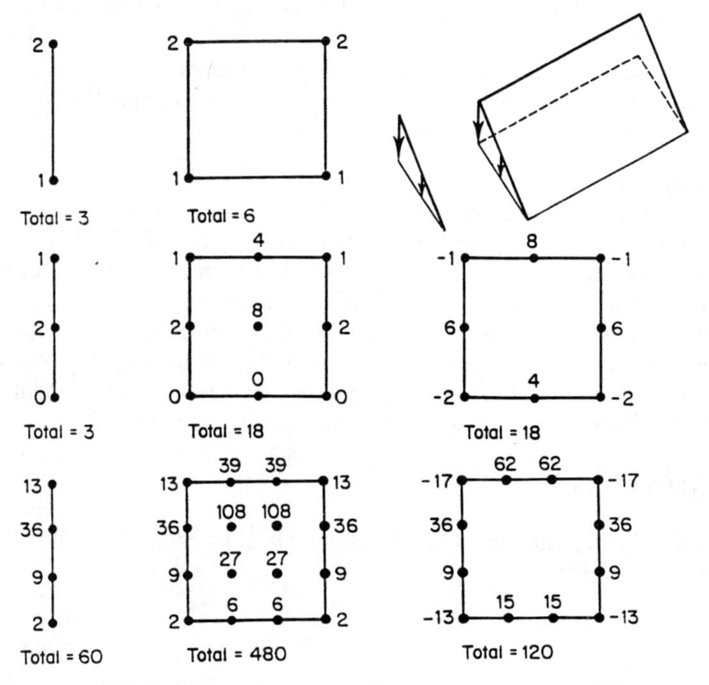

Figure 11.5.6 Nodal resultants for a linear source on a quadrilateral

$$\mathbf{D}^{eT}\,\mathbf{S}^e\,\mathbf{D}^e = 0$$

for $\mathbf{D}^e \neq 0$. Usually these zero energy modes, $\mathbf{D}^e$, are incompatible with the same modes in an adjacent element. Thus, the assembly of elements may have no zero energy modes (except for the standard *rigid* modes). Cook [3] illustrates that an eigen-analysis of the element can be used as a check since zero eigenvalues correspond to zero energy modes.

The integrand usually involves derivatives of the function of interest. Many solutions require the post-solution calculation of these derivatives for auxiliary calculations. Thus a related question is which points give the most accurate estimates for those derivatives. These points are often called *optimal points* or Barlow points. Their locations have been derived by Barlow [1, 2] and Moan [11]. The optimal points usually are the common quadrature points. For low order elements the optimal points usually correspond to the minimal integration points. This is indeed fortunate.

As discussed in Chap. 1, it is possible in some cases to obtain exact derivative estimates from the optimal points. Barlow considered line elements, quadrilaterals and hexahedra while Moan considered the triangular elements. The points were found by assuming that the p-th order polynomial solution, in a small element, is approximately equal to the $(p+1)$ order exact polynomial solution. The derivatives of the two forms were equated and the coordinates of points where the identity is satisfied were determined. For triangles the optimal rules are the symmetric rules involving 1, 4, 7, and 13 points. For machines with small word lengths the 4 and 13 point rules may require higher precision due to the negative centroid weights.

Generally, all interior point quadrature rules can be used to give accurate derivative estimates. The derivatives of the interpolation functions are least accurate at the nodes. Hinton and Campbell [6] have presented relations for extrapolating the optimal values to the nodal points.

For element formulations involving element constraints, or penalties, it is now considered best to employ selective integration rules [12]. For penalty formulations it is common to have equations of the form

$$(\mathbf{S}_1 + \alpha \mathbf{S}_2)\,\mathbf{D} = \mathbf{C}$$

where the constant $\alpha \to \infty$ in the case where the penalty constraint is exactly satisfied. In the limit as $\alpha \to \infty$ the system degenerates to $\mathbf{S}_2\,\mathbf{D} = 0$, where the solution approaches the trivial result, $\mathbf{D} = \mathbf{0}$. To obtain a non-trivial solution in this limit it is necessary for $\mathbf{S}_2$ to be singular. Therefore, the two contributing element parts $\mathbf{S}_1^e$ and $\mathbf{S}_2^e$ are *selectively* integrated. That is, $\mathbf{S}_2^e$ is under integrated so as to be rank deficient (singular) while $\mathbf{S}_1^e$ is integrated with a rule which renders $\mathbf{S}_1$ non-singular. Typical applications of selective integration were cited above and include problems such as plate bending where the bending contributions are in $\mathbf{S}_1^e$ while the shear contributions are in $\mathbf{S}_2^e$.

11.7 References

[1] Barlow, J., "Optimal Stress Locations in Finite Element Models," *Int. J. Num. Meth. Eng.*, **10**, pp. 243–251 (1976).

[2] Barlow, J., "More on Optimal Stress Points — Reduced Integration, Element Distortions and Error Estimation," *Int. J. Num. Meth. Eng.*, **28**, pp. 1487–1504 (1989).

[3] Cook, R.D., *Concepts and Applications of Finite Element Analysis*, New York: John Wiley (1974).

[4] Dunavant, D.A., "High Degree Efficient Symmetrical Gaussian Quadrature Rules for the Triangle," *Int. J. Num. Meth. Eng.*, **21**, pp. 1129–1148 (1985).

[5] Gellert, M. and Harbord, R., "Moderate Degree Cubature Formulas for 3-D Tetrahedral Finite-Element Approximations," *Comm. Appl. Num. Meth.*, **7**, pp. 487–495 (1991).

[6] Hinton, E. and Campbell, J., "Local and Global Smoothing of Discontinuous Finite Element Functions Using a Least Square Method," *Int. J. Num. Meth. Eng.*, **8**, pp. 461–480 (1974).

[7] Hughes, T.J.R., Cohen, M., and Haroun, M., "Reduced and Selective Integration Techniques in the Finite Element Analysis of Plates," *Nuclear Eng. Design*, **46**(1), pp. 203–222 (1978).

[8] Hughes, T.J.R., *The Finite Element Method*, Englewood Cliffs: Prentice-Hall (1987).

[9] Irons, B.M. and Ahmad, S., *Techniques of Finite Elements*, New York: John Wiley (1980).

[10] Malkus, D.S. and Hughes, T.J.R., "Mixed Finite Element Methods – Reduced and Selective Integration Techniques," *Comp. Meth. Appl. Mech. Eng.*, **15**(1), pp. 63–81 (1978).

[11] Moan, T., "Orthogonal Polynomials and Best Numerical Integration Formulas on a Triangle," *Zamm*, **54**, pp. 501–508 (1974).

[12] Zienkiewicz, O.C. and Hinton, E., "Reduced Integration Smoothing and Non-Conformity," *J. Franklin Inst.*, **302**(6), pp. 443–461 (1976).

[13] Zienkiewicz, O.C., *The Finite Element Method,* 3rd edition, New York: McGraw-Hill (1979).

Chapter 12

HEAT TRANSFER

12.1 Introduction

Earlier it was demonstrated that the basic relationships for the finite element formulation of elasticity problems could be obtained simply by minimizing the total potential energy of the system without any direct reference to the usual static equilibrium equations. In many areas of engineering and physics it is possible to obtain "exact" solutions to problems by minimizing some functional, subjected to certain boundary conditions. In the case of elasticity this functional was physically interpreted as the total potential energy of the system. For many areas of engineering and physics the functional may simply be a mathematically defined quantity. In other fields the physical interpretation may not be obvious. For example, in ideal fluid flow the functional may represent the rate of entropy production.

The physical behavior governing a variety of problems in engineering can be described by the well known Laplace and Poisson differential equations. The analytic solution of these equations in two- and three-dimensional field problems can present a formidable task, especially in the case where there are complex boundary conditions and irregularly shaped regions. The finite element formulation of this class of problems by using variational methods has proven to be a very effective and versatile approach to the solution. Previous difficulties associated with irregular geometry and complex boundary conditions are virtually eliminated. Some examples of problems frequently encountered in engineering practice falling into this category are: heat conduction, seepage through porous media, torsion of prismatic shafts, irrotational flow of ideal fluids, distribution of magnetic potential, etc. The following development will be concerned with the details of formulating the solution to the two-dimensional, steady-state heat conduction problem. The approach is general, however, and by redefining the physical quantities involved the formulation is equally applicable to other problems involving the Poisson equation.

12.2 Variational Formulation

We can obtain from any book on heat transfer the governing differential equation for steady and un-steady state heat conduction. The most general form of the heat conduction equation is the transient three-dimensional equation:

$$\frac{\partial}{\partial x}(k_x\frac{\partial T}{\partial x}) + \frac{\partial}{\partial y}(k_y\frac{\partial T}{\partial y}) + \frac{\partial}{\partial z}(k_z\frac{\partial T}{\partial z}) + Q = \frac{\partial}{\partial t}(\rho c T) \qquad (12.1)$$

where, k_x, k_y, k_z = thermal conductivity coefficients, T = temperature, Q = heat generation per unit volume, ρ = density, and c = specific heat. If we focus our attention to the two-dimensional $(\partial/\partial z = 0)$ steady-state $(\partial/\partial t = 0)$ problem, such as Fig. 12.2.1, the governing equation becomes

$$\frac{\partial}{\partial x}(k_x\frac{\partial T}{\partial x}) + \frac{\partial}{\partial y}(k_y\frac{\partial T}{\partial y}) + Q = 0 \tag{12.2}$$

in which k_x, k_y, and Q are known. Equations (12.1) or (12.2), together with the appropriate boundary conditions specify the problem completely. The two most commonly encountered boundary conditions are those in which the temperature, T, is specified on the boundary, i.e.,

$$T = T(s) \text{ on } \Gamma_s. \tag{12.3}$$

boundary heat flux is specified at a point, i.e.,

$$k_x\frac{\partial T}{\partial x}n_x + k_y\frac{\partial T}{\partial y}n_y + q + h(T - T_r) = 0 \text{ on } \Gamma_q.$$

or

$$k_n\frac{\partial T}{\partial n} + q + h(T - T_r) = 0, \tag{12.4}$$

where n_x and n_y are the direction cosines of the outward normal to the boundary surface, q represents the heat flux per unit of surface, and $h(T - T_r)$ is the convection heat loss.

As stated previously, an alternative formulation to the above heat conduction problem is possible using the calculus of variations. Euler's theorem of the calculus of variations states that if the integral

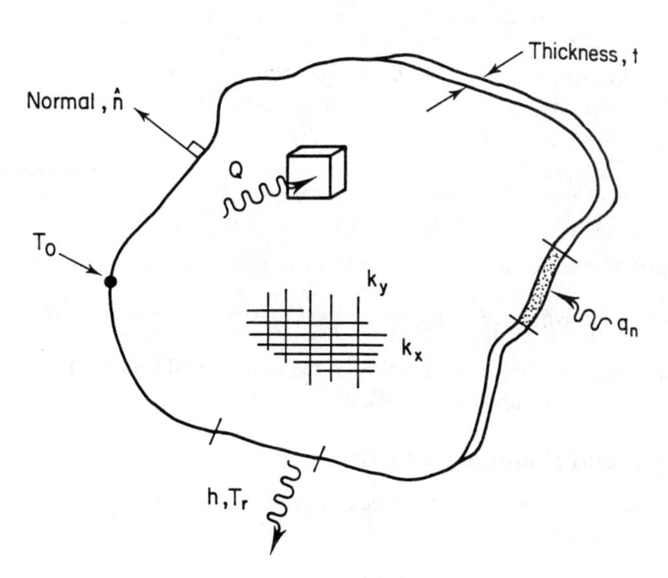

Figure 12.2.1 An anisotropic conduction region

$$I(u) = \int_V f(x, y, z, u, \frac{\partial u}{\partial x}, \frac{\partial u}{\partial y}, \frac{\partial u}{\partial z})\, dx\, dy\, dz + \int_S (qu + hu^2/2)\, da \tag{12.5}$$

is to be minimized, the necessary and sufficient condition for this minimum to be reached is that the unknown function $u(x, y, z)$ satisfy the following differential equation

$$\frac{\partial}{\partial x}\frac{\partial f}{\partial(\partial u/\partial x)} + \frac{\partial}{\partial y}\frac{\partial f}{\partial(\partial u/\partial y)} + \frac{\partial}{\partial z}\frac{\partial f}{\partial(\partial u/\partial z)} - \frac{\partial f}{\partial u} = 0 \tag{12.6}$$

within the region, provided u satisfies the essential boundary conditions in both cases. We can verify that the minimization of the volume integral

$$I = \int_V \left[\frac{1}{2}\left\{k_x(\frac{\partial T}{\partial x})^2 + k_y(\frac{\partial T}{\partial y})^2 + k_z(\frac{\partial T}{\partial z})^2\right\} - QT\right] dV + \int_S \left[qT + hT^2/2\right] da \tag{12.7}$$

leads directly to the formulation equivalent to Eq. (12.2) for the steady-state case. The functional volume contribution is

$$f = ½\left\{k_x(\frac{\partial T}{\partial x})^2 + k_y(\frac{\partial T}{\partial y})^2 + k_z(\frac{\partial T}{\partial z})^2\right\} - QT.$$

Thus, if f is to be minimized it must satisfy Eq. (12.6). Here

$$\frac{\partial f}{\partial(\partial T/\partial x)} = k_x\frac{\partial T}{\partial x}, \quad \frac{\partial f}{\partial(\partial T/\partial y)} = k_y\frac{\partial T}{\partial y}, \quad \frac{\partial f}{\partial(\partial T/\partial z)} = k_z\frac{\partial T}{\partial z}, \quad \frac{\partial f}{\partial T} = -Q.$$

Equation (12.6) results in

$$\frac{\partial}{\partial x}(k_x\frac{\partial T}{\partial x}) + \frac{\partial}{\partial y}(k_y\frac{\partial T}{\partial y}) + \frac{\partial}{\partial z}(k_z\frac{\partial T}{\partial z}) + Q = 0$$

verifying that the function f does lead to correct steady state formulation, if Eq. (12.3) is also satisfied. Euler also stated that the natural boundary condition associated with Eq. (12.5) on a surface with a unit normal vector $\vec{n}$ is

$$n_x\left\{\frac{\partial f}{\partial(\partial u/\partial x)}\right\} + n_y\left\{\frac{\partial f}{\partial(\partial u/\partial y)}\right\} + n_z\left\{\frac{\partial f}{\partial(\partial u/\partial z)}\right\} + g + hu = 0$$

on the boundary where the value of u is not forced. Here this equation simplifies to

$$n_x k_x\frac{\partial T}{\partial x} + n_y k_y\frac{\partial T}{\partial y} + n_z k_z\frac{\partial T}{\partial z} + g + hT = 0 = k_n\frac{\partial T}{\partial n} + g + hT$$

where k_n is the conductivity in the direction of the unit normal vector. This is the natural boundary condition given earlier in Eq. (12.4).

12.3 Element and Boundary Matrices

From Eqs. (12.1) and (12.7) it is clearly seen that the two-dimensional functional required for the steady-state analysis is

$$I = \int_A \left[½\left\{k_x(\frac{\partial T}{\partial x})^2 + k_y(\frac{\partial T}{\partial y})^2\right\} - QT\right] t\, dx\, dy + \int_\Gamma (gT + hT^2/2)t\, ds. \tag{12.8}$$

where t is the thickness of the domain. We will proceed in exactly the same manner as we did for the previous variational formulations. That is, we will assume that the area integral is the sum of the integrals over the element areas. Likewise, the boundary integral where the temperature is not specified is assumed to be the sum of the boundary segment integrals. Thus,

$$I = \sum_e I^e + \sum_b I^b$$

where the element contributions are

$$I^e = \int_{A^e} \left[\tfrac{1}{2} \left\{ k_x \left(\frac{\partial T}{\partial x}\right)^2 + k_y \left(\frac{\partial T}{\partial y}\right)^2 \right\} - QT \right] t \, da$$

and the boundary segment contributions are

$$I^b = \int_{\Gamma^b} (gT + hT^2/2) t \, ds.$$

If we make the usual interpolation assumptions in the element and on its typical edge then we can express these quantities as

$$I^e = \tfrac{1}{2} \mathbf{T}^{e^T} \mathbf{S}^e \mathbf{T}^e - \mathbf{T}^{e^T} \mathbf{C}^e, \qquad I^b = \tfrac{1}{2} \mathbf{T}^{b^T} \mathbf{S}^b \mathbf{T}^b - \mathbf{T}^{b^T} \mathbf{C}^b.$$

Here the element matrices are

$$\mathbf{S}^e = \int_{A^e} (k_x^e \, \mathbf{H}_x^{e^T} \mathbf{H}_x^e + k_y^e \, \mathbf{H}_y^{e^T} \mathbf{H}_y^e) t^e \, da \tag{12.9}$$

$$\mathbf{C}_Q^e = \int_{A^e} \mathbf{H}^{e^T} Q^e t^e \, da \tag{12.10}$$

$$\mathbf{S}_h^b = \int_{\Gamma^b} h^b \mathbf{H}^{b^T} \mathbf{H}^b t^b \, ds, \quad \mathbf{C}_h^b = \int_{\Gamma^b} T_r^b h^b \mathbf{H}^{b^T} t^b \, ds \tag{12.11}$$

$$\mathbf{C}_q^b = \int_{\Gamma^b} q^b \mathbf{H}^{b^T} t^b \, ds \tag{12.12}$$

where $\mathbf{H}$ denotes the shape functions and $\mathbf{H}_x = \partial \mathbf{H} / \partial x$, etc. For this class of problem there is only one unknown temperature per node. Once again, if $\mathbf{T}$ denotes all of these unknowns then $\mathbf{T}^e \subset \mathbf{T}$ and $\mathbf{T}^b \subset \mathbf{T}$.

12.3.1 Linear Triangle

If we select the three node (linear) triangle then the element interpolation functions, $\mathbf{H}^e$, are given in unit coordinates by Eq. (9.6) and in global coordinates by Eqs. (9.13) and (9.14). From either set of equations we note that

$$\begin{aligned} \mathbf{H}_x^e &= \partial \mathbf{H}^e / \partial x = [\, b_1 \ b_2 \ b_3 \,]^e / 2A^e = \mathbf{d}_x \\ \mathbf{H}_y^e &= \partial \mathbf{H}^e / \partial y = [\, c_1 \ c_2 \ c_3 \,]^e / 2A^e = \mathbf{d}_y . \end{aligned} \tag{12.13}$$

Since these are constant we can evaluate the integral by inspection if the conductivities are also constant:

$$\mathbf{S}^e = \frac{k_x^e t^e}{4A^e} \begin{bmatrix} b_1 b_1 & b_1 b_2 & b_1 b_3 \\ b_2 b_1 & b_2 b_2 & b_2 b_3 \\ b_3 b_1 & b_3 b_2 & b_3 b_3 \end{bmatrix}^e + \frac{k_y^e t^e}{4A^e} \begin{bmatrix} c_1 c_1 & c_1 c_2 & c_1 c_3 \\ c_2 c_1 & c_2 c_2 & c_2 c_3 \\ c_3 c_1 & c_3 c_2 & c_3 c_3 \end{bmatrix}^e . \tag{12.14}$$

This is known as the *element conductivity matrix*. Note that this allows for different conductivities in the x- and y- directions. Equations (12.14) show that the conduction in the x-direction depends on the size of the element in the y-direction, and vice versa. If the internal heat generation, Q, is also constant then Eq. (12.10) can be integrated via Eq. (11.3) to yield

$$\mathbf{C}^e = \frac{Q^e A^e t^e}{3} \begin{Bmatrix} 1 \\ 1 \\ 1 \end{Bmatrix}. \tag{12.15}$$

This internal source vector shows that a third of the internal heat generated, $Q^e A^e t^e$, is lumped to each of the three nodes as shown in Fig. 11.5.4. On a typical boundary segment the edge interpolation can be given by a linear form. The exact integrals can be evaluated for a constant Jacobian. For example, if the coefficient, h, is constant then the boundary segment square matrix is obtained from Eq. (12.11) as

$$\mathbf{S}^b = \frac{h^b L^b t^b}{6} \begin{bmatrix} 2 & 1 \\ 1 & 2 \end{bmatrix}. \tag{12.16}$$

Similarly if a constant normal flux, q, is given then the boundary flux vector is

$$\mathbf{C}^b = \frac{g^b L^b t^b}{2} \begin{Bmatrix} 1 \\ 1 \end{Bmatrix}. \tag{12.17}$$

In this case half the total normal flux is lumped at each of the two nodes on the segment. This condition is analogous to that in Fig. 11.5.1. That figure also shows similar nodal values for higher order segments.

12.3.2 Rectangular Bilinear Element *

When quadrilateral elements have a parallelogram shape in physical space, their Jacobian is constant. If it takes the form of a rectangle with sides parallel to the global axes, then the Jacobian matrix is a constant diagonal matrix that allows the analytic evaluation of the element matrices. Consider a Q4 element mapped into a rectangular element that is parallel to the global axes as shown in Fig. 12.3.1. Note that mapping is simple and gives a constant Jacobian:

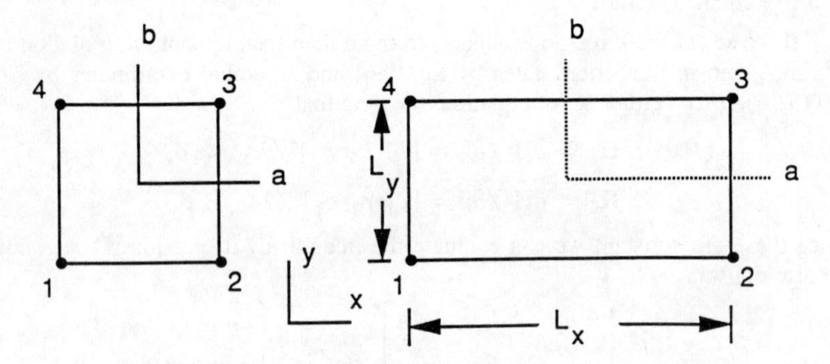

Figure 12.3.1 The rectangular conduction element

$$x = \bar{x} + a L_x/2\,, \qquad \frac{\partial x}{\partial a} = \frac{L_x}{2}\,, \qquad \frac{\partial x}{\partial b} \equiv 0\,,$$

$$y = \bar{y} + b L_y/2\,, \qquad \frac{\partial y}{\partial a} = 0\,, \qquad \frac{\partial y}{\partial b} = \frac{L_y}{2}\,,$$

$$\mathbf{J} = \tfrac{1}{2}\begin{bmatrix} L_x & 0 \\ 0 & L_y \end{bmatrix}, \qquad |J| = \frac{L_x L_y}{4} = \frac{A^e}{4}\,.$$

By inspection (or from the inverse Jacobian) we see

$$\frac{\partial}{\partial x} = \frac{\partial}{\partial a}\frac{\partial a}{\partial x} = \frac{2}{L_x}\frac{\partial}{\partial a}\,, \qquad \frac{\partial}{\partial y} = \frac{2}{L_y}\frac{\partial}{\partial b}$$

so that the typical term in the condition matrix due to k_x is

$$\mathbf{S}_{x_{i,j}} = \frac{4}{L_x^2}\int_{\Omega^e} k_x^e\, H_{i,a}\, H_{j,a}\, d\Omega\,, \tag{12.18}$$

but $d\Omega = |J|\, d\Box$ so that if k_x is constant

$$\mathbf{S}_{x_{i,j}} = \frac{k_x^e L_y^e}{L_x^e}\int_{\Box} H_{i,a}\, H_{j,a}\, d\Box\,.$$

The interpolation functions are

$$H_j = \frac{1}{4}(1 + a_j a)(1 + b_j b)\,, \qquad H_{j,a} = \frac{a_j}{4}(1 + b_j b)$$

and the integrand becomes

$$H_{i,a}\, H_{j,a} = a_i\, a_j\left[1 + (b_i + b_j)\, b + b_i\, b_j\, b^2\right]/\,16\,. \tag{12.19}$$

Invoking numerical integration in $\Box$ gives

$$\mathbf{S}_{x_{i,j}} = \frac{a_i\, a_j\, k_x^e L_y^e}{16\, L_x^e}\sum_{q=1}^{Q}\left[1 + (b_i + b_j)\, b_q + b_i\, b_j\, b_q^2\right] w_q\,.$$

Since this is quadratic in b, we need to pick only two points in the b direction. Likewise, for similar terms in the $\mathbf{S}_y$ matrix, we need two points in the a direction. Using the $Q = 4$ rule, we note that $w_q = 1$ and is constant, and that two $b_q = 1/\sqrt{3}$, while two are $-1/\sqrt{3}$. Thus, the linear terms cancel so that

$$\mathbf{S}_{x_{i,j}} = \frac{a_i\, a_j\, k_x^e L_y^e}{12\, L_x^e}\left[3 + b_i\, b_j\right]. \tag{12.20}$$

For the chosen local numbering, we have

j	a_j	b_j
1	−1	−1
2	1	−1
3	1	1
4	−1	1

so that

$$\mathbf{S}_x = \frac{k_x^e L_y^e}{6 L_x^e} \begin{bmatrix} 2 & -2 & -1 & 1 \\ -2 & 2 & 1 & -1 \\ -1 & 1 & 2 & -2 \\ 1 & -1 & -2 & 2 \end{bmatrix} \tag{12.21}$$

which agrees with the exact integration. Likewise, for a constant k_y^e:

$$\mathbf{S}_y = \frac{k_y^e L_x^e}{6 L_y^e} \begin{bmatrix} 2 & 1 & -1 & -2 \\ 1 & 2 & -2 & -1 \\ -1 & -2 & 2 & 2 \\ -2 & -1 & 1 & 2 \end{bmatrix}. \tag{12.22}$$

The typical convention square matrix term is

$$\begin{aligned} \mathbf{M}_{i,j} &= \int_{\Omega^e} \zeta^e H_i H_j \, d\Omega \\ &= \frac{L_x^e L_y^e}{64} \int_{\square^e} \zeta^e \left(1 + a_i a\right) \left(1 + a_j a\right) \left(1 + b_i b\right) \left(1 + b_j b\right) d\square \end{aligned} \tag{12.23}$$

so for constant ζ^e

$$\mathbf{M}_{i,j} = \frac{\zeta^e L_x^e L_y^e}{64} \int_{\square} \left[\left(1 + (a_i + a_j) a + a_i a_j a^2\right) \left(1 + (b_i + b_j) b + b_i b_j b^2\right) \right] d\square .$$

Again the four point rule is valid, and since $a_q = \pm 1/\sqrt{3}$, the linear terms cancel. Since $w_q \equiv 1$, we get

$$\mathbf{M}_{i,j} = \frac{\zeta^e L_x^e L_y^e}{64} \left[4 \left(1 + a_i a_j/3\right) \left(1 + b_j b_i/3\right) \right]$$

or

$$\mathbf{M} = \frac{\zeta^e L_x^e L_y^e}{36} \begin{bmatrix} 4 & 2 & 1 & 2 \\ 2 & 4 & 2 & 1 \\ 1 & 2 & 4 & 2 \\ 2 & 1 & 2 & 4 \end{bmatrix}. \tag{12.24}$$

The boundary matrices are the same as for the linear triangle, since both are linear.

12.3.3 General Elements

If we select a higher order element such as the isoparametric quadrilateral then some of the above integrations are more difficult to evaluate. In the notation of Chap. 9, Eq. (9.32) becomes

$$\mathbf{S}^e = \int_{A^e} \left[k_x^e \mathbf{d}_x^{e^T} \mathbf{d}_x^e + k_y^e \mathbf{d}_y^{e^T} \mathbf{d}_y^e \right] t^e \, da \, . \tag{12.25}$$

If we allow for general quadrilaterals and/or curved sides then we will need numerical integration. Thus, we write

$$\mathbf{S}^e = \sum_{i=1}^{NIP} W_i \left| J_i \right| \left[k_{x_i}^e \mathbf{d}_{x_i}^{e^T} \mathbf{d}_{x_i}^e + k_{y_i}^e \mathbf{d}_{y_i}^{e^T} \mathbf{d}_{y_i}^e \right] t_i^e \, . \tag{12.26}$$

Similar expressions are available for the source vectors. When the Jacobian is constant the nodal source resultants are again like those in Figs. 11.5.3 and 11.5.7.

12.4 Example Application

Consider a uniform square of material that has its exterior perimeter maintained at a constant temperature while its interior generates heat at a constant rate. We note that the solution will be symmetric about the square's centerlines as well as about its two diagonals. This means that we only need to utilize one-eighth of the region in the analysis. For simplicity we will assume that the material is homogeneous and $k_x = k_y = k$. The planes of symmetry have zero normal heat flux, $q = 0$. That condition is a natural boundary condition in a finite element analysis. That is true since $\mathbf{C}_q$ in Eq. (12.2) is identically zero when the normal flux, q, is zero. The remaining essential condition is that of the known external boundary temperature as shown in Fig. 12.4.1. For this model we have selected four elements and six nodes. The last three nodes have the known temperature and the first three are the unknown internal temperatures. For

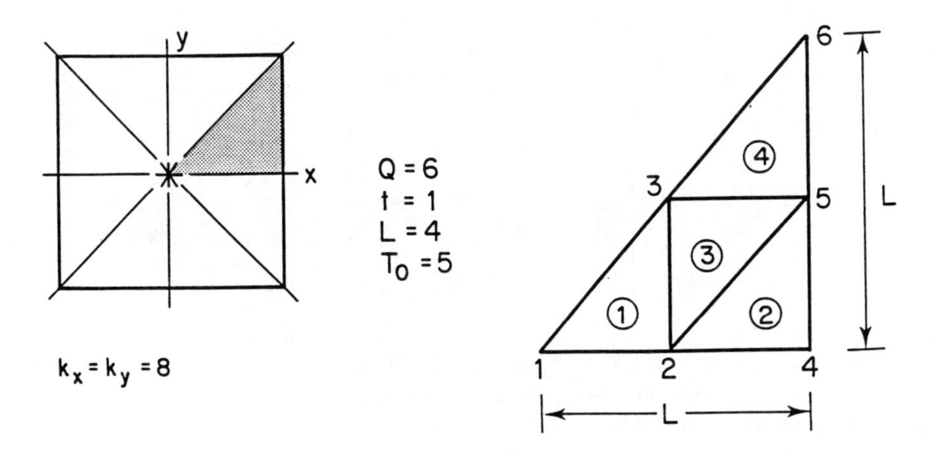

Figure 12.4.1 A one-eighth symmetry model of a square

this homogeneous region the data are:

Element	k^e	Q^e	*Topology*	t^e
1	8	6	1, 2, 3	1
2	8	6	2, 4, 5	1
3	8	6	5, 3, 2	1
4	8	6	3, 5, 6	1

From the geometry in the figure we determine that the element geometric properties from Eq. (9.14) are:

$$\begin{array}{ccc} & e=1,2,4 & e=3 \\ i & 1\ \ 2\ \ 3 & 1\ \ 2\ \ 3 \\ b_i & -2\ \ 2\ \ 0 & 2\ -2\ \ 0 \\ c_i & 0\ -2\ \ 2 & 0\ \ 2\ -2 \\ & A^e=2 & A^e=2 \end{array}$$

From Eq. (12.14) the conduction square matrix for elements 1, 2, and 4 are

$$\mathbf{S}^e = \frac{8(1)}{4(2)}\begin{bmatrix} 4 & -4 & 0 \\ -4 & 4 & 0 \\ 0 & 0 & 0 \end{bmatrix} + \frac{8(1)}{4(2)}\begin{bmatrix} 0 & 0 & 0 \\ 0 & 4 & -4 \\ 0 & -4 & 4 \end{bmatrix} = \begin{bmatrix} 4 & -4 & 0 \\ -4 & 8 & -4 \\ 0 & -4 & 4 \end{bmatrix} \quad (12.27)$$

Since element 3 results from a 180° rotation of element 1, it happens to have exactly the same S^e. Assembling the four element matrices gives the six system equations $\mathbf{ST} = \mathbf{C}$ where

$$\mathbf{S} = \begin{bmatrix} +4 & -4 & 0 & 0 & 0 & 0 \\ -4 & (+8+4+4) & (-4-4) & -4 & 0 & 0 \\ 0 & (-4-4) & (+4+8+4) & 0 & (-4-4) & 0 \\ 0 & -4 & 0 & +8 & -4 & 0 \\ 0 & 0 & (-4-4) & -4 & (+4+4+8) & -4 \\ 0 & 0 & 0 & 0 & -4 & +4 \end{bmatrix}$$

and

$$\mathbf{C} = \frac{QAt}{3}\begin{Bmatrix} 1 \\ 1+1+1 \\ 1+1+1 \\ 1 \\ 1+1+1 \\ 1 \end{Bmatrix} + \begin{Bmatrix} 0 \\ 0 \\ 0 \\ q_4 \\ q_5 \\ q_6 \end{Bmatrix} = \begin{Bmatrix} 4 \\ 12 \\ 12 \\ 4 \\ 12 \\ 4 \end{Bmatrix} + \begin{Bmatrix} 0 \\ 0 \\ 0 \\ q_4 \\ q_5 \\ q_6 \end{Bmatrix}.$$

In the above vector the q's are the nodal heat flux reactions required to maintain the specified external temperature. Since the last three equations have essential boundary conditions applied we can reduce the first three to

$$\begin{bmatrix} 4 & -4 & 0 \\ -4 & 16 & -8 \\ 0 & -8 & 16 \end{bmatrix} \begin{Bmatrix} T_1 \\ T_2 \\ T_3 \end{Bmatrix} = \begin{Bmatrix} 4 \\ 12 \\ 12 \end{Bmatrix} - T_4 \begin{Bmatrix} 0 \\ -4 \\ 0 \end{Bmatrix} - T_5 \begin{Bmatrix} 0 \\ 0 \\ -8 \end{Bmatrix} - T_6 \begin{Bmatrix} 0 \\ 0 \\ 0 \end{Bmatrix}. \quad (12.28)$$

Substituting the data that the exterior surface temperatures are $T_4 = T_5 = T_6 = 5$ yields the reduced source term

$$\mathbf{C}^* = \begin{Bmatrix} 4 \\ 12 \\ 12 \end{Bmatrix} + \begin{Bmatrix} 0 \\ 20 \\ 0 \end{Bmatrix} + \begin{Bmatrix} 0 \\ 0 \\ 40 \end{Bmatrix} = \begin{Bmatrix} 4 \\ 32 \\ 52 \end{Bmatrix}.$$

Solving for the interior temperatures using the inverse

$$\mathbf{S}^{*-1} = \frac{1}{512} \begin{bmatrix} 192 & 64 & 32 \\ 64 & 64 & 32 \\ 32 & 32 & 48 \end{bmatrix} \quad (12.29)$$

and multiplying by $\mathbf{C}^*$ yields:

$$\mathbf{T}^* = \begin{Bmatrix} 8.750 \\ 7.750 \\ 7.125 \end{Bmatrix} = \begin{Bmatrix} T_1 \\ T_2 \\ T_3 \end{Bmatrix}.$$

Substituting these values into the original system equations will give the exterior nodal heat flux values (*thermal reactions*) required by this problem. For example, the fourth equation yields:

$$-4T_2 + 8T_4 - 4T_5 = -4(7.75) + 8(5) - 4(5) = 4 + q_4, \quad \text{or} -15 = q_4.$$

The other two nodal fluxes are $q_5 = -29$, $q_6 = -4$ and the internal heat generated was

$$\sum_e Q^e A^e t^e = +48.$$

Thus, we have verified that the generated heat equals the heat outflow. Of course, this must be true for all steady state heat conduction problems. Note that while we started with six equations from the integral formulation only three were independent equations for the unknown temperatures. The other equations were independent equations for the thermal reactions necessary to maintain the essential boundary conditions on the temperature. One does not have to assemble and solve the reaction set but it is a recommended procedure. One must always modify the column vector for the temperatures to include essential boundary condition data. Coding for this example is included on the disk. Another example of thermal analysis will be given at the end of Chap. 20 to show its importance in computing thermal stress.

12.5 References

[1] Bass, J., *PHLEX User's Manual*, Austin, TX: Computational Mechanics Co. Inc. (1990).

[2] Desai, C.S., *Elementary Finite Element Method*, Englewood Cliffs: Prentice-Hall (1979).

[3] Hinton, E. and Owen, D.R.J., *Finite Element Programming*, London: Academic Press (1977).

[4] Hughes, T.J.R., *The Finite Element Method*, Englewood Cliffs: Prentice-Hall (1987).

[5] Meek, J.L., "Field Problems Solutions by Finite Element Methods," *Civil Eng. Trans., Inst. Eng. Aust.*, , pp. 173–180 (Oct. 1968).

[6] Myers, G.E., *Analytical Methods in Conduction Heat Transfer*, New York: McGraw-Hill (1971).

[7] Segerlind, L.J., *Applied Finite Element Analysis*, New York: John Wiley (1984).

[8] Silvester, P.P. and Ferrari, R.L., *Finite Elements for Electrical Engineers*, Cambridge: Cambridge University Press (1983).

[9] Zienkiewicz, O.C., *The Finite Element Method in Engineering Science*, New York: McGraw-Hill (1971).

Chapter 13

ELASTICITY

13.1 Introduction

The states of *plane stress* and *plane strain* are interesting and useful examples of stress analysis of a two-dimensional elastic solid (in the x-y plane). The assumption of plane stress implies that the component of all stresses normal to the plane are zero ($\sigma_z = \tau_{zx} = \tau_{zy} = 0$) whereas the plane strain assumption implies that the normal components of the strains are zero ($\varepsilon_z = \gamma_{zx} = \gamma_{zy} = 0$). The state of plane stress is commonly introduced in the first course of mechanics of materials. It was also the subject of some of the earliest finite element studies.

The assumption of plane stress means that the solid is very thin and that it is loaded only in the direction of its plane. At the other extreme, in the state of plane strain the dimension of the solid is very large in the z-direction. It is loaded perpendicular to the longitudinal (z) axis, and the loads do not vary in that direction. All cross-sections are assumed to be identical so any arbitrary x-y section can be used in the analysis. These two states are illustrated in Fig. 13.1.1. There are three common approaches to the variational formulation of the plane stress (or plane strain) problem: 1) Displacement formulation, 2) Stress formulation, and 3) Mixed formulation. We will select the common displacement method and utilize the total potential energy of the system. This can be proved to be equal to assuming a Galerkin weighted residual approach. In any event, note that it will be necessary to define all unknown quantities in terms of the displacements of the solid. Specifically, it will be necessary to relate the strains and stresses to the displacements as was illustrated in 1-D in Sec. 3.3.

The finite element form is based on the use of strain energy density, as discussed in Chap. 3. Since it is half the product of the stress and strain tensor components, we do not need, at this point, to consider either stresses or strains that are zero. From the first paragraph, this means that only three of six products will be used.

Our notation will follow that commonly used in mechanics of materials. The displacements components parallel to the x- and y-axes will be denoted by $u(x, y)$ and $v(x, y)$, respectively. The shear stress acting parallel to the x- and y-axes are σ_x and σ_y, respectively. The shear stress acting parallel to the y-axis on a plane normal to the x-axis is τ_{xy}, or simply τ. The corresponding components of strain are ε_x, ε_y, and γ_{xy}, or simply γ. Figure 13.1.2 summarizes these notations.

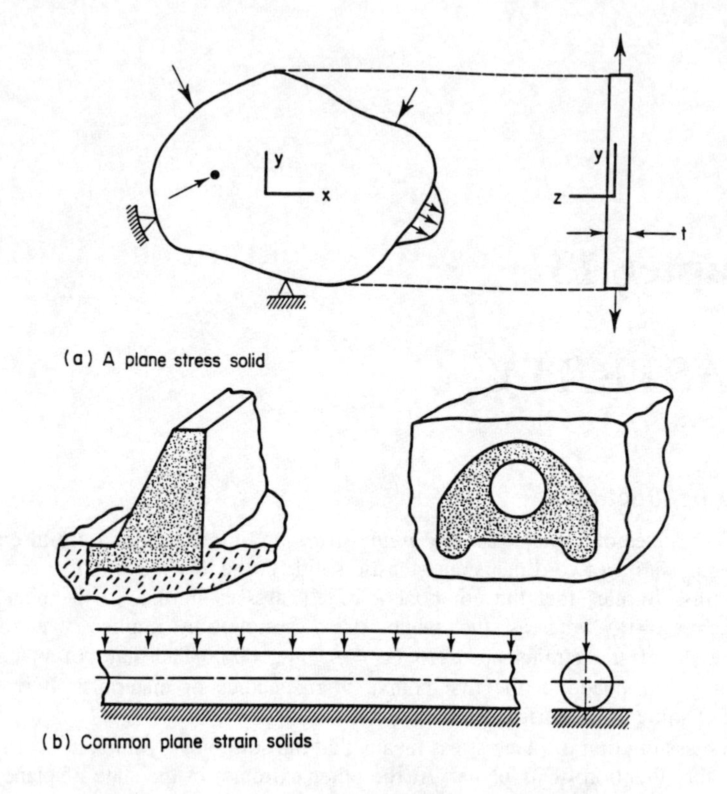

Figure 13.1.1 The states of plane stress and plane strain

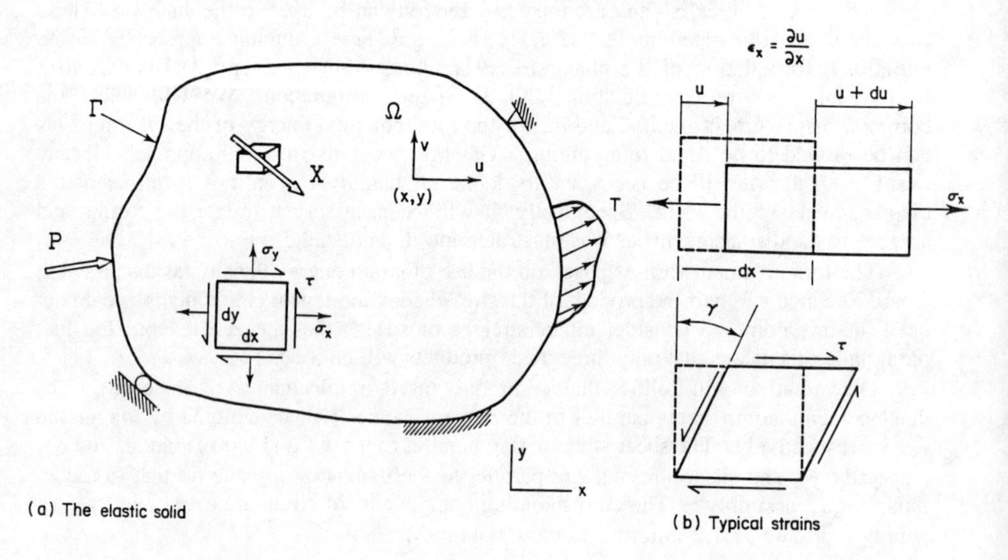

Figure 13.1.2 Notations for plane stress and plane strain

13.2 Minimum Total Potential Energy

Plane stress analysis, like other elastic stress analysis problems, is governed by the principle of minimizing the total potential energy in the system. These concepts were considered in Secs. 3.3 and 6.2. It is possible to write the generalized forms of the element matrices and boundary segment matrices defined in Sec. 4.5. The symbolic forms are:

(1) Stiffness matrix

$$\mathbf{S}^e = \int_{V^e} \mathbf{B}^{e^T} \mathbf{D}^e \mathbf{B}^e(x, y)\, dv \tag{13.1}$$

$\mathbf{B}^e$ = element strain-displacement matrix, $\mathbf{D}^e$ = material constitutive matrix;

(2) Body Force Matrix

$$\mathbf{C}_x^e = \int_{V^e} \mathbf{N}^e(x, y)^T\, \mathbf{X}^e(x, y)\, dv \tag{13.2}$$

$\mathbf{N}^e$ = generalized interpolation matrix, $\mathbf{X}^e$ = body force vector per unit volume;

(3) Initial Strain Load Matrix

$$\mathbf{C}_o^e = \int_{V^e} \mathbf{B}^{e^T}(x, y)\, \mathbf{D}^e \boldsymbol{\varepsilon}_o^e(x, y)\, dv \tag{13.3}$$

$\boldsymbol{\varepsilon}_o^e$ = initial strain matrix;

(4) Surface Traction Load Matrix

$$\mathbf{C}_T^b = \int_{A^b} \mathbf{N}^b(x, y)^T\, \mathbf{T}^b(x, y)\, da \tag{13.4}$$

$\mathbf{N}^b$ = boundary interpolation matrix, $\mathbf{T}^b$ = traction force vector per unit area;

and where V^e is the element volume, A^b is a boundary segment surface area, dv is a differential volume, and da is a differential surface area. Now we will specialize these relations for plane stress (or strain). At any point the two displacement components will be denoted by $\mathbf{u}^T = [\,u \quad v\,]$. Therefore, at each node there are two displacement components ($NG = 2$) to be determined. The total list of element degrees of freedom will be denoted by δ^e.

13.2.1 Displacement Interpolations

As before, it is necessary to define the spatial approximation for the displacement field. Consider the x-displacement, u, at some point in an element. The simplest approximation of how it varies in space is to assume a complete linear polynomial. In two-dimensions a complete linear polynomial contains three constants. Thus, we select a triangular element with three nodes and assume u is to be computed at each node. Then

$$u(x, y) = \mathbf{H}^e \mathbf{u}^e = H_1^e u_1^e + H_2^e u_2^e + H_3^3 u_3^e\,. \tag{13.5}$$

The interpolation can either be done in global (x, y) coordinates or in a local system. If global coordinates are utilized then, from Eq. (9.13), the form of the typical interpolation function is

$$H_i^e(x, y) = (a_i^e + b_i^e x + c_i^e y)/2A^e\,, \qquad 1 \le i \le 3 \tag{13.6}$$

where A^e is the area of the element and a_i^e, b_i^e, and c_i^e denote constants for node i that depend on the element geometry. Clearly, we could utilize the same interpolations for the y-displacement:

$$v(x, y) = \mathbf{H}^e \mathbf{v}^e . \quad (13.7)$$

To define the element dof vector a^e we chose to order these six constants such that

$$\delta^{e^T} = \begin{bmatrix} u_1 & v_1 & u_2 & v_2 & u_3 & v_3 \end{bmatrix}^e . \quad (13.8)$$

To refer to both displacement components at a point we employ a generalized element interpolation. Then

$$\mathbf{u}(x, y) = \begin{Bmatrix} u(x, y) \\ v(x, y) \end{Bmatrix} = \begin{bmatrix} H_1 & 0 & H_2 & 0 & H_3 & 0 \\ 0 & H_1 & 0 & H_2 & 0 & H_3 \end{bmatrix}^e \delta^e = \mathbf{N}^e \delta^e . \quad (13.9)$$

Of course, more advanced polynomials could be selected to define the **H** or **N** matrices. Note that the combined interpolation array, **N**, could be partitioned into typical partitions from a node. For example, from node j:

$$\mathbf{N}_j = \begin{bmatrix} H_j & 0 \\ 0 & H_j \end{bmatrix} .$$

13.2.2 Strain-Displacement Relations

From mechanics of materials we can define the strains in terms of the displacement. Order the three strain components so as to define $\varepsilon^T = [\, \varepsilon_x \;\; \varepsilon_y \;\; \gamma \,]$. These terms are defined as:

$$\varepsilon_x = \frac{\partial u}{\partial x}, \quad \varepsilon_y = \frac{\partial v}{\partial y}, \quad \gamma = \left(\frac{\partial u}{\partial y} + \frac{\partial v}{\partial x} \right)$$

if the common engineering form is selected for the shear strain, γ. Two of these terms are illustrated in Fig. 13.1.2. From Eqs. (13.5) and (13.7) we note

$$\varepsilon_x = \frac{\partial \mathbf{H}^e}{\partial x} \mathbf{u}^e, \quad \varepsilon_y = \frac{\partial \mathbf{H}^e}{\partial y} \mathbf{v}^e, \quad \gamma = \frac{\partial \mathbf{H}^e}{\partial y} \mathbf{u}^e + \frac{\partial \mathbf{H}^e}{\partial x} \mathbf{v}^e . \quad (13.10)$$

These can be combined into a single matrix identity to define

$$\begin{Bmatrix} \varepsilon_x \\ \varepsilon_y \\ \gamma \end{Bmatrix}^e = \begin{bmatrix} H_{1,x} & 0 & H_{2,x} & 0 & H_{3,x} & 0 \\ 0 & H_{1,y} & 0 & H_{2,y} & 0 & H_{3,y} \\ H_{1,y} & H_{1,x} & H_{2,y} & H_{2,x} & H_{3,y} & H_{3,x} \end{bmatrix}^e \delta^e \quad (13.11)$$

or symbolically, $\boldsymbol{\varepsilon}^e = \mathbf{B}^e(x, y)\boldsymbol{\delta}^e$, where the shorthand notation $H_x = \partial H / \partial x$, etc. has been employed. This defines the *element strain-displacement operator* $\mathbf{B}^e$ that would be used in Eqs. (13.1) and (13.3). Note that **B** could also be partitioned into 3×2 sub-partitions from each node on the element.

13.2.3 Stress-Strain Law

The stress-strain law (*constitutive relations*) between the strain components, ε, and the corresponding stress components, $\sigma^T = [\,\sigma_x \;\; \sigma_y \;\; \tau\,]$, is defined in mechanics of materials. For the case of an isotropic material in plane stress these are listed as

$$\sigma_x = \frac{E}{1-\nu^2}(\varepsilon_x + \nu\varepsilon_y), \quad \sigma_y = \frac{E}{1-\nu^2}(\varepsilon_y + \nu\varepsilon_x), \quad \tau = \frac{E}{2(1+\nu)}\gamma = G\gamma \quad (13.12)$$

where E is the elastic modulus, ν is Poisson's ration, and G is the shear modulus. In theory, G is not an independent property. In practice it is sometimes treated as independent. Some references list the inverse relations since the strains are usually experimentally determined from the applied stresses. In the alternate form the constitutive relations are

$$\varepsilon_x = \frac{1}{E}(\sigma_x - \nu\sigma_y), \quad \varepsilon_y = \frac{1}{E}(\sigma_y - \nu\sigma_x), \quad \gamma = \tau/G = 2\tau(1+\nu)/E. \quad (13.13)$$

We will write Eq. (13.13) in its matrix symbolic form

$$\boldsymbol{\sigma} = \mathbf{D}(\boldsymbol{\varepsilon} - \boldsymbol{\varepsilon}_o)\,. \quad (13.14)$$

Here we have allowed for the presence of initial strains, ε_o, that are not usually included in mechanics of materials. For plane stress

$$\mathbf{D} = \frac{E}{1-\nu^2}\begin{bmatrix} 1 & \nu & 0 \\ \nu & 1 & 0 \\ 0 & 0 & (1-\nu)/2 \end{bmatrix}. \quad (13.15)$$

Note that $\mathbf{D}$ is a symmetric matrix. This is almost always true. This observation shows that in general the element stiffness matrix, Eq. (13.1), will also be symmetric. For the sake of completeness the *constitutive matrix*, $\mathbf{D}$, for plane strain will also be given. It is

$$\mathbf{D} = \frac{E(1-\nu)}{(1+\nu)(1-2\nu)}\begin{bmatrix} 1 & a & 0 \\ a & 1 & 0 \\ 0 & 0 & b \end{bmatrix}, \quad (13.16)$$

where $a = \nu/(1-\nu)$ and $b = (1-2\nu)/2(1-\nu)$. Note that if one has an *incompressible material*, such as rubber where $\nu = 1/2$, then division by zero would cause difficulties for plane strain problems.

The most common type of initial strain, ε_o, is that due to temperature changes. For an isotropic material these *thermal strains* are

$$\varepsilon^{T_o} = \alpha\Delta\theta\,[\,1 \;\; 1 \;\; 0\,] \quad (13.17)$$

where α is the coefficient of thermal expansion and $\Delta\theta = (\theta - \theta_o)$ is the temperature rise from a stress free temperature of θ_o. Usually $\Delta\theta$ is supplied as data, or θ_o is given as data along with the nodal temperatures computed from the procedures in the previous chapter. Notice that thermal strains in isotropic materials do not include thermal shear strains. If the above temperature changes were present then the additional loading effects could be included via Eq. (13.3). An example of thermal stress loadings of T6 elements is given at the end of Chap. 20.

At this point, we do not know the nodal displacements, $\boldsymbol{\delta}^e$, of the element. Once we do know them, we will wish to use the above arrays to get postprocessing results for the stresses and, perhaps, for *failure criteria*. Therefore, for each element we usually store the arrays $\mathbf{B}^e$, $\mathbf{D}^e$, and $\mathbf{E}_o^e$ so that we can execute the products in Eqs. (13.11) and (13.14) after the displacements are known.

13.3 Matrices for the Constant Strain Triangle

Beginning with the simple linear displacement assumption of Eqs. (13.5) to (13.7) we note that for a typical CST interpolation function, H_i :

$$\frac{\partial H_i^e}{\partial x} = b_i^e / 2A^e, \quad \frac{\partial H_i^e}{\partial y} = c_i^e / 2A^e.$$

Therefore, from Eqs. (13.11) and (13.12), the strain components in the triangular element are constant. Specifically,

$$\mathbf{B}^e = \frac{1}{2A^e} \begin{bmatrix} b_1 & 0 & b_2 & 0 & b_3 & 0 \\ 0 & c_1 & 0 & c_2 & 0 & c_3 \\ c_1 & b_1 & c_2 & b_2 & c_3 & b_3 \end{bmatrix}^e . \tag{13.18}$$

For this reason this element is commonly known as the *constant strain triangle*, CST. If we also let the material properties, E and ν, be constant in a typical element then the stiffness matrix in Eq. (13.1) simplifies to

$$\mathbf{S}^e = \mathbf{B}^{e^T} \mathbf{D}^e \mathbf{B}^e V^e \tag{13.19}$$

where the element volume is

$$V^e = \int_{V^e} dv = \int_{A^e} t^e(x, y)\, dx\, dy \tag{13.20}$$

where t^e is the element thickness. Usually the thickness of a typical element is constant so that $V^e = t^e A^e$. Of course, it would be possible to define the thickness at each node and to utilize the interpolation functions to approximate $t^e(x, y)$. Similarly if the temperature change in the element is also constant within the element then Eq. (13.3) defines the thermal load matrix

$$\mathbf{C}_o^e = \mathbf{B}^{e^T} \mathbf{D}^e \boldsymbol{\varepsilon}_o^e \mathbf{t}^e A^e. \tag{13.21}$$

It would be possible to be more detailed and input the temperature at each node and integrate its change over the element.

It is common for plane stress problems to include body force loads due to gravity, centrifugal acceleration, etc. For simplicity, assume that the body force vector X^e, and the thickness, t^e, are constant. Then the body force vector in Eq. (13.2) simplifies to

$$\mathbf{C}_X^e = \mathbf{t}^e \int_{A^e} \mathbf{N}^{e^T}(x, y)\, dx\, dy\, \mathbf{X}^e. \tag{13.22}$$

From Eq. (13.9) it is noted that the non-zero terms in the integral typically involve scalar terms such as

$$I_i^e = \int_{A^e} H_i^e(x, y)\, da = \frac{1}{2A^e} \int_{A^e} (a_i^e + b_i^e x + c_i^e y)\, da. \tag{13.23}$$

These three terms can almost be integrated by inspection. The element geometric constants can be taken outside parts of the integrals. Then from the concepts of the first moment (centroid) of an area

$$a_i^e \int da = a_i^e A^e, \quad \int b_i^e x\, da = b_i^e \bar{x}^e A^e, \quad \int c_i^e y\, da = c_i^e \bar{y}^e A^e \tag{13.24}$$

where $\bar{x}$ and $\bar{y}$ denote the *centroid* coordinates of the triangle, that is,

$$\bar{x}^e = (x_1 + x_2 + x_3)^e/3, \quad \bar{y}^e = (y_1 + y_2 + y_3)^e/3 .$$

In view of Eq. (13.24), the integral in Eq. (13.23) becomes

$$I_i^e = A^e(a_i + b_i\bar{x} + c_i\bar{y})^e/2A^e.$$

Reducing the algebra to its simplest form, using Table 11.1, yields

$$I_i^e = A^e/3, \qquad 1 \le i \le 3. \tag{13.25}$$

Therefore, for the CST the expanded form of Eq. (13.22) is

$$\mathbf{C}_X^e = \frac{t^e A^e}{3}\begin{bmatrix} 1 & 0 \\ 0 & 1 \\ 1 & 0 \\ 0 & 1 \\ 1 & 0 \\ 0 & 1 \end{bmatrix} \begin{Bmatrix} X_x \\ X_y \end{Bmatrix}^e = \frac{t^e A^e}{3}\begin{Bmatrix} X_x \\ X_y \\ X_x \\ X_y \\ X_x \\ X_y \end{Bmatrix}^e \tag{13.26}$$

where X_x and X_y denote the components of the body force vector. To assign a physical meaning to this result note that $t^e A^e X_x^e$ is the resultant force in the x-direction. Therefore, the above calculation has replaced the distributed load with a statically equivalent set of three nodal loads. Each of these loads is a third of the resultant load. These *consistent loads* are illustrated in Figs. 11.5.4 and 13.3.1.

A body force vector, $\mathbf{X}$, can arise from several important sources. An example is one due to acceleration (and gravity) loads. We have been treating only the case of equilibrium. When the acceleration is unknown, we have a dynamic system. Then, instead of using Newton's second law,

$$\sum \mathbf{F} = m\,\mathbf{a}$$

where $\mathbf{a}(t)$ is the acceleration vector, we invoke the D'Alembert's principle and rewrite this as a pseudo-equilibrium problem

$$\sum \mathbf{F} - m\,\mathbf{a} = \mathbf{0}, \quad \text{or} \quad \sum \mathbf{F} + \mathbf{F}_I = \mathbf{0}$$

where we have introduced an inertial body force vector due to the acceleration, that is, we use $\mathbf{X} = -\rho\,\mathbf{a}$ for the equilibrium integral form. Since the acceleration is the second time derivative of the displacement vector, we can write

$$\mathbf{a}(t) = \mathbf{N}^e(x, y)\,\ddot{\boldsymbol{\delta}}^e$$

in a typical element. The typical element inertial contribution is, therefore,

$$-\,\mathbf{m}^e\,\mathbf{a}^e = -\int_{\Omega^e} \mathbf{N}^{e^T} \zeta\, \mathbf{N}^e\, d\Omega\, \ddot{\boldsymbol{\delta}}^e$$

where $\mathbf{m}^e$ is the element matrix. Since the acceleration vector is unknown, we move it to the LHS of the system equations:

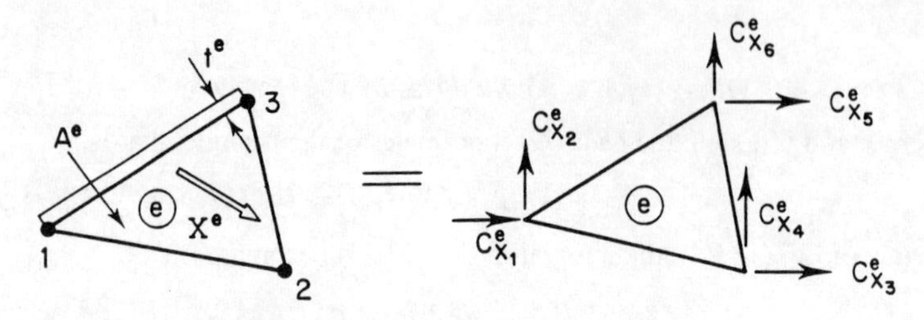

(a) Body force to element nodal load conversion

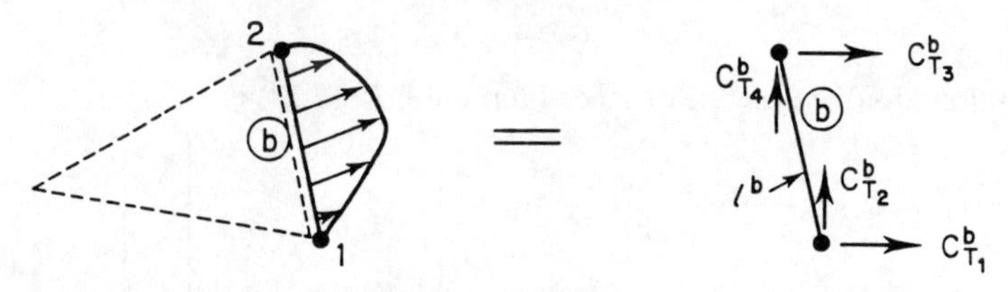

(b) Surface traction to boundary segment nodal load conversion

Figure 13.3.1 Element loads and consistent resultants

$$\mathbf{M}\ddot{\delta} + \mathbf{K}\delta = \mathbf{F}\,.$$

Here, **M** is the assembled system mass matrix, and the above are the *structural dynamic equations.* This class of problem will be considered in Chap. 17. If we had free ($\mathbf{F} = 0$) *simple harmonic motion,* so that

$$\delta(t) = \delta_j \operatorname{Sin}(\omega_j t),$$

then we get the alternate class known as the *eigenproblem,* where

$$\left[\mathbf{K} - \omega_j^2 \mathbf{M}\right] \delta_j = \mathbf{0}$$

where ω_j is the eigenvalue, or natural frequency, and δ_j is the mode shape, or eigenvector. The computational approaches to eigenproblems are covered in detail in the texts by Bathe [1], and by others.

Another type of body force that is usually difficult to visualize is that due to electromagnetic effects. In the past, they were usually small enough to be ignored. However, with the advances in superconducting materials, very high electrical current densities are possible, and they can lead to significant mechanical loads. Similar loads develop in medical scan devices, and in fusion energy reactors which are currently experimental. Recall from basic physics that the mechanical force, $\mathbf{F}$, due to a current density vector, $\mathbf{J}$, in a field with a magnetic flux density vector of **B** is the vector cross product $\mathbf{F} = \mathbf{J} \times \mathbf{B}$. For a thin wire conductor, **J** is easy to visualize since it is tangent to the conductor. However, the **B** field, like the earth's gravity field, is difficult to visualize. This could lead to important forces on the system which might be overlooked.

The final load to be considered is that acting on a typical boundary segment. As indicated in Fig. 13.3.1, such a segment is one side of an element being loaded with a traction. In plane stress problems these pressures or distributed shears act on the edge of the solid. In other words, they are distributed over a length L^b that has a known thickness, t. Those two quantities define the surface area, A^b, on which the tractions in Eq. (13.4) are applied. Similarly, the differential surface area is $da = t\,dL$. We observe that such a segment would have two nodes. We can refer to them as local boundary nodes 1 and 2. Of course, they are a subset of the three element nodes and also a subset of the system nodes. Before Eq. (13.4) can be integrated to define the consistent loads on the two boundary nodes it is necessary to form the boundary interpolation, $\mathbf{N}^b$. That function defines the displacements, u and v, at all points on the boundary segment curve. By analogy with Eq. (13.8) we can denote the dof of the boundary segment as

$$a^{b^T} = [u_1 \quad v_1 \quad u_2 \quad v_2]^b.$$

Then the requirement that $\mathbf{u} = \mathbf{N}^b \boldsymbol{\delta}^b$, for all points on L^b, defines the required $\mathbf{N}^b$. There are actually two ways that its algebraic form can be derived:

1) Develop a consistent (linear) interpolation on the line between the nodal dof.
2) Degenerate the element function $\mathbf{N}^e$ in Eq. (13.9) by restricting the x and y coordinates to points on the boundary segment.

If the second option is selected then all the H_i^e vanish except for the two associated with the two boundary segment nodes. Those two H^e are simplified by the restriction and thus define the two H_i^b functions. While the result of this type of procedure may be obvious, the algebra is tedious in global coordinates. (For example, let $y^b = mx^b + n$ in Eq. (13.6).) It is much easier to get the desired results if local coordinates are used. (For example, set $s^b = 0$ in Eq. (9.6). The net result is that one obtains a one-dimensional linear interpolation set for $\mathbf{H}^b$ that is analogous to Eq. (4.11). If we assume constant thickness, t^b, and constant tractions, $\mathbf{T}^b$, then Eq. (13.4) becomes

$$\mathbf{C}_T^b = t^b \int_{L^b} \mathbf{N}^{b^T} dL\, \mathbf{T}^b .$$

Repeating the procedure used for Eq. (5.5) a typical non-zero contribution to the integral is

$$I_i^b = \int_{L^b} H_i^b\, dL = L^b/2, \quad 1 \le i \le 2$$

and the final result for the four force components is

$$\mathbf{C}_T^b = \frac{t^b L^b}{2} \begin{Bmatrix} T_x \\ T_y \\ T_x \\ T_y \end{Bmatrix}^b \tag{13.27}$$

where T_x and T_y are the two components of T. Physically, this states that half of the resultant x-force, $t^b L^b T_x^b$ is lumped at each of the two nodes. The resultant y-force is lumped in the same way as illustrated in Fig. 13.3.1.

In the case of stress analysis there are times when it is desirable to rearrange the constitutive matrix, $\mathbf{D}$, into two parts. One part, $\mathbf{D}_n$, is due to normal strain effects, and the other, $\mathbf{D}_s$, is related to the shear strains. Therefore, in general it is possible to write Eq. (13.5) as

$$\mathbf{D} = \mathbf{D}_n + \mathbf{D}_s . \tag{13.28}$$

In this case such a procedure simply makes it easier to write the CST stiffness matrix in closed form. Noting that substituting Eq. (13.28) into Eq. (13.1) allows the stiffness to be separated into parts $\mathbf{S}^e = \mathbf{S}_n^e + \mathbf{S}_s^e$, where

$$\mathbf{S}_n^e = \frac{EV}{4A^2(1-\nu^2)} \begin{bmatrix} b_1^2 & & & & & \text{sym} \\ \nu b_1 c_1 & c_1^2 & & & & \\ b_1 b_2 & \nu c_1 b_2 & b_2^2 & & & \\ \nu b_1 c_2 & c_1 c_2 & \nu b_2 c_2 & c_2^2 & & \\ b_1 b_3 & \nu c_1 b_3 & b_2 b_3 & \nu c_2 b_3 & b_3^2 & \\ \nu b_1 c_3 & c_1 c_3 & \nu b_2 c_3 & c_2 c_3 & \nu b_3 c_3 & c_3^2 \end{bmatrix}$$

$$\mathbf{S}_s^e = \frac{EV}{8A^2(1-\nu)} \begin{bmatrix} c_1^2 & & & & & \text{sym} \\ c_1 b_1 & b_1^2 & & & & \\ c_1 c_2 & b_1 c_2 & c_2^2 & & & \\ c_1 b_2 & b_1 b_2 & c_2 b_2 & b_2^2 & & \\ c_1 c_3 & b_1 c_3 & c_2 c_3 & b_2 c_3 & c_3^2 & \\ c_1 b_3 & b_1 b_3 & c_2 b_3 & b_2 b_3 & c_3 b_3 & b_3^2 \end{bmatrix}$$

and where V is the volume of the element. As mentioned earlier, for constant thickness $V = At$.

The strain-displacement matrix $\mathbf{B}^e$ can always be partioned into sub-matrices associated with each node. Thus, the square stiffness matrix $\mathbf{S}$ can also be partitioned into square sub-matrices, since it is the product of $\mathbf{B}^T \mathbf{D}\, \mathbf{B}$. For local nodes j and k, they interact to give a contribution defined by:

$$\mathbf{S}_{jk} = \int_\Omega \mathbf{B}_j^T \mathbf{D}\, \mathbf{B}_k \, d\Omega .$$

If we choose to split $\mathbf{D}$ into two distinct parts, say $\mathbf{D} = \mathbf{D}_n + \mathbf{D}_s$, then we likewise have two contributions to the partitions of $\mathbf{S}$, namely

$$\mathbf{S}_{jk}^n = \int_\Omega \mathbf{B}_j^T \mathbf{D}_n \mathbf{B}_k \, d\Omega , \qquad \mathbf{S}_{jk}^s = \int_\Omega \mathbf{B}_j^T \mathbf{D}_s \mathbf{B}_k \, d\Omega .$$

Sometimes we may use different numerical integration rules on these two parts. For the constant strain triangle, CST, we have the nodal values of

$$\mathbf{B}_j^e = \frac{1}{2A^e} \begin{bmatrix} b_j & 0 \\ 0 & c_j \\ c_j & b_j \end{bmatrix}^e = \begin{bmatrix} d_{xj} & 0 \\ 0 & d_{yj} \\ d_{yj} & d_{xj} \end{bmatrix}$$

which involve the geometric constants defined earlier. For a constant isotropic $\mathbf{D}$, the integral gives the partitions

$$\mathbf{S}_{jk}^{s} = \frac{D_{33}}{4A}\begin{bmatrix} c_j c_k & b_j c_k \\ c_j b_k & b_j b_k \end{bmatrix}, \qquad \mathbf{S}_{jk}^{n} = \frac{1}{4A}\begin{bmatrix} D_{11} b_j b_k & D_{12} c_j b_k \\ D_{12} b_j c_k & D_{11} c_j d_k \end{bmatrix}$$

where D_{33} brings in the shear effects, and D_{11}, D_{12} couple the normal stress effects. If we allow j and k to range over the values 1, 2, 3, we would get the full 6×6 stiffness matrix

$$\mathbf{S} = \begin{bmatrix} \mathbf{S}_{11} & \mathbf{S}_{12} & \mathbf{S}_{13} \\ \mathbf{S}_{21} & \mathbf{S}_{22} & \mathbf{S}_{23} \\ \mathbf{S}_{31} & \mathbf{S}_{32} & \mathbf{S}_{33} \end{bmatrix}.$$

Since $\mathbf{D}$ is symmetric, it should be clear that $\mathbf{S}_{jk} = \mathbf{S}_{kj}^T$. A similar split can be made utilizing the constitutive law in terms of the Lamé constants,

$$\lambda = K - \frac{2G}{3} = \frac{E\nu}{(1+\nu)(1-2\nu)}, \qquad \mu = G = \frac{E}{2(1+\nu)},$$

where K and G are the bulk modulus and the shear modulus, respectively. Then the plane strain $\mathbf{D}$ matrix can be split as

$$\mathbf{D} = \lambda \begin{bmatrix} 1 & 1 & 0 \\ 1 & 1 & 0 \\ 0 & 0 & 0 \end{bmatrix} + \mu \begin{bmatrix} 2 & 0 & 0 \\ 0 & 2 & 0 \\ 0 & 0 & 1 \end{bmatrix}$$

$$\mathbf{D} = K \begin{bmatrix} 1 & 1 & 0 \\ 1 & 1 & 0 \\ 0 & 0 & 0 \end{bmatrix} + \frac{G}{3} \begin{bmatrix} 4 & -2 & 0 \\ -2 & 4 & 0 \\ 0 & 0 & 3 \end{bmatrix}.$$

Likewise, for the full three-dimensional case with six strains and six stresses, we have a similar form

$$\mathbf{D} = \lambda \begin{bmatrix} 1 & 1 & 1 & 0 & 0 & 0 \\ 1 & 1 & 1 & 0 & 0 & 0 \\ 1 & 1 & 1 & 0 & 0 & 0 \\ 0 & 0 & 0 & 0 & 0 & 0 \\ 0 & 0 & 0 & 0 & 0 & 0 \\ 0 & 0 & 0 & 0 & 0 & 0 \end{bmatrix} + \mu \begin{bmatrix} 2 & 0 & 0 & 0 & 0 & 0 \\ 0 & 2 & 0 & 0 & 0 & 0 \\ 0 & 0 & 2 & 0 & 0 & 0 \\ 0 & 0 & 0 & 1 & 0 & 0 \\ 0 & 0 & 0 & 0 & 1 & 0 \\ 0 & 0 & 0 & 0 & 0 & 1 \end{bmatrix}.$$

This type of split means that we have also split the strain energy into two corresponding parts: the distortional strain energy and the volumetric strain energy. We define an *incompressible material* as one that has no change in volume as it is deformed. For such a material $\nu = \frac{1}{2}$. For a nearly incompressible material, we note that as $\nu \rightarrow \frac{1}{2}$, we see that $\lambda \rightarrow \infty$ and $K \rightarrow \infty$. Since many rubber materials are nearly incompressible, we can expect to encounter this difficulty in several practical problems. Since incompressibility means no volume change, it also means there is no volumetric strain. In two-dimensions, this means that we have the *incompressibility constraint :*

$$\frac{\partial u}{\partial x} + \frac{\partial v}{\partial y} \equiv 0\,.$$

In such a case, we must either use an alternate variational form that involves the displacements and the mean stress (pressure), or we must undertake numerical corrections to prevent the solution from *locking*.

As an example of the use of the CST, consider the structure shown in Fig. 13.3.2. From Eq. (9.14) the element geometric constants are

$e=1$			$e=2$		
i	b_i	c_i	i	b_i	c_i
1	−2	−2	1	+2	+2
2	+2	0	2	−2	0
3	0	+2	3	0	−2

For the given data the constants multiplying the $\mathbf{S}_n$ and $\mathbf{S}_s$ matrices are 1×10^7 and 6×10^7, respectively. For the first element the two contributions to the element stiffness matrix are

$$\mathbf{S}_n^e = \frac{10^7}{2}\begin{bmatrix} +8 & & & & & \text{Sym.} \\ +2 & +8 & & & & \\ -8 & -2 & +8 & & & \\ 0 & 0 & 0 & 0 & & \\ 0 & 0 & 0 & 0 & 0 & \\ -2 & -8 & +2 & 0 & 0 & +8 \end{bmatrix}$$

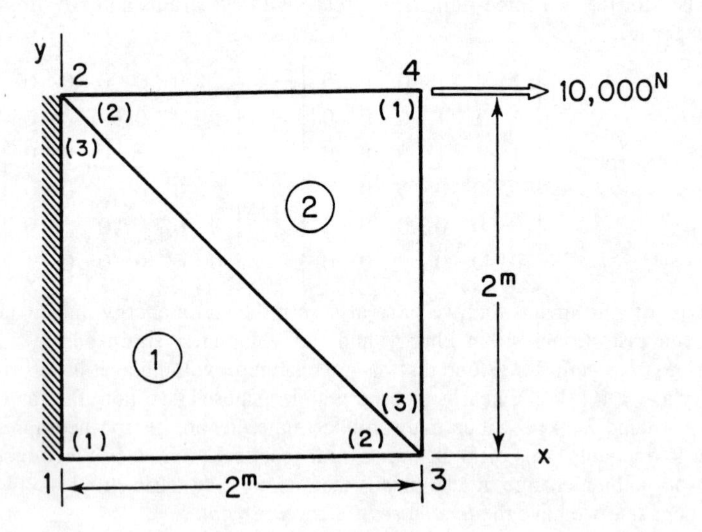

Figure 13.3.2 An example structure

$$\mathbf{S}_s^e = \frac{3 \times 10^7}{2} \begin{bmatrix} +1 & & & & & \text{Sym.} \\ +1 & +1 & & & & \\ 0 & 0 & 0 & & & \\ -1 & -1 & 0 & +1 & & \\ -1 & -1 & 0 & +1 & +1 & \\ 0 & 0 & 0 & 0 & 0 & 0 \end{bmatrix}$$

Thus, the first element stiffness is:

$$S^e = 5 \times 10^6 \begin{bmatrix} +11 & & & & & \text{Sym.} \\ +5 & +11 & & & & \\ -8 & -2 & +8 & & & \\ -3 & -3 & 0 & +3 & & \\ -3 & -3 & 0 & +3 & +3 & \\ -2 & -8 & +2 & 0 & 0 & +8 \end{bmatrix}$$

and its global and local degree of freedom numbers are the same. The second stiffness matrix happens to be the same due to its $180°$ rotation in space. Of course, its global dof numbers are different. That list is: 7, 8, 3, 4, 5, and 6. Since there are no body forces or surface tractions these matrices can be assembled to relate the system stiffness to the applied point load, P, and the support reactions. Applying the direct assembly procedure gives

$$\begin{array}{c} \quad 1 \quad 2 \quad 3 \quad 4 \quad 5 \quad 6 \quad 7 \quad 8 \quad \text{global} \end{array}$$

$$5 \times 10^6 \begin{bmatrix} +11 & & & & & & & \text{Sym.} \\ +5 & +11 & & & & & & \\ -3 & -3 & +11 & & & & & \\ -2 & -8 & 0 & +11 & & & & \\ -8 & -2 & 0 & +5 & +11 & & & \\ -3 & -3 & +5 & 0 & 0 & +11 & & \\ 0 & 0 & -8 & -3 & -3 & -2 & +11 & \\ 0 & 0 & -2 & -3 & -3 & -8 & +5 & +11 \end{bmatrix} \mathbf{u} = \begin{Bmatrix} R_1 \\ R_2 \\ R_3 \\ R_4 \\ 0 \\ 0 \\ 10^4 \\ 0 \end{Bmatrix}.$$

Applying the conditions of zero displacement at nodes 1 and 2 reduces this set to

$$5 \times 10^6 \begin{bmatrix} 11 & & & \text{Sym.} \\ 0 & 11 & & \\ -3 & -2 & 11 & \\ -3 & -8 & 5 & 11 \end{bmatrix} \begin{Bmatrix} u_3 \\ v_3 \\ u_4 \\ v_4 \end{Bmatrix} = \begin{Bmatrix} 0 \\ 0 \\ 10^4 \\ 0 \end{Bmatrix}.$$

Inverting the matrix and solving gives the required displacement vector, transposed: $10^5 \times u_r = [\,2.52 \quad -6.72 \quad 24.65 \quad -15.41\,]\, m$. Substituting to find the reactions yields $\mathbf{R}_g^T = [\,-0.002 \quad -756.3 \quad -10{,}000 \quad -756.3\,]\, N$. The deformed shape and resulting reactions are shown in Fig. 13.3.3. One should always check the equilibrium of the reactions and applied loads. Checking $\sum \mathbf{F}_x = 0$, $\sum \mathbf{M} = 0$ does show minor errors in about the sixth significant figure. Thus, the results are reasonable.

At this point we can recover the displacements for each element, and then compute the strains and stress. The element dof vectors (in meters) are, respectively

$$\boldsymbol{\delta}^{e^T} = \begin{bmatrix} 0 & 0 & 2.521 & -6.723 & 0 & 0 \end{bmatrix} \times 10^{-5}$$

$$\boldsymbol{\delta}^{e^T} = \begin{bmatrix} 24.650 & -15.406 & 0 & 0 & 2.521 & -6.723 \end{bmatrix} \times 10^{-5}$$

and the strain-displacement matrices, from Eq. (13.18) are

$$\mathbf{B}^e = \frac{1}{4}\begin{bmatrix} -2 & 0 & 2 & 0 & 0 & 0 \\ 0 & -2 & 0 & 0 & 0 & 2 \\ -2 & -2 & 0 & 2 & 2 & 2 \end{bmatrix}$$

for $e = 1$, while for $e = 2$

$$\mathbf{B}^e = \frac{1}{4}\begin{bmatrix} 2 & 0 & -2 & 0 & 0 & 0 \\ 0 & 2 & 0 & 0 & 0 & -2 \\ 2 & 2 & 0 & -2 & -2 & 0 \end{bmatrix}.$$

Recovering the element strains, $\boldsymbol{\varepsilon}^e = \mathbf{B}^e \boldsymbol{\delta}^e$ in meters/meter gives

$$e = 1, \quad \boldsymbol{\varepsilon}^{e^T} = 10^{-5}\begin{bmatrix} 1.261 & 0.000 & -3.361 \end{bmatrix}$$

$$e = 1, \quad \boldsymbol{\varepsilon}^{e^T} = 10^{-5}\begin{bmatrix} 12.325 & -4.342 & 3.361 \end{bmatrix}.$$

Utilizing the constitute law in Eq. (13.15), with $\boldsymbol{\varepsilon}_o = \mathbf{0}$, gives

$$\mathbf{D}^e = \frac{15 \times 10^9}{(15/16)}\begin{bmatrix} 1 & 1/4 & 0 \\ 1/4 & 1 & 0 \\ 0 & 0 & 3/8 \end{bmatrix} = 2 \times 10^9 \begin{bmatrix} 8 & 2 & 0 \\ 2 & 8 & 0 \\ 0 & 0 & 3 \end{bmatrix},$$

and the element stresses, in Newtons/meter^2, are

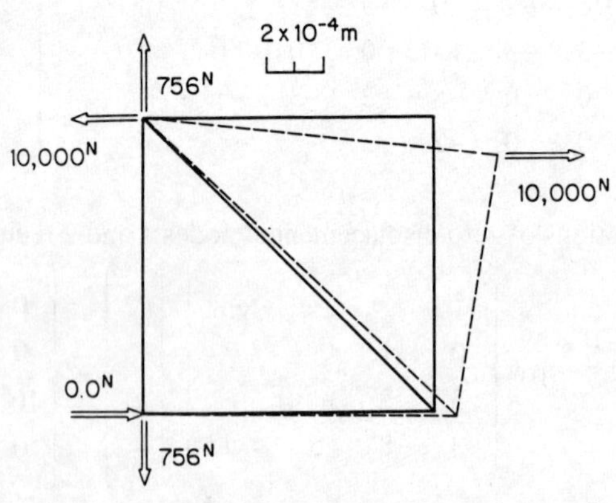

Figure 13.3.3 Deformed shape and reactions for a point load

$$e = 1, \quad \sigma^{e^T} = 10^4 \begin{bmatrix} 20.17 & 5.04 & -20.17 \end{bmatrix}$$

$$e = 2, \quad \sigma^{e^T} = 10^4 \begin{bmatrix} 179.83 & -20.17 & 20.17 \end{bmatrix}.$$

A good engineer should have an estimate of the desired solution before approaching the computer. For example, if the load had been at the center of the edge, then

$$\sigma_x = P/A = 10^4/(2)\,(5 \times 10^{-3}) = 10^6\,\text{N/m}^2,$$

and $\sigma_y = 0 = \tau$. The values are significantly different from the computed values. A better estimate would consider both the axial and bending effects so $\sigma_x = P/A \pm Mc/I$. At the centroid of these two elements ($y = 0.667$ and $y = 1.333$) the revised stress estimates are $\sigma_x = 0$ and $\sigma_x = 2 \times 10^6\,\text{N/m}^2$, respectively. The revised difference between the maximum centroidal stress and our estimate is only 10%. Of course, with the insight gained from the mechanics of materials our mesh was not a good selection. We know that while an axial stress would be constant across the depth of the member, the bending effects would vary linearly with y. Thus, it was poor judgement to select a single element through the thickness. Source code, data, and output for the above two examples are included on the .

To select a better mesh we should imagine how the stress would vary through the member. Then we would decide how many constant steps are required to get a good fit to the curve. Similarly, if we employed linear stress triangles (LST) we would estimate the required number of piece-wise linear segments needed to fit the curve. For example, consider a cantilever beam subjected to a bending load at its end. We know the exact normal stress is linear through the thickness and the shear stress varies quadratically through the thickness. Thus, through the depth we would need several CST, or a few LST, or a single cubic triangle (QST).

13.4 Stress and Strain Transformations *

Having computed the global stress components at a point in an element, we may wish to find the stresses in another direction. This can be done by employing the transformations associated with *Mohr's circle*. Mohr's circles of stress and strain are usually used to produce graphical solutions. However, here we wish to rely on automated numerical solutions. Thus, we will review the *stress transformation* laws. Refer to Fig. 13.4.1 where the quantities used in Mohr's transformation are defined. The alternate coordinate set (n, s) is used to describe the surfaces on which the normal stresses, σ_n and σ_s, and the shear stress, τ_{ns}, act. The n-axis is rotated from the x-axis by a positive (counter-clockwise) angle of β. By considering the equilibrium of the differential element, it is shown in mechanics of materials that

$$\sigma_n = \sigma_x \text{Cos}^2\beta + \sigma_y \text{Sin}^2\beta + 2\tau_{xy} \text{Sin}\,\beta\,\text{Cos}\,\beta\,. \tag{13.29}$$

Likewise the shear stress component is found to be

$$\tau_{ns} = -\sigma_x \text{Sin}\,\beta\,\text{Cos}\,\beta + \sigma_y \text{Sin}\,\beta\,\text{Cos}\,\beta + \tau_{xy}(\text{Cos}^2\beta - \text{Sin}^2\beta)\,. \tag{13.30}$$

For Mohr's circle only these two stresses are usually plotted in the $\sigma_n - \tau_{ns}$ space. However, for a useful analytical statement we also need to define σ_s. Again from equilibrium considerations it is easy to show that

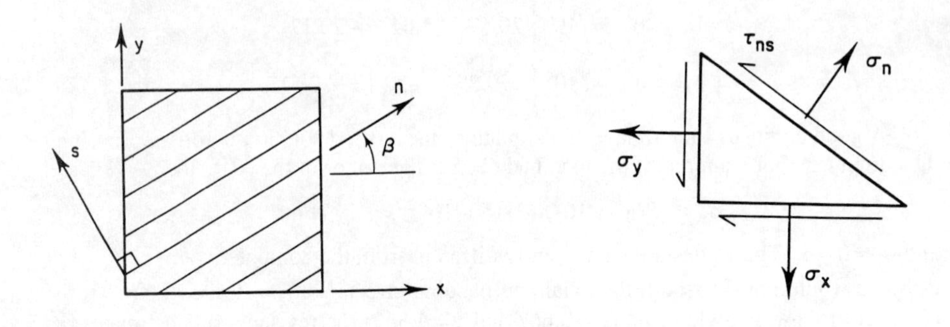

Figure 13.4.1 Local material or stress axes

$$\sigma_s = \sigma_x \text{Sin}^2\beta + \sigma_y \text{Cos}^2\beta - 2\tau_{xy} \text{Sin}\,\beta\,\text{Cos}\,\beta. \tag{13.31}$$

Prior to this point we have employed matrix notation to represent the stress components. Then we were considering only the global coordinates. But now when we refer to the stress components it will be necessary to indicate which coordinate system is being utilized. We will employ the subscripts *xy* and *ns* to distinguish between the two systems. Thus, our previous stress component array will be denoted by

$$\sigma^T = \sigma_{xy}^T = [\sigma_x \quad \sigma_y \quad \tau_{xy}]$$

while the new stress components will be ordered in a similar manner and denoted by

$$\boldsymbol{\sigma}_{ns}^T = [\sigma_n \quad \sigma_s \quad \tau_{ns}].$$

In this notation the stress transformation laws can be written as

$$\begin{Bmatrix} \sigma_n \\ \sigma_s \\ \tau_{ns} \end{Bmatrix} = \begin{bmatrix} +C^2 & +S^2 & +2SC \\ +S^2 & +C^2 & -2SC \\ -SC & +SC & (C^2 - S^2) \end{bmatrix} \begin{Bmatrix} \sigma_x \\ \sigma_y \\ \tau_{xy} \end{Bmatrix} \tag{13.32}$$

where $C \equiv \text{Cos}\,\beta$ and $S \equiv \text{Sin}\,\beta$ for simplicity. In symbolic matrix form this is

$$\boldsymbol{\sigma}_{ns} = \mathbf{T}(\beta)\,\boldsymbol{\sigma}_{xy} \tag{13.33}$$

where **T** will be defined as the stress transformation matrix. Clearly, if one wants to know the stresses on a given plane one specifies the angle β, forms **T**, and computes the results from Eq. (13.32).

A similar procedure can be employed to express Mohr's circle of strain as a strain matrix transformation law. If we denote the new strains as

$$\boldsymbol{\varepsilon}_{ns}^T = [\varepsilon_n \quad \varepsilon_s \quad \gamma_{ns}]$$

then the strain transformation law is

$$\begin{Bmatrix} \varepsilon_n \\ \varepsilon_s \\ \gamma_{ns} \end{Bmatrix} = \begin{bmatrix} +C^2 & +S^2 & +SC \\ +S^2 & +C^2 & -SC \\ -2SC & +2SC & (C^2 - S^2) \end{bmatrix} \begin{Bmatrix} \varepsilon_x \\ \varepsilon_y \\ \gamma_{xy} \end{Bmatrix} \tag{13.34}$$

or simply

$$\boldsymbol{\varepsilon}_{ns} = \mathbf{t}(\beta)\, \boldsymbol{\varepsilon}_{xy} . \tag{13.35}$$

Note that the two transformation matrices, **T** and **t**, are not identical. This is true because we have selected the engineering definition of the shear strain (instead of using the tensor definition). Also note that both of the transformation matrices are square. Therefore, the reverse relations can be found by inverting the transformations, that is,

$$\boldsymbol{\sigma}_{xy} = \mathbf{T}(\beta)^{-1} \boldsymbol{\sigma}_{ns}, \qquad \boldsymbol{\varepsilon}_{xy} = \mathbf{t}(\beta)^{-1} \boldsymbol{\varepsilon}_{ns} . \tag{13.36}$$

These two transformation matrices have the special property that the inverse of one is the transpose of the other, that is, it can be shown that

$$\mathbf{T}^{-1} = \mathbf{t}^T, \quad \mathbf{t}^{-1} = \mathbf{T}^T. \tag{13.37}$$

This property is also true when generalized to three-dimensional properties. Another generalization is to note that if we partition the matrices into normal and shear components, then

$$\mathbf{T} = \begin{bmatrix} \mathbf{T}_{11} & \mathbf{T}_{12} \\ \mathbf{T}_{21} & \mathbf{T}_{22} \end{bmatrix}, \quad \mathbf{t} = \begin{bmatrix} \mathbf{T}_{11} & \mathbf{T}_{12}/2 \\ 2\mathbf{T}_{21} & \mathbf{T}_{22} \end{bmatrix}.$$

In mechanics of materials it is shown that the principle normal stresses occur when the angle is given by

$$\text{Tan}(2\beta_p) = 2\tau_{xy} / (\sigma_x - \sigma_y).$$

Thus, if β_p were substituted into Eq. (13.31) one would compute the two principle normal stresses. In this case it may be easier to use the classical form that

$$\sigma_p = \frac{\sigma_x + \sigma_y}{2} \pm \left[\left(\frac{\sigma_x - \sigma_y}{2}\right)^2 + \tau_{xy}^2 \right]^{1/2}.$$

However, to illustrate the use of Eq. (13.31) we will use the results of the previous example to find the maximum normal stress at the second element centroid. Then

$$\text{Tan}(2\beta_p) = 2(20.17)/(179.83 - 20.17) = 0.2017$$

so $\beta_p = 5.70°$, $\text{Cos}\ \beta_p = 0.995$, $\text{Sin}\ \beta_p = 0.099$, and the transformation is

$$\begin{Bmatrix} \sigma_n \\ \sigma_s \\ \tau_{ns} \end{Bmatrix} = \begin{bmatrix} 0.9901 & 0.0099 & 0.1977 \\ 0.0099 & 0.9901 & -0.1977 \\ -0.0989 & 0.0989 & 0.9803 \end{bmatrix} \begin{Bmatrix} 179.83 \\ -20.17 \\ 20.17 \end{Bmatrix}$$

or

$$\boldsymbol{\sigma}_{ns}^T = [\ 181.84 \quad -22.18 \quad -0.00\ ]^{\text{N/m}^2}.$$

We should also recall that the maximum shear stress is

$$\tau_{\max} = (\sigma_n - \sigma_s)^2 / 2 = 102.01\ \text{N/m}^2.$$

These shear stresses occur on planes located at $(\beta_p \pm 45°)$. The classical form for $\tau_{\max}$ is

$$\tau_{max}^2 = (\frac{\sigma_x - \sigma_y}{2})^2 + \tau_{xy}^2 .$$

The designer needs to decide which failure criterion should be applied to the material to be utilized. The most common criterion are the Von Mises Effective Stress, the Maximum Stress, and the Maximum Shear Stress. There are discussed again in Sec. 15.5.

The above example is not finished at this point. In practice, we probably would want to check the failure criterion for this material, and obtain an error estimate to begin an adaptive solution. There are many failure criteria. The three most common ones are the Maximum Principal Stress, the Maximum Shear Stress, and the Von Mises Strain Energy of Distortion. The latter is most common for ductile materials. It can be expressed in terms of a scalar measure known as the *Effective Stress*, σ_E:

$$\sigma_E = \frac{1}{\sqrt{2}} \sqrt{(\sigma_x - \sigma_y)^2 + (\sigma_x - \sigma_z)^2 + (\sigma_y - \sigma_z)^2 + 6(\tau_{xy} + \tau_{xy} + \tau_{yz}^2)}$$

$$\sigma_E = \frac{1}{\sqrt{2}} \sqrt{(\sigma_1 - \sigma_2)^2 + (\sigma_1 - \sigma_3)^2 + (\sigma_2 - \sigma_3)^2}$$

in terms of the stress tensor components and principal stresses, respectively. For yielding in a simple tension test, $\sigma_x = \sigma_{yield}$, and all the other stresses are zero. Then,

$$\sigma_E = \sigma_{yield} \quad \leftrightarrow \quad \text{failure.}$$

This is the general test for ductile materials. For brittle materials, one may use the maximum stress criteria where

$$\sigma_1 = \sigma_{yield} \quad \leftrightarrow \quad \text{failure.}$$

The Tresca maximum shear stress criteria states

$$\tau_{max} = ½(\sigma_1 - \sigma_3) = ½\,\sigma_{yield} \quad \leftrightarrow \quad \text{failure.}$$

For the plane stress state, all the z-components of the stress tensor are zero. However, in the state of plane strain, σ_z is not zero and must be recovered using the Poisson ratio effect:

$$\sigma_z = \frac{E}{(1 + \nu)\,(1 - 2\nu)} \left[\nu\sigma_x + \nu\sigma_y \right] .$$

13.5 Anisotropic Materials *

A material is defined to be *isotropic* if its material properties do not depend on direction. Otherwise it is called *anisotropic*. Most engineering materials are considered to be isotropic. However, there are many materials that are anisotropic. Examples of anisotropic materials include reinforced concrete, plywood, and filament wound fiberglass. Several special cases of anisotropic behavior have been defined and are discussed in the Appendix. Probably the most common case is that of an *orthotropic material*. An orthotropic material has structural (or thermal) properties that can be defined in terms of two principal material axis directions. Let (*n, s*) be the principal material axis directions. For anisotropic materials it is usually easier to define the generalized constitutive law in the form:

$$\boldsymbol{\varepsilon}_{ns} = \mathbf{D}_{ns}^{-1}\boldsymbol{\sigma}_{ns} + \boldsymbol{\varepsilon}_{0\,ns}. \tag{13.38}$$

Note by way of comparison that Eq. (13.38) is written relative to the global coordinate axes. In Eq. (13.14) the square matrix contains the mechanical properties as experimentally measure relative to the principal material directions. For a two-dimensional orthotropic material the constitutive law is

$$\begin{Bmatrix} \varepsilon_n \\ \varepsilon_s \\ \gamma_{ns} \end{Bmatrix} = \begin{bmatrix} 1/E_n & -\nu_{sn}/E_s & 0 \\ -\nu_{ns}/E_n & 1/E_s & 0 \\ 0 & 0 & 1/G_{ns} \end{bmatrix} \begin{Bmatrix} \sigma_n \\ \sigma_s \\ \tau_{ns} \end{Bmatrix} + \varepsilon_{0\,ns}. \tag{13.39}$$

Here the moduli of elasticity in the two principal directions are denoted by E_n and E_s. The shear modulus, G_{ns}, is independent of the elastic moduli. The two Poisson's ratios are defined by the following notation:

$$\nu_{ij} = \varepsilon_i/\varepsilon_j \tag{13.40}$$

where i denotes the direction of the load, ε_i is the normal strain in the load directions, and ε_j is the normal strain in the transverse (orthogonal) direction. Symmetry considerations result in the additional requirement that

$$E_n\nu_{sn} = E_s\nu_{ns}. \tag{13.41}$$

Thus, four independent constants must be measured to define the orthotropic material mechanical properties. If the material is isotropic then $\nu = \nu_{ns} = \nu_{sn}$, $E = E_s = E_n$, and $G = G_{ns} = E/[2(1 + \nu)]$. In that case only two constants (E and ν) are required and they can be measured in any direction. When the material is isotropic then Eq. (13.39) reduces to Eq. (13.15).

Orthotropic materials also have thermal properties that vary with direction. If $\Delta\theta$ denotes the temperature change from the stress free state then the local initial thermal strain is

$$\varepsilon_{0\,ns}^T = \Delta\theta[\alpha_n \quad \alpha_s \quad 0]$$

where α_n and α_s are the principal coefficients of thermal expansion. If one is given the orthotropic properties it is common to numerically invert $\mathbf{D}_{ns}^{-1}$ to give the usual form

$$\boldsymbol{\sigma}_{ns} = \mathbf{D}_{ns}(\boldsymbol{\varepsilon}_{ns} - \boldsymbol{\varepsilon}_{0\,ns}). \tag{13.42}$$

This is done since the form of $\mathbf{D}_{ns}$ is algebraically much more complicated than its inverse. Due to experimental error in measuring the anisotropic constants there is a potential difficulty with this concept. For a physically possible material it can be shown that both $\mathbf{D}$ and $\mathbf{D}^{-1}$ must be positive definite. This means that the determinant must be greater than zero. Due to experimental error it is not unusual for this condition to be violated. When this occurs the occurs the program should be designed to stop and require acceptable data. Then the user must adjust the experimental data to satisfy the condition that

$$(E_n - E_s\nu_{ns}^2) > 0\,.$$

From the previous section on stress and strain transformations we know how to obtain $\boldsymbol{\varepsilon}_{xy}$ and $\boldsymbol{\sigma}_{xy}$ from given $\boldsymbol{\varepsilon}_{ns}$ and $\boldsymbol{\sigma}_{ns}$. But how do we obtain $\mathbf{D}_{xy}$ from $\mathbf{D}_{ns}$? We must have $\mathbf{D}_{xy}$ to form the stiffness matrix since it must be integrated relative to the x-y axes, that is,

$$\mathbf{K}_{xy} = \int \mathbf{B}_{xy}^T \mathbf{D}_{xy} \mathbf{B}_{xy} \, dv .$$

Thus, the use of the n–s axes, in Fig. 13.4.1, to define (input) the material properties requires that we define one more transformation law. It is the transformation from $\mathbf{D}_{ns}$ to $\mathbf{D}_{xy}$. There are various ways to derive the required transformation. One simple procedure is to recall that the strain energy density is a scalar. Therefore, it must be the same in all coordinate systems. The strain energy density at a point is

$$dU = \tfrac{1}{2}\, \sigma^T \varepsilon = \tfrac{1}{2}\, \varepsilon^T \sigma.$$

In the global axes it is

$$dU = \tfrac{1}{2}\, \sigma_{xy}^T \varepsilon_{xy} = \tfrac{1}{2}\, (\mathbf{D}_{xy} \varepsilon_{xy})^T \varepsilon_{xy}^T = \tfrac{1}{2}\, \varepsilon_{xy}^T \mathbf{D}_{xy} \varepsilon_{xy} .$$

In the principal material directions it is

$$dU = \tfrac{1}{2}\, \sigma_{ns}^T \varepsilon_{ns} = \tfrac{1}{2}\, (\mathbf{D}_{ns} \varepsilon_{ns})^T \varepsilon_{ns} = \tfrac{1}{2}\, \varepsilon_{ns}^T \mathbf{D}_{ns} \varepsilon_{ns} .$$

But from our Mohr's circle transformation for strain

$$\varepsilon_{ns} = \mathbf{t}_{ns} \varepsilon_{xy}$$

so in the n–s axes

$$dU = \tfrac{1}{2}\, (\mathbf{t}_{ns} \varepsilon_{xy})^T \mathbf{D}_{ns} (\mathbf{t}_{ns} \varepsilon_{xy}) = \tfrac{1}{2}\, \varepsilon_{xy}^T (\mathbf{t}_{ns}^T \mathbf{D}_{ns} \mathbf{t}_{ns}) \varepsilon_{xy} . \tag{13.43}$$

Comparing the two forms of dU gives the *constitutive transformation* law that

$$\mathbf{D}_{xy} = \mathbf{t}_{ns}^T \mathbf{D}_{ns} \mathbf{t}_{ns} . \tag{13.44}$$

The same concept holds for general three-dimensional problems.

Before leaving the concept of anisotropic materials we should review the initial thermal strains. Recall that for an isotropic material or for an anisotropic material in principal axes a change in temperature does not induce an initial shear strain. However, an anisotropic material does have initial thermal shear strain in other coordinate directions. From Eqs. (13.36) and (13.37) we have

$$\varepsilon_{0\,xy} = \mathbf{t}_{ns}^{-1} \varepsilon_{0\,ns} , \qquad \begin{Bmatrix} \varepsilon_x^0 \\ \varepsilon_y^0 \\ \gamma_{xy}^0 \end{Bmatrix} = \begin{bmatrix} +C^2 & +S^2 & -SC \\ +S^2 & +C^2 & +SC \\ +2SC & -2SC & (C^2 - S^2) \end{bmatrix} \begin{Bmatrix} \varepsilon_n^0 \\ \varepsilon_s^0 \\ 0 \end{Bmatrix} .$$

Thus, the thermal shear strain is $\gamma_{xy}^0 = 2 \,\mathrm{Sin}\, \beta \,\mathrm{Cos}\, \beta \,(\varepsilon_n^0 - \varepsilon_s^0)$. This is not zero unless the two axes systems are the same ($\beta = 0$ or $\beta = \pi/2$). Therefore, one must replace the previous null terms in Eq. (13.17).

13.6 Axisymmetric Solids*

There is another common elasticity problem class that can also be formulated as a two-dimensional problem involving two unknown displacement components. It is an axisymmetric solid subject to axisymmetric loads and axisymmetric supports. That is, the geometry, properties, loads, and supports do not have any variation around the circumference of the solid. This was sketched in Fig. 8.1.1. The problem is usually discussed in terms of axial and radial position, and axial and radial displacements. The solid is defined by the shape in the radial-axial plane as it is completely revolved about the axis. Let (R, Z) denote the coordinates in the plane of revolution, and u, v denote the corresponding radial and axial displacements at any point. This is an extension of the methods in Sec. 8.3 in that we now allow changes in the axial, Z, direction. The axisymmetric solid has four stress and strain components, three of them the same as those in the state of plane stress. We simply replace the x, y subscripts with R, Z.

The fourth strain is the so-called *hoop strain.* It arises because of the material around the circumference changes length as it moves radially. The circumferential strain at a radial position, R, is defined as

$$\varepsilon_\Theta = \frac{\Delta L}{L} = \frac{2\pi(R+u) - 2\pi R}{2\pi R} = \frac{u}{R}. \tag{13.45}$$

This is a normal strain, and it is usually placed after the other two normal strains. Note that on the axis of revolution, $R = 0$. It can be shown that both $u = 0$ and $\varepsilon_\Theta = 0$ on the axis of revolution. However, one can encounter numerical problems if numerical integration is employed with a rule that has quadrature points on the edge of an element.

We typically order the strains as $\boldsymbol{\varepsilon}^T = [\,\varepsilon_R \ \varepsilon_Z \ \varepsilon_\Theta \ \gamma_{RZ}\,]$ and the corresponding stresses as $\boldsymbol{\sigma} = [\,\sigma_R \ \sigma_Z \ \sigma_\Theta \ \tau_{RZ}\,]$ where σ_Θ is the corresponding *hoop stress.* From the above definition we now see that the contribution of a typical node j to the strain displacement $\mathbf{B}^e$ matrix is

$$\mathbf{B}_j^e = \begin{bmatrix} d_{Rj} & 0 \\ 0 & d_{Zj} \\ H_j/R & 0 \\ d_{Zj} & d_{Rj} \end{bmatrix}. \tag{13.46}$$

Therefore, we now see that in addition to the physical derivatives of the interpolation functions, we now must also include the actual interpolation functions and the radial coordinate. The corresponding stress-strain law for an isotropic material is given in terms of the constitutive matrix

$$\mathbf{D} = \frac{E}{1-\nu^2}\begin{bmatrix} 1 & \nu & \nu & 0 \\ \nu & 1 & \nu & 0 \\ \nu & \nu & 1 & 0 \\ 0 & 0 & 0 & \frac{1-\nu}{2} \end{bmatrix}.$$

For an orthotropic axisymmetric material, we utilize the material properties in the principal material axis (n, s, Θ) direction. In that case, the compliance matrix, $\mathbf{D}_{ns}^{-1}$, and initial strain matrix, $\varepsilon_{\Theta_{ns}}$ are

$$\begin{Bmatrix} \varepsilon_n \\ \varepsilon_s \\ \varepsilon_\Theta \\ \gamma_{ns} \end{Bmatrix} = \begin{bmatrix} \frac{1}{E_n} & \frac{-\nu_{sn}}{E_s} & \frac{-\nu_{\Theta n}}{E_\Theta} & 0 \\ \frac{-\nu_{ns}}{E_n} & \frac{1}{E_s} & \frac{-\nu_{\Theta s}}{E_\Theta} & 0 \\ \frac{-\nu_{n\Theta}}{E_n} & \frac{-\nu_{s\Theta}}{E_s} & \frac{1}{E_\Theta} & 0 \\ 0 & 0 & 0 & \frac{1}{G_{ns}} \end{bmatrix} + \Delta\Theta \begin{Bmatrix} \alpha_n \\ \alpha_s \\ \alpha_\Theta \\ 0 \end{Bmatrix}.$$

For the general anisotropic three-dimensional solid, there are nine independent material constants. However, due to the axisymmetry, there are only seven independent constants. When the material is also *transversely isotropic,* with the n - t plane being the plane of isotropic properties, then there are only five material constants and

$$E_n = E_\Theta, \quad \nu_{n\Theta} = \nu_{\Theta n}, \quad E_\Theta \nu_{s\Theta} = E_s \nu_{ns}.$$

In practical design problems with anisotropic materials, it is difficult to get accurate material constant measurements. To be a physically possible material, both $\mathbf{D}$ and $\mathbf{D}^{-1}$ should have a positive determinant. When that is not the case, the program should issue a warning and terminate the analysis. The possible values of Poisson's ratio can be bounded by

$$|\nu_{sn}| < \sqrt{E_s / E_n}, \quad -1 < \nu_{n\Theta} < 1 - 2 E_n \nu_{sn}^2 / E_s.$$

With the above changes and the observation that $d\Omega = 2\pi R\, da$, we note that the analytic integrals involve terms with $1/R$. These introduce logorithmic terms where we used to have only polynomial terms. Some of these become indeterminant at $R = 0$. For this and other practical considerations, one almost always employs numerical integration to form the element matrices. Clearly, one must interpolate from the given data to find the radial coordinate, R_q, at a quadrature point. Appendix II discusses other topics and notations for anisotropic materials.

13.7 General Solids *

For the completely general three-dimensional solid, there are three displacement components, $\mathbf{u}^T = [\, u \;\; v \;\; w \,]$, and the corresponding load vectors have three components. There are six stresses, $\boldsymbol{\sigma}^T = [\, \sigma_x \; \sigma_y \; \sigma_z \; \tau_{xy} \; \tau_{xz} \; \tau_{yz} \,]$, and six corresponding strain components, $\boldsymbol{\varepsilon}^T = [\, \varepsilon_x \; \varepsilon_y \; \varepsilon_z \; \gamma_{xy} \; \gamma_{xz} \; \gamma_{yz} \,]$. The enginering strains are defined by a partition at node j of:

$$\mathbf{B}_j^e = \begin{bmatrix} \frac{\partial H_j}{\partial x} & 0 & 0 \\ 0 & \frac{\partial H_j}{\partial y} & 0 \\ 0 & 0 & \frac{\partial H_j}{\partial z} \\ \frac{\partial H_j}{\partial y} & \frac{\partial H_j}{\partial x} & 0 \\ \frac{\partial H_j}{\partial z} & 0 & \frac{\partial H_j}{\partial x} \\ 0 & \frac{\partial H_j}{\partial z} & \frac{\partial H_j}{\partial y} \end{bmatrix}. \tag{13.47}$$

For an isotropic material, we have $D_{ij} = 0$, except for

$$D_{11} = D_{22} = D_{33} = (1-\nu)\,c, \qquad D_{12} = D_{13} = D_{23} = \nu c$$

$$D_{44} = D_{55} = D_{66} = G, \qquad c = E/(1+\nu)(1-2\nu), \qquad G = E/(1+\nu)\,2.$$

13.8 References

[1] Bathe, K.J., *Finite Element Procedures in Engineering Analysis*, Englewood Cliffs: Prentice-Hall (1982).

[2] Becker, E.B., Carey, G.F., and Oden, J.T., *Finite Elements – An Introduction*, Englewood Cliffs: Prentice-Hall (1981).

[3] Desai, C.S. and Abel, J.F., *Introduction to the Finite Element Method*, New York: Van Nostrand-Reinhold (1972).

[4] Gallagher, R.H., *Finite Element Analysis Fundamentals*, Englewood Cliffs: Prentice-Hall (1975).

[5] Hinton, E. and Owen, D.R.J., *Finite Element Programming*, London: Academic Press (1977).

[6] Hughes, T.J.R., *The Finite Element Method*, Englewood Cliffs: Prentice-Hall (1987).

[7] Meek, J.L., *Matrix Structural Analysis*, New York: McGraw-Hill (1972).

[8] Segerlind, L.J., *Applied Finite Element Analysis*, New York: John Wiley (1984).

[9] Szabo, B. and Babuska, I., *Finite Element Analysis*, New York: John Wiley (1991).

[10] Weaver, W.F., Jr. and Johnston, P.R., *Finite Elements for Structural Analysis*, Englewood Cliffs: Prentice-Hall (1984).

[11] Zienkiewicz, O.C., *The Finite Element Method*, 3rd edition, New York: McGraw-Hill (1979).

Chapter 14

ERROR MEASURES FOR ELLIPTIC PROBLEMS

14.1 Introduction

The Zienkiewicz and Zhu (or Z-Z) error estimator [23] has become popular for low order elements in adaptive applications. However, we now know tha it has some serious problems in general that have been overcome by its recent extension [20]. Here we will outline the basic method and notation of these types of approaches for error estimation. Consider a problem posed by the PDE written as

$$\mathbf{L}\,\phi + Q = 0 \quad in\,\Omega \tag{14.1}$$

with the essential boundary condition $\phi = \phi_o$ on boundary Γ_ϕ, and a prescribed traction $\mathbf{t} = \mathbf{t}_o$ on the boundary Γ_t with $\Gamma = \Gamma_\phi \cup \Gamma_t$. Here $\mathbf{L}$ is a linear differential operator that can be written in the symmetric form

$$\mathbf{L} \equiv \mathbf{S}^T \mathbf{D}\,\mathbf{S} \tag{14.2}$$

where $\mathbf{S}$ is a lower order operator and $\mathbf{D}$ contains material information. The gradient quantities of interest are denoted as

$$\boldsymbol{\varepsilon} \equiv \mathbf{S}\,\phi \tag{14.3}$$

and the flux quantities by

$$\mathbf{q} = -\mathbf{D}\,\boldsymbol{\varepsilon}. \tag{14.4}$$

On the boundary, Γ, of Ω we are often interested in a traction defined by

$$\mathbf{t} = \mathbf{G}\,\mathbf{q} \tag{14.5}$$

where $\mathbf{G}$ is usually the normal vector.

For example, in isotropic conduction ϕ is the temperature, Q, in internal volumetric heat source, $\mathbf{D} = k\,\mathbf{I}$, where k is the conductivity, and $\mathbf{S}$ is simply the gradient

$$\mathbf{S} = \nabla = \begin{Bmatrix} \dfrac{\partial}{\partial x} \\ \dfrac{\partial}{\partial y} \end{Bmatrix}$$

so that **L** becomes the Laplacian

$$\mathbf{L} = \nabla^T k \mathbf{I} \nabla = k \left[\frac{\partial^2}{\partial x^2} + \frac{\partial^2}{\partial y^2} + \frac{\partial^2}{\partial z^2} \right].$$

Here $\boldsymbol{\varepsilon}$ is the gradient vector

$$\boldsymbol{\varepsilon} = \nabla \phi, \quad \boldsymbol{\varepsilon}^T = \left[\frac{\partial \phi}{\partial x} \quad \frac{\partial \phi}{\partial y} \quad \frac{\partial \phi}{\partial z} \right]$$

and the Fourier Law defines the heat flux vector

$$\mathbf{q} = -k \mathbf{I} \nabla \phi = -k \nabla \phi, \quad q^T = [q_x \quad q_y \quad q_z].$$

Likewise, for $\mathbf{G} = \mathbf{n}$ the boundary traction is the normal heat flux:

$$\mathbf{t} = \mathbf{n}\mathbf{q} = k(q_x n_x + q_y n_y + q_z n_z) = q_n = -k \frac{\partial \phi}{\partial n}.$$

For the one-dimensional case of heat conduction these all reduce to scalars with

$$\mathbf{S} = \partial / \partial x, \quad \mathbf{D} = k, \quad \varepsilon = \partial \phi / \partial x, \quad \mathbf{q} = q_x = -k \, \partial \phi / \partial x$$

and $\mathbf{L}\phi + Q = 0$ becomes

$$\frac{\partial}{\partial x} \left[k \frac{\partial \phi}{\partial x} \right] + Q = 0$$

in Ω with $\phi = \phi_0$ on Γ_0. While on the boundary Γ_t, the traction $\mathbf{t} = q_n = -k n_x \partial \phi / \partial x$, and has an assigned value of $q_n = t_0$. In a finite element solution we seek an approximation $\hat{\phi}$ which, in turn, approximates the gradient and flux terms, $\hat{\boldsymbol{\varepsilon}}$ and $\hat{\boldsymbol{q}}$. The standard interpolation gives

$$\phi \approx \hat{\phi} = \mathbf{N}(x) \, \boldsymbol{\Phi}^e \quad \mathbf{x} \ in \ \Omega^e \tag{14.6}$$

with a corresponding gradient estimate

$$\varepsilon \approx \hat{\varepsilon} = \mathbf{S} \, \mathbf{N}(\mathbf{x}) \, \boldsymbol{\Phi}^e \tag{14.7}$$

$$\equiv \mathbf{B}^e(\mathbf{x}) \, \boldsymbol{\Phi}^e \tag{14.8}$$

for $\mathbf{x}$ in Ω^e. Likewise, the flux approximation is

$$\sigma \approx \hat{\sigma} = \mathbf{D}^e \, \mathbf{B}^e(\mathbf{x}) \, \boldsymbol{\Phi}^e. \tag{14.9}$$

In this notation the element square matrix and source vector are

$$\mathbf{K}^e = \int_{\Omega^e} \mathbf{B}^{e^T} \mathbf{D}^e \, \mathbf{B}^e \, d\Omega \tag{14.10}$$

and

$$\mathbf{F}_Q^e = \int_{\Omega^e} Q^e \, \mathbf{N}^{e^T} d\Omega \tag{14.11}$$

and the boundary traction contribution, if any, is

$$\mathbf{F}_{q_n}^b = \int_{\Gamma^b} q_n^b \, \mathbf{N}^{b^T} d\Gamma. \tag{14.12}$$

When the element degrees of freedom $\boldsymbol{\Phi}^e \subset \boldsymbol{\Phi}$ have been computed, the local errors in an element domain are

$$e_\phi(\mathbf{x}) \equiv \phi(\mathbf{x}) - \hat{\phi}(\mathbf{x}) \tag{14.13}$$

$$\mathbf{e}_\varepsilon(\mathbf{x}) \equiv \boldsymbol{\varepsilon}(\mathbf{x}) - \hat{\boldsymbol{\varepsilon}}(\mathbf{x}), \qquad \mathbf{x} \,\varepsilon\, \Omega^e \tag{14.14}$$

$$\mathbf{e}_\sigma(\mathbf{x}) \equiv \boldsymbol{\sigma}(\mathbf{x}) - \hat{\boldsymbol{\sigma}}(\mathbf{x}). \tag{14.15}$$

These quantities can be either positive or negative so we will mainly be interested in their absolute value or some normalized measure of them. We will employ integral *norms* for our error measures. On a linear space we can show that a norm has the properties given in Sec. 2.2. In finite elements we often use the *inner product* defined as

$$<u, v> \equiv \int_\Omega u(\mathbf{x})\, v(\mathbf{x})\, d\Omega \tag{14.16}$$

which passes a natural norm defined as

$$\|\phi\|^2 = <\phi, \phi> = \int_\Omega \phi(\mathbf{x})\, \phi(\mathbf{x})\, d\Omega. \tag{14.17}$$

This is also called the L_2 *norm,* since it involves the integral of the square of the argument. Error estimates commonly employ one of these norms which are

1. The energy norm $\| e \|$ defined as

$$\| e \| = \left[\int_\Omega (\varepsilon - \hat{\varepsilon})^T \mathbf{D} (\varepsilon - \hat{\varepsilon})\, d\Omega \right]^{1/2}. \tag{14.18}$$

$$= \left[\int_\Omega (\varepsilon - \hat{\varepsilon})^T (\sigma - \hat{\sigma})\, d\Omega \right]^{1/2} = \left[\int_\Omega \mathbf{e}_\varepsilon^T \mathbf{e}_\sigma\, d\Omega \right]^{1/2},$$

2. The L_2 flux or stress error norm

$$\| e_\sigma \|_{L_2} = \left[\int_\Omega (\sigma - \hat{\sigma})^T (\sigma - \hat{\sigma})\, d\Omega \right]^{1/2} = \left[\int_\Omega \mathbf{e}_\sigma^T \mathbf{e}_\sigma\, d\Omega \right]^{1/2}. \tag{14.19}$$

3. The root mean square stress error, $\Delta\sigma$, given by

$$\Delta\sigma = \| e_\sigma \|_{L_2}^2 / \Omega. \tag{14.20}$$

-4 In general, any of these norms is the sum of the corresponding individual element norms:

$$\|\phi\|^2 = \sum_e \|\phi\|_e^2 \tag{14.21}$$

where

$$\|\phi\|_e^2 = \int_{\Omega^e} \phi^2\, d\Omega \tag{14.22}$$

and $\Omega = \bigcup_e \Omega^e$. A relative percentage error can be defined as

$$\eta = \frac{\| e \|}{\| \phi \|} \times 100\% \tag{14.23}$$

which represents a weighted root mean square percentage error in the stresses. We can compute a similar estimate relative to the L_2 norms as

$$\eta_L = \frac{\| e \|_{L_2}}{\| \phi \|} \times 100\% . \tag{14.24}$$

In most of the papers on the subject of error estimators there is a discussion of the effectivity index, Θ. It is simply the ratio of the estimated error divided by the exact error. Usually an analtyical solution is employed to compute the exact error (and to assign the problem source and boundary conditions), but sometime very high precision numerical results are used. Clearly, one should search for methods where the effectivity index is very close to unity. Some methods employ a constant to increase their Θ to near unity for a specific element type.

14.2 Error Estimates

In general, we do not know the exact strain, ε, or stress values, σ, in Eqns. (14.3) and (14.4). We do have piecewise estimates for the element strains, $\hat{\varepsilon}$, and stresses, $\hat{\sigma}$. Unlike ϕ, these estimates are generally discontinuous between elements, as shown in Fig. 14.2.1. For homogeneous domains (homogeneous **D**), we expect the exact ε and σ to be continuous. At the interface of two different homogeneous materials ($\mathbf{D}_1$ and $\mathbf{D}_2$),

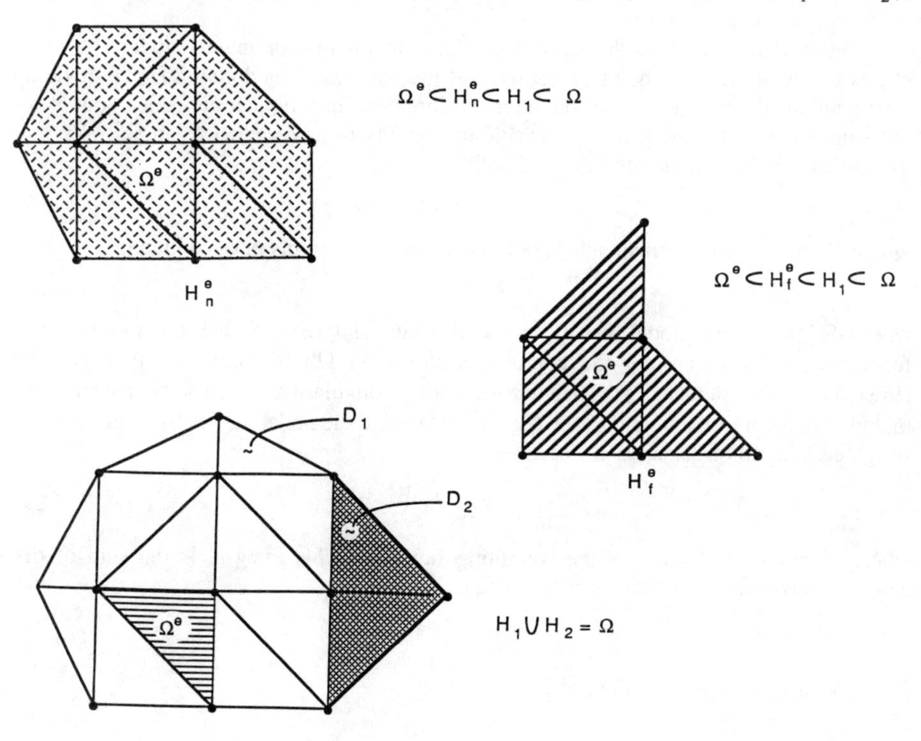

Figure 14.2.1 Patches for local error estimates

we expect $\boldsymbol{\varepsilon}$ to be discontinuous and $\boldsymbol{\sigma}$ to be continuous. In a homogeneous domain, $H \subseteq \Omega$ of Fig. 14.2.1, a continuous estimate of $\boldsymbol{\varepsilon}$ and $\boldsymbol{\sigma}$ should be more accurate than the piecewise continuous $\hat{\boldsymbol{\varepsilon}}$ and $\hat{\boldsymbol{\sigma}}$ would be. Denote such continuous approximations by $\boldsymbol{\varepsilon}^*$ and $\boldsymbol{\sigma}^*$, respectively. That is, $\hat{\boldsymbol{\sigma}}$ is discontinuous across element boundaries, while the $\boldsymbol{\sigma}^*$ are constructed to be continuous across those boundaries. Then within an element estimators with good accuracy are

$$\mathbf{e}_\varepsilon \approx \boldsymbol{\varepsilon}^*(\mathbf{x}) - \hat{\boldsymbol{\varepsilon}}(\mathbf{x}) \tag{14.25}$$

and

$$\mathbf{e}_\sigma \approx \boldsymbol{\sigma}^*(\mathbf{x}) - \hat{\boldsymbol{\sigma}}(\mathbf{x})\,. \tag{14.26}$$

There are various procedures for obtaining nodal values of the strains or stresses that will yield a continuous solution over the domain. Probably the most common is simply an averaging based on the number and/or size of elements contributing to a node. This is illustrated in Figs. 14.2.2 and 14.2.3, where continuous nodal stresses are obtained by averaging the values from surrounding elements. A precise mathematical procedure was given by Oden, et al. [5, 12, 13]. However, that "Conjugate Stress" approach required the assembly of element contributions and solving a system of equations equal in size to the number of nodes in the system. Approximations of that procedure still give useful results [13].

We will now outline the Z-Z procedures to obtain continuous nodal strains or stresses to be utilized in the error norms. Let the homogeneous domain, H, be made up of the union of some (or all) of the element domains. In a typical element domain, we have interpolated for the primary variable in Eqn. (14.6) as $\hat{\phi} = \mathbf{N}(\mathbf{x})\, \boldsymbol{\Phi}^e$ in Ω^e. We will interpolate for the continuous strains $\boldsymbol{\varepsilon}^*$ with

$$\boldsymbol{\varepsilon}^*(\mathbf{x}) = \mathbf{N}_\varepsilon(\mathbf{x})\mathbf{E}^e \quad \text{in } \Omega^e \tag{14.27}$$

where $\mathbf{E}^e$ denotes the element subset of the desired nodal strains in the domain:

$$\mathbf{E}^e \subset \mathbf{E}^H \tag{14.28}$$

where $\mathbf{E}^H$ is all the nodal values in the domain, and here $\mathbf{N}_\varepsilon$ are the interpolation functions used to represent the continuous strains in Ω^e. Usually, we pick $\mathbf{N}_\varepsilon = \mathbf{N}$. However, if the element is adjacent to a known singularity, $\mathbf{N}_\varepsilon$ may be modified to include the form of the singularity. We will determine the nodal values by requiring that weighted integral of the stress differences vanish:

$$I = \int_H \boldsymbol{\varepsilon}^{*T} (\boldsymbol{\varepsilon}^* - \hat{\boldsymbol{\varepsilon}})\, d\Omega = 0 \tag{14.29}$$

where $\boldsymbol{\varepsilon}^*$ has been chosen as the weighting function. This integral is the sum of the contributions from each element in H. Thus,

$$I = \sum_{\Omega^e \,\varepsilon\, H} I^e \tag{14.30}$$

where from Eqns. (14.8) and (14.27)

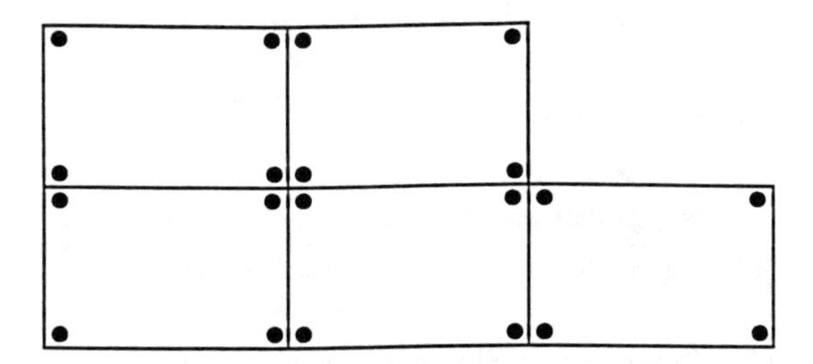

a) unaveraged flux on stress data

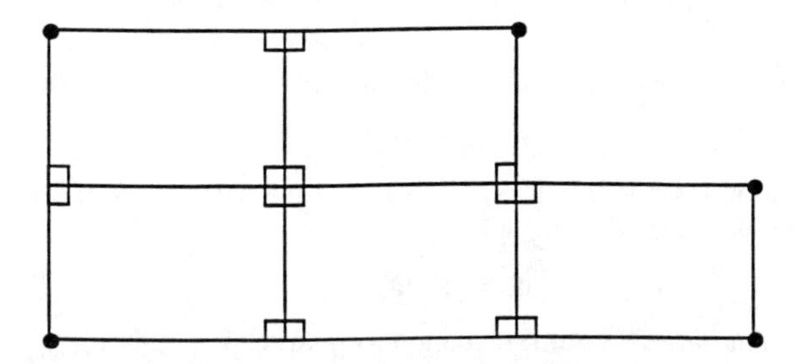

b) averaged data for post-processing

Figure 14.2.2 Forming nodal stress data

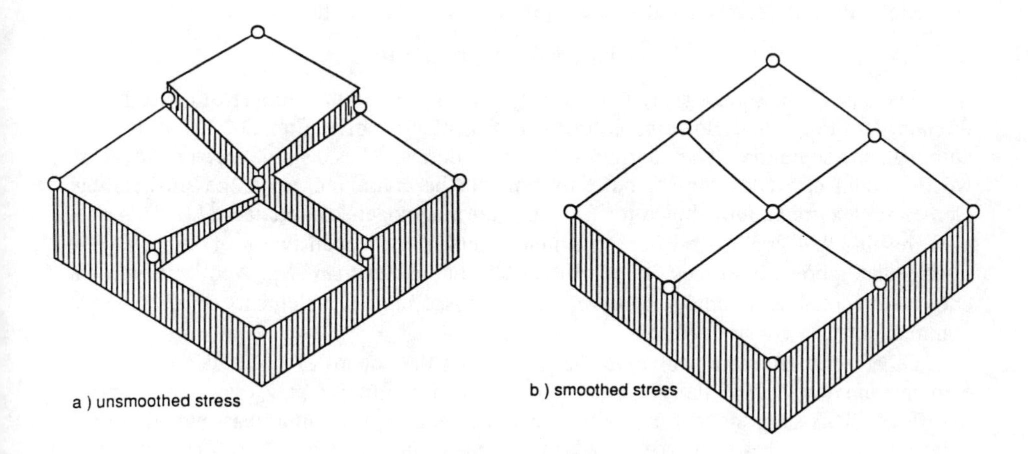

Figure 14.2.3 Local smoothing of stresses

$$I^e = \int_{\Omega^e} \boldsymbol{\varepsilon}^{*^T}(\boldsymbol{\varepsilon}^* - \hat{\boldsymbol{\varepsilon}})\, d\Omega = \int_{\Omega^e} \mathbf{E}^{e^T} \mathbf{N}_\varepsilon^T(\mathbf{N}_\varepsilon \mathbf{E}^e - \mathbf{B}^e \boldsymbol{\Phi}^e)\, d\Omega$$
$$= \mathbf{E}^{e^T} \mathbf{A}^e \mathbf{E}^e - \mathbf{E}^{e^T} \mathbf{P}^e \boldsymbol{\Phi}^e \qquad (14.31)$$

with the corresponding element square and rectangular matrices given by

$$\mathbf{A}^e = \int_{\Omega^e} \mathbf{N}_\varepsilon^T \mathbf{N}_\varepsilon\, d\Omega, \qquad \mathbf{P}^e = \int_{\Omega^e} \mathbf{N}_\varepsilon^T \mathbf{B}\, d\Omega\ . \qquad (14.32)$$

The assembly in Eqn. (14.30) is constructed in the standard way using the Boolean gather matrices, $\boldsymbol{\beta}^e$, $\boldsymbol{\lambda}^e$, such that $\boldsymbol{\Phi}^e = \boldsymbol{\beta}^e\, \boldsymbol{\Phi}$ and $\mathbf{E}^e = \boldsymbol{\lambda}^e\, \mathbf{E}$. Then, Eqn. (14.29) becomes

$$I = \sum_{\Omega^e \varepsilon H} \left[\mathbf{E}^T \boldsymbol{\lambda}^{e^T} \mathbf{A}^e \boldsymbol{\lambda}^e \mathbf{E} - \mathbf{E}^T \boldsymbol{\lambda}^{e^T} \mathbf{P}^e \mathbf{b}^e\, \boldsymbol{\Phi} \right] = \mathbf{E}^T(\mathbf{A}\,\mathbf{E} - \mathbf{P}\,\boldsymbol{\Phi}) = 0\,.$$

For a non-trivial solution, $\mathbf{E} \neq 0$, this requires that we solve the system

$$\mathbf{A}\,\mathbf{E} = \mathbf{p} \qquad (14.33)$$

to determine the continuous nodal values, $\mathbf{E}$, in sub-domain H. Here the vector $\mathbf{p}$ has been assembled from its element vector values

$$\mathbf{p}^e = \mathbf{P}^e\, \boldsymbol{\Phi}^e \qquad (14.34)$$

rather than storing the large system rectangular matrix $\mathbf{P}$. Solving the system in Eqn. (14.33) gives

$$\mathbf{E} = \mathbf{A}^{-1}\, \mathbf{p} \qquad (14.35)$$

where $\mathbf{A}$ is a sparse symmetric matrix obtained from the element $\mathbf{A}^e$. If H is large, then we may choose to employ a factorization of $\mathbf{A}$ rather than an inversion. Of course, if the $\mathbf{A}^e$ is a diagonal matrix, then the assembly and inversion is trivial and the storage requirements are minimal. A diagonal form for $\mathbf{A}^e$ can sometimes be obtained by special quadrature rules. Let $\mathbf{A}_d$ denote the diagonal approximation of $\mathbf{A}$. We can get a better estimate of the continuous nodal values by using n iterations of the form

$$\mathbf{E}_n = \mathbf{E}_{n-1} + \mathbf{A}_d^{-1}\,(\mathbf{A}\,\mathbf{E}_{n-1} - \mathbf{p})\,. \qquad (14.36)$$

The size of the square array A is directly proportional to the number of nodes in the domain H. For small domains it may be feasible to solve Eqn. (14.35) using the complete sparse matrix. The minimum choice for domain H is a single element Ω^e, then we get nodal estimates for $\mathbf{E}^e$, but they will not be continuous with the surrounding elements. The maximum choice for H is the entire homogeneous domain Ω_H. This will give results that are everywhere continuous but most expensive to compute. An intermediate choice is to pick H larger than Ω^e but smaller than Ω_H. Another common choice is to consider a nodal support region. That is, a region of elements connected to a common node, or constrained to it.

Other error estimators involve the integral of the square of residual error in the element and integral of the square of the jump discontinuity in $\boldsymbol{\varepsilon}$ over the element interfaces. This suggests that in addition to an element we want to at least include those adjacent elements that share a face with the element in question. That is probably the minimum domain to use for H, say H_F^e. Another choice is to use the element and all the element neighbors adjacent to it. That list of elements is generally available as a by-product of processing the elements for a frontal solution or for renumbering the system equations. We will denote the domain with all of the neighbors of the element as H_N^e.

A similar procedure is employed to obtain stress estimates. We replace the references to $\boldsymbol{\varepsilon}^*$ with $\boldsymbol{\sigma}^*$ and $\hat{\boldsymbol{\varepsilon}}$ with $\hat{\boldsymbol{\sigma}}$. The difference is that in Eqn. (14.32) the matrix $\mathbf{P}^e$ changes to

$$\mathbf{P}^e = \int_{\Omega^e} \mathbf{N}_\sigma^T \, \mathbf{D} \, \mathbf{B} \, d\Omega \,. \tag{14.37}$$

If the material properties $\mathbf{D}$ vary over the element, then this form would be a better approach. For constant properties we can simply compute $\boldsymbol{\sigma}^* = \mathbf{D} \, \boldsymbol{\varepsilon}^*$ and recover $\hat{\boldsymbol{\sigma}}$ from Eqn. (14.9).

Recent studies of the Z-Z method show that it is only reliable for quadratic elements applied to elliptic problems. For other elements, it does not act as a good error estimator. However, this approach can give smoothed values useful for postprocessing contour displays. The global nature of this method seems to lose local details that are important for accurate local error estimators. An alternate local procedure will be discussed later. The original Z-Z method has been found to be accurate mainly for the T3 and Q8 elements. It gives poor results for odd order elements. In some cases, its effectivity index converges to zero, and is not effective at all. It is implemented in some commercial codes so that the users of those programs are warned to use only the Q8 elements in the element library. They have developed a similar new hueristic approach, the super convergent patch recovery (SPR), that appears valid for most elements [26]. They have demonstrated numerically that one can generate superconvergence estimates for $\boldsymbol{\sigma}^*$ at a node by employing patches of elements surrounding the node. A local least squares fit is generated in the following way. Assume a polynomial approximation of the form

$$\boldsymbol{\sigma}^* = \mathbf{P}(\xi, n) \, \mathbf{a}$$

where $\mathbf{P}$ denotes a polynomial (in a local parametric coordinate system) that is of the same degree and completeness that was used to approximate the original solution, $\mathbf{u}_h$. That is, $\mathbf{P}$ is similar to $\mathbf{H}$. Recall that $\boldsymbol{\sigma}_h$ was computed using the derivatives of $\mathbf{H}$. To compute the estimate for $\boldsymbol{\sigma}^*$ at the nodes inside the patch, we minimize the function

$$F(a) = \sum_{j=1}^{n} \left[\boldsymbol{\sigma}_j^* - \hat{\boldsymbol{\sigma}}_j \right]^2 \quad \rightarrow \quad \min$$

where n is the number of integration points used in the elements that define the patch. Substituting the two different interpolation functions gives

$$F(\mathbf{A}) = \sum_{e=1} \sum_{j=1}^{n_e} \left[\mathbf{P}_j \mathbf{a} - \mathbf{D}^e \mathbf{B}_j^e \mathbf{u}^e \right]^2$$

where E denotes the number of elements in the patch and n_e is the number of integration points used to form $\hat{\boldsymbol{\sigma}}^e$. That is, we are seeking a least squares fit through the

$$n = \sum_{e=1}^{E} \sum_{j=1}^{n_e}$$

data points to compute the unknown coefficients, $\mathbf{a}$. The standard least squares minimization gives the local algebraic problem $\mathbf{S} \, \mathbf{a} = \mathbf{C}$ where

$$\mathbf{S} = \sum_{e=1}^{E} \sum_{j=1}^{n_e} \mathbf{P}^T(\xi_j, n_j)\, \mathbf{P}(\xi_j, n_j), \quad \mathbf{C} = \sum_{e=1}^{E} \sum_{j=1} n_e\, \mathbf{P}_j^T\, \mathbf{D}^e\, \mathbf{B}_j^e\, \mathbf{U}^e .$$

This is solved for the coefficients **a** of the local fit. The n stress sampling points are shown for interior edge, and corner nodes in Fig. 14.2.4. To avoid ill-conditioning common to least squares, the local patch fitting parametric space (ξ, n) is mapped to the square that encloses the patch of elements.

Zhu [21] has verified numerically that the derivatives estimated in this way have an accuracy of at least order $O(h^{p+1})$. There is a theorem that states that if the $\boldsymbol{\sigma}^*$ are super convergent of order $O(h^{p+\alpha})$ for $\alpha > 0$, then the error estimator will be asympotically exact. That is, the effectivity index should approach unit $\theta \rightarrow 1$. This means that we have the ability to get the maximum accuracy for a given number of degrees of freedom.

There is not yet a theoretical explaination for the "hyperconvergent" convergence (two orders higher) reported in the SPR numerical studies. It may be because the least square fit does not go exactly through the given Barlow points. Thus, they are really sampling nearby. In Sec. 3.8 we showed that derivative sampling points for a cubic are at ± 0.577, while those for the quartic are at ± 0.707. Therefore, patch may effectively be picking up those quartic derivative estimates and jumping to a higher degree of precision.

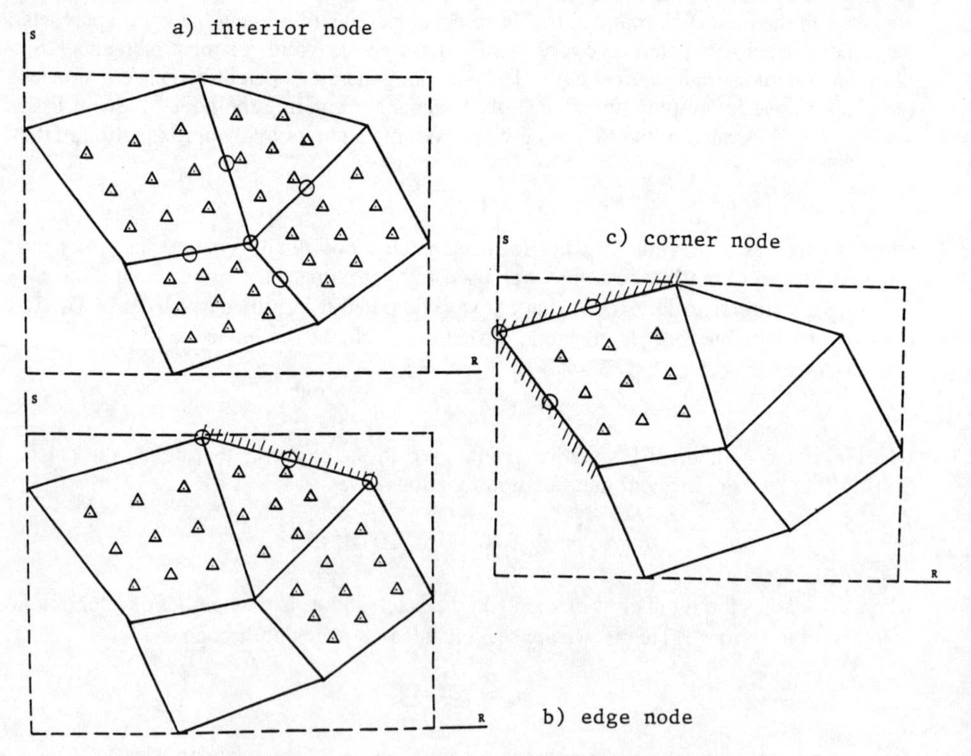

Figure 14.2.4 Local least square sampling points

We note that this, and other, local patch processing methods require some additional data from the database. In this case, we need the list of elements attached to a given node. This list can be calculated in various ways. For a fixed mesh, one could use a subroutine like ELTOPT (or LTOPV). If we are using adaptive solutions with constraint equations, we should add to this list those elements where the node is a master node in the constraint list. Other postprocessing operations may need similar lists. One can list the elements adjacent to a given element with subroutines ELTOEL and LTOLV, and the list of nodes connected to other nodes can be found from subroutines PTTOPT and PTOPV. Some adaptive programs keep only the list of elements that have only a common face with a given element. That is simply a subset of the complete element to element list. Several of these lists may already exist since they may be needed in an *equation re-ordering* system, such as REDUCE, or as part of pre-front or symbolic factorization process used in some more efficient equation solvers.

14.3 Adaptive Process

Theoretical and empirical studies show that the relative percentage error computed from Eqn. (14.23) or (14.24) will underestimate the true error. In practice a multiplying constant, $C^e > 1$, that accounts for the element type is employed to give the element estimate

$$\eta^* = C^e \eta_L . \tag{14.38}$$

For elasticity problems, Zienkiewicz [13] recommends that C^e be 1.1, 1.3, 1.6, and 1.4 for the Q4, T3, Q9 and T6 elements, respectively. To adaptively remesh the problem we seek to make the error percentage less than some specified value $\bar{\eta}$ that may be five percent or less. We want the energy norm to be the same in all the elements. The total allowed error per element is

$$T = \bar{\eta} \left[\frac{\| \phi \|^2 + \| e_\sigma \|^2}{m} \right]^{1/2}$$

for a system of m elements. We want each element to have a norm smaller than this value. For each element, i, we define the ratio $\xi_i = \| e \|_i / T$ to indicate needed refinement when $\xi_i > 1$ and de-refinement when $\xi_i < 1$. If h_i is the current element size, then asymptotic convergence rate estimates suggest that the new element size should be smaller than $h = h_i / \xi^{1/p}$ where p is the polynomial order of the interpolation for ϕ. Typically, a certain number of elements with the largest ξ are divided into smaller elements with each side reduced by a factor of two. This, in turn, requires that either constraint equations be introduced on the original sides to provide continuity to the extra nodes, or the surrounding elements must also be partially refined to yield continuity. A sample implementation of the above procedures are shown in Fig. 14.3.1 in subroutine ERROR.F . It was employed by Kyriacou for h-adaptive studies [9].

```
      SUBROUTINE ERROR (NE, N, NG, NELFRE, NDFREE, NODES, LNODE, INDEX,
     1                  DD, D, M, AVE, NS, NM, NQ, H, X, NSPACE, ELAVE,
     2                  B, EHAT, ESTAR, EEL, ENORM, GNORM, IADD, ERRAVE,
     3                  SYSNOR, NTAPE )
C     ----------------------------------------------------------------
C              ELEMENT ERROR ESTIMATES FOR ADAPTIVE SOLUTIONS
C     ----------------------------------------------------------------
      DIMENSION  DD(NDFREE), D(NELFRE), NODES(NE,N), AVE(M,NS+2), H(N),
     1           ELAVE (N,NS+2), B(NS,NELFRE), EHAT(NS), ESTAR(NS+2),
     2           EEL(NS+2), ENORM(NE), X(M,NSPACE), LNODE(N),
     3           INDEX(NELFRE)
C      THE FOLLOWING DIMENSIONS ARE FOR MOVIE.BYU PLOTS SUBROUTINE
      DIMENSION  IADD(M), ERRAVE(M)
C....  METHOD: ERROR ESTIMATE BY USING THE NORM OF DIFFERENCE
C....  BETWEEN  THE ELEMENT GRADIENT AND ITS NODAL AVERAGE GRADIENT.
C....  FOR STRESS ANALYSIS, GRADIENT MEANS STRESS "VECTOR".
C                           --- ARRAYS ---
C      AVE    = NODAL AVERAGED GRADIENT "VECTORS"
C      B      = B MATRIX IN STIFFNESS INTEGRAL
C      EEL    = VECTOR OF THE DIFFERENCE BETWEEN ESTAR AND EHAT
C      EHAT   = ARRAY CONTAINING THE VALUES OF THE DISCONTINUOUS
C               GRADIENTS AT THE QUADRATURE POINTS.
C      ELAVE = MATRIX CONTAINING THE VALUES OF THE NODAL
C               AVERAGE GRADIENTS FOR AN ELEMENT.
C      ENORM = THE ERROR L2 NORM FOR THE ELEMENT (RELATIVE)
C      ERRAVE= AVERAGE ERROR AT A NODE
C      ESTAR = ARRAY CONTAINING THE VALUES OF THE NODAL AVERAGE
C               GRADIENT AT THE QUADRATURE POINTS
C      GNORM = GLOBAL NORM
C      H      = ARRAY CONTAINING THE VALUES OF THE SHAPE FUNCTIONS
C               AT A QUADRATURE POINT
C      IADD   = NUMBER OF ELEMENTS CONNECTED TO A NODE
C      NS     = NUMBER OF STRAINS ( ROWS IN B , AND D )
C      SYSNOR= SYSTEM NORM. THE GRADIENT L2 NORM AS A ROOT MEAN
C               SQUARE OF ALL ELEMENTS
C
C....  INITIALIZING
      SYSNOR = 0.0
C....  REWIND THE FILE WITH THE DATA
      REWIND NTAPE
C....  LOOP OVER ELEMENTS TO FIND THE ERROR
      DO 50 IE = 1, NE
C....  EXTRACT D(NELFRE) ('PHI') AND ELAVE(N,NS) FOR THE ELEMENT
C....  FIND THE DEGREES OF FREEDOM FOR THIS ELEMENT
        CALL LNODES (IE, NE, N, NODES, LNODE)
        CALL INDXEL (N, NELFRE, NG, LNODE, INDEX)
        CALL ELFRE  (NDFREE, NELFRE, D, DD, INDEX)
C....  GET CONTRIBUTIONS FROM EACH COMPONENT OF THE GRADIENT 'AVE'
        DO 60 J = 1, NS
          CALL  ELFRE (M, N, ELAVE(1,J), AVE(1,J), LNODE)
   60   CONTINUE
C....  READ THE NUMBER OF POINTS (WRITTEN IN ISOPAR.F)
        READ (NTAPE) NIP
C....  LOOP OVER QUADRATURE POINTS FOR CURRENT ELEMENT
        ENORM(IE) = 0.0
        GNORM     = 0.0
```

Figure 14.3.1a Estimating element errors

```
        DO 70 IP = 1, NIP
C....    READ THE REQUIRED DATA :H,B,DETWT (WRITTEN IN ISOPAR.F)
          READ (NTAPE) ( H(I),I=1,N )
          READ (NTAPE) (( B(INS,I),INS=1,NS) ,I=1,NELFRE)
          READ (NTAPE) DETWT
C....    GET PRODUCTS  ESTAR= E(TRANSP) H(TRANSP)
C....      EHAT =   B D  FOR HEAT AND
C....      DB D  FOR STRESS ANALYSIS
C....      AND SUBTRACT THEM: ESTAR-EHAT
          DO 80 J = 1, NS
            SUM1 = 0.0
            SUM2 = 0.0
            DO  90 IN = 1, NELFRE
              SUM2 = SUM2 + B(J,IN) * D(IN)
 90         CONTINUE
            DO  91 IN = 1, N
              SUM1 = SUM1 + ELAVE(IN,J) * H(IN)
 91         CONTINUE
            EHAT(J)  = SUM2
            ESTAR(J) = SUM1
            EEL(J)   = ESTAR(J) - EHAT(J)
 80       CONTINUE
C....    FIND DOT PRODUCTS: SUM = EEL*EEL AND SUM2 = ESTAR*ESTAR
          SUM  = 0.0
          SUM2 = 0.0
          DO 100 J = 1, NS
            SUM2 = SUM2 + ESTAR(J)*ESTAR(J)
 100      SUM = SUM + EEL(J) * EEL(J)
C....    UPDATE NORM OF ERROR & NORM OF GRADIENT (NUMER INTEGR.)
          ENORM(IE) = ENORM(IE) + SUM * DETWT
          GNORM     = GNORM + SUM2 * DETWT
 70     CONTINUE
        ENORM(IE) = SQRT( ENORM(IE) )
C
C....   THE FOLLOWING APPLY THE EMPIRICAL CORRELATION FACTOR
C....   AS SUGGESTED BY ZIENKIEWICZ
C....   THIS IS 1.1 FOR BILINEAR,1.3 FOR LINEAR TRIANGLES,1.6
C....   FOR BIQUADRATIC AND 1.4 FOR QUADRATIC TRIANGLES
        IF ( N .EQ. 4 )  THEN
          FACTOR = 1.1
        ELSE IF ( N .EQ. 3 )  THEN
          FACTOR = 1.3
        ELSE IF ( N .EQ. 8 )  THEN
          FACTOR = 1.6
        ELSE IF ( N .EQ. 6 )  THEN
          FACTOR = 1.4
        ELSE
          FACTOR = 1.
        ENDIF
        ENORM(IE) = FACTOR*ENORM(IE)
C....   FINISHED APPLYING THE CORRELATION FACTOR
        SYSNOR = SYSNOR + GNORM
50    CONTINUE
C...  AVERAGE OVER ELEMENTS
      SYSNOR =  SYSNOR/NE
      PRINT *,'THE SYSTEM NORM (SYSNOR) IS',SYSNOR
```

Figure 14.3.1b Estimating element errors

```
C....   FIND THE RELATIVE ERRROR NORM ENORM FOR EACH ELEMENT
        DO 150 IE = 1, NE
          IF ( SYSNOR .NE. 0.0 ) THEN
            ENORM(IE) = (ENORM(IE)/SYSNOR)*100
            IF ( ENORM(IE) .GT. 0.05 ) THEN
              PRINT *,"THE ELEMENT #", IE," NEEDS ",
     1        "REFINEMENT. THE % ERROR EST. IS ", ENORM(IE)
            ELSE
              WRITE (*,*) 'THE % ERROR ESTIM. (ENORM) ',
     1                    'FOR EL #',IE,' IS', ENORM(IE)
            ENDIF
          ELSE
            PRINT *,'ERROR: SYSNOR IS EQUAL TO ZERO'
          ENDIF
150     CONTINUE
C....   CREATE A MOVIE.BYU FILE FOR VIEWING THE ELEMENTS ETC
C       CALL  MOVIE (N,M,NE,X,NODES,DD,ENORM,LNODE,IADD,ERRAVE)
        RETURN
        END
```

Figure 14.3.1c Estimating element errors

14.4 Hierarchical Error Indicator

Zienkiewicz and Morgan [22] have given a detailed study of how hierarchical interpolation functions can be employed to compute an error estimate. Here we will outline this approach in one-dimension. They define the error norm as

$$\| e \|_E^2 = -\int_{\Omega} e\, r\, d\Omega$$

where the error is $e = \phi - \hat{\phi}$ and r is the residual error on the interior of the domain

$$L\hat{\phi} + q = r \neq 0 .$$

Now we enrich the current approximate solution $\hat{\phi}$ to get a more accurate (higher degree) approximation by adding the next hierarchical bubble function

$$\phi^* = \hat{\phi} + H_b\, a_b$$

where a_b is the next unknown hierarchical degree of freedom. If we take this as representing the correction solution ($\phi \approx \phi^*$), then we have

$$e^e = H_b^e\, a_b$$

and

$$\| e^e \|_E = a_b \int_{\Omega^e} H_b^{e^T} r^e\, d\Omega .$$

If one can estimate the degree of freedom a_b, then we have an error indicator. If it is the only new dof, and if the hierarchical functions are orthogonal, the new system equations are

$$\begin{bmatrix} \mathbf{S} & \mathbf{0} \\ \mathbf{0} & s_{bb} \end{bmatrix} \begin{Bmatrix} \mathbf{a} \\ a_b \end{Bmatrix} = \begin{Bmatrix} \mathbf{C} \\ c_b \end{Bmatrix},$$

where $\mathbf{S}$ and $\mathbf{C}$ were the previous system matrices, and s_{bb} and c_b are the new element (and system) stiffness and source terms, respectively. From this diagonal system, we compute the new term $a_b = c_b / s_{bb}$, that is,

$$c_b = \int_{\Omega^e} H_b^T q^e \, d\Omega$$

or from the internal residual definition and the above orthogonality,

$$c_b = \int_{\Omega^e} H_b^T (r - L\hat{\phi}) \, d\Omega = \int_{\Omega^e} H_b^T r^e \, d\Omega .$$

Therefore, this error indicator simplifies to

$$\| e^e \|_E = a_b \, c_b = \frac{c_b^2}{s_{bb}} .$$

In the following we will use this approach on a one-dimensional sample problem. We will see that the effectivity index is only about one-half, which is unacceptably far from the desired value of unity. While we could introduce a "fudge factor" constant of two, it is wiser to search for a method that would yield an effectivity index that is always much closer to unity.

14.5 Example Error Calculations *

Consider the Zienkiewicz and Morgan (Z-M) error estimator for their Example 8.1 of [22] expanded to consider the local element errors and flux balances. The model problem is

$$-\frac{d^2\phi}{dx^2} + Q = 0, \qquad x \in \,]0, L[\, , \qquad \phi(0) = 0, \, \phi(L) = 0$$

with the exact solution $\phi = Q(x - L)\,x/2$, so $\phi' = Q(2x - L)/2$. Using the Galerkin approximation:

$$\int_L \phi Q \, dx - \int_L \phi \, \phi_{,xx} \, dx = \int_L \phi Q \, dx - \phi \, \phi_{,x} \Big|_0^L + \int_L \phi_{,x}^2 \, dx = 0$$

or finally

$$\int_L \phi_{,x}^2 \, dx = -\int_L \phi Q \, dx + \phi \, \phi_{,x} \Big|_0^L .$$

Splitting the domain into elements and using our interpolations $\phi_h = \mathbf{H}^e \mathbf{u}^e$ this reduces to the matrix form:

$$\sum_e \mathbf{u}^{e^T} \mathbf{K}^e \mathbf{u}^e = -\sum_e \mathbf{u}^{e^T} \mathbf{F}_Q^e + u(L) \, \phi_{,x}(L) - u(0) \phi_{,x}(0)$$

with the typical element matrices defined as

$$\mathbf{K}^e = \int_{L^e} \mathbf{H}_{,x}^{e^T} \mathbf{H}_{,x}^e \, dx, \qquad \mathbf{F}_Q^e = \int_{L^e} \mathbf{H}^{e^T} Q^e \, dx .$$

Recall for an initial linear interpolation with constant coefficients

$$\mathbf{K}^e = \frac{1}{L^e} \begin{bmatrix} 1 & -1 \\ -1 & 1 \end{bmatrix}, \qquad \mathbf{F}_Q^e = \frac{Q^e L^e}{2} \begin{Bmatrix} 1 \\ 1 \end{Bmatrix}.$$

First, consider a trivial single element solution. By inspection, $L^e = L$ so that

$$\frac{1}{L^e}\begin{bmatrix} 1 & -1 \\ -1 & 1 \end{bmatrix}\begin{Bmatrix} u_1 \\ u_2 \end{Bmatrix} = -\frac{QL^e}{2}\begin{Bmatrix} 1 \\ 1 \end{Bmatrix} + \begin{Bmatrix} -\phi_{,x}(0) \\ +\phi_{,x}(L) \end{Bmatrix}$$

but $u_1 = u_2 = 0$ from the boundary conditions. There are no unknown degrees of freedom to compute so we go directly to the flux recovery and error estimates. Solving for the flux

$$0 = \frac{-QL^e}{2}\begin{Bmatrix} 1 \\ 1 \end{Bmatrix} + \begin{Bmatrix} -\phi_{,x}(0) \\ +\phi_{,x}(L) \end{Bmatrix}$$

which gives $\phi_{,x}(0) = -QL^e/2$ and $\phi_{,x}(L) = QL^e/2$ as the two necessary nodal flux values. Checking we see that a useless solution has still give nodal fluxes that are exact as $L^e \equiv L$. The recovered nodal flux resultants are exact despite the fact that the single element solution is trivial, i.e., $\phi_h = \mathbf{H}^e\,\mathbf{u}^e = \mathbf{H}^e\,\mathbf{O}^e = 0$ (which is exact at nodes). For the single element solution we have 100% error on the interior of the element since $\phi_h = 0$. In the energy norm the error measure is

$$\| e \|^2 = -\int_L (\phi - \phi_h)\,(-\phi_{h,x}\,x + Q_h)\,dx = -\int_L e\,r\,dx$$

where r is the interior residual. For the single element case $\phi_h = 0$ and $\phi_{h,x}\,x = 0$ so $r = Q$ and $e = \phi$. Thus,

$$\| e \|^2_{exact} = -\int_L Q^2(x - L)\,x/2\,dx = -\frac{Q^2}{2}\int (x^2 - Lx)\,dx = \frac{Q^2 L^3}{12}.$$

To compute an error indicator, we add a quadratic hierarchical term to the linear element so

$$\phi_h^* = \phi_h + u_3^e\,H_3^e$$

where $H_3^e(x) = x(L - x)$ in global space, or $H_3^e(r) = r(1 - r)$ in a local unit coordinate space. The Z-M error indicator is

$$\| E^e \|^2 = \left[\int_{L^e} H_3\,r\,dx\right]^2 / K_{33}^e, \quad K_{33}^e = \int_{L^e} H_3\,[-H_3'']\,dx$$

is the new hierarchical stiffness term, and $E^e = \phi_h^* - \phi_h$. Here

$$K_{33}^e = \int_{L^e} r(1 - r)\left[-\frac{1}{L^{e^2}}(-2)\right] dx = \frac{1}{3L^e}$$

and

$$I^e = \int_{L^e} H_3^e R^e dx = \int_{L^e} r(1 - r)\,Q^e dx = Q^e L^e/6$$

so

$$\| E^e \|^2 = \frac{[Q^e L^e/6]^2}{(1/3L^e)} = \frac{3Q^{e^2} L^{e^3}}{36} = \frac{Q^{e^2} L^{e^3}}{12}$$

which happens to be exact for one element. We now repeat the solution and error indicators for two elements of equal size with $L^e = L/2$. The equilibrium equations are

$$\frac{1}{L^e}\begin{bmatrix} 1 & -1 & 0 \\ -1 & 2 & -1 \\ 0 & -1 & 1 \end{bmatrix}\begin{Bmatrix} u_1 \\ u_2 \\ u_3 \end{Bmatrix} = -\frac{QL^e}{2}\begin{Bmatrix} 1 \\ 2 \\ 1 \end{Bmatrix} + \begin{Bmatrix} -\phi_{,x}(0) \\ 0 \\ +\phi_{,x}(L) \end{Bmatrix}.$$

Setting $u_1 = 0 = u_3$ the remaining second equation yields $u_2 = -QL^{e^2}/2$, but $L^e = L/2$ so that $u_2 = -QL^2/8$ which is exact. Recovering the fluxes from equilibrium, we first check the global reactions. The first row gives

$$\frac{1}{L^e}\left[u_1 - u_2 + 0\right] = -\frac{QL^e}{2} - \phi_{,x}(0)$$

so that $\phi_{,x}(0) = -QL^e = -QL/2$ which is exact. Likewise, $\phi_{,x}(L) = +QL/2$, is also exact. Next, we find the fluxes on each element necessary for local equilibrium:

$$e = 1, \quad \frac{1}{L^e}\begin{bmatrix} 1 & -1 \\ -1 & 1 \end{bmatrix}\begin{Bmatrix} u_1 \\ u_2 \end{Bmatrix} = -\frac{QL^e}{2}\begin{Bmatrix} 1 \\ 1 \end{Bmatrix} + \begin{Bmatrix} -\phi_{,x}(x_1) \\ +\phi_{,x}(x_2) \end{Bmatrix}$$

$$-\frac{1}{L^e}\begin{Bmatrix} -QL^{e^2}/2 \\ +QL^{e^2}/2 \end{Bmatrix} + \frac{QL^e}{2}\begin{Bmatrix} 1 \\ 1 \end{Bmatrix} = QL^e\begin{Bmatrix} 1 \\ 0 \end{Bmatrix} = \begin{Bmatrix} -\phi_{,x}(x_1) \\ +\phi_{,x}(x_2) \end{Bmatrix}$$

which are exact since $L^e = L/2$. Likewise, for $e = 2$,

$$\begin{Bmatrix} -\phi_{,x}(x_2) \\ +\phi_{,x}(x_3) \end{Bmatrix} = QL^e\begin{Bmatrix} 0 \\ 1 \end{Bmatrix}.$$

The equilibrium of these global and local fluxes are sketched in Fig. 14.4.1. Note that the flux is zero at the symmetry point $(x = x_2)$ as expected. Since Q is a constant, the previously developed element error indicator,

$$\| E^e \|^2 = Q^{e^2} L^{e^3}/12,$$

is still valid for each element and the system error estimate is

$$\| E \|^2 = \sum_{e=1}^{NE=2} \| E^e \|^2 = Q^{e^2} L^{e^3}/6$$

and since $L^e = L/2$ we get $\| E \|^2 = Q^2 L^3/48$ compared to the exact value of $Q^2 L^3/24$. Thus, the total error is underestimated by a factor of two, but the indicator correctly shows each to have the same amount of error.

If we select two unequal elements, we still get exact values for the nodal values and fluxes. That is, if we let $L^e = L/4$ and $L^e = 3L/4$, respectively, we see the results in Fig. 14.4.2. There we note drastic differences in the local errors in each of the two elements. Checking our error indicators we get

$$e = 1, \quad \| E^e \|^2 = \frac{Q^{e^2} L^{e^3}}{12} = \frac{Q^2}{12}\left(\frac{L}{4}\right)^3 = \frac{Q^2 L^3}{768}$$

$$e = 2, \quad \| E^e \|^2 = \frac{Q^2}{12}\left(\frac{3L}{4}\right)^3 = \frac{27 Q^2 L^3}{768}.$$

Clearly, this indicates that the error in the second element is 27 times as large as for that in the first element. Thus, the second element would be selected for refinement. Of

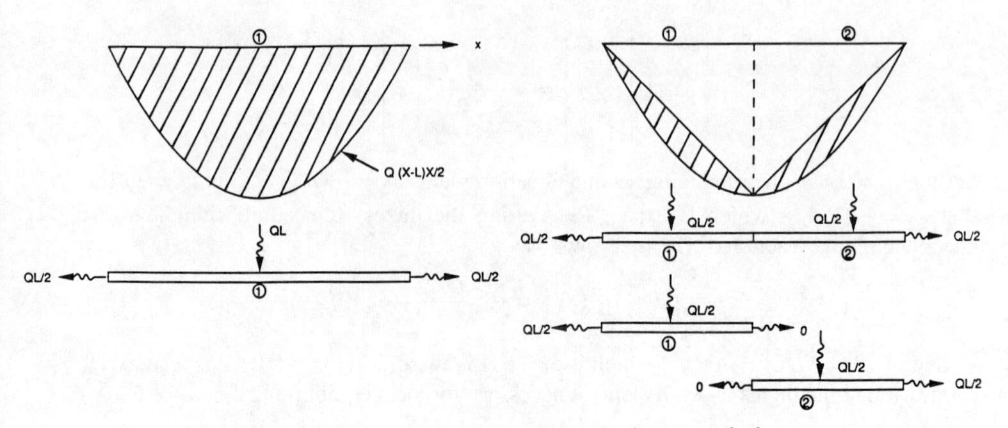

Figure 14.4.1 Sample one and two element solutions

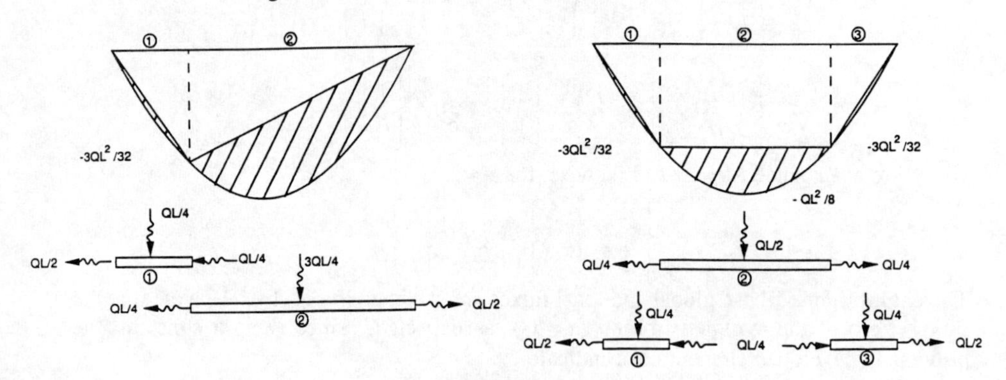

Figure 14.4.2 Sample two and three element solutions

course, the total error estimate for the two unequal elements is

$$\sum_{e=1}^{NE=2} \| E^e \|^2 = \frac{28Q^2L^3}{768}, \text{ and } \| E \|^2_{exact} = \frac{7}{96} Q^2L^3 = \frac{56}{768} Q^2L^3 .$$

Refining the second element by placing a new node at $x = 3L/4$ gives the results in Fig. 14.4.2. Clearly, the first and third elements have the same indicators $\| E^e \|^2 = Q^2L^3/768$ while the middle element has a value of $\| E^e \|^2 = Q^2L^3/96$. The total error estimate is

$$\| E \|^2 = \sum_{e=1}^{NE=2} \| E^e \|^2 = \frac{Q^2L^3}{768} [1 + 8 + 1] = \frac{10Q^2L^3}{768} .$$

Therefore, we notice that 10% of the error is in each of the two small elements and the remaining 80% is in the middle element. The exact error and effectivity measures are:

$$\| E \|^2_{exact} = \frac{10}{384} Q^2L^2, \qquad \frac{\| E \|^2}{\| E \|^2_{exact}} = 0.5 .$$

Finally, we observe the effects of four equally spaced elements on the error indicators. The system error indicator is the same for all elements and

$$\| E \|^2 = \sum_{e=1}^{NE=4} \| E^e \|^2 = \frac{1}{3} Q^{e^2} L^{e^3} = \frac{1}{192} Q^2 L^3,$$

and the exact value is $2Q^2L^3/192$, and once again we get an effectivity of only 50%.

14.6 Flux Balancing Methods

Recently, Ainsworth and Oden [1, 2] have developed a local patch error estimator that is very well justified through detailed functional analysis, is robust, easy and economical to implement, and gives very accurate local error estimates for any order interpolation functions. That is, it usually produces an efficivity index that is very close to unity and is much more reliable than other methods known to the author. By using a dual variational formulation, they have proved that this estimator provides an *upper bound estimate* of the true error. The Ainsworth-Oden flux balancing method uses a local patch of elements for each master node. A typical patch includes all elements connected, or constrained, to the master node. The goal is to choose a linear averaging function α_{KL} between each pair of adjacent elements, K and L, such that the residual internal error, r, and inter-element gradient jumps, R, are in equilibrium, that is,

$$\int_\Omega r \, d\Omega + \int_\Gamma R \, d\Gamma = 0.$$

They provide a detailed procedure for implementing this method, including pseudo-code for the flux-splitting algorithm. The equilibrium fluxes are used to compute the local error estimator. A summary of the method is as follows:

```
for each master node in patch A  do
      begin
            calculate a modified topology matrix, T
            factorize the matrix, L U ≡ T
            for every element e  in the patch  do
                  begin
                        calculate mean flux source, b^e
                        calculate inter-element weight, ζ_j^e
                        assemble patch source, b
                  end
            solve for patch constants λ ; L U λ = b
            for every inter-element edge Γ_KL in patch  do
                  begin
                        α_KL = ½ + (λ_K − λ_L)/ζ_KL
                  end
      end
```

with

$$T_{jk} = \begin{cases} 1 + \text{number of elements in patch,} & \text{if } j = k \\ 0, & \text{if } \Omega^j \text{ and } \Omega^k \text{ are neighbors in patch} \\ 1, & \text{otherwise}. \end{cases}$$

Letting Ψ be a piecewise linear function that is unity at the master node and zero on the patch boundary, the mean source is defined in terms of the model equation

$$-\nabla \cdot (k \nabla u) + \mathbf{b} \cdot \nabla u + cu = f$$

as

$$\mathbf{b}^e = L^e(\Psi) - B^e(\hat{u}, \Psi) + \int_{\Gamma^e \backslash \Gamma} \Psi \langle \mathbf{n}^e \cdot k \nabla \hat{u} \rangle_{1/2} \, d\Gamma$$

$$L^e(\Psi) = \int_{\Omega^e} f \Psi \, d\Omega + \int_{\Gamma_n} k \frac{\partial u}{\partial n} \Psi \, d\Gamma$$

$$B^e(\hat{u}, \Psi) = \int_{\Omega^e} \left[k \nabla \hat{u} \cdot \nabla \Psi + \Psi \mathbf{b} \cdot \nabla \hat{u} + c \hat{u} \Psi \right] d\Omega$$

$$\langle \mathbf{n}^e \cdot k \nabla \hat{u} \rangle_{1/2} = \mathbf{n}^e \cdot 1/2 \left(k^e \nabla \hat{u} \Big|_{\Omega^e} + k^j \nabla \hat{u} \Big|_{\Omega^j} \right)$$

and the inter-element weight is

$$\zeta_j^e = - \int_{\Gamma_j^e} \Psi \left(\mathbf{n}^e \cdot k^e \nabla \hat{u} \Big|_{\Omega^e} + \mathbf{n}^j \cdot k^j \nabla \hat{u} \Big|_{\Omega^j} \right) d\Gamma.$$

The actual flux-splitting function on the boundary between nodes K and L is

$$\alpha_{KL}(s) = \sum_A \alpha_{KL} \, \Psi(s)$$

where the sum has taken over all patches containing edge KL (and a non-zero Ψ). Once the fluxes are in equilibrium, the error, $e = u - \hat{u}$, is bounded above by the norm

$$\| e \|^2 \le \frac{1}{\beta^2} \sum_{e=1}^{NE} \| \phi^e \|^2$$

where $\beta > 0$ is a constant depending on the norm selected ($\beta = 1$ for the standard energy norm), and ϕ is obtained by solving the element local Neumann problem

$$a^e(\phi, w) = L^e(w) - B^e(\hat{u}, w) +$$

$$+ \int_{\Gamma^e} w \mathbf{n}^e \cdot \left[\left(1 - \alpha_{KL}(s) \right) k^e \nabla \hat{u} \Big|_{\Omega^e} + \alpha_{KL}(s) k^j \nabla \hat{u} \Big|_{\Omega^f} \right] d\Gamma.$$

The examples by Ainsworth and Oden show this procedure to be accurate and economical. The effectivity index, Θ, is usually very near unity as desired, and is usually above 0.9 for even crude initial mesh calculations.

14.7 References

[1] Ainsworth, M. and Oden, J.T., "A Procedure for *A Posteriori* Error Estimation for *h-p* Finite Element Methods," *Comp. Meth. Appl. Mech. Eng.*, **101**, pp. 73–96 (1992).

[2] Ainsworth, M. and Oden, J.T., "A Unified Approach to *A Posteriori* Error Estimation Using Element Residual Methods," *Comp. Meth. Appl. Mech. Eng.*, (to appear) (1993).

[3] Barnhill, R.E. and Whiteman, J.R., "Error Analysis of Finite Element Methods with Triangles for Elliptic Boundary Value Problems," in *The Mathematics of Finite Elements and Applications*, ed. J.R. Whiteman, London: Academic Press (1973).

[4] Bass, J., *PHLEX User's Manual*, Austin, TX: Computational Mechanics Co. Inc. (1990).

[5] Brauchli, H.J. and Oden, J.T., "On the Calculation of Consistent Stress Distribution in Finite Element Applications," *Int. J. Num. Meth. Eng.*, **3**, pp. 317–325 (1971).

[6] Hinton, E. and Campbell, J., "Local and Global Smoothing of Discontinuous Finite Element Functions Using a Least Square Method," *Int. J. Num. Meth. Eng.*, **8**, pp. 461–480 (1974).

[7] Hinton, E., Scott, F.C., and Ricketts, R.E., "Local Least Squares Stress Smoothing for Parabolic Isoparametric Elements," *Int. J. Num. Meth. Eng.*, **9**, pp. 235–238 (1975).

[8] Kelly, D.W., "The Self Equilibration of Residuals and Complementary *A Posteriori* Error Estimates in the Finite Element Method," *Int. J. Num. Meth. Eng.*, **20**, pp. 1491–1506 (1984).

[9] Kyriacou, S., "Adaptive Refinement and Related Topics for Finite Element Analysis of Linear Elliptic Problems," M.S. Thesis, Rice University, Houston, TX (1991).

[10] Ladeveze, D. and Leguillon, D., "Error Estimate Procedure in the Finite Element Method and Applications," *SIAM J. Num. Anal.*, **20**(3), pp. 485–509 (1983).

[11] Liusternik, L.A. and Sobolev, V.J., *Elements of Functional Analysis*, New York: Frederick Ungar (1961).

[12] Oden, J.T., *Finite Elements of Nonlinear Continua*, New York: McGraw-Hill (1972).

[13] Oden, J.T. and Reddy, J.N., "Note on Approximation Method for Computing Consistent Conjugate Stresses in Finite Elements," *Int. J. Num. Meth. Eng.*, **6**, pp. 55–63 (1973).

[14] Oden, J.T., *Applied Functional Analysis*, Englewood Cliffs: Prentice-Hall (1979).

[15] Oden, J.T. and Demkowicz, L., "Adaptive Finite Element Methods for Complex Problems in Solid and Fluid Mechanics," pp. 1, 3–13 in *Finite Elements in Computational Mechanics*, ed. T. Kant, Oxford: Pergamon Press (1985).

[16] Oden, J.T., Demkowicz, L., Rochowicz, W., Westermann, T.A., and Hardy, O., "Toward a Universal *h-p* Adaptive Finite Element Strategy – Parts I, II, III," *Comp. Meth. Appl. Mech. Eng.*, **77**, pp. 79–212 (1989).

[17] Oden, J.T., "The Best FEM," *Finite Elements in Analysis and Design*, **7**, pp. 103–114 (1990).

[18] Shephard, M.S., "Approaches to the Automatic Generation and Control of Finite Element Meshes," *Appl. Mech. Rev.*, **41**(4), pp. 169–190 (Apr. 1988).

[19] Whiteman, J.R., ed., *The Mathematics of Finite Elements and Applications*, London: Academic Press (1973).

[20] Zhu, J.Z. and Zienkiewicz, O.C., "Superconvergence Recovery Techniques and *A Posteriori* Error Estimators," *Int. J. Num. Meth. Eng.*, **30**, pp. 1321–1339 (1990).

[21] Zhu, J.Z., "Derivative Recovery Techniques and *a Posteriori* Error Estimation in the Finite Element," *SIAM J. Appl. Num. Anal.*, (to appear) (1992).

[22] Zienkiewicz, O.C. and Morgan, K., *Finite Elements and Approximation*, Chichester: John Wiley (1983).

[23] Zienkiewicz, O.C. and Zhu, J.Z., "A Simple Error Estimator and Adaptive Procedure for Practical Engineering Analysis," *Int. J. Num. Meth. Eng.*, **24**, pp. 337–357 (1987).

[24] Zienkiewicz, O.C., Kelley, D.W., Gago, J., and Babuska, I., "Hierarchal Finite Element Approaches Error Estimates and Adaptive Refinement," pp. 313–346 in *The Mathematics of Finite Elements and Applications, VI*, ed. J.R. Whiteman, London: Academic Press (1988).

[25] Zienkiewicz, O.C., Zhu, J.Z., Craig, A.W., and Ainsworth, M., "Simple and Practical Error Estimation and Adaptivity," pp. 100–114 in *Adaptive Methods for Partial Differential Equations*, ed. J.E. Flaherty et al., SIAM (1989).

[26] Zienkiewicz, O.C. and Zhu, J.Z., "Superconvergent Patch Recovery Techniques and Adaptive Finite Element Refinement," *Comp. Meth. Appl. Mech. Eng.*, **101**, pp. 207–224 (1992).

Chapter 15

SENSITIVITY ANALYSIS *

15.1 Introduction

The optimal design of structural and thermal systems are usually obtained by employing finite element models. Usually these are viewed as either a shape optimization for minimum weight or fully stressed designs, or the determination of design parameters, like thickness, for a given shape. Sensitivity analysis is an important aspect of these optimal design problems. Various sensitivity formulations are available but we must be concerned about which is computationally most efficient when used in conjunction with a finite element analysis. Most of the literature addresses structural problems. Thus, we will begin with a review of the notation for finite element structural analysis. Next, we will introduce the notation for the optimal design problem. Then we will introduce the sensitivity analysis for structures and introduce a new method for addressing stress failure criteria sensitivity. In some cases the details for a specific element type are presented.

15.2 Structural Models

In our finite element models of linear solid mechanics problems we obtained the algebraic equilibrium equations

$$\mathbf{K}\boldsymbol{\delta} = \mathbf{F} \tag{15.1}$$

where $\boldsymbol{\delta}$ is the system displacement vector, $\mathbf{F}$ is the resultant force vector from all sources, and $\mathbf{K}$ is the symmetric positive definite stiffness matrix of the system. For a typical element there is a subset of δ associated with that element. The subset is defined by a theoretical Boolean assembly matrix such that

$$\boldsymbol{\delta}^e \equiv \boldsymbol{\beta}^e \boldsymbol{\delta}. \tag{15.2}$$

Here the Boolean bookkeeping operation "gathers" the element $\boldsymbol{\delta}^e$ from the rows of $\boldsymbol{\delta}$. Sometimes we need the reverse operation to "scatter" the $\boldsymbol{\delta}^e$ back to the system level where they become the only non-zero terms in a subset of $\boldsymbol{\delta}$ defined as

$$\boldsymbol{\delta}_s^e \equiv \boldsymbol{\beta}^{e^T} \boldsymbol{\delta}^e \tag{15.3}$$

where

$$\boldsymbol{\delta} = \bigcup_e \boldsymbol{\delta}_s^e . \tag{15.4}$$

Similar operations are valid for other vectors. For example, if $\mathbf{F}^e$ is an element force vector then the corresponding system resultant is

$$\mathbf{F} = \sum_e \boldsymbol{\beta}^{e^T} \mathbf{F}^e \tag{15.5}$$

and this is contained in the resultant vector $\mathbf{P}$. The system stiffness matrix is the assembly of all the element stiffness matrices, $\mathbf{K}^e$:

$$\mathbf{K} = \sum_e \boldsymbol{\beta}^{e^T} \mathbf{K}^e \boldsymbol{\beta}^e . \tag{15.6}$$

The direct assembly operator, $\underset{e}{\mathbf{A}}$, can be thought of as using the Boolean array, $\boldsymbol{\beta}^e$, to convert each element local subscript to the corresponding system subscript before adding the element coefficient to the corresponding system coefficient. We often see the last two equations replaced by the notation

$$\mathbf{F} \equiv \underset{e}{\mathbf{A}}\, \mathbf{F}^e, \quad \mathbf{K} \equiv \underset{e}{\mathbf{A}}\, \mathbf{K}^e$$

to represent the statement that the system matrices are the assembly of all the element matrices.

The domain of analysis is denoted as Ω and its boundary is Γ. Thus, the closure of the domain is $\overline{\Omega} = \Omega \cup \Gamma$. A typical element domain is $\Omega^e \subset \Omega$ and has a boundary Γ^e and a closure $\overline{\Omega}^e = \Omega^e \cup \Gamma^e$. At times we may want to consider a boundary segment $\Gamma^b \subset \Gamma$ which is usually also a subset of an element boundary, $\Gamma^b \subset \Gamma^e$. Any scalar quantity, q , can be interpolated in an element from its nodal values, $\mathbf{q}^e$, as

$$q = \mathbf{H}^e \mathbf{q}^e \qquad \varepsilon\, \Omega^e$$

where $\mathbf{H}^e$ are the interpolation functions (or shape functions) of the element. On a typical boundary segment this may degenerate to $q = \mathbf{H}^b \mathbf{q}^b \ \varepsilon\, \Gamma^b$. Here the displacement vector, $\mathbf{f}$, at a point in Ω^e is similarly interpolated as

$$\mathbf{f} = \mathbf{N}^e \boldsymbol{\delta}^e \qquad \varepsilon\, \Omega^e$$

where $\mathbf{N}^e$ contains zeros and the H_i terms. The mechanical strain obtained from the strain-displacement relations as

$$\boldsymbol{\varepsilon}^e = \mathbf{B}^e \boldsymbol{\delta}^e \qquad \varepsilon\, \Omega^e .$$

We employ a constitutive relation defined by a generalized Hooke's Law given as

$$\boldsymbol{\sigma}^e = \mathbf{D}(\boldsymbol{\varepsilon} - \boldsymbol{\varepsilon}_o) + \boldsymbol{\sigma}_o$$

where $\mathbf{D}$ is the symmetric positive definite matrix of material property constants, and $\boldsymbol{\varepsilon}_o$ and $\boldsymbol{\sigma}_o$ denote initial strains and initial stresses, respectively. The later two terms, when present, contribute to element force vectors, $\mathbf{F}_\varepsilon^e$ and $\mathbf{F}_\sigma^e$. Usually, we discuss the mechanical strains in a element

$$\boldsymbol{\sigma}^e = \mathbf{D}^e \boldsymbol{\varepsilon}^e = \mathbf{D}^e \mathbf{B}^e \boldsymbol{\delta}^e \qquad \varepsilon\, \Omega^e \tag{15.7}$$

The element stiffness matrix is defined as

$$\mathbf{K}^e = \int_{\Omega^e} \mathbf{B}^{e^T} \mathbf{D}^e \mathbf{B}^e \, d\Omega \,. \tag{15.8}$$

Likewise, the element load vectors due to body forces $\mathbf{X}^e$, initial strains $\boldsymbol{\varepsilon}_0$, and initial stresses $\boldsymbol{\sigma}_0$ are defined as

$$\mathbf{F}_X^e = \int_{\Omega^e} \mathbf{N}^{e^T} \mathbf{X}^e \, d\Omega \,, \quad \mathbf{F}_\varepsilon^e = \int_{\Omega^e} \mathbf{B}^{e^T} \mathbf{D} \varepsilon_o \, d\Omega \,, \quad \mathbf{F}_\sigma^e = \int_{\Omega^e} \mathbf{B}^{e^T} \sigma_o \, d\Omega$$

and the force from a boundary traction vector, $\mathbf{T}$ is

$$\mathbf{F}_T^b = \int_{\Gamma^b} \mathbf{N}^{b^T} \mathbf{T}^b \, d\Omega \,.$$

The resultant force vector is the sum of all the sources of loads, including external concentrated nodal loads $\mathbf{P}_c$,

$$\mathbf{P} = \mathbf{P}_c + \mathbf{F}_X + \mathbf{F}_\varepsilon + \mathbf{F}_\sigma \,.$$

The strain energy of the system, and the external work are, respectively:

$$U = \frac{1}{2} \boldsymbol{\delta}^T \mathbf{K} \boldsymbol{\delta} \,, \quad W = \boldsymbol{\delta}^T \mathbf{P} \,. \tag{15.9}$$

15.3 Optimal Design Problem

The finite element analysis is used as a tool to help solve constrained optimal design problem. These are usually stated as a minimization problem:

$$\text{Minimize } W(\mathbf{a}) \tag{15.10}$$

where $\mathbf{a}$ denotes the design variable, and W is some merit function such as the weight of a system. In optimal shape design problems $\mathbf{a}$ represents the coordinates at certain control points. Often the design process is subjected to a functional inequality constraint

$$g_i(\mathbf{a}) \le 0, \quad i = 1, m \tag{15.11}$$

and regional constraints on the design variables

$$L_j \le a_j \le U_j, \quad j = 1, n. \tag{15.12}$$

Here, g_i may represent a constraint on the displacement or the stress at a point. In solving the design problem we need to know how some of the quantities in the finite element analysis depend on the design variables. This is called the design sensitivity analyses. The constraints can often be written as

$$g_i = g_i(\boldsymbol{\delta}, \mathbf{a}), \ \boldsymbol{\delta} = \boldsymbol{\delta}(\mathbf{a}) \,.$$

15.4 Sensitivity Relations

We almost always need to know how the equilibrium equations depend on the design variables. Taking the partial derivative of the equilibrium equation, Eq. (15.1) with respect to a typical design variable a_j in $\mathbf{a}$ gives

$$\frac{\partial \mathbf{K}}{\partial a_j} \boldsymbol{\delta} + \mathbf{K} \frac{\partial \boldsymbol{\delta}}{\partial a_j} = \frac{\partial \mathbf{P}}{\partial a_j}$$

which could be solved for the displacement sensitivity

$$\frac{\partial \boldsymbol{\delta}}{\partial a_j} = \mathbf{K}^{-1}\left[\frac{\partial \mathbf{P}}{\partial a_j} - \frac{\partial \mathbf{K}}{\partial a_j}\boldsymbol{\delta}\right] = \mathbf{K}^{-1}\hat{\mathbf{P}}. \tag{15.13}$$

The quantity in brackets is often referred to as the force derivative. We will be interested in relatively economical ways to compute the stiffness sensitivity, $\partial \mathbf{K}/\partial a_j$. If we have solved Eq. (15.1) by a Crout factorization with lower and upper triangular matrices $\mathbf{K} \equiv \mathbf{L}\mathbf{L}^T$ then the displacement sensitivity is obtained by a cheap forward and backward substitution of the force derivative.

We are also interested in the sensitivities of the constraint equations. They usually depend on a set of *analysis functions*, or *state variables*, $\mathbf{Z}$. Applying the chain rule

$$\frac{dg_i}{da_j} = \frac{\partial g_i}{\partial a_j} + \left[\frac{\partial g_i}{\partial \mathbf{Z}}\right]^T \frac{d\mathbf{Z}}{da_j}. \tag{15.14}$$

Since the equilibrium equations depend on the vector $\boldsymbol{\delta}$ we could likewise write

$$\frac{d}{da_j}(\mathbf{K}\boldsymbol{\delta} - \mathbf{P} = \mathbf{O}) = \frac{d\mathbf{K}}{da_j}\boldsymbol{\delta} + \mathbf{K}\frac{d\boldsymbol{\delta}}{da_j} - \frac{d\mathbf{P}}{da_j} = \mathbf{O}$$

or

$$\left[\frac{\partial \mathbf{K}}{\partial a_j} + \frac{\partial \mathbf{K}}{\partial \boldsymbol{\delta}}\frac{\partial \boldsymbol{\delta}}{\partial a_j}\right]\boldsymbol{\delta} + \mathbf{K}\frac{d\boldsymbol{\delta}}{da_j} = \left[\frac{\partial \mathbf{P}}{\partial a_j} + \frac{\partial \mathbf{P}}{\partial \boldsymbol{\delta}}\frac{\partial \boldsymbol{\delta}}{\partial a_j}\right].$$

For linear problems the partials with respect to $\boldsymbol{\delta}$ vanish. If the partial derivatives of $\boldsymbol{\delta}$ also vanish then

$$\frac{\partial \mathbf{K}(\mathbf{a})}{\partial a_j}\boldsymbol{\delta} + \mathbf{K}(\mathbf{a})\frac{d\boldsymbol{\delta}}{da_j} = \frac{\partial \mathbf{P}(\mathbf{a})}{\partial a_j}$$

or

$$\mathbf{K}(\mathbf{a})\frac{d\boldsymbol{\delta}}{da_j} = \frac{\partial}{\partial a_j}(\mathbf{P}(\mathbf{a}) - \mathbf{K}(\mathbf{a})\boldsymbol{\delta}) \equiv \mathbf{E}(a_j), \qquad \mathbf{E} = \underset{e}{\mathbf{A}}\, \mathbf{E}^e. \tag{15.15}$$

Note that $\mathbf{E}$ could be assembled from its element values. In the special case where the displacements are the only state variables ($\mathbf{Z} = \boldsymbol{\delta}$) we could rewrite Eq. (15.14) as

$$\frac{dg_i}{da_j} = \frac{\partial g_i}{\partial a_j} + \left[\frac{\partial g_i}{\partial \boldsymbol{\delta}}\right]^T \left[\mathbf{K}^{-1}\mathbf{E}(a_j)\right]. \tag{15.16}$$

At this point it may be useful to review the size of the arrays in use here. Let the system have M degrees of freedom, L loadcases, while an element has N degrees of freedom. Then $\mathbf{K}$ is $M \times M$, $\mathbf{K}^e$ is $N \times N$, $\boldsymbol{\delta}$ and $\mathbf{P}$ are $M \times L$, and the various $\mathbf{F}^e$ are $N \times L$. If D is the number of design variables then $\mathbf{a}$ is $D \times 1$, the $\delta\mathbf{K}/\delta\mathbf{a}$ is $M \times M \times D$, the $d\boldsymbol{\delta}/d\mathbf{a}$ is $M \times L \times D$, and so forth. If the number of strain components is S then ε, ε_o, and σ_o are $S \times 1$, $\mathbf{D}$ is $S \times S$, and the strain-displacement matrix $\mathbf{B}^e$ is $S \times N$.

The stress sensitivity in an element (with constant $\mathbf{D}$ and no initial effects) is obtained from Eq. (15.14) as

$$\frac{\partial \sigma^e}{\partial a_j} = \mathbf{D}^e\left[\frac{\partial \mathbf{B}^e}{\partial a_j}\boldsymbol{\delta}^e + \mathbf{B}^e\frac{\partial \boldsymbol{\delta}^e}{\partial a_j}\right]. \tag{15.17}$$

Here the first term represents the change of the shape of the element and the second is due to a change in the elements displacement. Later we will consider isoparametric elements and the details that go into formulating $\mathbf{B}^e$ in general.

15.5 Yield Criteria Sensitivity

According to the distortional energy theory for ductile materials, failure occurs when the *Von Mises Effective Stress*, σ_E, equals the yield stress, σ_Y. Actually we are computing an energy criteria that states that failure occurs when the distortional energy equals its value at yield. The effective stress can be expressed in several forms. In terms of the Principal Stresses it is:

$$\sigma_E = \frac{1}{\sqrt{2}}\sqrt{(\sigma_1 - \sigma_2)^2 + (\sigma_2 - \sigma_3)^2 + (\sigma_3 - \sigma_1)^2}$$

(note that for a uniaxial tension yield $\sigma_1 = \sigma_Y$, $\sigma_2 = \sigma_3 = 0$ so $\sigma_E = \sigma_Y$). For general three-dimensional stresses

$$2\sigma_E^2 = (\sigma_x - \sigma_y)^2 + (\sigma_y - \sigma_z)^2 + (\sigma_z - \sigma_x)^2 + 6(\tau_{xy}^2 + \tau_{xz}^2 + \tau_{zy}^2)\,,$$

while in the axisymmetric case ($\tau_{rc} = 0 = \tau_{cz}$, $r \sim$ radial, $z \sim$ axial, $c \sim$ circumference)

$$2\sigma_E^2 = (\sigma_r - \sigma_z)^2 + (\sigma_z - \sigma_c)^2 + (\sigma_c - \sigma_r)^2 + 6\tau_{rz}^2\,,$$

and for the common plane stress state

$$2\sigma_E^2 = 2(\sigma_x^2 + \sigma_y^2 - \sigma_x\sigma_y) + 6\tau_{xy}^2\,,$$

and for plane strain we must include the terms from $\sigma_z = -\nu(\sigma_x + \sigma_y)$ which comes from the condition that $\varepsilon_z = 0$. A related quantity is the *Effective Strain*

$$\varepsilon_E = \frac{\sqrt{2}}{3}\sqrt{(\varepsilon_1 - \varepsilon_2)^2 + (\varepsilon_2 - \varepsilon_3)^2 + (\varepsilon_3 - \varepsilon_1)^2}.$$

Note at a tension yield $\varepsilon_1 = \varepsilon_Y$, $\varepsilon_2 = \varepsilon_3 = -\nu\varepsilon_Y$ so that $\varepsilon_E = 2(1+\nu)\varepsilon_Y/3$ where $\varepsilon_Y = \sigma_Y/E$. The *distortional energy* at yield is

$$U_D = \frac{1}{2}\varepsilon_E\sigma_E = \frac{1}{2}\frac{2}{3}(1+\nu)\frac{\sigma_Y}{E}\sigma_y = \frac{1}{3}(1+\nu)\frac{\sigma_Y^2}{E}.$$

Most ductile materials are predicted to fail when the Von Mises effective stress reaches a critical value. Other materials may be subjected to the Tresca failure criterion. The Von Mises stress is defined in terms of the principal stresses as

$$2\sigma_E^2 = [(\sigma_1 - \sigma_2)^2 + (\sigma_1 - \sigma_3)^2 + (\sigma_2 - \sigma_3)^2]$$

or in terms of the full stress tensor

$$2\sigma_E^2 = \left[2(\sigma_x^2 + \sigma_y^2 + \sigma_z^2 - \sigma_x\sigma_y - \sigma_x\sigma_z - \sigma_y\sigma_z) + 6(\tau_{xy}^2 + \tau_{xz}^2 + \tau_{yz}^2)\right].$$

The allowed value, σ_a, is usually the tension yield point. The *stress constraint* is often written as

$$g = \sigma_E - \sigma_a \le 0\,.$$

Note that the Von Mises effective stress can be written as a matrix identity

$$2\sigma_E^2 = \sigma^T \begin{bmatrix} 2 & -1 & -1 & 0 & 0 & 0 \\ -1 & 2 & -1 & 0 & 0 & 0 \\ -1 & -1 & 2 & 0 & 0 & 0 \\ 0 & 0 & 0 & 6 & 0 & 0 \\ 0 & 0 & 0 & 0 & 6 & 0 \\ 0 & 0 & 0 & 0 & 0 & 6 \end{bmatrix} \sigma \equiv \sigma^T \mathbf{V} \sigma \tag{15.18}$$

where in general

$$\sigma^T = [\, \sigma_x \;\; \sigma_y \;\; \sigma_z \;\; \tau_{xy} \;\; \tau_{xz} \;\; \tau_{yz} \,] .$$

For the plane stress problem $\sigma^T = [\sigma_x \;\; \sigma_y \;\; \tau_{xy}]$ and $\mathbf{V}$ reduces to

$$\mathbf{V} = \begin{bmatrix} 2 & -1 & 0 \\ -1 & 2 & 0 \\ 0 & 0 & 6 \end{bmatrix} .$$

This has some computational advantages if we define the stress constraint at any point as

$$g = \sigma_E^2 - \sigma_a^2 \le 0 , \quad \text{or} \quad g = \sigma_E^2 - \sigma_T \sigma_c \le 0$$

where σ_c and σ_T are the absolute values of the yield points in compression and tension, respectively. Then we could write the sensitivity as

$$\frac{\partial g}{\partial a_j} = \frac{\partial \sigma^T}{\partial a_j} \mathbf{V} \sigma + \sigma^T \mathbf{V} \frac{\partial \sigma}{\partial a_j} .$$

At a point in a typical element one can substitute Eq. (15.7) to define

$$\sigma_E^2 = \frac{1}{2} \delta^{e^T} \mathbf{B}^{e^T} \mathbf{D}^{e^T} \mathbf{V} \, \mathbf{D}^e \mathbf{B}^e \delta^e = \frac{1}{2} \delta^{e^T} \mathbf{B}^{e^T} \mathbf{D}_E^e \mathbf{B}^e \delta^e , \quad \varepsilon \, \Omega^e ,$$

where we have defined define a pseudo-constitutive matrix for effective stress: $\mathbf{D}_E \equiv \mathbf{D}^{e^T} \mathbf{V} \, \mathbf{D}^e$.

In optimization procedures the point where a constraint is violated usually jumps around from one site to the next as the design iterations proceed. Thus, the merit function tends to oscillate and slow the convergence. This is especially true for shape optimization problems. There are an infinite number of points where the design constraints might be checked. Most analysts restrict such checks to the nodes and/or to the Gauss points. Since we have only a finite number of elements it makes sense to try to develop a constraint check that can be applied at the element level, at least in the early stages of the design iterations. The last equation is very similar to Eqs. (15.8 and 15.9), and suggests that an integral measure could be useful. Define

$$\Omega^e \times \bar{\sigma}_E^2 = \frac{1}{2} \delta^{e^T} \int_{\Omega^e} \mathbf{B}^{e^T} \mathbf{D}_E^e \mathbf{B}^e \, d\Omega \, \delta^e \equiv \frac{1}{2} \delta^{e^T} \mathbf{k}_E^e \delta^e . \tag{15.19}$$

Note that $\mathbf{k}_E^e$ has the same form as $\mathbf{K}^e$ in Eq. (15.8) and would easily be computed in the same subroutines. Rather than check σ_E or a stress constraint at all points in Ω it may be useful to scan for a related scalar quantity in each element. Thus, it is proposed that a useful stress constraint would be

$$g = \int_{\Omega} \sigma_E^2 \, d\Omega - \alpha \int_{\Omega} \sigma_T \sigma_c \, d\Omega \tag{15.20}$$

where $0 < \alpha \leq 1$. If the element domain, Ω^e, were infinitesimal then we would be making a pointwise check and we would be justified to set $\alpha = 1$. However, since the integral in Eq. (15.20) smooths out the peak value we should use a smaller α, say $\alpha = 0.9$, or have it depend on the element measure, Ω^e. This smoothed constraint measure can be evaluated as the sum of the element contributions:

$$g = \sum_e g^e, \quad g^e = \frac{1}{2} \boldsymbol{\delta}^{e^T} \mathbf{k}_E^e \boldsymbol{\delta}^e - \alpha \sigma_T^e \sigma_c^e \Omega^e \leq 0 \,.$$

These matrices follow the usual assembly relations so one could write

$$\mathbf{k}_E = \mathop{\mathbf{A}}_e \mathbf{k}_E^e, \quad g^e = \frac{1}{2} \boldsymbol{\delta}^T \mathbf{k}_E \boldsymbol{\delta}^T - \alpha \sum_e \sigma_T^e \sigma_c^e \, \Omega^e \leq 0 \,. \tag{15.21}$$

If the constraint on a typical element, Eq. (15.21), is violated then we may wish to activate the a_j or the $\boldsymbol{\delta}^e$ associated with it. If an element is activated by the $\boldsymbol{\delta}^e$ then the surrounding elements making up a regional set, R, also would become active. Assuming we know the displacement sensitivity from Eq. (15.13) then we may need to find

$$\frac{\partial g^e}{\partial \boldsymbol{\delta}^e} = \mathbf{k}_E^e \boldsymbol{\delta}^e$$

or the system value, see Eq. (15.3),

$$\frac{\partial g}{\partial \boldsymbol{\delta}^e} = \mathbf{k}_E \boldsymbol{\delta}_S^e = \sum_{e \subset R} \mathbf{k}_E^e \boldsymbol{\delta}_R^e, \quad \boldsymbol{\delta}_R^e \subset \boldsymbol{\delta}^e \,. \tag{15.22}$$

This corresponds to multiplying the assembled columns in $\mathbf{k}_E$, associated with $\boldsymbol{\delta}^e$, by the values in $\boldsymbol{\delta}^e$.

A similar approach could be used for other design or failure criteria. The *maximum shear stress criterion* at any point for plane stress is

$$g = \sqrt{[(\sigma_x - \sigma_y)/2]^2 + \tau_{xy}^2} - \tau_{\max} \leq 0.$$

If we apply a similar concept and use the square of each quantity so

$$g = [\,(\,[\sigma_x - \sigma_y]/2)^2 + \tau_{xy}^2\,] - \tau_{\max}^2 \leq 0 \tag{15.23}$$

then we can define a new $\mathbf{V}$ for maximum shear given by

$$\mathbf{V}_s = \begin{bmatrix} 1/2 & -1/2 & 0 \\ -1/2 & 1/2 & 0 \\ 0 & 0 & 1 \end{bmatrix}$$

to define a similar element criterion given as

$$g^e = \frac{1}{2} \boldsymbol{\delta}^{e^T} \mathbf{k}_s^e \boldsymbol{\delta}^e - \alpha \tau_{\max}^{e2} \leq 0,$$

which is amendable to the same approach. For either failure criteria the constraint sensitivity given by Eq. (15.22) is quite economical to compute.

15.6 Computing Stiffness Sensitivity

15.6.1 Exact Sensitivity

For many common simple elements the stiffness and load matrices can have their partial derivatives given in closed form. To illustrate the concept recall the elastic bar linear element matrices

$$\mathbf{K}^e = \frac{EA}{L}\begin{bmatrix} 1 & -1 \\ -1 & 1 \end{bmatrix}, \quad \mathbf{F}^e = \frac{\gamma AL}{2}\begin{Bmatrix} 1 \\ 1 \end{Bmatrix}.$$

If the area, A, is a design parameter then we have

$$\frac{\partial \mathbf{K}^e}{\partial A} = \frac{E}{L}\begin{bmatrix} 1 & -1 \\ -1 & 1 \end{bmatrix} = \frac{1}{A}\mathbf{K}^e, \quad \frac{\partial \mathbf{F}^e}{\partial A} = \frac{\gamma L}{2}\begin{Bmatrix} 1 \\ 1 \end{Bmatrix} = \frac{1}{A}\mathbf{F}^e$$

while if we are concerned with the length, L,

$$\frac{\partial \mathbf{K}^e}{\partial L} = -\frac{EA}{L^2}\begin{bmatrix} 1 & -1 \\ -1 & 1 \end{bmatrix} = -\frac{1}{L}\mathbf{K}^e, \quad \frac{\partial \mathbf{F}^e}{\partial L} = \frac{\gamma A}{2}\begin{Bmatrix} 1 \\ 1 \end{Bmatrix} = \frac{1}{L}\mathbf{F}^e.$$

The linear triangle (T3) is common in many two-dimensional applications. For heat transfer with constant properties it has conduction and source matrices of the form

$$\mathbf{K}^e = tA\mathbf{B}^T\mathbf{D}\mathbf{B}, \quad \mathbf{F}^e = \frac{tAQ}{3}\begin{Bmatrix} 1 \\ 1 \\ 1 \end{Bmatrix}$$

where t is the thickness, A is the area, Q is the source per unit volume, and $\mathbf{D}$ is the constant constitutive matrix. As above, we note

$$\frac{\partial \mathbf{K}}{\partial t} = \frac{1}{t}\mathbf{K}, \quad \frac{\partial \mathbf{F}}{\partial t} = \frac{1}{t}\mathbf{F}$$

The area, A, and global derivative matrix, $\mathbf{B}$, depend on the nodal coordinates of the three nodes. Specifically, $2A = (x_2 - x_1)(y_3 - y_1) - (x_3 - x_1)(y_2 - y_1)$, and

$$\mathbf{B} = \frac{1}{2A}\begin{bmatrix} (y_2 - y_3) & (y_3 - y_1) & (y_1 - y_2) \\ (x_3 - x_2) & (x_1 - x_3) & (x_2 - x_1) \end{bmatrix} = \frac{1}{A}\mathbf{B}^*.$$

To evaluate the change in $\mathbf{B}$ with respect to a nodal coordinate, say x_2, we compute

$$\frac{\partial \mathbf{B}}{\partial x_i} = -\frac{1}{A^2}\frac{\partial A}{\partial x_i}\mathbf{B}^* + \frac{1}{A}\frac{\partial \mathbf{B}^*}{\partial x_i}$$

with

$$\frac{\partial A}{\partial x_2} = (y_3 - y_1)/2, \quad \frac{\partial \mathbf{B}^*}{\partial x_2} = \frac{1}{2}\begin{bmatrix} 0 & 0 & 0 \\ -1 & 0 & 1 \end{bmatrix}$$

so that

$$\frac{\partial \mathbf{B}^e}{\partial x_2} = \frac{(y_1 - y_3)^e}{2A^e}\mathbf{B}^e + \frac{1}{2A^e}\begin{bmatrix} 0 & 0 & 0 \\ -1 & 0 & 1 \end{bmatrix}$$

$$\frac{\partial \mathbf{B}^e}{\partial y_2} = \frac{(-x_1 + x_3)^e}{2A^e} \mathbf{B}^e + \frac{1}{2A^e} \begin{bmatrix} 1 & 0 & -1 \\ 0 & 0 & 0 \end{bmatrix}.$$

15.6.2 Finite Difference Approximations

As we have seen in Eq. (15.13), we often wish to calculate the stiffness sensitivity, $\partial \mathbf{K}/\partial a_j$. This can be done numerically. Various numerical methods are available but some, like the finite difference method, are extremely expensive when done in the usual way and one must exercise care when executing this calculation within a finite element system. Most authors compute the finite difference approximation as

$$\frac{\partial \mathbf{K}}{\partial a_j} \approx \frac{\mathbf{K}(a_j + \Delta a_j) - \mathbf{K}(a_j)}{\Delta a_j}$$

where Δa_j is some small change in parameter a_j, and where the numerator represents the difference in two different full square matrices. This is a grossly inefficient procedure because a small change in a_j usually affects only a very small number of entries in $\mathbf{K}$. Thus, most of the effort is spent in computing terms that are zero. In fact, Δa_j may affect only a small sub-region, $R \subseteq \Omega$, or possibly a single element. In that case, we need only assemble the changes in the element matrices:

$$\frac{\partial \mathbf{K}}{\partial a_j} \approx \frac{\underset{e}{A} \left[\mathbf{K}^e(a_j + \Delta a_j) - \mathbf{K}(a_j) \right]}{\Delta a_j}, \quad e \in R$$

which is much cheaper.

In optimal shape design the variable, a_j, is often a nodal coordinate, say $a_j = x_j$ and $a_j = y_j$. That is, in shape analysis we are interested in the change of stiffness due to the relocation of a node in the mesh. This class of problem is frequently addressed in linear fracture mechanics. Thus, the literature in that field can provide some useful experience in selecting an efficient numerical method for forming the $\partial \mathbf{K}/\partial x_j$. A numerical method for computing such changes in linear *fracture mechanics* was developed independently by a number of people [2, 7, 12]. Today this is most commonly referred to as the method of virtual crack extension.

Since the formation of $\mathbf{K}^e$ always involves using its nodal coordinates, it should be obvious that moving the location of a node will change the stiffness of any and all elements attached to that node. It should be clear that such a change would affect *only* those elements. Thus, in the case of shape optimization, the region $R \subseteq \Omega$ consists only of those elements connected (or constrained) to the node being moved to a new location. Changing the location of a node in shape optimization is a vector operation, that is, one must find as many a_j sensitivities for node j as there are spatial dimensions. Experiences with the method of *virtual crack extension* suggests that one should be able to economically compute a preferred direction of relocation, and reduce the effective search for the a_j to one per node. What would be the preferred direction for a *virtual shape extension?* It logically would be the direction that gives the maximum change in energy at the node. A small shape change has a very small effect on the displacements. Thus, the change in energy due to a relocation Δa_k in the amount a in the direction of Θ_k would be

$$\frac{\partial U}{\partial a_k} \approx \frac{1}{2} \frac{\sum_e \delta^{e^T} \left[\mathbf{K}^e(\Delta a_k) - \mathbf{K}^e \right] \delta^e}{a} \approx \frac{\frac{1}{2} \sum_e \delta^{e^T} \mathbf{K}^e(\Delta a_k)\, \delta^e - \sum_e U_0^e}{a}, \quad e \in R$$

where U_0^e is the strain energy in the element before its change of shape. Such a calculation is a simple and fast postprocessing operation for any node of interest. For any small reference value of a, it gives the change of energy as a function of the direction of the node movement in the change of shape. The direction giving the maximum rate of change of energy with shape relocation is easily output for any node. Usually one would only consider nodes on the free surface of the solution domain.

15.7 Isoparametric Elements

For isoparametric elements the Jacobian is generally not constant. This can complicate the calculation of the sensitivities of the stiffness matrix and load vector. Recall that the stiffness matrix is

$$\mathbf{K}^e = \int_{\Omega^e} \mathbf{B}^{e^T} \mathbf{D}^e \mathbf{B}^e \, d\Omega = \int_{\Box} \mathbf{B}^{e^T} \mathbf{D}^e \mathbf{B}^e \, |J^e| \, d\Box$$

where $\mathbf{D}^e$ is the symmetric constitutive matrix and $\mathbf{J}^e$ is the Jacobian of the coordinate transformation. If a_i is a design variable then the stiffness derivative is denoted as $(\ )' = \partial(\ )/\partial a_i$, then

$$\mathbf{K}^{e'} = \int_{\Box} [\, \mathbf{B}^{e'^T} \mathbf{D}^e \mathbf{B}^e \, |\mathbf{J}^e| + \mathbf{B}^{e^T} \mathbf{D}^{e'} \mathbf{B}^e \, |\mathbf{J}^e| \tag{15.24}$$

$$+ \mathbf{B}^{e^T} \mathbf{D}^e \mathbf{B}^{e'} \, |J^e| + \mathbf{B}^{e^T} \mathbf{D}^e \mathbf{B}^e \, |\mathbf{J}^e|' \,] \, d\Box.$$

Usually a_i is not a constitutive term, such as E or ν. Thus, we usually have $\mathbf{D}^{e'} = \mathbf{0}$. Recall that the definition of the element stresses at any point in Ω^e is $\boldsymbol{\sigma}^e = \mathbf{D}^e \mathbf{B}^e \boldsymbol{\delta}^e$. This allows the product of Eq. (15.24) and the element displacement vector $\boldsymbol{\delta}^e$, to be expressed as

$$\mathbf{P}_K^e = \mathbf{K}^{e'} \boldsymbol{\delta}^e = \int_{\Box} [\, (\mathbf{B}^{e'^T} |\mathbf{J}^e| + \mathbf{B}^{e^T} |\mathbf{J}^e|')\, \boldsymbol{\sigma}^e + \mathbf{B}^{e^T} \mathbf{D}^e \mathbf{B}^{e'} \, |\mathbf{J}^e| \, \boldsymbol{\delta}^e \,] \, d\Box.$$

Likewise, a typical element load vector is

$$\mathbf{P}_q^e = \int_{\Omega^e} \mathbf{N}^{e^T} \mathbf{q} \, d\Omega = \int_{\Box} \mathbf{N}^{e^T} \mathbf{q} \, |\mathbf{J}^e| \, d\Box \tag{15.25}$$

and its derivative is

$$\mathbf{P}_q^{e'} = \int_{\Box} [\, \mathbf{N}^{e'^T} \mathbf{q} \, |\mathbf{J}^e| + \mathbf{N}^{e^T} \mathbf{q} \, |J^e|' \,] \, d\Box \tag{15.26}$$

since we usually assume that $\mathbf{q}$ does not depend on the a_i. Note that Eqs. (15.25 and 15.26) can be used to compute the element force derivative, $\hat{\mathbf{P}}^e$, needed in Eq. (15.13). These element quantities can be formed while computing the element stresses. At the same time these vectors are assembled to give the system result

$$\hat{\mathbf{P}} = \underset{e}{A} \, (\mathbf{P}_q^{e'} - \mathbf{P}_K^e).$$

This approach does not require the assembly of $\mathbf{K}'$.

15.8 Shape Sensitivity

When the design parameter, a_j, is a spatial coordinate of one of the nodes then we are attempting to change the shape of the mesh that describes the object. This is called shape optimization. Some implementations allow all elements to be considered in moving the nodes but most define subregions of elements near selected boundaries and allow only the subregions to move and redefine the shape. There have been a large number of procedures developed for optimizing structural shapes. Most are very expensive and as a result more approximate methods are being utilized in commercial finite element systems [7, 10]. Despite the merit of the several procedures in use today there is still a need for better shape optimization approaches. The "swelling" procedure outlined here is equally valid for two-dimensional and three-dimensional continuum stress analysis systems. Isotropic versions of it have been successfully used in the literature. It effectively utilizes the computational investment previously made in the assembling and solving of the stiffness equations. An analogous procedure should be practical for shape optimization for general field problems.

We will have the object change shape in response to an "anisotropic swelling" of the original elements. For a given shape, the system equilibrium equations for the assembled continuum elements are $\mathbf{K}\,\boldsymbol{\delta} = \mathbf{F}$ where $\mathbf{K}$ and $\mathbf{F}$ are the assembled stiffness and force vectors, respectively, and $\boldsymbol{\delta}$ is the vector of nodal displacements to be determined. We denote the current shape by an array of nodal coordinates, $\mathbf{R}$, that is the same size as $\boldsymbol{\delta}$. Solving this system gives the displacements, $\boldsymbol{\delta}$, and the associated subset of element displacement vectors, $\boldsymbol{\delta}^e \subset \boldsymbol{\delta}$.

Assume that we have numerically integrated isoparametric elements. Then at every Gauss quadrature point, q, in any element we can recover the strains and stresses

$$\varepsilon^e(\mathbf{x}_q) = \mathbf{B}^e(\mathbf{x}_q)\,\delta^e, \quad \sigma_q^e = \mathbf{D}_q^e\,\varepsilon^e$$

where $\mathbf{x}_q$ denotes the coordinates of the quadrature point. Having the stress tensor at the point we compute the principle stresses and their principle axes. To induce a shape change that responds to these stresses, we introduce a pseudo anisotropic swelling strain ε_o^e that varies over the element volume. This introduces an additional resultant element load $\mathbf{F}_o^e$ defined in the standard way as the volume integral

$$\mathbf{F}_o^e = \int_{V^e} \mathbf{B}^{e^T}\,\mathbf{D}^e\,\varepsilon_o^e\,dV. \tag{15.27}$$

The swelling strain, $\boldsymbol{\varepsilon}_o$, will be assigned anisotropic components that vary at each point in the element and correspond to the directions of the principle stresses. Given a principle stress magnitude and direction, we want the material to either expand, or contract, in a direction perpendicular to that stress component. The amount of expansion should be proportional to the stress level. If the stress is too high we need to increase the area normal to the stress direction so as to reduce the stress in the next step. If it is too low we want to decrease the normal area.

The components of orthotropic swelling strains (in plane stress or strain) at a quadrature point q would be defined as,:

$$\boldsymbol{\varepsilon}_{oq}^e = \begin{Bmatrix} \varepsilon_1 \\ \varepsilon_2 \\ 0 \end{Bmatrix}_o^e . \tag{15.28}$$

That is, the shear component of the local strain is zero. This is analogous to an orthotropic thermal strain which has no local (principle) shear strain component but contains all the components when transformed to the physical axes. The integration in Eq. (15.27) involves the full physical components of $\boldsymbol{\varepsilon}_o$. These are obtained from the local orthotropic components by the Mohr's Circle transformation for strain that was discussed in Chap. 13:

$$\boldsymbol{\varepsilon}_o^e = \begin{Bmatrix} \varepsilon_x \\ \varepsilon_y \\ \varepsilon_{xy} \end{Bmatrix}_o^e = \boldsymbol{t}(\,\beta(\boldsymbol{x})\,) \begin{Bmatrix} \varepsilon_1 \\ \varepsilon_2 \\ 0 \end{Bmatrix}_o^e .$$

Here the square transformation matrix, $\mathbf{t}$, depends on the local principle stress direction, β. Note that this approach is anisotropic, that is, it allows a shear swelling effect whereas most applications in the literature have included only swelling from normal strains. Returning to Eq. (15.28), we wish to define the local shape-changing swelling strains. Let the difference between the principle stress σ_1 and its desired value be denoted by $\Delta\sigma_1$. We can normalize this with respect to the material modulus of the element. This stress difference should cause a shape change in the direction perpendicular to it. Thus, we set the swelling strain in that direction to be $\varepsilon_2 = \alpha\,\Delta\sigma_1 / E$, where α is a weighting constant to be selected later. In a similar manner we set

$$\boldsymbol{\varepsilon}_{oq} = \frac{\alpha}{E} \begin{Bmatrix} \Delta\sigma_2 \\ \Delta\sigma_1 \\ 0 \end{Bmatrix} = \frac{\alpha}{E}\,\boldsymbol{\sigma}_o$$

for plane stress or plane strain. A similar rule could be used in three-dimensions. For example

$$\boldsymbol{\varepsilon}_{oq} = \frac{\alpha}{E} \begin{Bmatrix} \Delta\sigma_2 + \Delta\sigma_3 \\ \Delta\sigma_1 + \Delta\sigma_3 \\ \Delta\sigma_1 + \Delta\sigma_2 \\ 0 \\ 0 \\ 0 \end{Bmatrix} .$$

Having defined the swelling strains, one computes the resultant effect on the element by numerical integration:

$$\boldsymbol{F}_o^e = \frac{\alpha}{E} \sum_q \boldsymbol{B}_q^{e^T} \boldsymbol{D}_q^e \, \boldsymbol{t}_q^e \, \boldsymbol{\sigma}_{oq}^e \, |J_q| \, W_q .$$

All of the above quantities but $\mathbf{t}$ and $\boldsymbol{\varepsilon}_o$ were generated previously (and stored) during the generation of the element stiffness matrix. Thus, this is a relatively computationally efficient procedure.

Assembling all the elements gives a new global load case for the system forces, $\boldsymbol{F}_o$, that cause increments of the displacements, say $\Delta\boldsymbol{\delta}$ for the new load case. Equilibrium gives

$$K(\delta + \Delta\delta) = F + F_o, \quad \text{or} \quad K\,\Delta\delta = F_o. \tag{15.29}$$

Here $\Delta\delta$ for this new load case is efficiently found from the previous inverse (actually factorization) of **K**. Adjacent elements will tend to have similar element resultants, F_o^e. An interior node, completely surrounded by elements, will have an assembled component, in F_o, that may be relatively small or zero. Conversely, a high incremental stress, $\Delta\sigma$, in an element near the surface will cause an assembled entry, in F_o, that may be relatively large. Thus F_o has entries that are expected to be larger nearer the object's boundary (or surface). The corresponding displacement increments, $\Delta\delta$, would likely have a similar magnitude distribution.

The components of $\Delta\delta$ indicate the directions in which the object would like to move in order to relieve the unbalanced optimal stress differences, $\Delta\sigma$. These components can be scaled to compute a corresponding change in shape of the nodal coordinates: $\Delta R = \gamma\,\Delta\delta$ to get the new mesh locations, $R + \Delta R$. The shape scaling factor could be computed by various criterion. If we are going to follow this shape change with a new computation and assembly of the global stiffnesses, then we want the element shape change, ΔR^e, to be large enough to actually make the calculation necessary. For example, we may want the maximum component of ΔR to be half the size, h^e, of the largest element attached to that node. Then we would have $\gamma = h^e/(2\,\Delta R_{\max})$. Another approach would be to consider the rate of material growth, per load increment. By integrating the normal displacement, $\Delta\delta \cdot n$, over the surface, Γ, of the object the net volume and mass change, Δm, is computed. If ΔM is the desired mass change for this step then we could set

$$\gamma = \frac{\Delta M}{\Delta m} = \frac{\Delta M}{\int_\Gamma \Delta\delta \cdot n \, d\Gamma}.$$

Other heuristic rules can be studied for these scaling processes. For example, one may wish to break Γ into regions of positive and negative material flows and scaling one material flow to cancel the other.

One can proceed with the above calculation in an alternate iterative manner. Let Eq. (15.29) be viewed as the k-th step in the process:

$$K\,\Delta\delta_k = F_{ok}.$$

We update the total displacement $(\delta + \sum \Delta\delta)$ and its element subsets. Then we loop again through the swelling calculations and update them:

$$\varepsilon_{k+1}^e = B^e\,\delta_k^e, \quad \sigma_{k+1}^e = D^e\,\varepsilon_{k+1}^e, \quad F_{o_{k+1}}^e = \int_{V^e} B^{e^T} D^e\,\varepsilon_{o_{k+1}}^e\,dv.$$

We would stop this process when $F_{o_{k+1}}^e$ is sufficiently small, or after N terms. Then, scale the net swelling displacement correction to get a shape change

$$\Delta R = \gamma\left[\sum_k^N = 1\,\Delta\,\delta_k\right].$$

15.9 References

[1] Adelman, H.M. and Haftka, R.T., "Sensitivity Analysis of Discrete Structural Systems," *AIAA J.*, **24**(5), pp. 823–832 (May 1986).

[2] Akin, J.E., Younan, M.Y., and Dewey, B.R., "Fracture Mechanics of Weldments Using Finite Elements," 3rd Intern. Conf. Structural Mechanics in Reactor Technology, Vol. 3 (G 5/2), Amsterdam, pp. 1–10, North Holland (1975).

[3] Atrek, E., Gallagher, R.H., Ragsdell, K.M., and Zienkiewicz, O.C., *New Directions in Optimum Structural Design*, New York: John Wiley (1984).

[4] Braibant, V. and Fleury, C., "Shape Optimal Design – A Performing CAD Oriented Formulation," AIAA 25th Structures Structural Dynamics and Materials Conference, AIAA-84-0857 (1984).

[5] Gallagher, R.H. and Zienkiewicz, O.C., *Optimum Structural Design Theory and Application*, New York: John Wiley (1973).

[6] Haftka, R.T. and Kamat, M.P., eds., *Elements of Structural Optimization*, Martinus Nijhoff (1985).

[7] Hellen, T.K., "On the Method of Virtual Crack Extensions," *Int. J. Num. Meth. Eng.*, **9**, pp. 187–207 (1975).

[8] Imgrund, M.C., "Design Improvements Using Finite Elements and Optimizing with Approximate Functions," Proceedings of 9th ASCE Conf. Electronic Computation, pp. 508–518 (Feb. 1986).

[9] Imgrund, M.C., "Using the Approximate Optimization Algorithm in ANSYS for the Solution of Optimum Convective Surfaces and Other Thermal Design Problems," Fifth Intern. Conf. on Thermal Problems, Montreal (1986).

[10] Kirsch, U., *Optimal Structural Design*, New York: McGraw-Hill (1981).

[11] Oda, J., "On a Technique to Obtain an Optimum Strength Shape by the Finite Element Method," *Bulletin of JSME*, **20**(140), pp. 160–167 (1977).

[12] Parks, D.M., "A Stiffness Derivative Finite Element Technique for Determination of Crack Tip Stress Intensity Factors," *Intern. J. of Fracture*, **10**(4), pp. 487–501 (Dec. 1974).

[13] Vanderplaats, G.N., "Structural Optimization – Past, Present, and Future," *AIAA J.*, **20**(7), pp. 992–1000 (1982).

[14] Vanderplaats, G.N., Miura, H., and Chargin, M., "Large Scale Structural Synthesis," *Finite Elements in Analysis and Design*, **2**(2), pp. 117–130 (1985).

[15] Vanderplaats, G.N. and Sugimoto, H., "A General Purpose Optimization Program for Engineering Design," *Intern. J. Computers and Structures*, **24**, p. 1 (1986).

[16] Yang, R.J. and Botkin, M.E., "A Modular Approach for Three-dimensional Shape Optimization of Structures," *AIAA J.*, **25**(3), pp. 492–497 (Mar. 1987).

[17] Yang, R.J. and Fiedler, M.J., "Design Modeling for Large-scale Three-dimensional Shape Optimization Problems," *ASME Computers in Engineering*, **2**, pp. 177–182 (1987).

Chapter 16

SPECIAL ELEMENTS *

16.1 Introduction

Current applications of the finite element method of analysis show a trend towards the enhancement of solution accuracy through the utilization of elements with *special interpolation* functions. A survey of these types of special elements, developed with the aid of physical insight, is presented. This class includes point and line singularity elements, semi-infinite elements, and elements with *upwind* weighting. The relative advantages and disadvantages of these special elements are also discussed.

16.2 Interpolation Enhancement for C^0 Transition Elements

The procedures in the previous sections were designed for elements having the same order of interpolation on each edge or face. If one wishes to go from one order of interpolation to another then it is necessary to supply transition elements between the two groups of elements or to define generalized constraint equations on the boundaries between the two different groups. The former method is not difficult to implement and will be illustrated here. Several special *transition* elements have been developed by various authors using trial and error methods. Special elements can be generated by enhancing low order elements, usually linear, or by degrading high order elements. Here a typical enhancement procedure will be utilized. Assume that the element to be enriched satisfies the usual conditions

$$u(r, s) = \sum_{i=1}^{n} H_i(r, s)\, u_i\,, \quad \sum_{i=1}^{n} H_i(r, s) = 1\,, \quad H_i(r_j, s_j) = \delta_{ij}\,. \tag{16.1}$$

Here the H_i are the original unenhanced interpolation functions, the u_i are nodal values, δ_{ij} is the Kronecker delta, and n is the original number of nodes. Let the enriching functions, say $M_j(r, s)$ be defined at m additional points. These satisfy the similar conditions that

$$M_i(r_j, s_j) = \delta_{ij}\,, \qquad i > n\,, \qquad 1 \le j \le (m+n)\,. \tag{16.2}$$

Thus, the enriched elements have values defined by

$$u(r, s) = \sum_{i=1}^{(m+n)} H_i(r, s)\, u_i , \tag{16.3}$$

where the enriched interpolation functions, H_i, are given by

$$H_i(r, s) \leftarrow H_i(r, s) - \sum_{j=n+1}^{n+m} H_i(r_j, s_j)\, M_j(r, s) , \qquad i \le n \tag{16.4}$$

$$H_i(r, s) \leftarrow M_i(r, s) , \qquad i > n .$$

where "$\leftarrow$" reads "is replaced by". These functions satisfy analogue relations to the second and third of Eq. (16.1). That can be seen by summing the enriched function so that

$$\sum_{i=1}^{m+n} H_i \leftarrow \sum_{i=1}^{n} H_i - \sum_{i=1}^{n} \sum_{j=1+n}^{m+n} H_i(r_j, s_j)\, M_j + \sum_{j=1+n}^{m+n} M_j$$

$$= 1 - \sum_{j=1+n}^{m+n} M_j(r, s) \left[1 - \sum_{i=1}^{n} H_i(r_j, s_j) \right] = 1 - 0 = 1$$

as required. Similarly if $i > n$ $H_i(r_k, s_k) \leftarrow M_i(r_k, s_k) = \delta_{ik}$, from Eqs. (16.2) and (16.4) while if $i \le n$

$$H_i(r_k, s_k) \leftarrow H_i(r_k, s_k) - \sum_{j=1+n}^{m+n} H_i(r_j, s_j)\, M_j(r_k, s_k)$$

$$= H_i(r_k, s_k) - \sum_{j=1+n}^{m+n} H_i(r_j, s_j)\, \delta_{jk} = \begin{cases} 0 & \text{if } k > n \\ \delta_{ik} & \text{if } k \le n ; \end{cases}$$

thus, as required, $H_i(r_k, s_k) = \delta_{ik}$. As the simplest example of enhancement, consider a line element to which a center node is to be added, see Fig. 16.2.1. First, assume that the enriching function is to be quadratic. The interpolation functions are

$$H_1(r) = 1 - r, \quad H_2(r) = r, \quad M_3(r) = 4r - 4r^2 . \tag{16.5}$$

Then the enriched interpolation function for Node 1 is

$$\begin{aligned} H_1(r) &= H_1(r) - H_1(r_3)\, M_3(r) \\ &= [1 - r] - (\tfrac{1}{2})\, [4r - 4r^2] = 1 - 3r + 2r^2 . \end{aligned} \tag{16.6}$$

This, of course, is the standard function for a full quadratic line element. The value of M_2 is obtained in a similar manner.

Figure 16.2.1 also shows a standard bilinear quadrilateral to which a fifth node is to be added. Again, only interpolation functions for Nodes 1 and 2 need be enriched. These functions are

$$H_1(r, s) = 1 - r - s + rs , \quad H_2(r, s) = r - rs . \tag{16.7}$$

If the enriching function is quadratic, then it is $M_5(r, s) = 4r(1 - r)(1 - s)$, so that

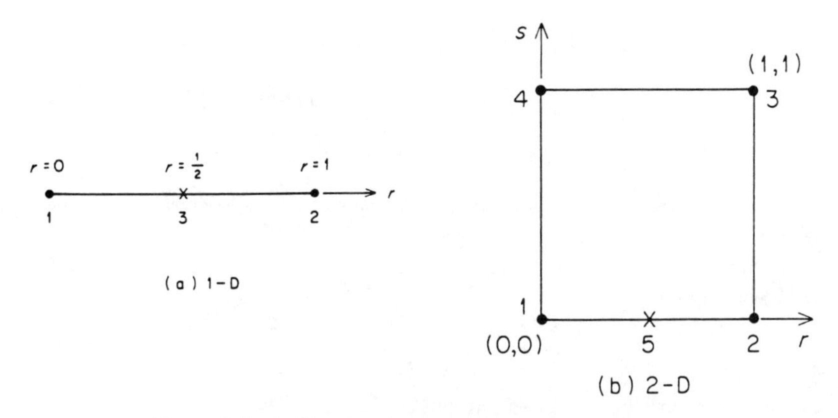

Figure 16.2.1 Simple element enhancements

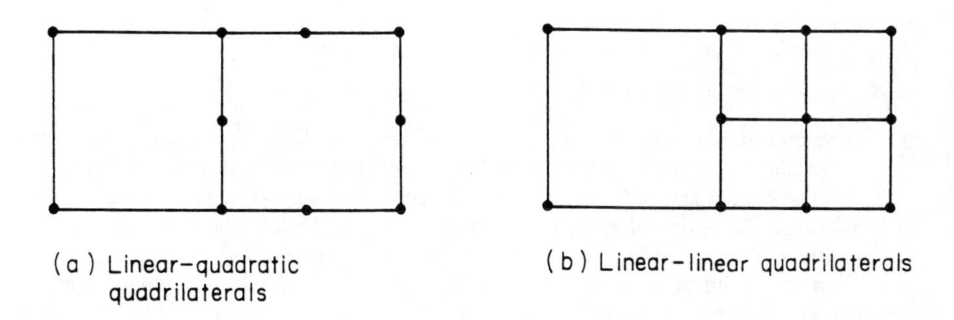

Figure 16.2.2 Typical transition elements

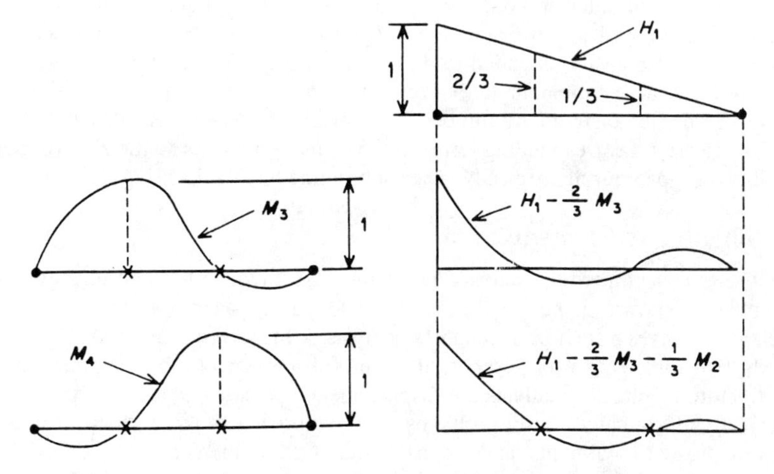

Figure 16.2.3 Enrichment of a line element to cubic degree

$$H_1(r,s) = H_1(r,s) - H_1(r_5,\ s_5)\,M_5(r,s)$$
$$= (1-r)(1-s) - (\tfrac{1}{2})\,[\,4r(1-r)(1-s)\,] \qquad (16.8)$$
$$= (1-2r)(1-r)(1-s)\,.$$

This corresponds to the linear-quadratic transition quadrilateral given by Bathe and Wilson [3].

One may also wish to have a linear-linear transition element for grid refinement. Then we may take

$$M_5(r,s) = 2r(1-s)\,, \quad 0 \le r \le \tfrac{1}{2} \qquad (16.9)$$
$$= 2(1-r)(1-s)\,, \quad \tfrac{1}{2} \le r \le 1\,,$$

and the corresponding modified interpolation functions are

$$H_1 = (1-r)(1-s) - (\tfrac{1}{2})\,[\,2r(1-s)\,]\,, \quad r \le \tfrac{1}{2}\,,$$
$$H_1 = (1-r)(1-s) - (\tfrac{1}{2})\,[\,2(1-r)(1-s)\,]\,, \quad r > \tfrac{1}{2}\,,$$

which reduce to

$$H_1 = (1-2r)(1-s)\,, \quad 0 \le r \le \tfrac{1}{2}\,, \qquad (16.10)$$
$$= 0 \quad \tfrac{1}{2} \le r \le 1\,.$$

These correspond to the functions given by Whiteman [20] for a linear-linear transition quadrilateral. The above elements, and corresponding triangles, are shown in Fig. 16.2.2.

The enrichment of higher elements proceeds in a similar manner. A linear to cubic enrichment for a line element is summarized in Fig. 16.2.3. The results for a series of quadrilateral elements are given in Fig. 9.2.8. That figure presents interpolation functions which can be linear, quadratic or cubic on any of the four sides of a quadrilateral. It is generated by enhancing the bilinear procedure. These operations can be programmed in different ways. Figure 9.2.8 illustrates only one way of generating such an element. That procedure is implemented as subroutine SHP412 which could easily be extended to include one or four interior nodes so as to generate Lagrangian elements. A similar procedure can be used to generate a 3 to 10 node enhanced triangle. Of course, these procedures are also easily extended to three-dimensional elements. Since such elements allow for local node numbers to be zero it is necessary to avoid zero or negative subscripts as degree of freedom numbers. For example, during assembly if a local node is zero, there are no corresponding terms in the element matrices so the zero (or negative) subscript is skipped during assembly. (See subroutine STORCL in Fig. 1.3.2.)

16.3 Singularity Elements

There are numerous analysis problems that involve the presence of singular derivatives at known points in the solution domain. The most common singularity application involves a fracture mechanics analysis of the stresses near a crack tip or crack edge. In two dimensions let ρ denote the radial distance from the singular point. In a linear fracture mechanics analysis the displacements and stresses are $0(\rho^{1/2})$ and $0(\rho^{-1/2})$, respectively. Originally these problems were approached by a direct solution that employed standard elements and an extensive grid refinement near the singularity. Engineers expected that if they employed their physical insight and used special elements around the singularity, they would obtain more accurate results. Thus they attempted to

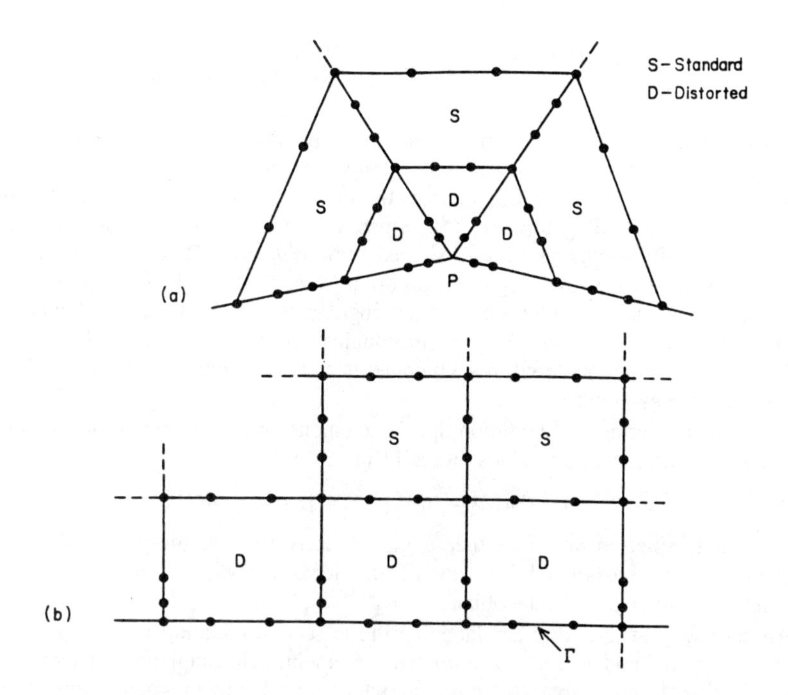

Figure 16.3.1 Distorted elements for singularities

```
      SUBROUTINE  SINGLR (P, H, DH, N, NSPACE)
C     * * * * * * * * * * * * * * * * * * * * * * * *
C      CONVERT STANDARD FUNCTIONS TO SINGULAR FUNCTIONS
C     WITH DERIV SINGULARITIES AT NODE 1 OF O(R**(-P))
C     * * * * * * * * * * * * * * * * * * * * * * * *
CDP   IMPLICIT REAL*8 (A-H,O-Z)
      DIMENSION  H(N), DH(NSPACE,N)
C     H      = SHAPE FUNCTION ARRAY (STANDARD)
C     DH     = LOCAL DERIVATIVES OF H (STANDARD)
C     N      = NUMBER OF SHAPE FUNCTIONS
C     NSPACE = DIMENSION OF SPACE
C      REQUIRES SUM OF H(I) = 1, & CONST JACOBIAN
      IF ( P .EQ. 0.0 )  RETURN
      W = 1.0 - H(1)
      R = W**P
      DO 20  I = 1,NSPACE
        DO 10  J = 2,N
   10   DH(I,J) = DH(I,J)/R + P*DH(I,1)*H(J)/R/W
        DH(I,1) = (1.0 - P)*DH(I,1)/R
   20 CONTINUE
      DO 30  J = 2,N
   30 H(J) = H(J)/R
      H(1) = 1.0 - W/R
      RETURN
      END
```

Figure 16.3.2 Singularity interpolation modifications

modify standard interpolation functions to include the $\rho^{1/2}$ displacement behavior near the crack tip. Many special two-dimensional elements have been proposed for this type of application.

There are numerous other problems that involve point singularities. For example, in the solution of the Poisson equation in a domain with a re-entrant corner there is a singularity at the corner point. If b denotes the interior obtuse corner angle, the solution near the corner has leading term $0(\rho^{\pi/b})$. Since $\pi < b \le 2\pi$ one can encounter a wide range of singularity strengths. For L-shaped regions $b = 3\pi/2$ so that the solution is $0(\rho^{2/3})$ and its derivatives are singular with order $0(\rho^{-1/3})$. Only if the region contains a slit, $b = 2\pi$, does one encounter the stronger singularity for which many special elements have been proposed by the fracture mechanics' researchers. The above variable singularity could be encountered in such contexts as heat conduction, potential flow, and transverse electromagnetic lines.

Before considering the common special elements we will review the relationship between global and local derivatives. Recall from Sect. 9.4:

$$\partial_g u = \mathbf{J}^{-1}\,\partial_l u,\ \mathbf{J} = \partial_l \mathbf{H}\,[X^e\!: Y^e],\ \partial_l u = \partial_l \mathbf{H}\mathbf{U}^e.$$

From the first equation we note that a global derivative singularity in the physical coordinate can be introduced by way of the local derivatives of the interpolation functions or the inverse of the Jacobian.

An easy way of affecting the Jacobian in the second equation is to vary the global coordinates of the nodes, i.e., to *distort* the element. Henshell and Shaw [10] were among the first to note that the point at which the Jacobian goes singular can be controlled through the use of distorted elements. Generally, the order of the geometric interpolation functions is quadratic or higher. One can control the location of the singular point by moving nodes in the global coordinates. The singular point usually lies outside the element and one must force it to occur on an element boundary or node. To illustrate this concept consider a one-dimensional quadratic element, $0 \le x \le h$, of length h, as discussed in Sect. 3.9. The coordinate transformation of the standard unit element, $0 \le \rho \le 1$, with an interior node at ah, is

$$x(\rho) = h(4a-1)\,\rho + 2h(1-2a)\,\rho^2 \tag{16.11}$$

and the Jacobian is

$$J = \frac{\partial x}{\partial \rho} = h(4a-1) + 4h(1-2a)\,\rho\ . \tag{16.12}$$

Thus, if one wants a singularity to occur at $\rho = 0$, i.e., node 1, one must set $a = 1/4$. That means that the global coordinates of node 3 must be specified so that $x_3 = h/4$. One then has $x = h\rho^2$ so that $\rho = (x/h)^{1/2}$ and the inverse Jacobian becomes

$$J^{-1} = \frac{\partial \rho}{\partial x} = \frac{1}{2}\,(hx)^{-1/2}$$

which has a singularity of order $0(x^{-1/2})$ at $x = 0$. The derivative singularity can also be seen by noting that for quadratic interpolation on the unit interval

$$u(r) = \alpha_1 + \alpha_2\,\rho + \alpha_3\,\rho^2$$

becomes, in physical space,

$$u(x) = \alpha_1 = \beta_2\,x^{1/2} + \beta_3 x\ .$$

Although the above one-dimensional example is simple, the algebra involved in this type of procedure can be rather involved in higher-dimensional elements. In two and three dimensions, respectively, the point or line of singularity is surrounded with the special distorted elements as shown in Fig. 16.3.1. The locations of nodes on the element sides are usually the same as for the one-dimensional case. Any interior nodes usually lie on lines connecting the shifted side nodes. The above procedure is easily generalized, and typical nodal locations for function interpolations of order $0(r^{1/m})$ are given in Table 16.1.

Table 16.1 Node Locations for Interpolations of Order $\rho^{1/m}$

	Standard node				Mapped node				
Number of nodes $n > m$	r_i $1 \le i \le n$				$(x/l)_i = r_i^m$ $1 \le i \le n$				m
2	0			1	0			1	1
3	0	1/2		1	0	1/2		1	1
					0	1/4		1	2
4	0	1/3	2/3	1	0	1/3	2/3	1	1
					0	1/9	4/9	1	2
					0	1/27	8/27	1	3

The second alternative of introducing the singularity through the local derivative requires special element interpolation functions. A number of special elements of this type have been developed for the form $u = 0(\rho^{1/2})$. However, most such elements cannot be generalized to other singularities. A conforming element having the general form of $u = 0(\rho^{1-a})$ for $0 < a < 1$ was given by Akin [1]. This procedure is valid for, and compatible with, an element that satisfies the condition that $\sum_i N_i = 1$. This easily programmed procedure is outlined here. If the singularity occurs at node K in an element one defines a function $W(p, q) = 1 - H_k(r, s)$ which is zero at node k and equal to unity at all other nodes. To obtain the power singularity form one can divide by a function $R = W_a$. If N_i denote the standard functions, then the modified singular functions, H_i, are defined as

$$H_i = [1 - 1/R]\,\delta_{ki} + N_i/R\ .$$

These also satisfy Eq. (16.1) and near node K give a solution of the form $\phi(\rho^{1-a})$ so that the first derivative is $O(\rho^{-a})$. Thus, since $a > 0$ the first derivative is singular. This procedure has been extended to line singularities by Akin [2]. The programming of these operations for point singularities is given in subroutine SINGLR in Fig. 16.3.2. The above procedure, summarized in subroutine SINGLR, assumes that the geometric Jacobian is constant. Otherwise, the net value of the introduced singularity is changed. Therefore, it is recommended that this procedure be used with straight-sided triangles. The modification is intended to primarily enhance the response in the radial direction. The analytic solutions often also have rapid angular variations. Thus, one should utilize several elements to allow for such angular response.

The examples of Sect. 16.2 illustrated how elements can be enriched to give higher order functions on edges and grid refinement on edges. The method can also be utilized to incorporate known singularities. For example, consider the line element again as in

Eq. (16.5) and let the center node have a more general enrichment of the form:

$$M_3(r) = K(r^a - r^b), \quad a \neq b \tag{16.13}$$

where $K = 1/[1/2)^a - (1/2)^b]$. This can yield the standard quadratic function, $a = 1$ and $b = 2$, or a singularity function. For example, if one desires to have terms or order $0(r^{1/2})$ enriched into the previous linear element one could set $a = 1$ and $b = 1/2$. Then

$$M_3 = 2K(r - r^{1/2}), \qquad K = 1/(1 - \sqrt{2})$$

and the modified functions are

$$H_1(r) = 1 - (1 + K)\, r + K r^{1/2}, \qquad H_2(r) = (1 + K)\, r - K r^{1/2}.$$

If standard interpolation is utilized for the geometry then this element gives derivative singularities at node 1 of order $0(\rho^{1/2})$ where ρ is the distance from the node.

The integration of the singularity interpolation functions is often done with special quadrature rules. The author recommends rules that lie on radial lines through the singular point. For triangular elements the Radau quadrature data have this property. It is a rule that yields points that are symmetric with respect to the line $r = s$. Their values are given in subroutine RADAU.

Hughes and Akin [12] have presented another interpolation algorithm that can generate singularities of various orders at once. For example, $\phi(r) = c_1 r^{\alpha} + c_2 r^{\beta}$ where α and β can have any values. Many analytic solutions are actually series. The crack tip solution is of the form

$$\phi(\theta, r) = c_1(\theta)\, r^{1/2} + c_2(\theta)\, r^{3/2} + \cdots$$

where some boundary conditions make $c_1(\theta) = 0$. Thus, if we build an interpolation with $\alpha = 1/2$ and $\beta = 3/2$, then the process can better model a wider range of boundary conditions. Their form also includes the "rigid body modes" that are missing in most other singularity treatments. The procedure is easily implemented numerically, but is difficult to express analytically except for one-dimensional problems. One begins the procedure with a set of standard functions that include the necessary rigid body modes. One then adds an additional node for each new singularity to be included. The old functions are modified to create a new set of *rational polynomial* interpolation functions that include the desired singularity. The algorithm is:

Step 1: Provide an initial set of m standard interpolation functions, and a group of p functions of the form r^{α}. Thus, there will be a total of $n = m + p$ functions. Note that $H_i(r_j) = \delta_{ij}$ only for $i, j \leq m$.

Step 2: Modify H_{m+1} so it satisfies the interpolation property for all n nodes:

$$H_{m+1}(r) \leftarrow \frac{H_{m+1}(r) - \sum_{k=1}^{m} H_{m+1}(r_k)\, H_k(r)}{H_{m+1}(r_{m+1}) - \sum_{k=1}^{m} H_{m+1}(r_k)\, H_k(r_{m+1})}.$$

Step 3: Update the preceding interpolations:

$$H_k(r) \leftarrow H_k(r) - H_k(r_{m+1})\, H_{m+1}(r), \quad 1 \leq k \leq m.$$

Step 4: If $m + 1 < n$, replace m by $m + 1$, repeat Steps 2 to 4. If $m + 1 = n$, stop.

The derivatives could, and usually would, be computed at the same time. If needed, they would be computed from:

Step 2b:

$$\frac{\partial H_{m+1}(r)}{\partial r} \leftarrow \frac{\dfrac{\partial H_{m+1}(r)}{\partial r} - \sum_{k=1}^{m} H_{m+1}(r_k) \dfrac{\partial H_k(r)}{\partial r}}{H_{m+1}(r_{m+1}) - \sum_{k=1}^{m} H_{m+1}(r_k) H_k(r_{m+1})} .$$

Step 3b:

$$\frac{\partial H_k(r)}{\partial r} \leftarrow \frac{\partial H_k(r)}{\partial r} - H_k(r_{m+1}) \frac{\partial H_{m+1}(r)}{\partial r} .$$

As an example of this procedure, consider a one-dimensional element in unit coordinates that is to be enhanced from linear interpolation to include an r^α singularity. The initial functions are

$$H_1(r) = 1 - 2r, \quad H_2(r) = 2r, \quad H_3(r) = r^\alpha .$$

Note that the linear interpolations are through the left $(r = 0)$ and middle $(r = 1/2)$ nodes. Invoking the above algorithm gives the singularity functions

$$H_1(r) = 1 - 2r + H_3(r), \quad H_2(r) = 2r - 2H_3(r), \quad H_3(r) = \frac{r^\alpha - 2r(1/2)^\alpha}{1 - 2(1/2)^\alpha}$$

which satisfy the desired condition that $\sum_j H_j(r) \equiv 1$.

16.4 Semi-infinite Elements

A number of practical applications involve infinite or semi-infinite domains. They include soil mechanics, external flows, and electromagnetic fields. The most common method of solution has been to extend the *standard elements* out as far as one can afford and thereto approximate the boundary conditions at infinity. Recently, special spatial coordinate interpolation functions have been suggested to map standard elements into a semi-infinite domain. Thus, in a typical problem one would surround the region of interest with standard elements. The outer layer of the elements would then be mapped to infinity and the corresponding boundary conditions applied at the points at infinity. Semi-infinite elements have been successfully employed in several classes of static, transient, and dynamic problems. Here we will consider some of the elementary concepts. Analysts faced with such problems should consult the extensive review articles on the subject by Bettess [5, 6]. They include about 170 references on this subject.

In a standard element one defines the global coordinates in terms of the local coordinates by the standard mapping; for example, for a one-dimensional quadratic element $x(r) = H_1(r)\, x_1 + H_2(r)\, x_2 + H_3(r)\, x_3$. Thus, $x_1 \le x(r) \le x_3$ and the element is a finite domain. To generate a semi-infinite domain, one could replace the standard interpolation functions, $H_i(r)$, with a special set, say $N_i(r)$, that lead to the proper physical behavior as $x \to \infty$ (i.e., $r \to 1$). Two approaches to this problem have been suggested. Gartling and Becker [9] proposed the use of rational functions for the N_i while Bettess [4] suggested a combined exponential and Lagrangian interpolation for the N_i. Both approaches lead to functions involving a characteristic length, say L, that is related to the input coordinates. The results obtained from their analyses can vary significantly with L. As an example, consider the mapping which takes the local coordinates, $r = 0$, $r = 1/2$, and $r = 1$, into the semi-infinite domain x_1, x_2, ∞. The lower limit, x_1, is necessary to maintain compatibility with the standard element mesh. One possible mapping is $x(r) = x_1 + Lr/(1 - r)$ where $L = (x_2 - x_1)$. That is, where $N_1(r) = 1 - r/(1 - r)$,

$N_2(r) = r/(1-r)$ and $N_3 = 0$. Using the standard local interpolation of the dependent variable, $u(r) = H_1(r)\, u_1 + H_2(r)\, u_2 + H_3(r)\, u_3$, together with the spatial mapping, the gradient in the semi-infinite element is $\partial u / \partial x = (\partial r / \partial x)\ (\partial u / \partial r) = [\,(1-r)\,(1-r)/L\,]\, \partial u / \partial p$. Thus, as $r \to 1$, $x \to \infty$, $u \to u_3$, and $\partial u / \partial x \to 0$. The value of L affects the 2nd and 4th of the limits. To find the optimum L one wolud have to use physical insite. If one arbitrarily selects L but knows a physical condition such as $\partial u / \partial x \to 1/x^2$ as $x \to \infty$, then the above form implies a multi-point constraint between u_1 and u_2. The generation of semi-infinite elements and procedures for finding optimum parameters, such as L, still merits attention.

An alternative procedure for treating semi-infinite domains was suggested by Silvester et al [18]. It does not introduce arbitrary constants as most mappings do and is computationally efficient. Consider a finite region R containing a standard finite element mesh. The region has been selected such that there exists at least one point, P, which can see all points on the boundary, Γ. Assume that this region needs to be extended to infinity to satisfy a boundary condition. Define a mapping of Γ onto a series of temporary interface boundaries, γ_k, such that for each pair of points on γ_k and γ_{k+1}: $\mathbf{r}_{k+1} = K\mathbf{r}_k$, where $\mathbf{r}$ is the position vector from P_k, $K > 1$ is a constant and $\gamma_0 \equiv \Gamma$. Thus at the exterior boundary, say γ_n:

$$\mathbf{r}_n = K^n r_0 .$$

Between two interfaces, say γ_k and γ_{k+1}, an angular segment defines an element e_{k+1}. The volumes of the extended elements e_{k+1} and e_k will have a ratio of K^{NSPACE} and the elements are geometrically similar. Assume the exterior domain satisfies a homogeneous differential equation and that the material properties are constant. Then, if the square matrix involves an inner product of the first global derivatives the element matrices in e_k and e_{k+1} are given by

$$\mathbf{S}^{k+1} = \lambda \mathbf{S}^k$$

where $\lambda = K^{-NSPACE}$. This observation allows a recursion relation to be developed so that only contributions to nodes on Γ and γ_n need be retained. Since the function of interest on γ_n, say ϕ_n, is usually zero it may only be necessary to consider the net contributions to nodes of Γ.

16.5 Upwind Elements

The use of physical insight to generate an *upwind* difference procedure for the finite difference analysis of flows is known to be very useful. Thus it is logical to develop a similar procedure for finite elements. For one-dimensional applications the special elements seem quite promising. However, the programming difficulties associated with the extension to higher-dimensional problems may limit the usefulness of the procedure. The upwind procedure recommended by Hughes [13] seems more practical from the computational point of view. A recent review of this approach has been given by Hughes [14].

A newer computational procedure has been successfully applied by Rice and Schnipke [16] for the Q4 quadrilaterals and H8 brick elements. Their method is summarized here for the Q4 element. Consider a Q4 element in a flow field. We can identify one of its nodes as the most "downwind node". From that node we proceed in opposite directions along the element boundary to find the point on a side where the edge inflow and outflow balance. In other words, we find where the streamline through the downwind node intersects the other side of the element. The quantity being convected is computed at that point and at the downwind node. The difference in those values divided by the distance between the two points is taken as the quantitie's upwind gradient. That value is assumed to be constant and is used as such in all the integrals involving upwind gradient terms. Those procedures can be generalized for upwinding any type of element (see GUS3D). Kondo, et al, [15] have demonstrated another higher order accuracy method for Q4 elements.

16.6 Gap-Contact Elements

The contact element or gap element may be used to impose constraints between nodes on elastic or rigid solids. Either perfect friction (i.e., "stick") or frictionless (i.e., "slip") conditions may be achieved. A contact element is defined by two nodes, one of which is the master; a spring constant, k ; and a fixed direction vector, $\mathbf{n}$, that is normal to the contact plane. The present location of node A is given by $\mathbf{x}_A + \mathbf{d}_A$, where $\mathbf{x}_A$ is the initial position vector and $\mathbf{d}_A$ is its displacement vector. The contact plane passes through the point $\mathbf{x}_A + \mathbf{d}_A$ and is perpendicular to $\mathbf{n}$ (see Fig. 16.6.1). The contact/release condition is defined as follows:

$$\delta > 0 \quad \textit{release}, \qquad \delta \le 0 \quad \textit{contact}$$

where $\delta = \rho \cdot \mathbf{n}$ and ρ is the relative position of node B with respect to the position of master node A, that is,

$$\rho = \mathbf{x}_B + \mathbf{d}_B - \mathbf{x}_A - \mathbf{d}_A .$$

The quantity δ is a measure of the signed distance between $\mathbf{x}_B + \mathbf{d}_B$ and the contact

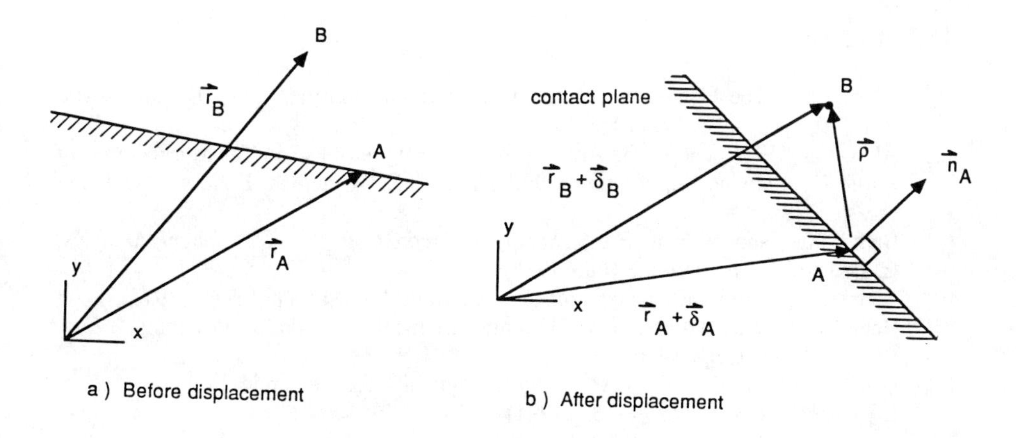

Figure 16.6.1 Gap geometry

plane. When contact is noted, a contact element stiffness and out-of-balance force are added to the global equations. These arrays are defined as follows:

$$\mathbf{k}^{(stick)} = k \begin{bmatrix} 1 & 0 & -1 & 0 \\ & 1 & 0 & -1 \\ & & 1 & 0 \\ sym & & & 1 \end{bmatrix}$$

$$\mathbf{k}^{(slip)} = k \begin{bmatrix} n_1^2 & n_2 n_2 & -n_1^2 & -n_1 n_2 \\ & n_2^2 & -n_1 n_2 & -n_2^2 \\ & & n_1^2 & n_1 n_2 \\ sym & & & n_2^2 \end{bmatrix} = k \begin{Bmatrix} \mathbf{n} \\ -\mathbf{n} \end{Bmatrix} \begin{Bmatrix} \mathbf{n} \\ -\mathbf{n} \end{Bmatrix}^T .$$

Note that the stick case is simply a repeated freedom (or Type 2 constraint) between the components at both nodes. The slip case simply means that there is a spring parallel to the normal vector and perpendicular to the normal vector that are attached to these two nodes. The out-of-balance force between the two nodes is

$$\mathbf{f} = -\mathbf{k}^{(slip)} \begin{Bmatrix} \mathbf{x}_A + \mathbf{d}_A \\ \mathbf{x}_B + \mathbf{d}_B \end{Bmatrix} = k\delta \begin{Bmatrix} \mathbf{n} \\ -\mathbf{n} \end{Bmatrix} .$$

If $k > 0$ is sufficiently large, the point $\mathbf{x}_B + \mathbf{d}_B$ will be forced to lie (approximately) on the contact plane. The problem must be solved iteratively. Both $\mathbf{k}$ and $\mathbf{f}$ are used in the first step. In subsequent steps, only the contact stiffness is assembled and the decision to remain in contact, or release, is made on the basis of the sign of δ, as above. Practical guidelines for selecting the gap stiffness have been given by Rizzo [17].

16.7 References

[1] Akin, J.E., "The Generation of Elements with Singularities," *Int. J. Num. Meth. Eng.*, **10**, pp. 1249–1259 (1976).

[2] Akin, J.E., "Elements for the Analysis of Line Singularities," in *The Mathematics of Finite Elements and Applications, Vol. III*, ed. J.R. Whiteman, London: Academic Press (1978).

[3] Bathe, K.J. and Wilson, E.L., *Numerical Methods for Finite Element Analysis*, Englewood Cliffs: Prentice-Hall (1976).

[4] Bettess, P., "Infinite Elements," *Int. J. Num. Meth. Eng.*, **11**, pp. 53–64 (1977).

[5] Bettess, P. and Bettess, J.A., "Infinite Elements for Static Problems," *Eng. Computations*, **1**, pp. 4–16 (1984).

[6] Bettess, P. and Bettess, J.A., "Infinite Elements for Dynamic Problems," *Eng. Computations*, **8**(2), pp. 99–152 (1991).

[7] Blackburn, W.S., "Calculation of Stress Intensity Factors at Crack Tips Using Special Finite Elements," pp. 327–336 in *The Mathematics of Finite Elements and Applications*, ed. J.R. Whiteman, London: Academic Press (1973).

[8] Daly, P., "Singularities in Transmission Lines," pp. 337–350 in *The Mathematics of Finite Elements and Applications*, ed. J.R. Whiteman, London: Academic Press (1973).

[9] Gartling, D. and Becker, E.B., "Computationally Efficient Finite Element Analysis of Viscous Flow Problems," in *Computational Methods in Nonlinear Mechanics*, ed. J.T. Oden, et al., Austin, Texas: T.I.C.O.M. (1974).

[10] Henshell, R.D. and Shaw, K.G., "Crack Tip Elements are Unnecessary," *Int. Num. Meth. Eng.*, **9**, pp. 495–509 (1975).

[11] Hughes, T.J.R., "A Simple Scheme for Developing 'Upwind' Finite Elements," *Int. J. Num. Meth. Eng.*, **12**, pp. 135–167 (1978).

[12] Hughes, T.J.R. and Akin, J.E., "Techniques for Developing Special Finite Element Shape Functions with Particular Reference to Singularities," *Intern. J. Num. Meth. Eng.*, **15**, pp. 733–751 (1980).

[13] Hughes, T.J.R., *The Finite Element Method*, Englewood Cliffs: Prentice-Hall (1987).

[14] Hughes, T.J.R., "Recent Progress in the Development and Understanding of SUPG Methods with Special Reference to the Compressible Euler and Navier-Stokes Equations," pp. 273–287 in *Finite Elements in Fluids* – Volume 7, ed. R.H. Gallagher, R. Glowinski, P.M. Gresho, J.T. Oden and O.C. Zienkiewicz, New York: John Wiley (1987).

[15] Kondo, N., Tosaka, N., and Nishimura, T., "High Reynolds Solutions of the Navier-Stokes Equations Using the Third-Order Upwind Finite Element Method," in *Computational Methods in Flow Analysis*, ed. H. Niki and M. Kawahara, Okayama University of Science Press (1988).

[16] Rice, J.G. and Schnipke, R.J., "A Monotone Streamline Upwind Finite Element Method for Convection-Dominated Flows," *Comp. Meth. Appl. Mech. Eng.*, **48**, pp. 313–327 (1985).

[17] Rizzo, A.R., "FEA Gap Elements: Choosing the Right Stiffness," *Mechanical Engineering*, **113**(6), pp. 57–59 (June 1991).

[18] Silvester, P.P., Lowther, D.A., Carpenter, C.J., and Wyatt, E.A., "Exterior Finite Elements for 2-Dimensional Field Problems with Open Boundaries," *Proc. IEE*, **124**(12), pp. 1267–1270 (1977).

[19] Whiteman, J.R., "Numerical Solution of Steady State Diffusion Problems Containing Singularities," pp. 101–120 in *Finite Elements in Fluids II*, ed. R.H. Gallagher, New York: John Wiley (1975).

[20] Whiteman, J.R., "Some Aspects of the Mathematics of Finite Elements," pp. 25–42 in *The Mathematics of Finite Elements and Applications, Vol. II*, ed. J.R. Whiteman, London: Academic Press (1976).

[21] Whiteman, J.R. and Akin, J.E., "Finite Elements, Singularities and Fracture," in *The Mathematics of Finite Elements and Applications, Vol. III*, ed. J.R. Whiteman, London: Academic Press (1978).

[22] Zienkiewicz, O.C., *The Finite Element Method,* 3rd edition, New York: McGraw-Hill (1979).

Chapter 17

TIME DEPENDENT PROBLEMS

17.1 Introduction

Many problems require the solution of time-dependent equations. In this context, there are numerous theoretical topics that an analyst should investigate before selecting a computational algorithm. These include the stability limits, amplitude error, phase error, etc. A large number of implicit and explicit procedures have been proposed. Several of these have been described in the texts by Bathe and Wilson [1], Chung [3], and Zienkiewicz [14]. It is even possible to combine these procedures as suggested by Hughes and Liu [6]. A recent review of the stability considerations was given by Park [11].

The purpose here is to cite typical additional computational procedures that arise in the time integration problems. Both simple explicit and implicit algorithms will be illustrated. Applications involving second order spatial derivatives and first order time derivatives will be referred to as parabolic or transient, while those with second order time derivatives will be referred to as hyperbolic or dynamic. For these classes of problems it is common to select various temporal operators to approximate the time derivatives. They approximate the velocity, $v = dR/dt$, by the forward difference, $v \approx [R(t+\Delta t) - R(t)]/\Delta t$, backward difference, $v \approx [R(t) - R(t-\Delta t)]/\Delta t$, or central difference $v \approx [R(t+\Delta t) - R(t-\Delta t)]/(2\Delta t)$, where Δt denotes a small time difference. Similar expressions can be developed to estimate the second derivative or acceleration, $a = d^2/dt^2$. The actual time integration algorithm is determined by the choices of the difference operators and the ways that they are combined.

17.2 Parabolic Equations

We saw in Sec. 2.7 (and will again in the next chapter) that the governing system equations for a transient application are generally ordinary differential equations of the form

$$\mathbf{M}\,\dot{\mathbf{R}}(t) + \mathbf{K}\,\mathbf{R}(t) = \mathbf{P}(t) \tag{17.1}$$

where $(\dot{\ }) = d(\ )/dt$ denotes the derivative with respect to time. Generally, one or more of the coefficients in $\mathbf{R}$ say R_i, will be defined in a boundary condition as a function of

Table 17.1 System matrices for linear transients

$$\mathbf{A}\dot{\mathbf{R}} + \mathbf{B}\mathbf{R} = \mathbf{P} \rightarrow \mathbf{S}\,\mathbf{R}(t) = \mathbf{F}$$

1. Euler (forward difference), $k = \Delta t$

$$\mathbf{S} = \mathbf{A}/k$$
$$\mathbf{F} = \mathbf{P}(t-k) + (\mathbf{A}/k - \mathbf{B})\,\mathbf{R}(t-k)$$

2. Crank-Nicolson (mid-difference), $h = \Delta t/2$

$$\mathbf{S} = \mathbf{A}/k + \mathbf{B}/2$$
$$\mathbf{F} = \mathbf{P}(t-h) + (\mathbf{A}/k - \mathbf{B}/2)\,\mathbf{R}(t-k)$$

3. Linear velocity

$$\mathbf{S} = 2\mathbf{A}/k + \mathbf{B}$$
$$\mathbf{F} = \mathbf{P}(t) + \mathbf{A}\left[2\mathbf{R}(t-k)/k - \mathbf{R}(t-k)\right] \text{ and}$$
$$\dot{\mathbf{R}}(t) = 2\left[\mathbf{R}(t) - \mathbf{R}(t-k)\right]/k + \dot{\mathbf{R}}(t-k)$$

4. Galerkin

$$\mathbf{S} = \mathbf{A}/k + 2\mathbf{B}/3$$
$$\mathbf{F} = (\mathbf{A}/k - \mathbf{B}/3)\,\mathbf{R}(t-k) + 2\int_0^k P(\tau)\,\tau d\tau/k^2$$

5. Least-squares

$$\mathbf{S} = \mathbf{B}^T\mathbf{B}\,k/3 + (\mathbf{B}^T\mathbf{A} + \mathbf{A}^T\mathbf{B})/2 + \mathbf{A}^T\mathbf{A}/k$$
$$\mathbf{F} = \left[\mathbf{A}^T\mathbf{A}/k + (\mathbf{B}^T\mathbf{A} - \mathbf{A}^T\mathbf{B})/2 - \mathbf{B}^T\mathbf{B}\,k/6\right]\mathbf{R}(t-k) + \mathbf{B}^T\int_0^k P(\tau)\,\tau d\tau + \mathbf{A}^T\int_0^k P(\tau)\,d\tau/k$$

time, i.e., $R_i = g(t)$. Also, the initial values $\mathbf{R}(0)$, must be known to start the transient solution. Note that the governing equations now involve two square matrices, $\mathbf{M}$ and $\mathbf{K}$, at the system level. Thus, in general it will be necessary to apply the previously discussed square matrix assembly procedure twice. This is,

$$\mathbf{M} = \sum_{e=1}^{NE} \mathbf{M}^e, \qquad \mathbf{K} = \sum_{e=1}^{NE} \mathbf{K}^e \tag{17.2}$$

where $\mathbf{M}^e$ and $\mathbf{K}^e$ are generated from the Boolean assembly matrices and the corresponding element contributions, say $\mathbf{m}^e$ and $\mathbf{k}^e$. In a nonlinear problem the system matrices usually depend on the values of $\mathbf{R}(t)$ and an iterative solution is required. For example, a heat transfer problem may involve material conductivities which are temperature-dependent. For the sake of simplicity, such nonlinear applications will not be considered at this point. Only the direct step-by-step time integration of Eq. (17.1) will be considered. There are many such procedures published in the literature. When selecting a computational procedure from the many available algorithms one must consider the relative importance of storage requirements, stable step size, input-output operations, etc. The text by Myers [9] examines in detail many of the aspects of simple time integration procedures for linear transient applications. He utilizes both finite difference and finite element spatial approximations and illustrates how their transient solutions differ. Bathe and Wilson [1] also give details of several methods of direct time integration and their extension to nonlinear time dependent problems.

17.2.1 Simple Approximations

The accuracy, stability, and relative computational cost of a transient integration scheme depend on how one approximates the velocity, $d\mathbf{R}/dt$, during the time step. For example, one could assume that the velocity during the time step is (a) constant, (b) equal to the average value at the beginning and end of the step, (c) varies linearly during the step, etc. To illustrate cases (a) and (c) consider a time interval of $k = \Delta t$ and assume a Taylor series for $\mathbf{R}(t)$ in terms of the value at the previous time step, $\mathbf{R}(t-k)$:

$$\mathbf{R}(t) = \mathbf{R}(t-k) + k\dot{\mathbf{R}}(t-k) + k^2/2\,\ddot{\mathbf{R}}(t-k) + \cdots \tag{17.3}$$

Then, as illustrated in Fig. 17.2.1, the first assumption gives $\ddot{\mathbf{R}} = 0$ and the above equation yields $\mathbf{R}(t)$. The standard Euler integration procedure is obtained by multiplying by $\mathbf{M}$:

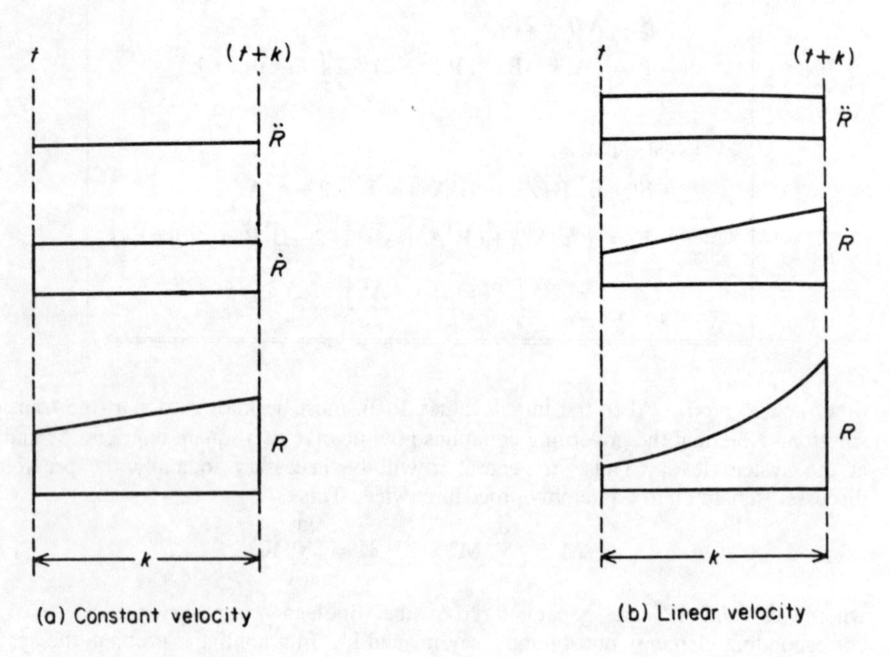

Figure 17.2.1 Typical integration assumptions

$$\mathbf{M}\,\mathbf{R}(t) = \mathbf{M}\,\mathbf{R}(t-k) + k\mathbf{M}\dot{\mathbf{R}}(t-k) \tag{17.4}$$

and substituting Eq. (17.1) at $(t-k)$

$$\mathbf{M}\dot{\mathbf{R}}(t-k) = \mathbf{P}(t-k) - \mathbf{K}\,\mathbf{R}(t-k)$$

to obtain the final result that

$$\mathbf{M}\,\mathbf{R}(t) = k\mathbf{P}(t-k) + \{\,\mathbf{M} - k\mathbf{K}\,\}\,\mathbf{R}(t-k)\,. \tag{17.5}$$

One can make the general observation that the governing ordinary differential equations have been reduced to a new set of algebraic equations of the form

$$\mathbf{S}\,\mathbf{R}(t) = \mathbf{F}(t) \tag{17.6}$$

which must be solved at each time step. In the present case of the Euler method one has system matrices $\mathbf{S} = \mathbf{M}$, and

$$\mathbf{F}(t) = k\mathbf{P}(t-k) + (\mathbf{M} - k\mathbf{K})\,\mathbf{R}(t-k)\,. \tag{17.7}$$

As shown in Table 17.1, all integrations can be reduced to the form of Eq. (17.6). When the problem is linear and the time step, k, is held constant, the system square matrix does not change with time. Thus is need be assembled and 'inverted' only once. Then at each time step is is only necessary to evaluate $\mathbf{F}(t)$ and solve for $\mathbf{R}(t)$.

Before considering the practical significance of the alternate forms of Eq. (17.6), let us return to the assumption that $\dot{\mathbf{R}}$ is linear during the time step. From Fig. 17.2.1 one notes that

$$\dot{\mathbf{R}}(t) = \dot{\mathbf{R}}(t-k) + k\ddot{\mathbf{R}}(t-k)\,. \tag{17.8}$$

Solving Eq. (17.3) for $\ddot{\mathbf{R}}$ and substituting into the above equation leads to

$$\dot{\mathbf{R}}(t) = 2\,[\,\mathbf{R}(t) - \mathbf{R}(t-k)\,]/k - \dot{\mathbf{R}}(t-k)\,. \tag{17.9}$$

Substituting into Eq. (17.1) at time t yields the system equations $\mathbf{S}\,\mathbf{R}(t) = \mathbf{F}(t)$, where now

$$\mathbf{S} = \mathbf{K} + 2\mathbf{M}/k\,,\quad \mathbf{F}(t) = \mathbf{P}(t) + \mathbf{M}\,[\,2\mathbf{R}(t-k)/k + \dot{\mathbf{R}}(t-k)\,]\,. \tag{17.10}$$

This is referred to as the linear velocity algorithm. A comparison of Eqs. (17.5) and (17.10) is useful. The Euler method is known as an explicit method since $\mathbf{R}(t)$ is obtained explicitly from $\mathbf{R}(t-k)$. The linear velocity formulation involves an implicit dependence on $\mathbf{v}(t-k)$ and is one of many implicit algorithms. Note that the Euler form requires no additional storage while the linear velocity algorithm must store the array $\mathbf{v}(t-k)$, and perform the calculations necessary to update its value at each time step. The necessary recurrence relation which utilizes the above calculated values for $\mathbf{R}(t)$ is obtained from Eq. (17.9). Also note that the implicit procedure requires one to have initial starting values for the velocity $\mathbf{v}(0)$. These can be obtained from Eq. (17.1) as

$$\dot{\mathbf{R}}(0) = \mathbf{M}^{-1}\,[\,\mathbf{P}(0) - \mathbf{K}\,\mathbf{R}(0)\,]\,. \tag{17.11}$$

However, this is a practical approach only so long as $\mathbf{M}$ is a diagonal matrix. The system matrices [14] resulting from Crank-Nicolson, Galerkin, and least-squares approximations in the time interval are summarized in Table 17.1. A typical subroutine, EULER, for executing the simple Euler intergration algorithm is given in Fig. 17.2.2.

```
      SUBROUTINE  EULER (NDFREE, IBW, A, B, P, R, DT, NSTEPS,
     1                   IPRINT, NBC, IBC)
C     * * * * * * * * * * * * * * * * * * * * * * * * * * *
C        STEP BY STEP INTEGRATION OF MATRIX EQUATIONS
C                 A*DR(T)/DT + B*R(T) = P(T)
C                   BY THE EULER METHOD
C     * * * * * * * * * * * * * * * * * * * * * * * * * * *
CDP   IMPLICIT REAL*8 (A-H,O-Z)
      DIMENSION A(NDFREE,IBW), B(NDFREE,IBW), P(NDFREE),
     1          R(NDFREE), IBC(NBC)
      PARAMETER ( NPRT = 6, ZERO = 0.0 )
C     INITIAL VALUES OF R ARE PASSED THRU ARGUMENT LIST
C     NBC    = NO. D.O.F. WITH SPECIFIED VALUES OF ZERO
C     IBC    = ARRAY OF NBC DOF NUMBERS WITH ZERO BC
C     NDFREE = TOTAL NUMBER OF SYSTEM DOF
C     IBW    = MAX HALF BANDWIDTH, INCLUDING DIAGONAL
C     NSTEPS = NO. OF INTEGRATION STEPS
C     IPRINT = NO. OF INTEGRATION STEPS BETWEEN PRINTING
C               *** INITIAL CALCULATIONS ***
      WRITE (NPRT,*) 'EULER STEP BY STEP INTEGRATION'
      IF ( IPRINT .LT. 1 )  IPRINT = 1
      NSTEPS = (NSTEPS/IPRINT)*IPRINT
      IF ( NSTEPS .EQ. 0 )  NSTEPS = IPRINT
      IF ( NBC .LT. 1 )  STOP 'NO CONSTRAINTS IN EULER'
      DO 30  J = 1, IBW
        DO 20  I = 1, NDFREE
   20   B(I,J) = -DT*B(I,J) + A(I,J)
   30 CONTINUE
      ICOUNT = IPRINT - 1
C      PRINT INITIAL VALUES OF R
      ISTEP = 0
      T      = ZERO
      WRITE (NPRT,5020)  ISTEP,T
 5020 FORMAT (/,' STEP NUMBER = ',I5,5X,' TIME = ',1PE12.5,/,
     1 '         I          R(I)')
      WRITE (NPRT,5030) ( K, R(K), K = 1, NDFREE)
 5030 FORMAT (I10, 2X, 1PE12.5)
C         *** APPLY BOUNDARY CONDITIONS TO A ***
      DO 40  I = 1, NBC
        CALL  MODFY1 (NDFREE, IBW, IBC(I), ZERO, B, P)
   40 CALL  MODFY1 (NDFREE, IBW, IBC(I), ZERO, A, P)
C         *** TRIANGULARIZE A ***
      CALL  FACTOR (NDFREE, IBW, A)
      IBWL1 = IBW - 1
C         *** END INITIAL CALCULATIONS ***
C         *** CALCULATE SOLUTION AT TIME T ***
      DO 80  ISTEP = 1, NSTEPS
        ICOUNT = ICOUNT + 1
        T      = DT*FLOAT(ISTEP-1)
        TLESS  = T - DT
        IF ( ICOUNT .EQ. IPRINT )  WRITE (NPRT, 5020) ISTEP, T
C       FORCER DEFINES THE FORCING FUNCTION P(T)
        CALL  FORCER (T, P, NDFREE)
```

Figure 17.2.2a Explicit Euler integration

```
C       FORM MODIFIED FORCING FUNCTION
        DO 60  I = 1, NDFREE
          SUM = ZERO
C           CONSIDER ONLY MATRIX BAND
          J1 = I - IBWL1
          J2 = I + IBWL1
          J1 = MAX0 (1, J1)
          J2 = MIN0 (J2, NDFREE)
          DO 50  J = J1, J2
C             CONVERT SUBSCRIPTS FOR COMPRESSED STORAGE
            CALL  BANSUB (I, J, IB, JB)
            BIJ = B(IB, JB)
            IF ( BIJ .NE. ZERO )  SUM = SUM + BIJ*R(J)
   50     CONTINUE
        P(I) = P(I)*DT + SUM
   60   CONTINUE
C        *** APPLY BOUNDARY CONDITIONS TO P ***
        DO 70  I = 1, NBC
          IN = IBC(I)
   70   P(IN) = ZERO
C        SOLVE FOR R AT TIME T
        CALL SOLVE (NDFREE, IBW, A, P, R)
C        OUTPUT RESULTS FOR TIME T
        IF ( ICOUNT .EQ. IPRINT) THEN
          WRITE (NPRT, 5030) ( K, R(K), K = 1, NDFREE)
          ICOUNT=0
        ENDIF
   80 CONTINUE
      RETURN
      END
```

Figure 17.2.2b Explicit Euler integration

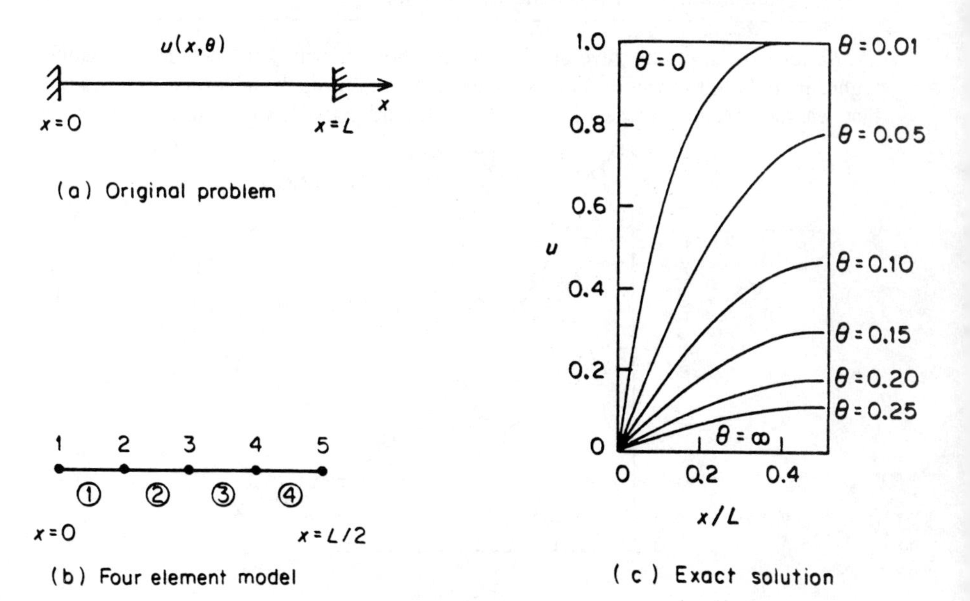

Figure 17.2.3 Typical linear transient mesh

17.2.2 Examples

The one-dimensional problem presented by Myers [9] will serve as a useful example application of subroutine EULER and the effects of condensed matrices. Myers presents finite difference, finite element, and exact analytic solutions for the following transient problem. Consider a rod having a uniform initial temperature of unity. Suddenly both ends of the rod have their temperature reduced to zero. The object is to determine the time history of the temperature at interior points. By utilizing half symmetry one can apply a simpler finite element model and use the natural boundary of zero thermal gradient at the center (see Fig. 17.2.3). Myers considered finite element solutions involving 1, 2, 3, and 4 elements. The 1, 2, and 3 element models yielded consistent matrices which could be integrated analytically, and these results are also given in reference [9].

Table 17.2 Stable solution comparisons at various steps

	$\theta = 0.04$				$\theta = 0.96$			
	100 C	100 D	75 D	50 D	100 C	100 D	75 D	50 D
u_2	0.550	0.594	0.585	0.565	0.042	0.045	0.045	0.044
u_3	0.878	0.903	0.902	0.898	0.078	0.083	0.082	0.081
u_4	1.016	0.988	0.990	1.000	0.102	0.109	0.108	0.106
u_5	1.004	1.000	1.000	1.000	0.111	0.118	0.117	0.114

C = Consistent, D = Diagonalized, θ = Time

The latter two solutions gave small time solutions that had the center-line temperature higher than its initial value. This is a clear violation of the basic laws of thermodynamics that was not present in the finite difference solutions. This response is due to the off-

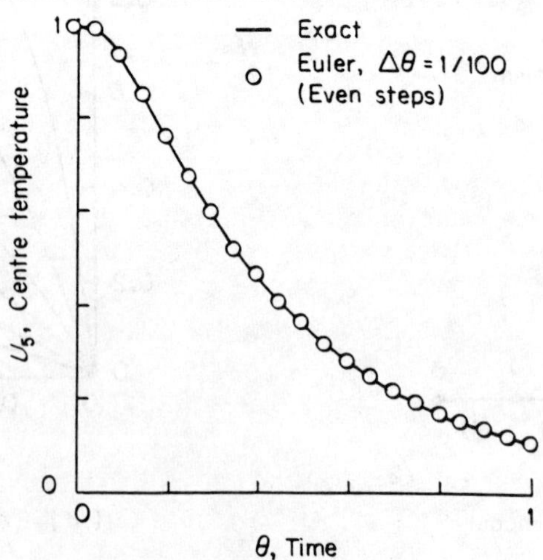

Figure 17.2.4 Transient response at center point

element solution. The matrices are

$$\mathbf{M} = \begin{bmatrix} 2 & 1 & 0 & 0 & 0 \\ & 4 & 1 & 0 & 0 \\ & & 4 & 1 & 0 \\ & & & 4 & 1 \\ \text{sym} & & & & 2 \end{bmatrix}, \quad \mathbf{K} = \frac{6k}{\rho c l^2} \begin{bmatrix} 1 & -1 & 0 & 0 & 0 \\ & 2 & -1 & 0 & 0 \\ & & 2 & -1 & 0 \\ & & & 2 & -1 \\ \text{sym} & & & & 1 \end{bmatrix}$$

where in this problem $6k/(\rho c l^2) = 96$, and the forcing function is zero, i.e., $\mathbf{P}(t) = \mathbf{0}$. These equations were stored in upper half-bandwidth form and integrated by subroutine EULER. The results at the center point from $t = 0$ to $t = 1$ for the consistent and condensed **M** matrices for a step size of $k = 1/100$ are shown in Fig. 17.2.4. To the scale shown they are in agreement with each other and the exact solution. However, Myers shows that other integration algorithms can give good results with $k = 1/16$. Thus, it is useful to observe what happens to the Euler procedure when k is increased. The consistent formulation was found to be unstable for $k = 1/75$. The condensed form was stable at $k = 1/75$ and $k = 1/50$ but was unstable at $k = 1/30$. A comparison of these results at selected small and large times in presented in Table 17.2. Since **P** was zero the problem-dependent forcing function, subroutine FORCER, simply set **P** to zero on each call. The version of EULER shown in Fig. 17.2.2 is written for zero boundary values for one or more parameters. Only minor changes are required for time-dependent boundary values.

In the above example the lumped and diagonalized matrices were identical. As an example of a problem where the results are different the following equations were obtained for a two degree of freedom *cylindrical conduction* problem

$$\mathbf{M}^e = \frac{1}{12} \begin{bmatrix} 9 & 5 \\ 5 & 24 \end{bmatrix}, \quad \mathbf{K}^e = \begin{bmatrix} 9 & -5 \\ -5 & 12 \end{bmatrix}. \tag{17.12}$$

Both condensed forms yielded the exact steady state results, but as shown in Fig. 17.2.5 the small time results were different. The consistent EULER was unstable for the same step size.

17.3 Hyperbolic Equations

Many solid mechanics problems and wave problems involve the solution of dynamic equations. The text by Bathe and Wilson [1] presents detailed discussions of both linear and nonlinear dynamic problems. They present both direct integration and model (eigen) algorithms. In this section a typical direct integration procedure will be illustrated. The governing system equations for dynamic applications are generally ordinary differential equations of the form

$$\mathbf{M}\ddot{\mathbf{R}} + \mathbf{C}\dot{\mathbf{R}} + \mathbf{K}\mathbf{R} = \mathbf{P}(t) \tag{17.13}$$

where **M**, **C**, and **K** are square banded matrices, $\mathbf{P}(t)$ is a time-dependent forcing function, and **R** is the vector of unknown nodal parameters. The **C** matrix is usually called the *damping matrix*. In many cases it is a linear combination of the **M** and **K** matrices. The system square matrices are again assembled from corresponding element matrices. Generally the initial values of **R** and velocity are known at time $t = 0$, and it is necessary to solve for the initial values of the acceleration from Eq. (17.13). That is,

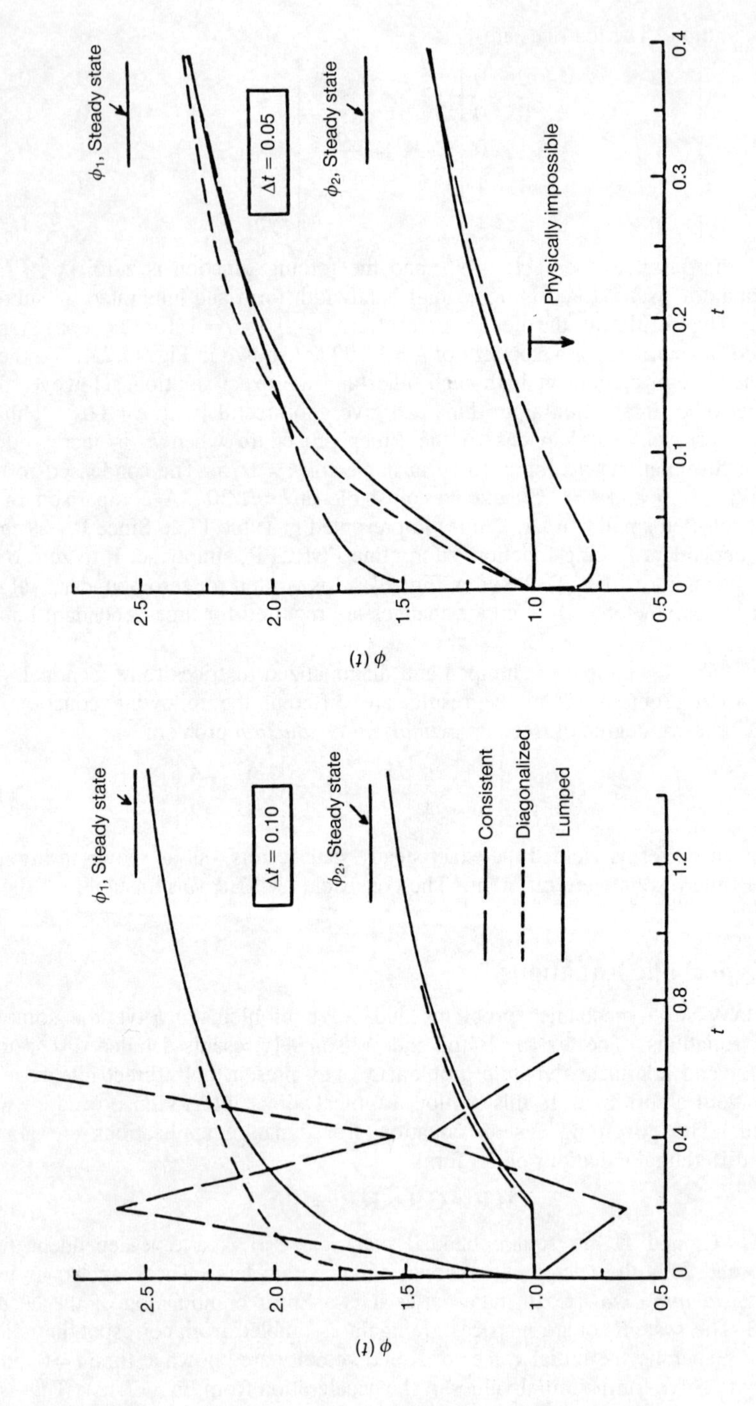

Figure 17.2.5 Euler solution of single element examples

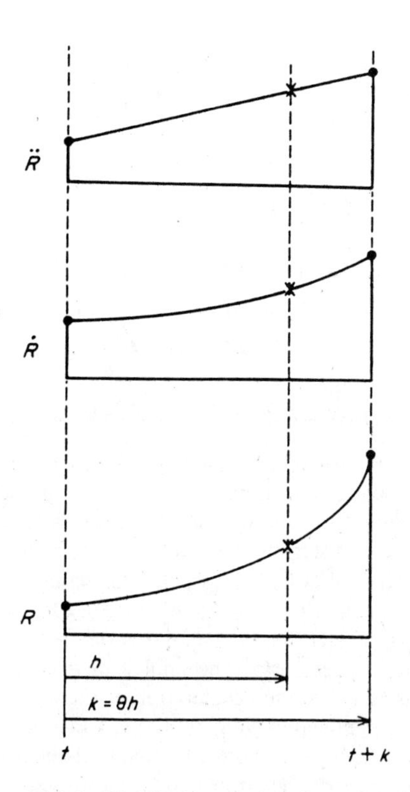

Figure 17.3.1 The linear acceleration procedure

$$\ddot{\mathbf{R}}(0) = \mathbf{M}^{-1} [\, \mathbf{P}(0) - \mathbf{K}\,\dot{\mathbf{R}}(0) - \mathbf{K}\,\mathbf{R}(0) \,]. \tag{17.14}$$

If $\mathbf{M}$ is a diagonal matrix this is easily evaluated. Otherwise, it may not be computationally efficient to exactly calculate the values of the initial acceleration. In the algorithm to be presented the initial values of the acceleration will be approximated by a simple technique. Assume that $\mathbf{R}(t)$, and velocity and acceleration are the known values of the above quantities at some time t. We desire to obtain the solution of Eq. (17.13) at time $(t+h)$ where h represents the size of the time increments to be used. There are numerous procedures for accomplishing this goal. The procedure to be presented herein has been successfully applied to many problems. It can be programmed so as to reduce the storage requirements and solution time.

17.3.1 Simple Approximations

The basic assumption of the algorithm is that within a time increment, $0 \le t \le k = \alpha h$, the second time derivative (acceleration) varies with t^n, that is,

$$\ddot{\mathbf{R}}(t) \equiv \mathbf{f}_1(t) = \ddot{\mathbf{R}}(0) + [\, \ddot{\mathbf{R}}(k) - \ddot{\mathbf{R}}(0) \,]\, t^n / k^n. \tag{17.15}$$

The first time derivative is obtained by integrating with respect to time:

$$\dot{\mathbf{R}}(t) \equiv \mathbf{f}_2(t) = \dot{\mathbf{R}}(0) + \int \mathbf{f}_1(t)\, dt$$

or

$$\dot{\mathbf{R}}(t) = \dot{\mathbf{R}}(0) + \ddot{\mathbf{R}}(0)t + [\, \ddot{\mathbf{R}}(k) - \ddot{\mathbf{R}}(0) \,]\, t^{n-1} / [\, (n+1)\, k^n \,]. \tag{17.16}$$

In a similar manner the value of $\mathbf{R}(t)$ is obtained from:

$$\mathbf{R}(t) = \mathbf{f}_3(t) = \mathbf{R}(0) + \int \mathbf{f}_2(t)\, dt \tag{17.17}$$

$$= \mathbf{R}(0) + \dot{\mathbf{R}}(0)t + \mathbf{R}(0)t^2/2 + [\, \mathbf{R}(k) - \mathbf{R}(0) \,]\, t^{n+2} / [\, (n+1)(n+2)k^n \,].$$

It is useful to note that if one evaluates Eqs. (17.16) and (17.17) at time $t = k$ one obtains:

$$\dot{\mathbf{R}}(k) = \dot{\mathbf{R}}(0) + (1-\gamma)\, k\, \ddot{\mathbf{R}}(0) + \gamma k \ddot{\mathbf{R}}(k) \tag{17.18}$$

and

$$\mathbf{R}(k) = \mathbf{R}(0) + k\, \dot{\mathbf{R}}(0) + (\tfrac{1}{2} - \beta)\, k^2\, \ddot{\mathbf{R}}(0) + \beta k^2\, \ddot{\mathbf{R}}(k) \tag{17.19}$$

where the constants γ and β have been defined as $\gamma = 1/(n+1)$, and $\beta = \gamma/(n+2)$. These equations represent the basic equations of the algorithm and they will be briefly discussed before proceeding with the details of the algorithm. Of course, Eqs. (17.18) and (17.19) could be taken as a starting point with β and γ being arbitrary constants.

Note that Eqs. (17.16) and (17.17) imply that the values of **R**, **v** and **a** are known at time $t = 0$ and that **a** is known at time $t = k$. As a point of fact, the latter quantity is not known and must be estimated. Of course, iteration could be used to improve the accuracy of the estimate if necessary. The choice of n in Eq. (17.15) determines the relative weight that is assigned to the acceleration at the beginning and end of the time step. If $n = 1$ they are assigned equal weights; if $n < 1$ then the acceleration $\mathbf{a}(k)$ is more important; and if $n > 1$ then $\mathbf{a}(0)$ is more important. This is clearly illustrated by the limiting cases ($n = 0$ and $n = \infty$). The most commonly used values are $n = 1$ and $n = 0$. Newmark [10] has considered the application of Eqs. (17.18) and (17.19) to the solution of Eq. (17.13) in some detail. Starting with these equations he assumed that the parameters γ and β were independent. That assumption does not apply in the present analysis; nevertheless, portions of his analysis can be utilized here. By considering the known difference solution for the simple harmonic motion of a single degree of freedom system, Newmark concluded that to avoid the introduction of erroneous damping in the solution one must set $\gamma = \tfrac{1}{2}$. For the present assumptions this result requires that $n = 1$, which in turn implies that the second derivative is linear during the time interval, k. Of course, if $n = 1$ then $\beta = 1/6$. For $\beta = 1/6$ and $k = h(\alpha = 1)$, Newmark shows that for the above problem the solution will theoretically be stable and converge if $h/T \leq 0.389$ where T represents the smallest period of the system. For multi degree of freedom systems subjected to forced motion a much smaller ratio must be used in practice. The actual algorithm will be developed using $n = 1$. For $n = 1$ Eqs. (17.18) and (17.19) can be generalized to

$$\dot{\mathbf{R}}(t+k) = \dot{\mathbf{R}}(t) + [\, \ddot{\mathbf{R}}(t+k) + \ddot{\mathbf{R}}(t) \,]\, k/2 \tag{17.20}$$

$$\mathbf{R}(t+k) = \mathbf{R}(t) + k\, \dot{\mathbf{R}}(t) + k^2\, [\, \ddot{\mathbf{R}}(t+k) + 2\, \ddot{\mathbf{R}}(t) \,]/6. \tag{17.21}$$

For $t = h(\alpha = 1)$ the above equations correspond to the standard "linear acceleration" method. This technique, which requires iterations within each time step to establish the acceleration $\mathbf{a}(t+h)$, has been outlined in detail (including a flow chart) by Fenves [4]. A similar study, with examples, has been given by Biggs [2].

To avoid these iterations, and at the same time maintain numerical stability, modified integration schemes are usually preferred. A typical modified *linear acceleration* integration scheme will now be considered. This extrapolation algorithm utilizes Eqs. (17.20) and (17.21). These equations can be arranged to make any one of the quantities $\mathbf{a}(t+k)$, $\mathbf{v}(t+k)$, or $\mathbf{R}(t+k)$ the independent unknown. The best choice of the three is not yet known; however, the "displacement", $\mathbf{R}(t+k)$, formulation is the most common. Since the eventual objective is to solve for $\mathbf{R}(t+k)$ in terms of $\mathbf{R}(t)$, and its velocity and acceleration, it is desirable to solve Eqs. (17.20) and (17.21) simultaneously for $\mathbf{v}(t+k)$ and $\mathbf{a}(t+k$,. that is, (for $n=1$),

$$\ddot{\mathbf{R}}(t+k) = 6[\,\mathbf{R}(t+k) - \mathbf{R}(t)\,]/k^2 - 6\,\dot{\mathbf{R}}(t)k - 2\,\ddot{\mathbf{R}}(t) \tag{17.22}$$

and

$$\dot{\mathbf{R}}(t+k) = 3[\,\mathbf{R}(t+k) - \mathbf{R}(t)\,]/k - k\,\ddot{\mathbf{R}}(t)/2. \tag{17.23}$$

To establish the relations for $\mathbf{R}(t+k)$ we return to Eq. (17.13). Evaluating that equation at time $t+k$ yields

$$\mathbf{M}\,\ddot{\mathbf{R}}(t+k) + \mathbf{C}\,\dot{\mathbf{R}}(t+k) + \mathbf{K}\,\mathbf{R}(t+k) = \mathbf{P}(t+k). \tag{17.24}$$

Substituting Eqs. (17.20) and (17.21) into Eq. (17.24) and collecting like terms gives

$$[\,6\mathbf{M}/k^2 + 3\,\mathbf{C}/k + \mathbf{K}]\,\mathbf{R}(t+k) = \mathbf{P}(t+k) + [\,6\mathbf{M}/k^2 + 3\,\mathbf{C}/k\,]\,\mathbf{R}(t) \tag{17.25}$$

$$+ [\,6\mathbf{M}/k + 2\,\mathbf{C}\,]\,\dot{\mathbf{R}}(t) + [\,2\,\mathbf{M} + k\mathbf{C}/2\,]\,\mathbf{R}(t)$$

where the only unknown is $\mathbf{R}(t+k)$. Experience indicates that if k is to be held constant throughout the integration then the computational efficiency can be increased by defining a square matrix $\mathbf{D} = 2\,\mathbf{M}/k + \mathbf{C}$, and rewriting Eq. (17.25) as

$$[\,3\mathbf{D}/k + \mathbf{K}\,]\,\mathbf{R}(t+k) =$$

$$\left[\mathbf{P}(t+k) + \mathbf{D}[\,3\,\mathbf{R}(t)/k + 3\,\dot{\mathbf{R}}(t) + k\,\mathbf{R}(t)\,] - \mathbf{C}[\,\dot{\mathbf{R}}(t) + k\,\mathbf{R}(t)/2\,]\right]. \tag{17.26}$$

One notes that the right hand side of Eq. (17.26) is a column matrix so that the final result is analogous to Eq. (17.6), i.e.,

$$\mathbf{S}(t+k)\,\mathbf{R} = \mathbf{F}(t+k) \tag{17.27}$$

where $\mathbf{S}$ is the resultant square matrix and $\mathbf{F}$ is the resultant forcing function. In the following programs, $\mathbf{D}$ is stored in the original location of $\mathbf{M}$ while $\mathbf{S}$ is stored in the original location of $\mathbf{K}$. If $\mathbf{K}$ is zero, and/or $\mathbf{M}$ is diagonal, then one would use alternative storage schemes. The above equations would yield the values of $\mathbf{R}$ at time $t+k$ but we desire the values at a smaller time $t+h$ where $k = \theta h$, $\theta \geq 1$. Once the above equation has been solved the required parameters are obtained by interpolating from the old t and new $t+k$ values. This is accomplished by utilizing Eqs. (17.15), (17.16), and (17.17) evaluated at the intermediate time, that is,

$$\ddot{\mathbf{R}}(t+h) = \ddot{\mathbf{R}}(t) + [\,\ddot{\mathbf{R}}(t+k) - \ddot{\mathbf{R}}(t)\,]/\theta$$

$$\dot{\mathbf{R}}(t+h) = \dot{\mathbf{R}}(t) + h\,\ddot{\mathbf{R}}(t) + h[\,\ddot{\mathbf{R}}(t+k) - \ddot{\mathbf{R}}(t)\,]/(2\theta) \tag{17.28}$$

$$\mathbf{R}(t+h) = \mathbf{R}(t) + h\,\dot{\mathbf{R}}(t) = h^2\,\ddot{\mathbf{R}}(t)/2 + h^2[\,\ddot{\mathbf{R}}(t+k) - \ddot{\mathbf{R}}(t)\,]/(6\theta),$$

where $\ddot{\mathbf{R}}(t+k)$ and $\dot{\mathbf{R}}(t+k)$ are given by Eqs. (17.22) and (17.23), respectively. These concepts are illustrated in Fig. 17.3.1. The linear acceleration algorithm is implemented in subroutine DIRECT and is shown in Fig. 17.3.2. Since the forcing function $\mathbf{P}(t)$ is problem-dependent in general, it must be supplied to subroutine DIRECT by the function program FORCER. Subroutine DIRECT assumes that $\mathbf{M}$ is not a diagonal matrix; thus Eq. (17.14) is not utilized to initialize $\mathbf{a}(0)$. Instead, this routine starts the solution at $t=-h$ using $\mathbf{R}(-h)=\mathbf{R}(0)$, $\mathbf{v}(-h)=\mathbf{0}$, and $\mathbf{a}(-h)=\mathbf{0}$. It solves the standard equations and interpolates for the value of the acceleration $\mathbf{a}(0)$. At this point the entire integration

```
      SUBROUTINE  DIRECT (NDFREE, MBW, A, B, C, DRP, R, DR, D2R,
     1                    P, OMEGA, DT, NSTEPS, IPRINT, NBC, IBC)
C     * * * * * * * * * * * * * * * * * * * * * * * * * * * * * *
C     STEP BY STEP INTEGRATION OF MATRIX EQUATIONS:
C        A*D2R(T)/DT2 + B*DR(T)/DT + C*R(T) = P(T)
C     * * * * * * * * * * * * * * * * * * * * * * * * * * * * * *
CDP   IMPLICIT  REAL*8(A-H,O-Z)
      DIMENSION  A(NDFREE,MBW), B(NDFREE,MBW), IBC(NBC),
     1        R(NDFREE), DR(NDFREE), D2R(NDFREE), P(NDFREE),
     2        DRP(NDFREE), C(NDFREE,MBW)
      PARAMETER ( NPRT = 6, ZERO = 0.0 )
C     INITIAL VALUES OF R AND DR ARE PASSED THRU ARGUMENTS
C     NBC       = NO. D.O.F. WITH SPECIFIED VALUES OF ZERO
C     IBC       = ARRAY CONTAINING THE NBC DOF NOS WITH ZERO BC
C     R,DR,D2R = 0,1,2 ORDER DERIV. OF R W.R.T. T AT TIME=T
C     DRP       = VALUE OF DR AT TIME = T + DELT
C     NSTEPS    = NO. OF INTEGRATION STEPS
C     IPRINT    = NO. OF INTEGRATION STEPS BETWEEN PRINTING
C     OMEGA     = 1.25 IS SUGGESTED
C                ** INITIAL  CALCULATIONS **
      WRITE (NPRT,*) 'DIRECT STEP BY STEP INTEGRATION'
      IF ( IPRINT .LT. 1 )  IPRINT = 1
      NSTEPS = (NSTEPS/IPRINT)*IPRINT
      IF ( NSTEPS .EQ. 0 )  NSTEPS = IPRINT
      DELT   = (OMEGA-1.)*DT
      TAU    = OMEGA*DT
      MBWL1 = MBW - 1
      IF ( NBC .LT. 1 )  STOP 'NO CONSTRAINTS IN DIRECT'
      DO 30  I = 1, NDFREE
        P(I)   = ZERO
        DRP(I) = ZERO
        D2R(I) = ZERO
        DO 20  J = 1, MBW
   20   B(I,J) = B(I,J) + 2.*A(I,J)/TAU + TAU*C(I,J)/3.
   30 CONTINUE
      ICOUNT = IPRINT - 1
C              ** APPLY BOUNDARY CONDITIONS ( TO B ) **
      DO 40  I = 1,NBC
   40 CALL MODFY1 (NDFREE, MBW, IBC(I), ZERO, B, P)
C                 ** TRIANGULARIZE B **
      CALL  FACTOR (NDFREE, MBW, B)
C      APPROXIMATE THE INITIAL VALUE OF D2R
      TPLUS = ZERO
      ISTEP = 0
C        ( THE FOLLOWING IS AN EXTRA LEGAL STATEMENT )
      GO TO 60
C             ** END OF INITIAL CALCULATIONS **
```

Figure 17.3.2a Direct integration procedure

```
C             ** CALCULATE SOLUTION AT TIME T **
   50 DO 120  ISTEP = 1, NSTEPS
        ICOUNT = ICOUNT + 1
        T      = DT*FLOAT(ISTEP-1)
        TPLUS  = T + DELT
        IF ( ISTEP .EQ. 1 )  TPLUS = ZERO
        IF ( ICOUNT .EQ. IPRINT )  WRITE (NPRT,5020)  ISTEP,T
 5020   FORMAT ( //,' STEP NUMBER = ',I5,5X,'TIME = ',1PE13.5,/,
     1  '     I          R(I)                DR/DT              ',
     2  'D2R/DT2',/)
C        FORCER IS A SUBR TO DEFINE THE FORCING FUNCTION P(T)
   60   CALL  FORCER ( TPLUS, P, NDFREE)
C        FORM MODIFIED FORCING FUNCTION AT T + DELT
        DO 90  I = 1, NDFREE
          SUM = 0.
          J1 = I - MBWL1
          J2 = I + MBWL1
          J1 = MAX0 (1, J1)
          J2 = MIN0 (J2, NDFREE)
          DO 80  J = J1, J2
            CALL  BANSUB (I, J, IB, JB)
            AIJ = A(IB,JB)
            IF ( AIJ .NE. ZERO )
     1        SUM = SUM + AIJ*( 2.*DR(J)/TAU+D2R(J) )
            CIJ = C(IB,JB)
            IF ( CIJ .NE. ZERO )
     1        SUM = SUM - CIJ*( R(J) + 2.*TAU*DR(J)/3.
     2              + TAU*TAU*D2R(J)/6. )
   80     CONTINUE
          P(I) = P(I) + SUM
   90   CONTINUE
C             ** APPLY BOUNDARY CONDITIONS ( TO P ) **
        DO 100  I = 1, NBC
          IN = IBC(I)
  100   P(IN) = ZERO
C        SOLVE FOR DRP AT TIME T+DELT
        CALL  SOLVE (NDFREE, MBW, B, P, DRP)
C        USING DATA AT T-DT AND T+DELT CALCULATE VALUES AT T
        DO 110  I = 1, NDFREE
          DRDT = (1.-1./OMEGA)*DR(I) + DRP(I)/OMEGA
          D2RDT2 = D2R(I)*(1.-2./OMEGA)+2.*(DRDT-DR(I))/OMEGA/DT
C          APPROXIMATE THE INITIAL VALUE OF D2R
          IF ( ISTEP.LT.2 )  GO TO 110
            R(I) = R(I) + DT*(2.*DR(I)+DRDT)/3. + DT*DT*D2R(I)/6.
            DR(I) = DRDT
  110   D2R(I) = D2RDT2
        IF ( ISTEP .EQ. 0 )  GO TO 50
C        OUTPUT RESULTS FOR TIME T
        IF ( ICOUNT .NE. IPRINT )  GO TO 120
          WRITE (NPRT,5030) (K, R(K), DR(K), D2R(K), K=1, NDFREE)
 5030     FORMAT ( I10, 2X, 1PE13.5, 2X, 1PE13.5, 2X, 1PE13.5 )
          ICOUNT = 0
  120 CONTINUE
      RETURN
      END
```

Figure 17.3.2b Direct integration procedure

solution is begun. The above approximation gives the exact value of the initial acceleration for many forcing functions, $\mathbf{P}(t)$.

17.3.2 Lumped Mass Forced Vibration Example

To illustrate the step by step integration procedure, consider the forced vibration of the lumped mass system given in Fig. 17.3.3. The exact solution of this problem is given by Biggs [2]. The loads are linearly decreasing ramp functions. The function FORCER for these loads is shown in Fig. 17.3.4. The calculated displacements and velocities of node three are compared with the exact values in Figs. 17.3.5 and 17.3.6, respectively. Fig. 17.3.7 shows the effect of increased time step size.

17.4 Diagonal "Mass" Matrices *

When one uses a finite difference spatial formulation, the system mass matrix **M** is a diagonal matrix. However, if one utilizes a consistent finite element formulation it is not a diagonal matrix. Clearly, converting **M** to a diagonal matrix would also save storage and make the evaluation of products and inversions much more economical. Some engineering approaches for modifying **M** have been shown to be successful. To illustrate these, consider the form of a typical element contribution, say **m**. The consistent definition for constant properties is

$$\mathbf{m}^e \equiv q \int_{V^e} \mathbf{H}^{e^T} \mathbf{H}^e \, dv ,$$

where q is some constant property per unit volume and $\mathbf{H}^e$ denotes the element interpolation functions. This generally can be written as $\mathbf{m}^e = Q\mathbf{M}^e$, where the total property is $Q = qV^e$, and $\mathbf{M}^e$ is a symmetric full matrix. In some cases it is possible to obtain a diagonal form by using a nodal quadrature rule. Since that procedure often leads to negative terms other procedures will be considered here. Let the sum of the coefficients of the matrix $\mathbf{M}^e$ be T, that is

$$T \equiv \sum_i \sum_j M_{ij}^e . \tag{17.28}$$

In most cases T will be unity, but this is not true for axisymmetric integrals. Another quantity of interest is the sum of the diagonal terms of $\mathbf{M}^e$, i.e.,

$$d \equiv \sum_i M_{ii}^e . \tag{17.29}$$

The most common engineering solution to defining a diagonal matrix is to *lump*, or sum, all the terms in a given row onto the diagonal of the row and then set the off-diagonal terms to zero. That is, the *lumped matrix* $\mathbf{M}_L^e$ is defined such that

$$\mathbf{M}_{L_{ij}}^e = 0 \quad \text{if } i \neq j, \; M_{L_{ii}}^e = \sum_j M_{ij}^e . \tag{17.30}$$

Note that doing this does not alter the value of T. Another diagonal matrix, $\mathbf{M}_D^e$, with the same value of T can be obtained by simply extracting the diagonal of **M** and scaling it by a factor of T/d, that is,

$$\mathbf{M}_{D_{ij}}^e = 0 \quad \text{if } i \neq j, \; \mathbf{M}_{D_{ii}}^e = M_{ii}^e T/d .$$

The matrix $\mathbf{M}_D^e$ will be called the diagonalized matrix and $\mathbf{M}_L^e$ the lumped matrix. The corresponding assembled matrices $\mathbf{M}_D$ and $\mathbf{M}_L$ are referred to as "condensed" matrices. For linear simplex elements in two and three dimensions both procedures yield identical diagonal matrices. However, for axisymmetric problems and higher order elements they yield different results and the diagonalized matrix appears to be best in general. This is

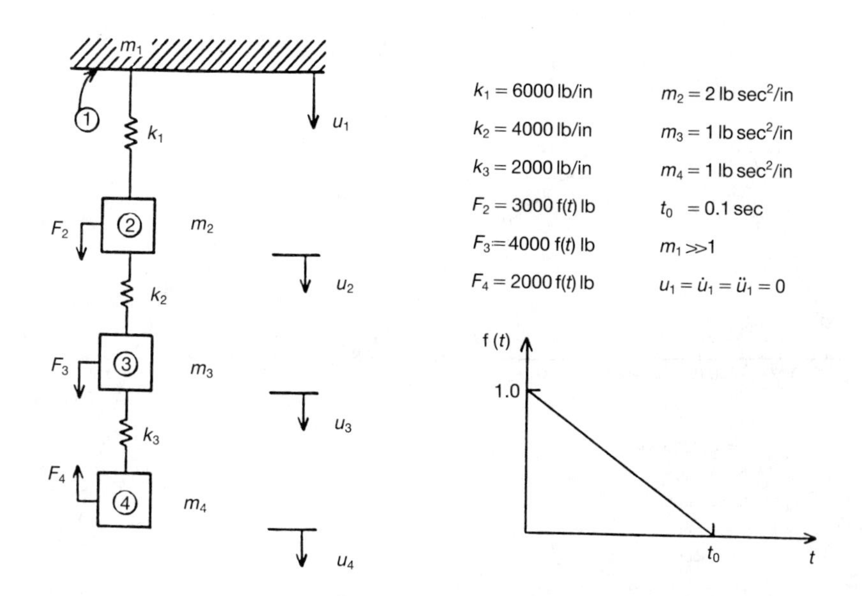

Figure 17.3.3 Dynamic test problem

```
      PROGRAM  BIGGS
C     ** DIRECT INTEGRATION FOR DECREASING RAMP LOAD **
      DIMENSION  AK(4,2), AM(4,2), U(4), F(4), VEL(4),
     1           ACC(4), IBC(4), AC(4,2), UP(4)
      PARAMETER ( MBW=2, NDFREE=4, NBC=1, KPRINT=1,
     1            NSTEPS=120, DT=0.0025, OMEGA=1.4 )
      DATA  AC, AM / 8*0., 100., 2., 1., 1., 4*0./
      DATA  AK / 6000., 10000., 6000., 2000.,
1               -6000., -4000.,-2000., 0.   /
      DATA U, VEL, IBC / 4*0., 4*0., 1, 3*0 /
      CALL DIRECT (NDFREE, MBW, AM, AC, AK, UP, U, VEL, ACC,
     1             F, OMEGA, DT, NSTEPS, KPRINT, NBC, IBC)
      RETURN
      END
      SUBROUTINE  FORCER (T, P, NDFREE)
C     * * DEFINE FORCING FUNCTION FOR INTEGRATION **
      DIMENSION  P(NDFREE)
      PARAMETER ( F2=3000., F3=4000., F4=-2000., TONE=0.1 )
      FOFT = 0.0
      IF ( T .LT. TONE )  THEN
        FOFT = 1.-T/TONE
      ELSE
        P(1) = 0.
        P(2) = F2*FOFT
        P(3) = F3*FOFT
        P(4) = F4*FOFT
      ENDIF
      RETURN
      END
```

Figure 17.3.4 Data and forcing routines

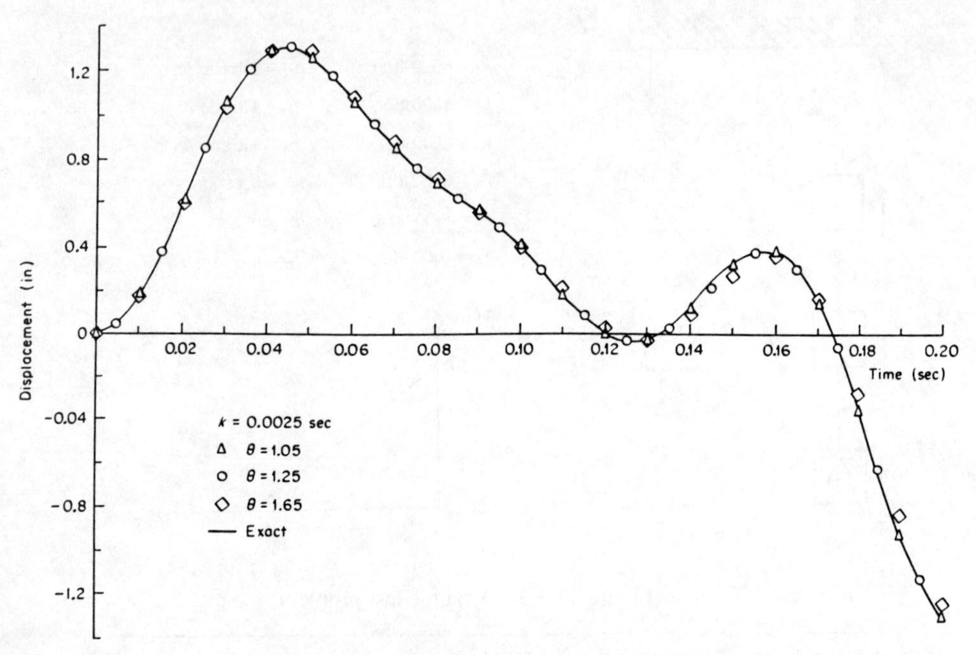

Figure 17.3.5 Displacement response at node three

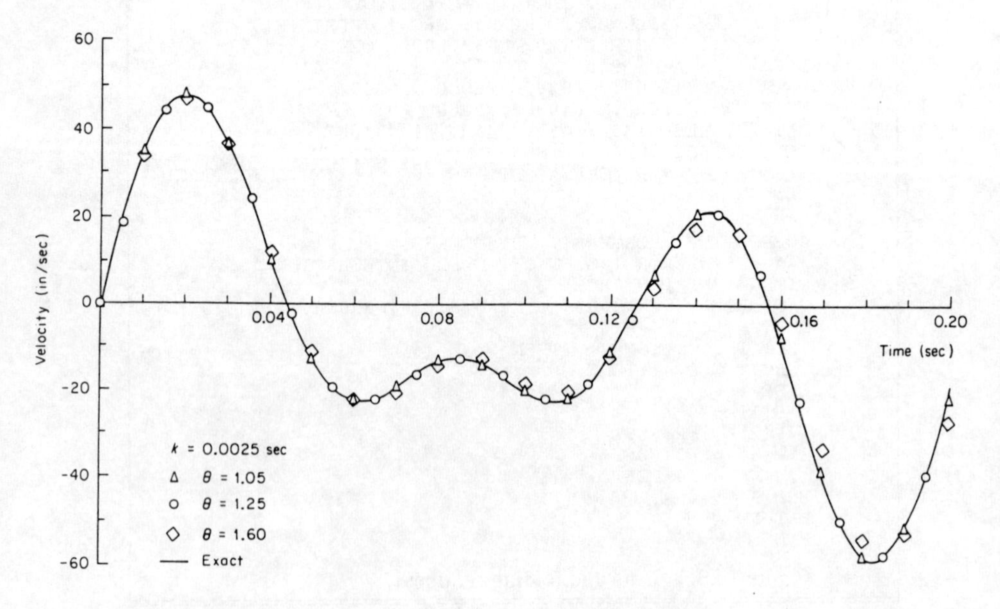

Figure 17.3.6 Velocity response at node three

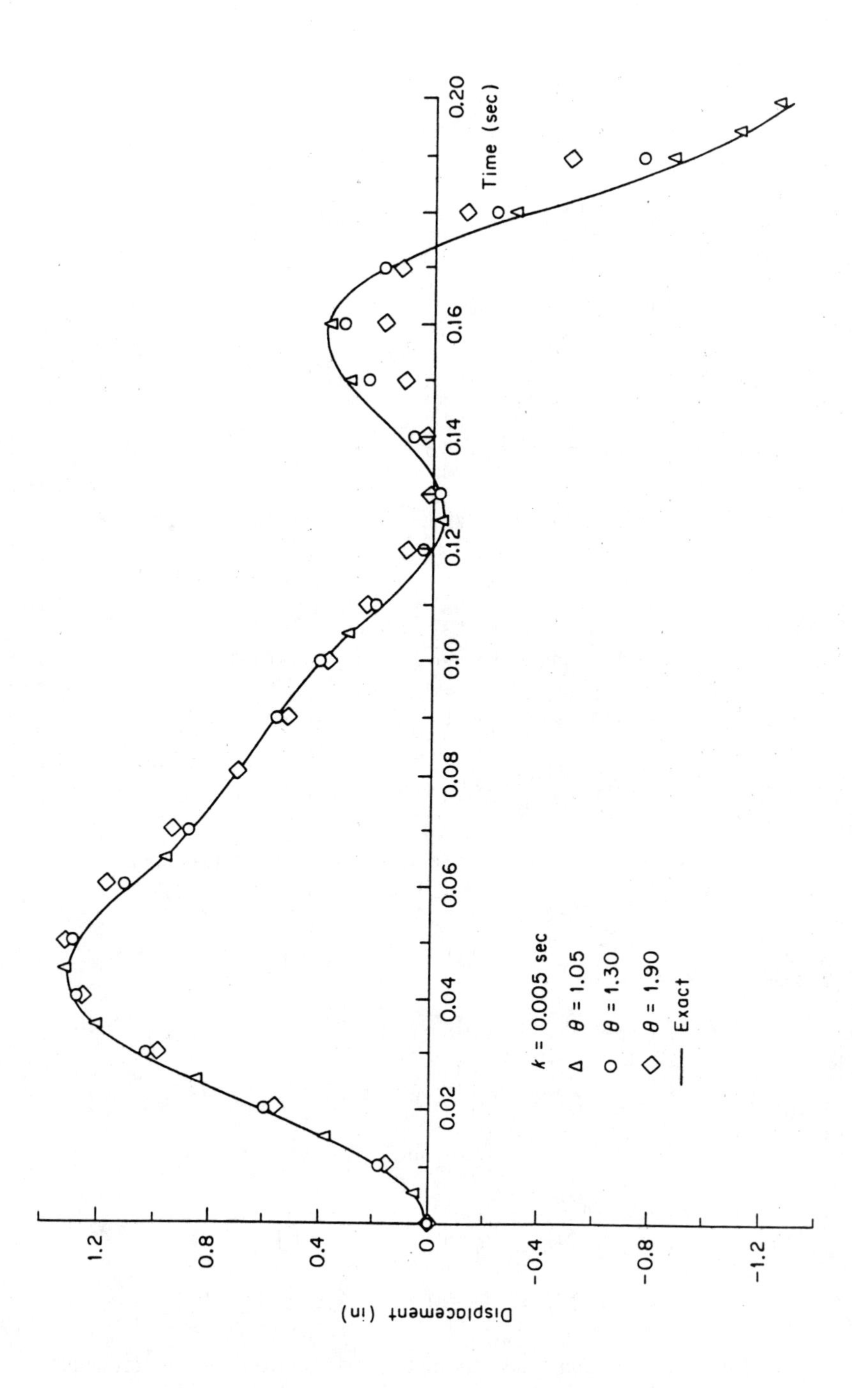

Figure 17.3.7 Effect of time step on displacement

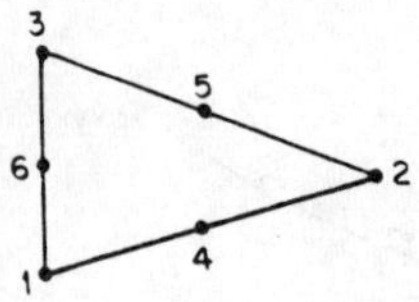

(a) Consistent

$$\mathbf{M} = \frac{1}{180}\begin{bmatrix} 6 & -1 & -1 & 0 & -4 & 0 \\ & 6 & -1 & 0 & 0 & -4 \\ & & 6 & -4 & 0 & 0 \\ & & & 32 & 16 & 16 \\ & & & & 32 & 16 \\ \text{sym} & & & & & 32 \end{bmatrix}$$

(b) Lumped

$$\mathbf{M}_L = \frac{1}{180}\lceil 0 \quad 0 \quad 0 \quad 60 \quad 60 \quad 60 \rfloor$$

(c) Diagonalized

$$\mathbf{M}_D = \frac{1}{114}\lceil 6 \quad 6 \quad 6 \quad 32 \quad 32 \quad 32 \rfloor$$

Figure 17.4.1 Matrices for a quadratic triangle

(a) Consistent

$$\mathbf{M} = \frac{2\pi}{12}\begin{bmatrix} (3r_1 + r_2) & (r_1 + r_2) \\ \text{sym.} & (r_1 + 3r_2) \end{bmatrix}$$

(b) Lumped

$$\mathbf{M}_L = \frac{2\pi}{6}\lceil (2r_1 + r_2) \quad (r_1 + 2r_2) \rfloor$$

(c) Diagonalized

$$\mathbf{M}_D = \frac{2\pi}{8}\lceil (3r_1 + r_2) \quad (r_1 + 3r_2) \rfloor$$

Figure 17.4.2 Matrices for an axisymmetric line element

(a) Linear: 1 -------- 2

$$\mathbf{M}_L^e = \frac{\rho L^e}{2}\lceil 1 \quad 1 \rfloor$$

$$\mathbf{M}_D^e = \frac{\rho L^e}{2}\lceil 1 \quad 1 \rfloor$$

(b) Quadratic: 1 ---- 2 ---- 3

$$\mathbf{M}_L^e = \frac{\rho L^e}{6}\lceil 1 \quad 4 \quad 1 \rfloor$$

$$\mathbf{M}_D^e = \frac{\rho L^e}{6}\lceil 1 \quad 4 \quad 1 \rfloor$$

(c) Cubic: 1 --- 2 --- 3 --- 4

$$\mathbf{M}_L^e = \frac{\rho L^e}{8}\lceil 1 \quad 4 \quad 4 \quad 1 \rfloor$$

$$\mathbf{M}_D^e = \frac{\rho L^e}{1552}\lceil 128 \quad 648 \quad 648 \quad 128 \rfloor$$

Figure 17.4.3 Diagonal Lagrangian mass matrices for line elements

because for higher order elements the lumped form can introduce zeros or negative numbers on the diagonal. The matrices $\mathbf{M}^e$, $\mathbf{M}_L^e$, and $\mathbf{M}_D^e$ are illustrated for a quadratic triangle in Fig. 17.4.1, for an axisymmetric line element in Fig. 17.4.2, and for line elements in Fig. 17.4.3. The effects of the condensed matrices on transient calculations will be considered in a later section.

17.5 Units Diagonal Scaling *

The diagonalization of the mass matrix is a commonly utilized approximation. The simple lumping procedure of adding the terms on a row works for linear elements but fails for higher order elements because it renders the diagonal zero or negative. For isoparametric elements a scaling of the diagonal of the mass matrix has been successfully used. This procedure has been described in detail by several authors. That scaling method fails, or is incorrect, for hierarchical interpolation and for Hermite interpolation. Here we employ a scaling method for use with these later elements. It is called the Unit Diagonal Scaling (UDS) method because it is founded on the basic principle that all terms in a valid equation must have the same units or dimensions. The act of taking a derivative changes the units (or dimensions) by introducing a new term in the denominator. The new unit there is that of the unit of the quantity with respect to which the derivative was taken. Thus, it is clear that the Hermite and hierarchical families have terms of different units in the higher order parts.

We begin by forming the consistent mass matrix, **m**, which always has a diagonal containing positive terms. We next group the diagonal terms into subsets that have the same units. Then the diagonal values of that subset are added to give a total, say d. For that subset partition of **m** there are other non-diagonal terms that have the same units. These are summed to a value n. The resulting total of coefficients in the subset partition of **m** that have the same units is therefore given by $t = d + n$. For this partition of the diagonal we scale it by multiplying the diagonal by the ration T/d. This scaled diagonal is retained. The same procedure is applied to the remaining subsets in turn. Note that for two and three dimensional spaces this means that derivatives of equal order (and thus equal units) will be scaled together. For example, in two dimensions, the hierarchical terms $\partial^2/\partial x^2$, $\partial^2/\partial y^2$, and $\partial^2/\partial x\partial y$ would be scaled in a single subset. Since isoparametric families always involves zero order derivatives the usual scaling algorithm is included here as a special case.

To illustrate these concepts we begin with the common beam in cubic Hermite form in unit coordinates, given in Fig. 4.5.1, $x = L^e\, r$,

$$U = H_1\, U_1 + H_2\, \theta_1 + H_3\, U_2 + H_4\, \theta_2\,.$$

Here we have chosen the derivative degrees of freedoms, $\theta = du/dx$, relative to the global coordinate x because they have a useful physical meaning (the slope) to most analysts. Here it is clear that the derivative dof's bring in the units of the length, L. The mass matrix for this element is well known:

$$\mathbf{m} = \frac{m}{420}\begin{bmatrix} 156 & 22L & 54 & -13L \\ 22L & 4L^2 & 13L & -3L^2 \\ 54 & 13L & 156 & -22L \\ -13L & -3L^2 & -22L & 4L^2 \end{bmatrix}.$$

In this case it is clear that we have three sets of units involved: m, mL, and mL^2.

However, the diagonal terms contain only two sets of units: m and mL^2. The first and third diagonal terms involve zero derivatives so they are our first subset. Its terms are

$$\mathbf{m}^o = \frac{m}{420}\begin{bmatrix} 156 & 54 \\ 54 & 156 \end{bmatrix}$$

so $d = 312$ and $n = 108$ so $T = 420$ and our diagonal scale term is $r = 420/312$, and the subset partition becomes

$$\mathbf{m}_D^o = \frac{m}{2}\lceil 1 \quad 1 \rfloor .$$

The second and fourth rows involve the integrals of products of first derivatives (and the units of mL^2). Their subset for this algorithm is

$$\mathbf{m}^1 = \frac{mL^2}{420}\begin{bmatrix} 4 & -3 \\ -3 & 4 \end{bmatrix}$$

with $d = 8$, $n = -6$, $t = 2$, and $r = 8/2$ so that

$$\mathbf{m}_D^1 = \frac{4mL^2}{105}\lceil 1 \quad 1 \rfloor .$$

No higher derivative products exist on the diagonal so we stop here. Note that the eight terms with units not on the diagonal, i.e., mL, were not considered. Picking a common denominator we get the scaled result

$$\mathbf{m}_D = \frac{m}{210}\lceil 105 \quad 4L^2 \quad 105 \quad 4L^2 \rfloor = m\lceil 0.5 \quad 0.019L^2 \quad 0.5 \quad 0.019L^2 \rfloor$$

which differs from the Hinton, Rock, and Zienkiewicz [5] procedure, which gives

$$\mathbf{m}_D = m\lceil 0.5 \quad 0.013L^2 \quad 0.5 \quad 0.013L^2 \rfloor .$$

Their procedure determines the scaling factor of the zero derivative subset only and applies it to all diagonal terms. As high order derivative products are considered the difference between these two procedures increases.

For hierarchical elements the changes in units are less obvious since they are usually taken with respect to non-dimensional coordinate systems. Consider a cubic hierarchical family given by Zienkiewicz and Morgan [15] as

$$H_1 = (1 - \xi)/2 \qquad H_2 = (1 + \xi)/2$$

$$H_3 = \xi^2 - 1 \qquad H_4 = 2(\xi^3 - \xi)$$

for $-1 \le \xi \le 1$, where H_1 and H_2 are associated with the unknown function (its zeroth derivative), while H_3 and H_4 are associated with nodal parameters at the midside, that are the second and third local derivatives of the unknown, respectively. The linear, quadratic, and cubic partitions to the mass matrix are

$$\mathbf{m}^e = \frac{L^e}{210}\begin{bmatrix} 70 & 35 & -70 & 28 \\ 35 & 70 & -70 & -28 \\ -70 & -70 & 112 & 0 \\ 28 & -28 & 0 & 64 \end{bmatrix}.$$

The products of the derivative orders involved here are

$$\begin{matrix} 0,0 & 0,0 & 2,0 & 3,0 \\ 0,0 & 0,0 & 2,0 & 3,0 \\ 0,2 & 0,2 & 2,2 & 3,2 \\ 0,3 & 0,3 & 2,3 & 3,3 \end{matrix}$$

which sum to

$$\begin{matrix} 0 & 0 & 2 & 3 \\ 0 & 0 & 2 & 3 \\ 2 & 2 & 4 & 5 \\ 3 & 3 & 5 & 6 \end{matrix}.$$

The off diagonal sums (units) match only in the linear partition. Thus, only its subset needs scaling to

$$\frac{L^e}{2}\lceil 1 \quad 1 \rfloor$$

and the third and fourth diagonals remain the same

$$\frac{L^e}{210}\lceil 112 \quad 64 \rfloor.$$

So if we pick a common denominator we get

$$\mathbf{m}_D^e = \frac{L^e}{210}\lceil 105 \quad 105 \quad 112 \quad 64 \rfloor.$$

which is the same result given by the HRZ method since there is no coupling of the units at this low hierarchical order. Here, simple lumping gives

$$\mathbf{m}_L^e = \frac{L^e}{210}\lceil 105 \quad 105 \quad -28 \quad 64 \rfloor$$

which has a negative term on the third row. The UDS algorithm begins to differ from that of the HRZ at the next level of hierarchy. For example, if we add the fourth, fifth, and sixth derivative contributions we get derivative orders that sum to

$$\begin{matrix} 0 & 0 & 2 & 3 & 4 & 5 & 6 \\ 0 & 0 & 2 & 3 & 4 & 5 & 6 \\ 2 & 2 & 4 & 5 & 6 & 7 & 8 \\ 3 & 3 & 5 & 6 & 7 & 8 & 9 \\ 4 & 4 & 6 & 7 & 8 & 9 & 10 \\ 5 & 5 & 7 & 8 & 9 & 10 & 11 \\ 6 & 6 & 8 & 9 & 10 & 11 & 12 \end{matrix}.$$

At this stage we get five terms to scale the third diagonal, seven for the fourth diagonal, five for the fifth, three for the sixth, and only one for the seventh. If we go to a full eighth order polynomial by adding a seventh derivative the only changes are three couplings for the seventh diagonal and one for the eight. For tensor products in 2-D and 3-D we get many more terms to consider, and the final UDS form would differ greatly from that of HRZ scaling and simple lumping. The above comments provide enough information to test the algorithm for one-dimensional formulations. For a hierarchical code in 2-D or 3-D with a general mixture of derivative orders it would be necessary to generate a code to identify the even order derivative terms to be associated with the corresponding hierarchical diagonal coefficients. This appears to be easy to implement, since the order of each diagonal term is known before selecting an integration rule. Note that this procedure would also mean that, except for the linear contributions, the integrals of derivatives of odd order sum could be skipped since they are not included in the UDS.

17.6 References

[1] Bathe, K.J. and Wilson, E.L., *Numerical Methods for Finite Element Analysis*, Englewood Cliffs: Prentice-Hall (1976).

[2] Biggs, J.M., *Introduction to Structural Dynamics*, New York: McGraw-Hill (1964).

[3] Chung, T.J., *Finite Element Analysis in Fluid Dynamics*, New York: McGraw-Hill (1978).

[4] Fenves, S.J., *Computer Methods in Civil Engineering*, Englewood Cliffs: Prentice-Hall (1967).

[5] Hinton, E., Rock, T., and Zienkiewicz, O.C., "A Note on Mass Lumping and Related Processes in the Finite Element Method," *Earthquake Eng. Struct. Dyn.*, **4**(3), pp. 245–249 (1976).

[6] Hughes, T.J.R. and Liu, W.K., "Implicit-Explicit Finite Elements in Transient Analysis: Implementations and Numerical Examples," *J. Appl. Mech.*, **45**(2), pp. 375–378 (1978).

[7] Hughes, T.J.R., *The Finite Element Method*, Englewood Cliffs: Prentice-Hall (1987).

[8] Malkus, D.S. and Plesha, M.E., "Zero and Negative Masses in Finite Element Vibration and Transient Analysis," *Comp. Meth. Appl. Mech. and Eng.*, **59**, pp. 281–306 (1986).

[9] Myers, G.E., *Analytical Methods in Conduction Heat Transfer*, New York: McGraw-Hill (1971).

[10] Newmark, N.M., "A Method of Computation for Structural Dynamics," *ASCE J. Eng. Mech. Div.*, **85**(EM3), pp. 67–94 (July 1959).

[11] Park, K.C., "Practical Aspects of Numerical Time Integration," *Computers and Structures*, **7**(2), pp. 343–354 (1977).

[12] Weaver, W.F., Jr. and Johnston, P.R., *Structural Dynamics by Finite Elements*, Englewood Cliffs: Prentice-Hall (1987).

[13] Zienkiewicz, O.C. and Lewis, R.W., "An Analysis of Various Time-Stepping Schemes for Initial Value Problems," *Earthquake Eng. Struct. Dynamics*, **1**, pp. 407–408 (1973).

[14] Zienkiewicz, O.C., *The Finite Element Method,* 3rd edition, New York: McGraw-Hill (1979).

[15] Zienkiewicz, O.C. and Morgan, K., *Finite Elements and Approximation*, Chichester: John Wiley (1983).

Chapter 18

COMPUTATIONAL FLUID DYNAMICS *

With Dr. Bala Ramaswamy

18.1 Introduction

The fundamental equations for incompressible flows of Newtonian fluids are the momentum (Navier-Stokes) and the continuity equations. These equations are written in the primitive variables of velocity components and pressure, with reference to an Eulerian frame, i.e., a space-fixed system of coordinates through which the fluid flows. In primitive variable formulations, methods based on a consistent mass representation and an implicit time integration procedure have been extensively studied in the literature.

The development of finite element methods (FEM) for the numerical simulation of viscous, incompressible laminar flows has received considerable attention in the last decade, particularly the treatment of pressure in the primitive variable formulation. Finite element methods applied to the Navier-Stokes (NS) equations, using the velocity and pressure primitive variables, can be categorized into three groups. These are the mixed interpolation methods, the penalty methods, and the segregated velocity-pressure solution methods [24, 25].

Fully implicit finite element methods for solving transient fluid flow problems are often stable for any positive value of the time increment. At the present time, however, these "integrated" solution techniques are not always economically feasible. Alternatively, a few explicit and semi-implicit finite element analyses based on the segregated velocity-pressure solution method (velocity and pressure are solved sequentially at each iteration or time step) have appeared. In this chapter, we discuss different segregated approaches with emphasis on the solution implementation in the finite element context. These iterative schemes generally require much less execution time and storage than the classical velocity-pressure integrated and mixed interpolation methods, particularly for three-dimensional problems. The most critical step is, of course, to choose integration methods that combine efficiency and accuracy.

In recent years, a few finite element analyses based on the segregated velocity-pressure solution method (velocity and pressure are solved sequentially at each iteration or time step) have appeared. Various modifications to the Chorin's projection algorithm [4] have been investigated, particularly with respect to their effects on stability and accuracy. These iterative schemes generally require much less execution time and storage than the mixed-Galerkin methods and the penalty methods, particularly for three-dimensional problems. In the present study, the solution procedure consists of a semi-implicit approach using the explicit Lax-Wendroff type scheme for the nonlinear convective terms.

The classic velocity-pressure integrated and mixed interpolation methods are based on the quadratic velocity and linear pressure interpolation. Here, a finite element procedure for the NS equations is developed. The numerical scheme employs a semi-implicit, time-splitting method to integrate the two-dimensional full NS equations satisfying continuity to machine accuracy. The efficient use of direct solvers for the uncoupled momentum and pressure equations is demonstrated. The Poisson equation system for the pressure correction is computed only once, is assembled, modified for the pressure boundary conditions, and factored (only once). Then at each time step, or iteration, it is only necessary to perform a forward and backward substitution to obtain the pressures. The element mass, convection, diffusion, gradient, and divergence matrices are computed only once (if storage permits) and are used again at each time step. Often it is possible to identify *geometrically equivalent* elements (i.e., the same shape and geometric Jacobian) that allows one element matrix set to be used for hundreds of others of the same geometric group. A driven cavity flow is calculated, and it is shown that the present method needs fewer iterations and has an accuracy as good as existing methods. It does not have the checkerboard pressure oscillations of the other equal order interpolation methods.

18.2 Mathematical Formulation

The fluid is considered to be viscous, incompressible, with constant properties, and the flow is laminar. The governing equations are derived from the basic physical principles of conservation of mass, and momentum. Let Ω be a bounded domain in R^2 and T be a positive real number. The spatial and temporal coordinates are denoted by $\mathbf{x} \in \overline{\Omega}$ and $t \in [0, T]$, where a superposed bar indicates the set closure. The equations of interest are the 2D time-dependent incompressible NS equations for the velocity $\mathbf{u}$, and kinematic pressure (p, pressure divided by density) in the bounded domain Ω:

$$\frac{\partial \mathbf{u}}{\partial t} + (\mathbf{u} \cdot \nabla)\,\mathbf{u} = -\nabla p + \frac{1}{\mathrm{Re}} \nabla^2 \mathbf{u} \quad \text{in} \quad \Omega, \tag{18.1}$$

$$\nabla \cdot \mathbf{u} = 0 \quad \text{in} \quad \Omega, \tag{18.2}$$

where Re is the Reynolds number. These governing equations are elliptic in spatial coordinates and parabolic in space and time. Therefore, in order to formulate a well-posed problem, the boundary conditions must be specified for all boundaries and an initial condition for the velocities in the domain of investigation is needed. Let Γ be a piecewise smooth boundary of the domain Ω. The following boundary conditions are specified:

1. On wall, Γ_1, both velocity components are zero (no-slip condition) :

$$\mathbf{u} = \hat{\mathbf{u}} = 0 \quad \text{on} \quad \Gamma_1 . \tag{18.3}$$

2. Natural boundary conditions are specified on Γ_2:

$$-p + \frac{1}{\text{Re}} \frac{\partial u_n}{\partial n} = f_n \tag{18.4}$$

$$\frac{1}{\text{Re}} \frac{\partial u_r}{\partial n} = f_\tau , \tag{18.5}$$

where n and τ are the unit normal and tangential vectors to the boundary Γ. The initial conditions specify the value of velocity at the initial time $t = 0$:

$$\mathbf{u}(\mathbf{x}, 0) = \mathbf{u}_0(\mathbf{x}) \quad \text{in} \quad \overline{\Omega} \tag{18.6}$$

where

$$\mathbf{n} \cdot \mathbf{u}_0 = \mathbf{n} \cdot \hat{\mathbf{u}}(x, 0) \quad \text{on} \quad \Gamma_1 \tag{18.7}$$

and

$$\nabla \cdot \mathbf{u}_0 = 0 \quad \text{in} \quad \Gamma \tag{18.8}$$

in order that a solution exists.

18.3 Solution Procedure Outline

Here, the FEM is used to obtain algebraic representation of the governing equations. Let us discretize the time interval $(0, T)$ by a time increment Δt. Let $\mathbf{u}^0$ be the velocity field at time t^0. From $\mathbf{u}^0$, and the boundary specifications, the fields $\mathbf{u}^{n+1}$ and p^{n+1} are calculated at $t^n = t^0 + \Delta t$ as follows. At first we consider only the convective term in the momentum equation:

$$\frac{\partial \mathbf{u}}{\partial t} + (\mathbf{u} \cdot \nabla)\, \mathbf{u} = 0 . \tag{18.9}$$

Equation (18.9) is discretized in time by considering a Taylor-series expansion in the time step Δt, up to second-order, as

$$\mathbf{u}^c = \mathbf{u}^0 + \Delta t \frac{\partial \mathbf{u}^0}{\partial t} + \frac{\Delta t^2}{2} \frac{\partial^2 \mathbf{u}^0}{\partial t^2} + O(\Delta t^3) . \tag{18.10}$$

The first- and second-order time derivative terms are expressed as

$$\frac{\partial \mathbf{u}^0}{\partial t} = -(\mathbf{u}^0 \cdot \nabla)\, \mathbf{u}^0 \tag{18.11}$$

$$\frac{\partial^2 \mathbf{u}^0}{\partial t^2} = \left(\left[(\mathbf{u}^0 \cdot \nabla)\, \mathbf{u}^0 \right] \cdot \nabla \right) \mathbf{u}^0 + (\mathbf{u}^0 \cdot \nabla)\,(\mathbf{u}^0 \cdot \nabla)\, \mathbf{u}^0 . \tag{18.12}$$

The convection step allows us to determine an intermediate velocity field $\mathbf{u}^c$ from $\mathbf{u}^0$, starting with $\mathbf{u}_0$, by means of Eqs. (18.10) – (18.12)

$$\begin{aligned} \frac{\mathbf{u}^c - \mathbf{u}^0}{\Delta t} = {} & -(\mathbf{u}^0 \cdot \nabla)\, \mathbf{u}^0 + \frac{\Delta t}{2} ([(\mathbf{u}^0 \cdot \nabla)\, \mathbf{u}^0] \cdot \nabla)\, \mathbf{u}^0 \\ & + \frac{\Delta t}{2} (\mathbf{u}^0 \cdot \nabla)\,(\mathbf{u}^0 \cdot \nabla)\, \mathbf{u}^0 . \end{aligned} \tag{18.13}$$

The weak form of this stage would be to solve

$$\int_\Omega \mathbf{w} \left[\frac{\partial \mathbf{u}}{\partial t} + (\mathbf{u} \cdot \Delta)\, \mathbf{u} \right] d\Omega = 0$$

with appropriate integration by parts, subject to the boundary conditions that $\mathbf{u}^c = \mathbf{g}$ on Γ_g. On Γ_h we typically have boundary source integrals to evaluate as a result of the integration by parts. After integrating the second order spatial derivatives terms by parts, and using the divergence theorem, the weak form is :

$$\begin{aligned} \langle \mathbf{u}^*, \frac{\mathbf{u}^c - \mathbf{u}^0}{\Delta t} \rangle &= -\langle \mathbf{u}^*, (\mathbf{u}^0 \cdot \nabla)\, \mathbf{u}^0 \rangle - \frac{\Delta t}{2} \langle (\mathbf{u}^0 \cdot \nabla)\, \mathbf{u}^*, (\mathbf{u}^0 \cdot \nabla)\, \mathbf{u}^0 \rangle \\ &+ \frac{\Delta t}{2} \langle \mathbf{u}^*, \left[[(\mathbf{u}^0 \cdot \nabla)\, \mathbf{u}^0] \cdot \nabla \right] \mathbf{u}^0 \rangle - \frac{\Delta t}{2} \langle \mathbf{u}^*(\nabla \cdot \mathbf{u}^0), (\mathbf{u}^0 \cdot \nabla)\, \mathbf{u}^0 \rangle \\ &+ \frac{\Delta t}{2} \langle \mathbf{u}^*(\mathbf{u}^0 \cdot \mathbf{n}), (\mathbf{u}^0 \cdot \nabla)\, \mathbf{u}^0 \rangle_\Gamma \end{aligned} \tag{18.14}$$

with

$$\mathbf{u}^c = \hat{\mathbf{u}} \quad \text{on} \quad \Gamma_1 . \tag{18.15}$$

The second term on the right hand side of Eq. (18.14) is the most important one. It is present even for a uniform convection field, and has a tensorial structure which enables it to correct the velocity field only in the direction of the convection velocity and not transversely. Therefore, it has a streamline character, which is provided directly by this method without introducing any special upwind mechanism or the introduction of special parameters. This formulation represents an implementation of a Lax-Wendroff type scheme for the convective phase. The scheme is second order accurate in time. It is important to emphasize that the second order terms are not to be interpreted as numerical diffusion or viscosity inherent to the scheme, but the correction terms are an element of the improved difference approximation to the time derivative, with respect to the explicit Euler algorithm.

Now we consider the viscosity term of the Navier-Stokes equations:

$$\frac{\partial \mathbf{u}}{\partial t} - \frac{1}{\mathrm{Re}} \nabla^2 \mathbf{u} = 0 . \tag{18.16}$$

A first order implicit Euler time integration scheme is used here and a new intermediate velocity $\mathbf{u}^v$ is determined from $\mathbf{u}^c$ by:

$$\mathbf{u}^v = \mathbf{u}^c + \frac{\Delta t}{\mathrm{Re}} \nabla^2 \mathbf{u}^v . \tag{18.17}$$

The weak formulation,

$$\int_\Omega \mathbf{w} \left[\frac{\partial \mathbf{u}}{\partial t} - \frac{1}{\mathrm{Re}} \nabla^2 \mathbf{u} \right] d\Omega = 0$$

after integration by parts, is

$$\langle \mathbf{u}^*, \frac{\mathbf{u}^v - \mathbf{u}^c}{\Delta t} \rangle = \frac{1}{\mathrm{Re}} \langle \nabla \mathbf{u}^*, \nabla \mathbf{u}^v \rangle \tag{18.18}$$

with

$$\mathbf{u}^{v} = \hat{\mathbf{u}} \quad \text{on} \quad \Gamma_1 . \tag{18.19}$$

In some cases, the above two phases are combined.

The computed velocity $\mathbf{u}^v$ does not generally satisfy the continuity constraint, therefore we must account for this in the pressure calculation. The auxiliary velocity field $\mathbf{u}^v$ can be decomposed into the sum of a vector field with zero divergence and a vector field with zero curl. The zero-divergence component is the end-of-step velocity vector $\mathbf{u}^n$, whereas the irrotational one is related to the gradient of the pressure field p^n. The final velocity field $\mathbf{u}^n$ for the cycle is obtained by combining the temporary velocity $\mathbf{u}^v$ with the pressure acceleration. To compute the pressure, we employ the pair of equations

$$\nabla \cdot \mathbf{u} \equiv 0$$

and

$$\frac{\partial \mathbf{u}^p}{\partial t} + \nabla p = 0 .$$

When we take the gradient of the second equation and substitute the incompressibility constant, we obtain a Poisson equation which is subject to essential boundary conditions of $p = p_h$ on Γ_h, and on Γ_g one has a normal gradient given by

$$\frac{\partial p}{\partial n} = -\bar{\mathbf{n}} \cdot \frac{\partial \mathbf{u}}{\partial t} .$$

This is solved in the standard fashion for an elliptic BVP. Our approximation here is

$$\frac{(\mathbf{u}^n - \mathbf{u}^v)}{\Delta t} + \nabla p^n = 0 \tag{18.20}$$

$$\nabla \cdot \mathbf{u}^n = 0 . \tag{18.21}$$

The pressure in Eq. (18.20) ensures the satisfaction of the incompressibility constraint, that is, the final velocity field must possess a zero-velocity divergence in every element. The following Poisson equation, with respect to pressure, is obtained by eliminating $\mathbf{u}^n$ from Eqs. (18.20) and (18.21):

$$\nabla^2 p = \frac{1}{\Delta t} \nabla \cdot \mathbf{u}^v . \tag{18.22}$$

We solve Eqs. (18.20) – (18.21) by first solving the Poisson equation and then using (18.20) to compute $\mathbf{u}^n$. To solve Eq. (18.22), the following boundary conditions are applied:

$$p = \hat{p} \quad \text{on} \quad \Gamma_2 \tag{18.23}$$

$$\frac{\partial p}{\partial n} = \mathbf{n} \cdot \frac{1}{\Delta t} \left(\mathbf{u}^v - \mathbf{u}^n \right) \quad \text{on} \quad \Gamma_1 . \tag{18.24}$$

Once the pressure has been determined from Eq. (18.22), nodal velocities $\mathbf{u}^n$ can be computed from the weak form of Eq. (18.20):

$$\langle \mathbf{u}^*, \frac{\mathbf{u}^{n+1} = \mathbf{u}^v}{\Delta t} \rangle = \langle \nabla \cdot \mathbf{u}^*, \ p^{n+1} \rangle \tag{18.25}$$

with

$$\mathbf{u}^n = \hat{\mathbf{u}} \quad \text{on} \quad \Gamma_1 . \tag{18.26}$$

In the preceding discussion, the explanation of the algorithm was given in a continuum space. In the finite element method, the variables **u** and p are discretized by the following finite element approximations:

$$\mathbf{u} = \Phi_\alpha \mathbf{u}_\alpha \tag{18.27}$$

$$p = \Phi_\alpha p_\alpha \tag{18.28}$$

where $\mathbf{u}_\alpha$ and p_α are the discretized velocity components, and pressure at node α, respectively; Φ_α is the interpolation function for velocity, and pressure. Here, the bilinear quadrilateral elements are used for each. The finite element method described in the previous section is amenable to all types of elements.

18.4 Algorithm Details

Stage I : Taylor-Galerkin Convective Approximation: Recall our starting velocity prediction

$$\frac{\partial \mathbf{u}}{\partial t} + (\mathbf{u} \cdot \nabla)\, \mathbf{u} = 0$$

so that

$$\frac{\partial^2 \mathbf{u}}{\partial t^2} = \left[[\,(\mathbf{u} \cdot \nabla)\, \mathbf{u}\,] \cdot \nabla \right] u + (\mathbf{u} \cdot \nabla)\,(\mathbf{u} \cdot \nabla)\, \mathbf{u} .$$

To better understand the development of the definitions, we will switch to Cartesian tensor form. Thus,

$$\frac{\partial u_i}{\partial t^2} = -u_j \frac{\partial u_i}{\partial x_j}$$

and

$$\frac{\partial^2 u_i}{\partial t^2} = 2u_k \frac{\partial u_j}{\partial x_k} \frac{\partial u_i}{\partial x_j} + u_j\, u_k \frac{\partial^2 u_i}{\partial x_j\, \partial x_k} .$$

Using the Taylor series,

$$\mathbf{u}^n = \mathbf{u}^o + \Delta t \frac{\partial \mathbf{u}^o}{\partial t} + ½ \Delta t^2 \frac{\partial^2 \mathbf{u}^o}{\partial t^2} + 0(\Delta t^3)$$

or

$$u_i^n = u_i^o + \Delta t \left[-u_j^o \frac{\partial u_i^o}{\partial x_j} \right] + ½ \Delta t^2 \left[2u_k^o \frac{\partial u_j^o}{\partial x_k} \frac{\partial u_i^o}{\partial x_j} + u_j^o\, u_k^o \frac{\partial^2 u_i^o}{\partial x_j\, \partial x_k} \right] + 0(\Delta t^3) \tag{18.29}$$

where the superscript o denotes the result from the previous time step. Multiply this by the test functions w_i and integrate over the solution domain:

$$\int_\Omega w_i\, u_i^n\, d\Omega = \int_\Omega w_i\, u_i^o\, d\Omega - \Delta t \int_\Omega w_i\, u_j^o\, \frac{\partial u_i^o}{\partial x_j}\, d\Omega$$

$$+ \frac{2}{2}\,\Delta t^2 \int_\Omega w_i\, u_k^o\, \frac{\partial u_j^o}{\partial x_k}\, \frac{\partial u_i^o}{\partial x_j}\, d\Omega + \frac{\Delta t^2}{2} \int_\Omega w_i\, u_j^o\, u_k^o\, \frac{\partial^2 u_i^o}{\partial x_j\, \partial x_k}\, d\Omega\,.$$

However,

$$\left[w_i\, u_j\, u_k\, \frac{\partial^2 u_i}{\partial x_j\, \partial x_k} \right] = \frac{\partial}{\partial x_k} \left[w_i\, u_j\, u_k\, \frac{\partial u_i}{\partial x_j} \right]$$

$$- \frac{\partial w_i}{\partial x_k}\, u_j\, u_k\, \frac{\partial u_i}{\partial x_j} - w_i\, \frac{\partial u_j}{\partial x_k}\, u_k\, \frac{\partial u_i}{\partial x_j} - w_i\, u_j\, \frac{\partial u_k}{\partial x_k}\, \frac{\partial u_i}{\partial x_j}$$

and

$$\int_\Omega \frac{\partial}{\partial x_k} \left[w_i\, u_j\, u_k\, \frac{\partial u_i}{\partial x_j} \right] d\Omega = \int_\Gamma w_i\, u_j\, u_k\, \frac{\partial u_i}{\partial x_j}\, n_k\, d\Gamma$$

so that

$$\int_\Omega w_i\, u_i^n\, d\Omega =$$

$$\int_\Omega w_i \left\{ u_i^o - \Delta t\, u_j^o\, \frac{\partial u_i^o}{\partial x_j} + \frac{\Delta t^2}{2} \left[(2-1)\, u_k^o\, \frac{\partial u_j^o}{\partial x_k}\, \frac{\partial u_i^o}{\partial x_j} - u_j^o\, \frac{\partial u_k^o}{\partial x_k}\, \frac{\partial u_i^o}{\partial x_j} \right] \right\} d\Omega$$

$$- \frac{\Delta t^2}{2} \int_\Omega \frac{\partial w_i}{\partial x_k} \left\{ u_j^o\, u_k^o\, \frac{\partial u_i^o}{\partial x_j} \right\} d\Omega + \frac{\Delta t^2}{2} \int_\Gamma w_i\, u_j^o\, u_k^o\, \frac{\partial u_i^o}{\partial x_j}\, n_k\, d\Gamma\,. \tag{18.30}$$

This weak form can be solved directly to obtain the convective approximation. The consistent mass matrix on the LHS is usually approximated by a diagonal form. The RHS source vector uses the previous velocities and their gradients.

Stage II : Viscous Prediction : Including the body force, f_i ,we have the viscous phase PDE

$$\frac{\partial u_i}{\partial t} - \nu\, \frac{\partial}{\partial x_j} \left(\frac{\partial u_i}{\partial x_j} + \frac{\partial u_j}{\partial x_i} \right) = f_i \tag{18.31}$$

where $\nu = 1/\mathrm{Re}$. Here we denote the previous prediction as u_i^o and assume

$$\begin{aligned} u_i = u_i^o &+ \alpha\, \Delta t \left[\nu\, \frac{\partial}{\partial x_j} \left(\frac{\partial u_i}{\partial x_j} + \frac{\partial u_j}{\partial x_i} \right) + f_i \right] \\ &+ (1-\alpha)\, \Delta t \left[\nu\, \frac{\partial}{\partial x_j} \left(\frac{\partial u_i^o}{\partial x_j} + \frac{\partial u_j^o}{\partial x_i} \right) + f_i^o \right]. \end{aligned} \tag{18.32}$$

Here we refer to α as the implicitness parameter. The Euler method is obtained with $\alpha = 0$ while $\alpha = 1/2$ gives the *Crank-Nickelson* method. Multiplying by a test function w_i and integrating over Ω gives :

$$\int_\Omega w_i\, u_i\, d\Omega = \int_\Omega w_i\, u_i^o\, d\Omega + \alpha\,\Delta t \int_\Omega w_i\, f_i\, d\Omega$$

$$+ \alpha\,\Delta t v \int_\Omega w_i \frac{\partial}{\partial x_j}\left[\frac{\partial u_i}{\partial x_j} + \frac{\partial u_j}{\partial x_i}\right] d\Omega + (1-\alpha)\,\Delta t \int_\Omega w_i\, f_i^o\, d\Omega$$

$$+ (1-\alpha)\,\Delta t\, v \int_\Omega w_i \frac{\partial}{\partial x_j}\left[\frac{\partial u_i^o}{\partial x_j} + \frac{\partial u_j^o}{\partial x_i}\right] d\Omega .$$

However,

$$w_i \frac{\partial}{\partial x_j}\left[\frac{\partial u_i}{\partial x_j} + \frac{\partial u_j}{\partial x_i}\right] = \frac{\partial}{\partial x_j}\left[w_i\left[\frac{\partial u_i}{\partial x_j} + \frac{\partial u_j}{\partial x_i}\right] - \frac{\partial w_i}{\partial x_j}\left[\frac{\partial u_i}{\partial x_j} + \frac{\partial u_j}{\partial x_i}\right]\right]$$

so that from Green's Theorem,

$$\int_\Omega w_i \frac{\partial}{\partial x_j}\left[\frac{\partial u_i}{\partial x_j} + \frac{\partial u_j}{\partial x_i}\right] d\Omega = \int_\Gamma w_i\left[\frac{\partial u_i}{\partial x_j} + \frac{\partial u_j}{\partial x_i}\right] n_j\, d\Gamma$$

$$- \int_\Omega \frac{\partial w_i}{\partial x_j}\left[\frac{\partial u_i}{\partial x_j} + \frac{\partial u_j}{\partial x_i}\right] d\Omega$$

and our final weak form for the viscous predictions is

$$\int_\Omega w_i\, u_i\, d\Omega = \int_\Omega w_i\left[u_i^o + \Delta t\left\{\alpha f_i + (1-\alpha) f_i^o\right\}\right] d\Omega$$

$$+ \alpha\, v\, \Delta t \int_\Gamma w_i\left[\frac{\partial u_i}{\partial x_j} + \frac{\partial u_j}{\partial x_i}\right] n_j\, d\Gamma - \alpha\, v\, \Delta t \int_\Omega \frac{\partial w_i}{\partial x_j}\left[\frac{\partial u_i}{\partial x_j} + \frac{\partial u_j}{\partial x_i}\right] d\Omega \tag{18.33}$$

$$+ (1-\alpha)\, v\, \Delta t \int_\Gamma w_i\left[\frac{\partial u_i^o}{\partial x_j} + \frac{\partial u_j^o}{\partial x_i}\right] n_j\, d\Gamma - (1-\alpha)\, v\, \Delta t \int_\Omega \frac{\partial w_i}{\partial x_j}\left[\frac{\partial u_i^o}{\partial x_j} + \frac{\partial u_j^o}{\partial x_i}\right] d\Omega .$$

In the notation of the subroutines given later this is expressed in matrix form as

$$(\mathbf{M} - v\,\Delta t\,\mathbf{S})\,\mathbf{u}^v = \mathbf{M}\mathbf{u}^c + \Delta t\,\mathbf{F} .$$

Stage III : Pressure Correction: Recall that our approximation is

$$\frac{\partial \mathbf{u}^n}{\partial t} = -\nabla p^n .$$

Taking the gradient of both sides

$$\nabla \cdot \left[\frac{\mathbf{u}^n - \mathbf{u}^v}{\Delta t} = -\nabla p^n\right]$$

$$\frac{\nabla \cdot \mathbf{u}^n}{\Delta t} - \frac{\nabla \cdot \mathbf{u}^v}{\Delta t} = -\nabla^2 p^n$$

but $\nabla \cdot \mathbf{u}^n \equiv 0$ since $\mathbf{u}^n$ satisfies the incompressibility condition, so that

$$-\nabla^2 p^n = -\frac{1}{\Delta t}\,\nabla \cdot \mathbf{u}^v \tag{18.34}$$

is our governing PDE. Multiplying by a test function q and integrating over Ω:

$$\int_\Omega q\,\frac{\partial}{\partial x_j}\left[\frac{\partial p}{\partial x_j}\right] d\Omega = \frac{1}{\Delta t}\int_\Omega q\,\frac{\partial u_j^v}{\partial x_j}\,d\Omega$$

but

$$q\,\frac{\partial}{\partial x_j}\left[\frac{\partial p}{\partial x_j}\right] = \frac{\partial}{\partial x_j}\left[q\,\frac{\partial p}{\partial x_j}\right] - \frac{\partial q}{\partial x_j}\,\frac{\partial p}{\partial x_j}$$

so that

$$\int_\Gamma q\,\frac{\partial p}{\partial x_j}\,n_j\,d\Gamma - \int_\Omega \frac{\partial q}{\partial x_j}\,\frac{\partial p}{\partial x_j}\,d\Omega = \frac{1}{\Delta t}\int_\Omega q\,\frac{\partial u_j^v}{\partial x_j}\,d\Omega \tag{18.35}$$

which is the weak form. The first term vanishes since we have $\partial p/\partial n = 0$ on Γ for the correction term. In matrix form this is

$$\mathbf{A}\mathbf{p} = -\mathbf{H}^T\mathbf{u}^v\ .$$

Stage IV : Velocity Correction: The pressure, p^n, is now known. We seek the final velocity field from

$$\frac{\partial \mathbf{u}^n}{\partial t} = -\nabla p^n \tag{18.36}$$

or

$$\mathbf{u}^n = \mathbf{u}^v - \Delta t\,\nabla p^n\ .$$

Multiply by test function w_i and integrate over Ω to get the weak form

$$\int_\Omega w_i\,u_i^n\,d\Omega = \int_\Omega w_i\left[u_i^v - \Delta t\,\frac{\partial p^n}{\partial x_i}\right] d\Omega\,. \tag{18.37}$$

A typical controlling subroutine, NSFLOW, is shown in Fig. 18.4.1. The integrals given above are identified as array names beginning with E to denote element level, and end with X or Y to denote the component to which they contribute. The subroutine to form these typical matrices at a Gauss point is FORM, which is shown in Fig. 18.4.2. Other element routines, such as FORMH, FORMKS, and LAPLAC, are simply sub-sets of that subroutine. For added efficiency the interpolation functions and their derivatives are computed only once at each Gauss point (in subroutine FILL), and simply stored with an additional subscript to denote the quadrature point number. Since the nonlinear element matrices are relatively expensive to compute, the program also uses the equivalent element type concept discussed in Chap. 9 to reduce duplicate calculations. Element arrays are given an additional subscript to denote the geometry kind number. The most common kinds are computed once and used over at each time step. The last kind number is used to flag elements that always require re-evaluation by numerical integration at each time step. Therefore, for groups of similar elements, this code runs quite fast. However, if the mesh is a highly curvilinear and almost no elements have the same shape, the efficiency drops drastically to the standard approach of numerically integrating every element in the mesh at every time step.

```
      SUBROUTINE  NSFLOW (NMAX, NE, NL, NS, NQP, MAXK, RHO, FNU, G,
     1                   ISTF, ISTL, KOUT, DELT, NBU, NBV, NBW, NBP,
     2                   NBS, XYZ, UB, VB, WB, PB,
     3                   SB, UN, VN, WN, PN,
     4                   UBAR, VBAR, WBAR, H, DLH,
     5                   B, COORD, XJ, XJI, GP,
     6                   GW, PT, WT, EDIAG, EHX,
     7                   EHY, EHZ, EKX, EKY, EKZ,
     8                   EMASS, ESS, DMASS, RHS, VSKY,
     9                   NODES, NVU, NVV, NVW, NVP,
     1                   NVS, LEK, LNODE, NK1, NK2,
     2                   NTABLE, ITOP, LPVT, NB,
     3                   RN, IN, JPT, KPT, MAXR, MAXI, NUMR, NUMI,
     4                   MAXSKY, LASTI, LASTR, JPTSKY, NIP, NEXTK,
     5                   NEXTI, R, LTYPE )
C     * * * * * * * * * * * * * * * * * * * * * * * * * * * * * * * *
C                        --- NSFLOW ---
C     FINITE ELEMENT PROGRAM FOR NAVIER_STOKES EQUATIONS USING
C              ISOPARAMETRIC ELEMENTS WITH SPLITTING
C                        COPYRIGHT 1990
C
C      PROGRAMMED BY: Dr. B. RAMASWAMY, Dr. J.E. AKIN
C                     RICE UNIVERSITY, HOUSTON, TX 77251-1892
C     * * * * * * * * * * * * * * * * * * * * * * * * * * * * * * * *
      COMMON / UNITS  /  ICTL,ILPR,IMSH,IBC,IBUG,IOUT,IRST,IUSR
      COMMON / BUGS   /  MBUG
      CHARACTER*8  RN, IN
      PARAMETER  ( PSTOP = 1.E6 )
      DIMENSION  R(MAXR), RN(NUMR), IN(NUMI), JPT(NUMR), KPT(NUMI)
      DIMENSION  XYZ(NMAX,NS), UB(1), VB(1), WB(1), PB(1), SB(1),
     1    UN(NMAX), VN(NMAX), WN(NMAX), PN(NMAX), UBAR(NMAX),
     2    VBAR(NMAX), WBAR(NMAX), H(NL,NIP), DLH(NS,NL,NIP),
     3    B(NS,NL), COORD(NL,NS), XJ(NS,NS), XJI(NS,NS),
     4    GP(NQP), GW(NQP), PT(NS,NIP), WT(NIP), EDIAG(NL,MAXK),
     5    EHX(NL,NL,MAXK), EHY(NL,NL,MAXK), EHZ(NL,NL,MAXK),
     6    EKX(NL,NL,NL,MAXK), EKY(NL,NL,NL,MAXK), EKZ(NL,NL,NL,MAXK),
     7    EMASS(NL,NL), ESS(NL,NL,MAXK), DMASS(NMAX),
     8    RHS(NMAX), VSKY(1)
     DIMENSION  NODES(NE,NL), NVU(1), NVV(1), NVW(1), NVP(1), NVS(1),
     1    LEK(NE), LNODE(NL), NK1(NMAX), NK2(NMAX), NTABLE(NMAX),
     2    ITOP(NMAX), LPVT(NMAX+1), NB(1)
C                 ---- Notation ----
C  B      = Global derivatives of H
C  COORD = Spatial coordinates of nodes on element
C  DELT   = Time increment
C  DET    = Determinant of Jacobian matrix, XJ
C  DLH    = Local derivatives of H
C  DMASS = Diagonal system mass matrix
C  EHX, EHY = Gradient matrix in X- and Y- directions
C  EKX, EKY = Advection matrix for X- and Y-directions
C  EMASS = Element consistent mass matrix
C  ESS    = Laplacian matrix for pressure (or stream function)
C  FNU    = Viscosity, NU
C  G      = Gravity
C  GP     = 1D Gaussian quadrature coordinates
C  GW     = 1D Gaussian weights
```

Figure 18.4.1a Incompressible Navier-Stokes control

```
C  H     = Interpolation function for geometry
C  HP    = Interpolation for pressure
C  HV    = Interpolation for velocity component
C  IN    = Names of integer sub-arrays
C  JPT   = Pointers to real sub-arrays
C  KPT   = Pointers to integer sub-arrays
C  RHS   = RHS for pressure
C  KOUT  = Desired output step
C  IBUG  = Unit for error messages
C  ILPR  = Unit of standard Line PRinter
C  ISTF  = Starting step
C  ISTL  = Ending step
C  LASTI = Location of last used position in array I
C  LASTR = Location of last used position in array R
C  LNODE = List of nodes on an element
C  MBUG  = Flag for debug list. 0 =none, IBUG =max
C  MAXI  = Maximum available position in array I
C  MAXR  = Maximum available position in array R
C  NBP   = Number of boundary conditions for P-pressure
C  NBS   = Number of boundary conditions for stream function
C  NBU   = Number of boundary conditions for U-velocity
C  NBV   = Number of boundary conditions for V-velocity
C  NE    = Number of elements
C  NEK   = Number of element kinds in the problem
C  NEXTI = Number of next sub-array in array I
C  NEXTK = Number of next sub-array in array R
C  NIP   = Total number of integration points
C  NL    = Number of nodes per element
C  NMAX  = Total number of nodal points
C  NODES = Element connectivities
C  NPL   = Number of pressure nodes per element
C  NQP   = Number of 1D quadrature points
C  NS    = Dimension of space
C  NVL   = Number of velocity nodes per element  = NL
C  NVP   = Node numbers for specified Pressure
C  NVS   = Node numbers for specified Streamline
C  NVU   = Node numbers for specified U-velocity
C  NVV   = Node numbers for specified V-velocity
C  PB    = Specified values for P-pressure on the boundary
C  PN    = Final pressure
C  PT    = Local quadrature point coordinates
C  RN    = Names of real sub-arrays
C  RHO   = Density
C  SB    = Specified values for Streamline on the boundary
C  U1    = Final U-Velocity
C  UB    = Specified values for U-velocity on the boundary
C  UBAR  = Intermediate U-velocity
C  UN    = Final U-velocity
C  VSKY  = Vector with upper sq matrix in skyline storage
C  V1    = Final V-Velocity
C  VB    = Specified values for V-velocity on the boundary
C  VBAR  = Intermediate V-velocity
C  VN    = Final V-velocity
C  XJ    = Jacobian matrix, (Inverse is XJI)
C  XYZ   = Spatial coordinates of all nodes
C  WBAR  = Intermediate W-velocity
C  WN    = Final W-velocity
C  WT    = Local quadrature point weights
C *******************************************************
C
```

Figure 18.4.1b Incompressible Navier-Stokes control

```
      WRITE(ILPR,*) 'Available integer & real storage:', MAXI, MAXR
      WRITE(IBUG,*) 'Available integer & real storage:', MAXI, MAXR
      CALL IZERO1 (LEK, NE)
      IF ( ISTF .EQ. 1)  THEN
C         Zero solution if this is first step
        CALL ZERO1 (PN, NMAX)
        CALL ZERO1 (UN, NMAX)
        CALL ZERO1 (VN, NMAX)
        IF ( NS .EQ. 3 )  CALL ZERO1 (WN, NMAX)
      ELSE
C         Read the restart values
        OPEN (UNIT=IRST, FILE='RST')
        READ (IRST, *)  ISTEP, TIME
        READ (IRST, *)  (I, UN(I), VN(I), PN(I), J = 1, NMAX)
        DO 20  J = 1, NMAX
   20   READ (IRST,*)  I, ( XYZ(I,K), K = 1,NS )
        CLOSE (IRST)
      ENDIF
C       Read problem data
      CALL INPUT (XYZ, NMAX, NODES, NE, NL, UB, NBU, VB, NBV, WB, NBW,
     1            PB, NBP, SB, NBS, NVU, NVV, NVW, NVP, NVS, NS)
C       Compute required skyline storage
      CALL  PRESKY (NE, NL, NODES, NMAX, NMAX+1, LPVT, MAXSKY, ITOP)
C         MAXSKY is really final now update LASTR
      JPT(JPTSKY + 1) = JPT(JPTSKY) + MAXSKY
      RN(JPTSKY) = "VSKY run"
C-->    --- check skyline and limits ---
      LASTR = JPT(JPTSKY+1)
      WRITE (ILPR,*) ' Integers used vs max: ', LASTI, MAXI
      WRITE (IBUG,*) ' Integers used vs max: ', LASTI, MAXI
      IF ( LASTI .GT. MAXI ) THEN
        WRITE (ILPR,*) ' ERROR, MAXI MUST EXCEED ', LASTI
        WRITE (IBUG,*) ' ERROR, MAXI MUST EXCEED ', LASTI
        IERR = 1
      ENDIF
      WRITE (ILPR,*) ' Reals used vs max:      ', LASTR, MAXR
      WRITE (IBUG,*) ' Reals used vs max:      ', LASTR, MAXR
      IF ( LASTR .GT. MAXR ) THEN
        WRITE (ILPR,*) ' ERROR, MAXR MUST EXCEED ', LASTR
        WRITE (IBUG,*) ' ERROR, MAXR MUST EXCEED ', LASTR
        IERR = 1
        CALL  SIZER (JPTSKY, JPTSKY+1, RN, JPT)
      ENDIF
      IF ( MBUG .GT. 0 )  CALL  SIZER (JPTSKY, JPTSKY+1, RN, JPT)
      IF ( IERR .GT. 0 )  STOP
C       Assign element kind based on relative shape
      CALL  ELKIND (NMAX, NE, NL, NS, XYZ, COORD,
     1              NODES, LNODE, LEK, NEK, MAXK)
C       Get integration data for element type
      IF ( LTYPE .EQ. 0 )  THEN
C         Quad, use Gauss quadratures
        IF ( NS .EQ. 1 )  CALL  GAUS1D (NQP, GP, GW, NIP, PT, WT)
        IF ( NS .EQ. 2 )  CALL  GAUS2D (NQP, GP, GW, NIP, PT, WT)
      ELSE
C         Triangle, use symmetric rule
        CALL  SYMRUL (NIP, PT, WT)
      ENDIF
```

Figure 18.4.1c Incompressible Navier-Stokes control

```
      IF ( MBUG .GT. 0 )  THEN
        WRITE (IBUG,*) 'NIP PTS', PT
        WRITE (IBUG,*) 'NIP WTS', WT
      ENDIF
C      Fill interpolation arrays at each Gauss point
      CALL  FILL (NIP, NL, NS, PT, WT, H, DLH, LTYPE)
C      Generate matrices for each kind
      IK = 0
C-->   Loop over elements
      DO 101 IE = 1, NE
C        Extract element nodes
      CALL  LNODES (IE, NE, NL, NODES, LNODE)
C        Extract element coordinates
      CALL  ELCORD  (NMAX, NL, NS, XYZ, COORD, LNODE)
C        Get geometry kind number
      ILEK = LEK(IE)
      IF ( IK .GE. ILEK ) GO TO 101
C     warning, kinds may no longer be in order
      IK = IK + 1
C      MAXK is reserved for numerical integration
      IF ( IK .GE. MAXK )  GO TO 101
        CALL  FORM (EMASS, EHX, EHY, EHZ, ESS, EKX, EKY, EKZ,
     1              COORD, MAXK, NE, NL, IE, NS, IK,
     2              XJ, XJI, H, DLH, B, PT, WT, NIP)
C         Build diagonal element mass matrix
        CALL  DIAGSQ (NL, EMASS, EDIAG(1, IK))
C<--   End element loop
  101 CONTINUE
C      Carry out the upper skyline assembly mode
      CALL  SKY (ESS, XYZ, COORD, NODES, LNODE, MAXK, NE, NL,
     1           NMAX, NS, LEK, XJ, XJI, H, DLH, B, PT, WT,
     2           NIP, MAXSKY, LPVT, VSKY)
      DO 90 I = 1, NBP
   90 NB(I) = 1
      DO 100 I = 1, NMAX
      NK1(I) = 1
      NK2(I) = I
  100 NTABLE(I) = I
C      Apply essential boundary conditions
      NNN = NMAX + 1
      CALL ABOUN (VSKY, MAXSKY, LPVT, RHS, NMAX, NNN, NVP, NB,
     1            PB, NTABLE, NK1, NK2, NBP, NMAX, 1)
C      Reduce mass matrix to diagonal form
      CALL  LUMP (MAXK, NE, NIP, NL, NMAX, NS, NODES, LNODE,
     1            LEK, COORD, DLH, DMASS, EMASS, H, PT, WT,
     2            XJ, XJI, XYZ, EDIAG)
C      Factor skyline system once
      CALL AOSOL (VSKY, MAXSKY, RHS, LPVT, NMAX, 1)
C      Invert diagonal mass matrix
      DO 110 I = 1, NMAX
  110 DMASS(I) = 1.E0/DMASS(I)
      RODT  = RHO/DELT
      RODTI = 1.0E0/RODT
C
```

Figure 18.4.1d Incompressible Navier-Stokes control

```
C==>> Loop through time steps
C
      DO 320 ISTEP = ISTF, ISTL
C       Zero velocity vector
      CALL ZERO1 (UBAR, NMAX)
      CALL ZERO1 (VBAR, NMAX)
      IF ( NS .EQ. 3 )  CALL ZERO1 (WBAR, NMAX)
C
C-->   Loop over all elements
C
      DO 150 IELEM = 1, NE
C        Extract element nodes
      CALL  LNODES (IELEM, NE, NL, NODES, LNODE)
C        Get the element kind flag
      IK = LEK(IELEM)
C        Does this one have to be numerically integrated?
      IF ( IK .GE. MAXK )  THEN
        IK = MAXK
C          Extract element coordinates
        CALL  ELCORD (NMAX, NL, NS, XYZ, COORD, LNODE)
C          Build element EK and ESS arrays
        CALL FORMKS (ESS, EKX, EKY, EKZ, COORD, MAXK, NE, NL,
     1          IELEM, NS, IK, XJ, XJI, H, DLH, B, PT, WT, NIP)
      ENDIF
C        Recover EKX, EKY, ESS for kind and multiply with UN, VN
C        Ui bar @ n+1 = -nu*(Ui,j + Uj,i),j - UjUi,j @ n
      DO 140 I = 1, NL
        II = LNODE(I)
        DO 130 J = 1, NL
          JJ = LNODE(J)
          UNJ = UN(JJ)
          VNJ = VN(JJ)
          UBAR(II) = UBAR(II) - ESS(I,J,IK)*UNJ*FNU
          VBAR(II) = VBAR(II) - ESS(I,J,IK)*VNJ*FNU
          DO 120 K = 1, NL
            KK = LNODE(K)
            UBAR(II) = UBAR(II) - EKX(I,K,J,IK)*UN(KK)*UNJ
     *                          - EKY(I,K,J,IK)*VN(KK)*UNJ
            VBAR(II) = VBAR(II) - EKX(I,K,J,IK)*UN(KK)*VNJ
     *                          - EKY(I,K,J,IK)*VN(KK)*VNJ
  120     CONTINUE
  130   CONTINUE
  140 CONTINUE
  150 CONTINUE
C<--    End element loop
C
C       Ui bar @ n+1 = Ui @ n + dt*Ui bar @ n+1
C
C-->    Begin nodal loops
      DO 160 I = 1, NMAX
      DTDM = DELT*DMASS(I)
      UBAR(I) = UN(I) + UBAR(I)*DTDM
  160 VBAR(I) = VN(I) + VBAR(I)*DTDM
C       Insert velocity essential boundary values
      DO 170 I = 1, NBU
  170 UBAR(NVU(I)) = UB(I)
      DO 180 I = 1, NBV
  180 VBAR(NVV(I)) = VB(I)
C<--     End nodal loops
C
C      Build net forcing vector, RHS, at this time
C       Del dot U bar @ n+1
C
      CALL ZERO1 (RHS, NMAX)
C-->    Begin element loop
```

Figure 18.4.1e Incompressible Navier-Stokes control

```
      DO 210 IELEM = 1, NE
C        Extract element nodes
      CALL  LNODES (IELEM, NE, NL, NODES, LNODE)
C       Get kind of element
      IK = LEK(IELEM)
C         Does this one have to be numerically intergrated?
      IF ( IK .GE. MAXK )  THEN
        IK = MAXK
C          Extract element coordinates
        CALL  ELCORD (NMAX, NL, NS, XYZ, COORD, LNODE)
C          Build element EH array
        CALL FORMH (EHX, EHY, EHZ, COORD, MAXK, NE, NL, IELEM,
     1               NS, IK, XJ, XJI, H, DLH, B, PT, WT, NIP)
      ENDIF
C       Nodal loops
      DO 200 I = 1, NL
        II = LNODE(I)
        DO 190 J = 1, NL
          JJ = LNODE(J)
          RHS(II) = RHS(II) + EHX(J,I,IK)*UBAR(JJ)
     *                      + EHY(J,I,IK)*VBAR(JJ)
  190   CONTINUE
  200 CONTINUE
  210 CONTINUE
C<--    End element loop
C
C       Source for Del^2 P is Del dot U bar * rho / dt @ n+1
C
C-->    Begin nodal loops
      DO 220 I = 1, NMAX
  220 RHS(I) =   - RODT*RHS(I)
C       Pressure boundary condition
      DO 230 IBP = 1, NBP
  230 RHS(NVP(IBP)) = PB(IBP)
C<--     End nodal loops
C       Forward and backward substitution for Pressure solution.
      CALL AOSOL (VSKY, MAXSKY, RHS, LPVT, NMAX, 2)
C       Transfer Pressure into PN
      DO 240 I = 1, NMAX
      PN(I) = RHS(I)
      IF ( ABS( PN(I) ) .GE. PSTOP )  THEN
        WRITE (IBUG,*) ' STOP PRESSURE EXCEEDED AT NODE =', I
        WRITE (ILPR,*) ' STOP PRESSURE EXCEEDED AT NODE =', I
        STOP
      ENDIF
 240  CONTINUE
C
C     Build final velocity vector
C
      CALL ZERO1 (UN, NMAX)
      CALL ZERO1 (VN, NMAX)
C-->    Loop over all elements
      DO 270 IELEM = 1, NE
C       Extract element nodes
      CALL  LNODES (IELEM, NE, NL, NODES, LNODE)
C       Recover element kind
      IK = LEK(IELEM)
```

Figure 18.4.1f Incompressible Navier-Stokes control

```
C       Does this one have to be numerically intergrated ?
      IF ( IK .GE. MAXK )  THEN
        IK = MAXK
C        Extract element coordinates, build element EH array
        CALL  ELCORD (NMAX, NL, NS, XYZ, COORD, LNODE)
        CALL FORMH (EHX, EHY, EHZ, COORD, MAXK, NE, NL,
     1       IELEM, NS, IK, XJ, XJI, H, DLH, B, PT, WT, NIP)
      ENDIF
C      Set U @ n+1 = P,i @ n+1
      DO 260 I = 1, NL
        II = LNODE(I)
        DO 250 J = 1, NL
          JJ = LNODE(J)
          UN(II) = UN(II) + EHX(J,I,IK)*PN(JJ)
          VN(II) = VN(II) + EHY(J,I,IK)*PN(JJ)
  250   CONTINUE
  260 CONTINUE
  270 CONTINUE
C<--     End element loop
C     Update to current velocity vector
C      U @ n+1 = U bar - P,i / rho * M inverse @ n+1
C-->     Begin node loops
      DO 280 K = 1, NMAX
      UN(K) = UBAR(K) - RODTI*UN(K)*DMASS(K)
  280 VN(K) = VBAR(K) - RODTI*VN(K)*DMASS(K)
C      Insert velocity essential values
      DO 290 I = 1, NBU
  290 UN(NVU(I)) = UB(I)
      DO 300 I = 1, NBV
  300 VN(NVV(I)) = VB(I)
C<--    End node loops, Update time steps
      TIME = ISTEP*DELT
      JL = MOD(ISTEP, KOUT)
C      Print results
      IF ( JL .EQ. 0 ) THEN
        WRITE (ILPR,2005) ISTEP, TIME
        WRITE (IBUG,*)    ISTEP, TIME
 2005   FORMAT (' Time' , I10, 1PE14.6, /,
     1      ' Node          U              V                P' )
        DO 24  I = 1, NMAX
   24   WRITE (ILPR,2010) I, UN(I), VN(I), PN(I)
 2010   FORMAT ( I5, 3(1PE 14.5) )
      ENDIF
C<==   End time step loop, write to restart unit
  320 CONTINUE
      WRITE (ILPR,2005) ISTEP, TIME
      WRITE (IOUT,2006) ISTEP, TIME
 2006 FORMAT ( I10, 1PE14.6 )
      DO 25 I = 1, NMAX
      WRITE (ILPR,2010) I, UN(I), VN(I), PN(I)
   25 WRITE (IOUT,2010) I, UN(I), VN(I), PN(I)
      WRITE (ILPR,*) ' Node         X              Y                Z'
      DO 23  I = 1, NMAX
      WRITE (ILPR,2010) I, (XYZ(I,K), K = 1,NS)
 23   WRITE (IOUT,2010) I, (XYZ(I,K), K = 1,NS)
      WRITE (ILPR,*) ' NORMAL EXIT OF NSFLOW'
      WRITE (IBUG,*) ' NORMAL EXIT OF NSFLOW'
      WRITE (IOUT,*) ' NORMAL EXIT OF NSFLOW'
      RETURN
      END
```

Figure 18.4.1g Incompressible Navier-Stokes control

```
      SUBROUTINE FORM (EMASS, EHX, EHY, EHZ, ESS, EKX, EKY, EKZ,
     1                 COORD, MAXK, NE, NL, IE, NS, IK,
     2                 XJ, XJI, H, DLH, B, PT, WT, NIP)
C
C     Form element matrices for element kind IK
C
      COMMON / UNITS / ICTL,ILPR,IMSH,IBC,IBUG,IOUT,IRST,IUSR
      DIMENSION EHX(NL,NL,MAXK), EHY(NL,NL,MAXK), EHZ(NL,NL,MAXK),
     1          EMASS(NL,NL), ESS(NL,NL,MAXK),
     2          EKX(NL,NL,NL,MAXK), EKY(NL,NL,NL,MAXK),
     3          EKZ(NL,NL,NL,MAXK), COORD(NL,NS),
     4          XJ(NS,NS), XJI(NS,NS), H(NL,NIP),
     5          DLH(NS,NL,NIP), B(NS,NL), PT(NS,NIP), WT(NIP)
C      Zero matrices
      CALL ZERO2 (EMASS, NL, NL)
      CALL ZERO2 (ESS(1,1,IK), NL, NL)
      CALL ZERO2 (EHX(1,1,IK), NL, NL)
      CALL ZERO2 (EHY(1,1,IK), NL, NL)
      CALL ZERO3 (EKX(1,1,1,IK), NL, NL, NL)
      CALL ZERO3 (EKY(1,1,1,IK), NL, NL, NL)
C-->    Loop over quadrature points
      DO 90 J = 1, NIP
C      Compute Jacobian, its inverse, and determinant
      CALL  JACOB (NL, NS, DLH(1,1,J), COORD, XJ)
      CALL  INVDET (XJ, XJI, DET, NS)
      IF ( DET .LE. 0.0 )  THEN
        WRITE (ILPR, *) ' WARNING: JACOBIAN BAD FOR ELEMENT ', IE
        DET = ABS( DET )
      ENDIF
      DETWT = DET*WT(J)
C      Convert to global derivatives, B
      CALL  GDERIV (NS, NL, XJI, DLH(1,1,J), B)
C      Add contributions at current quadrature point
      DO 70 M = 1, NL
        HMDW = H(M,J)*DETWT
        B1M = B(1,M)
        B2M = B(2,M)
        DO 60 L = 1, NL
          HL = H(L,J)
C          Build consistent mass matrix
          EMASS(L,M) = EMASS(L,M) + HL*HMDW
C          Form EHX, EHY, EKX, EKY, ESS
          EHX(L,M,IK) = EHX(L,M,IK) + B(1,L)*HMDW
          EHY(L,M,IK) = EHY(L,M,IK) + B(2,L)*HMDW
          ESS(L,M,IK) = ESS(L,M,IK) + B(1,L)*B1M*DETWT
     *                              + B(2,L)*B2M*DETWT
C          Form EKX, EKY
          DO 50 K = 1, NL
            EKX(K,L,M,IK) = EKX(K,L,M,IK) + H(K,J)*HL*B1M*DETWT
            EKY(K,L,M,IK) = EKY(K,L,M,IK) + H(K,J)*HL*B2M*DETWT
   50     CONTINUE
   60   CONTINUE
   70 CONTINUE
C<--  End quadrature loops
   90 CONTINUE
      RETURN
      END
```

Figure 18.4.2 Forming nonlinear element integrals

18.5 Examples

18.5.1 Lid-Driven Cavity Problem

The driven cavity problem presented here concerns the steady-state solution obtained by the time integration procedure described in the preceding sections. Corner singularities for two-dimensional fluid flows are very important, since most problems of physical interest have corners. Like most elliptic problems, singularities will develop where the boundary contour is not smooth. A typical example is driven flow in a rectangular cavity where the top surface moves with constant velocity along its length. The upper corners where the moving surface meets the stationary walls are singular points of the flow at which the vorticity becomes unbounded and the horizontal velocity is multi-valued. The lower corners are also weakly singular points. Flow near the sharp corners is dominated by viscous terms. We consider a domain $\Omega = \{ (x,y)/0 \le x \le 1, 0 \le y \le 1 \}$ with non-slip boundary conditions, i.e., $u = v = 0$ in all boundaries except $y = 1$ where $u = \bar{u} = 1$. A mesh of 40×40 elements was used in this example. The flow is completely determined by the Reynolds number at $y = 1$, and we chose the cases $Re = 0$, 100, 200, 400, and 1000. A pressure datum $p = 0$ was specified at the middle of the bottom wall. The solution of Stokes flow ($Re = 0$) in a driven cavity was symmetric with respect to the x-axis, as expected. The streamlines and the pressure contours at steady state are plotted in Fig. 18.5.1 for Re = 1, 100, 400, 1000, 5,000 and 10,000. Values of pressure contours are given in Table 18.1.

Table 18.1 Pressure contour values for cavity flow

Contour Number	Reynolds number (Re) 1	100	400	1000	5000	10000
1	–9.6000	–0.0800	–0.1000	–0.1000	–0.1000	–0.0800
2	–6.4000	–0.0600	–0.0800	–0.0800	–0.0800	–0.0600
3	–4.0000	–0.0400	–0.0600	–0.0600	–0.0600	–0.0400
4	–1.5000	–0.0200	–0.0400	–0.0400	–0.0400	–0.0200
5	0.0000	0.0000	–0.0200	–0.0200	–0.0200	0.0000
6	1.5000	0.0200	0.0000	0.0000	0.0000	0.0200
7	4.0000	–	–	–	0.0200	0.0400
8	6.4000	–	–	–	0.0400	0.0600
9	9.6000	–	–	–	0.0600	–

The stream function value of the corner eddies is in good agreement with those predicted by higher-order finite difference methods with more refined grids [10]. As the Reynolds number increased to this value, the vortex moved away from the driving wall toward the geometric center of the cavity, while the streamlines became more circular near the center. The profiles of the dimensionless horizontal velocity along the vertical centerline of the driven cavity ($x_1 = 0.5$) and the dimensionless vertical velocity along the horizontal centerline of the driven cavity ($x_2 = 0.5$) are illustrated for various Reynolds numbers in Fig. 18.5.2. These results are compared with results of Ghia

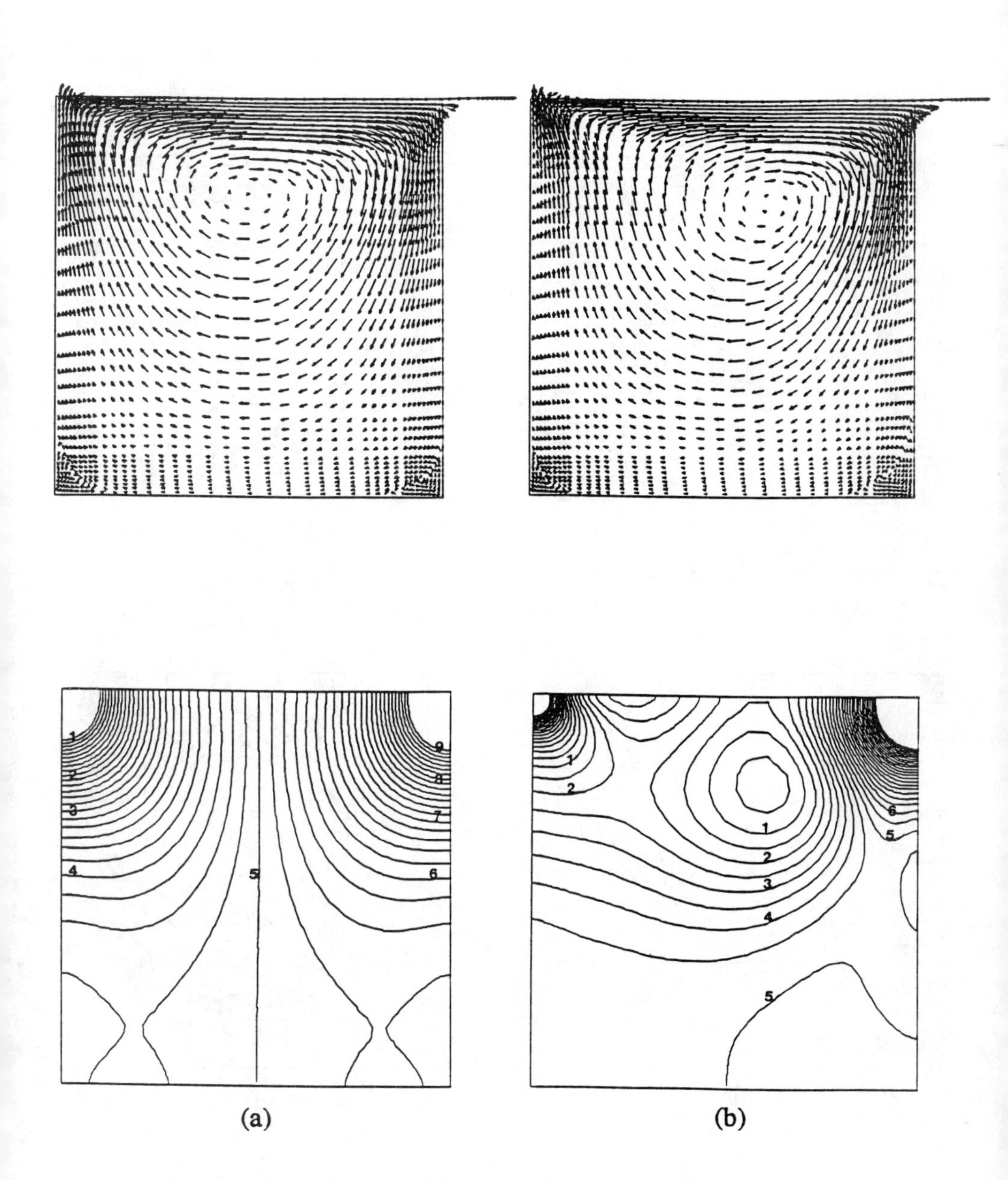

Figure 18.5.1a Steady velocity and pressure for cavity flow
a) Re = 1, b) Re = 100

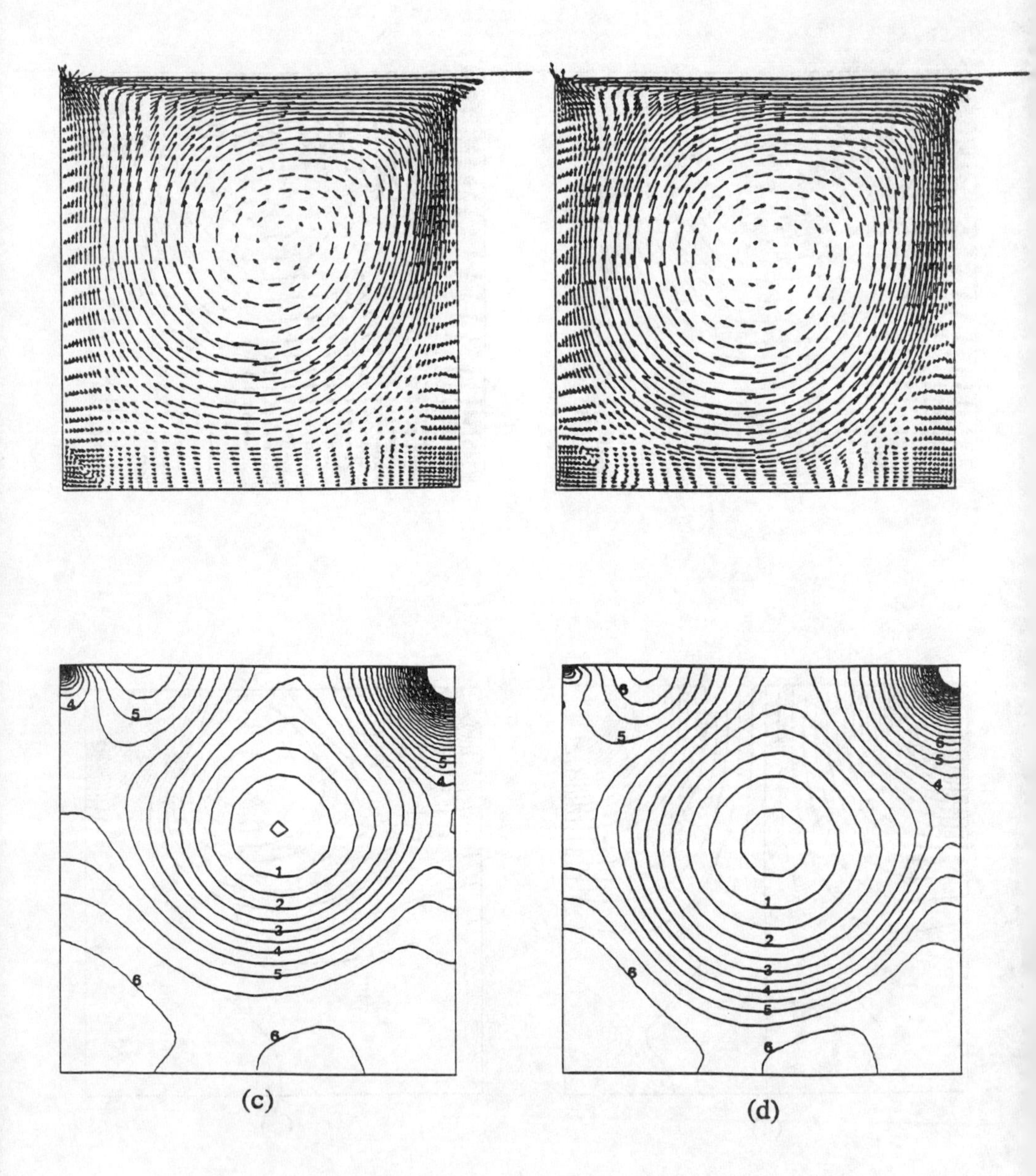

Figure 18.5.1b Steady velocity and pressure for cavity flow
c) Re = 400, d) Re = 1000

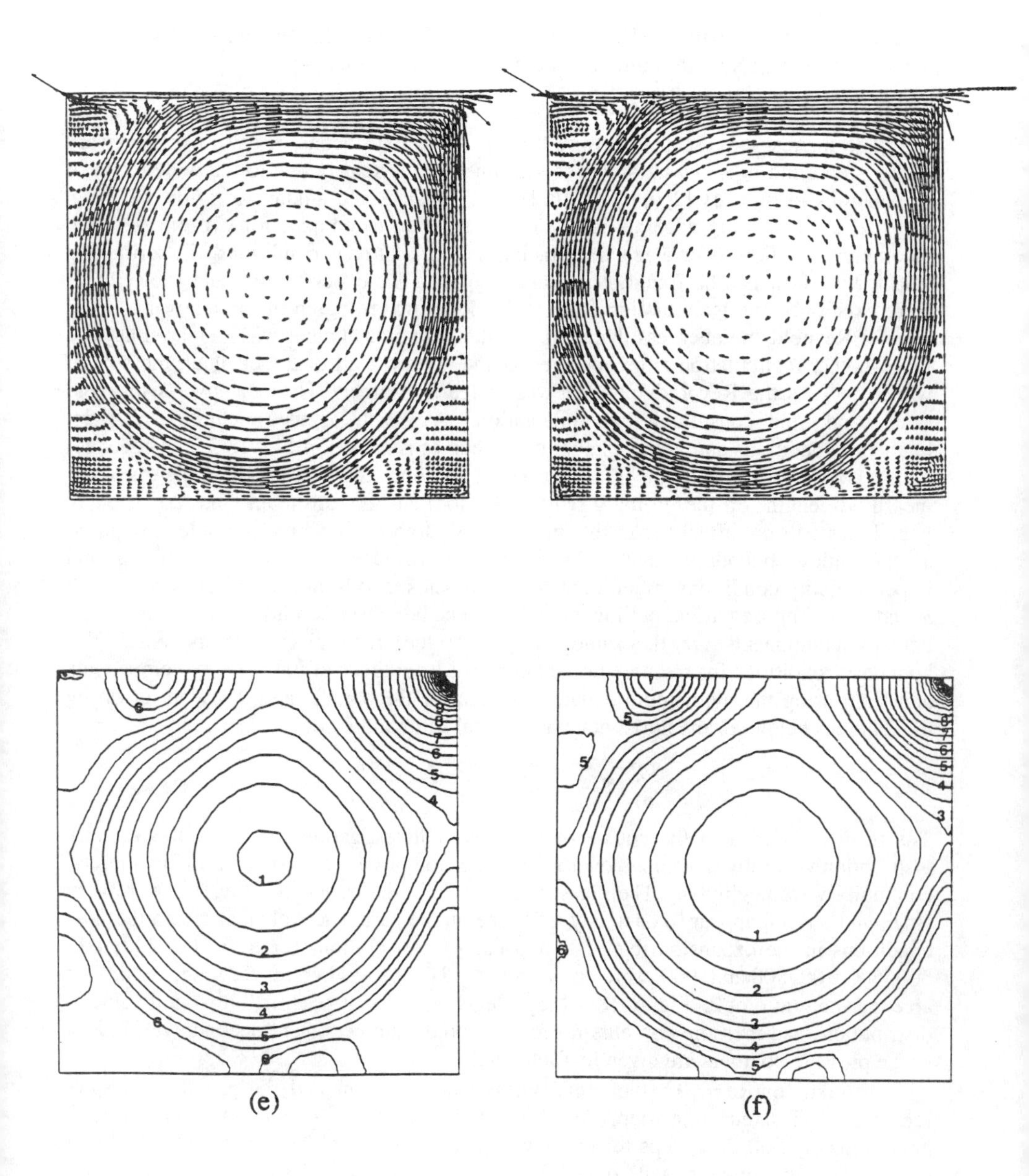

Figure 18.5.1c Steady velocity and pressure for cavity flow
e) Re = 5000, f) Re = 10000

et al. [10] obtained with a 129×129 grid via a multigrid technique. Very good agreement can be found for both maximum velocity position and value.

18.5.2 Sudden Expansion Problem

The interest expressed in flows through abrupt expansions, rearward or backward facing steps, and other variations in this geometry is primarily due to the wide range of applications in industry [2, 9, 11, 14-16, 19]. A considerable amount of effort has been invested in developing accurate and versatile solution techniques for two-dimensional, incompressible flow over a rearward-facing step for laminar conditions. Experimental data for the flow can be found in Armaly et al [1]. The experimental data showed that there exist only one recirculation zone at the downstream region of the rearward facing step for Reynolds number less than 450. As the Reynolds number was increased beyond 450, another recirculation zone appeared at the bottom wall of the channel, the size of which grew as the Reynolds number was increased further. The objective is to avoid costly and time consuming experimentation. A sketch of the integration domain, boundary conditions and nomenclature of the eddies is shown in Fig. 18.5.3. The length of the computational domain is chosen equal to about three times the experimentally measured length of the primary vortex [1] so that the upstream and downstream boundaries will not affect the results implied. A parabolic u velocity profile is imposed at the inflow boundary together with a zero vertical velocity v. Non-slip and impermeability conditions are enforced on solid walls, and Neumann conditions for both u and v are imposed at the outflow boundary. The boundary conditions for the auxiliary velocity components are the same as that for the true velocity components. The boundary conditions for the pressure are deduced from those of the velocity components and from using the momentum equation written for the boundaries. They are given by the relations below for the horizontal and vertical solid boundaries, respectively

$$\frac{\partial p}{\partial y} = \nu \frac{\partial^2 v}{\partial y^2}, \quad \frac{\partial p}{\partial x} = \nu \frac{\partial^2 u}{\partial x^2}.$$

The results obtained in the course of the present investigation show that the velocity distribution was fully developed over a major portion of the flow domain. Errors in mass continuity were negligible. The pressure gradients in the transverse direction were very small and did not appear to have any influence on the parameters characterizing the flow conditions in the expansion region. The steady state solutions for Re = 100, 200, 400, 500, 600, 700, 800 and 900 were obtained with 81×41 grid points (see Fig. 18.5.3) for an expansion ratio of 2. Figure 18.5.4 represents the velocity vectors, horizontal velocity distributions and pressure contours for few Reynolds numbers mentioned above. Values of the pressure contours are given in Table 18.2.

At first, for Re = 100, the initial guess was quiescent and the calculations were repeated at a fixed time step, $\Delta t = 0.1$, until steady flow was obtained. It took approximately 500 time steps for the flow to become steady. The numerical solution at time $t = 50.0$ was obtained with only 1.5 min. of CPU time on a CRAY X-MP. This flow was used as the initial guess for Re = 200, and so on. As the Reynolds number increases, the time step Δt also increased. Maximum time increment $\Delta t = 0.5$ is used for Re = 900. At low Reynolds numbers, the recirculating region is quite small. The predicted results agree quite well with both the predictions and measurements by Armaly et al [1]. The re-attachment length is a key parameter and appears to within a few percent of the measured data. The pressure distributions in the vicinity of the step is shown in Figs. 18.5.4a-d. In this region the pressure distribution is strongly two-dimensional while

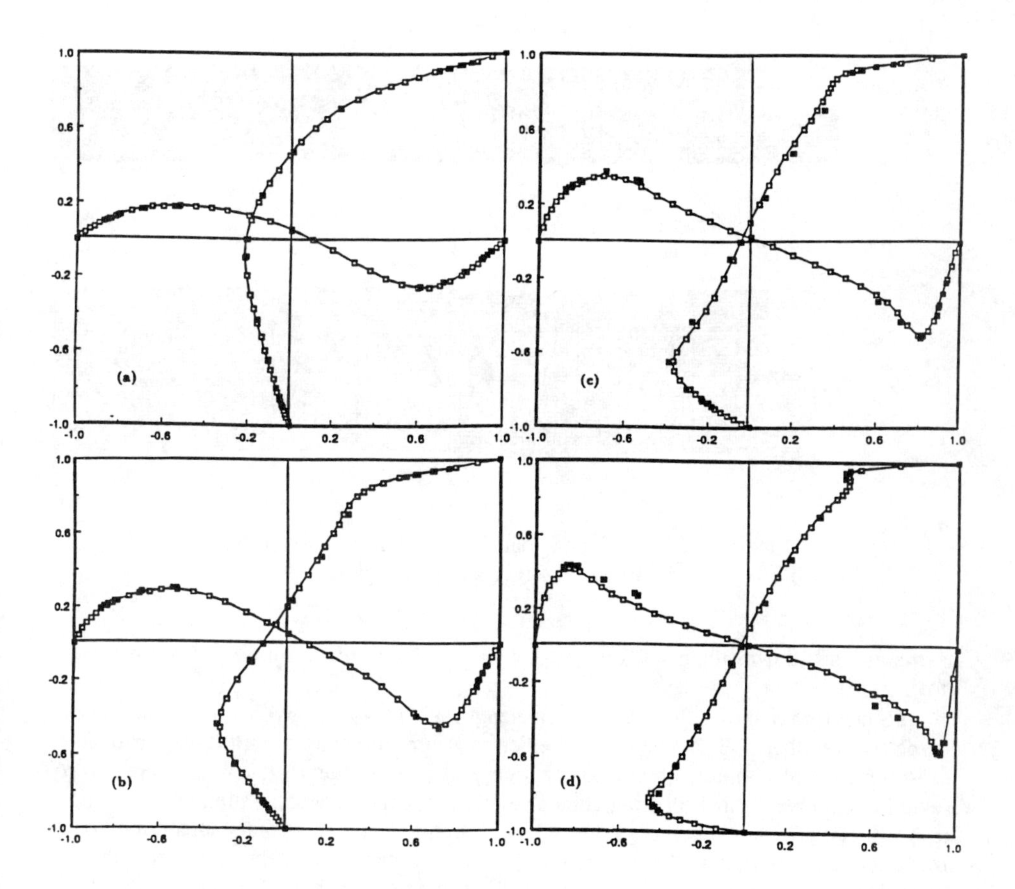

Figure 18.5.2 Velocity profiles for cavity flow. a) Re = 100, b) Re = 400, c) Re = 1000, d) Re = 5000

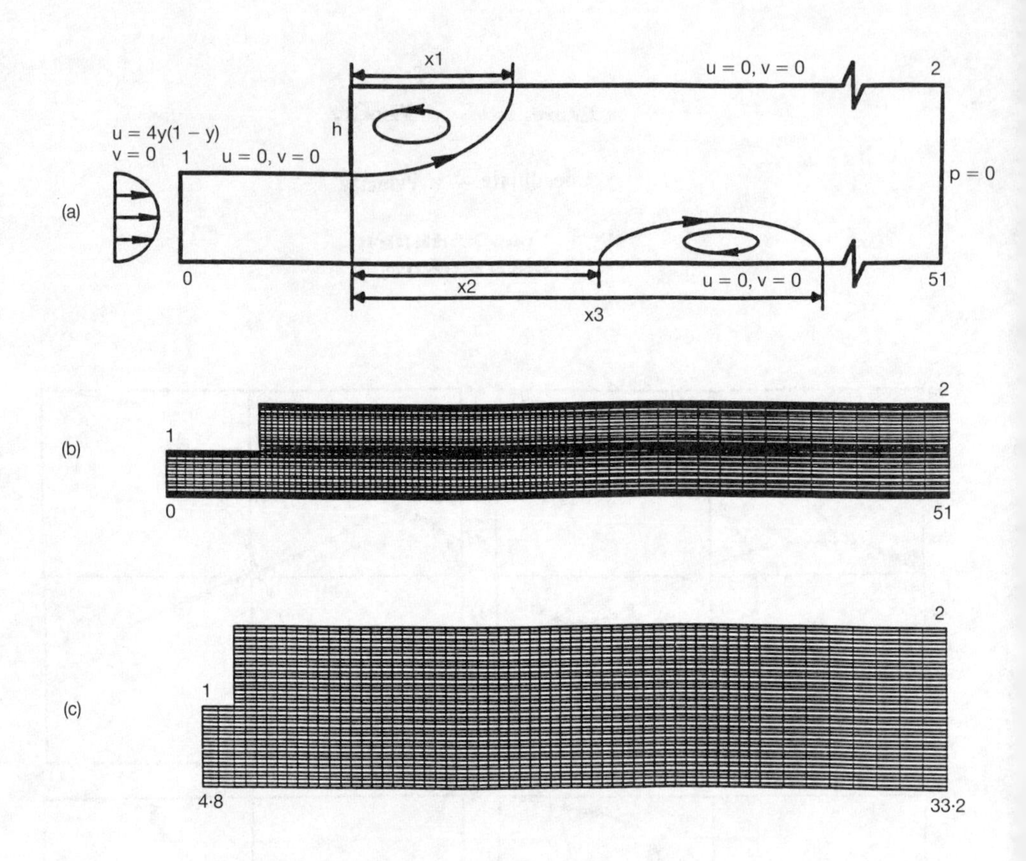

Figure 18.5.3 Sudden expansion problem: a) notation, b) mesh, c) corner plot region

upstream and downstream of this region the pressure distribution is nearly one-dimensional.

Figure 18.5.5 shows the reattachment length of the primary vortex as a function of Reynolds number. The present predictions are compared with experimental measurements of Armaly et al [1]. The dependence of the reattachment length on Reynolds number is in good agreement with the experimental results up to about Re = 500. For Reynolds number higher than 500, the computed primary separation zone size starts to fall below the measured data. This may be attributed to the three-dimensionality of the experimental flow at this Reynolds number. The general trend of the curve of x_1 (the re-attachment length of the recirculating zone behind a rearward-facing step) is the same as that of Kim and Moin [18], who used the finite difference method with 101×101 grid points. Also, predictions of the beginning and end points of the recirculating zone near the upper wall compare satisfactorily with experimental data.

This chapter describes the development and evaluation of a second-order accurate explicit finite element scheme based on a Taylor-Galerkin projection algorithm for

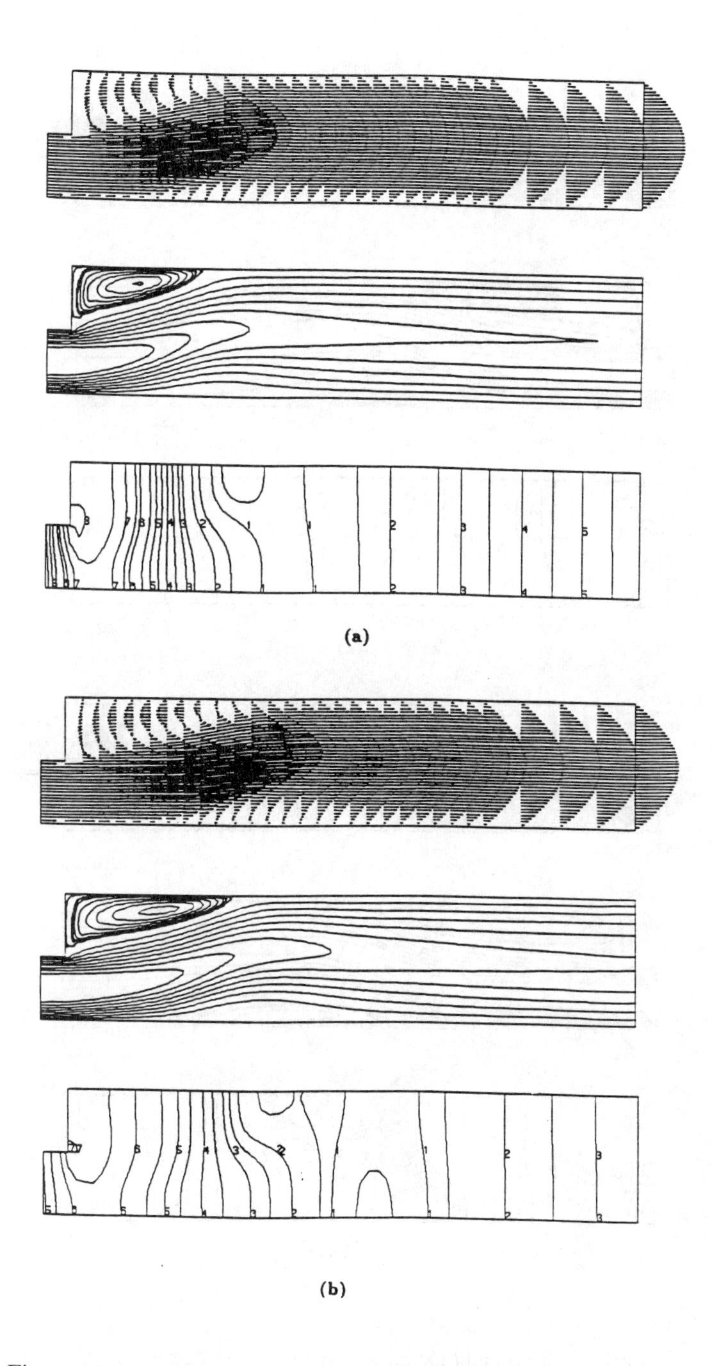

Figure 18.5.4a Steady expansion velocities, streamlines, pressure
a) Re = 200, b) Re = 400

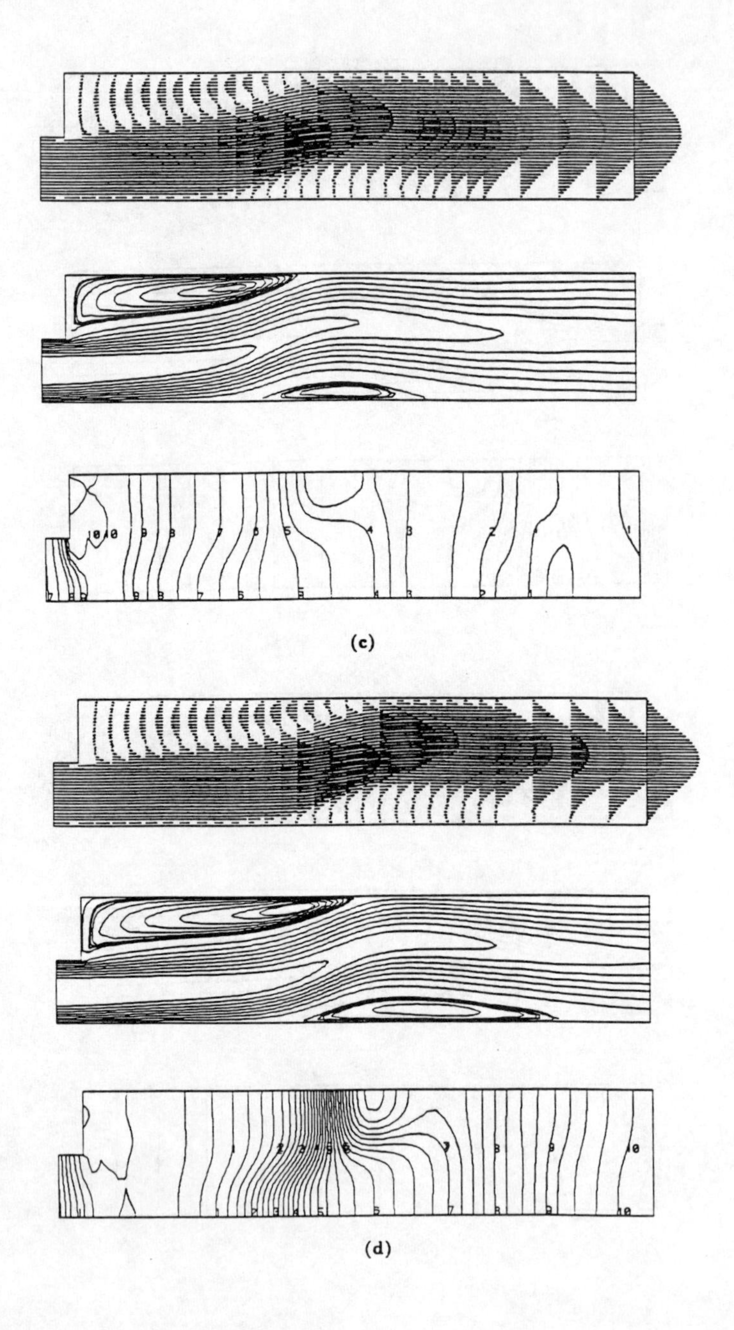

Figure 18.5.4b Steady expansion velocities, streamlines, pressure
c) Re = 600, d) Re = 800

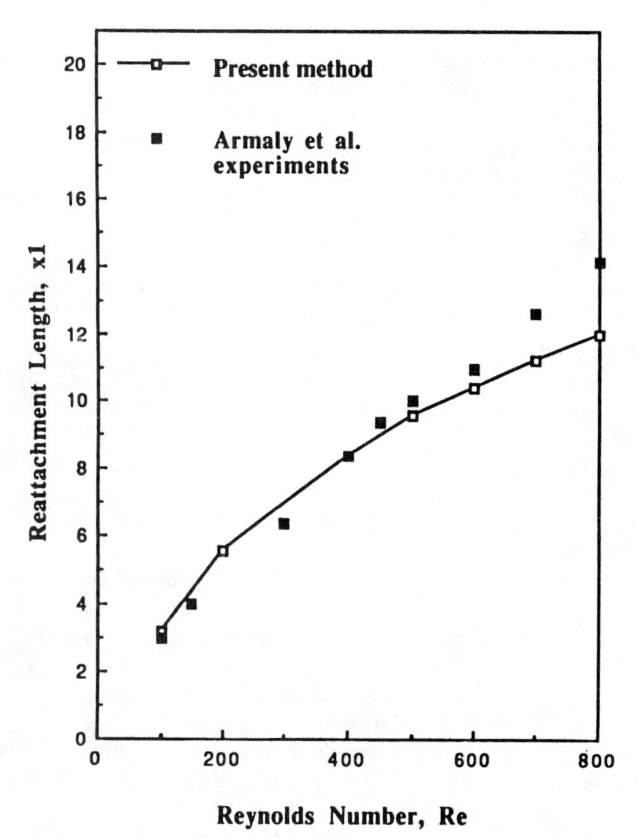

Figure 18.5.5 Sudden expansion, size of the primary vortex

Table 18.2 Values of pressure contours for sudden expansion

Contour Number	Reynolds number (Re)			
	200	400	600	800
1	0.1230	0.0600	0.0140	−0.0927
2	0.1100	0.0524	0.0120	−0.0829
3	0.0950	0.0406	0.0000	−0.0732
4	0.0810	0.0190	−0.0083	−0.0634
5	0.0670	−0.0065	−0.0240	−0.0537
6	0.0500	−0.0300	−0.0480	−0.0439
7	0.0390	−0.0400	−0.0680	−0.0341
8	0.0250	–	−0.0800	−0.0244
9	–	–	−0.0840	−0.0146
10	–	–	−0.0862	−0.0480

the transient incompressible Navier-Stokes equations. From the results presented here, it is clear that a segregated finite element solution scheme with equal-order interpolation for the velocity components and the pressure can be used. With the pressure correction derived here, a checkerboard solution for pressure should not occur. The numerical examples discussed in the chapter indicate that in addition to its simplicity, the method is capable of producing reasonably accurate results.

18.6 References

[1] Armaly, B.F., Durst, F., Pereira, J.C.F., and Schonung, B., "Experimental and Theoretical Investigation of Backward-Facing Step Flow," *J. Fluid Mech.*, **127**, pp. 473–496 (1983).

[2] Aung, W., Baron, A., and Tsou, F., "Wall Independency and Effect of Initial Shear-layer Thickness in Separated Flow and Heat Transfer," *Int. J. Heat Mass Transfer*, **28**, pp. 1757–1771 (1985).

[3] Baker, A.J., *Finite Element Computational Fluid Mechanics*, New York: Hemisphere Pub. (1983).

[4] Chorin, A.J., "Numerical Solution of the Navier-Stokes Equations," *Mathematics of Computation*, **22**, pp. 745–762 (1968).

[5] Chung, T.J., *Finite Element Analysis in Fluid Dynamics*, New York: McGraw-Hill (1978).

[6] Connor, J.C. and Brebbia, C.A., *Finite Element Techniques for Fluid Flow*, London: Butterworth (1976).

[7] Gallagher, R.H., et al., eds., *Finite Elements in Fluids*, New York: John Wiley (Vol. 1, 1973; Vol. 2, 1975; Vol. 3, 1978).

[8] Gartling, D. and Becker, E.B., "Computationally Efficient Finite Element Analysis of Viscous Flow Problems," in *Computational Methods in Nonlinear Mechanics*, ed. J.T. Oden, et al., Austin, Texas: T.I.C.O.M. (1974).

[9] Ghia, K.N., Hankey, W.L., and Hodge, J.K., "Use of Primitive Variables in the Solution of Incompressible Navier-Stokes Equations," *AIAA J.*, **17**, pp. 298–301 (1979).

[10] Ghia, U., Ghia, K.N., and Shin, C.T., "High-Re solutions for incompressible flow using the Navier-Stokes equations and a multigrid method," *J. Comp. Phys.*, **48**, pp. 387–411 (1982).

[11] Goldstein, R.J., Eriksen, V.L., Olson, R.M., and Eckert, E.R.G., "Laminar Separation, Reattachment, and Transition of the Flow over a Downstream-Facing Step," *J. Basic Eng., Trans. of the ASME*, , pp. 732–741 (1970).

[12] Gresho, P.M., Chan, S., Upson, C., and Lee, R.L., "A Modified Finite Element Method for Solving the Time-Dependent, Incompressible Navier-Stokes Equations," *Int. J. Numer. Meth. Fluids*, **14**, pp. 557–598 (1984).

[13] Gresho, P.M. and Chan, S., "Solving the Incompressible Navier-Stokes Equations Using Consistent Mass and a Pressure Poisson Equation," Proceedings of ASME Symposium on Recent Advances in Computational Fluid Dynamics, Chicago, IL (Nov. 28–Dec. 2, 1988).

[14] Guj, G. and Stella, F., "Numerical Solutions of High-*Re* Recirculating Flows in Vorticity-Velocity Form," *Int. J. Num. Meth. Fluids*, **8**, pp. 405–416 (1988).

[15] Hackman, L.P., Raithby, G.D., and Strong, A.B., "Numerical Prediction of Flow over a Backward-Facing Step," *Int. J. Num. Meth. Fluids.*, **4**, pp. 711–724 (1984).

[16] Honji, H., "The Starting Flow Down a Step," *J. Fluid. Mech.*, **69**, pp. 229–240 (1975).
[17] Hughes, T.J.R., Liu, K., and Brooks, A., "Finite Element Analysis of Incompressible Viscous Flows by the Penalty Function Formulation," *J. Comp. Phys.*, **30**, pp. 1–60 (1979).
[18] Kim, J. and Moin, P., "Application of a Fractional-Step Method to Incompressible Navier-Stokes Equations," *J. Comp. Phys.*, **59**, pp. 308–323 (1985).
[19] Morgan, K., Periaux, J., and Thomasset, F., eds., "Analysis of Laminar Flow over a Backward Facing Step," A GAMM–Workshop, Germany, Friedr Vieweg and Son (1984).
[20] Oden, J.T. and Demkowicz, L., "Adaptive Finite Element Methods for Complex Problems in Solid and Fluid Mechanics," pp. 1, 3–13 in *Finite Elements in Computational Mechanics*, ed. T. Kant, Oxford: Pergamon Press (1985).
[21] Oden, J.T., Demkowicz, L., Strouboulis, T., and Devloo, P., "Adaptive Methods for Problems in Solid and Fluid Mechanics," pp. 249–280 in *Accuracy Estimates and Adaptive Refinements in Finite Element Computations*, ed. I. Babuska, O.C. Zienkiewicz, J. Gago, and E.R. de A. Oliveira, Chichester: John Wiley (1986).
[22] Pironneau, O., *Finite Element Methods for Fluids*, New York: John Wiley (1991).
[23] Ramaswamy, B. and Kawahara, M., "Lagrangian Finite Element Analysis Applied to Viscous Free Surface Fluid Flow," *Int. J. Num. Meth. Fluids*, **7**, pp. 953–984 (1987).
[24] Ramaswamy, B., "Finite Element Solution for Advection and Natural Convection Flows," *Comp. Fluids*, **16**, pp. 349–388 (1988).
[25] Ramaswamy, B., "Efficient Finite Element Method for Two-Dimensional Fluid Flow and Heat Transfer Problems," *Num. Heat Transfer*, **17**, pp. 123–154 (1990).
[26] Schneider, G.E., Raithby, G.D., and Yavanovich, M.M., "Finite Element Solution Procedures for Solving the Incompressible Navier-Stokes Equations Using Equal-Order Interpolation," *Numer. Heat. Transfer*, **1**, pp. 435–451 (1978).
[27] Stevens, W.N.R., "Finite Element, Stream Function-Vorticity Solution of Steady Laminar Natural Convection," *Int. J. Num. Meth. Fluids*, **2**, pp. 349–366 (1982).

Chapter 19

AUTOMATIC MESH GENERATION

19.1 Introduction

As we have seen, the practical application of the finite element method requires extensive amounts of input data. Analysts quickly learned that special mesh generation programs could help reduce this burden. Today the most powerful commercial codes offer extensive mesh generation and data supplementation options. Most mesh generation techniques are divided into two general classes: automatic generation and manual generation. In manual generation, the user at least determines how many elements will be used in each section of the mesh, and how many element neighbors each will have. In the extreme case, the manual method means the analyst supplies all the nodal and element data. It is now common for manual procedures to employ an isoparametric mapping or an IJK mapping [7]. A typical mesh generation program locates and numbers the nodal points, numbers the elements, and determines the element incidences. In addition, it usually allows for assignment of codes to each node and each element. Thus, it also has a capability to generate boundary condition codes and element material codes. Many such codes use a mapping, or transformation, method to generate the mesh data.

Tessellation methods for mesh generation begin with the definition of the boundary of the domain [3, 5]. They then work from the boundary to the interior and fill the interior with elements. Usually triangles and tetrahedra are used in two- and three-dimensions, respectively. The user typically has control of the size of the elements on the boundary. Some procedures also allow for size control of interior elements.

Certain tree structures are used in *solid modeling* to represent design parts [1]. They therefore provide a beginning point for automatic mesh generation [8, 9]. The most common of these methods are the bintree, quadtree, and octree for one-, two-, and three-dimensional objects, respectively. In the following sections, we will illustrate some of these techniques for mesh generation.

19.2 Automatic Mesh Generation

There is a continuing trend toward the use of automatic mesh generators. They are often tied to the use of *solid modeling* systems, or *computer-aided design* systems. The designer defines the surfaces and volumes of the various material regions that make up the object. The automatic mesh generator then employs those data to generate the nodes and elements for that volume. The number of elements in a region of the solid is not known before the mesh generation is complete. However, the user can sometimes specify a minimum element size at a given location. If the generator creates a larger element, it can split it into smaller elements until the size restriction has been met. Usually the size of the element is governed by the shape of the object. They tend to get smaller when there are rapid changes in the geometry. That is often consistent with the distribution needed for accurate analysis results. Adaptive methods selectively divide the original elements into smaller elements. Thus, the automatic mesh generators can be stopped with relatively large element sizes. However, if the large elements lie on a curved boundary, one must provide the adaptive code with enough information to ensure that the new elements match the actual boundary. Otherwise, the new elements will simply match the shape of the larger parent element.

There have been some recent review articles (see [4] and [2]) on automatic mesh generation to which the reader is referred. Here we will discuss the main concepts of the most common methods. These include Tesselation methods and the octree methods. If one is given a collection of node points, one can pick a starting location and create a simplex element. To generate its neighbor, one selects the node point that gives the least distorted element shape. As shown in Fig. 19.2.1, one proceeds in this fashion until all points are utilized to form a complete mesh. A mesh generated in this fashion is shown in Fig. 19.2.2. Another approach is to define the boundary of the object by a series of points. Selected boundary points are connected to simplex elements, as illustrated in Fig. 19.2.3. The interior face of the elements and the unattached portions of the original boundary define a new boundary. This continues until no well-shaped elements can be connected to points on the boundary. Then, a new interior point is inserted to form a new nearly equilateral simplex element. The result of that operation is to define a new interior boundary, and the process begin anew.

19.3 Tree Structure Methods

The *octree mesh generation* and other tree structure mesh generation methods are historically and practically tied to solid modeling methods and computer graphics display methods. An early approach to solid modeling was to divide the space into a uniform grid of cells or elements to locate the object in the space. The object was then approximated by those cells that lay completely inside the object. Boolean operations, such as union and intersection, are made simple by the use of uniform cells. For example, to find the intersection of objects A and B, we scan over the cells of grid A. If a cell is inside A, we then check the corresponding cell in grid B. If it is occupied by B, then the two objects intersect at that cell and we can flag that occurrence in a new grid that represents the result of that operation. Similar procedures are employed for the union and difference operators. This approach employs uniform accuracy everywhere and is somewhat wasteful, since many of the uniform cells may be empty.

For common engineering shapes the boundaries will not match a course grid, and we would like a finer resolution on the edges and a coarser one on the interior. This observation leads to an organization structure that has several levels. Each cell level is a factor of two smaller, in each spatial dimension, than the parent cell from which it was obtained. The structures for representing these forms in one dimension, two dimensions, and three dimensions are called bintrees, quadtrees, and octrees, respectively.

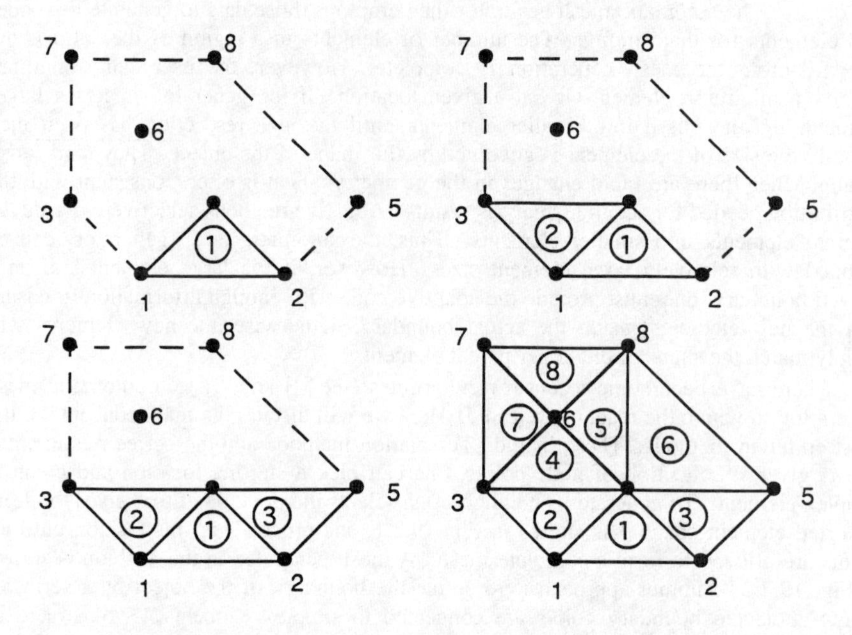

Figure 19.2.1 Surface tesselation concepts

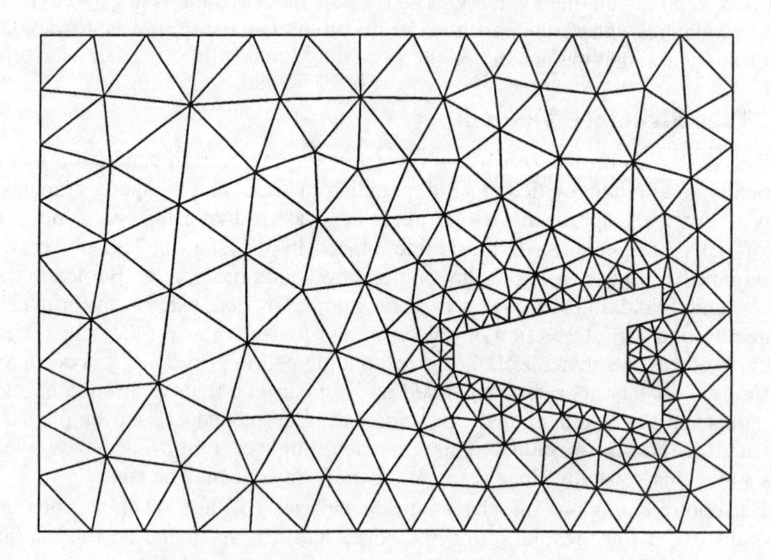

Figure 19.2.2 Example tesselation result

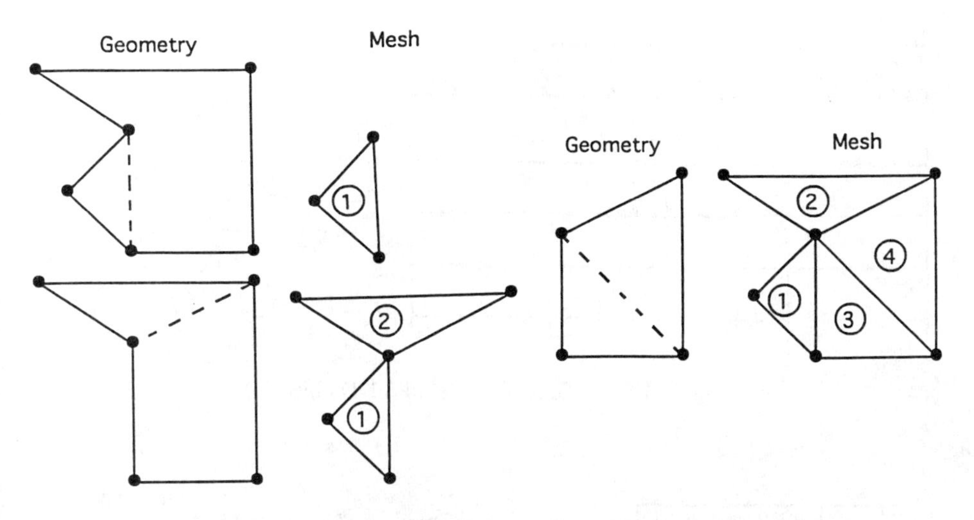

Figure 19.2.3 Mesh generation by removing domains

Figure 19.3.1 shows a bintree model of a 1-D space. We begin with a single cell that completely encloses the object. This cell will be partially filled by the object. Thus, to get more detail we divide it into two equal cells. If one of these cells is completely full or empty, we stop its division into child cells. A partially occupied cell is flagged for additional division. This continues until all cells have stopped dividing (are full or empty) or until we reach the final level of division. The latter limit is usually governed by the amount of available storage. The preceding three cell options are also referred to as inside, outside, and boundary cells, or as black, white, and gray cells. This alternate procedure of tree division can give much more accurate boundary resolutions than a uniform grid. The relative sizes at selected levels are given below.

Level	0	1	2	4	8	16	32
Cell Width	1.0	0.5	0.25	6.25E-2	3.9E-3	1.5E-5	2.3E-10

A quadtree model of a two-dimensional solid is shown in Fig. 19.3.2. Note that it is easy to select either the region inside the solid or the region outside. They have the same tree structure, but the nature of the final leafs reverse, that is, a completely empty node in representing the solid is reversed and becomes completely filled when viewed as the external region. This is significant since we may want to use the solid model to define an interior mesh for stress analysis, or an exterior mesh for a fluid flow analysis.

Once a tree structure has been completed, it is simple to determine if a specified point is interior or exterior to an object or on its surface. This fundamental test is often needed when processing a solid model. The answer is obtained by traversing down the tree by always selecting the branch that contains the coordinates of the point. When the final leaf node is reached its type (e.g., inside or outside) specifies the status of the point. If it is a partially full, or boundary cell, then we can consider the point to be on the surface, within a tolerance equal to the cell size. Regardless of how small the cells are, most practical shapes will still have some cells that are only partially filled by the object.

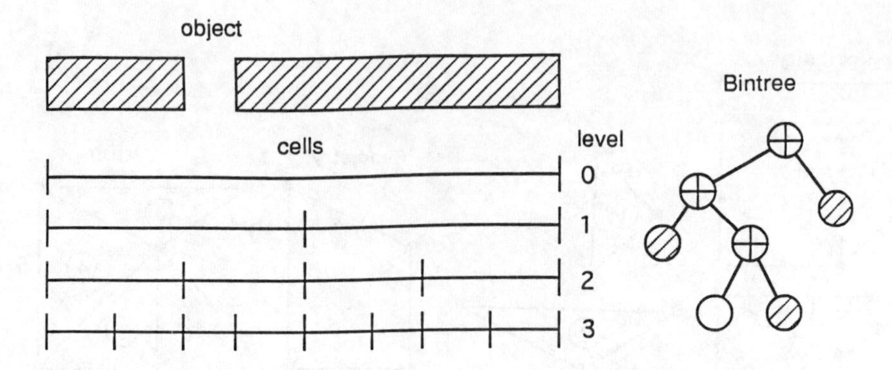

Figure 19.3.1 Bintree model of a 1-D object

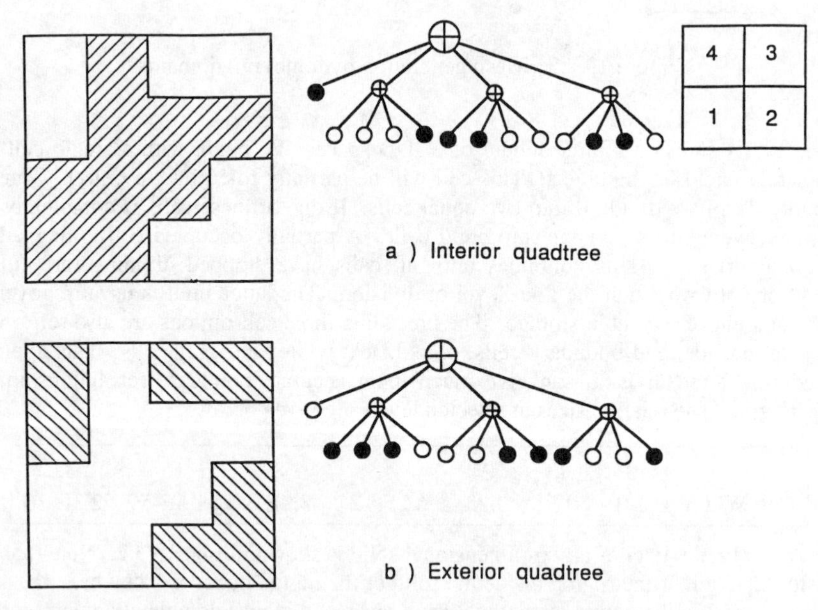

Figure 19.3.2 Quadtree model representation

These are the boundary (partial) cells. The others are either full (inside) or empty (outside) cells. The extension to octrees for solids is illustrated in Fig. 19.3.3.

The cells used in a solid modeler are often used in a finite element analysis. The trend is toward having the quadtree or octree automatically generate a FEA mesh. As shown in Fig. 19.3.4a, tree structure solid modelers can be used to generate a one-level difference in the size of adjacent elements. Adaptive codes can readily accept such a mesh since they have the necessary constraint equations to enforce the continuity between the large and small elements. However, the majority of FEA systems will not allow the level-one (size factor of 2) difference in adjacent elements. The original tree

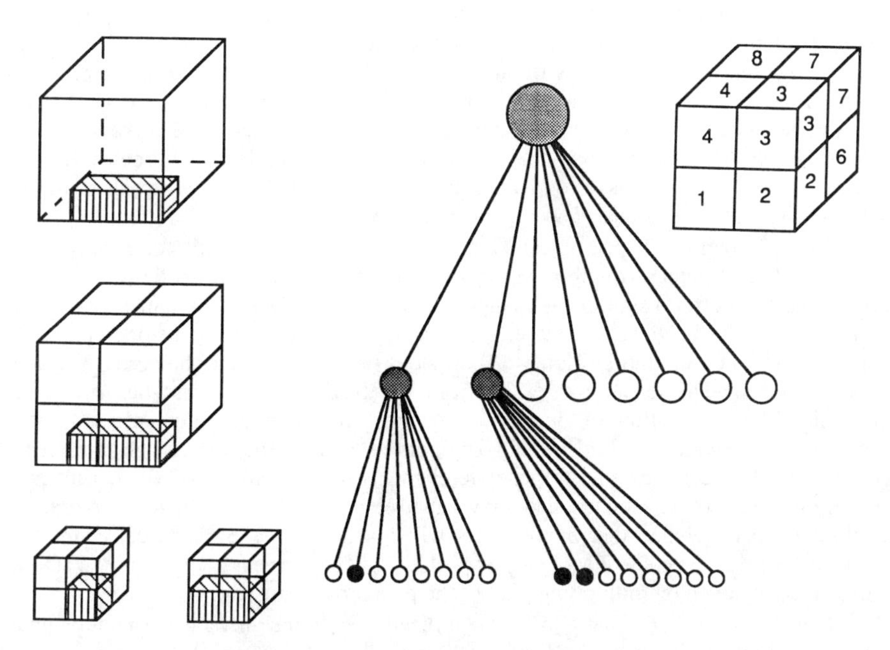

Figure 19.3.3 Octree model of a solid

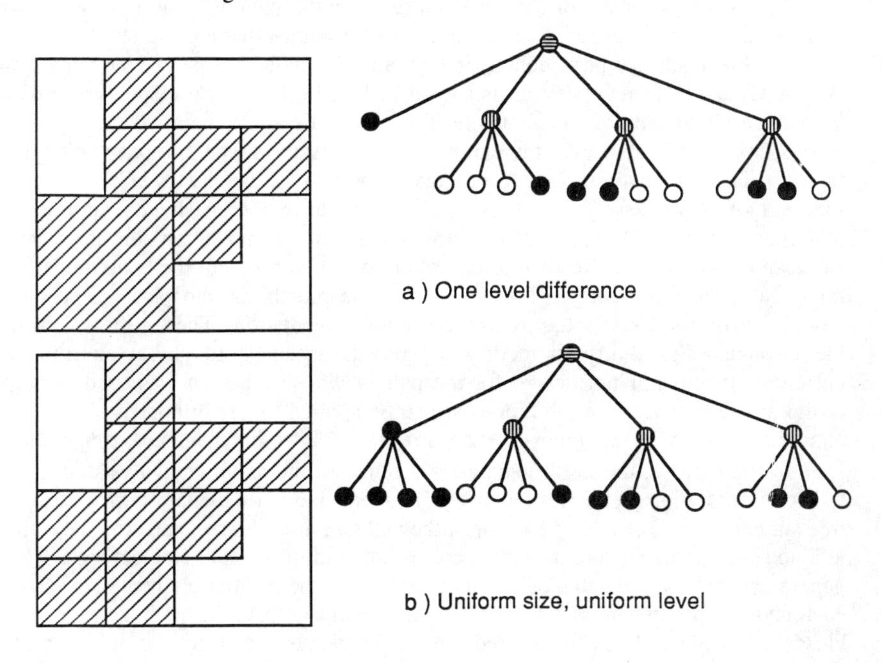

Figure 19.3.4 Converting geometry tree to a mesh tree

structure is easily extended to create meshes made up of uniform-sized elements. First, the level of the smallest uniform mesh size is selected. Then, all completely full leaf nodes above that level are simply divided into equal-sized children until the depth of the desired level is reached. This is shown in Fig. 19.3.4b, where the finest element size is only one level deeper than the tree shown in part a. The full leaf node at the left of part a must be run (at least) one level deeper, and split into four uniform children as shown in part b. If one desired to have four times as many uniform elements, then all of the full bottom level leaf nodes in part b would be split into their four children.

For real irregular shaped objects, the boundary of the object will not exactly match that of the tree structure. In other words, the bottom leaf nodes are partially filled cells. The partial cells offer a significant challenge if we are also generating a mesh. We must decide how to place elements in these cells in a way that avoids badly distorted elements. There are limits on the ratio of long to short sides (aspect ratio) and the corner angles of these elements. This economy results from the fact that the parent has a constant Jacobian. The application of quadtree structures for automatic finite element mesh generation is illustrated in Fig. 19.3.5. The bottom level of the tree can be modified to give nonstandard cells in order to generate elements that exactly lie on the perimeter of the original object. The figure also shows that elements can be split to overcome the level-one size difference that many codes will not accept. The inside cells are often identical for homogeneous linear analysis problems. The element matrices of the smaller cells can be obtained by multiplying that of the parent by a constant.

Reuben, Akin, and Chang [6] have utilized the octree method to model human bones and medical implants. A *Computer Tomography* (CT) scan of the patient is used to define the mass density of the bone and tissue in the volume. These data are provided on a uniform grid in space. There are ranges of densities that are used to define the bone volume. Each cell that has a density in this range is retained as part of the bone structure. A typical result of this operation is shown in Fig. 19.3.6, which displays a portion of a human proximal femur. In this application, the cells are actually rectangular parallel-pipeds, instead of cubes. All the children have the same shape, but have an edge length that is a factor of two smaller. Since the modulus of elasticity of bone is a nonlinear function of bone density, it is also possible to have the mesh generator create the materials property file. A similar octree model of the metal prosthesis is available. Boolean operations are executed to determine which portions of the bone model must be removed (by the surgeon) to insert and glue the prosthesis into place. The combined models are then solved for the stress states after implantation. These results, in turn, give the physician the ability to modify the prosthesis shape to obtain an improved or optimum custom implant design for the patient whose CT scan was used to begin the design process. Clearly, such a process can be applied to any orthopedic joint structure, and it has recently been employed to construct models of the knee joint components.

In some cases, the mesh generators are tied to the adaptive analysis code and can receive information on the desired size of elements based on the latest error estimator. A tree structure mesh generator can check the cell size at each level and stop the refinement at a node on the tree where the cell is consistent with the desired element size. The mesh generators that use Tessellation methods will have the desired element size provided at each node used in the previous solution. As it works in from the boundary, it can find the closest old node and use the desired size of the element to select the points to generate the new element. When the Tessellator is made fast enough, some analysts create a completely new mesh after each of the error estimation procedures has been completed.

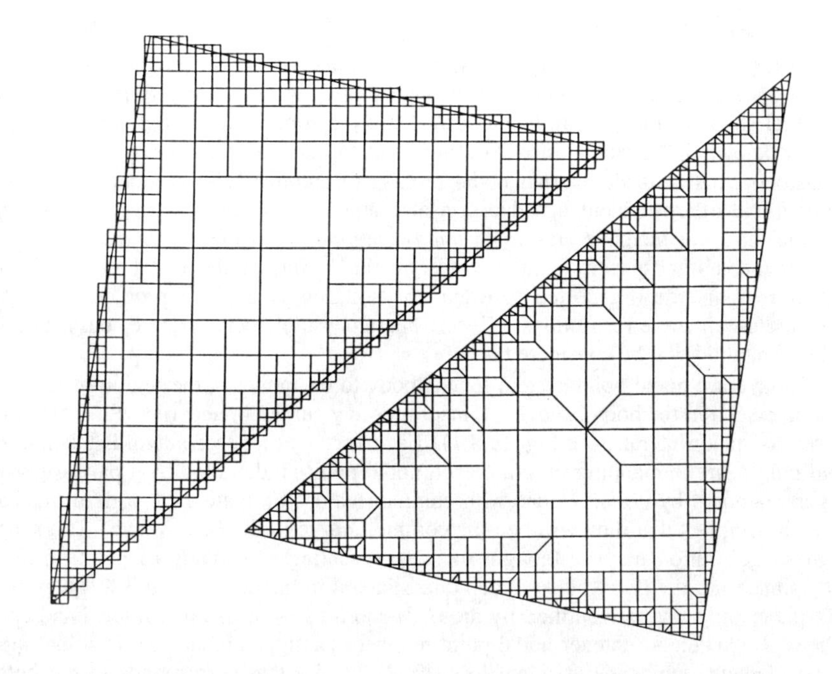

Figure 19.3.5 Examples of standard and modified quadtree meshes

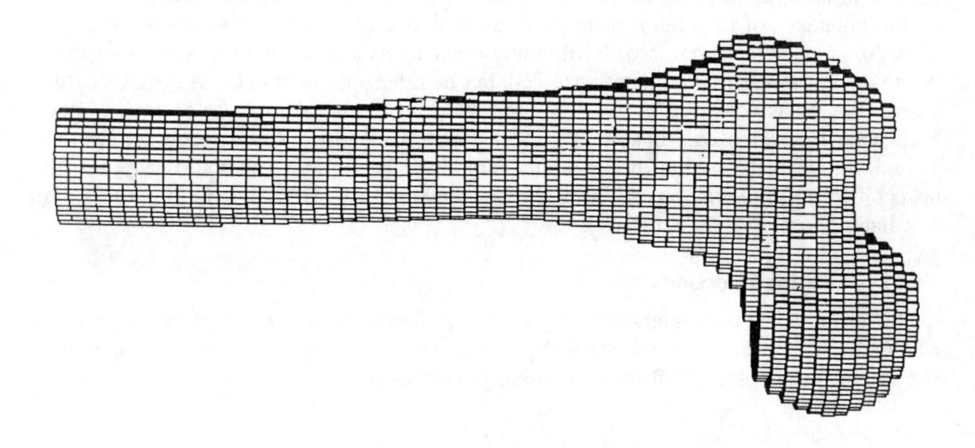

Figure 19.3.6 Octree mesh of the proximal femur region

19.4 Mapping Methods*

Mesh generation by mapping is a technique where the element connectivity (topology) is simplified to a square or triangular grid system which is then mapped into the actual shape of the domain of interest. Only the coordinates of the perimeter points must be known in advance since they are used to execute the mapping process. The spatial coordinates of all internal nodal points are automatically computed. The node numbers and element incidence data are also automatically generated. The two most common mapping methods are *isoparametric mapping* and the *I-J mapping*. Both of these methods impose certain restrictions on the layout of the nodal points, but the reduction in the effort required to effect the solution of a given problem more than compensates for these restrictions. Since isoparametric elements were discussed earlier, the latter method will be considered here.

To layout a nodal point system for the body to be analyzed, the region of the (x, y)-plane intersection the body is covered, insofar as any curved boundaries will permit, with an array of quadrilaterals (see Fig. 19.4.1). Each vertex of a quadrilateral is taken to be a nodal point. In the mapping program each nodal point is identified by a pair of positive integers, denoted by (I, J). Nodes with common values of J are said to lie in the same row. This implies that they will be on a common curve in the (x, y)-plane. The scheme for mesh generation may be thought of as representing a mapping of points from the (I, J)-plane into the (x, y)-plane. Each quadrilateral in the (x, y)-plane is a square in the (I, J)-plane and may be identified by the (I, J) coordinates of its bottom left-hand corner in the (I, J)-plane. An integer nodal point number is assigned to each (I, J) point, and an integer element number is assigned to each (I, J) point that corresponds to the bottom left-hand corner of the squares in the (I, J)-plane. Another typical mapping is illustrated in Fig. 19.4.2. In this figure, the body to be analyzed is shown in part (a), the points in the (I, J)-plane are in (b), and their mapped locations in the (x, y)-plane are shown in (c).

The mesh generation is accomplished in the following manner. Data containing (among other things) the values of I, J and the x- and y-coordinates are input for each node whose coordinates are to be specified. Such nodes must include at least all nodes on the boundary of the region of interest, as well as on any interfaces between regions with different element properties. As many other interior nodal points as the user may desire may have their coordinates specified, but no others are required. As the data cards are read, a list is compiled of the minimum and maximum values of I for each J, and each node for which coordinates have been input is identified and the coordinates are stored. After all the desired nodal points cards have been input, the coordinates for all unspecified nodes which have I in the interval $IMIN \le I \le IMAX$ for the proper J, are calculated for all J $(1 \le J \le JMAX)$.

19.4.1 Mapping Functions

The first method generates a mesh by approximately solving Laplace's equation using the given perimeter (and interior) points as boundary conditions. For example, the x-coordinates are obtained from an approximate solution of

$$\frac{\partial^2 x}{\partial I^2} + \frac{\partial^2 x}{\partial J^2} = 0,$$

where x is specified on the perimeter of the (I, J) domain. First the x-coordinates of the boundary points are used as boundary values, and the values obtained in the interior, $x(I, J)$, are taken as the x-coordinates of the corresponding points in the (x, y)-plane. A

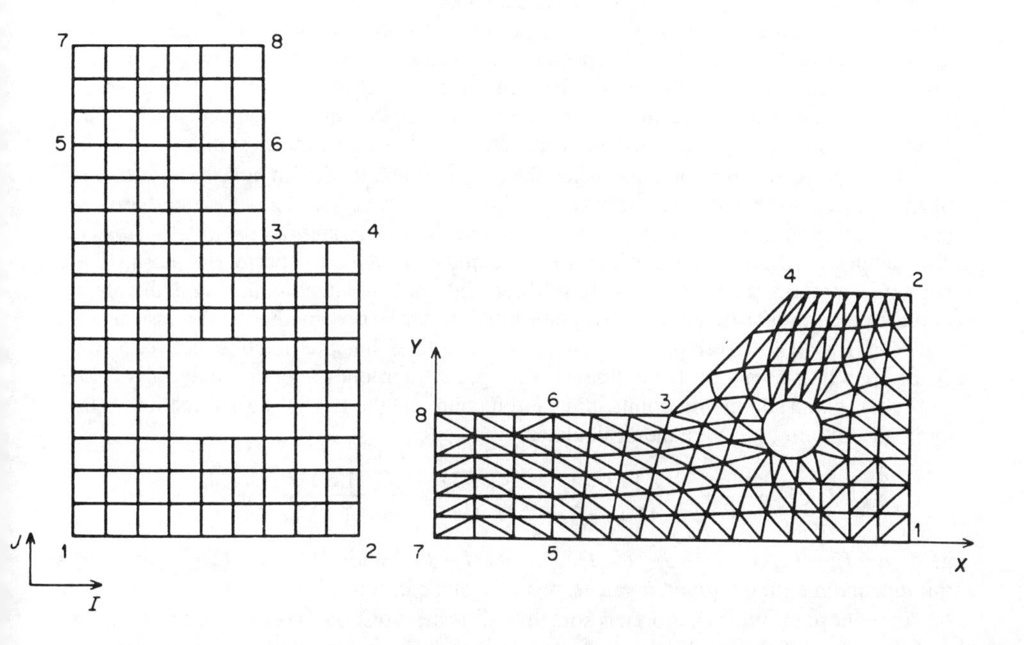

Figure 19.4.1 Mapping from (I,J) to (x,y) space

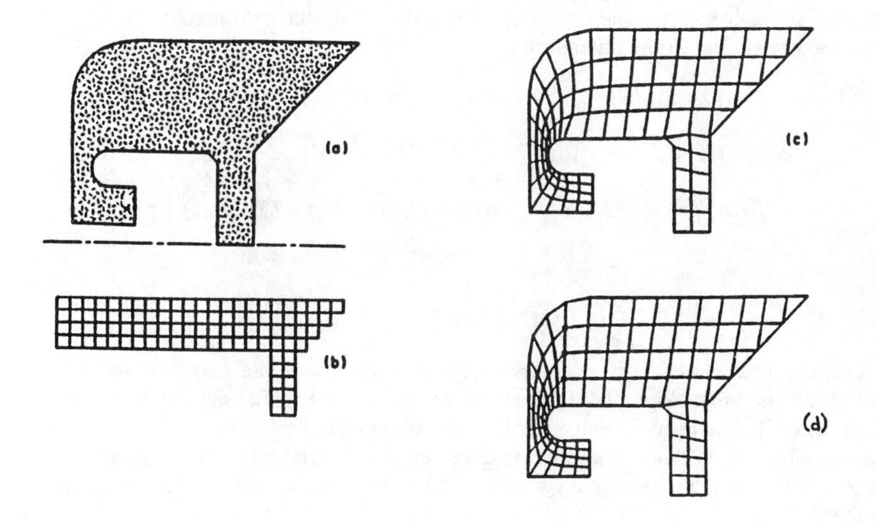

Figure 19.4.2 A typical domain mapping

similar procedure yields the y-coordinates. This method tends to yield nodal points with uniform curvilinear spacing. If this is not deemed desirable, some points interior to the region may have their coordinates specified to control the remaining points. This last point is illustrated by comparing Fig. 19.4.2d with Fig. 19.4.2c. In Fig. 19.4.2d some interior nodes have been specified to make desired changes in the mapped mesh. The interior nodes which have been specified are shown with the blocked-in symbols.

The Laplace mapping method takes the largest amount of computer time, but it is still quite fast and usually the preferred method. However, for regions with sharp re-entrant regions the interior mesh lines may cross the boundary and fall outside the object. If that happens additional interior lines or points must be added to control the mesh. The other three methods to be discussed are direct interpolation techniques, and therefore require little computer time. The two point linear interpolation method is the fastest but it can easily lead to overlapping mesh lines. It utilizes an interpolation between two (given) bounding points on the J line. A four point interpolation option determines the two closest (given) bounding points in the I direction and the two in the J direction. The coordinates of point (I, J) are then calculated by:

$$X(I,J) = \frac{abcd}{ac+bd}\left\{\frac{X(I2,J)}{a(a+c)} + \frac{X(I,J2)}{b(b+d)} + \frac{X(I1,J)}{c(a+c)} + \frac{X(I,J1)}{d(b+d)}\right\}$$

where $a = I2 - I$, $b = J2 - J$, $c = I - I1$, $d = I - J1$, and $I1$ and $I2$ denote the minimum and maximum given I values, respectively, etc. for J.

An eight point interpolation has some minor restrictions on the shape of the mesh in the (I, J)-plane. If the mesh is not a rectangle then it *must* consist of sub-areas, defined by interior points, that are all rectangular. The two closest bounding points in both the I and J directions are determined. These four points are used to define the four *corner* points of the rectangular sub-area. Then the coordinates of point (I, J) are calculated by nonlinear interpolation. The eight node interpolation leads to smooth variations in nodal locations. In addition, it tends to make the calculated lines orthogonal to the four boundary curves. The Coon's blending is:

$$\begin{aligned} X(I,J) &= f_1(u)\,X(I1,J) + f_0(u)\,X(I2,J) + f_1(v)\,X(I,J1) \\ &\quad + f_0(v)\,X(I,J2) - f_1(u)\,f_1(v)\,X(I1,J1) \\ &\quad - f_1(u)\,f_1(v)\,X(I2,J1) - f_1(u)\,f_0(v)\,X(I1,J2) - f_0(u)\,f_0(v)\,X(I2,J2) \end{aligned}$$

where

$$f_0(z) \equiv 10z^3 - 15z^4 + 6z^5, \qquad f_1(z) \equiv 1 - f_0(z), \qquad u = \frac{I - I1}{I2 - I1}, \qquad v = \frac{J - J1}{J2 - J1}.$$

The iterative mapping by Laplace's equation can cause the elements to become distorted, as shown in Fig. 19.4.3c, that is, the corner angles for the element become unacceptable. It has been shown that this can be corrected by solving the *biharmonic equation.* The results from that solution gives elements and mesh lines that are nearly orthogonal, as well as having a smooth change in element sizes. The biharmonic equation is

$$\frac{\partial^4 x}{\partial I^4} + 2\,\frac{\partial^4 x}{\partial I^2\,\partial J^2} + \frac{\partial^4 x}{\partial J^4} = 0$$

with the corresponding essential boundary conditions on x everywhere on the boundary, as well as at selected interior points.

Usually the program calculates the location of each point and assigns a nodal number to each point (except those that are interior to a hole). After this, the program forms quadrilateral or pairs of triangular elements (except those interior to a hole) by assigning the proper nodal point numbers to the corners of the elements. After the elements (and any element codes) have been generated, they are checked for possible errors involving negative areas. The output of the mesh generation program is designed so that it may be fed, as input data, directly to the finite element analysis program MODEL, or similar codes.

All of these mapping methods work well for convex domains. However, for concave domains the nodes may become too crowded, and it is usually best to divide concave domains into nearly convex by specifying various interior lines. It is possible to extend all of the above techniques to surface generation problems that might be required in the analysis of general shell structures. They may also be generalized to three-dimensional solids, but other methods appear to be the simpler for such problems.

19.4.2 Higher Order Elements

For the higher order elements there are additional considerations. One is that for the Serendipity elements, some internal nodes are omitted. The simple procedure is to map all points in the usual manner but omit them in the output state while they are being numbered. The second problem is that of identifying the element incidences. As usual, the elements are identified by the (I, J) values of their bottom left-hand corner. For higher order elements the element spans more than one row and column. It is necessary to store the (I, J) values of all nodes of a typical element type. This is done by building two lists with the numbers to be added to (I, J) to identify each of the nodes.

For example, for the simple four node quadrilateral one could define $IQ = [0, 1, 1, 0]$ and $JQ = [0, 0, 1, 1]$ to identify the four relative positions $(I + IQ, J + JQ)$, i.e., (I, J), $(I + 1, J)$ $(I + 1,\ J + 1)$, and $(I, J + 1)$. Similar arrays exist for each type but they range from zero to LINC, the number of increments between corner nodes. Of course, for the triangles it is necessary to have two sets to allow for the different diagonal directions which can be selected. Once these relative positions are known one simply loops over each I, J pair and extracts the corresponding node numbers. Figure 19.4.3 shows the typical relative node positions and the omitted internal nodes. It illustrates only half of the triangular options. MESH2D can generate all the linear to fourth degree isoparametric triangles, Serendipity quadrilaterals, and Lagrangian quadrilaterals. Each element has an assigned code (material number) for use in the output. Consider an element associated with point (I, J). The element's nodes not lying on $I +$ LINC or $J +$ LINC are assigned the same code as (I, J). It is important in non-homogeneous problems to verify that one has correctly input the various element codes. As an aid the example program prints a sketch of the input element codes. The codes are initialized to unity, and negative values imply elements in a hole to be omitted from both the point and element output.

It is also possible to use a mesh generator to generate constraint or flux data. However, this tends to become rather specialized. It is relatively easy to do since after the mesh generation is complete one has the coordinate arrays, boundary codes, and node numbers for all (I, J) points in the mapped space. Thus it is just a matter of combining these data in a particular form. For example, to generate flux data one could input two

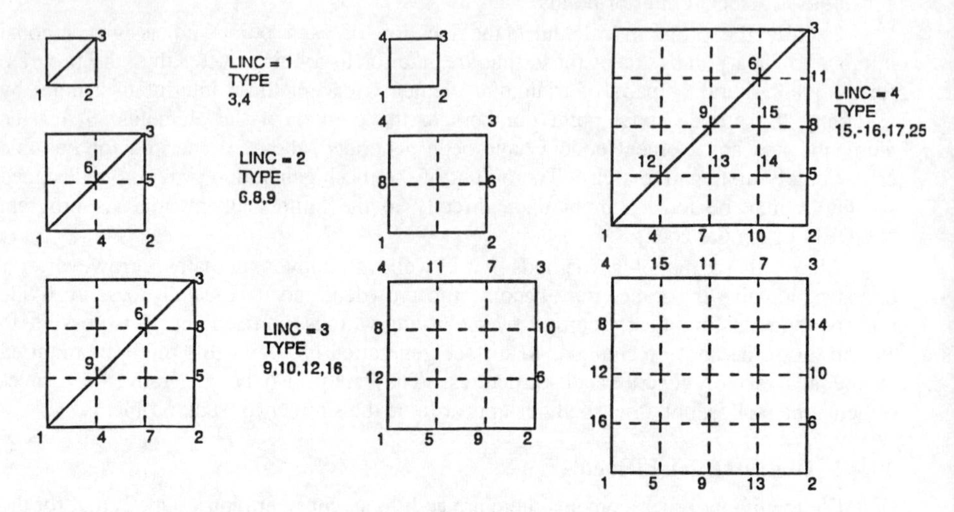

Figure 19.4.3 Node locations relative to element (I,J) point

(I, J) points and their corresponding flux components, and then all nodes along the line could be recovered and the flux obtained by linear interpolation.

19.4.3 MESH2D Program

The previously outlined procedures have been implemented in a program, MESH2D, which is discussed here and is included on the disk. The program allows the user the option of selecting elements that are linear, quadratic, cubic, or quartic. Either quadrilaterals or triangles can be generated. The triangles are obtained by dividing the quadrilaterals across their shortest diagonal. Points being used to define the physical shape can be input in either cartesian or polar coordinates. The minimum data consists of a title, control card, the nodal definition cards, and any material code changes (default value is unity). These data are explained in the table of input formats on the disk. Lines or arcs are generated by linear (or quadratic) interpolation between the given two (or three) values.

The points in the integer space are defined by their (I, J) coordinates, and a generation flag, INTEST. Data can be for a single point, two new points, three new points, two previously given points, or three old points. Holes in the mesh and any changes in element codes are defined by the element code generation cards. These assign element codes for rectangular regions defined by their lower left and upper right I, J coordinates. Negative codes are interpreted to denote holes which are omitted from the point and element output phases. In practical problems one must usually carefully define any interior lines that divide material regions.

The output of the code usually consists of print, punch, and plot files. For punched data the MODEL code formats are default but a provision has been made for a user supplied set of formats. The plotting logic is quite simple. First, any existing J-lines are plotted, then the I-lines are plotted, and if the elements are triangles, one must then insert

```
      SUBROUTINE  MESH2D (MAXR, MAXC, MINNPT, MINEL, NPTCD,
     1                    NMTCD, NBCCD, ECHO, IOPT, LINC,
     2                    LTYPE, MAP, OUTEL, OUTPT, LPLOT,
     3                    INNER, MAT, IMAX, IMIN, LIMIN,
     4                    LIMAX, IGIVEN, NPTNUM, R, Z, CODE,
     5                    ALINE, IJ2JI, IJFLAG)
C     * * * * * * * * * * * * * * * * * * * * * * * * * * * *
C     FEM MESH GENERATION FOR TRIANGLES AND QUADRILATERALS
C     * * * * * * * * * * * * * * * * * * * * * * * * * * * *
      INTEGER  OUTPT, OUTEL, ECHO
      COMMON /IOUNIT/ NCRD, NPRT, NPUN, NPLT
      COMMON / BUGS / MAXBUG
      DIMENSION  ALINE(MAXR), CODE(MAXC,MAXR), R(MAXC,MAXR),
     1           Z(MAXC,MAXR), IGIVEN(MAXC,MAXR), IMIN(MAXR),
     2           IMAX(MAXR), LIMAX(MAXR), LIMIN(MAXR),
     3           MAT(MAXC,MAXR), NPTNUM(MAXC,MAXR),
     4           INNER(MAXC,MAXR)
C     MAXR  - MAX. NUMBER OF INPUT ROWS (J MAX).
C     MAXC  - MAX. NUMBER OF INPUT COLUMNS (I MAX).
C     MINNPT- MINIMUM NODE POINT NUMBER FOR THIS MESH
C     MINEL - MINIMUM ELEMENT NUMBER FOR THIS MESH
C     NPTCD - NUMBER OF NODAL POINT CARD GENERATION SETS.
C     NMTCD - NUMBER OF MATERIAL CODE CHANGE SETS.
C     LTYPE - ELEMENT TYPE. THAT IS, NODES PER ELEMENT:
C               3 - LINEAR TRIANGLE
C               4 - BILINEAR QUADRILATERAL
C               6 - QUADRATIC TRIANGLE
C               8 - SERENDIPITY BIQUADRATIC QUADRILATERAL
C               9 - LAGRANGIAN BIQUADRATIC QUADRILATERIAL
C              10 - FULL CUBIC TRIANGLE ( PAFEC 9T )
C              12 - SERENDIPITY BICUBIC QUADRILATERAL
C              15 - FULL QUARTIC TRIANGLE
C              16 - LAGRANGIAN BICUBIC QUADRILATERAL
C             -16 - SERENDIPITY BIQUARTIC QUADRILATERAL
C              17 - SERENDIPITY BIQUARTIC QUADRILATERAL W CENTER
C              25 - LAGRANGIAN BIQUARTIC QUADRILATERAL
C     LPLOT - ELEMENT PLOTTING OPTION TO FILE POLTTER.
C               0 = NO PLOTS
C               1 = PLOT PHYSICAL MESH
C               2 = PLOT PARENT MESH
C               3 = PLOT BOTH MESHES
C     MAP   -  MESH MAPPING OPTION.
C               0 = ITERATIVE LAPLACE SMOOTHING (DEFAULT).
C               1 = LINEAR INTERPOLATION ON J = CONSTANT.
C               2 = FOUR POINT BILINEAR INTERPOLATION.
C               3 = EIGHT POINT BLENDING (COONS) INTERPOLATION.
C     OUTP  -   NODAL OUTPUT OPTIONS.
C               0 = LIST ONLY
C               1 = LIST AND PUNCH FOR IOPT PROGRAM
C               2 = PUNCH FOR IOPT PROGRAM ONLY
C               3 = NO OUTPUT
C              THIS ALSO APPLIES TO DEFAULT ESSENTIAL BC VALUES.
C     OUTL  -   ELEMENT OUTPUT OPTION, SAME AS OUTP.
C     IOPT  -   FORMAT OPTION
C               0 = MODEL CODE
C               1 = FINITE CODE
C               2 = PAFEC CODE
C               3 = FLOWNS NAVIER-STOKES
C     ECHO  -   ECHO INPUT DATA IF 0.
```

Figure 19.4.4a   Mesh generator control

```
C     NBCCD -  NUMBER OF ESSENTIAL BC VALUE GENERATION SETS.
C     ALINE -  A LINE FOR PRINT-PLOT
C     CODE  -  BOUNDARY CONDITION CODE FOR EACH I, J NODE
C     R     -  X COORDINATE ARRAY
C     Z     -  Y COORDINATE ARRAY
C     IGIVEN-  FLAGS ALL INPUT NODES ON THE GRID
C     INNER -  ARRAY DEFINEING INTERIOR NODES AND HOLES
C     MAT   -  MATERIAL CODES AT I, J
C     IMAX  -  MAX. I FOR EACH J NODE
C     IMIN  -  MIN. I FOR EACH J NODE
C     LIMAX -  MAX. I FOR EACH J ELEMENT
C     LIMIN -  MIN. I FOR EACH J ELEMENT
C     NPTNUM-  GENERATED NODE POINT NUMBERS ARRAY
C             *** INITIALIZE ARRAYS ***
      CALL  INITAL  (MAXR, MAXC, MAT, IMIN, LIMIN)
C
C-->          *** READ INPUT POINTS ***
C
      CALL  INPTS (MAXC, MAXR, R, Z, CODE, INNER, IMIN, IMAX,
     1            LIMIN, LIMAX, ECHO, NPTCD, MAP, IJ2JI)
      WRITE (NPRT, *) ' END OF GEOMETRY INPUT'
C             *** PRINT MAP OF INPUT POINTS ***
      CALL IJMESH (MAXC, MAXR, ALINE, INNER, IMAX, IMIN, 1)
C             *** PRINT MAP OF INPUT BC FLAGS ***
      CALL IJMESH (MAXC, MAXR, ALINE, CODE, IMAX, IMIN, 2)
C
C-->        *** READ CHANGES IN ELEMENT CODES ***
C
      IF ( NMTCD .GT. 0 )  THEN
        CALL INEL (MAXC, MAXR, MAT, LIMIN, LIMAX, ECHO, NMTCD,
     1            IJ2JI, IJFLAG, LINC, IMIN, IMAX)
        WRITE (NPRT, *) ' END OF MATERIAL CHANGES'
      ENDIF
C
C-->              *** MAP POINTS ***
C
      IF ( LPLOT .EQ. 2 .AND. OUTPT .EQ. 3 )  GO TO 15
      IF ( MAP .EQ. 0 )  CALL LAPLAC (MAXC, MAXR, R, Z, INNER,
     1                              IMIN, IMAX, IGIVEN, N1)
      IF ( MAP .EQ. 1 )  CALL TWOPT  (MAXC, MAXR, R, Z, INNER,
     1                              IMIN, IMAX, N1)
      IF ( MAP .EQ. 2 )  CALL FOURPT (MAXC, MAXR, R, Z, INNER,
     1                              IMIN, IMAX, IGIVEN, N1)
      IF ( MAP .EQ. 3 )  CALL EIGHTP (MAXC, MAXR, R, Z, INNER,
     1                              IMIN, IMAX, IGIVEN, N1)
      IF ( OUTPT .LT. 2 )  WRITE (NPRT, *)
     1    'COORDINATE ITERATIONS USED =', N1
C
C-->               *** NUMBER POINTS ***
C
   15 CONTINUE
      IF ( LTYPE .EQ. 8 .OR. LTYPE .EQ. 12 .OR.
     1    LTYPE .EQ. 17 )  CALL OMITP (MAXC, MAXR, MAT, LTYPE,
     2                                LINC, IJFLAG, R, Z)
      CALL  PCOUNT (MAXC, MAXR, R, Z, CODE, MAT, NPTNUM, IMIN,
     1             IMAX, IJFLAG, IOPT, OUTPT, MINNPT, LINC)
```

Figure 19.4.4b Mesh generator control

```
C                   DEBUG LISTS
      IF ( MAXBUG .NE. 0 )  THEN
        WRITE (NPRT, *) ' Material Numbers'
        CALL IPRINT (MAT, MAXC, MAXR)
        WRITE (NPRT, *) ' Nodal Numbers'
        CALL IPRINT (NPTNUM, MAXC, MAXR)
      ENDIF
C              *** PRINT MAP OF ACTUAL NODES ***
      CALL IJNODE (MAXC, MAXR, ALINE, NPTNUM, IMAX, IMIN)
      CALL IJMESH (MAXC, MAXR, ALINE, NPTNUM, IMAX, IMIN, 3)
C
C-->                  *** GENERATE ELEMENTS ***
C              (AND OVERWRITE ARRAY IGIVEN WITH LEMNUM)
      IF ( OUTEL .LT. 4 )  CALL  LCOUNT (MAXC, MAXR, R, Z, LIMAX,
     1                       LIMIN, MAT, NPTNUM, LTYPE, LINC, MINEL,
     2                       OUTEL, IOPT, IGIVEN)
C          *** PRINT MAP OF ACTUAL ELEMENT NUMBERS***
      CALL IJELEM (MAXC, MAXR, ALINE, IGIVEN, LIMAX, LIMIN)
C            *** PRINT MAP OF MATERIAL CODES ***
      CALL IJMAT (MAXC, MAXR, ALINE, IGIVEN, LIMAX, LIMIN, MAT, LINC)
C
C-->               *** GENERATE BOUNDARY CONDITIONS ***
C
C              *** PUNCH DEFAULT ESSENTIAL BC FLAGS ***
      IF (  OUTPT .EQ. 1 .OR. OUTPT .EQ. 2 )
     1      CALL  PBCOUT (MAXC, MAXR, NPTNUM, CODE, IOPT)
C-->         *** PUNCH GIVEN ESSENTIAL BC VALUES ***
C               ( AND OVERWRITE INNER )
      IF (  NBCCD.GT.0 .AND. OUTPT .EQ. 1 .OR. OUTPT .EQ. 2 )
     1      CALL  VALUE  (MAXC, MAXR, ALINE, INNER, IMIN, IMAX,
     2                      NPTNUM, ECHO, NBCCD, IJ2JI)
C
C-->                     *** PLOT RESULTS ***
C
      IF ( LPLOT .EQ. 0 )  GO TO 20
C        INITIALIZE PLOT FILE AND SET FIRST ORIGIN
      CALL  STARTP
C        PLOT ACTUAL R-Z MESH
      IF ( LPLOT .EQ. 1 .OR. LPLOT .EQ. 3 )
     1     CALL  PLOTIJ (MAXC, MAXR, R, Z, NPTNUM, MAT, IMAX, IMIN,
     2                   LIMIN, LIMAX, LINC, LTYPE, LPLOT)
      IF ( LPLOT .LT. 2 )  GO TO 20
C        ADVANCE PLOTTER FOR NEXT FRAME
      IF (  LPLOT .EQ. 3 )  CALL ADVANC
C        REPLACE R, Z BY I, J VALUES AND PLOT
      CALL  IJTORZ (MAXC, MAXR, R, Z)
      CALL  PLOTIJ (MAXC, MAXR, R, Z, NPTNUM, MAT, IMAX,
     1              IMIN, LIMIN, LIMAX, LINC, LTYPE, LPLOT)
   20 CONTINUE
C        CLOSE PLOT FILE
      IF (  LPLOT .GT. 0 )  CALL  FINSHP
      WRITE (NPR, *) ' NORMAL ENDING OF MESH/2D'
      RETURN
      END
```

Figure 19.4.4c Mesh generator control

the required diagonals on each J-row of elements. The print file echoes the given data and lists data generated by the point and element definition cards. As a debugging aid it also prints a list of given integer points and the element codes. The controlling program is given in Fig. 19.4.7. The full source is available on the disk for this book. After the coordinates have been found, subroutine OMITP identifies any points that will be omitted from the higher order elements. Then the nodal points are numbered and output by PCOUNT. The element incidences, material number, etc., are established by subroutine LCOUNT. It calls IJQUAD or IJTRI to obtain the list of relative (I, J) coordinates of the element nodes. Subroutine LCOUNT also outputs the element data. Now that the points and elements are completely defined in the (I, J) data base they can be plotted using PLOTIJ. Hardware-dependent plotter calls have been hidden in the calls to dummy routines. These include WINDO, MOVETO, LINETO, STARTP, and FINSHP.

Useful print-plots, relative to the I–J space, show the input points, points with boundary condition flags, material numbers, the node numbers, element numbers, generated nodes for the element type, and points where boundary conditions were generated. These are useful in debugging meshes for higher order elements, or regions with multiple materials. For CFD problems the generation of constant, linear, or quadratic boundary values is very helpful since the entire boundary must be assigned constraints for internal flows.

19.5 References

[1] Akin, J.E., *Computer Assisted Mechanical Design*, Englewood Cliffs: Prentice-Hall (1990).

[2] Armstrong, C.G., "Special Issue: Automatic Mesh Generation," *Advances in Eng. Software*, **13**(516), pp. 217–337 (1991).

[3] Cavendish, J.C., Field, D.A., and Frey, W.B., "An Approach to Automatic Three Dimensional Finite Element Mesh Generation," *Int. J. Num. Meth. Eng.*, **21**, pp. 329–347 (1985).

[4] George, P.L., *Automatic Generation of Meshes*, New York: John Wiley (1991).

[5] Hall, M. and Warren, J., "Adaptive Tessellation of Implicitly Defined Surfaces," Technical Report COMP TR88-84, Department of Computer Science, Rice University (1988).

[6] Reuben, J.D., Akin, J.E., and Chang, C-H., "A Computer-aided Design and Manufacturing System for Total Hip Replacement," Seminars in Orthpaedics (Mar. 1992).

[7] Stillman, D.W. and Hallquist, J.O., "INGRID: A Three-Dimensional Mesh Generator for Modeling Non-Linear Systems," Lawrence Livermore National Laboratory, Report UCID-20506 (July 1985).

[8] Yerry, M.A. and Shephard, M.S., "A Modified Quadtree Approach to Finite Element Mesh Generation," *IEEE Trans. Computer Graphics and Applications*, **3**(1), pp. 39–46 (Jan. 1983).

[9] Yerry, M.A. and Shephard, M.S., "Automatic Three Dimensional Mesh Generation by the Modified Octree Technique," *Int. J. for Num. Mech. Eng.*, **20**, pp. 1965–1990 (1984).

Chapter 20

COMPUTATIONAL PROCEDURES

20.1 Introduction

The practical application of the finite element method requires the use of a computer and a sizable amount of programming effort. If a potential user of the finite element method were forced to program the entire code from scratch, he would be faced with a long, difficult, and often impractical task. It is therefore essential that the programming effort and knowledge of previous workers be utilized as much as possible. A number of large finite element computer codes have been developed. Several of the early codes were developed at a great cost in both money and labor, yet they were highly specialized, machine dependent, and difficult to modify; thus, they rapidly became obsolete.

The most desirable types of general purpose finite element codes are those that are designed for comprehension, modification, and updating by the user. It is also desirable for the user to be familiar with the internal functions of the computer code if full advantage is to be taken of its basic potential and if inappropriate applications are to be avoided. These desirable objectives are most easily met if the program consists of several *building blocks* or *modules*. Generally, any specific function performed with some degree of repetition (or having a particular educational value) is separated into a distinct subroutine (or module) with a conscious effort on the part of the original programmer to make obvious the subroutine's function. By applying this concept one can establish a program *library* that contains many such building blocks. Then, for a given application, it is possible for the user to assemble the appropriate subroutines from the library plus any that he has written to meet his special needs. Of course, such a system of finite element codes generally requires a simple main program whose function is to call the various required subroutines in the appropriate order.

This modular approach to programming a finite element code has a number of important advantages. From the educational point of view, it is useful in that the student of the finite element method has only to master one elementary concept at a time. Therefore, it is easier to master the general programming concepts associated with the finite element method. The modular approach also makes it easier to apply the finite element technique to various fields of analysis such as heat transfer, fluid flow, and solution of differential equations. It also reduces to a minimum the amount of computer codes that the user must write when undertaking an analysis in a new field of application. Of still greater importance, however, is the fact that the modularity of such a code allows someone other than the original author to keep a program from becoming obsolete by incorporating new types of elements, solution techniques, input/output routines and other facilities as they become available. The various building block programs to be presented herein were developed by the author with these objectives in mind. These programs have been kept as short and independent as is practical in order to make their application as general as possible. The author has also attempted to reduce the solution time and storage requirements.

A typical finite element program will require the storage of a large number of matrices. These matrices include the spatial coordinates of the nodes, the element incidences, the element matrices, the system equations, etc. In general, the size of a typical matrix can be defined by the control parameters (number of nodes, number of elements, etc.) and/or the input data (element nodal incidences, etc.). This fact allows the programmer to utilize efficiently the available in-core storage. Unfortunately, most programs do not take advantage of this option. Instead they specify, once and for all, the maximum size of each array, that is, the program sets limits on the maximum number of nodal points, elements, boundary conditions, etc. Usually, a typical problem will at most reach one or two of these limits so that the sizes of the other matrices are larger than required and storage is wasted. Of course, the size of the limits can be changed, but this usually involves changing several dimension statements in the program and its various subroutines. A procedure is described which allows for the semi-dynamic allocation of storage for these matrices. The first step is to dimension a single vector of a size large enough to contain all the matrices needed in an average problem. Next, the size of each matrix is calculated, for the problem of interest, by using the control parameters and the input data. Then, as illustrated in Fig. 20.1.1, a set of *pointers* is defined to locate the

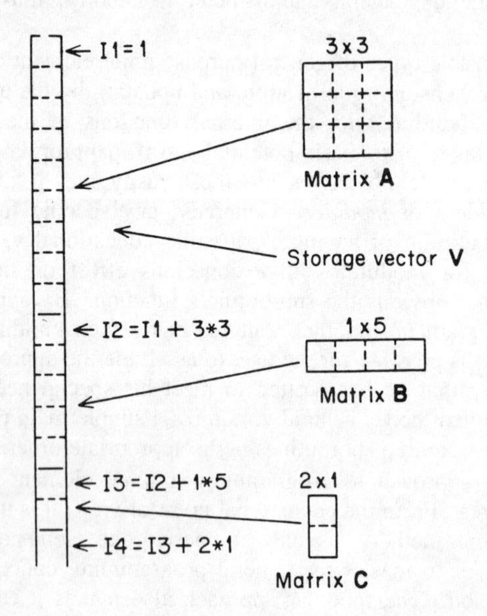

Figure 20.1.1 An efficient matrix storage technique

position, in the storage vector, of the first coefficient of each matrix of interest. By using this approach, the user has only one limit about which to worry. That limiting number is the dimension of the single storage vector and that limit in turn is usually governed by the size of the computer system to which the user has access. Only one dimension card has to be changed to modify the in-core storage capacity of the present building block program. The disadvantage is that one must do the extra "bookkeeping" necessary to define the pointers to locate the matrices.

In some cases, an additional storage saving is possible through overwriting. Referring to Fig. 20.1.1 for three representative matrices, $\mathbf{A}\,(3 \times 3)$, $\mathbf{B}\,(1 \times 5)$, $\mathbf{C}\,(2 \times 1)$, assume that the matrix **B** is generated, stored and completely utilized before matrix **C** is generated and stored. Thus, to save storage space one could define the pointer for matrix **C** as $I3 = I2$. This would result in the coefficients of matrix **C**, and the following matrices, being stored in the now unneeded space that matrix **B** occupied. It is possible to write a FORTRAN code so as to obtain true dynamic storage allocation abilities (see [4]). However, these extra details will not be considered here. The controlling main program and the data input routines are very important to the user. They significantly affect the ease of program use as well as its range of analysis capabilities. However, they vary greatly from one program to the next. For example, large commercial packages tend to have special control macros, free-format input, and extensive graphics to simplify the ease of use. Other small systems may use special macros to simplify the order of data input. An example of this type of code is that presented by Taylor [15] in the appendix of the text by Zienkiewicz. Other programs often use the more classical type of fixed format data input. That is the procedure utilized here in the MODEL program.

The major point of the discussions of the routines in MODEL is to illustrate the typical calculations that go on within a finite element analysis system. The details of interfacing with the user are given lesser importance here. However, when considering a large commercial code a user should probably reverse the order of the above priorities. Figure 20.1.2 presents a flow chart that illustrates the major steps whichf all under the control of the main program, MODEL. It is important that one understand the relationships between these operations. The first half of the steps, down to point 1, deal with initializing the system, determining the input options to be utilized, reading and printing the input data, and assigning storage locations that will hold the information to be generated in later phases. For future reference, Fig. 20.1.3 lists the names of the major subroutines that control the corresponding segments listed in Fig. 20.1.2.

The steps from points 1 to 2, in Fig. 20.1.2, deal with the generation of element matrices, setting up data to be used for later secondary calculations, assemblying the system matrices, and applying essential boundary conditions before solving the equations. Between points 2 and 3 the system equations are solved (factorized) to yield the unknown nodal parameters. If additional calculations are desired, then the required auxiliary data are recovered from storage, and the post-solution calculations are executed. A problem with a material or geometric nonlinearity would have an additional loop such as that shown between points 1 and 3 in Fig. 20.1.2. For example, if a temperature calculation involves temperature, dependent thermal conductivities, then it is necessary to estimate initially the conductivities in each element. After the nodal temperatures have been computed, a more accurate value of the conductivity is obtained and compared with the previous estimate. If all element conductivities are reasonably close to the estimate, then the calculated temperatures are acceptable, otherwise a new estimate must be made and the solution repeated. A linear time dependent problem would involve a loop through points 2 and 3 at each time step. Other combinations are possible such as a nonlinear dynamic analysis that requires material property iterations at the end of each time step. Thus, the main controlling program can have complicated logic if it is designed to be very general.

The major subroutines ASYMBL, APLYBC, FACTOR, SOLVE, and POST that are cited in Fig. 20.1.3 will be considered in more detail elsewhere. Their names and comparison of Figs. 20.1.2 and 20.1.3 should suggest their major functions. The assembly segment control program, ASYMBL, uses several subroutines mentioned in Fig. 20.1.4. It is included here to outline the major operations of ASYMBL and to

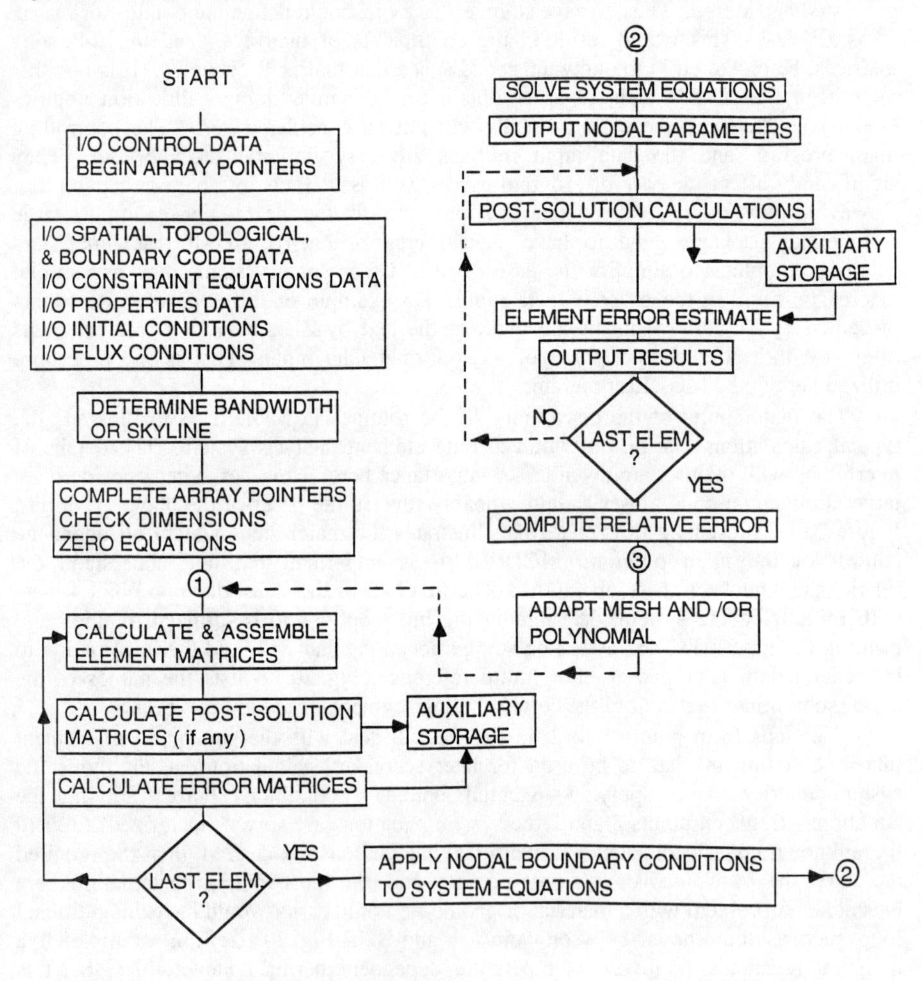

Figure 20.1.2 Typical analysis sequence flowchart

identify the relative functional positions of some of the routines presented in this text. The following sections will discuss typical procedures utilized in the generation of element matrices. Figure 20.1.5 illustrates the flow of operations in the element generation control segment, GENELM, and identifies some of the subroutines it utilizes. Since the generation of the popular isoparametric elements requires special programming considerations, they are given detailed attention in Sec. 9.4. The major steps that occur in a program, say ISOPAR, that controls the generation of isoparametric elements is shown in Figs. 20.1.6 and 9.4.3. It also identifies typical routines to be presented later. From Fig. 20.1.6 it is noted that isoparametric element generation usually requires the knowledge of element interpolation, or shape, equations and numerical integration.

A flow chart for the operations in a controlling routine, say APLYBC in Fig. 20.1.3, is given in Fig. 20.1.7. The type of constraint is defined by the number of dof occurring in the linear constraint equation given in the input data. Later sections will give specific details of the equation solution process and the output of the calculated nodal dof, that is the operations of FACTOR, SOLVE, WRTPT, and WRTELM cited in Fig. 20.1.3. They will also give the typical procedures for executing any post-solution calculations defined at the element level and stored at assembly time. These are listed in Fig. 20.1.8.

Very detailed examples of common applications of finite element analysis methods are given in Chap.s 21 and 22. These include the element matrices, typical post-solution quantities of practical importance, and specific input and output data. To aid in generating larger and more practical meshes, Chap. 19 was included to present an algorithm for two-dimensional mesh generation. Chapter 17 was included to demonstrate the implementation aspects of selected time dependent finite element solutions. However, since there are at least 30 time integration algorithms those presented may not be typical of the procedures employed in the more advanced applications.

As the reader proceeds through the following discussions, it will probably be useful to refer back to flow charts given previously. If the reader is interested in searching in the text or disk for specific programming details, then the Index of Subroutines along with these flow charts and the summary of subroutine purposes should allow the text to also serve as a quick reference book.

The basic data dealing with the nodal points and elements are read by subroutine INPUT. This routine reads the node number, boundary code indicator and the spatial coordinates for every node in the system. Next it reads the element number and the element incidences (node numbers of all nodal points associated with the element) for every element in the system. These data are also printed for the purpose of checking the input data. Given the number of nodes, number of elements, the number of nodes per element, number of parameters per node, and the dimensions of the space, this subroutine sets aside storage for the spatial coordinates, element incidences, and the nodal point boundary code indicator. This routine and its input instructions are included on the disk.

The nodal point boundary code indicator serves the important task of indicating the number and type of nodal parameter boundary conditions occurring in the system. Several pieces of data are included in this code. Subroutine PTCODE extracts these data. This program takes the integer boundary code indicator (consisting of NG digits, right justified) and extracts NG single digit integer codes. These single digit codes range from zero to nine and they identify the type of constraint equation (defined by the analyst) that the parameter must satisfy. A zero code implies no constraint. This allows the analyst to write programs to handle up to nine types of boundary constraint conditions. The present

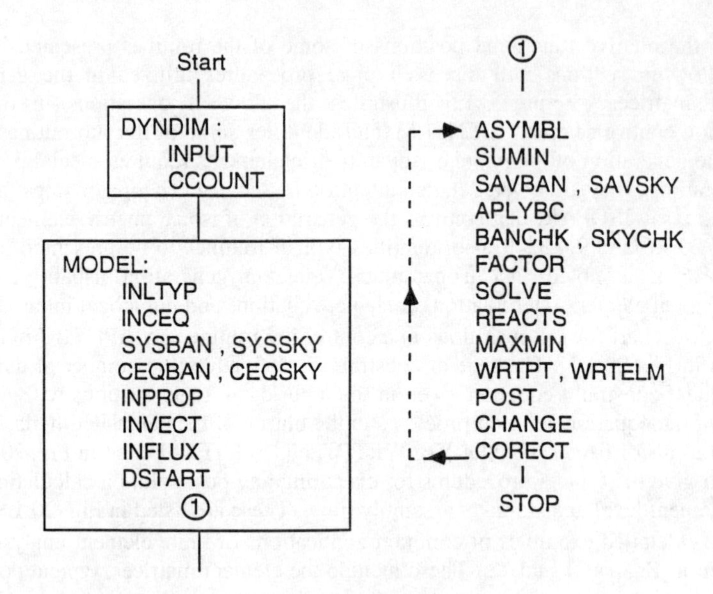

Figure 20.1.3 Subroutines in the major segments

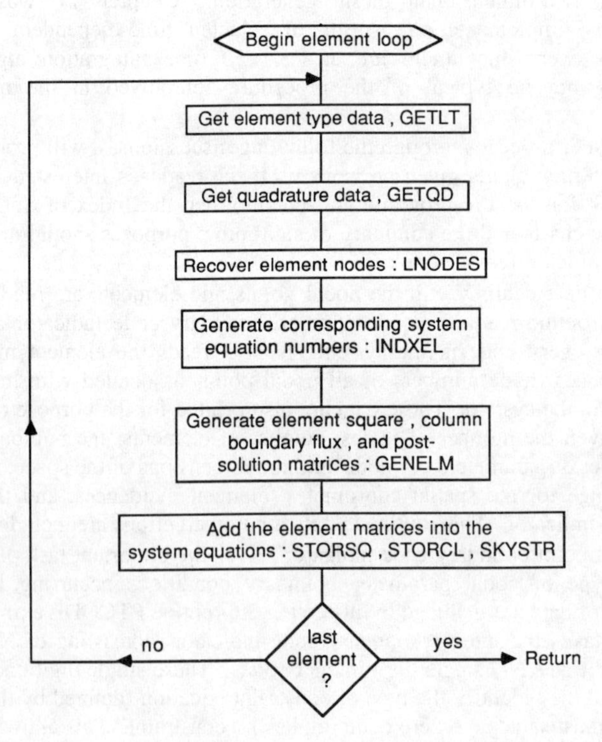

Figure 20.1.4 Assembly flowchart

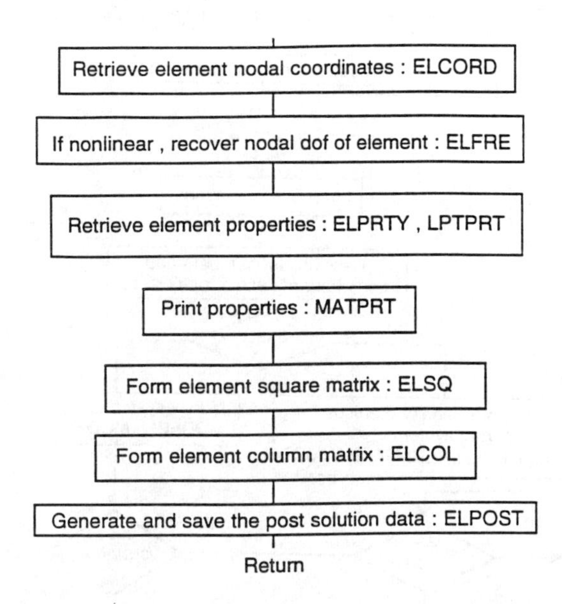

Figure 20.1.5 Element matrix generation flowchart

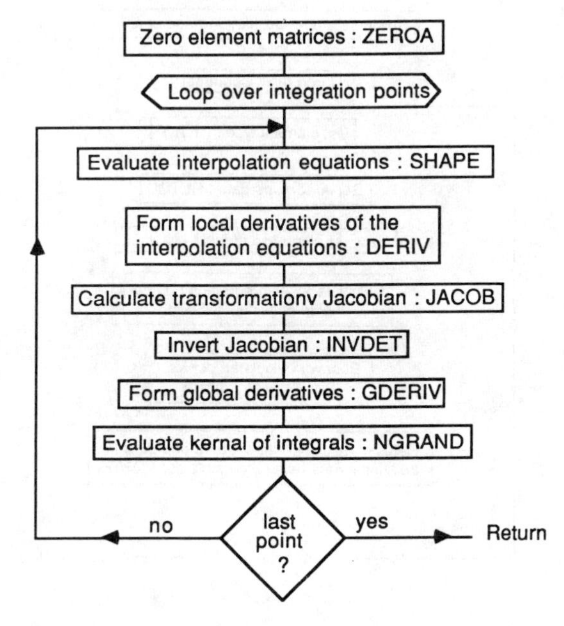

Figure 20.1.6 Programming isoparametric elements

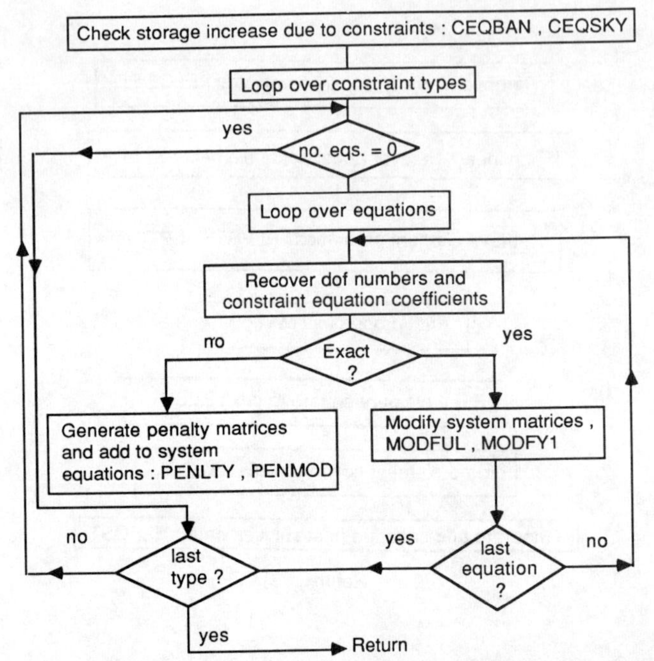

Figure 20.1.7 Applying essential boundary conditions and constraints

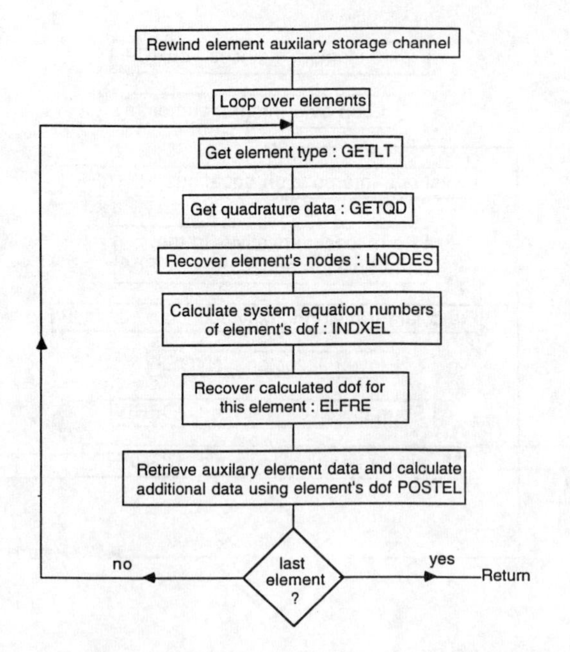

Figure 20.1.8 Post-solution calculations

set of building block programs use only the single digit values of 0, 1, 2, and 3. After subroutine PTCODE has extracted the NG single digit integer codes associated with the NG parameters at a node, these codes must be scanned to count the number of different types of nodal parameter boundary conditions encountered in the system. This is done by subroutine CCOUNT which utilizes the nodal point indicators, KODES, to count the number of non-zero constraints of each type that occur in the system. Subroutine CCOUNT calls subroutine PTCODE, and both are given in the text disk. For the example shown in Fig. 20.1.9, the constraint indicator counts are nineteen zeros, nine ones and two twos. These data imply nine Type 1 nodal parameter constraints and one Type 2 constraint.

Once the number and type of nodal parameter boundary conditions have been determined, it is necessary to supply the constraint equation data. These nodal parameter constraint data are read by subroutine INCEQ (see disk). The nodal parameter constraint equations will be discussed in detail in Sec. 8.1. At this point it is only necessary to note that the present program allows the following three types of linear equations:

$$\text{TYPE 1} \quad D_j = A\,, \quad \text{TYPE 2} \quad AD_j + BD_k = C\,, \qquad (20.1)$$

$$\text{TYPE 3} \quad AD_j + BD_k + CD_l = E$$

where A, B, C, and E are constants and j, k, and l denote system degree of freedom numbers associated with the nodal parameters (D). The program can easily be expanded to allow for nine types of linear constraint equations. Regardless of which type of constraint equation one has, it is still necessary for the analyst to input the constant(s) and identify the degree of freedom number(s). The degree of freedom numbers are identified indirectly by specifying the node where the parameter is located and its local parameter number. The coefficients in the equation are specified directly. Subroutine INCEQ reads two sets of data for each constraint equation of each type. These data include identification (node number, corresponding coefficients of the constraint equation). The input formats for this routine are shown at the beginning of Appendix II.

As outlined in the above section, the present program specifies three allowable types of linear constraint equations. It can easily be expanded to allow for nine (or more) types of linear constraint equations. For any given type of linear constraint equation it is necessary to define an integer array, NDX, and a floating point array, CEQ. Both of these arrays will have the same number of rows. The number of rows is, of course, equal to the maximum number of degrees of freedom in the constraints. Each row of the integer array contains the system degree of freedom numbers of the nodal parameters

		Boundary code for nodal parameter number, J					
Node	IBC	J = 1	2	3	4	5	
1	111111	1	1	1	1	1	1
2	0	0	0	0	0	0	0
3	220000	2	2	0	0	0	0
4	100	0	0	0	1	0	0
5	11000	0	1	1	0	0	0

Figure 20.1.9 Interpretation of essential boundary condition code

occurring in a particular constraint equation. The floating point array contains the corresponding set of coefficients occurring in the constraint equation. The number of columns equals the total number of nodal constraint equations. The degree of freedom numbers will be calculated from the input data by using

$$NDF = NG*(I-1)+J \qquad 1 \le I \le M, \qquad 1 \le J \le NG, \tag{20.2}$$

and the input coefficients will be modified so that the coefficient associated with the first parameter in the equation is defined to be unity. For example, a TYPE 2 constraint equation $AD_j + BD_k = C$ becomes $D_j + aD_k = b$, where $a = B/A$ and $b = C/A$. Thus, it will be necessary to store only two integers (*j* and *k*) and two floating point numbers (*a* and *b*) for each equation of this type. In general, the number of rows in these two arrays equals the constraint type number.

Subroutine INPROP inputs the various properties necessary to define the problem. If there are nodal point properties, then the node number and corresponding properties are read for each nodal point. Similarly, if element properties are to be input, the element number and corresponding properties are read for each element. Finally, any miscellaneous system properties are also read. The user is allowed the option of specifying the nodal point and element properties to be homogeneous. This greatly reduces the required amount of data in several problems. Later applications will illustrate the use of these properties.

In some applications, the initial forcing vector (system column matrix) terms will be known to be non-zero. In such cases, the non-zero initial contributions must be input by the analyst. Subroutine INVECT reads these contributions and then prints the entire initial load vector. The subroutine is also included on the disk, as are the corresponding input instructions. In subroutine INVECT each specified coefficient has an identifying parameter number and system node number associated with it. It is necessary to convert the latter two quantities to the corresponding system degree of freedom number. This calculation is defined by Eq. (20.2), which is included in subroutine DEGPAR for possible uses elsewhere.

There are numerous applications with boundary conditions which do not directly involve the nodal parameters. Usually these conditions involve a flux across, or traction on, a specified segment of an element boundary. For example, in a heat transfer problem one may specify the components of the heat flux across a portion of the boundary, while in stress analysis one may have an applied pressure acting on a portion of the structure's surface. Generally, these types of boundary conditions lead to contributions to the system matrices. The calculations of these contributions are closely related to the calculation of the element's square matrix and column matrix. The latter two matrices were discussed in the early chapters and will be considered in detail in Chap.s 21 and 22. The addition of the flux contributions to the system matrices is closely related to the procedures to be given in Sec. 22.2.

Subroutine INFLUX reads the flux type boundary condition data for each boundary segment. The actual calculation of the flux contributions is carried out by a problem dependent subroutine called BFLUX, which is discussed later. If the specified segment has LBN nodes associated with it and NG parameters per node, then subroutine BFLUX would utilize the NG components of FLUX to calculate the contributions of the flux to the NFLUX = LBN*NG system degrees of freedom on that segment. Since the actual calculations in BFLUX would vary from one application to another, the analyst could choose to interpret the specified flux components to be normal and tangential to the segment instead of being components in each of the spatial directions.

20.2 Element Calculations

After the problem control parameters and properties have been input, some useful preliminary calculations can be executed. One important step is the calculation of the half-bandwidth size of the system equations. This information is necessary to calculate the amount of storage required by the system equations. It is used to test the feasibility of proceeding with the calculations. This type of calculation requires the identification of the node numbers (element incidences) connected to a typical element. The latter operation is performed by subroutine LNODES. This subroutine extracts the N element incidences of a specified element from the element incidences list for the entire system. As shown in Fig. 20.2.1, the N element incidences are stored in a column matrix, called LNODE.

This elementary operation is also very important to the bookkeeping involved in various parts of the program. The operation is illustrated in Fig. 20.2.2. Once the element incidences of the specified element have been extracted (by subroutine LNODES) from the input data, they are scanned to determine the maximum difference in these node numbers. This difference in node numbers is then substituted into

$$IBW = NG*(I_{max} - I_{min} + 1) \tag{20.3}$$

to determine the half-bandwidth associated with that particular element, without regard

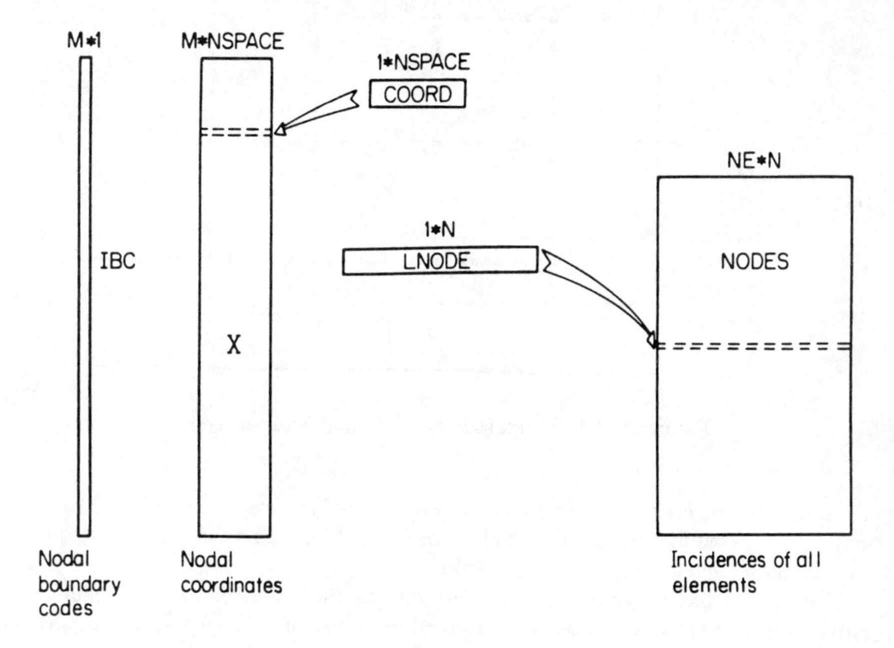

Figure 20.2.1 Element connectivity storage

```
      SUBROUTINE   LNODES (LID, NE, N, NODES, LNODE)
C     * * * * * * * * * * * * * * * * * * * * * * * *
C          EXTRACT NODES ASSOCIATED WITH ELEMENT LID
C     * * * * * * * * * * * * * * * * * * * * * * * *
      DIMENSION  NODES(NE,N), LNODE(N)
C     NE    = NUMBER OF ELEMENTS IN SYSTEM
C     N     = NUMBER OF NODES PER ELEMENT
C     LID   = ELEMENT NUMBER
C     NODES = NODAL INCIDENCES OF ALL ELEMENTS
C     LNODE = THE N NODAL INCIDENCES OF THE ELEMENT
      DO 10  I = 1, N
   10 LNODE(I) = NODES(LID,I)
      RETURN
      END
```

Figure 20.2.2 Extracting the element topology

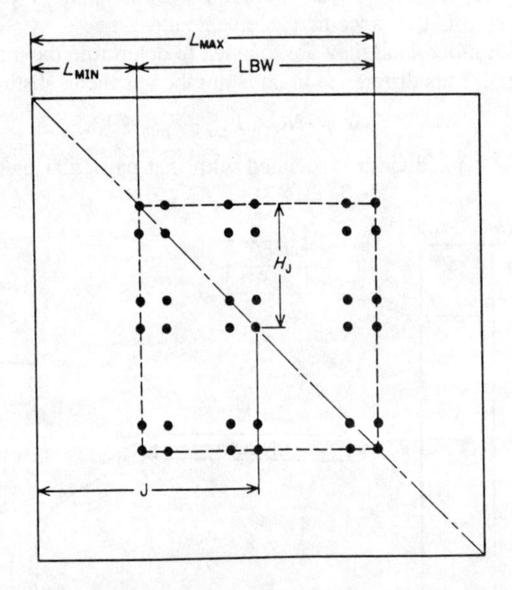

Figure 20.2.3 Element bandwidth and column sizes

to the effect of constraints. These elementary operations are are based on Fig. 20.2.3 and executed by subroutine ELBAND, which is shown in Fig. 20.2.4. Similar operations are used for column heights or element wavefronts.

The simple but important task of determining the value of the maximum half-bandwidth of the system equations is performed by subroutine SYSBAN, see Fig. 20.2.5. The half-bandwidth of each element is checked and the largest value found in the system is retained. The system equations usually occupy a very large percentage of the total storage. Knowing the system equation's half-bandwidth, determined above, and the total number of degrees of freedom in the system, the storage requirements for these banded, symmetric equations are easily determined. It is possible for nodal parameter constraint equations to increase the half-bandwidth of the system equations.

Recall that the input data include the element properties for each element in the system. Thus it is a simple matter to scan these data and extract the properties for any particular element. This operation is executed by subroutine ELPRTY. After the element properties have been extracted they can be passed to the subroutine(s) that calculate the element matrices. The above discussion assumed that the element properties do not vary spatially over the element. If this is not the case then it would be necessary to introduce an alternative procedure that could define a spatial variation of element properties. This could be accomplished by inputting the properties at the nodal points and using the element interpolation functions to also define the spatial variations of the properties. Subroutine LPTPRT can be used to establish an array containing the nodal point properties of all nodes connected to a particular element. The routine uses the element incidence list to guide the extraction of the proper data from the input nodal properties array. The properties data associated with the element incidences are stored in array PRTLPT and are passed to the element matrices routines. Subroutines ELPRTY and LPTPRT are shown in Fig. 20.2.8 and Fig. 20.2.9, respectively. The ability of the finite element methods to treat many different types of properties is one of their major practical advantages. There are certain basic operations that must be carried out before one actually generates the element matrices for the particular problem under study. For simplicity it will be assumed that the problem under study is linear and involves a quadratic functional. One must always use the spatial coordinates of the nodes

```
      SUBROUTINE  ELBAND (N, NG, IBW, LNODE)
C     * * * * * * * * * * * * * * * * * * * * * * * * *
C             ELEMENT BANDWIDTH CALCULATION
C     * * * * * * * * * * * * * * * * * * * * * * * * *
      DIMENSION  LNODE(N)
C     LNODE = ELEMENT INCIDENCES
C     N     = NUMBER OF NODES PER ELEMENT
C     NG    = NUMBER OF PARAMETERS PER NODE
C     IBW   = UPPER HALF BANDWIDTH, INCLUDING DIAGONAL
      IBW = 1
      NLESS = N - 1
      DO 20 I = 1, NLESS
        II = I + 1
        LNI = LNODE(I)
C        ALLOW FOR OMITTED NODES
        IF ( LNI .GT. 0 )  THEN
          DO 10  J = II, N
            LNJ = LNODE(J)
            IF ( LNJ .GT. 0 )  THEN
              NEW = NG*( IABS( LNJ-LNI ) + 1)
              IF ( NEW .GT. IBW ) IBW = NEW
            ENDIF
 10       CONTINUE
        ENDIF
 20   CONTINUE
      RETURN
      END
```

Figure 20.2.4 Element half bandwidth calculation

```
      SUBROUTINE  SYSBAN (NE,N,NG,IBW,NODES,LNODE,LMAX)
C     * * * * * * * * * * * * * * * * * * * * * * * *
C        DETERMINE UPPER HALF BAND WIDTH OF SYSTEM
C     * * * * * * * * * * * * * * * * * * * * * * * *
      DIMENSION  NODES(NE,N), LNODE(N)
C     NE    = NUMBER OF ELEMENTS IN SYSTEM
C     N     = NUMBER OF NODES PER ELEMENT
C     NG    = NUMBER OF PARAMETERS (DOF) PER ELEMENT
C     IBW   = MAXIMUM HALF BANDWIDTH = LBW MAX
C     NODES = NODAL INCIDENCES OF ALL ELEMENTS
C     LNODE = ELEMENT INCIDENCES LIST
C     LBW   = ELEMENT HALF BANDWIDTH
C     LMAX  = LAST ELEMENT CAUSING LBW
      LMAX = 1
      IBW  = 1
      DO 10  I = 1, NE
        CALL  LNODES (I, NE, N, NODES, LNODE)
        CALL  ELBAND (N, NG, LBW, LNODE)
        IF ( LBW .GT. IBW )  THEN
          IBW  = LBW
          LMAX = I
        ENDIF
   10 CONTINUE
      RETURN
      END
```

Figure 20.2.5 System upper half bandwidth

```
      SUBROUTINE  ELPRTY (LID, LHOMO, NE, NLPFIX, NLPFLO,
     1                    LPFIX, FLTEL, LPROP, ELPROP)
C     * * * * * * * * * * * * * * * * * * * * * * * * *
C     GET PROPERTIES OF ELEMENT LID FROM TOTAL PROPERTIES
C     * * * * * * * * * * * * * * * * * * * * * * * * *
      DIMENSION  FLTEL(NE,0:NLPFLO), ELPROP(0:NLPFLO),
     1           LPFIX(NE,0:NLPFIX), LPROP(0:NLPFIX)
C     LPFIX  = SYSTEM ARRAY OF FIXED PT ELEM PROPERTIES
C     LPROP  = ELEM FIXED PT PROPERTIES ARRAY
C     FLTEL  = SYS ARRAY OF FLOATING PT NODAL PROP
C     ELPROP = ELEM FLOATING PT PROPERTIES ARRAY
C     LHOMO  = 1, IF PROPERTIES ARE SAME IN ALL ELEMENTS
C     NLPFIX = NUMBER OF INTEGER ELEMENT PROPERTIES
C     NLPFLO = NUMBER OF REAL ELEMENT PROPERTIES
      IF ( LHOMO .EQ. 1 )  THEN
        I = 1
      ELSE
        I = LID
      ENDIF
C       REAL PROPERTIES, THEN INTEGER PROPERTIES
      DO 10  J = 1, NLPFLO
   10   ELPROP(J) = FLTEL(I,J)
      DO 20  J = 1, NLPFIX
   20   LPROP(J) = LPFIX(I,J)
      RETURN
      END
```

Figure 20.2.6 Recover integer and real element properties

```
      SUBROUTINE  LPTPRT (N, M, NNPFLO, FLTNP, PRTLPT,
     1                NNPFIX, NPFIX, LPPROP, LNODE, NHOMO)
C     * * * * * * * * * * * * * * * * * * * * * * * * * * *
C         GET REAL PROPERTIES AT NODES OF AN ELEMENT
C     * * * * * * * * * * * * * * * * * * * * * * * * * * *
      DIMENSION  FLTNP(M,0:NNPFLO), PRTLPT(N,0:NNPFLO),
     1     NPFIX(M,0:NNPFIX), LPPROP(N,0:NNPFIX), LNODE(N)
C     FLTNP  = FLOATING POINT PROP ARRAY OF SYSTEM NODES
C     PRTLPT = FLOATING POINT PROP ARRAY OF ELEMENT NODES
C     NPFIX  = INTEGER PROPERTY ARRAY OF SYSTEM NODES
C     LPPROP = INTEGER PROPERTY ARRAY OF ELEMENT NODES
C     NNPFIX = NUMBER OF INTEGER PROPERTIES PER NODE
C     NNPFLO = NUMBER OF REAL PROPERTIES PER NODE
      DO 20  I = 1, N
        IROW = LNODE(I)
C         ALLOW FOR OMITTED NODES
        IF ( IROW .GT. 0 )  THEN
          IF ( NHOMO .EQ. 1 )  IROW = 1
          IF ( NNPFLO .GT. 0 )  THEN
            DO 10  J = 1, NNPFLO
 10           PRTLPT(I,J) = FLTNP(IROW,J)
          ENDIF
          IF ( NNPFIX .GT. 0 )  THEN
            DO 30  J = 1, NNPFIX
 30           LPPROP(I,J) = NPFIX(IROW,J)
          ENDIF
        ENDIF
 20   CONTINUE
      RETURN
      END
```

Figure 20.2.7 Properties for each element node

associated with the element. Recall that the spatial coordinates of all the nodal points and the element incidences of all elements are available as input data. Thus to define an array containing the spatial coordinates of the nodes associated with an element one simply extracts its element incidences with subroutine LNODES and then extracts the spatial coordinates of each node in the list of incidences. The present program for this purpose, ELCORD, is shown in Fig. 20.2.6. A similar program, PTCORD, for extracting the spatial coordinates of a single node is shown in Fig. 20.2.7. Examples of the types of properties (or differential equation coefficients) that one commonly encounters are the thermal conductivity in a heat conduction analysis and the nodal point temperatures in a thermal stress analysis. Several specific examples of typical property data will be given in the example applications beginning in Chap. 21.

Consider a typical nodal point in the system and recall that there are NG parameters associated with each node. Thus, at a typical node there will be NG local degree of freedom numbers ($1 \leq J \leq NG$) and a corresponding set of system degree of freedom numbers. If I denotes the system node number of the point, then the NG corresponding system degree of freedom numbers are calculated by utilizing Eq. (20.2). These elementary calculations are carried out by subroutine INDXPT. The program assigns NG storage locations for the vector, say INDEX, containing the system degree of freedom numbers associated with the specified nodal point — see Tables 20.1 and 20.2 for the related details. Storage locations are established for the N element incidences (extracted by subroutine LNODES) and the N*NG system degree of freedom numbers, in vector INDEX, associated with the element.

```
      SUBROUTINE ELCORD (M,N,NSPACE,X,COORD,LNODE)
C     * * * * * * * * * * * * * * * * * * * * * * * * *
C     DETERMINE COORDINATES OF NODES ON ELEMENT
C     * * * * * * * * * * * * * * * * * * * * * * * * *
      DIMENSION  X(M,NSPACE), COORD(N,NSPACE), LNODE(N)
C     M      = NUMBER OF NODES IN SYSTEM
C     NSPACE = DIMENSION OF SPACE
C     N      = NUMBER OF NODES PER ELEMENT
C     X      = COORDINATES OF SYSTEM NODES
C     COORD  = COORDINATES OF ELEMENT NODES
C     LNODE  = N ELEMENT INCIDENCES OF ELEMENT
      DO 20  K = 1, NSPACE
        DO 10  I = 1, N
C          ALLOW FOR OMITTED NODES
          IF ( LNODE(I) .GT. 0 )
     1      COORD(I,K) = X(LNODE(I),K)
 10     CONTINUE
 20   CONTINUE
      RETURN
      END
```

Figure 20.2.8 Coordinates for a single element

```
      SUBROUTINE  PTCORD (IPT, M, NSPACE, X, COORD)
C     * * * * * * * * * * * * * * * * * * * * * * * * *
C     EXTRACT COORDINATES OF POINT NUMBER IPT
C     * * * * * * * * * * * * * * * * * * * * * * * * *
      DIMENSION  X(M,NSPACE), COORD(1,NSPACE)
C     X      = SPATIAL COORDINATES OF ALL SYSTEM NODES
C     COORD  = SPATIAL COORDINATES OF THE NODE
C     M      = TOTAL NUMBER OF NODES IN SYSTEM
C     N      = NUMBER OF NODES PER ELEMENT
C     NSPACE = DIMENSION OF THE SPACE
      DO 10  J = 1, NSPACE
 10   COORD(1,J) = X(IPT,J)
      RETURN
      END
```

Figure 20.2.9 Coordinates at a single node

If the properties depend on the nodal parameters, as they often do in an iterative solution, then one may need to recover the previous nodal parameters associated with an element. Subroutine ELFRE, Fig. 20.2.10, uses the above element INDEX vector to extract those data. We will see typical uses of the above pre-element calculations in later sections.

Up to this point we have been assuming that a banded storage mode will be used. While this is used in most of the discussions for the sake of simplicity, here it should be noted that other important storage modes can also be dealt with easily. For example, the *skyline* or profile storage is suggested by Taylor [16] and Bathe and Wilson [2]. It is the current default in MODEL. To use these procedures, we must know the maximum height (out to the last non-zero term) of each column of the assembled system equations. One easy way to find these data is to find the contribution of each element to the column

Table 20.1 Degree of freedom numbers at node I_s

Local	System[†]
1	INDEX(1)
2	INDEX(2)
.	.
.	.
.	.
J	INDEX(J)
.	.
.	.
.	.
NG	INDEX(NG)

[†] INDEX(J) = NG*(I_s – 1) + J

Table 20.2 Relating local and system equation numbers

Local node I_L	Parameter number J	System node I_s = LNODE (I_L)	Element degree of freedom numbers: Local NG*(I_L – 1) + J	Element degree of freedom numbers: System NG*(I_s – 1) + J
1	1	LNODE(1)	1	NG*(LNODE(1) – 1) + 1
1	2	LNODE(1)	2	
.	.	.		
.	.	.		
.	.	.	.	.
1	NG	LNODE(1)	.	.
2	1	LNODE(2)	.	.
.	.	.		
.	.	.		
.	.	.		
K	IG	LNODE(K)	NG*(K – 1) + IG	NG*(LNODE(K) – 1) + IG
.	.	.		
.	.	.		
.	.	.	.	.
N	1	LNODE(N)	.	.
.	.	.	.	.
.	.	.		
.	.	.		
N	NG	LNODE(N)	N*NG	NG*(LNODE(N) – 1) + NG

```
      SUBROUTINE  ELFRE (NDFREE, NELFRE, D, DD, INDEX)
C     * * * * * * * * * * * * * * * * * * * * * * * * * * *
C     EXTRACT ELEMENT DEGREES OF FREEDOM FROM SYSTEM DOF
C     * * * * * * * * * * * * * * * * * * * * * * * * * * *
      DIMENSION  D(NELFRE), DD(NDFREE), INDEX(NELFRE)
C     D      = NODAL PARAMETERS ASSOCIATED
C     DD     = SYSTEM ARRAY OF NODAL PARAMETERS
C     INDEX  = ARRAY OF SYSTEM DEGREE OF FREEDOM NUMBERS
C     NELFRE = NUMBER OF DEGREES OF FREEDOM PER ELEMENT
C     NDFREE = TOTAL NUMBER OF SYSTEM DEGREES OF FREEDOM
      DO 10  I = 1, NELFRE
C        ALLOW FOR OMITTED NODES
        IF ( INDEX(I) .GT. 0 ) THEN
          D(I) = DD(INDEX(I))
        ELSE
          D(I) = 0.0
        ENDIF
 10   CONTINUE
      RETURN
      END
```

Figure 20.2.10 Extract element parameters from solution

```
      SUBROUTINE  ELHIGH (NELFRE,INDEX,LHIGH)
C     * * * * * * * * * * * * * * * * * * * * * * * *
C      FIND SYSTEM COLUMN HEIGHTS OF AN ELEMENT
C     * * * * * * * * * * * * * * * * * * * * * * * *
      DIMENSION  INDEX(NELFRE), LHIGH(NELFRE)
C     NELFRE   = NO OF DEGREES OF FREEDOM OF ELEMENT
C     INDEX    = SYSTEM DOF NOS OF ELEMENT PARAMETERS
C     LHIGH(I) = COLUMN HEIGHT FOR EQUATION INDEX(I)
      MIN = INDEX(1)
C      FIND MINIMUM INDEX
      DO 10  I = 1, NELFRE
        LHIGH(I) = 0
        NDX = INDEX(I)
C        ALLOW FOR OMITTED NODES
        IF ( NDX .GT. 0 .AND. NDX .LT. MIN ) MIN = NDX
 10   CONTINUE
C      CONVERT TO COLUMN HEIGHTS
      MIN = MIN - 1
      DO 20  I = 1, NELFRE
        NDX = INDEX(I)
        IF ( NDX .GT. 0 )  LHIGH(I) = NDX - MIN
 20   CONTINUE
      RETURN
      END
```

Figure 20.2.11 Element matrix column heights

heights and retain the maximum height encountered for each column in the system square matrix. Thus, we can have routines similar to ELBAND and SYSBAN. From Fig. 20.2.4 we note that the typical column height, H_J, of an element is $H_J = J - L_{min} + 1$. These simple calculations are executed by subroutine ELHIGH and the system maximums are tabulated by routine SKYHI which loops over all of the elements and calls ELHIGH.

```
      SUBROUTINE  SKYHI (NDFREE, NE, N, NG, NELFRE, NODES,
     1                 LNODE, INDEX, LHIGH, IDOFHI)
C     * * * * * * * * * * * * * * * * * * * * * * * * * * *
C           FIND COLUMN HEIGHTS OF SYSTEM EQUATIONS
C              SYMMETRIC SKYLINE STORAGE MODE
C     * * * * * * * * * * * * * * * * * * * * * * * * * * *
      DIMENSION  NODES(NE,N), LNODE(N), INDEX(NELFRE),
     1           LHIGH(NELFRE), IDOFHI(NDFREE)
C     NDFREE     = TOTAL NO OF SYSTEM DOF
C     NELFRE     = NO OF ELEMENT PARAMETERS (DOF)
C     NE         = NUMBER OF ELEMENTS
C     N          = NUMBER OF NODES PER ELEMENT
C     NG         = NO OF PARAMETERS PER NODE
C     NODES      = NODAL INCIDENCES OF ALL ELEMENTS
C     LNODE      = ELEMENT NODAL INCIDENCES
C     INDEX(I)   = SYS DOF NO OF ELEMENT DOF I
C     IDOFHI(I) = COL HEIGHT OF SYS DOF I
C      ZERO HEIGHTS
      CALL  ZEROI (NDFREE,IDOFHI)
C      LOOP OVER ELEMENTS
      DO 20  IE = 1, NE
C        EXTRACT NODES, FIND DOF NOS
        CALL LNODES (IE, NE, N, NODES, LNODE)
        CALL INDXEL (N, NELFRE, NG, LNODE, INDEX)
C        FIND ELEMENT COLUMN HEIGHTS
        CALL ELHIGH (NELFRE, INDEX, LHIGH)
C        COMPARE WITH CURRENT MAXIMUMS
        DO 10  J = 1,NELFRE
          NDX = INDEX(J)
C          ALLOW OMITTED NODES
          IF ( NDX .GT. 0 .AND. IDOFHI(NDX) .LT. LHIGH(J) )
     1         IDOFHI(NDX) = LHIGH(J)
 10     CONTINUE
 20   CONTINUE
      RETURN
      END
```

Figure 20.2.12 Assembled system matrix column heights

```
      SUBROUTINE  SKYDIA (NDFREE, IDOFHI, IDIAG)
C     * * * * * * * * * * * * * * * * * * * * * * * * * *
C     USE COLUMN HEIGHTS TO FIND DIAGONAL COEFFICIENTS
C     FOR SYMMETRIC SKYLINE IN VECTOR STORAGE MODE
C     * * * * * * * * * * * * * * * * * * * * * * * * * *
C      ASSUMING SYMMETRIC COLS STORED FROM TOP DOWN
      DIMENSION  IDOFHI(NDFREE), IDIAG(NDFREE)
C     NDFREE    = TOTAL NO OF SYSTEM EQUATIONS
C     IDOFHI(I) = COL HEIGHT OF EQ I, WITH DIAG
C     IDIAG(I)  = LOCATION OF DIAG OF I-TH EQ NUMBER
C                 COEFF IN UPPER TRIANGLE
C      TOTAL NUMBER OF SQ MATRIX TERMS = IDIAG(NDFREE)
      IPOINT = 0
      DO 10  I = 1, NDFREE
        IPOINT = IPOINT + IDOFHI(I)
 10   IDIAG(I) = IPOINT
      RETURN
      END
```

Figure 20.2.13 Locate diagonal terms in vector

Other information of importance to skyline solution procedures is the location of the diagonal elements and the total number of coefficients to be stored (which is always less than or equal to the band storage). If we decide to store the columns from the top down, then these calculations are simple, as shown in subroutine SKYDIA. Note that the location of the last diagonal coefficient also corresponds to the total number of coefficients to be stored. The last three programs are shown in Figs. 20.2.11 to 20.2.13. An example of typical equation heights due to various element contributions is shown in Fig. 20.2.14. Also shown is the assembled form of the system equations for the mesh in the figure. Figure 20.2.15 shows a typical system square matrix. It indicates which columns would be stored, and the location of the diagonal coefficients.

The above discussion has shown how to reduce storage requirements and problem costs by using efficient storage techniques. More importantly, we have a better understanding of how to relate the element and system equation numbers, and how to recover basic input data needed for calculating the element matrices. In the next chapter we will outline additional important concepts associated with the generation of various element matrices. Very specific and extensive details of these concepts will be presented in the chapters on typical applications.

Next the typical computations related to problem, dependent element calculations will be outlined. These correspond to the boxes after point 1 in the flowchart of Fig. 20.2.2. Much more detail will be given when example applications are considered later. The major subjects of interest here are typical data requirements for the element square matrix, element column matrix, auxiliary element calculations, and boundary flux contributions. The absolute minimum problem-dependent calculation involves the generation of an element square matrix. Thus, our discussions begin with this.

A typical definition of an element square matrix $\mathbf{S}$, frequently called the *element stiffness matrix*, is often of the form

$$\mathbf{S}^e = \int_{\Omega^e} \left[P_x \frac{\partial \mathbf{H}^T}{\partial x} \frac{\partial \mathbf{H}}{\partial x} + P_y \frac{\partial \mathbf{H}^T}{\partial y} \frac{\partial \mathbf{H}}{\partial y} + P_z \frac{\partial \mathbf{H}^T}{\partial z} \frac{\partial \mathbf{H}}{\partial z} \right] d\Omega .$$

This can be considered as a three-dimensional extension of Eq. (1.11). The integral is over the volume of the element, thus the necessity of defining the spatial orientations of the element; i.e., the nodal coordinates. The integrand involves typical element properties (P_x, P_y, P_z) and the element interpolation functions, $\mathbf{H}$, and/or their spatial derivatives, $\partial\mathbf{H}/\partial x$. The evaluation of such an integral can be difficult or impossible in closed form, with the result that numerical integration is often used. In the present program the element square matrix is calculated in subroutine ELSQ, which is shown in Fig. 20.2.16. It also has the option to calculate the column matrix. The element column matrix can be calculated in ELCOL from the same arguments. These subroutines are two of the few programs that change from one typical problem to another. Several specific examples are given in Chap.s 21 and 22.

Most of the parameters included in the argument lists to the routines ELSQ and ELCOL, which are used to calculate the problem dependent element square and column matrices, are defined in the comments in Fig. 20.2.16. Some of the arrays are often needed to form element matrices, while others are often useful in storing data for post processing. Thus, the user can opt to store some data here rather than in subroutine ELPOST. Remember setting NTAPE1 to a value greater than zero is the flag used turn on the calls to the postprocessing routines ELPOST, POST, and POSTEL. Some specific items of interest are: B(NRB,NELFREE) = global $\mathbf{B}$ matrix, BODY(NSPACE) = volume

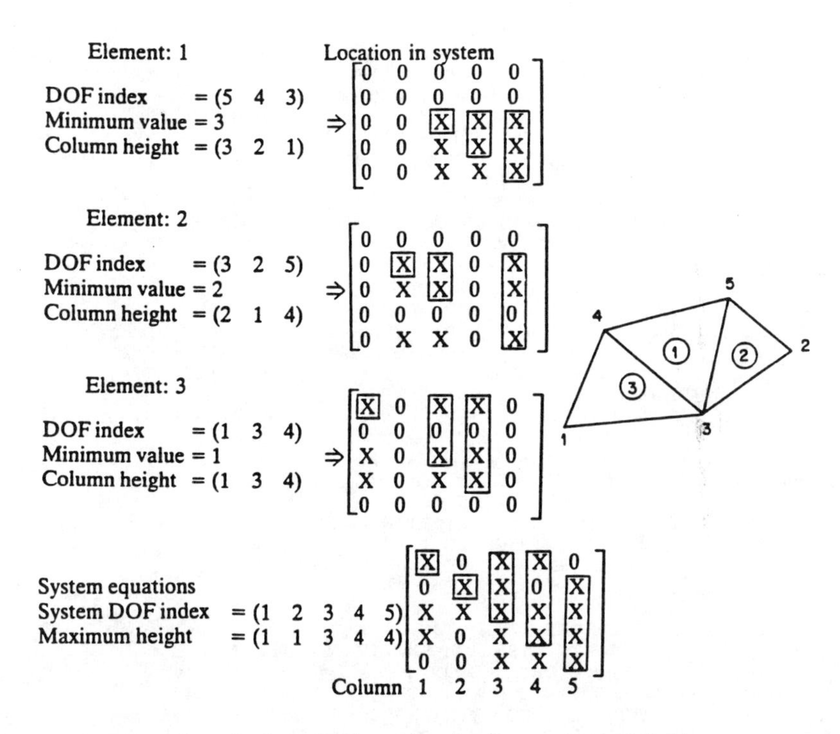

Figure 20.2.14 Sample element and system column heights

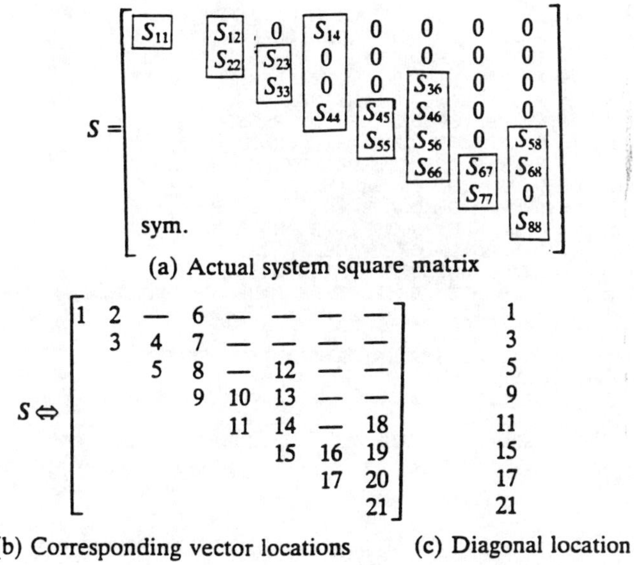

$$S = \begin{bmatrix} S_{11} & S_{12} & 0 & S_{14} & 0 & 0 & 0 & 0 \\ & S_{22} & S_{23} & 0 & 0 & 0 & 0 & 0 \\ & & S_{33} & 0 & 0 & S_{36} & 0 & 0 \\ & & & S_{44} & S_{45} & S_{46} & 0 & 0 \\ & & & & S_{55} & S_{56} & 0 & S_{58} \\ & & & & & S_{66} & S_{67} & S_{68} \\ & & & & & & S_{77} & 0 \\ \text{sym.} & & & & & & & S_{88} \end{bmatrix}$$

(a) Actual system square matrix

$$S \Leftrightarrow \begin{bmatrix} 1 & 2 & - & 6 & - & - & - & - \\ & 3 & 4 & 7 & - & - & - & - \\ & & 5 & 8 & - & 12 & - & - \\ & & & 9 & 10 & 13 & - & - \\ & & & & 11 & 14 & - & 18 \\ & & & & & 15 & 16 & 19 \\ & & & & & & 17 & 20 \\ & & & & & & & 21 \end{bmatrix} \qquad \begin{matrix} 1 \\ 3 \\ 5 \\ 9 \\ 11 \\ 15 \\ 17 \\ 21 \end{matrix}$$

(b) Corresponding vector locations (c) Diagonal location

Figure 20.2.15 Skyline storage of the system square matrix

```
      SUBROUTINE  ELSQ (N, NSPACE, NELFRE, NRB, NQP, NGEOM,
     1     NPARM, NNPFIX, NNPFLO, MISCFX, MISCFL, NLPFIX,
     2     NLPFLO, COORD, S, C, H, DGH, B, E, EB, STRAIN,
     3     STRAN0, STRESS, BODY, PT, WT, XYZ, DLH, G, DLG,
     4     AJ, AJINV, HINTG, D, PRTLPT, FLTMIS, ELPROP, PRTMAT,
     5     MISFIX, LSHAPE, LPROP, LPPROP, NTAPE1, NTAPE2,
     6     NTAPE3, NTAPE4, NTAPE5, LNODE, NG, IE )
C     * * * * * * * * * * * * * * * * * * * * * * * * * * * * * *
C     GENERATE ELEMENT SQUARE MATRIX, OPTIONAL COLUMN MATRIX
C     * * * * * * * * * * * * * * * * * * * * * * * * * * * * * *
C        ALWAYS USED
      DIMENSION  COORD(N,NSPACE), S(NELFRE,NELFRE)
C        USUALLY USED
      DIMENSION  C(NELFRE), H(N), DGH(NSPACE,N), B(NRB,NELFRE),
     1           E(NRB,NRB), EB(NRB,NELFRE), STRAIN(NRB+2),
     2           STRAN0(NRB), STRESS(NRB+2), BODY(NSPACE)
C        OPTIONAL FOR NUMERICAL INTEGRATION
      DIMENSION  PT(NPARM,0:NQP), WT(0:NQP), DLH(NSPACE,N),
     1           G(NGEOM), DLG(NPARM,NGEOM), AJ(NSPACE,NSPACE),
     2           AJINV(NSPACE,NSPACE), HINTG(N), LNODE(N),
     3           XYZ(NSPACE)
C        OPTIONAL PROPERTY AND SOLUTION VALUES
      DIMENSION  D(NELFRE), PRTLPT(N,0:NNPFLO), FLTMIS(0:MISCFL),
     1           ELPROP(0:NLPFLO), PRTMAT(0:NLPFLO),
     2           MISFIX(0:MISCFX), LPROP(0:NLPFIX),
     3           LPPROP(0:NNPFIX)
C
C        VARIABLES: (SEE TEXT ABOVE)
C     B        = STRAIN-DISPLACEMENT (GRADIENT) MATRIX
C     BODY     = BODY FORCE VECTOR
C     C        = ELEMENT COLUMN MATRIX
C     COORD    = SPATIAL COORDINATES OF ELEMENT'S NODES
C     D        = NODAL PARAMETERS ASSOCIATED WITH AN ELEMENT
C     DGH      = GLOBAL DERIVATIVES INTERPOLATION FUNCTIONS
C     ELPROP   = ELEMENT ARRAY OF FLOATING PT PROPERTIES
C     FLTMIS   = SYSTEM STORAGE OF FLOATING PT MISC PROP
C     H        = SOLUTION INTERPOLATION FUNCTIONS
C     LPPROP   = INTEGER PROPERTIES AT EACH ELEMENT NODE
C     LPROP    = ARRAY INTEGER POINT ELEMENT PROPERTIES
C     MISFIX   = MISCELLANEOUS INTEGER SYSTEM PROPERTIES
C     NELFRE   = NUMBER OF DEGREES OF FREEDOM PER ELEMENT
C     NGEOM    = NUMBER OF GEOMETRY NODES
C     NMAT     = NUMBER OF MATERIAL TYPES
C     NPARM    = DIMENSION OF PARAMETRIC SPACE
C     NQP      = NUMBER OF QUADRATURE POINTS
C     NRB      = NUMBER OF ROWS IN B AND E MATRICES
C     NSPACE   = DIMENSION OF SPACE
C     PRTLPT   = REAL PROPERTIES AT ELEMENT NODES
C     PRTMAT   = REAL ELEM PROPERTIES BASED ON MATERIAL NUMBER
C     S        = ELEMENT SQUARE MATRIX
C     STRAIN   = STRAIN OR GRADIENT VECTOR
C     STRAN0   = INITIAL STRAIN OR GRADIENT VECTOR
C     STRESS   = STRESS VECTOR
C     XYZ      = SPACE COORDINATES AT A POINT
C     ..........................................................
C      ***  ELSQ PROBLEM DEPENDENT STATEMENTS FOLLOW ***
C     ..........................................................
      RETURN
      END
```

Figure 20.2.16 Available arguments for element square matrix

source term, C(NELFRE) = element column matrix (output), D(NELFRE) = nodal parameters from the previous iteration (if any), E(NRB,NRP) = constitutive matrix, EB(NRB,NELFREE) = product of $\mathbf{E} \times \mathbf{B}$, S(NELFRE,NELFRE) = element square matrix (output), STRAIN(NS+2) = strain or gradient vector, STRAN0(NS) = initial strain or gradient, and STRESS(NS+2) = stress or flux vector, where NRB = number of rows in **B**. The array **D** is usually utilized only in problems requiring iterations. Other problem-dependent data could be included in additional dimension or common statements within the application-dependent subroutines ELSQ, ELCOL, ELPOST, and POSTEL. All of these routines also have in their argument lists four auxiliary storage unit numbers, NTAPE1, NTAPE2, NTAPE3, and NTAPE4. A typical column matrix, **C**, frequently called the load or force vector, is often defined by an integral of the form

$$\mathbf{C}^e = \int_{\Omega^e} g(x, y, z)\, \mathbf{H}^T d\Omega\,,$$

where g is some known quantity (e.g., rate of heat generation) that varies over the element. In the present program the calculation of the element column matrix is carried out in either subroutine ELCOL or in ELSQ. Several specific examples will be presented in Chap.s 21 and 22. In some cases the programming will be most efficient if both the element square and column matrices are generated in a single subroutine. This is particularly true if the integrals are evaluated numerically. Therefore, ELSQ has both **S** and **C** as return arguments, whereas only **C** can be returned from ELCOL. There are two other problem-dependent subroutines that are closely related to ELSQ. They are ELPOST, Fig. 20.2.17, and POSTEL, and they are both related to any post-solution calculations to be performed once the nodal parameters have been calculated. Subroutines ELPOST and POSTEL, respectively, generate and process element data that are required in any post-solution calculations. POSTEL has the same arguments as ELSQ plus data for possible numerical integration and the element matrices **S** and **C** that may be desired for reaction calculations. It also has iteration counters for nonlinear solutions. For example, if the nodal parameters represent temperatures, these routines could perform the calculations for the temperatures and temperature gradients at specific points within the elements once the temperatures (nodal parameters) have been calculated. Subroutine ELPOST may share selected element data with subroutine ELSQ by means of a problem-dependent COMMON statement. It utilizes these data and/or other information not supplied in the standard arguments to relate "secondary" quantities of interest to the, as yet unknown, nodal parameters. Subroutine ELPOST contains the input parameter NTAPE1 in its argument list. It is called only if NTAPE1 is greater than zero. NTAPE1 usually represents the external unit number on which any element data associated with the post-solution calculations are to be stored. Of course, the parameter NTAPE1 must be greater than zero or this subroutine is not called. Subroutine POSTEL is called (if NTAPE1 > 0) after the nodal parameters have been calculated. This program reads, from NTAPE1, the data that were generated by ELPOST. POSTEL combines the data stored by ELPOST and the calculated nodal parameters, **D**, to determine secondary quantities of interest within element IE. Its argument list contains most of the same arguments as ELSQ and is shown in Fig. 20.2.18.

The element square and column matrices must be generated for each and every element in the system. These calculations can represent a large percentage of the total computing time and so these calculations should be programmed as efficiently as possible. Some economy considerations associated with generating the element matrices will be presented in Sec. 20.5. In an integral formulation involving a Neumann-type

```
      SUBROUTINE  ELPOST (N, NSPACE, NELFRE, NRB, NQP, NNPFIX,
     1            NNPFLO, MISCFX, MISCFL, NLPFIX, NLPFLO, H,
     2            DGH, B, E, EB, STRAIN, STRAN0, STRESS, BODY,
     3            HINTG, D, PRTLPT, FLTMIS, ELPROP, PRTMAT,
     4            MISFIX, LSHAPE, LPROP, LPPROP, NTAPE1,
     5            NTAPE2, NTAPE3, NTAPE4, NTAPE5, LNODE, NG )
C     * * * * * * * * * * * * * * * * * * * * * * * * * * * * *
C     GENERATE OR STORE DATA FOR ELEMENT POST-SOLUTION USE
C     * * * * * * * * * * * * * * * * * * * * * * * * * * * * *
      DIMENSION  H(N), DGH(NSPACE,N), B(NRB,NELFRE), HINTG(N),
     1           E(NRB,NRB), EB(NRB,NELFRE), STRAIN(NRB+2),
     2           STRAN0(NRB), STRESS(NRB+2), BODY(NSPACE)
      DIMENSION  D(NELFRE), PRTLPT(N,0:NNPFLO), FLTMIS(0:MISCFL),
     1           ELPROP(0:NLPFLO), PRTMAT(0:NLPFLO),
     2           MISFIX(0:MISCFX), LPROP(0:NLPFIX),
     3           LPPROP(0:NNPFIX), LNODE(N)
C     ....................................................
C      *** ELPOST PROBLEM DEPENDENT STATEMENTS FOLLOW ***
C     ....................................................
      RETURN
      END
```

Figure 20.2.17 Arguments for storing auxiliary items

```
      SUBROUTINE  POSTEL (N, NSPACE, NELFRE, NRB, NQP, NGEOM,
     1    NPARM, NNPFIX, NNPFLO, MISCFX, MISCFL, NLPFIX,
     2    NLPFLO, COORD, S, C, H, DGH, B, E, EB, STRAIN,
     3    STRAN0, STRESS, BODY, PT, WT, XYZ, DLH, G, DLG,
     4    AJ, AJINV, HINTG, D, PRTLPT, FLTMIS, ELPROP, PRTMAT,
     5    MISFIX, LTSHAP, LPROP, LPPROP, NTAPE1, NTAPE2,
     6    NTAPE3, NTAPE4, NTAPE5, IT, NITER, IE, NE, LNODE, NG,
     7    USEREL, USERPT )
C     * * * * * * * * * * * * * * * * * * * * * * * * * * * * *
C          ELEMENT LEVEL POST-SOLUTION CALCULATIONS
C     * * * * * * * * * * * * * * * * * * * * * * * * * * * * *
      DIMENSION COORD(N,NSPACE), S(NELFRE,NELFRE), C(NELFRE)
      DIMENSION H(N), DGH(NSPACE,N), B(NRB,NELFRE),
     1          E(NRB,NRB), EB(NRB,NELFRE), STRAIN(NRB+2),
     2          STRAN0(NRB), STRESS(NRB+2), BODY(NSPACE)
C        OPTIONAL FOR NUMERICAL INTEGRATION
      DIMENSION PT(NPARM,0:NQP), WT(0:NQP), DLH(NSPACE,N),
     1          G(NGEOM), DLG(NPARM,NGEOM), AJ(NSPACE,NSPACE),
     2          AJINV(NSPACE,NSPACE), HINTG(N), LNODE(N),
     3          XYZ(NSPACE)
C        OPTIONAL PROPERTY AND SOLUTION VALUES
      DIMENSION D(NELFRE), PRTLPT(N,0:NNPFLO), FLTMIS(0:MISCFL),
     1          ELPROP(0:NLPFLO), PRTMAT(0:NLPFLO),
     2          MISFIX(0:MISCFX), LPROP(0:NLPFIX),
     3          LPPROP(0:NNPFIX)
C        OPTIONAL USER APPLICATION AT NODE OR ELEMENT
      DIMENSION USERPT(NG), USEREL(NG,N)
C      USEREL, USERPT  = (USER) ELEMENT OR POINT SCRATCH AREA
C     ..................................................
C      *** POSTEL PROBLEM DEPENDENT STATEMENTS FOLLOW ***
C     ..................................................
      RETURN
      END
```

Figure 20.2.18 Arguments for element postprocessing

boundary condition the boundary flux contributions usually define a column matrix, say $\mathbf{C}^b$, with NFLUX coefficients and a corresponding square matrix. These matrices are often defined by integrals of the form

$$\mathbf{C}^b = \int_{\Gamma^b} \mathbf{H}_b^T \, \mathbf{f} \, d s, \qquad \mathbf{S}^b = \int_{\Gamma^b} \mathbf{H}_b^T \, \mathbf{H}_b \, P \, ds,$$

where $\mathbf{f}$ denotes the specified flux components, $P = P(\mathbf{f})$ is a coefficient defined on the boundary, and $\mathbf{H}_b$ represents the element interpolation functions along that boundary segment. If the flux components and the parameter of interest both vary linearly along the element boundary segment, one obtains a boundary column matrix contribution such as the one shown in Figs. 11.5.7, and 20.2.19, where l_{ij} is the line segment and there is a linear variation in the flux (f_x, f_y). Similar contributions for two flux components on a quadratic segment are shown in Fig. 20.2.20. The calculation of boundary segment matrices for typical segments, such as these are carried out by the problem-dependent subroutine BFLUX using the arguments shown in Fig. 20.2.21. Alternate approaches for face and edge flux effects are used in FFLUX and EFLUX, respectively.

Typical examples of the problem-dependent arguments in ELSQ, ELCOL, ELPOST, POSTEL, and BFLUX are given in Chap.s 21 and 22. Future references to these routines will show only the specific problem-dependent calculations. Most of the previous operations that are required to generate the element problem-dependent calculations can be combined into a single routine. Figure 20.2.22 shows subroutine GENELM which is typical of an element generation control routine. This was illustrated in Fig. 20.1.5. Some programs also store the element matrices on auxiliary storage for later use.

20.3 Condensation of Internal Degrees of Freedom *

As one becomes more experienced with the finite element method one finds it desirable at times to consider elements with internal degrees of freedom. Here the word internal is used to mean that the quantity of interest is not shared with any other element (or master element). In high order hierarchical elements, the number of degrees of freedom at the centroid can be quite significant. Therefore, it is desirable to condense out the element's internal degrees of freedom. To do this it is necessary, in effect, to require the governing functional to also be minimized over the element. This allows one to write the element equations as:

$$\begin{bmatrix} S_{AA} & S_{AB} \\ S_{BA} & S_{BB} \end{bmatrix} \begin{Bmatrix} D_A \\ D_B \end{Bmatrix} = \begin{Bmatrix} C_A \\ C_B \end{Bmatrix}, \tag{20.4}$$

where the element equations have been partitioned to distinguish between shared degrees of freedom, $\mathbf{D}_A$, and internal degrees of freedom, $\mathbf{D}_B$. From these equations it is possible to relate $\mathbf{D}_B$ to $\mathbf{D}_A$; that is

$$\mathbf{D}_B = \mathbf{S}_{BB}^{-1} (\mathbf{C}_B - \mathbf{S}_{BA} \mathbf{D}_A). \tag{20.5}$$

Therefore, the element equations can be rewritten in condensed form as

$$\mathbf{S}_{AA}^* \mathbf{D}_A = \mathbf{C}_A^*, \tag{20.6}$$

where $\mathbf{S}^*$ and $\mathbf{C}^*$ are defined in terms of the original matrices $\mathbf{S}$ and $\mathbf{C}$. The use of internal degrees of freedom has been shown to improve accuracy and reduce the number of system equations that must be solved in many cases. The operations required to

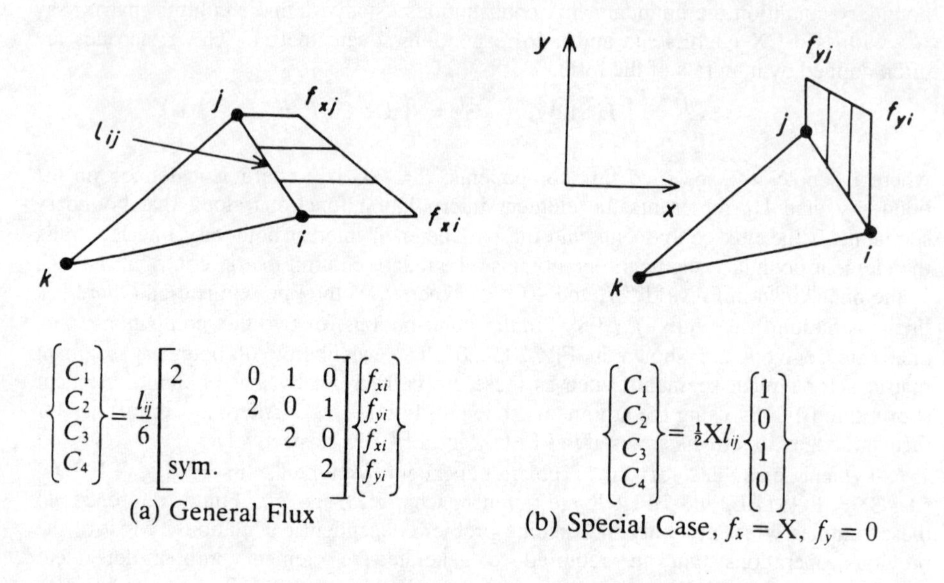

$$\begin{Bmatrix} C_1 \\ C_2 \\ C_3 \\ C_4 \end{Bmatrix} = \frac{l_{ij}}{6} \begin{bmatrix} 2 & 0 & 1 & 0 \\ & 2 & 0 & 1 \\ & & 2 & 0 \\ \text{sym.} & & & 2 \end{bmatrix} \begin{Bmatrix} f_{xi} \\ f_{yi} \\ f_{xi} \\ f_{yi} \end{Bmatrix} \qquad \begin{Bmatrix} C_1 \\ C_2 \\ C_3 \\ C_4 \end{Bmatrix} = \tfrac{1}{2} X l_{ij} \begin{Bmatrix} 1 \\ 0 \\ 1 \\ 0 \end{Bmatrix}$$

Figure 20.2.19 Linear flux boundary segments

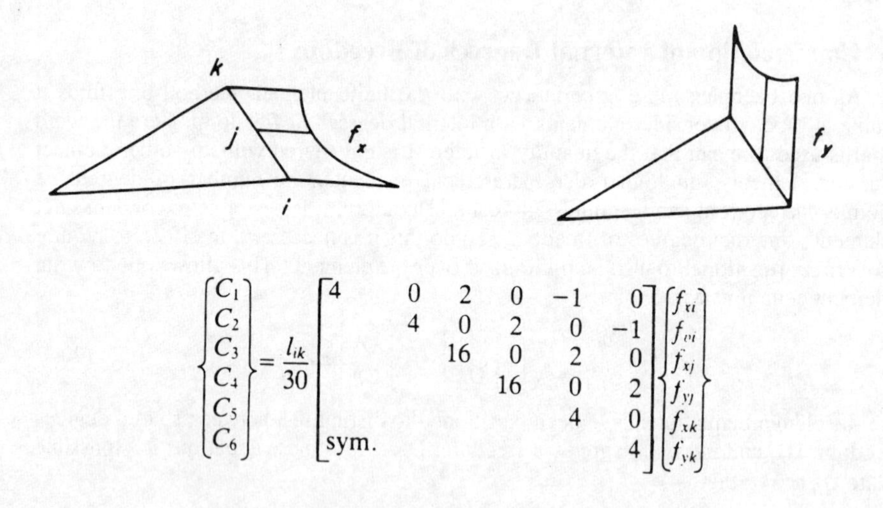

$$\begin{Bmatrix} C_1 \\ C_2 \\ C_3 \\ C_4 \\ C_5 \\ C_6 \end{Bmatrix} = \frac{l_{ik}}{30} \begin{bmatrix} 4 & 0 & 2 & 0 & -1 & 0 \\ & 4 & 0 & 2 & 0 & -1 \\ & & 16 & 0 & 2 & 0 \\ & & & 16 & 0 & 2 \\ & & & & 4 & 0 \\ \text{sym.} & & & & & 4 \end{bmatrix} \begin{Bmatrix} f_{xi} \\ f_{yi} \\ f_{xj} \\ f_{yj} \\ f_{xk} \\ f_{yk} \end{Bmatrix}$$

Figure 20.2.20 Quadratic flux boundary segments

```
      SUBROUTINE  BFLUX (FLUX, COORD, LBN, N, NSPACE, NFLUX,
     1            NG, C, S, IOPT, NQP, NPARM, H, DGH, PT, WT,
     2            XYZ, DLH, G, DLG, AJ, AJINV, LHOMO, ISEG, NSEG,
     3            NBSFIX, NBSFLO, NBSPFX, FLTBS, GPT, GWT, NGF)
C     * * * * * * * * * * * * * * * * * * * * * * * * *
C        PROBLEM DEPENDENT BOUNDARY FLUX CONTRIBUTIONS
C     * * * * * * * * * * * * * * * * * * * * * * * * *
C       ALWAYS USED
      DIMENSION  COORD(LBN,NSPACE), FLUX(LBN,NGF), C(NFLUX),
     1           S(NFLUX,NFLUX)
C
C       OPTIONAL FOR NUMERICAL INTEGRATION
      DIMENSION   H(N), DGH(NSPACE,N), PT(NPARM,0:NQP),
     1            WT(0:NQP), XYZ(NSPACE), DLH(NSPACE,N),
     2            G(LBN), DLG(NPARM,LBN), AJ(NSPACE,NSPACE),
     3            AJINV(NSPACE,NSPACE), GPT(0:NQP), GWT(0:NQP)
C       OPTIONAL SEGMENT PROPERTIES
      DIMENSION   FLTBS(0:NSEG,0:NBSFLO), NBSPFX(0:NSEG,0:NBSFIX)
C
C     AJ      = JACOBIAN
C     C       = BOUNDARY FLUX COLUMN MATRIX CONTRIBUTIONS
C     COORD   = SPATIAL COORDINATES OF SEGMENT NODES
C     DGH     = GLOBAL DERIVATIVES INTERPOLATION FUNCTIONS
C     DLG     = LOCAL DERIVATIVES GEOMETRIC INTERPOLATION
C     DLH     = LOCAL DERIVATIVES INTERPOLATION FUNCTIONS
C     FLTBS   = REAL PROPERTIES ON THE SEGMENTS
C     FLUX    = SPECIFIED BOUNDARY FLUX COMPONENTS
C     G       = GEOMETRIC INTERPOLATION FUNCTIONS
C     GPT,GWT = GAUSSIAN QUADRATURE 1-D DATA
C     H       = SOLUTION INTERPOLATION FUNCTIONS
C     IOPT    = PROBLEM MATRIX REQUIREMENTS, MUST BE SET.
C             = 1, CALCULATE C ONLY
C             = 2, CALCULATE S ONLY
C             = 3, CALCULATE BOTH C AND S
C     ISEG    = SEGMENT NUMBER
C     LBN     = NO. OF NODES ON AN ELEMENT BOUNDARY SEGMENT
C     LHOMO   = 1 IF SEGMENT PROPERTIES ARE HOMOGENEOUS
C     N       = NUMBER OF ELEMENT NODES, N > LBN USUALLY
C     NBSFIX  = NUMBER OF INTEGER PROPERTIES PER SEGMENT
C     NBSFLO  = NUMBER OF REAL PROPERTIES PER SEGMENT
C     NBSPFX  = INTEGER PROPERTIES ON THE SEGMENTS
C     NFLUX   = LBN*NG = MAXIMUM NUMBER OF SEGMENT DOF
C     NG      = NUMBER OF PARAMETERS PER NODE POINT
C     NGF     = NUMBER OF GENERALIZED FLUX COMPONENTS PER NODE
C     NPARM   = PARAMETRIC GEOMETRY NODES, = NSPACE USUALLY
C     NQP     = NUMBER OF QUADRATURE POINTS
C     NSEG    = MAX NUMBER OF SEGMENTS
C     NSPACE  = DIMENSION OF SOLUTION SPACE
C     PT      = QUADRATURE COORDINATES
C     S       = BOUNDARY FLUX SQUARE MATRIX
C     WT      = QUADRATURE WEIGHTS
C     XYZ     = SPACE COORDINATES AT A POINT
C     ...................................................
C       ** BFLUX PROBLEM DEPENDENT STATEMENTS FOLLOW **
C     ...................................................
      IOPT = 0
      RETURN
      END
```

Figure 20.2.21 Matrices from boundary flux sources

```
      SUBROUTINE  GENELM ( IE, M, NE, NDFREE, NITER, LPTEST, LHOMO,
     1    NHOMO, NULCOL, N, NSPACE, NELFRE, NRB, NQP, NGEOM,
     2    NPARM, NNPFIX, NNPFLO, MISCFX, MISCFL, NLPFIX, NLPFLO,
     3    LNODE, INDEX, X, DDOLD, COORD, S, C, H, DGH, B, E, EB,
     4    STRAIN, STRAN0, STRESS, BODY, PT, WT, XYZ, DLH, G, DLG,
     5    AJ, AJINV, HINTG, D, PRTLPT, FLTNP, FLTEL, FLTMIS,
     6    ELPROP, PRTMAT, MISFIX, NPFIX, LPFIX, LPROP, LPPROP,
     7    NTAPE1, NTAPE2, NTAPE3, NTAPE4, NTAPE5, LT,
     8    LSHAPE, LTUSER, NG )
C     * * * * * * * * * * * * * * * * * * * * * * * * * * * * * * * *
C          GENERATE ELEMENT MATRICES AND POST SOLUTION DATA
C     * * * * * * * * * * * * * * * * * * * * * * * * * * * * * * * *
CDP   IMPLICIT REAL*8 (A-H,O-Z)
C       SYSTEM DATA
      DIMENSION  X(M,NSPACE), DDOLD(NDFREE), LNODE(N), INDEX(NELFRE)
C       SYSTEM PROPERTIES
      DIMENSION  PRTLPT(N,0:NNPFLO), FLTNP(M,0:NNPFLO),
     1           FLTEL(NE,0:NLPFLO), NPFIX(M,0:NNPFIX),
     2           LPFIX(NE,0:NLPFIX)
C       FOR USE IN ELSQ, ELCOL, OR ELPOST:
      DIMENSION  COORD(N,NSPACE), S(NELFRE,NELFRE),C(NELFRE), H(N),
     1           DGH(NSPACE,N), B(NRB,NELFRE), E(NRB,NRB),
     2           EB(NRB,NELFRE), STRAIN(NRB+2), STRAN0(NRB),
     3           STRESS(NRB+2), BODY(NSPACE), PT(NPARM,0:NQP),
     4           WT(0:NQP), DLH(NSPACE,N), G(NGEOM), DLG(NPARM,NGEOM),
     5           AJ(NSPACE,NSPACE), AJINV(NSPACE,NSPACE), HINTG(N),
     6           XYZ(NSPACE), D(NELFRE), FLTMIS(0:MISCFL),
     7           ELPROP(0:NLPFLO), PRTMAT(0:NLPFLO), MISFIX(0:MISCFX),
     8           LPROP(0:NLPFIX), LPPROP(0:NNPFIX)
C          VARIABLES: (SEE TEXT ABOVE)
C     ELPROP  = ELEMENT ARRAY OF REAL PROPERTIES
C     FLTMIS  = MISCELLANEOUS REAL PROPERTIES OF SYSTEM
C     LPPROP  = INTEGER PROPERTIES AT EACH ELEMENT NODE
C     LPROP   = ARRAY OF ELEMENT INTEGER PROPERTIES
C     PRTLPT  = REAL PROPERTIES AT ELEMENT NODES
C     PRTMAT  = REAL ELEM PROPERTIES BASED ON MATERIAL NUMBER
C
C-->    EXTRACT NODAL COORDINATES
      CALL  ELCORD (M, N, NSPACE, X, COORD, LNODE)
C       EXTRACT NODAL PARAMETERS FROM LAST ITERATION (IF ANY)
      IF ( NITER .GT. 1 )
     1     CALL  ELFRE (NDFREE, NELFRE, D, DDOLD, INDEX)
C       EXTRACT NODAL POINT PROPERTIES (IF ANY)
      IF ( NNPFLO .GT. 0 )
     1     CALL  LPTPRT (N, M, NNPFLO, FLTNP, PRTLPT, NNPFIX, NPFIX,
     2                   LPPROP, LNODE, NHOMO)
C-->    EXTRACT ELEMENT PROPERTIES (IF ANY)
      IF ( LPTEST .GT. 0 )
     1     CALL ELPRTY (IE, LHOMO, NE, NLPFIX, NLPFLO, LPFIX, FLTEL,
     2                  LPROP, ELPROP)
C-->    EXTRACT MATERIAL PROPERTIES (IF ANY)
      IF ( NMAT .GT. 0 )
     1     CALL MATPRT (NMAT, NLPFLO, MISCFL, FLTMIS, PRTMAT)
```

Figure 20.2.22a Generate element level matrices and data

```
C-->   GENERATE ELEMENT SQUARE AND COLUMN MATRICES
      CALL  ELSQ (N, NSPACE, NELFRE, NRB, NQP, NGEOM, NPARM, NNPFIX,
     1            NNPFLO, MISCFX, MISCFL, NLPFIX, NLPFLO, COORD, S,
     2            C, H, DGH, B, E, EB, STRAIN, STRAN0, STRESS, BODY,
     3            PT, WT, XYZ, DLH, G, DLG, AJ, AJINV, HINTG, D,
     4            PRTLPT, FLTMIS, ELPROP, PRTMAT, MISFIX,
     5            LSHAPE, LPROP, LPPROP, NTAPE1, NTAPE2, NTAPE3,
     6            NTAPE4, NTAPE5, LNODE, NG, IE )
      IF ( NULCOL .EQ. 0 )
     1     CALL  ELCOL (N, NSPACE, NELFRE, NRB, NQP, NGEOM, NPARM,
     2           NNPFIX, NNPFLO, MISCFX, MISCFL, NLPFIX, NLPFLO,
     3           COORD, C, H, DGH, B, E, EB, STRAIN, STRAN0, STRESS,
     4           BODY, PT, WT, XYZ, DLH, G, DLG, AJ, AJINV, HINTG,
     5           D, PRTLPT, FLTMIS, ELPROP, PRTMAT, MISFIX,
     6           LSHAPE, LPROP, LPPROP, NTAPE1, NTAPE2, NTAPE3,
     7           NTAPE4, NTAPE5, LNODE, NG )
C-->   STORE DATA FOR POST SOLUTION CALCULATIONS (IF ANY)
      IF ( NTAPE1 .GT. 0 )
     1     CALL   ELPOST (N, NSPACE, NELFRE, NRB, NQP, NNPFIX,
     2            NNPFLO, MISCFX, MISCFL, NLPFIX, NLPFLO, H,
     3            DGH, B, E, EB, STRAIN, STRAN0, STRESS, BODY,
     4            HINTG, D, PRTLPT, FLTMIS, ELPROP, PRTMAT,
     5            MISFIX, LSHAPE, LPROP, LPPROP, NTAPE1,
     6            NTAPE2, NTAPE3, NTAPE4, NTAPE5, LNODE, NG )
C      NOTE:  SYSTEM PROPERTIES UPDATE COULD BE DONE HERE
      RETURN
      END
```

Figure 20.2.22b Generate element level matrices and data

condense out the internal degrees of freedom are executed by subroutine CONDSE. Since matrix $\mathbf{S}_{BB}$ can be relatively large for some elements, it is desirable to avoid the actual inversion of this matrix. Therefore, the routine uses Gaussian elimination, which is computationally much more efficient. The elimination begins with the last degree of freedom to be condensed and works backwards towards the first one.

20.4 Assembly into the System

20.4.1 Introduction

An important but often misunderstood topic is the procedure for *assemblying* the system equations from the element equations. Here assemblying is defined as the operation of adding the coefficients of the element equations into the proper locations in the system equations. There are various methods for accomplishing this but, as will be illustrated later, most are numerically inefficient. The numerically efficient *direct assembly* technique will be described here in some detail. First, it is desirable to review the simple but important relationship between a set of local (nodal point, or element) degree of freedom numbers and the corresponding system degree of freedom numbers. These simple calculations were described earlier in the discussions of INDXPT and INDXEL (Fig. 1.3.3).

Once the system degree of freedom numbers for the element have been stored in a vector, say INDEX, then the subscripts of a coefficient in the element equation can be directly converted to the subscripts of the corresponding system coefficient to which it is to be added. This correspondence between local and system subscripts is illustrated in Fig. 20.4.1. The expressions are generally of the form

Columns						
Index (1)	...	Index (i)	...	Index (j)	...	Index (N*NG)
1	...	i	...	j	...	N*NG

$$\begin{bmatrix} S^e_{11} & \cdots & S^e_{1i} & \cdots & S^e_{1j} & \cdots & S^e_{1,N\cdot NG} \\ \vdots & & \vdots & & \vdots & & \vdots \\ S^e_{i1} & \cdots & S^e_{ii} & \cdots & S^e_{ij} & \cdots & S^e_{i,N\cdot NG} \\ \vdots & & \vdots & & \vdots & & \vdots \\ S^e_{j1} & \cdots & S^e_{ji} & \cdots & S^e_{jj} & \cdots & S^e_{j,N\cdot NG} \\ \vdots & & \vdots & & \vdots & & \vdots \\ S^e_{N\cdot NG,1} & \cdots & S^e_{N\cdot NG,i} & \cdots & S^e_{N\cdot NG,j} & \cdots & S^e_{N\cdot NG,N\cdot NG} \end{bmatrix}$$

Local	System
1	Index (1)
⋮	⋮
I	Index (i)
⋮	⋮
J	Index (j)
⋮	⋮
N*NG	Index (N*NG)

$$\begin{Bmatrix} C^e_1 \\ \vdots \\ C^e_i \\ \vdots \\ C^e_j \\ \vdots \\ C^e_{N\cdot NG} \end{Bmatrix}$$

Figure 20.4.1 Local and system degrees of freedom for a typical element, $\mathbf{S}^e\,\mathbf{D}^e = \mathbf{C}^e$

```
      SUBROUTINE  STORSQ (NDFREE, IBW, NELFRE, INDEX, S, SS)
C     * * * * * * * * * * * * * * * * * * * * * * * * * * *
C      ADD ELEMENT SQUARE MATRIX TO UPPER HALF BANDWIDTH
C          OF THE SYMMETRIC SYSTEM SQUARE MATRIX
C     * * * * * * * * * * * * * * * * * * * * * * * * * * *
      DIMENSION  S(NELFRE,NELFRE), SS(NDFREE,IBW),
     1           INDEX(NELFRE)
C     INDEX  = SYSTEM DOF NOS OF THE ELEMENT DOF
C     S      = SQUARE ELEMENT MATRIX
C     SS     = SQUARE SYSTEM MATRIX
C     NDFREE = DEGREES OF FREEDOM IN THE SYSTEM
C     IBW    = HALF BAND WIDTH INCLUDING THE DIAGONAL
C     NELFRE = NUMBER OF PARAMETERS (DOF) PER ELEMENT
C     I,J    = ROW AND COLUMN POSITIONS IN THE
C              UNPACKED SYSTEM MATRIX, RESPECTIVELY
C     JJ     = COLUMN POSITION IN PACKED SYSTEM MATRIX
      DO 20 L = 1,NELFRE
        I = INDEX(L)
C        ALLOW FOR OMITTED NODES
        IF ( I .GT. 0 )  THEN
          DO 10 K = 1,NELFRE
            J = INDEX(K)
            IF ( J .GT. 0 )  THEN
C             STORE UPPER BAND ONLY
              IF ( I .LE. J )  THEN
                JJ        = J - I + 1
                SS(I,JJ) = SS(I,JJ) + S(L,K)
              ENDIF
            ENDIF
 10       CONTINUE
        ENDIF
 20   CONTINUE
      RETURN
      END
```

Figure 20.4.2 Assembly for a banded storage mode

$$C_I = C_I + C_i^e, \quad S_{I,J} = S_{I,J} + S_{i,j}^e \tag{20.7}$$

where i and j are the local subscripts of a coefficient in the element square matrix, and I, J are the corresponding subscripts of the system equation coefficient to which the element contribution is to be added. The direct conversions are given by $I = \text{INDEX}(i)$, $J = \text{INDEX}(j)$, where the INDEX array for element, e, is generated from Eq. (20.2) by subroutine INDXEL.

20.4.2 Assembly Programs

Utilizing the vector containing the system degree of freedom numbers for the element, the coefficients of the element column matrix are easily added to the coefficients of the system column matrix. This simple operation is performed by subroutine STORCL, which is illustrated in Fig. 1.3.2. Today this is often referred to as a scatter operation. The reverse procedure (in ELFRE) is a gather operation. The addition of the NELFRE*NELFRE square element matrix to the "square" matrix of the system is accomplished by using the system degree of freedom number array, INDEX, to covert both the element subscripts to system subscripts. This is known as the *direct method* of assemblying the system equations. A simple Fortran routine, STRFUL, to accomplish this direct assembly is illustrated in Fig. 1.3.2. Although this type of routine can be used on small problems, most practical problems have a large number of unknowns and are not economical if the full system equations are retained and solved. Fig. 1.3.1 gives a graphical example of full assembly. The more practical procedure for storing the element equations in the upper half-bandwidths of the system equations is illustrated in subroutine STORSQ, Fig. 20.4.2. The procedure utilized in STORSQ is slightly more involved since, due to the banded symmetric nature of the equations, the complete square system matrix is not stored in the computer. As mentioned before, the upper half-bandwidth (including the diagonal) of the system equations is stored as a

```
      SUBROUTINE  BANSUB (I, J, K, L)
C     * * * * * * * * * * * * * * * * * * * * * *
C     CONVERT SUBSCRIPTS (I,J) OF SYMMETRIC MATRIX
C     TO SUBSCRIPTS (K,L) IN UPPER HALF BANDWIDTH
C     * * * * * * * * * * * * * * * * * * * * * *
      ITEST = I - J
      IF ( ITEST )  10, 20, 30
C      BELOW DIAGONAL
 10   K = I
      L = 1 - ITEST
      RETURN
C      ON DIAGONAL
 20   K = I
      L = 1
      RETURN
C      ABOVE DIAGONAL
 30   K = J
      L = 1 + ITEST
      RETURN
      END
```

Figure 20.4.3 Convert from full to banded mode

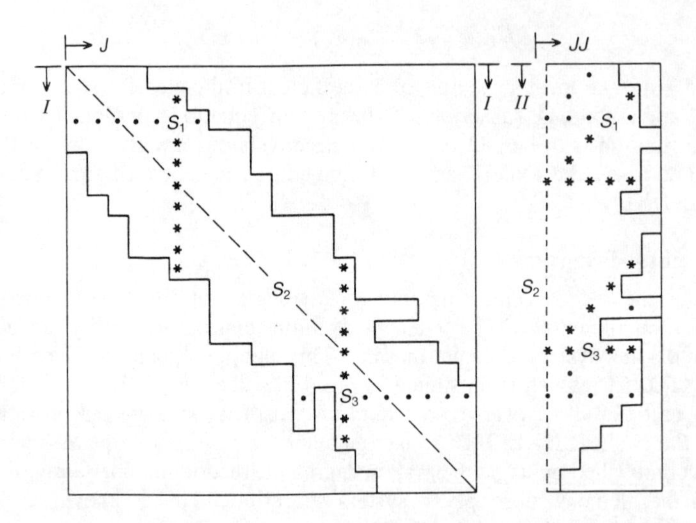

Figure 20.4.4 Locations of coefficients in a banded mode

rectangular array. If the element coefficient corresponds to a system coefficient that is below the main diagonal then it is ignored, but if it corresponds to a coefficient in the upper half-bandwidth then it is stored in the rectangular array. The program defines NELFRE*NELFRE storage locations for the element matrix, NDFREE*MAXBAN locations for the coefficient of the system upper half-bandwidth, and NELFRE storage locations for the system degree of freedom numbers associated with the element.

As illustrated in Figs. 20.4.3 and 20.4.4, the main diagonal of the square array becomes the first column of the rectangular array. In general, a row in the square array becomes a row in the rectangular array. A column in the upper half-bandwidth of the square array becomes a diagonal in the rectangular array. Thus, if a coefficient belongs in the I-th row and J-th column of the square array ($J \geq 1$) then that coefficient is stored in the I-th row and the JJ-th column of the rectangular array where

$$JJ = J - I + 1 . \tag{20.8}$$

Note that while the banded node requires less storage space we pay a price in extra bookkeeping to find terms now. This is true of all sparse storage nodes. Again referring to Fig. 20.4.4, one should note for the future reference that, due to symmetry, the coefficients of the lower half-bandwidth portion of the J-th column are also contained in the upper half-bandwidth portion of the J-th row. Subroutine BANSUB, Fig. 20.4.3, allows the novice to write code in terms of a full matrix, and then convert to the band storage mode.

As shown in Fig. 20.1.5, the generation and assembly of the element matrices often takes place in a single element loop. A typical subroutine, ASYMBL, for controlling these operations is shown in Fig. 20.4.5. It employs GENELM, given in Fig. 20.2.22, to generate the element data. Subroutine STORCL is used to assembly the force vector, **C**, and STORSQ is used to assembly **S** in a banded storage mode. Of course, an alternative storage mode like the skyline procedure (now the default) could have been used to assembly **S**. Both options require the use of a few extra calculations. If the skyline

```
      SUBROUTINE  ASYMBL (NG, NCOEFF, MODE, IDIAG, NODES, SS, CC,
     1    M, NE, NDFREE, NITER, LPTEST, LHOMO, NHOMO, NULCOL, N,
     2    NSPACE, NELFRE, NRB, NQP, NGEOM, NPARM, NNPFIX, NNPFLO,
     3    MISCFX, MISCFL, NLPFIX, NLPFLO, LNODE, INDEX, X, DDOLD,
     4    COORD, S, C, H, DGH, B, E, EB, STRAIN, STRAN0, STRESS,
     5    BODY, PT, WT, XYZ, DLH, G, DLG, AJ, AJINV, HINTG, D,
     6    PRTLPT, FLTNP, FLTEL, FLTMIS, ELPROP, PRTMAT,
     7    MISFIX, NPFIX, LPFIX, LPROP, LPPROP, NTAPE1, NTAPE2,
     8    NTAPE3, NTAPE4, NTAPE5, LTYPE, NLTYPE, LTDATA, LSHAPE,
     9    GPT, GWT )
C     * * * * * * * * * * * * * * * * * * * * * * * * * * * * * * *
C              ASSEMBLE SYSTEM EQUATIONS AND STORE POST
C                       SOLUTION ELEMENT DATA
C     * * * * * * * * * * * * * * * * * * * * * * * * * * * * * * *
CDP   IMPLICIT REAL*8 (A-H,O-Z)
      DATA  LASTLT, LTFREE  / 0, 0 /
      DIMENSION  CC(NDFREE), SS(NCOEFF), IDIAG(NDFREE), NODES(NE,N),
     1           LTYPE(NE), LTDATA(6,NLTYPE)
C       JUST PASSING THROUGH: SYSTEM DATA
      DIMENSION  X(M,NSPACE), DDOLD(NDFREE), LNODE(N), INDEX(NELFRE)
C       SYSTEM PROPERTIES
      DIMENSION  PRTLPT(N,0:NNPFLO), FLTNP(M,0:NNPFLO),
     1           FLTEL(NE,0:NLPFLO), NPFIX(M,0:NNPFIX),
     2           LPFIX(NE,0:NLPFIX)
C      FOR USE IN ELSQ, ELCOL, OR ELPOST:
      DIMENSION  COORD(N,NSPACE), S(NELFRE,NELFRE), C(NELFRE),
     1           H(N,0:NQP), DGH(NSPACE,N), B(NRB,NELFRE),
     2           E(NRB,NRB), EB(NRB,NELFRE), STRAIN(NRB+2),
     3           STRAN0(NRB), STRESS(NRB+2), BODY(NSPACE),
     4           PT(NPARM,0:NQP), WT(0:NQP), DLH(NSPACE,N,0:NQP),
     5           G(NGEOM,0:NQP), DLG(NPARM,NGEOM,0:NQP),
     6           AJ(NSPACE,NSPACE), AJINV(NSPACE,NSPACE),
     7           HINTG(N,0:NQP+1), GPT(0:NQP), GWT(0:NQP),
     8           XYZ(NSPACE), D(NELFRE), FLTMIS(0:MISCFL),
     9           ELPROP(0:NLPFLO), PRTMAT(0:NLPFLO),
     1           MISFIX(0:MISCFX), LPROP(0:NLPFIX),
     2           LPPROP(0:NNPFIX)
C        VARIABLES: (SEE TEXT ABOVE)
C     CC      = SYSTEM EQUATIONS COLUMN MATRIX
C     IDIAG   = DIAGONAL LOCATION IN SKYLINE VECTOR
C     INDEX   = SYSTEM DOF NUMBERS ASSOCIATED WITH ELEMENT
C     LNODE   = THE N ELEMENT INCIDENCES OF THE ELEMENT
C     LSHAPE  = SHAPE FLAG FOR QUADRATURE RULE SELECTION
C     LTQP    = NUMBER OF QUADRATURE PTS FOR ELEMENT TYPE
C     LTYPE   = ELEMENT TYPE NUMBER
C     MODE    = MODE OF STORAGE, 0-SKYLINE, 1-BANDED
C     NCOEFF  = TOTAL NUMBER OF TERMS IN SS
C     NDFREE  = TOTAL NUMBER OF SYSTEM DEGREES OF FREEDOM
C     NE      = NUMBER OF ELEMENTS
C     NELFRE  = NUMBER OF DEGREES OF FREEDOM PER ELEMENT
C     NG      = NUMBER OF PARAMETERS PER NODE
C     NGEOM   = NUMBER OF GEOMETRY NODES
C     NITER   = NO. OF ITERATIONS TO BE RUN (USUALLY 1)
C     NODES   = NODAL INCIDENCES OF ALL ELEMENTS
C     NPARM   = DIMENSION OF PARAMETRIC SPACE
C     NQP     = NUMBER OF QUADRATURE POINTS, >= LTQP
C     NRB     = NUMBER OF ROWS IN B AND E MATRICES
C     NSPACE  = DIMENSION OF SPACE
```

Figure 20.4.5a Typical equation assembly procedure

```
C     NTAPE1  = UNIT FOR POST SOLUTION MATRICES STORAGE
C     NULCOL  > 0, IF ELEMENT COLUMN MATRIX IS ALWAYS ZERO
C     S       = ELEMENT SQUARE MATRIX
C     SS      = SYSTEM EQUATIONS SQUARE MATRIX
C     STRAIN  = STRAIN OR GRADIENT VECTOR
C     STRAN0  = INITIAL STRAIN OR GRADIENT VECTOR
C     STRESS  = STRESS VECTOR
C     X       = COORDINATES OF SYSTEM NODES
C
C     GENERATE ELEMENT EQUATIONS & POST SOLUTION MATRICES
      DO 10  IE = 1, NE
C-->      GET ELEMENT TYPE NUMBER
        LT = 1
        IF ( NLTYPE .GT. 1 )  LT = LTYPE(IE)
C         SAME AS LAST TYPE ?
        IF ( LT .NE. LASTLT )  THEN
          LASTLT = LT
C           GET CONTROLS FOR THIS TYPE
          CALL  GETLT (LT, NLTYPE, LTDATA, LTN, LTQP, LTGEOM,
     1                 LTPARM, LTSHAP, LTUSER )
          LTFREE = LTN*NG
C-->        GET QUADRATURE RULE FOR ELEMENT TYPE AND SHAPE
          IF ( LTQP .GT. 0 )  THEN
            IF ( LTQP .GT. NQP ) STOP 'LTQP > NQP IN ASYMBL'
            CALL  GETQD (LTSHAP, LTQP, NSPACE, GPT, GWT, PT, WT)
          ENDIF
        ENDIF
C-->      EXTRACT ELEMENT NODE NUMBERS
        CALL  LNODES (IE, NE, LTN, NODES, LNODE)
C-->      CALCULATE DEGREE OF FREEDOM NUMBERS
        CALL  INDXEL (LTN, LTFREE, NG, LNODE, INDEX)
C-->      GENERATE ELEMENT PROBLEM DEPENDENT MATRICES
        CALL  GENELM ( IE, M, NE, NDFREE, NITER, LPTEST, LHOMO,
     1       NHOMO, NULCOL, LTN, NSPACE, LTFREE, NRB, LTQP,
     2       LTGEOM, LTPARM, NNPFIX, NNPFLO, MISCFX, MISCFL,
     3       NLPFIX, NLPFLO, LNODE, INDEX, X, DDOLD, COORD, S,
     4       C, H, DGH, B, E, EB, STRAIN, STRAN0, STRESS, BODY,
     5       PT, WT, XYZ, DLH, G, DLG, AJ, AJINV, HINTG, D,
     6       PRTLPT, FLTNP, FLTEL, FLTMIS, ELPROP, PRTMAT,
     7       MISFIX, NPFIX, LPFIX, LPROP, LPPROP, NTAPE1,
     8       NTAPE2, NTAPE3, NTAPE4, NTAPE5, LT,
     9       LTSHAP, LTUSER, NG )
C-->      STORE THE MATRICES IN SYSTEM EQUATIONS
        IF ( MODE .EQ. 0 )  THEN
C           SKYLINE VECTOR STORAGE MODE
          CALL  SKYSTR (NCOEFF, NDFREE, LTFREE, INDEX, IDIAG,
     1                  S, SS)
        ELSE
C           BANDED STORAGE
          MAXBAN = NCOEFF/NDFREE
          CALL  STORSQ (NDFREE, MAXBAN, LTFREE, INDEX, S, SS)
        ENDIF
        IF ( NULCOL .EQ. 0 )
     1      CALL STORCL (NDFREE, LTFREE, INDEX, C, CC)
   10 CONTINUE
C       ASSEMBLY COMPLETED
      RETURN
      END
```

Figure 20.4.5b   Typical equation assembly procedure

```
      SUBROUTINE  SKYSTR (NOCOEF, NDFREE, NELFRE, INDEX,
     1                    IDIAG, S, SS)
C     * * * * * * * * * * * * * * * * * * * * * * * * *
C       STORE ELEMENT SQUARE MATRIX TO SYSTEM SQUARE
C        MATRIX STORED IN SYMMETRIC SKYLINE MODE
C     * * * * * * * * * * * * * * * * * * * * * * * * *
      DIMENSION  SS(NOCOEF), S(NELFRE,NELFRE),
     1           INDEX(NELFRE), IDIAG(NDFREE)
C     NOCOEF   = NO COEFF IN SQ MATRIX = IDIAG(NDFREE)
C     NDFREE   = TOTAL NO OF DOF IN SYSTEM
C     NELFRE   = NUMBER OF ELEMENT DEGREES OF FREEDOM
C     INDEX(I) = SYS DOF NO OF ELEMENT DOF I
C     IDIAG(I) = LOCATION OF DIAGONAL OF I-TH EQ
C     S        = ELEMENT SQUARE MATRIX
C     SS       = SYS SQ MATRIX IN SKYLINE VECTOR MODE
C
C      LOOP OVER ELEMENT COEFFICIENTS
      DO 20  J = 1, NELFRE
        NDXJ = INDEX(J)
C        ALLOW FOR OMITTED NODES
        IF ( NDXJ .GT. 0 )  THEN
          JTEMP = IDIAG(NDXJ) - NDXJ
          DO 10  I = 1, NELFRE
            NDXI = INDEX(I)
            IF ( NDXI .LE. NDXJ .AND. NDXI .GT. 0 ) THEN
C              FIND SYSTEM COEFF IN VECTOR S
              NDXV = JTEMP + NDXI
C             NDXV = IDIAG(AMAX0(NDXI,NDXJ))
C                    - IABS(NDXJ-NDXI)
              SS(NDXV) = SS(NDXV) + S(I,J)
            ENDIF
 10       CONTINUE
        ENDIF
 20   CONTINUE
      RETURN
      END
```

Figure 20.4.6 Assemble into skyline vector storage mode

```
      SUBROUTINE  SKYSUB (NDFREE, IDIAG, I, J, IJV)
C     * * * * * * * * * * * * * * * * * * * * * * * *
C     CONVERT (I,J) FULL SYMMETRIC MATRIX SUBSCRIPTS
C     TO IJV SUBSCRIPT OF VECTOR SKYLINE STORAGE MODE
C     * * * * * * * * * * * * * * * * * * * * * * * *
C     ASSUMING SYMM EQS, COLS STORED FROM TOP DOWN
      DIMENSION  IDIAG(NDFREE)
C     NDFREE   = TOTAL NO OF SYSTEM EQUATIONS
C     IDIAG(I) = LOCATION OF DIAG OF I-TH EQ
      ID  = MAX0 (I,J)
      IJV = IDIAG(ID) - IABS(I-J)
      RETURN
      END
```

Figure 20.4.7 Convert from full to skyline vector storage

method is used then STORSQ would be replaced by a procedure such as SKYSTR in Fig. 20.4.6. The goal in both cases is the same. It is to take a coefficient that is to be added to the term $S_{I,J}$ in the full matrix and to determine where it must actually be placed when the sparse matrix storage method is considered. In the skyline mode the vector

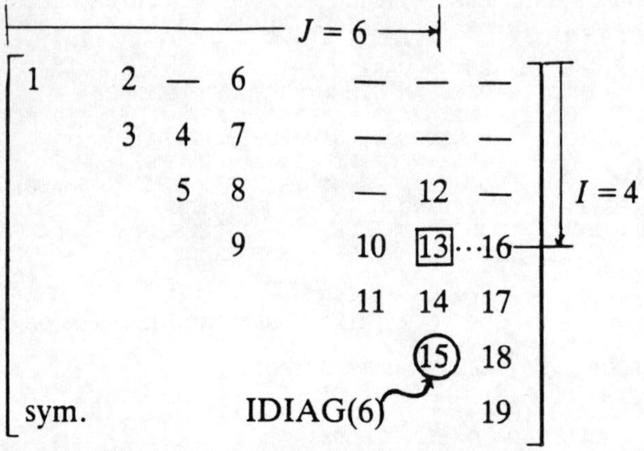

$$\overbrace{}^{J=6}$$

$$
\left[\begin{array}{ccccccc}
1 & 2 & - & 6 & - & - & - \\
 & 3 & 4 & 7 & - & - & - \\
 & & 5 & 8 & - & 12 & - \\
 & & & 9 & 10 & \boxed{13} \cdots & 16 \\
 & & & & 11 & 14 & 17 \\
 & & & & & (15) & 18 \\
\text{sym.} & & & \text{IDIAG}(6) & & & 19
\end{array}\right] \quad I = 4
$$

$$[S_{IJ}] \leftrightarrow \{S_K\} \quad K = \text{IDIAG}(\text{MAXO}(I, J)) - |I - J|$$
$$I = 4, J = 6 \rightarrow K = \text{IDIAG}(6) - |4 - 6|$$
$$K = 15 - 2 = 13$$

Figure 20.4.8 Locating skyline terms

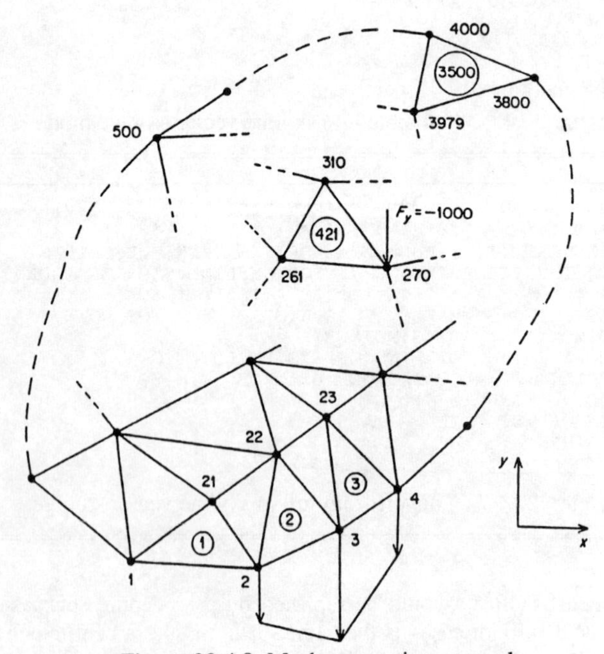

Figure 20.4.9 Mesh generation example

position, IDIAG(J), of the diagonal term, $S_{I,J}$, is found and then steps are taken backwards (up the column) to locate where $S_{I,J}$ is stored. The latter is summarized in in Figs. 20.4.7 and 20.4.8. In the bandwidth mode, Fig. 20.4.2, the diagonal term $S_{I,J}$ is found in the first column of row I of the rectangular storage array. Then Eq. (20.9) is employed to count over the correct number of columns to the location of $S_{I,J}$. In both cases symmetry is assumed. In a non-symmetric problem both the upper and lower band segments are needed. Then the same pointers could be used with two arrays (upper and lower parts of **S**). Then only the diagonal is duplicated in storage and the same routines are still valid.

20.4.3 Example

Consider a two-dimensional problem (NSPACE = 2) involving 4000 nodal points (M = 4000) and 3500 elements (NE = 3500). Assume two parameters per node (NG = 2) and let these parameters represent the horizontal and vertical components of some vector. In a stress analysis problem, the vector could represent the displacement vector of the node, whereas in a fluid flow problem it could represent the velocity vector at the nodal point. Assume the elements to be triangular with three corner nodes (N = 3). The local numbers of these nodes will be defined in some consistent manner, e.g., by numbering counter-clockwise from one corner. This mesh is illustrated in Fig. 20.4.9. For simplicity assume that two floating point material properties are associated with each element, and that the elements are homogeneous throughout the domain. (This implies that NLPFLO = 2, LHOMO = 1, NNPFIX = NNPFLO = NLPFIX = MISCFX = MISCFL = NHOMO = 0.) In addition, assume that the nodal parameter boundary conditions specify the following conditions: (1) both parameters are zero at nodal point 1; (2) at nodal point 500 the first parameter equals zero; and (3) at nodal point 4000 the two parameters are related by the equation $3/5\, P_1 - 4/5\, P_2 = 0$, where P_1 and P_2 denote the first and second parameters, respectively (at that node). Define the two material properties of the elements to have the values of 30×10^6 and 0.25, respectively. Assume that in the y-direction at node 270 there is a known initial contribution to the load vector (system column matrix). Let this coefficient be associated with the second nodal parameter at the node and have a value of -1000. Finally, assume that there are specified flux components in the y-direction along the two element segments (NSEG = 2 and LBN = 2) between nodes 2 and 3, and 3 and 4. Let the magnitudes of the fluxes be -100, -150, and -25 at nodes 2, 3, and 4, respectively. The input data to, and the arrays stored by the building block codes for this example, are shown below. The resulting arrays stored by the program are identified by their names in capital letters and/or are surrounded by boxes in Tables 20.3 through 20.9.

Table 20.3 Nodal data

Node	Boundary condition code	Spatial coordinates	
1	11	0.00	0.00
2	00	0.25	0.00
3	00	0.50	0.10
⋮	⋮	⋮	⋮
261	00	2.25	1.50
⋮	⋮	⋮	⋮
270	00	2.75	1.50
⋮	⋮	⋮	⋮
310	00	2.60	1.90
⋮	⋮	⋮	⋮
500	10	0.00	3.40
⋮	⋮	⋮	⋮
4000	22	9.80	5.60
	Array IBC	Array X	

Table 20. 4 Connectivity

Element	Element indices		
1	1	2	21
2	2	3	22
3	3	4	23
⋮	⋮	⋮	⋮
421	261	270	310
⋮	⋮	⋮	⋮
3500	3979	3980	4000
	Array NODES		

Table 20.5 Type 1 Constraint

Node	Parameter	Value	Calculated dof
1	1	0.0	1
1	2	0.0	2
500	1	0.0	999
		Array CEO1	Array NDX1

Table 20.6 Type 2 Constraint

First dof		*Second dof*		*Equation coefficients*		
Node	Parameter	Node	Parameter	c_1	c_2	c_3
4000	1	4000	2	0.6	−0.8	0.0
7999					−1.33	
8000					0.00	
Array NDX2					Array CEQ2	

Table 20.7 Element properties

Element no.	Properties	
1	30.E6	0.25
	Array FLTEL	

Table 20.8 Nodal sources

Node	Parameter	Value
270	2	−1000
4000	2	0

$J = 2(270 - 1) + 2 = 540$

CC(540) = −1000 , plus 7999 zero terms

Table 20.9 Flux data

	Segment 1			
Nodes	2	3		
Flux components	0.0	−100	0.0	−150
	Segment 2			
Nodes	3	4		
Flux components	0.0	−100	0.0	−25

By utilizing the above control parameters, it is easy to determine the total number of degrees of freedom in the system, NDFREE, and associated with a typical element, LEMFRE: NDFREE = $M*NG$ = 4000*2 = 8000, LEMFRE = $N*NG$ = 3*2 = 6. In addition to the total number of degrees of freedom in the system, it is important to be able to identify the system degree of freedom number that is associated with any parameter in the system. Equation (20.2) or subroutine DEGPAR, provides this information. This relation has many practical uses. For example, when one specifies that the first parameter ($J = 1$) at node 500 ($I = 500$) has some given value what one is indirectly saying is that system degree of freedom number NDF = 2*(500 − 1) + 1 = 999 has a given value. In a similar manner, we often need to identify the system degree of freedom numbers that correspond to the ELMFRE local degrees of freedom of the element. In order to utilize Eq. (20.2) to do this, one must be able to identify the N node numbers associated with the element of interest. This is relatively easy to accomplish since these data are part of the input data (element incidences). For example, for element number 421 we find the three element incidences (row 421 of NODES) to be

System	261	270	310	← Array LNODE
Local	1	2	3	(1∗N)

Therefore, by applying Eq. (20.2), we find the degree of freedom numbers in Table 20.10.

Table 20.10 Element indices

Node local i	Number system I	Parameter number J	Degree of Freedom System NDF_s	Degree of Freedom Local NDF_L
1	261	1	521	1
		2	522	2
2	270	1	539	3
		2	540	4
3	310	1	619	5
		2	620	6
	Array LNODE		Array INDEX	

The element array INDEX has many programming uses. Its most important application is to aid in the assembly (scatter) of the element equations to form the governing system equations. Recall from Fig. 20.4.1 that the element equations are expressed in terms of local degree of freedom numbers. In order to add these element coefficients into the system equations one must identify the relation between the local degree of freedom numbers and the corresponding system degree of freedom numbers. Array INDEX provides this information for a specific element. In practice, the assembly procedure is as follows. First the system matrices are set equal to zero. Then a loop over all the elements if performed. For each element, the element matrices are generated in terms of the local degrees of freedom. The coefficients of the element matrices are added to the corresponding coefficients in the system matrices. Before the addition is carried out, the element array INDEX is used to convert the local subscripts of the coefficient to the system subscripts of the term in the system equations to which the coefficient is to be added. That is, we scatter

$$S^e_{i,j} \overset{+}{\rightarrow} S_{I,J}, \quad C^e_i \overset{+}{\rightarrow} C_I \tag{20.9}$$

where $I_s = \text{INDEX}(i_L)$ and $J_s = \text{INDEX}(j_L)$ are the corresponding row and column numbers in the system equations, i_L, j_L are the subscripts of the coefficients in terms of the local degrees of freedom, and $\overset{+}{\rightarrow}$ reads "is added to". Considering all of the the terms in the element matrices for element 421 in the previous example, one finds

$$\begin{aligned}
S^e_{1,1} &\overset{+}{\rightarrow} S_{521,521} & C^e_1 &\overset{+}{\rightarrow} C_{521} \\
S^e_{2,3} &\overset{+}{\rightarrow} S_{522,539} & C^e_2 &\overset{+}{\rightarrow} C_{522} \\
S^e_{3,4} &\overset{+}{\rightarrow} S_{539,540} & C^e_3 &\overset{+}{\rightarrow} C_{539} \\
S^e_{4,5} &\overset{+}{\rightarrow} S_{540,620} & C^e_4 &\overset{+}{\rightarrow} C_{540} \\
S^e_{5,6} &\overset{+}{\rightarrow} S_{619,620} & C^e_5 &\overset{+}{\rightarrow} C_{619} \\
S^e_{1,6} &\overset{+}{\rightarrow} S_{521,620} & C^e_6 &\overset{+}{\rightarrow} C_{620}\,.
\end{aligned}$$

Let L_{min} and L_{max} denote, respectively, the maximum and minimum system degree of freedom numbers associated with a particular element. Then one can make the general observation that the coefficients of the square element matrix will be included in a square subregion of the system matrix. As illustrated in Fig. 20.2.4, the limits of this square region are defined by L_{min} and L_{max}. From this figure, one can also note that the half-bandwidth associated with a typical element (including the diagonal) is given by $\text{HBW} = L_{max} - L_{min} + 1$, For example, for element 421, the $\text{HBW}_{421} = 620 - 521 + 1 = 100$. The equation for the half-bandwidth (including the diagonal) may be expressed in a much more useful form. Note that from Eq. (20.2) $L_{max} = \text{NG}*(I_{max} - 1) + J_{max}$, $L_{min} = \text{NG}*(I_{min} - 1) + J_{min}$ where $J_{min} = 1$ and $J_{max} = \text{NG}$ and where I_{max} and I_{min} denote the largest and smallest system node numbers associated with the element, respectively. Substituting these two relations into the above expression for the half-bandwidth yields:

$$\text{HBW} = \text{NG}*(I_{max} - I_{min} + 1) \tag{20.10}$$

which was stated earlier in Eq. (20.3). As a check, we can return to element 421 of the example to determine that $\text{HBW}_{421} = 2*(310 - 261 + 1) = 100$ which agrees with the previous calculations. After the equations have been completely assembled, it is necessary to apply the boundary conditions. Searching array IBC, one determines that there are three Type 1 constraint equations and one (2/2) Type 2 constraint equation. Again, Eq. (20.2) is applied to determine the system degree of freedom that must be modified to include these constraint equations. The programming concepts associated with applying the boundary conditions will be discussed in the next section.

While on the subject of assembly processes, it may be useful to cite the *assembled normal algorithm* of Belytischko, et al [3]. This is a very useful procedure for identifying the elements and/or nodes that lie on the boundary of the mesh. Those data can greatly reduce the graphical display time of some output results. They are also important in problems where surfaces may move into contact with each other. One begins by computing the unit normal vector on each face of an element. These are assigned to each node on the face. The sum of the normal vectors from the faces yields the nodal normal vectors on that element. If a global assembly of these nodal vectors is carried out over all the elements, the result will be the net normal vector at every node. If the node is on the interior of the object, then the vectors from its surrounding elements cancel, and the result is (approximately) a null vector. For nodes on the surface of the domain, only the components normal to the surface do not cancel. Thus, those nodes have a vector whose resultant is not zero. We flag those nodes as boundary nodes and re-scale each of the corresponding vectors to unit vectors so that they can be used to provide light shading for image display. They could also be employed to convert normal pressures to the global components needed in the resultant source integral.

20.5 Essential Boundary Conditions and Constraints

20.5.1 Application of Essential Boundary Conditions

After the system equations have been fully assembled, it is necessary to apply the *essential boundary conditions* before the equations can be solved for the unknown nodal parameters. There are at least three different formal procedures for applying these boundary constraints. In this section, three types of nodal parameter boundary constraints will be defined. The most common boundary conditions (Type 1) are of the form

$$D_j = b\,, \tag{20.11}$$

where D_j represents system degree of freedom number, j, and b is the specified value assigned to that degree of freedom. Obviously, this boundary condition has many practical applications. For example, in a heat transfer problem this can be used to assign a specific temperature to a particular nodal parameter. Also, this type of condition occurs very frequently in structural analysis problems. The next boundary constraint (Type 2) to be considered is of the form

$$AD_j + BD_k = C\,, \tag{20.12}$$

where D_j and D_k represent two system degrees of freedom, and A, B, and C are arbitrary constants ($A \neq 0$). This restraint contains the first condition as a special case. The second type of constraint also has various practical applications. For example, in a two-dimensional inviscid fluid flow problem, this can be used to require that the two velocity components at a node be tangential to a solid boundary at that node. It is also known as a coupled node condition when $A = 1$, $B = -1$, and $C = 0$. The third and final boundary constraint (Type 3) to be considered is of the form

$$AD_j + BD_k + CD_l = E. \tag{20.13}$$

This type of constraint can occur in several three-dimensional problems. For example, in a structural analysis problem one could have a node constrained to displace tangential to an arbitrary plane in space (general roller support). Such a support condition can be described by Eq. (20.13). These three constraints are probably the most commonly used (nodal parameter) boundary conditions found in practical finite element problems. It should be clear that this process can be extended to any number of linear constraints. The analyst should observe that if there are m Type 2 constraint equations (and corresponding input data), then there are $2m$ nodal parameters that will have Type 2 boundary code indicators. Similarly, if there are m Type 3 constraint equations then there are $3m$ nodal parameters with Type 3 indicator codes, etc.

As mentioned earlier, there are two common methods used to apply the boundary conditions. The first method is to modify the assembled equations to reflect the boundary constraints and then to solve the complete set of equations. The second method is to eliminate the restrained parameters and then to solve the reduced set of equations. The first method has the advantage that it can more easily handle general forms of restraint. However, it requires all the nodal parameters to be calculated; even those with specified values. The second method is often used in structural analysis programs and has the advantage that only the unrestrained parameters have to be calculated. Its major disadvantage is that it is difficult to program for constraints other than the first type.

20.5.2 Matrix Manipulations

The first method will be illustrated by outlining the modifications necessary to incorporate a Type 2 constraint. To simplify the procedure, Eq. (20.12) will be rewritten as $D_j + aD_k = b$ where $a = B/A$ and $b = C/A$. Since the procedure is not restricted to banded matrices the assumption of a banded matrix will temporarily be lifted. The assumption of symmetry is retained. The original form of the system equations is

$$\begin{array}{c} \\ 1 \\ j \\ k \\ p \end{array}\begin{array}{c} \begin{array}{cccc} 1 & j^* & k^* & p \end{array} \\ \begin{bmatrix} S_{11} & S_{1j} & S_{1k} & S_{1p} \\ S_{j1} & S_{jj} & S_{jk} & S_{jp} \\ S_{k1} & S_{kj} & S_{kk} & S_{kp} \\ S_{p1} & S_{pj} & S_{pk} & S_{pp} \end{bmatrix} \end{array} \begin{Bmatrix} D_1 \\ D_j \\ D_k \\ D_p \end{Bmatrix} = \begin{Bmatrix} C_1 \\ C_j \\ C_k \\ C_p \end{Bmatrix}.$$

Recalling that $D_j = b - aD_k$ one can replace D_j times the j-th column by $(b - aD_k)$ times the j-th column. This substitution results in the modification

$$\begin{array}{c} \\ 1 \\ j \\ k \\ p \end{array}\begin{array}{c} \begin{array}{cccc} 1 & j & k & p \end{array} \\ \begin{bmatrix} S_{11} & 0 & (S_{1k} - aS_{1j}) & S_{1p} \\ S_{j1} & 0 & (S_{jk} - aS_{jj}) & S_{jp} \\ S_{k1} & 0 & (S_{kk} - aS_{kj}) & S_{kp} \\ S_{p1} & 0 & (S_{pk} - aS_{pj}) & S_{pp} \end{bmatrix} \end{array} \begin{Bmatrix} D_1 \\ D_j \\ D_k \\ D_p \end{Bmatrix} = \begin{Bmatrix} C_1 - bS_{1j} \\ C_j - bS_{jj} \\ C_k - bS_{kj} \\ C_p - bS_{pj} \end{Bmatrix}.$$

Note that these changes destroy the symmetry of the equations and result in a determinant of zero. To restore the symmetry, subtract 'a' times the j-th row from the

k-th row and then multiply the j-th row by zero. This second modification is

$$\begin{array}{c} \\ 1 \\ j \\ k \\ p \end{array}\begin{array}{c} \begin{array}{cccc} 1 & j & k & p \end{array} \\ \left[\begin{array}{cccc} S_{11} & 0 & (S_{1k}-aS_{1j}) & S_{1p} \\ 0 & 0 & 0 & 0 \\ (S_{k1}-aS_{j1}) & 0 & (S_{kk}-2aS_{jk}+a2S_{jj}) & (S_{kp}-aS_{jp}) \\ S_{p1} & 0 & (S_{pk}-aS_{pj}) & S_{pp} \end{array}\right] \end{array}\begin{Bmatrix} D_1 \\ D_j \\ D_k \\ D_p \end{Bmatrix} =$$

$$= \begin{Bmatrix} C_1 - bS_{1j} \\ 0 \\ C_k - aC_j - bS_{kj} + abS_{jj} \\ C_p - bS_{pj} \end{Bmatrix}.$$

These operations restore the symmetry of the problem, but the equations are still singular since the dependence on D_j has been removed. Either the equations must be reduced in size by one or modified still further. To be consistent, the latter approach will be used. expressed in a symmetric matrix form as

$$\begin{bmatrix} 1 & a \\ a & a^2 \end{bmatrix} \begin{Bmatrix} D_j \\ D_k \end{Bmatrix} = \begin{Bmatrix} b \\ ab \end{Bmatrix}. \tag{20.14}$$

Combining these relations with the last modification results in the final form necessary to include the second type of constraint. This final form is

$$\begin{array}{c} \\ 1 \\ j \\ k \\ p \end{array}\begin{array}{c} \begin{array}{cccc} 1 & j & k & p \end{array} \\ \left[\begin{array}{cccc} S_{11} & 0 & (S_{1k}-aS_{1j}) & S_{1p} \\ 0 & 1 & a & 0 \\ (S_{k1}-aS_{j1}) & a & (S_{kk}-2aS_{jk}+a2S_{jj}+a^2) & (S_{kp}-aS_{jp}) \\ S_{p1} & 0 & (S_{pk}-aS_{pj}) & S_{pp} \end{array}\right] \end{array}\begin{Bmatrix} D_1 \\ D_j \\ D_k \\ D_p \end{Bmatrix}$$

$$= \begin{Bmatrix} C_1 - bS_{1j} \\ b \\ C_k - aC_j - bS_{kj} + abS_{jj} + ab \\ C_p - bS_{pj} \end{Bmatrix}.$$

The solution of this complete set of symmetric equations will exactly satisfy the given constraint equation. It should be noted that the modification necessary to incorporate the first type of boundary constraint is obtained as a special case ($a = 0$) of the above procedure. The final modification for this type of constraint is

$$\begin{array}{c} 1 \\ j \\ k \\ p \end{array}\begin{bmatrix} S_{11} & 0 & S_{1k} & S_{1p} \\ 0 & 1 & 0 & 0 \\ S_{k1} & 0 & S_{kk} & S_{kp} \\ S_{p1} & 0 & S_{pk} & S_{pp} \end{bmatrix}\begin{Bmatrix} D_1 \\ D_j \\ D_k \\ D_p \end{Bmatrix} = \begin{Bmatrix} C_1 - bS_{1j} \\ b \\ C_k - bS_{kj} \\ C_p - bS_{pj} \end{Bmatrix}.$$

(columns labelled 1, j, k, p)

Note that if the original matrix is banded then the modification for constraint Types 2 and 3 will increase the bandwidth size. The Type 1 constraint, which is most common in practice, does not increase the bandwidth. In practice, the increase in bandwidth for Types 2 and 3 will be small since the constrained parameters are usually at the same node. The increase in bandwidth can be minimized by making programming changes (see REVISE and CONVRT) that will partition, during assembly, the system equations so that the parameters are separated according to their constraint type number. Figure 2.6.1 illustrated a simple Fortran subroutine, MODFUL, for applying a Type 1 boundary

```
      SUBROUTINE  MODFY1 (NDFREE, MBW, L1, C1, SS, CC)
C     * * * * * * * * * * * * * * * * * * * * * * * * * * *
C     APPLY TYPE 1 CONSTRAINT EQUATION MODIFICATIONS TO
C     UPPER HALF BANDWIDTH,      SS*DD = CC, DD(L1) = C1
C     * * * * * * * * * * * * * * * * * * * * * * * * * * *
      DIMENSION  SS(NDFREE,MBW), CC(NDFREE)
C     SS      = UPPER HALF BANDWIDTH OF SYMMETRIC SYSTEM
C     CC      = SYSTEM COLUMN MATRIX
C     L1      = SPECIFIED SYSTEM DEGREE OF FREEDOM NUMBER
C     C1      = SPECIFIED CONSTRAINT EQUATION COEFFICIENT
C     MBW     = MAX. HALF BANDWIDTH OF SYSTEM
C     NDFREE = TOTAL DEGREES OF FREEDOM OF SYSTEM
      PARAMETER  ( ZERO = 0.0 )
      M1 = MIN0 (L1,MBW) - 1
      IF ( M1 .GT. 0 )  THEN
        DO 10  I = 1,M1
          IROW = L1 - I
          ICOL = I + 1
          IF ( C1 .NE. ZERO )  THEN
            CC(IROW) = CC(IROW) - C1*SS(IROW,ICOL)
          ENDIF
 10     SS(IROW,ICOL) = ZERO
      ENDIF
      M1 = MIN0 ( (NDFREE + 1 - L1),MBW )
      DO 20  I = 1,M1
        IROW = L1 - 1 + I
        ICOL = I
        IF ( C1 .NE. ZERO )  THEN
          CC(IROW) = CC(IROW) - C1*SS(L1,ICOL)
        ENDIF
 20   SS(L1,ICOL) = ZERO
      SS(L1,1) = 1.0
      CC(L1) = C1
      RETURN
      END
```

Figure 20.5.1 Boundary condition change of a banded matrix

constraint to a full system (or element) square matrix. Again it should be noted that these types of routines would only be practical for problems involving a small number of unknowns.

It was pointed out above that, except for Type 1 constraint equations (which fortunately are the most common), the modification of the system equations will result in an increase in the size of the half-bandwidth of the equations. The increased bandwidth results from the addition of two columns (or rows) that takes place during the modification procedure and the size of the new bandwidth depends on the original size of the bandwidth, due to the element topology, and the difference in the column (or row) numbers that are being added. Once the constraint equation data and the system list of element incidences have been input, it is possible to calculate the increase (if any) in the bandwidth. First, the original half-bandwidth is determined (by subroutine SYSBAN). Next, the degree of freedom numbers of the parameters in the constraint equations are determined (by subroutine INDXPT). The differences in these (column and row) numbers are combined with the original bandwidth to determine the new bandwidth. These calculations (executed by subroutine MODBAN) take place early in the solution process so that adequate storage space can be set aside for the upper half-bandwidth of the system equations. The increased bandwidth (if any) results in many additional zero coefficients being introduced into the equations to be solved. This is serious since the initial zeros become non-zeros during the factorization of the system square matrix. Thus, the exact modifications are undesirable for banded modes with other than Type 1 nodal constraints. The banded storage mode modifications are executed by MODFY1, MODFY2 and MODFY3. Figure 20.5.1 shows subroutine MODFY1. These modifications would usually be under the control of a routine (such as APLYBC) which recovers the input constraint equation degree of freedom numbers and coefficients. Its flow chart was given earlier, while the actual details are given in the subroutines on the disk. The generalization to any number of linear constraints for a full system are in MODLFL, and in SKYLCE for a skyline vector storage mode.

The above procedure is clearly correct when there is only one linear constraint equation. If there is more than one constraint equation, there is a danger that the zeroing operations in one constraint will destroy data in later constraint equations. This is usually not going to happen unless two constraint equations involve a common degree of freedom. Many programs do not allow that case as a valid input condition, and thus, forces the user to avoid the danger. However, adaptive analysis programs can generate a very large number of constraint equation sets, and it is common for the master corner nodes to occur in several constraint equations at one time. To remove the potential danger of loosing interacting coefficients, the coding should simply be changed to do all the zeroing loops after all the constraint equations row and column operations have been completed. If a skyline solution procedure is utilized then the exact manipulation procedure could be used with little additional cost if high accuracy in multi-point constraints is required. Then one uses routines like CEQSKY and MODSKY to invoke the essential boundary conditions. Of course, there are other procedures, such as transformation methods, that can be employed to enforce constraints. In problems with a high percentage of constrained degrees of freedom an iterative solver with constraints may be best.

The second method for applying Type 1 boundary conditions will be outlined at this point. This procedure has been described in detail in several texts. Note that the system equations could be arranged in the following partitioned form

$$\begin{bmatrix} \mathbf{S}_{\alpha\alpha} & \mathbf{S}_{\alpha\beta} \\ \mathbf{S}_{\beta\alpha} & \mathbf{S}_{\beta\beta} \end{bmatrix} \begin{Bmatrix} \mathbf{D}_{\alpha} \\ \mathbf{D}_{\beta} \end{Bmatrix} = \begin{Bmatrix} \mathbf{C}_{\alpha} \\ \mathbf{C}_{\beta} \end{Bmatrix}$$

where $\mathbf{D}_\alpha$ represents the unknown nodal parameters, and $\mathbf{D}_\beta$ represents the boundary values of the prescribed Type 1 nodal parameters. The sub-matrices $\mathbf{S}_{\alpha\alpha}$ and $\mathbf{S}_{\beta\beta}$ are square, whereas $\mathbf{S}_{\alpha\beta}$ and $\mathbf{S}_{\beta\alpha}$ are rectangular, in general. It has been assumed that the equations have been assembled in a manner that places the prescribed Type 1 parameters at the end of the system equations. The above matrix relations can be rewritten as

$$\begin{aligned} \mathbf{S}_{\alpha\alpha}\mathbf{D}_{\alpha} + \mathbf{S}_{\alpha\beta}\mathbf{D}_{\beta} &= \mathbf{C}_{\alpha} \\ \mathbf{S}_{\beta\alpha}\mathbf{D}_{\alpha} + \mathbf{S}_{\beta\beta}\mathbf{D}_{\beta} &= \mathbf{C}_{\beta} \end{aligned} \tag{20.15}$$

so that the unknown nodal parameters are obtained by inverting the matrix $\mathbf{S}_{\alpha\alpha}$, that is

$$\mathbf{D}_{\alpha} = \mathbf{S}_{\alpha\alpha}^{-1}(\mathbf{C}_{\alpha} - \mathbf{S}_{\alpha\beta}\mathbf{D}_{\beta}). \tag{20.16}$$

Also, if desired, the value of $\mathbf{C}_\beta$ is determined from

$$\mathbf{C}_{\beta} = \mathbf{S}_{\beta\alpha}\mathbf{S}_{\alpha\alpha}^{-1}\mathbf{C}_{\alpha} - (\mathbf{S}_{\beta\alpha}\mathbf{S}_{\alpha\alpha}^{-1}\mathbf{S}_{\alpha\beta} - \mathbf{S}_{\beta\beta})\,\mathbf{D}_{\beta}. \tag{20.17}$$

20.5.3 Constraints Applied at the Element Level *

The previous sections on the application of boundary nodal constraints have assumed that the constraints would be applied at the system level after the equations have been fully assembled. This is generally the most efficient way of applying the nodal constraints. However, there are situations where it is necessary or useful to apply the nodal constraints at the element level. The above procedures for introducing the nodal constraints can also be applied to the element matrices, $\mathbf{S}^e$ and $\mathbf{C}^e$. This is true so long as all the degrees of freedom given in the linear constraint equation occur at a common node. For example, consider a problem with one parameter per node and assume that the value at a particular node is defined to be V, i.e., a Type 1 constraint. Further, assume that there are g elements connected to that node. Then the matrices of each of the g elements would be modified such that the corresponding square matrix diagonal term and column matrix terms are one and V respectively. When the g elements are assembled, the system matrix has a corresponding diagonal term of g and the column matrix term has a value of $g \times V$. Thus, the final result gives the same as would be obtained if the modifications were carried out at the system level. Since the element matrices usually are stored in their full form, instead of in upper half-bandwidth form, one would need a subroutine such as MODFUL (Fig. 2.6.1) to apply the Type 1 modification at the element level.

20.5.4 Penalty Modifications for Nodal Constraints *

Nodal parameter constraints can also be accomplished by the use of an artifice such as the following. Let 'i' denote the degree of freedom to be assigned a given value, say V. This trick consists of modifying two terms of the system matrix and column vector, i.e., $S_{i,i}$ and C_i. These terms are re-defined to be

$$C_i = \beta V S_{i,i}, \quad S_{i,i} = \beta S_{i,i} \tag{20.18}$$

where β is a very large number. This yields the i-th system equation

$$S_{i,1} D_1 + \cdots + \beta S_{i,i} D_i + \cdots + S_{i,n} D_n = \beta V S_{i,i},$$

which is a good approximation of the boundary condition $D_i = V$, if β is sufficiently large. This artifice is the fastest method for introducing Type 1 constraints, but it can lead to numerical ill-conditioning. There is also a problem of how large a value should be assigned to β. Commonly used values range from 10^{12} to 10^{28}. The higher order constraints could be introduced in a similar manner. For example, a Type 2 constraint such as that in Eq. (20.5.4) could be also added to the system equations after being multiplied by some penalty weight. Such a procedure has the advantage that it adds very little to skyline or frontal storage procedures. However, there are numerical difficulties. The multiplication by the weighting factor introduces large numbers into off-diagonal terms and may cause ill-conditioning of the equations. There is also a problem of how to select the penalty weight. If a very large penalty is used, as in the Type 1 case, the system tends to *lock-up,* i.e., yield a zero solution, and the constraints approach a set of Type 1 constraints. To prevent this one must retain some information about element contributions to the corresponding equations. One way of doing this is to find the average value of the original diagonal terms to be modified by the constraint and then the constraint equation penalty could be selected to be 100 or 1000 times larger. This gives reasonable accuracy, but the exact matrix manipulation procedure is better.

The symmetric constraints in Eq. (20.5.4) can be generalized to any number of terms. Recall that the normalized coefficients, CEQ, and the degree of freedom numbers, NDXC, are stored for each constraint of Type n, where

$$\begin{aligned} 1 \cdot D_{NDXC(1)} + CEQ(1) \cdot D_{NDXC(2)} + \cdots + CEQ(i) \cdot D_{NDXC(i+1)} + \cdots \\ + CEQ(n-1) \cdot D_{NDXC(n)} = CEQ(n) . \end{aligned} \tag{20.19}$$

Thus, a least squares procedure can be used to form the penalty square and column matrices, **SP** and **CP**. Form a temporary vector, $\mathbf{V}(n+1)$, such that $\mathbf{V}(1) = 1$ and $\mathbf{V}(i+1) = CEQ(i)$ for $1 \le i \le n$. Calculate the product $\mathbf{V}^T \mathbf{V}$ and delete the last row. Then the last column contains **CP** while the remainder of the expression is **SP**. Several other types of penalty constraints have been given by Campbell [5]. Constraints via the penalty method could be invoked with PENLTY and PENMOD which are on the disk.

20.6 Economical Solution of the System Equations

Modern computer methods of finite element analysis present us with the problem of having to solve large systems of simultaneous algebraic equations. If the analysis is to be performed economically, it is critical that we give careful consideration to the equation solving process. Present experience indicates that from one-third to one-half of the computer time involved in a linear analysis is associated with solving simultaneous equations. In nonlinear and dynamic analysis, as much as 80% of the computer time is associated with solving the system equations.

Certain properties of the square system matrix in the finite element method allow us to employ a number of techniques that reduce not only the solution time, but the amount of computer storage required to perform the analysis. As mentioned before, the system equations are *symmetric, positive definite, sparse,* and are often *banded.* Symmetry allows us to economize on computer storage since only the elements on the main diagonal and those above (or below) the diagonal need be stored. Because the matrix is positive definite, the process of *pivoting* is never required in order to ensure a stable solution. This allows us to save on computational steps. Banding, however, depends

(a) Original equations

$$\begin{bmatrix} S_{11} & S_{12} & S_{1j} & S_{1n} \\ S_{21} & S_{22} & & S_{2n} \\ \vdots & & & \vdots \\ S_{j1} & & S_{jj} & S_{jn} \\ \vdots & & & \vdots \\ S_{n1} & & & S_{nn} \end{bmatrix} \begin{Bmatrix} D_1 \\ D_2 \\ \vdots \\ D_j \\ \vdots \\ D_n \end{Bmatrix} = \begin{Bmatrix} C_1 \\ C_2 \\ \vdots \\ C_j \\ \vdots \\ C_n \end{Bmatrix}$$

(b) Equations after factorization

$$\begin{bmatrix} S'_{11} & S'_{11} & S'_{1j} & S'_{1n} \\ & S'_{22} & & S'_{2n} \\ & & & \vdots \\ & & S'_{jj} & \vdots \\ & & & \vdots \\ \text{zero} & & & S'_{nn} \end{bmatrix} \begin{Bmatrix} D_1 \\ D_2 \\ \vdots \\ D_j \\ \vdots \\ D_n \end{Bmatrix} = \begin{Bmatrix} C'_1 \\ C'_2 \\ \vdots \\ C'_j \\ \vdots \\ C'_n \end{Bmatrix}$$

Figure 20.6.1 Result of Gauss factorization

directly on the way the node numbers are assigned and the analyst's ability to prepare an efficient numbering scheme. Banding provides us with the capability to further save on computer storage and computational steps in that only those elements within the band will be stored and involved in the operations. Figures 20.2.4, 20.2.15, and 20.4.2 are illustrative of a banded, symmetric matrix. Since the solution time varies approximately as the square of the bandwidth, the nodal numbering producing the minimum bandwidth results in the most economical solution. By way of comparison, it should be mentioned here that the solution time for equation solvers which operate on all elements in a square array varies cubically with the number of equations.

20.6.1 Efficient Node Numbering

The previous section has again highlighted the importance of the half-bandwidth on the storage requirements and the solution time. There are more advanced programming techniques that can be utilized to reduce the significance of the half-bandwidth, but these methods will not be considered here. Since the solution procedures described here are banded techniques, the effects of the nodal numbering scheme will be briefly reviewed in the hope that the user will exercise the proper care when numbering nodes. Recall that the half-bandwidth caused by a typical element is given by:

$$HBW = NG \times (I_{max} - I_{min} + 1)$$

and is illustrated graphically in Figs. 20.2.4 and 20.4.2. Note that the analyst assigns the node numbers. Thus, he has control over largest and smallest node numbers (I_{max} and I_{min}, respectively) that are assigned to any element. He, therefore, has direct control over the maximum half-bandwidth that occurs in the system of elements. Usually the analyst can number the nodes so that the minimum half-bandwidth is achieved by inspection. However, for three-dimensional problems and multiple connected regions, one may find it necessary to rely on algorithms (see REDUCE Sec. 20.7) that automatically renumber the nodes so as to reduce (but not necessarily minimize) the bandwidth.

20.6.2 Equation Solving Techniques

Before discussing specific equation solving algorithms, it is necessary to discuss briefly those capabilities that are essential for efficient and economical solutions of the system equations. The equation solver should be able to take advantage of the sparse, symmetric, banded nature of the equations. It should also be capable of handling multiple solution vectors economically. In addition, solution algorithms should be compared on the basis of calculation and data handling efficiencies.

There are three principle direct solution algorithms in wide use today. These are (1) Gauss elimination, (2) Choleski factorization, and (3) modified Gauss-Choleski factorization. From the standpoint of calculation efficiency, it has been proven that no algorithm for equation solving can involve fewer calculations than Gauss elimination. The original Choleski algorithm requires only $10N$ (where N is the number of equations) more calculations than Gauss elimination. This results from the square-root operation in the original Choleski algorithm. It has been shown that there is much less error associated with the Gauss process than the original Choleski procedure. This is attributed to the fact that the repeated square-root operation severely degrades the accuracy of the process on some computers. This is the fault of the computer and not the algorithm. Despite the disadvantages of the original Choleski process, it has important data storage advantages. It is possible to retain the best features of both the Gauss and Choleski algorithms within a modified Gauss-Choleski algorithm.

For simplicity we first consider full matrices. Basically, the Gauss elimination procedure consists of two parts: a factorization procedure; and a back-substitution process. The effect of the factorization procedure on a full matrix is illustrated in Figs. 20.6.1 to 20.6.5. The coefficients below the diagonal are reduced to zero (i.e., *eliminated*) by performing a sequence of row operations. For example, to eliminate the terms below the diagonal in column i, one multiplies row i by $(-k_{ji}/k_{ii})$ and adds the result to row j, where $(i+1) \leq j \leq 1$. After the factorization is completed, as shown in Fig. 20.6.5, the last row contains one equation with one unknown. At this point the back-substitution process may be started, that is, the last equation, $S_{nn} D_n = C_n$, is solved for the value of D_n. Once D_n is known, it is substituted into the next to the last equation, which is then solved for D_{n-1}. The back-substitution continues in a similar manner until all the D_i are known. Algorithms for the Gaussian elimination technique can be found in many texts on numerical analysis. However, most finite element codes do not use this method since other methods, such as the Choleski method, result in more efficient storage utilization and simpler handling of multiple solution vectors, **C**.

The (original) Choleski factorization method factors the system square matrix **S** into the product of a lower triangular matrix (i.e., coefficients above the diagonal are zero), **L**,

$$
\begin{aligned}
S_{11} &= d_{11} \\
S_{21} &= d_{11} L_{21} \\
S_{22} &= d_{11} L_{21} L_{21} + d_{22} \\
S_{31} &= d_{11} L_{31} \\
S_{32} &= d_{11} L_{31} L_{21} + d_{22} L_{32} \\
S_{33} &= d_{11} L_{31} L_{31} + d_{22} L_{32} L_{32} + d_{33} \\
S_{41} &= d_{11} L_{41} \\
S_{42} &= d_{11} L_{41} L_{21} + d_{22} L_{42} \\
S_{43} &= d_{11} L_{41} L_{31} + d_{22} L_{42} L_{32} + d_{33} L_{43} \\
S_{44} &= d_{11} L_{41} L_{41} + d_{22} L_{42} L_{42} + d_{33} L_{43} L_{43} + d_{44}
\end{aligned}
$$

(a) The $\mathbf{S} = \mathbf{L}\mathbf{d}\mathbf{L}^T$ identity

$$
\begin{aligned}
d_{11} &= S_{11} \\
L_{21} &= S_{21}/d_{11} \\
d_{22} &= S_{22} - d_{11} L_{21} L_{21} \\
L_{31} &= S_{31}/d_{11} \\
L_{32} &= (S_{32} - d_{11} L_{31} L_{21})/d_{22} \\
d_{33} &= S_{33} - d_{11} L_{31} L_{31} - d_{22} L_{32} L_{32} \\
L_{41} &= S_{41}/d_{11} \\
L_{42} &= (S_{42} - d_{11} L_{41} L_{21})/d_{22} \\
L_{43} &= (S_{43} - d_{11} L_{41} L_{31} - d_{22} L_{42} L_{32})/d_{33} \\
d_{44} &= S_{44} - d_{11} L_{41} L_{41} - d_{22} L_{42} L_{42} - d_{33} L_{43} L_{43}
\end{aligned}
$$

(b) Solution for coefficients of **L** and **d**

Figure 20.6.2 A factorization of four equations

and the transpose of this lower triangular matrix, that is,

$$\mathbf{S}\,\mathbf{D} = \mathbf{C} \tag{20.19}$$

is factored into

$$\mathbf{L}\,\mathbf{L}^T\mathbf{D} = \mathbf{C}. \tag{20.20}$$

Then defining

$$\mathbf{G} = \mathbf{L}^T\mathbf{D}, \tag{20.21}$$

the previous equation can be written as

$$\mathbf{L}\,\mathbf{G} = \mathbf{C}. \tag{20.22}$$

$$\begin{aligned}
G_1 &= C_1 \\
L_{21}\,G_1 + G_2 &= C_2 \\
L_{31}\,G_1 + L_{32}\,G_2 + G_3 &= C_3 \\
L_{41}\,G_2 + L_{42}\,G_2 + L_{43}\,G_3 &= C_4
\end{aligned}$$

(a) The $\mathbf{L}\mathbf{G} = \mathbf{C}$ identity

$$\begin{aligned}
G_1 &= C_1 \\
G_2 &= C_2 - L_{21}\,G_1 \\
G_3 &= C_3 - L_{31}\,G_1 - L_{32}\,G_2 \\
G_4 &= C_4 - L_{41}\,G_1 - L_{42}\,G_2 - L_{43}\,G_3
\end{aligned}$$

(b) Solution for $\mathbf{G}$

Figure 20.6.3 Forward substitution for four equations

$$\begin{aligned}
G_4 &= d_{44}\,D_4 \\
G_3 &= d_{33}\,(D_3 + L_{43}\,D_4) \\
G_2 &= d_{22}\,(D_2 + L_{32}\,D_3 + L_{42}\,D_4) \\
G_1 &= d_{22}\,(D_1 + L_{21}\,D_2 + L_{31}\,D_3 + L_{41}\,D_4)
\end{aligned}$$

(a) The (reverse) $\mathbf{G} = \mathbf{d}\mathbf{L}^T\mathbf{D}$ identity

$$\begin{aligned}
D_4 &= G_{41}/d_{44} \\
D_3 &= G_{31}/d_{33} - L_{43}\,D_4 \\
D_2 &= G_{21}/d_{22} - L_{32}\,D_3 - L_{42}\,D_4 \\
D_1 &= G_1/d_{11} - L_{21}\,D_2 - L_{31}\,D_3 - L_{41}\,D_4
\end{aligned}$$

(b) Solution for the coefficients of $\mathbf{D}$

Figure 20.6.4 Back-substitution for four equations

Since **L** and **C** are known and **L** is a triangular matrix (with one equation and one unknown in the first row), the intermediate vector, **G**, can be determined by a forward-substitution process. After **G** has been determined from Eq. (20.22), the unknown nodal parameters, **D**, can be obtained by a back-substitution into Eq. (20.21). These concepts are shown graphically in Fig. 2.4.2.

The attractive features of the Choleski algorithm should now be more apparent. First, note that it is only necessary to store **L** or $\mathbf{L}^T$, whichever is more convenient. Secondly, once the system square matrix, **S**, has been factored the unknown nodal parameters, **D**, can be evaluated by a simple forward- and backward-substitution utilizing the given solution vector(s), **C** (and the intermediate vector, **G**). In most problems the equations only need be solved for one value of **C**. However, in dynamic or transient problems and in structural problems with several load conditions one may require the solutions involving many different values of **C**. With the Choleski scheme the time-consuming operations of assembling and factoring **S** are only performed once in linear problems. For every given **C** only the relatively fast forward- and back-substitutions need be repeated. The difficulty with the original Choleski algorithm was that the calculation of the diagonal terms of **L** required a square-root operation. This operation required an undesirable amount of time and also led to relatively inaccurate results. Thus, a modified Gauss-Choleski algorithm was developed.

A modified Gauss-Choleski algorithm retains the better features of both techniques and at the same time eliminates the undesirable square-root operations in the factorization. The algorithm is outlined below in symbolic form. The system equations, $\mathbf{S}\,\mathbf{D} = \mathbf{C}$, are factored into

$$(\mathbf{L}\,\mathbf{d}\,\mathbf{L}^T)\,\mathbf{D} = \mathbf{C} \tag{20.23}$$

in which **L** is a lower triangular matrix with ones on the diagonal matrix (see Figs. 2.4.2

(a) Original system, $\mathbf{SD} = \mathbf{C}$, $\begin{bmatrix} 1 & -1 & 1 & 2 \\ & 5 & -3 & 0 \\ & & 3 & 0 \\ & \text{sym} & & 7 \end{bmatrix} \begin{Bmatrix} D_1 \\ D_2 \\ D_3 \\ D_4 \end{Bmatrix} = \begin{Bmatrix} 2 \\ -4 \\ 4 \\ 1 \end{Bmatrix}$

(b) Factorization, $\mathbf{L}^T = \begin{bmatrix} 1 & -1 & 1 & 2 \\ 0 & 1 & -1/2 & 1/2 \\ 0 & 0 & 1 & -1 \\ 0 & 0 & 0 & 1 \end{bmatrix}$, $\mathbf{d} = \begin{bmatrix} 1 & & & \\ & 4 & & \\ & & 1 & \\ & & & 1 \end{bmatrix}$,

$\mathbf{S} = \mathbf{L}\mathbf{d}\mathbf{L}^T$

(c) Forward-substitution, $\mathbf{L}\,\mathbf{G} = \mathbf{C}$, $\mathbf{G}^T = [\,2 \quad -2 \quad 1 \quad -1\,]$

(d) Scaling, $\mathbf{d}\,\mathbf{H} = \mathbf{G}$, $\mathbf{H}^T = [\,2 \quad -1/2 \quad 1 \quad -1\,]$

(e) Back-substitution, $\mathbf{L}^T\mathbf{D} = \mathbf{H}$, $\mathbf{D}^T = [\,4 \quad 0 \quad 0 \quad -1\,]$

Figure 20.6.5 Sample factorization and solution

and 20.6.5). It is the introduction of the diagonal matrix that eliminates the square-root operations. Proceeding as before, let

$$\mathbf{G} = \mathbf{d}\,\mathbf{L}^T\mathbf{D}, \tag{20.24}$$

then

$$\mathbf{L}\,\mathbf{G} = \mathbf{C}. \tag{20.25}$$

Given the value of **C** one obtains **G** by a forward-substitution into Eq. (20.27), and then **G** is back-substituted into Eq. (20.24) to yield the nodal parameters, **D**. The last operation can be divided into two operations; first, a scaling calculation, using the diagonal matrix **d**

$$\mathbf{L}^T\mathbf{D} = \mathbf{d}^{-1}\mathbf{G} \equiv \mathbf{H},$$

and then the back substitution of **H** to find the **D**.

20.6.3 Programming the Solution

Clearly, the first step in programming the solution of the system equations is to identify the algorithm for calculating the arrays **L** and **d**. To illustrate these operations consider a system with only four equations. Carrying out the products and equating the lower (or upper) triangle gives the identities of Fig. 20.6.2(a). As shown in Fig. 20.6.2(b) the identities can be solved to define the necessary terms in **L** and **d**. The above operations can be generalized for any number of equations by the expressions:

$$d_{11} = S_{11}, \quad d_{ii} = S_{ii} - \sum_{k=1}^{i-1} d_{kk} L_{ik}^2, \qquad i > 1, \tag{20.26}$$

$$L_{ii} = 1, \quad L_{ij} = \left[S_{ij} - \sum_{k=1}^{j-1} d_{kk} L_{ik} L_{jk} \right] / d_{jj}, \quad i > j, L_{ij} = O, \quad i < j.$$

Of course, one only calculates the L_{ij} terms when $i > j$. The L_{ij} terms are stored in the original locations of the S_{ij}. The d_{ii} are stored on the diagonal in place of the $L_{ii} = 1$ terms. Since the above factorization operations can represent a significant part of the total computational costs every effort is usually made to program Eqs. (20.26) as efficiently as possible. Thus, it is often difficult to compare the software with Eqs. (20.26). This is in part due to the fact that large finite element matrices are not stored in their full mode as was assumed in Eqs. (20.26). If they are banded the summations must be restricted to the bandwidth. Conversely, if a skyline storage is used the summations are carried out only with the stored columns.

Since both **L** and **C** are known, the identity of Eq. (20.25) gives the forward-substitution algorithm for the determination of **G**. The procedure is shown in Figs. 20.6.3 and 20.6.5. The general expression is

$$G_i = C_i - \sum_{k=1}^{i-1} G_k L_{ik}. \tag{20.27}$$

Note that **G** could be stored in either the old **C** locations or the, as yet unneeded, **D** locations. Here we use the latter. The backward-substitution is defined by Eq. (20.24). The operations are illustrated (in reverse order) in Fig. 20.6.5, and the resulting expressions are

```
      SUBROUTINE  FACTOR (NDFREE, IBW, S)
C     * * * * * * * * * * * * * * * * * * * * * * * * * * *
C       LDLT FACTOR OF BANDED SYMMETRIC SQUARE MATRIX
C     * * * * * * * * * * * * * * * * * * * * * * * * * * *
      DIMENSION  S(NDFREE,IBW)
C     NDFREE = MAX. DEGREES OF FREEDOM OF SYSTEM
C     IBW    = MAXIMUM HALF BANDWIDTH OF SYSTEM EQS
C     S      = RECT MATRIX WITH UPPER HALF BAND OF SYS EQS
      TEMP = 1.0/S(1,1)
      DO 10 J = 2,IBW
 10   S(1,J) = S(1,J)*TEMP
      DO 40  I = 2,NDFREE
        LL = I - 1
        NN = NDFREE - LL
        IF (NN .GT. IBW) NN = IBW
        DO 30 J = 1,NN
          L = IBW - J
          SUM = 0.0
C           ALLOW FOR OMITTED NODES
          IF ( L .GT. 0 ) THEN
            IF ( LL .LT. L ) L = LL
            DO 20 K = 1,L
              K1 = I - K
              K2 = 1 + K
              K3 = J + K
 20         SUM = SUM + S(K1,K2)*S(K1,K3)*S(K1,1)
          ENDIF
        S(I,J) = S(I,J) - SUM
        IF ( J .GT. 1 )  S(I,J) = S(I,J) / S(I,1)
 30     CONTINUE
 40   CONTINUE
      RETURN
      END
```

Figure 20.6.6 Banded matrix factorization

$$D_i = G_i/d_{ii} - \sum_{k=1}^{n-1} D_{i+k}\, L_{i+k,\,i} \tag{20.28}$$

for n equations. Again note that **D** could be stored in **G**, that is in the old **C**. Many codes utilize this storage efficiency. In the actual subroutines in use here, **D** and **C** are given as different arrays. Of course, they could be assigned the same storage pointer for added storage efficiency. A sample solution is shown in Fig. 20.6.5. The actual subroutines used for solutions, FACTOR and SOLVE, are given in Figs. 20.6.6 and 20.6.7, respectively; for linear problems FACTOR need be called only once while SOLVE would be called once for each loading condition. For a skyline storage mode the programming becomes more involved. Typical programs have been given by Taylor [15] and Bathe and Wilson [2]. Programs such as SKYSOLVE illustrate one approach. They employ a function program, DOT, to evaluate the dot (or scaler) product of two vectors. They are compatible with the previously discussed skyline subroutines.

In 1992, the National Aeronautics and Space Administration released to the public an efficient Choleski direct equation solver designed for large (> 100,000) linear systems. This code, PVSOLVE, is vectorized and also designed for parallel computers, as well as sequential computers [14]. It is a column solver, but has extra zeros included to make the vectorization and loop unrolling efficient. Test problems with about 60,000 unknowns were solved by NASA in less than four seconds on a supercomputer. PVSOLVE employs special sparse storage arrays and therefore the assembly procedures must use those features. Programs ELWIDE, SKYPAD, SKYPVD, SKYROW, STORPV, and SKYRMV were developed to allow MODEL to be interfaced with it.

20.7 Frontal Assembly and Solution Procedures *

In addition to banded and skyline or profile methods of storage and solution, many finite element codes utilize *frontal methods*. Usually an element by element frontal assembly and elimination procedure is employed. Such a procedure can utilize a relatively small memory. Unlike the procedures considered to this point, the efficiency of frontal methods is independent of node numbers and is dependent on the element processing order. Usually the element matrices are written to a sequential file. They are read from that file, assembled and factored simultaneously. A very detailed description of the frontal method, along with the supporting software, was given in the text by Hinton and Owen [10]. Here only a limited insight into the frontal assembly procedure will be presented. This involves the definition between the element degree of freedom numbers and the current system dof number destination number in memory.

The frontal method stores only equations that are currently active. Equations that have received all of their element (and constraint) contributions are eliminated to provide storage for equations that may become active with the next element. To illustrate these points consider a mesh with eight nodes, 14line elements, two nodes per element and one degree of freedom per node (M = 8, NE = 12, N = 2, NG = 1, NDFREE = 8). If assembled in full form before elimination takes place, storage must be provided for the total NDFREE = 8 equations. The frontal method generally requires less storage. This is illustrated in Figs. 20.7.1 and 20.7.2, which show the mesh and the storage location (or destination) of each system equation as the elements are read. Note that after the first four elements are read, node 3 (destination 2) has made its last contribution. Thus equation 3 can be eliminated and its storage locations freed for the next new equation. When the next element, 5, is read, node 6 makes its first appearance and can be stored where equation 3 used to be located. Note that proceeding in this manner requires storage for 7 equations instead of 8. Note that equations 5 and 7 are the last two to be eliminated.

The adaptive methods of solution discussed in Chap. 10 have degrees of freedom that appear and/or disappear at each solution stage. To conserve space one must place newly appearing equations in the spaces vacated by those no longer active. Thus, it is not possible to employ a simple expression like Eq. (20.2) to find the INDEX terms needed in the assembly process. It becomes necessary to use two arrays to keep up with the equation numbers. Therefore, the examples in Fig. 20.7.2 may also be useful in planning the assembly software for adaptive solutions. It shows that if the elements are input in a different order only a maximum of 4 equations must be stored instead of 8. The maximum number of equations required in memory is usually called the front size or wavefront. Algorithms for reordering elements to yield reduced fronts have been given in reference [1]. An improved procedure, REDUCE, is included on the disk. The above discussion shows that frontal solvers (and multi-front solvers on parallel machines) depend only on the element processing order, and is independent of the node numbers. Conversely, the column solvers and band solvers depend only on the node numbering order, and is independent of the element numbers. Various algorithms have been developed to reduce the maximum and/or *rms* wavefront or column height. The

```
      SUBROUTINE  SOLVE (NDFREE, IBW, S, P, D)
C     * * * * * * * * * * * * * * * * * * * * * * * * * * *
C      FORWARD AND BACK SUBSTITUTION OF SYSTEM EQUATIONS
C          PART TWO OF CHOLESKY-GAUSSIAN SOLUTION
C     * * * * * * * * * * * * * * * * * * * * * * * * * * *
      DIMENSION  S(NDFREE,IBW), P(NDFREE), D(NDFREE)
C     NDFREE = MAX. DEGREES OF FREEDOM IN SYSTEM
C     IBW    = MAXIMUM HALF BANDWIDTH OF THE SYSTEM
C     S      = FACTORED SYS SQ MATRIX FROM SUBR FACTOR
C     D      = SYSTEM DEGREES OF FREEDOM TO BE DETERMINED
C     P      = SYSTEM COLUMN MATRIX (KNOWN)
      D(1) = P(1) / S(1,1)
C---  FORWARD SUBSTITUTION
      DO 20  I = 2,NDFREE
        II = I + 1
        J = II - IBW
        IF(II .LE. IBW ) J = 1
        IK = I - 1
        SUM = 0.0
        DO 10 K = J,IK
          KK = II - K
 10     SUM = SUM + S(K,KK)*S(K,1)*D(K)
 20   D(I) = ( P(I) - SUM ) / S(I,1)
C---  BACK SUBSTITUTION
      DO 40  NN = 2,NDFREE
        I = NDFREE + 1 - NN
        LL = I - 1
        J = LL + IBW
        IF ( J.GT.NDFREE )  J = NDFREE
        L = I + 1
        SUM = 0.0
        DO 30 K = L,J
          KK = K - LL
 30     SUM = SUM + S(I,KK)*D(K)
 40   D(I) = D(I) - SUM
      RETURN
      END
```

Figure 20.6.7 Forward and back substitution of banded matrix

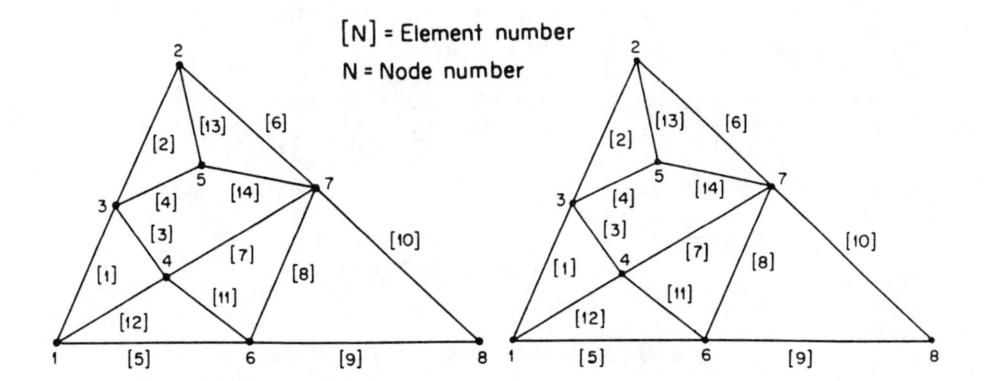

Figure 20.7.1 Example mesh topologies, [N] = element number

	Destination number			
	1	2	3	4
Element	Node active			
1	1	3	–	–
2	1	3	2	–
3	1	3	2	4
4	1	3	2	4
5	1	6	2	4
6	1	6	2	4
7	1	6	2	4
8	1	6	2	4
9	1	6	2	4
10	1	6	2	4
11	1	6	2	4
12	1	–	2	4
13	–	–	2	–
14	–	–	–	–

a) Original

	Destination number			
	1	2	3	4
Element	Node active			
9 → 1	6	8	–	–
10 → 2	6	8	7	–
5 → 3	6	1	7	–
8 → 4	6	1	7	–
11 → 5	6	1	7	4
12 → 6	–	1	7	4
1 → 7	3	1	7	4
7 → 8	3	–	7	4
3 → 9	3	–	7	4
14 → 10	3	5	7	–
6 → 11	3	5	7	2
4 → 12	3	5	–	2
13 → 13	3	5	–	2
2 → 14	3	–	–	2

b) New element order

Figure 20.7.2 Active front size during solution

program REDUCE provided on the disk is designed primarily to reduce wavefront sizes, but can also be employed to reduce column heights. The main re-ordering logic is in subroutine LRESEQ in the REDUCE program. This re-ordering program uses the complete list of elements connected to other elements. This is a large and expensive array to build, so that it would be desirable to save this list and pass it to the analysis program to be used in the adaptivity process and/or in a pre-front equation solving stage. The program can be re-started so as to avoid the large expense of rebuilding this list.

20.8 Iterative Solvers *

Iterative equation solvers let us avoid the assembly of the system square matrix, but they usually do assemble to system column matrix. The element square matrix and its degree of freedom numbers are usually written to a sequential file during the generation phase. Then the products necessary to obtain the iterative solution are carried out in loops over all the elements. The terms in the estimated solution vector are gathered to the element level, an element matrix-vector product is computed, and that resulting addition to the correction vector is scattered to the system level.

The conjugate gradient (CG) method of Hestenes and Stiefel [9] is an algorithm for solving the system of equations $\mathbf{Ax} = \mathbf{b}$ in the case that $\mathbf{A}$ is symmetric and positive definite. The method can be recommended over simple factorization methods if $\mathbf{A}$ is very large and sparse. Theoretically, the conjugate gradient algorithm will yield the solution of the system $\mathbf{Ax} = \mathbf{b}$ in n steps if $\mathbf{A}$ is $n \times n$. In practice, however, the algorithm is used as an iterative method to produce a sequence of vectors converging to the solution. This is necessary because roundoff errors often prevent the algorithm from furnishing a sufficiently precise solution at the n-th step. If $\mathbf{A}$ is symmetric positive definite then the problem of solving $\mathbf{Ax} = \mathbf{b}$ is equivalent to the problem of minimizing the quadratic form $Q(x) = ½\mathbf{x}^T\mathbf{AX} - \mathbf{x}^T\mathbf{b}$. We begin by finding out how the function Q behaves along a one-dimensional ray from $\mathbf{x}$ in the direction $\mathbf{v}$. Straightforward calculation reveals that for a step size of t:

$$\begin{aligned} Q(\mathbf{x} + t\,\mathbf{v}) &= \frac{1}{2}\,(\mathbf{x} + t\,\mathbf{v})^T\,\mathbf{A}(\mathbf{x} + t\,\mathbf{v}) - (\mathbf{x} + t\,\mathbf{v})^T\,\mathbf{b} \\ &= Q(\mathbf{x}) + t\,\mathbf{v}^T\mathbf{Ax} - t\,\mathbf{v}^T\mathbf{b} + \frac{1}{2}\,t^2\,\mathbf{v}^T\mathbf{Av}\,. \end{aligned} \tag{20.29}$$

From this we compute the derivative with respect to the distance, t, traveled in the direction $\mathbf{v}$ as

$$\frac{d}{dt}\,Q(\mathbf{x} + t\,\mathbf{v}) = \mathbf{v}^T(\mathbf{Ax} - \mathbf{b}) + t\,\mathbf{v}^T\mathbf{Av}\,. \tag{20.30}$$

The minimum of Q along the ray occurs when this derivative is zero. The value of t which yields the minimum point is, therefore,

$$t^* = \mathbf{v}^T(\mathbf{b} - \mathbf{Ax})/\mathbf{v}^T\mathbf{Av}\,.$$

Using this value, t^*, we compute the minimum of Q at that point on the ray:

$$Q(\mathbf{x} + t^*\,\mathbf{v}) = Q(\mathbf{x}) - t^*\,\mathbf{v}^T(\mathbf{b} - \mathbf{Ax}) = Q(\mathbf{x}) - [\mathbf{v}^T(\mathbf{b} - \mathbf{Ax})]^2/\mathbf{v}^T\mathbf{Av}\,.$$

Since the last term is always positive or zero this calculation shows that a reduction of the value of Q always occurs in passing from $\mathbf{x}$ to $\mathbf{x} + t^*\mathbf{v}$ unless $\mathbf{v}$ is orthogonal to the residual, $\mathbf{b} - \mathbf{Ax}$. If $\mathbf{x}$ is not a solution of the system $\mathbf{Ax} = \mathbf{b}$, then many vectors $\mathbf{v}$ exist satisfying $\mathbf{v}^T(\mathbf{b} - \mathbf{Ax}) \neq 0$. Hence, if $\mathbf{Ax} \neq \mathbf{b}$, then $\mathbf{x}$ does not minimize Q. On the other hand, if $\mathbf{Ax} = \mathbf{b}$, then there is no ray emanating from $\mathbf{x}$ on which Q takes a lesser value than $Q(\mathbf{x})$. Hence such an $\mathbf{x}$ is a minimum point of Q. These observations suggest an iterative method for solving $\mathbf{Ax} = \mathbf{b}$. We proceed by minimizing Q along a succession of rays. At the k-th step in such an algorithm $\mathbf{x}_k$ would be available. Then a suitable search direction $\mathbf{v}_k$ is chosen. The next point in our sequence is

$$\mathbf{x}_{k+1} = \mathbf{x}_k + t_k \mathbf{v}_k ,$$

where the current estimate of the optimal step size is

$$t_k = \mathbf{v}_k^T (\mathbf{b} - \mathbf{Ax}_k) / \mathbf{v}_k^T \mathbf{Av}_k .$$

The steepest decent direction could be chosen for the direction $\mathbf{v}_k$, but in practice we find that it results in lots of little steps and the convergence to the minimum is very slow. The conjugate gradient method employs the residual vector $\mathbf{r}_k \equiv \mathbf{b} - \mathbf{Ax}_k$ to pick the search direction. The sequence of residuals are specially chosen so that they are non-interfering or conjugate directions. This means that the sequence $[\, \mathbf{r}_0 , \mathbf{r}_1 , \ldots , \mathbf{r}_{k+1} \,]$ is an orthogonal set. The outline of such a conjugate gradient iterative solver is:

$\mathbf{x}_0$ is arbitrary choice
Enforce constraints on $\mathbf{x}_0$
$\mathbf{r}_0 = \mathbf{Ax}_0 - \mathbf{b}$
$\mathbf{v}_0 = -\mathbf{r}_0$
Correct $\mathbf{r}_0$ for constraints on $\mathbf{x}_0$

For $k = 0,\ 1, 2, \ldots, k_{\text{max}}$

Stop if $\mathbf{v}_k = \mathbf{0}$. Otherwise, compute
$t_k = \mathbf{r}_k^T \mathbf{r}_k / \mathbf{v}_k^T \mathbf{Av}_k$
$\mathbf{x}_{k+1} = \mathbf{x}_k + t_k \mathbf{v}_k$
Enforce constraints on $\mathbf{x}_{k+1}$
$\mathbf{r}_{k+1} = \mathbf{r}_k + t_k \mathbf{Av}_k$
Correct $\mathbf{r}_{k+1}$ for $\mathbf{x}$ constraints

$s_k = \mathbf{r}_{k+1}^T \mathbf{r}_{k+1} / \mathbf{r}_k^T \mathbf{r}_k$
$\mathbf{v}_{k+1} = -\mathbf{r}_{k+1} + s_k \mathbf{v}_k$

End for.

In practice, for finite element applications we do not store the system square matrix $\mathbf{A}$ so its products with the vectors $\mathbf{x}$ or $\mathbf{v}$ must be evaluated as a vector, say $\mathbf{g}$, which is summed over element products of $\mathbf{A}^e$ and $\mathbf{x}^e$ or $\mathbf{v}^e$. Forming the work space vector $\mathbf{g}$ is the most computationally intensive parts of the algorithm. In reading the above algorithm one should read $\mathbf{Ax}_0$ as $\mathbf{g}_0$ and $\mathbf{Av}_k$ as $\mathbf{g}_k$. Fried [8] has discussed the use of CG methods in detail for finite element applications. Subroutine CGPROD is utilized to evaluate the product $\mathbf{g}$ with element level products. It utilizes routine ELFRE to pull out the element level vector $\mathbf{g}^e$ to multiply the element square matrix, $\mathbf{S}^e$. Since we do not have the assembled system square matrix we must either apply the constraints at the element level, or evoke them in a weak sense with pseudo element penalty arrays, or modify the iterative solver to satisfy the essential boundary conditions and/or the linear constraints. It also makes sense to start the iteration with a guess, $\mathbf{x}_0$, that satisfies them. In general, we can rewrite our linear constraints discussed earlier to put the redundant or slave degree of freedom on the LHS of the equation

$$D_{\text{NDXC}(1)} = \text{CEQ}(n) - \sum_{j=1}^{n-1} \text{CEQ}(j)\, D_{\text{NDXC}(j+1)}$$

which degenerates to an essential Dirichlet condition if $n = 1$. In a matrix symbolic form for each constraint equation set we would write this as

$$x_s = \mathbf{C}\,\mathbf{x}_m + f_s$$

where $\mathbf{x}_m$ is the master degrees of freedom for the constraint set that define the single slave value, $\mathbf{x}_s$. Subroutine CGCEQ carries out this operation on the current trial solution, $\mathbf{x}$. When a slave has been constrained in this manner, or assigned a specific value, then we want it to remain at that value. Thus, its residual should be zero and its pointer vector should be zero since we do not want it to move away. In other words, the residual and pointer must also be corrected for each constraint set. The correction is

$$\mathbf{r}_m = \mathbf{C}^T r_s\,, \qquad r_s = 0$$

and the same for the pointer vector. Both correction operations are carried out in CGCEQC. These three tools give us what we need to execute the CG iterative solution. The controlling program for this, CGITER, is given on the disk. Note that in addition to the system source vector and solution vector we must provide storage for the residual and pointer vectors.

Before leaving the topics of efficient assembly and solution methods the reader is reminded that the advent of parallel computers is having an important impact on finite element analysis techniques. For examples of this new important area the reader is referred to the work of Farhat [6, 7], Johnsson and Mathur [11], Law [12], Owen [13], and Zois [17].

20.9 Programming Parametric Elements

In Sec. 9.4, we saw that the element matrices, $\mathbf{S}^e$ or $\mathbf{C}^e$, are usually defined by integrals of the symbolic form

$$\mathbf{A}^e = \iiint_{\Omega^e} \mathbf{B}^e(x, y, z)\, dx\, dy\, dz = \int_{-1}^{1}\int_{-1}^{1}\int_{-1}^{1} \hat{\mathbf{B}}^e(r, s, t)\, \left|J(r, s, t)\right|\, dr\, ds\, dt\,.$$

where $\mathbf{B}^e$ is usually the sum of products of other matrices involving the element interpolation functions and problem properties. With the element formulated in terms of the local (r, s, t) coordinates, where $\mathbf{B}^e(x, y, z)$ is transformed into $\hat{\mathbf{B}}^e(r, s, t)$, one would use numerical integration to obtain

$$\mathbf{A}^e = \sum_{i=1}^{NIP} W_i\, \hat{\mathbf{B}}^e(r_i, s_i, t_i)\, \left|J(r_i, s_i, t_i)\right|$$

where $\mathbf{B}^e$ and $|J|$ are evaluated at each of the NIP integration points, and where (r_i, s_i, t_i) and W_i denote the tabulated abscissae and weights, respectively. The general procedure for programming an isoparametric element is outlined below:

(A) Initial Programming Steps

1. Define a subroutine (such as SHP4Q) to evaluate the N interpolation functions $\mathbf{H}(r, s, t)$ for any given values of the local coordinates (r, s, t).
2. Define a subroutine (such as DER4Q) to evaluate the NSPACE local derivatives of the interpolation functions $\Delta(r, s, t)$ for any given values of the local coordinates.

3. Tabulate or calculate the number of integration points, NIP, and the weight coefficients W_i, and local coordinate abscissae (r_i, s_i, t_i) of each integration point.

(B) Steps Preceding the Element Matrix Subroutines

1. Extract the global coordinates, $[\mathbf{x}^e : \mathbf{y}^e : \mathbf{z}^e]$, **COORD**, of the nodes of the element.
2. Extract any constant element properties (ELPROP and/or LPROP), and tabulate the nodal point values (LPTPRT) of any properties which vary with local coordinates.

(C) Steps Within the Element Matrix Routine

1. Establish storage for any problem-dependent variables ($\mathbf{H}$, $\mathbf{\Delta}$, $\mathbf{d}$, $\mathbf{J}$, $\mathbf{J}^{-1}$, etc.).
2. Zero the element matrices.
3. Perform the numerical integration (see Part D).

(D) Steps for Each Integration Point

1. Extract, or calculate, the weight and abscissae of the integration point.
2. For the given abscissae (r_i, s_i, t_i) evaluate the interpolation functions, $[\mathbf{H}]$, (call SHP4Q) and the local derivatives of the interpolation functions, $[\Delta]$, (call DER4Q).
3. Calculate the Jacobian matrix at the point (call JACOB):

$$\mathbf{J}(r_i, s_i, t_i) = \mathbf{\Delta}(r_i, s_i, t_i)\,\mathbf{COORD}.$$

4. Calculate the inverse, $\mathbf{J}(r_i, s_i, t_i)^{-1}$, and determinant $|J|$ of the Jacobian matrix (e.g., call I2BY2).
5. Calculate the first order global derivatives, $\mathbf{d}$, of the interpolation functions (call GDERIV):

$$\mathbf{d} = \mathbf{J}^{-1}\,\mathbf{\Delta}.$$

6. If the problem properties vary with the local coordinates, use the nodal values of the properties and the interpolation functions to calculate the properties at the integration point (call CALPRT, Fig. 20.9.1).
7. Execute the matrix operations defining the matrix integrand. In general, this involves the sum of products of element properties, $\mathbf{H}$, and $\mathbf{d}$.
8. Multiply the resulting matrix by the weighting coefficient and the determinant of the Jacobian and add it to the previous contributions to the element matrix.
9. If any of the above data are to be utilized in post-solution calculations, write them on auxiliary storage.

A typical arrangement of the latter operations is illustrated in subroutine ISOPAR, which is shown in Fig. 9.4.3. Note that the global coordinates of the integration point are also determined since they are often used for later output of the derivatives. Of course, in an axisymmetric analysis the global radial coordinate is required in the integrand. For most practical problems the Jacobian inverse can be easily obtained with a routine such as INVDET. The interpolation function routine, SHAPE, and the corresponding local derivative routine, DERIV, will be considered later. Numerical integration is also reviewed in more detail in a later section. The control section of subroutine NGRAND is used to evaluate the actual problem-dependent integrands. As we saw in Sec. 11.6, it is sometimes desirable to use different integration rules for different parts of the integrand. Thus, two or more versions of NGRAND might be required in a typical application. Therefore, in calling ISOPAR the name NGRAND is treated as a dummy EXTERNAL variable so that the user passes in the name of the subroutine that is actually required for the problem. Specific examples will be given in Chap.s 21 and 22.

```
      SUBROUTINE  CALPRT (N, NNPFLO, H, PRTLPT, VALUES)
C     * * * * * * * * * * * * * * * * * * * * * * * * * *
C       CALCULATE NNPFLO PROPERTIES AT A LOCAL PT USING
C        ELEMENT'S NODAL PROPERTIES, PRTLPT, AND THE N
C          INTERPOLATION FUNCTIONS, H, AT THE POINT
C     * * * * * * * * * * * * * * * * * * * * * * * * * *
      DIMENSION  H(N), PRTLPT(N,0:NNPFLO), VALUES(0:NNPFLO)
C     N      = NUMBER OF NODES PER ELEMENT
C     NNPFLO = NO. OF FLOATING POINT NODAL PROPERTIES
C     H      = INTERPOLATION FUNCTIONS FOR AN ELEMENT
C     PRTLPT = FLOATING PT PROPS OF ELEMENT'S NODES
C     VALUES = LOCAL VALUES OF PROPERTIES
      IF ( NNPFLO .LT. 1 ) STOP 'NNPFLO = 0, CALPRT'
      DO 20  I = 1, NNPFLO
        SUM = 0.0
        DO 10  J = 1, N
          SUM = SUM + H(J)*PRTLPT(J,I)
 10     CONTINUE
        VALUES(I) = SUM
 20   CONTINUE
      RETURN
      END
```

Figure 20.9.1 Interpolating for spatial properties

20.10 Output of Results

Once the nodal parameters of the system have been calculated it is necessary to output these quantities in a practical form. The number of nodal parameters (degrees of freedom) can be quite large. Common problems often involve 2000 parameters but some special problems have required as many as 50,000 (or more) parameters. Also, the reader should recall that these nodal parameters are often utilized to calculate an equally large set of secondary quantities. Obviously, a simple printed list of these nodal parameters is not the optimum form of output for the analyst. Data presented in a graphical form are probably the most practical. Several computer graphics programs have been developed for use with certain specialized finite element codes. However, these graphics packages are often expensive and usually highly hardware-dependent. Thus it is not feasible here to delve into the subject of computer graphics as a mode of data presentation. Instead the following sections will discuss the utilization of printed output.

The most straightforward way to list the calculated nodal parameters is to list each of the M nodal points in the system and the values of the NG parameters that correspond to each of these nodal points. It is common practice to also list the NSPACE spatial coordinates of the nodal points so as to save the analyst the inconvenience of referring back to the input data to determine the location of the point. The above format for printing the calculated nodal parameters is utilized by subroutine WRTPT. It determines the system degree of freedom to be printed by substituting the system node number (I) of the point and the local parameter (J) into Eq. (1.1).

It is desirable at times to know the values of the calculated parameters that are associated with some particular element. This can be accomplished by listing the data of interest associated with each of the NE elements in the system. These data usually include a list of all nodal points (element incidences) associated with the element, and a corresponding list of the NG nodal parameters that occur at each of these nodes. When using this approach it is obvious that several of the nodal parameters will be listed more than once since most of the nodal points are shared by more than one element. The element incidences are available as input data and thus can be used with Eq. (20.2) to identify which system degrees of freedom should be printed with a particular element. These elementary operations are executed by subroutine WRTELM, which is included on the disk with WRTPT. Once the nodal parameters, **D**, have been calculated, the analyst may desire to know the extreme values of each of the NG nodal parameters and the node numbers where each extreme value occurred. These simple operations are performed by subroutine MAXMIN. This returns two arrays, RANGE and NRANGE. Both arrays are NG*2 in size. The first column of RANGE contains the NG maximum values of **D** and the second column contains the corresponding minimum values. Array NRANGE contains the node numbers where the corresponding extreme values in RANGE were encountered.

It is desirable to also have information on the contours of the nodal dof. One would prefer to have the computer plot the contour lines illustrating the spatial variation of each of the NG types of nodal parameters. Lacking this ability one would like the computer to supply data that would make plotting by hand as simple as possible. As discussed in the previous section, the extreme values of any particular type of parameter (there are NG) are readily determined. The procedure utilized here for each parameter type is to print the spatial coordinates of points lying on specified contour curves. The analyst specifies the number of contours (> 2) to range from 5% above the minimum value to 5% below the maximum value for each parameter. The contour curves can be generated in several ways. One simple procedure (not the most efficient) is as follows. Examine, in order, each element in the system to ascertain if the contour passes through that element. That is, determine if the specified contour value is bounded by the nodal parameters associated with that element. If so, assume that the contour passes through the element only once and calculate the spatial coordinates of the points were it pierces the boundary of the element. For simplicity, the present program, subroutine CONPLT, utilizes linear interpolation in these calculations, and plots the results. Subroutine CONTUR simply prints the contour coordinates.

Once the intersection points have been located, their NSPACE spatial coordinates are printed along with an identifying integer point number. For printed output this procedure is inefficient since for NSPACE > 1 most points will be listed (and numbered) twice. (This repetition could be reduced by calculating, printing, and numbering only the first piercing point.) Nevertheless, this procedure is much more efficient than doing the calculations by hand. If the analyst has access to a plotter, the above procedure is easily modified to yield plotted contours (for NSPACE ≥ 2). Once the two intersection points for an element have been identified, one simply makes the plotter draw a straight line through the two points. Of course, it would be desirable to have a number or symbol plotted that would clearly identify the contour curve to which the segment belongs. After all the elements have been surveyed, the completed contour, consisting of many straight line segments, will have been plotted. Subroutines MAXMIN and CONPLT have other possible applications. For example, if properties have been defined at each nodal point (in array FLTNP) then these properties values could also be contoured, if NNPFLO $\leq$ NG. This would clearly identify regions of different materials as a visual data check. However, since array FLTNP is double subscripted and array **D** in CONPLT has only a single subscript, a few minor changes would be required in subroutine CONPLT.

```
      SUBROUTINE  POST ( M, NE, NG, NDFREE, NODES, LNODE, INDEX,
     1    DD, N, NSPACE, NELFRE, NRB, NQP, NGEOM, NPARM, NNPFIX,
     2    NNPFLO, MISCFX, MISCFL, NLPFIX, NLPFLO, COORD, S, C, H,
     3    DGH, B, E, EB, STRAIN, STRAN0, STRESS, BODY, PT, WT,
     4    XYZ, DLH, G, DLG, AJ, AJINV, HINTG, D, PRTLPT, FLTMIS,
     5    ELPROP, PRTMAT, MISFIX, MISFLO, LPROP, LPPROP, NTAPE1,
     6    NTAPE2, NTAPE3, NTAPE4, NTAPE5, IT, NITER )
C     * * * * * * * * * * * * * * * * * * * * * * * * * * *
C            ELEMENT LEVEL POST-SOLUTION CALCULATIONS
C     * * * * * * * * * * * * * * * * * * * * * * * * * * *
C        ALWAYS USED
      DIMENSION  COORD(N,NSPACE), S(NELFRE,NELFRE), DD(NDFREE),
     1           D(NELFRE), NODES(NE,N), LNODE(N), INDEX(NELFRE)
C        USUALLY USED
      DIMENSION  C(NELFRE), H(N), DGH(NSPACE,N), B(NRB,NELFRE),
     1           E(NRB,NRB), EB(NRB,NELFRE), STRAIN(NRB+2),
     2           STRAN0(NRB), STRESS(NRB+2), BODY(NSPACE)
C        OPTIONAL FOR NUMERICAL INTEGRATION
      DIMENSION  PT(NPARM,NQP), WT(NQP), XYZ(3), DLH(NSPACE,N),
     1           G(NGEOM), DLG(NPARM,NGEOM), AJ(NSPACE,NSPACE),
     2           AJINV(NSPACE,NSPACE), HINTG(N)
C        OPTIONAL PROPERTY AND SOLUTION VALUES
      DIMENSION  D(NELFRE), PRTLPT(N,0:NNPFLO), FLTMIS(0:MISCFL),
     1           ELPROP(0:NLPFLO), PRTMAT(0:NLPFLO),
     2           MISFIX(0:MISCFX), LPROP(0:NLPFIX),
     3           LPPROP(0:NNPFIX)
C                    VARIABLES:
C       ... SEE NOTATION APPENDIX OR TEXT ABOVE
      REWIND  NTAPE1
C-->   LOOP OVER ELEMENTS
      DO 10  IE = 1, NE
C        EXTRACT ELEMENT'S NODES
        CALL LNODES (IE, NE, N, NODES, LNODE)
C        EXTRACT NODAL PARAMETERS OF THE ELEMENT
        CALL INDXEL (N, NELFRE, NG, LNODE, INDEX)
        CALL ELFRE (NDFREE, NELFRE, D, DD, INDEX)
C-->     PERFORM PROBLEM DEPENDENT CALCULATIONS AND OUTPUT
        CALL   POSTEL (N, NSPACE, NELFRE, NRB, NQP, NGEOM,
     1    NPARM, NNPFIX, NNPFLO, MISCFX, MISCFL, NLPFIX,
     2    NLPFLO, COORD, S, C, H, DGH, B, E, EB, STRAIN,
     3    STRAN0, STRESS, BODY, PT, WT, XYZ, DLH, G, DLG,
     4    AJ, AJINV, HINTG, D, PRTLPT, FLTMIS, ELPROP,
     5    PRTMAT, MISFIX, MISFLO, LPROP, LPPROP, NTAPE1,
     6    NTAPE2, NTAPE3, NTAPE4, NTAPE5, IT, NITER )
 10   CONTINUE
      RETURN
      END
```

Figure 20.11.1 Controlling the post solution calculations

20.11 Post-solution Calculations Within the Elements

As stated Sec. 4.3, if NTAPE1 is greater than zero this implies that problem-dependent calculations are to be performed in each element after the nodal parameters have been calculated. These post-solution calculations are under the control of subroutine POST, which is shown in Fig. 20.11.1. Its argument includes those in ELSQ plus the following quantities: M = total number of nodes, NE = total number of elements, NELFRE = number of element degrees of freedom, NDFREE = total number of system degrees of freedom, NODES(NE,N) = element incidences lists for all elements, DD(NDFREE) = calculated array of nodal parameters, and INDEX(NELFRE) = element degree of freedom numbers. For every element in the system, subroutine POST extracts the list of nodes on that element, and the nodal parameters, **D**, associated with the element. It then calls the problem-dependent subroutine POSTEL which was given earlier. Subroutine POSTEL reads the necessary used defined problem-dependent data (generated by ELPOST in Fig. 20.2.17) from NTAPE1, combines these with the nodal parameters, **D**, and calculates additional quantities of interest within the element. These problem-dependent quantities are usually printed along with the identifying element number, IE. Application-dependent examples are given in the next two chapters.

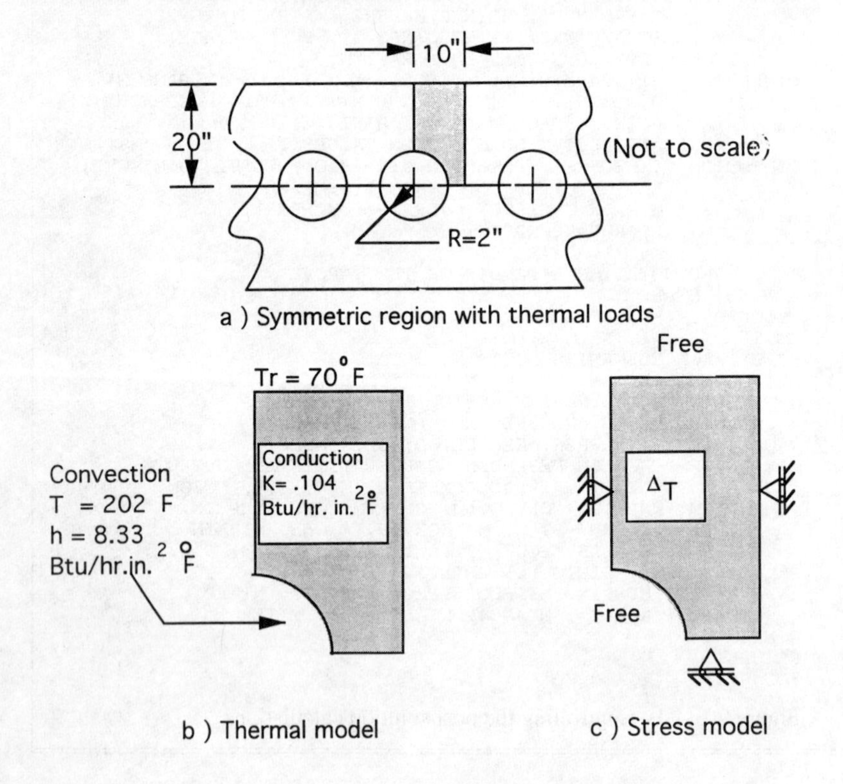

Figure 20.12.1 Structure with cooling holes

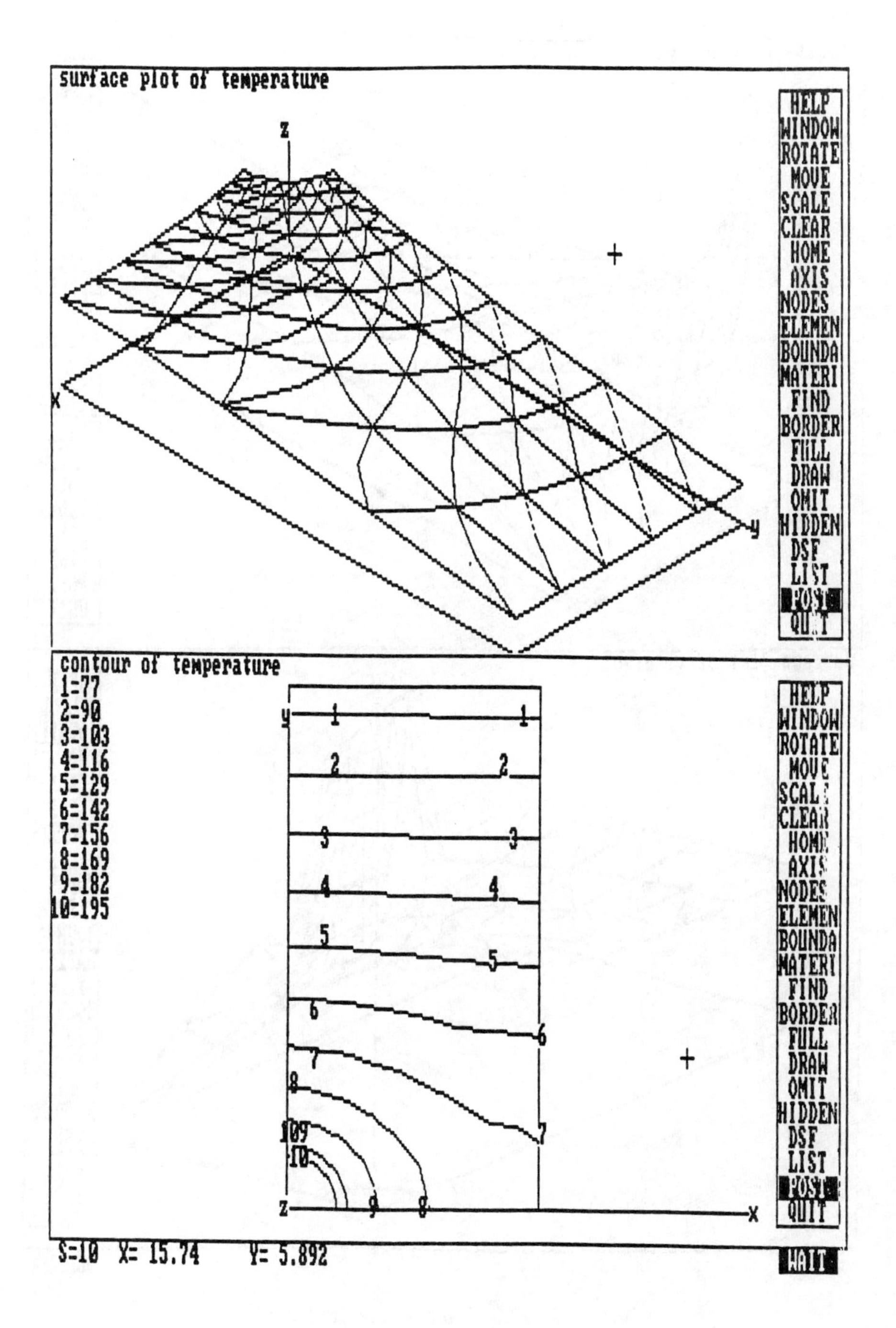

Figure 20.12.2 Temperature results for cooling hole

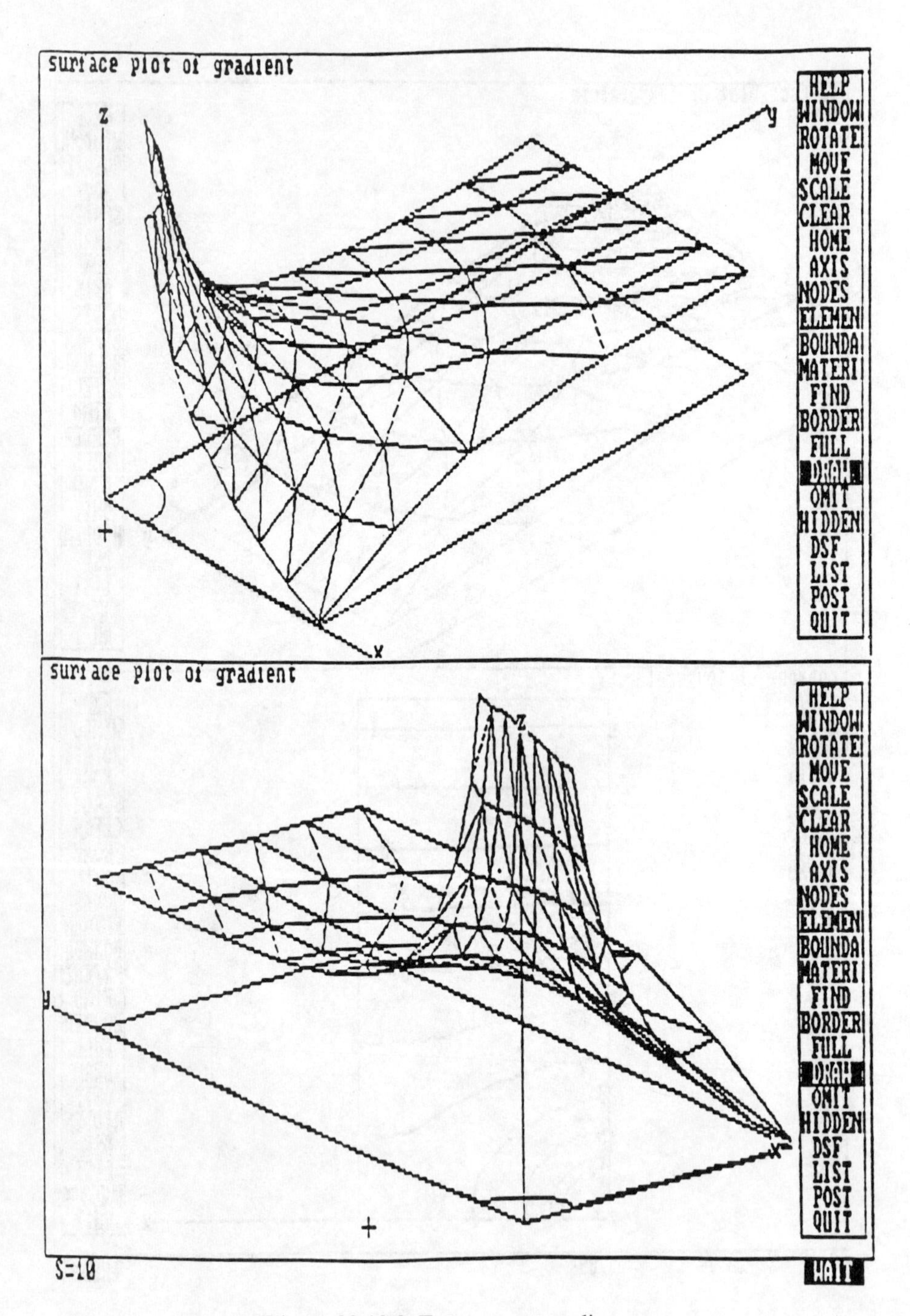

Figure 20.12.3 Temperature gradients

Table 20.11 Computed Stress Values for Element at Point *B*

Element No.	Node No.	SX S1	SY S2	SXY β	SVME
13	13	37.36	1666.	−135.0	1664.
		1677.	26.25	−4.707	
13	15	139.4	2050.	−114.7	1994.
		2057.	132.5	−3.424	
13	45	−171.4	−195.9	241.6	457.5
		58.31	−425.6	−43.55	
13	14	88.38	1858.	−124.9	1828.8
		1677.	26.25	3.977	
13	30	−15.98	927.1	63.45	941.6
		2057.	132.5	−3.798	
13	29	−67.00	735.0	53.31	776.2
		58.31	−425.6	−3.754	

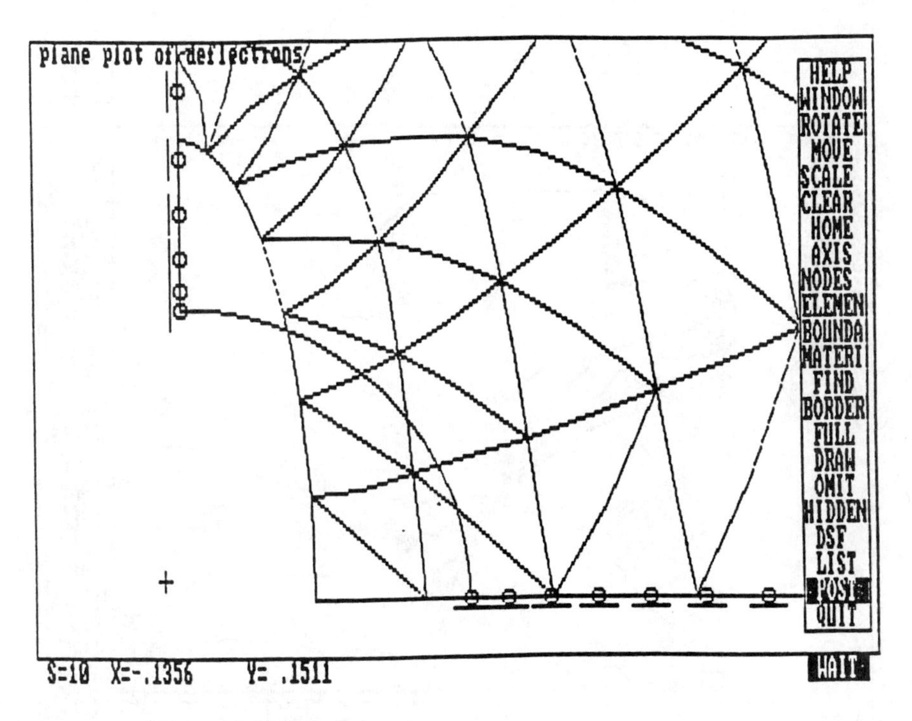

Figure 20.12.4 Relative displacements around cooling hole

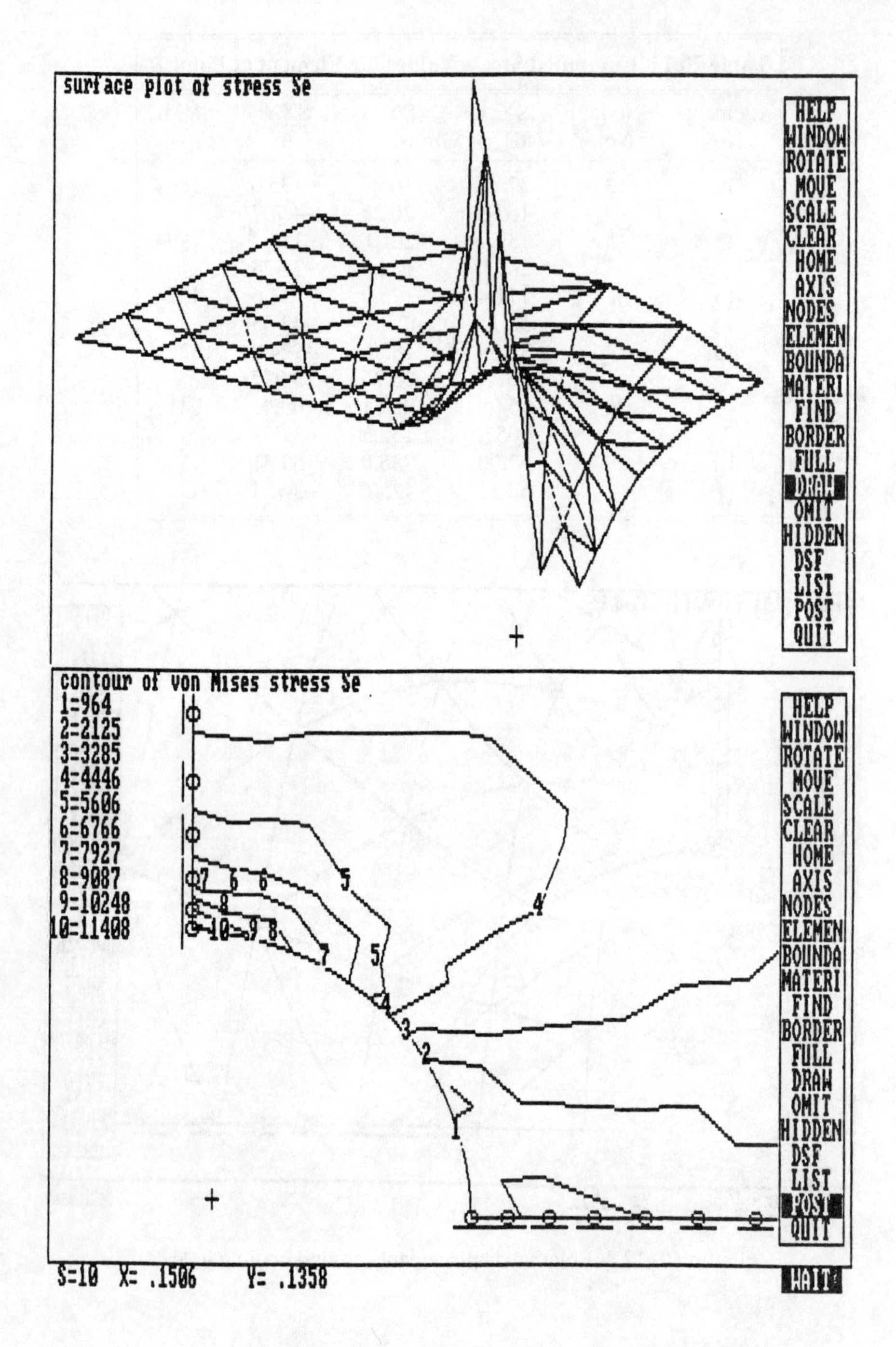

Figure 20.12.5 Von Mises stress due to cooling

20.12 Sample Graphical Outputs

To illustrate some typical graphical output, an example of running MODEL on a personal computer will be presented. The mesh plots and *carpet plots* used programs DRAWSP and VIEW3D (included on the disk) as well as a window manager. Consider a sheet containing symmetrically placed heating channels as shown in Fig. 20.12.1. We wish to determine the thermal stresses that result in the material. This requires that a thermal analysis (from Chap. 12) be completed to determine the temperature distribution. Then this non-uniform temperature is passed to the stress analysis option (in Chap. 13) to form a thermal load state. In the thermal analysis and stress analysis, the same mesh is utilized to model a typical region with three planes of symmetry. In the thermal analysis, these planes have zero heat flux crossing them. The exterior plane is held at room temperature while the interior channel convects heat from the heating fluid in the channel. In the corresponding stress model, the material must be constrained to move with displacements that are tangent to the symmetry planes. The room surface and channel surface are free to expand as necessary. The mesh generator MESH2D was utilized to create the mesh.

For the thermal problem, the parameters were: interior conductivity K = 0.104 Btu/hr.in.°F; channel wall convection coefficient $h = 8.33$ Btu/hr.in^2.°F with a fluid temperature $T = 202$°F. The top (room) surface was held at $T_f = 70$°F. The computed temperature contours, and their carpet plot, are shown in Fig. 20.12.2. These temperatures are automatically saved for later use as thermal loads. The magnitude of the temperature gradient is plotted as a surface in Fig. 20.12.3. This serves as a visual check of the zero flux on the symmetry planes. Note that the flux magnitude goes to zero at the intersection of two of the symmetry planes.

The stress analysis assumed a homogeneous material with an elasticity of $E = 30 \times 10^6$ psi, Poisson's ratio of $\nu = 0.3$, and a stress free temperature of $T_f = 40$°F. The symmetric displacement boundary conditions are partially shown in Fig. 20.12.4. The sole loading was from thermal effects. However, other loads such as gravity, point loads, etc. could have also been included. The nodal temperature distributions were recovered from the previous thermal analysis. The resulting displacements and stresses were computed and stored. Figure 20.12.4 also shows the deformed shape in the region around the heating channel hole. The deformed shape plot is superimposed on the undeformed boundary, with the displacement scale exaggerated. The Von Mises' effective stress was computed and displayed in Fig. 20.12.5. All of the stress data were stored in a file that could be reviewed by the user. Table 20.11 lists the global stresses and principal stresses for the element at point B at the edge of the hole.

20.13 References

[1] Akin, J.E. and Pardue, R.M., "Element Resequencing for Frontal Solutions," pp. 535–541 in *The Mathematics of Finite Elements and Applications, Vol. II*, ed. J.R. Whiteman, London: Academic Press (1976).

[2] Bathe, K.J. and Wilson, E.L., *Numerical Methods for Finite Element Analysis*, Englewood Cliffs: Prentice-Hall (1976).

[3] Belytischko, T., Chiapetta, R.L., and Bartel, H.D., "Efficient Large Scale Non-Linear Transient Analysis by Finite Elements," *Int, J. Num. Meth. Eng.*, **10**, pp. 579–596 (1976).

[4] Berztiss, A.T., *Data Structures – Theory and Practice,* 2nd Edition, New York: Academic Press (1975).

[5] Campbell, J.S., "A Penalty Function Approach to the Minimization of Quadratic Functionals in Finite Element Analysis," pp. 33–54 in *Finite Element Methods in Engineering*, Univ. of NSW, Australia (1977).

[6] Farhat, C. and Wilson, E., "A Parallel Active Column Equation Solver," *Computers & Structures*, **28**(2), pp. 289–304 (1988).

[7] Farhat, C., Pramono, E., and Felippa, C., "Towards Parallel I/O in Finite Element Simulations," *Int. J. Num. Meth. Eng.*, **28**, pp. 2541–2553 (1989).

[8] Fried, I., "A Gradient Computational Procedure for the Solution of Large Problems Arising from the Finite Element Discretization Method," *Int. J. Num. Meth. Eng.*, **2**, pp. 477–494 (1970).

[9] Hestenes, M.R. and Stiefel, E.L., "Methods of Conjugate Gradients for Solving Linear Systems," *Nat. Bur. Std. J. Res.*, **49**, pp. 409–436 (1952).

[10] Hinton, E. and Owen, D.R.J., *Finite Element Programming*, London: Academic Press (1977).

[11] Johnsson, S.L. and Mathur, K.K., "Data Structures and Algorithms for the Finite Element Method on a Data Parallel Supercomputer," *Int. J. Num. Meth. Eng.*, **29**, pp. 881–908 (1990).

[12] Law, K.H., "A Parallel Finite Element Solution Method," *Computers & Structures*, **23**(6), pp. 845–858 (1986).

[13] Owen, D.R.J. and Alves F., J.S.R., "Parallel Finite Element Algorithms for Transputer Systems," in *The Mathematics of Finite Elements and Applications VII (MAFELAP 1990)*, ed. J.R. Whiteman, London: Academic Press (1991).

[14] Storaasli, O.O., Nguyen, D.T., and Agarwal, T.K., "A Parallel-Vector Algorithm for Rapid Structural Analysis on High-Performance Computers," Technical Memorandum 102614, NASA Langley Research Center, 18 pp. (Apr. 1990).

[15] Taylor, R.L., "Computer Procedures for Finite Element Analysis," in *The Finite Element Method*, ed. O.C. Zienkiewicz, London: McGraw-Hill (1977).

[16] Zienkiewicz, O.C., *The Finite Element Method,* 3rd edition, New York: McGraw-Hill (1979).

[17] Zois, D., "Parallel Processing Techniques for FE Analysis: Stiffnesses, Loads and Stresses Evaluation," *Computers & Structures*, **28**(2), pp. 247–260 (1988).

Chapter 21

MODEL APPLICATIONS IN 1-D

21.1 Introduction

The simplest applications of the finite element method involve problems which are independent of time. The general programming considerations of this class of problems were presented in detail in the first four chapters. At this point one is ready to consider the programming of the problem- dependent subroutines associated with the application of interest. The following sections will present several representative examples of the application of the finite element method to steady state problems.

Generally, an element which can be defined by a single spatial variable in the local coordinates is relatively simple to implement. Thus, the typical illustrative applications begin with this class of element. This, of course, does not mean that the global coordinates must also be one-dimensional. For example, a set of one-dimensional *truss* elements can be utilized to construct a three-dimensional structure.

21.2 Conductive and Convective Heat Transfer

The text by Myers [6] presents a detailed study of heat transfer analysis from both the analytic and numerical points of view. Several examples are solved by finite difference and finite element methods as well as by analytic techniques. All three methods were applied to the problem to be illustrated in this section. Consider the heat transfer in a slender rod, illustrated in Fig. 21.2.1, that is insulated at $x = L$, has a specified temperature, t_0, at $x = 0$, and is surrounded by a medium with a temperature of $t_\infty = 0$. The temperature, $t(x)$, in this one-dimensional problem is governed by the differential equation

$$KA \frac{d^2 t}{dx^2} - hPt = 0 \tag{21.1}$$

and the boundary conditions $t(0) = t_0$, and $dt/dx\,(L) = 0$ where A = cross-sectional area of rod, h = convective transfer coefficient, K = thermal conductivity, L = length of rod, P = perimeter of cross-section, t = temperature, and x = spatial coordinate. The exact solution for this problem is known to be $t(x) = \cosh[\,m(L-x)\,]/\cosh(mL)$ where $m^2 = hP/KA$. Myers shows that an integral formulation can be obtained by rendering stationary the functional

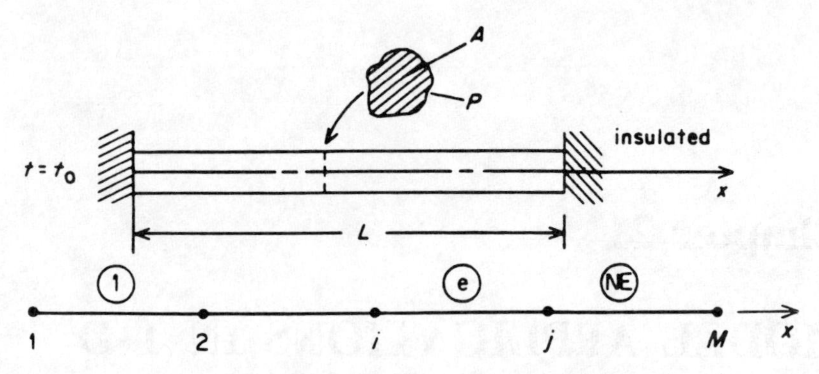

Figure 21.2.1 Rod with conduction and convection

$$I = \frac{1}{2}\int_0^L \left[KA \left(\frac{dt}{dx}\right)^2 + hPt^2 \right] dx, \tag{21.2}$$

subject to the boundary condition $t(0) = t_0$. The boundary condition imposed at $x = 0$ is an *essential* boundary condition. The second boundary condition is automatically satisfied, and is known as a *natural boundary condition*. Of course, if a second temperature had been specified at $x = L$, it would have taken precedence over the natural boundary condition at that end. The reader may wish to verify that applying the weighted residual method of Galerkin and integrating the first term by parts yields an identical formulation. Assuming a linear element the interpolation functions are known to be

$$t^e(x) = \mathbf{M}(x)\,\mathbf{g}^{e^{-1}}\,\mathbf{T}^e$$

where

$$\mathbf{M}(x) = [\,1 \quad x\,], \quad \mathbf{g}^{e^{-1}} = \frac{1}{l}\begin{bmatrix} x_j & -x_i \\ -1 & 1 \end{bmatrix}, \quad \mathbf{T}^{eT} = [\,T_i \quad T_j\,]$$

and $l = (x_j - x_i)$ denotes the length of the element. That is, $t^e(x)$ is a function of an assumed spatial variation, $\mathbf{M}(x)$, the element geometry, $\mathbf{g}^{e^{-1}}$, and the element's nodal temperatures, $\mathbf{T}^e$. This can be expressed in a simpler form here as

$$t^e(x) = \mathbf{H}^e(x)\,\mathbf{T}^e, \quad \mathbf{H}^e(x) = \mathbf{M}(x)\,\mathbf{g}^{e^{-1}} = \left[\frac{(x_j - x)}{l} \quad \frac{(x - x_i)}{l} \right] \tag{21.3}$$

is called the *element interpolation matrix*. Myers shows that the element square matrix for this element, $\mathbf{S}^e$, can be defined as

$$\mathbf{S}^e = \int_{x_i}^{x_j} \left(K^e A^e \frac{\partial \mathbf{H}^{eT}}{\partial x} \frac{\partial \mathbf{H}^e}{\partial x} + h^e P^e \mathbf{H}^{eT} \mathbf{H}^e \right) dx \tag{21.4}$$

which reduces to

```
C       .......................................................
C        ***  ELSQ PROBLEM DEPENDENT STATEMENTS FOLLOW ***
C       .......................................................
C       APPLICATION: ONE-DIM. STEADY STATE HEAT TRANSFER
C       REFER: MYERS, ANAL. METH. IN COND. HEAT TRANSFER
C            NSPACE=1, N=2, NG=1, NELFRE=2
        COMMON /ELARG/  PH, XIJ, TOTAL, QC(1)
C       ELPROP(1) = CROSS-SECTIONAL AREA
C       ELPROP(2) = PERIMETER
C       ELPROP(3) = THERMAL CONDUCTIVITY
C       ELPROP(4) = CONVECTIVE LOSS COEFFICIENT
        AK  = ELPROP(1)*ELPROP(3)
        PH  = ELPROP(2)*ELPROP(4)
        XIJ = COORD(2,1)  - COORD(1,1)
        S(1,1) =  AK/XIJ + PH*XIJ/3.
        S(1,2) = -AK/XIJ + PH*XIJ/6.
        S(2,1) = S(1,2)
        S(2,2) = S(1,1)
        RETURN
        END
```

Figure 21.2.2 Conduction and convection in 1-D

$$\mathbf{S}^e = \begin{bmatrix} \left[1/3P^eh^el + A^eK^e/l\right] & \left[1/6P^eh^el - A^eK^e/l\right] \\ sym. & \left[1/3P^eh^el + A^eK^e/l\right] \end{bmatrix} \tag{21.5}$$

which is the form that is utilized in subroutine ELSQ which is shown in Fig. 21.2.2. This simple expression is all that is required to generate the equations necessary to solve for the nodal temperatures.

Generally, one is also interested in other quantities that can be evaluated once the nodal temperatures have been calculated. For example, one may need to perform a post-solution calculation of the heat lost through convection along the length of the bar. Myers shows that the convective heat loss, dq, from a segment of length dx is

```
C       .......................................................
C        *** ELPOST PROBLEM DEPENDENT STATEMENTS FOLLOW ***
C       .......................................................
C-->    GENERATE CONVECTIVE HEAT LOSS DATA
C       APPLICATION: ONE-DIM. STEADY STATE HEAT TRANSFER
C       REFER: MYERS, ANAL. METH. IN COND. HEAT TRANSFER
        COMMON /ELARG/  PH, XIJ, TOTAL, Q(1)
C       HINTG = QC HERE
        HINTG(1) = 0.5*PH*XIJ
        HINTG(2) = 0.5*PH*XIJ
        WRITE (NTAPE1)  HINTG
        RETURN
        END
```

Figure 21.2.3 Generate convective loss matrix

$dq = HPt\,dx$. Thus, the loss from a typical element (with constant properties) is

$$\Delta Q^e = \int_{x_i}^{x_j} dq = h^e P^e \int_{x_i}^{x_j} t^e\,dx\,,\ \ \Delta Q^e = \mathbf{QC}^e\mathbf{T}^e \tag{21.6}$$

where

$$\mathbf{QC}^e = \frac{1}{2} h^e P^e l \begin{bmatrix} 1 & 1 \end{bmatrix}.$$

The simple matrix, $\mathbf{QC}^e$, can be generated at the same time as $\mathbf{S}^e$ and stored on auxiliary storage for use after the nodal temperatures, $\mathbf{T}^e$, are known. It is very common for an application to use the integral of the interpolation functions for postprocessing. Therefore that array, HINTG, is a standard argument to those routines. Thus, $\mathbf{QC}^e$ is stored in it. The matrix $\mathbf{QC}^e$ is generated and stored by subroutine ELPOST. It is retrieved and multiplied by $\mathbf{T}^e$ in subroutine POSTEL which calculates and prints ΔQ^e for each element. The latter two problem-dependent routines are shown in Figs. 21.2.3 and 21.2.4, respectively.

The above equations were derived by assuming constant properties within each element. In some practical applications, one or more of the properties might vary along the length of the bar. The above element could still be utilized to obtain a good approximation by assuming average properties over the element. This could be accomplished by defining the variable properties at each node and inputting them as nodal properties. For each element the program would then automatically extract the properties associated with the element's nodal points (array PRTLPT) and pass those data to subroutine ELSQ. One could then use the average of the two nodal properties,

```
C     ..............................................................
C      *** POSTEL PROBLEM DEPENDENT STATEMENTS FOLLOW ***
C     ..............................................................
C     APPLICATION: ONE-DIM. STEADY STATE HEAT TRANSFER
C     CALCULATE CONVECTIVE HEAT LOSS BY ELEMENT
C     REFER: MYERS, ANAL. METH. IN COND. HEAT TRANSFER
      COMMON /ELARG/  PH, XIJ, TOTAL, Q(1)
      DATA KALL / 1 /
      IF ( KALL .EQ. 1 )  THEN
C-->       SET CONST, PRINT TITLES ON FIRST CALL
 10     WRITE (6,5000)
 5000   FORMAT ('** CONVECTIVE HEAT LOSS**',//,
     1          'ELEMENT          LOSS          TOTAL')
        KALL  = 0
        TOTAL = 0.0
      ENDIF
C-->   FIND ELEMENT LOSS
      READ (NTAPE1) HINTG
      CALL  MMULT (HINTG, D, Q, 1, N, 1)
      TOTAL = TOTAL + Q(1)
      WRITE (6,5010)  IE, QC, TOTAL
 5010 FORMAT ( I7, 2(1PE16.6) )
      RETURN
      END
```

Figure 21.2.4 Computing convection loss

e.g. instead of A = ELPROP(1) set A = 0.5 [PRTLPT(1,1) + PRTLPT(2,1)].

A more accurate approach would be to integrate Eq. (21.2) numerically with due consideration for the variable properties. The variation of the properties could be approximated by assuming, for example, that **VALUES**(x) = $\mathbf{H}^e$(x)**PRTLPT**, where **VALUES** denotes the four properties of interest within the element. Recall that the latter operation is carried out by subroutine CALPRT, which was presented in Sec. 5.3.

As a numerical example consider a problem where $A = 0.01389\,ft^2$, $h = 2.0$ BTU/hr ft^{2}°F, $K = 120$ BTU/hr ft °F, $L = 4\,ft$, $P = 0.5\,ft$ and $t_0 = 10°\,F$. Select a mesh with eight nodes (M = 8), seven elements (NE = 7), one temperature per node (NG = 1), two nodes per element (N = 2), and four floating point properties per element (NLPFLO = 4). This one-dimensional problem, shown in Fig. 3.7.1, involves homogeneous element properties and a null element column matrix (NSPACE = 1, LHOMO = 1, NULCOL = 1). Since data are to be stored by subroutine ELPOST on NTAPE1, assign unit eight for that purpose (NTAPE1 = 8). The above four properties could also have been stored as miscellaneous system properties. Since node 1 is at $x = 0$, it is necessary to define the first (and only) parameter at node 1 to have a value of 10.0. The output data from the MODEL code are shown in Fig. 21.2.5. Table 21.1 lists the configuration options of the MODEL code used in the present problem.

Table 21.1 Parameter definitions for one-dimensional heat transfer

CONTROL:

NSPACE = 1	NLPFLO = 4	N = 2
NULCOL = 1	NG = 1	NTAPE1 = 8

DEPENDENT VARIABLES:
1 = temperature, t

PROPERTIES:
Element level:
Floating point:
1 = cross-sectional area, A
2 = perimeter of cross-section, P
3 = thermal conductivity, K
4 = convection coefficient ($t_\infty \equiv 0$), h

CONSTRAINTS:
Type 1: specified nodal temperatures

POST-SOLUTION CALCULATIONS:
1 = e lement convective heat loss, q

```
NUMBER OF NODAL POINTS IN SYSTEM =..........     8
NUMBER OF ELEMENTS IN SYSTEM =..............     7
NUMBER OF NODES PER ELEMENT =...............     1
NUMBER OF PARAMETERS PER NODE =.............     2
DIMENSION OF SPACE =........................     1
NUMBER OF ITERATIONS TO BE RUN =............     1
NUMBER OF ROWS IN B MATRIX =................     1
ELEMENT SHAPE: LINE, TRI, QUAD, HEX, TET =...    1
NUMBER OF DIFFERENT ELEMENT TYPES =.........     1
STIFFNESS STORAGE MODE: SKY, BAND =.........     1
NUMBER OF REAL PROPERTIES PER ELEMENT =.....     4
OPTIONAL UNIT NUMBERS: NTAPE1 =  8
*** NODAL POINT DATA ***
 NODE, CONSTRAINT FLAG, 1 COORDINATES
         1         1    0.0000
         2         0    0.2500
         3         0    0.5000
         4         0    1.0000
         5         0    1.5000
         6         0    2.0000
         7         0    3.0000
         8         0    4.0000
*** ELEMENT CONNECTIVITY DATA ***
 ELEMENT NO., 2 NODAL INCIDENCES.
    1    1    2
    2    2    3
    3    3    4
    4    4    5
    5    5    6
    6    6    7
    7    7    8
***  NODAL PARAMETER CONSTRAINT LIST  ***
TYPE            EQUATIONS
  1                 1
*** CONSTRAINT EQUATION DATA ***
CONSTRAINT TYPE ONE
EQ. NO.    NODE1   PAR1          A1
     1       1      1   1.00000E+01
***  ELEMENT  PROPERTIES   ***
ELEMENT  PROPERTY       VALUE
    1         1    1.38900E-02
    1         2    5.00000E-01
    1         3    1.20000E+02
    1         4    2.00000E+00
***  OUTPUT OF RESULTS  ***
NODE, 1 COORDINATES, 1 PARAMETERS.
    1  0.00000E+00  1.00000E+01
    2  2.50000E-01  8.23848E+00
    3  5.00000E-01  6.78781E+00
    4  1.00000E+00  4.61384E+00
    5  1.50000E+00  3.14964E+00
    6  2.00000E+00  2.16995E+00
    7  3.00000E+00  1.13219E+00
    8  4.00000E+00  8.49165E-01
```

Figure 21.2.5 Conduction convection solution output

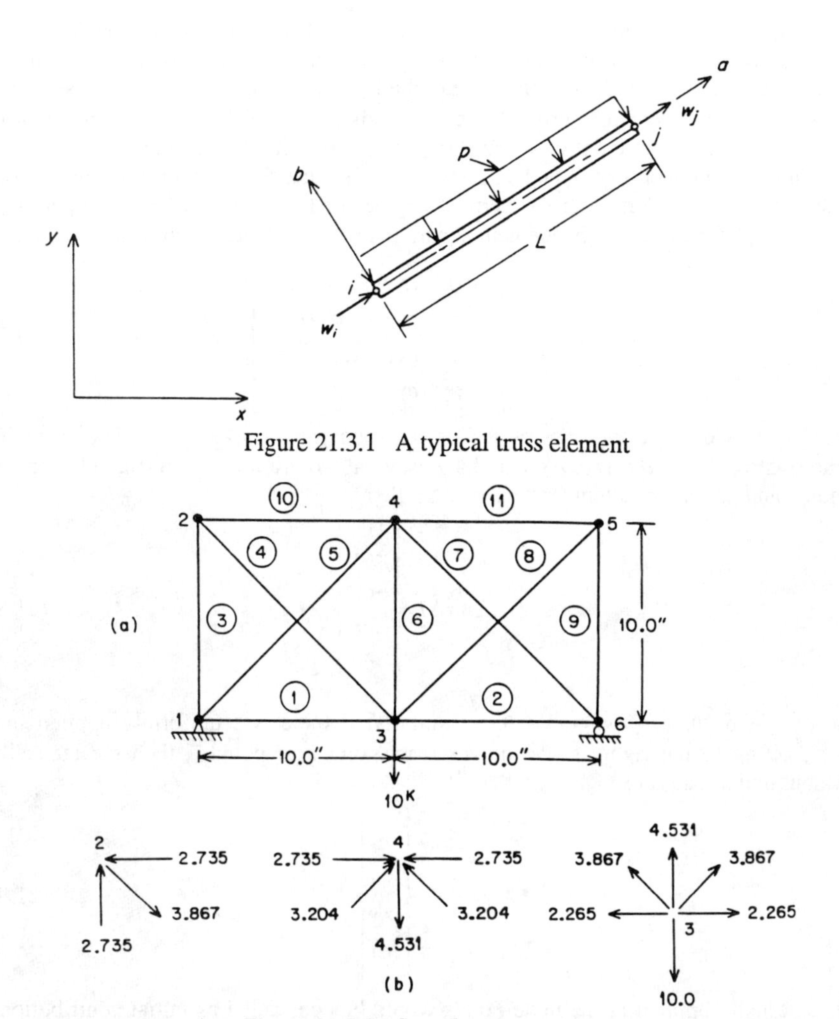

Figure 21.3.1 A typical truss element

Figure 21.3.2 Sample truss structure

21.3 Plane Truss Structures *

A truss structure is an assemblage of bar members connected at their end points by smooth pins. Most texts on structural analysis present a detailed analysis of the simple truss element. The element equations presented here were taken from the first finite element text by Zienkiewicz [10] and will be checked against a simple example presented in the text by Meek [5]. The truss element is a one-dimensional element that can be utilized in a one-dimensional global space to solve problems involving collinear force systems. However, this is of little practical interest and one is usually more interested in two- and three-dimensional structures.

Consider a two-dimensional truss constructed from the one-dimensional bar elements. Let the bar go from point i to point j and thus have direction angles from the x- and y-axes of θ_x and θ_y, respectively. Let the bar, of length L, have a cross-sectional area, A, with a moment in inertia, I, and a depth of $2d$. The bar has a modulus of elasticity, E, and a coefficient of thermal expansion, α. The displacement at each node has x- and y-components of u and v, respectively, so that the element has four degrees of freedom. If one orders the four element degrees of freedom as $\delta^{eT} = [\, u_1 \; v_1 \; u_2 \; v_2 \,]$, Zienkiewicz [10] shows that the element square matrix (or *stiffness matrix*) is

$$\mathbf{S}^e = \frac{EA}{L}\begin{bmatrix} cxx & cxy & -cxx & -cxy \\ cxy & cyy & -cxy & -cyy \\ -cxx & -cxy & cxx & cxy \\ -cxy & -cyy & cxy & cyy \end{bmatrix} \tag{21.7}$$

where $cxx = \text{Cos}^2\,\theta_x$, $cxy = \text{Cos}\,\theta_x \text{Cos}\,\theta_y$, and $cyy = \text{Cos}^2\,\theta_y$. The element column matrix (or *load vector*) can have several contributions. If the element is restrained and undergoes a temperature rise ΔT then

$$\mathbf{C}^e = EA\alpha\Delta T \begin{Bmatrix} -cx \\ -cy \\ cx \\ cy \end{Bmatrix}, \tag{21.8}$$

where $cx = \text{Cos}\,\theta_x$ and $cy = \text{Cos}\,\theta_y$. Similarly, if there is a uniformly applied line load, p, acting to the right as one moves from point i to point j, then the statically equivalent nodal loads are

$$\mathbf{C}^e = \frac{pL}{2}\begin{Bmatrix} cy \\ -cx \\ cy \\ -cx \end{Bmatrix}. \tag{21.9}$$

Loads externally applied at the node points would be considered as initial contributions to the system column matrix. If only external loads are in effect, then the element column matrix is null. In summary, the problem under consideration is two-dimensional (NSPACE = 2) and involves an element with two nodes (N = 2), two displacements per node (NG = 2) and seven $(A, E, \alpha, \Delta T, p, I, d)$ possible element properties (NLPFLO = 7). A typical truss element is shown in Fig. 21.3.1.

After the displacements, δ^e, have been calculated, the post-solution stress calculations can be performed. The stresses at the mid-section on the right (σ_1) and left (σ_2) sides (going from i to j) are

$$\begin{Bmatrix} \sigma_1 \\ \sigma_2 \end{Bmatrix} = \frac{E}{L}\begin{bmatrix} -cx & -cy & cx & cy \\ -cx & -cy & cx & cy \end{bmatrix} \delta^e + \frac{pL^2 d}{8I}\begin{Bmatrix} 1 \\ -1 \end{Bmatrix} - E\alpha\Delta T \begin{Bmatrix} 1 \\ 1 \end{Bmatrix}, \tag{21.10}$$

where a negative stress denotes compression. A word of caution should be mentioned in the presentation of this element. A derivation of the truss element considers only the

```
C      .......................................................
C       ***  ELSQ PROBLEM DEPENDENT STATEMENTS FOLLOW ***
C      .......................................................
C      APPLICATION:  ANALYSIS OF A TWO-DIMENSIONAL TRUSS
C      ZIENKIEWICZ, F.E.M. IN STRUCTURAL & CONTINUUM MECH.
C           NG = 2, N = 2, NSPACE = 2, NLPFLO = 7
C      ELPROP(1) = AREA, ELPROP(2) = MODULUS OF ELASTICITY
C      ELPROP(3) = TEMP RISE, ELPROP(4) = COEF THERMAL EXP
C      ELPROP(5) = LINE LOAD, ELPROP(6) = MOMENT OF INERTIA
C      ELPROP(7) = HALF DEPTH OF BAR
       COMMON  /ELARG/  CX, CY, A, E, T, AL, P, BI, BD, BARL
CDP    SQRT(Z) = DSQRT(Z)
       XI = COORD(1,1)
       XJ = COORD(2,1)
       YI = COORD(1,2)
       YJ = COORD(2,2)
       A  = ELPROP(1)
       E  = ELPROP(2)
       T  = ELPROP(3)
       AL = ELPROP(4)
       P  = ELPROP(5)
       BI = ELPROP(6)
       IF ( BI .LE. 0.0 )  BI = 1.0
       BD = ELPROP(7)
       IF ( BD .LE. 0.0 )  BD = 1.0
C-->   FIND BAR LENGTH AND DIRECTION COSINES
       DX   = XJ-XI
       DY   = YJ-YI
       BARL = SQRT( DX*DX + DY*DY )
       CX   = DX/BARL
       CY   = DY/BARL
C-->   FIND 1-D STIFFNESS, K=E*A/L
       STIF = E*A/BARL
C-->   TRANSFORM TO 2-D STIFFNESS
       CXX = CX*CX
       CXY = CX*CY
       CYY = CY*CY
       S(1,1) =  STIF*CXX
       S(1,2) =  STIF*CXY
       S(1,3) = -STIF*CXX
       S(1,4) = -STIF*CXY
       S(2,1) =  STIF*CXY
       S(2,2) =  STIF*CYY
       S(2,3) = -STIF*CXY
       S(2,4) = -STIF*CYY
       S(3,1) = -STIF*CXX
       S(3,2) = -STIF*CXY
       S(3,3) =  STIF*CXX
       S(3,4) =  STIF*CXY
       S(4,1) = -STIF*CXY
       S(4,2) = -STIF*CYY
       S(4,3) =  STIF*CXY
       S(4,4) =  STIF*CYY
       RETURN
       END
```

Figure 21.3.3 Truss stiffness calculation

```
C     ......................................................
C      ***  ELCOL PROBLEM DEPENDENT STATEMENTS FOLLOW ***
C     ......................................................
C     APPLICATION:  ANALYSIS OF A TWO-DIMENSIONAL TRUSS
      COMMON  /ELARG/  CX, CY, A, E, T, AL, P, BI, BD, BARL
      DO 10  I = 1,4
   10 C(I) = 0.0
C      CHECK FOR P=0 AND DT=0  (NO ELEMENT LOADS)
      IF ( P .EQ. 0.0 .AND. T .EQ. 0.0 )  RETURN
C-->    INITIAL STRAIN EFFECTS    F = E*A*ALPHA*DT
      IF ( T .NE. 0.0 )  THEN
        F     = E*A*AL*T
        C(1) = -CX*F
        C(2) = -CY*F
        C(3) = CX*F
        C(4) = CY*F
      ENDIF
C-->    LINE LOAD EFFECTS  F = 0.5*P*L (POS TO R , I TO J )
      IF ( P .EQ .0.0 )  RETURN
        F = 0.5*P*BARL
        C(1) = C(1) + CY*F
        C(2) = C(2) - CX*F
        C(3) = C(3) + CY*F
        C(4) = C(4) - CX*F
        RETURN
      END
```

Figure 21.3.4 Joint load calculations

```
C     ......................................................
C      *** ELPOST PROBLEM DEPENDENT STATEMENTS FOLLOW ***
C     ......................................................
C     APPLICATION:  ANALYSIS OF A TWO-DIMENSIONAL TRUSS
      COMMON /ELARG/  CX, CY, A, E, T, AL, P, BI, BD, BARL
C-->    FORM STRESS-DISPLACEMENT MATRIX
      EDL = E/BARL
      EB(1,1) = -EDL*CX
      EB(1,2) = -EDL*CY
      EB(1,3) =  EDL*CX
      EB(1,4) =  EDL*CY
      EB(2,1) = -EDL*CX
      EB(2,2) = -EDL*CY
      EB(2,3) =  EDL*CX
      EB(2,4) =  EDL*CY
C-->    BENDING EFFECTS     = P*L*L*D/8*I
      STRAIN(1) = 0.125*P*BARL*BARL*BD/BI
      STRAIN(2) = -STRAIN(1)
C-->    FORM THERMAL EFFECT = -E*ALPHA*DT
      STRAN0(1) = -E*AL*T
      STRAN0(2) = STRAN0(1)
C-->    STORE DATA ON NTAPE1 FOR USE BY POSTEL
      WRITE (NTAPE1)  EB, STRAIN, STRAN0
      RETURN
      END
```

Figure 21.3.5 Saving stress data

```
C      ..................................................
C       *** POSTEL PROBLEM DEPENDENT STATEMENTS FOLLOW ***
C      ..................................................
C      APPLICATION:  ANALYSIS OF A TWO-DIMENSIONAL TRUSS
       COMMON /ELARG/  CX, CY, A, E, T, AL, P, BI, BD, BARL
       DATA  KODE / 1 /
       IF ( KODE .EQ. 1 )  THEN
         KODE = 0
C-->       WRITE STRESS HEADINGS
         WRITE (6,1000)
 1000    FORMAT ('E L E M E N T    S T R E S S E S',//,
     A   ' ELEMENT       MID SECTION STRESS AT:',/,
     B   ' NUMBER      RIGHT                LEFT',/)
       ENDIF
C-->     READ STRESS DATA OFF NTAPE1 FROM ELPOST
       READ (NTAPE1)  EB, STRAIN, STRAN0
C--->    CALCULATE STRESS, STRESS = EB*D
       CALL  MMULT (EB, D, STRESS, NRB, NELFRE, 1)
C        ADD THERMAL AND PRESSURE EFFECTS
       STRESS(1) = STRESS(1) + STRAN0(1) + STRAIN(1)
       STRESS(2) = STRESS(2) + STRAN0(2) + STRAIN(2)
       WRITE (6,1010)  IE, (STRESS(K), K = 1, 2)
 1010 FORMAT (I5, 7X, 2E20.7)
       RETURN
       END
```

Figure 21.3.6 Calculate and print stresses

axial effects of applied loads. The effects of the transverse line load, p, are handled in a special way. First, the force effects are simply lumped at the nodes by inspection. Secondly, the calculated bending stresses are obtained from elementary beam theory and not from finite element theory. These bending effects are available in finite element models which utilize six degree of freedom frame elements (see Sec. 21.6).

As a specific example, consider the problem in Fig. 21.3.2 which was solved by Meek [5]. It has six nodes (M = 6), eleven elements (NE = 11) with homogeneous element properties (LHOMO = 1), and no element loads due to pressures or temperature changes (NULCOL = 1). At node 1 both displacements are zero and at node 6 the y-displacement is zero. There is a negative externally applied y-load of 10 kips at node 3 (thus INRHS = 1). The properties are assumed to be $A = 1\ in^2$, $E = 30{,}000$ ksi, $\alpha = \Delta T = p = 0.0$, $I = 1/12 in^4$ and $d = 0.5$ in. The problem-dependent subroutines are shown in Figs. 21.3.3 to 21.3.6. The input data are shown in Fig. 21.3.7 and the MODEL output in Fig. 21.3.8. The calculated displacements and stresses are much more accurate than the solution given by Meek.

If one wanted to check joint force equilibrium and determine the support reactions, one would have to multiply the original system stiffness matrix by the calculated nodal displacements. To accomplish this, it would be necessary to store the original system square matrix on auxiliary storage and recall it after the displacements are known. A subroutine, BANMLT, for executing the multiplication to find the resultant joint forces is included on the disk. One could easily obtain the member forces in subroutine POSTEL by multiplying the area by the stress resulting from the thermal effects and deformation of the member. Referring to Table 21.2, one finds the list of options used for this element.

```
NUMBER OF NODAL POINTS IN SYSTEM =..........     6
NUMBER OF ELEMENTS IN SYSTEM =..............    11
NUMBER OF NODES PER ELEMENT =...............     2
NUMBER OF PARAMETERS PER NODE =.............     2
DIMENSION OF SPACE =........................     2
NUMBER OF ROWS IN B MATRIX =................     2
NUMBER OF REAL PROPERTIES PER ELEMENT =.....     7
*** NODAL POINT DATA ***
NODE, CONSTRAINT FLAG, 2 COORDINATES
        1        11    0.0000    0.0000
       ...      ...    ...       ...
*** ELEMENT CONNECTIVITY DATA ***
ELEMENT NO., 2 NODAL INCIDENCES.
    1    1    3
   ...  ...  ...
*** CONSTRAINT EQUATION DATA ***
CONSTRAINT TYPE ONE
EQ. NO.   NODE1    PAR1          A1
      1       1       1   0.00000E+00
     ...     ...     ...     ...
***  ELEMENT  PROPERTIES   ***
ELEMENT  PROPERTY        VALUE
    1          1    1.00000E+00
    1          2    3.00000E+04
   ...        ...       ...
    1          6    8.33330E-02
    1          7    5.00000E-01
*** INITIAL FORCING VECTOR DATA ***
NODE   PARAMETER        VALUE       DOF
    3     2      -1.00000E+01       6
    6     2       0.00000E+00      12
***  OUTPUT OF RESULTS  ***
NODE, 2 COORDINATES, 2 PARAMETERS.
    1  0.00000E+00  0.00000E+00  0.00000E+00  0.00000E+00
    2  0.00000E+00  1.00000E+01  1.66667E-03 -9.11530E-04
    3  1.00000E+01  0.00000E+00  7.55136E-04 -4.40126E-03
    4  1.00000E+01  1.00000E+01  7.55136E-04 -2.89098E-03
    5  2.00000E+01  1.00000E+01 -1.56394E-04 -9.11530E-04
    6  2.00000E+01  0.00000E+00  1.51027E-03  0.00000E+00
** E L E M E N T     S T R E S S E S **
ELEMENT       MID SECTION STRESS AT:
NUMBER      RIGHT                LEFT
    1      2.26541E+00      2.26541E+00
    2      2.26541E+00      2.26541E+00
    3     -2.73459E+00     -2.73459E+00
    4      3.86729E+00      3.86729E+00
    5     -3.20377E+00     -3.20377E+00
    6      4.53082E+00      4.53082E+00
    7     -3.20377E+00     -3.20377E+00
    8      3.86729E+00      3.86729E+00
    9     -2.73459E+00     -2.73459E+00
   10     -2.73459E+00     -2.73459E+00
   11     -2.73459E+00     -2.73459E+00
*** EXTREME VALUES OF THE NODAL PARAMETERS ***
PARAMETER      MAXIMUM, NODE         MINIMUM, NODE
     1     1.6667E-03,    2    -1.5639E-04,    5
     2     0.0000E+00,    6    -4.4013E-03,    3
```

Figure 21.3.7 Example truss results output

Table 21.2 Parameter definitions for a plane truss (* Default, ** Optional)

CONTROL:
NSPACE = 2 NLPFLO = 7 NG = 2
N = 2 ISOLVT = 0*, or 1**

DEPENDENT VARIABLES:
1 = x-component of displacement, u
2 = y-component of displacement, v

PROPERTIES:
Element level:
Floating point:
1 = cross-sectional area, A
2 = modulus of elasticity, E
3 = temperature rise from unstressed state, Δt
4 = coefficient of thermal expansion, α
5 = uniform line load (to the right when moving from node 1 to node 2 of the element)
6 = moment of inertia, I
7 = half depth of the bar, d

CONSTRAINTS:
Type 1: specified nodal displacement components
Type 2: inclined roller supports, rigid members, etc.

INITIAL VALUES OF COLUMN VECTOR:
Specified external nodal forces**

POST-SOLUTION CALCULATIONS:
1 = principal element stresses

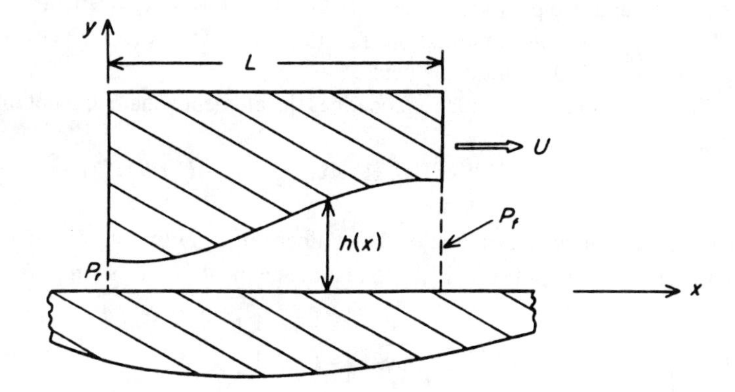

Figure 21.4.1 Slider bearing notation

21.4 Slider Bearing Lubrication

Several references are available on the application of the method to lubrication problems. These include the early work of Reddi [7], a detailed analysis and computer program for the three node triangle by Allan [2], and a presentation of higher order elements by Wada and Hayashi [8]. The most extensive discussion is probably found in the text by Huebner [3]. These formulations are based on the Reynolds equation of lubrication. For simplicity a one-dimensional formulation will be presented here. Consider the slider bearing shown in Fig. 21.4.1, which is assumed to extend to infinity out of the plane of the figure. It consists of a rigid bearing and a slider moving relative to the bearing with a velocity of U. The extremely thin gap between the bearing and the slider is filled with an incompressible lubricant having a viscosity of ν. For the one-dimensional case the governing Reynolds equation reduces to

$$\frac{d}{dx}\left(\frac{h^3}{6\nu}\frac{dP}{dx}\right) = \frac{d}{dx}(Uh), \tag{21.11}$$

where $P(x)$ denotes the pressure and $h(x)$ denotes the thickness of the gap. The boundary conditions are that P must equal the known external pressures (usually zero) at the two ends of the bearing. It can be shown that the variational equivalent of the one-dimensional Reynolds equation requires the minimization of the functional

$$I = \int_0^L \left[\frac{h^3}{12\nu}\left(\frac{dP}{dx}\right)^2 + hU\left(\frac{dP}{dx}\right)\right] dx. \tag{21.12}$$

As a word of warning, it should be noted that, while the pressure P is continuous, the film thickness h is often discontinuous at one or more points on the bearing. Another related quantity of interest is the load capacity of the bearing. From statics one finds the resultant normal force per unit length in the z-direction, F_y, is

$$F_y = \int_0^L P\,dx.$$

This is a quantity which would be included in a typical set of post-solution calculations. As a specific example of a finite element formulation, consider a linear element with two nodes (N = 2) and one pressure per node (NG = 1). Thus, $P(x) = \mathbf{H}^e(x)\mathbf{P}^e$ where as before $\mathbf{P}^{eT} = [P_i \quad P_j]$ and the interplation functions are $\mathbf{H}^e = [\,(x_j - x)\ (x - x_i)\,]/l$, where $l = (x_j - x_i)$ is the length of the element.

Minimizing the above functional defines the element square and column matrices as

$$\mathbf{S}^e = \frac{1}{6\nu}\int_{x_i}^{x_j} h^3 \mathbf{H}_{,x}^{eT}\,\mathbf{H}_{,x}^e\,dx, \quad \mathbf{C}^e = -U\int_{x_i}^{x_j} h\,\mathbf{H}_{,x}^{eT}\,dx.$$

For the element under consideration $\mathbf{H}^e$ is linear in x so that its first derivative will be constant. That is, $\mathbf{H}_{,x}^e = [-1 \quad 1]/l$ so that the element matrices simplify to

$$\mathbf{S}^e = \frac{1}{6\nu l^2}\begin{bmatrix} 1 & -1 \\ -1 & 1 \end{bmatrix}\int_{x_i}^{x_j} h^3\,dx \tag{21.13}$$

```
C      ..................................................
C       ***  ELSQ PROBLEM DEPENDENT STATEMENTS FOLLOW ***
C      ..................................................
C      APPLICATION: LINEAR SLIDER BEARING
C      NSPACE = 1, N = 2, NG = 1, NELFRE = 2, MISCFL = 2
C      FLTMIS(1) = VISCOSITY, FLTMIS(2) = VELOCITY
C      ELPROP(1) OR PRTLPT(K,1) = FILM THICKNESS
       COMMON /ELARG/  SP(2, 2), CP(2), VIS, VEL, DL, THICK
       DATA  KALL / 1 /
       IF ( KALL .EQ. 1 )  THEN
C-->      ON FIRST CALL GENERATE CONSTANT MATRICES
         KALL = 0
         VIS = FLTMIS(1)
         VEL = FLTMIS(2)
         V6I = 1./(6.*VIS)
         SP(1,1) = V6I
         SP(2,2) = V6I
         SP(1,2) = -V6I
         SP(2,1) = -V6I
         CP(1) = VEL
         CP(2) = -VEL
       ENDIF
C-->     DEFINE ELEMENT PROPERTIES
       DL    = COORD(2,1) - COORD(1,1)
       THICK = ELPROP(1)
C       CHECK FOR ALTERNATE DEFINITION
       IF ( THICK .LE. 0.0 )  THICK = 0.5*( PRTLPT(1,1)
      1                             + PRTLPT(2,1) )
       H3L = THICK**3/DL
C-->   GENERATE ELEMENT SQUARE MATRIX
       S(1,1) = H3L*SP(1,1)
       S(1,2) = H3L*SP(1,2)
       S(2,2) = H3L*SP(2,2)
       S(2,1) = H3L*SP(2,1)
       RETURN
       END
```

Figure 21.4.2 Slider bearing square matrix

```
C      ..................................................
C       ***  ELCOL PROBLEM DEPENDENT STATEMENTS FOLLOW ***
C      ..................................................
C      APPLICATION: LINEAR SLIDER BEARING
C      NSPACE = 1, N = 2, NG = 1, NELFRE = 2, MISCFL = 2
C      FLTMIS(1) = VISCOSITY, FLTMIS(2) = VELOCITY
C      ELPROP(1) OR PRTLPT(K,1) = FILM THICKNESS
       COMMON /ELARG/  SP(2, 2), CP(2), VIS, VEL, DL, THICK
C-->     FORM ELEMENT COLUMN MATRIX
       C(1) = THICK*CP(1)
       C(2) = THICK*CP(2)
       RETURN
       END
```

Figure 21.4.3 Slider source term

```
C      .....................................................
C        *** ELPOST PROBLEM DEPENDENT STATEMENTS FOLLOW ***
C      .....................................................
C      APPLICATION: LINEAR SLIDER BEARING
C      NSPACE = 1, N = 2, NG = 1, NELFRE = 2, MISCFL = 2
       COMMON /ELARG/  SP(2, 2), CP(2), VIS, VEL, DL, THICK
C-->     GENERATE DATA FOR LOAD CALCULATIONS AND STORE
       HINTG(1) = 0.5*DL
       HINTG(2) = 0.5*DL
       WRITE (NTAPE1)  HINTG
       RETURN
       END
```

Figure 21.4.4 Matrix for load resultant

```
C      .....................................................
C        *** POSTEL PROBLEM DEPENDENT STATEMENTS FOLLOW ***
C      .....................................................
C      APPLICATION: LINEAR SLIDER BEARING
C       NSPACE = 1, N = 2, NG = 1, NELFRE = 2, MISCFL = 2
       COMMON /ELARG/  SP(2, 2), CP(2), TOT, VEL, DL, THICK
       DATA  KALL / 1 /
       IF ( KALL .EQ. 1 )  THEN
C-->        PRINT TITLES ON THE FIRST CALL
          KALL = 0
          WRITE (6,5000)
 5000     FORMAT ('***   E L E M E N T   L O A D S  ***',//,
     1           ' ELEMENT          LOAD           TOTAL')
          TOT = 0.D0
       ENDIF
C-->     CALCULATE LOADS ON THE ELEMENTS, F = HINTG*D
       READ  (NTAPE1)  HINTG
       CALL  MMULT (HINTG, D, F, 1, N, 1)
       TOT = TOT + F
       WRITE (6,5010)  IE, F, TOT
 5010 FORMAT (I5, 1PE16.5, 3X, 1PE16.5)
       RETURN
       END
```

Figure 21.4.5 Resultant element loads

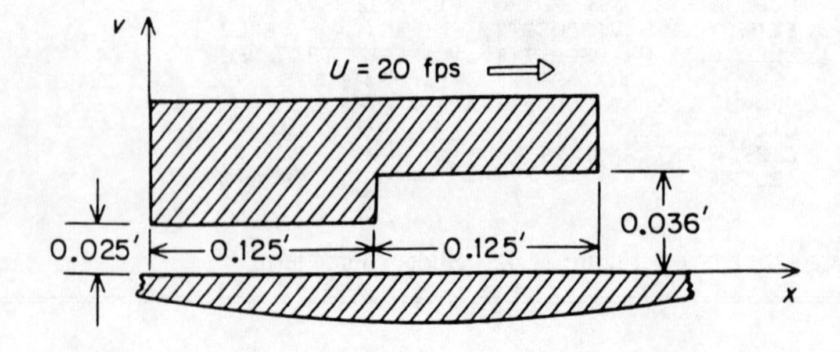

Figure 21.4.6 Example slider bearing

```
LINEAR SLIDER BEARING

NUMBER OF NODAL POINTS IN SYSTEM =..........      3
NUMBER OF ELEMENTS IN SYSTEM =..............      2
NUMBER OF NODES PER ELEMENT =...............      1
NUMBER OF PARAMETERS PER NODE =.............      2
DIMENSION OF SPACE =........................      1
NUMBER OF ITERATIONS TO BE RUN =............      1
NUMBER OF ROWS IN B MATRIX =................      1
ELEMENT SHAPE: LINE, TRI, QUAD, HEX, TET =...     1
NUMBER OF DIFFERENT ELEMENT TYPES =.........      1
STIFFNESS STORAGE MODE: SKY, BAND =.........      1
NUMBER OF REAL PROPERTIES PER ELEMENT =......     1
NUMBER OF REAL MISCELLANEOUS  PROPERTIES =...     2
OPTIONAL UNIT NUMBERS: NTAPE1 =  8
*** NODAL POINT DATA ***
NODE, CONSTRAINT FLAG, 1 COORDINATES
         1         1    0.0000
         2         0    0.1250
         3         1    0.2500
*** ELEMENT CONNECTIVITY DATA ***
ELEMENT NO., 2 NODAL INCIDENCES.
    1    1    2
    2    2    3
***  NODAL PARAMETER CONSTRAINT LIST   ***
TYPE      EQUATIONS
   1         2
*** CONSTRAINT EQUATION DATA ***
CONSTRAINT TYPE ONE
EQ. NO.    NODE1    PAR1          A1
     1       1       1    0.00000E+00
     2       3       1    0.00000E+00
***  ELEMENT  PROPERTIES    ***
ELEMENT  PROPERTY      VALUE
    1         1    2.50000E-02
    2         1    3.60000E-02
***  MISCELLANEOUS SYSTEM PROPERTIES    ***
PROPERTY       VALUE
       1      2.00000E-03
       2      2.00000E+01

***  OUTPUT OF RESULTS  ***
NODE, 1 COORDINATES, 1 PARAMETERS.
    1  0.00000E+00  0.00000E+00
    2  1.25000E-01  5.29857E+00
    3  2.50000E-01  0.00000E+00

***   E L E M E N T   L O A D S  ***
ELEMENT        LOAD                TOTAL
    1      3.31160E-01         3.31160E-01
    2      3.31160E-01         6.62321E-01

*** EXTREME VALUES OF THE NODAL PARAMETERS ***
PARAMETER     MAXIMUM, NODE        MINIMUM, NODE
      1    5.2986E+00,    2     0.0000E+00,    3
```

Figure 21.4.7 Slider bearing results

$$\mathbf{C}^e = \frac{U}{l} \begin{Bmatrix} 1 \\ -1 \end{Bmatrix} \int_{x_i}^{x_j} h \, dx .$$

Thus, it is clear that the assumed thickness variation within the element has an important effect on the complexity of the element matrices. It should also be clear that the nodal points of the mesh must be located such that any discontinuity in h occurs at a node. The simplest assumption is that h is constant over the length of the element. In this case the latter two integrals reduce to $h^3 l$ and hl, respectively. One may wish to utilize this element to approximate a varying distribution of h by a series of constant steps. In this case, one could use an average thickness of $h = (h_i + h_j)/2$, where h_i and h_j denote the thickness at the nodal points of the element. Subroutines ELSQ and ELCOL for this element are shown in Figs. 21.4.2 and 21.4.3. Note that these subroutines allow for two methods of defining the film thickness, h, in each element. In the default option (NLPFLO = 1, NNPFLO = 0) the value of h is input as a floating point element property, i.e., H = ELPROP(1). In the second option (NLPFLO = 0, NNPFLO = 1) the thickness is specific at each node as a floating point property. Note that for the latter option NLPFLO = 0, which always causes ELPROP(1) = 0.0 and thus $h = 0$. The program checks for this occurrence and then skips to the second definition of h.

The programs for performing the post-solution calculations, ELPOST and POSTEL, are shown in Figs. 21.4.4 and 21.4.5. These two routines evaluate the force, F_y^e, carried by each element. The load on a typical element is $F_y^e = \mathbf{Q}^e \mathbf{P}^e$ where $\mathbf{Q}^e = [l/2 \;\; l/2]$. Subroutine ELPOST generates and stores $\mathbf{Q}^e$ for each element. Subroutine POSTEL carries out the multiplication once the nodal pressures, $\mathbf{P}^e$, are known. In addition, it sums the force on each element to obtain the total load capacity of the bearing. It prints the element number and its load and the total load on the bearing. With the addition of a few extra post-solution calculations, one could also output the location of the resultant bearing force. In closing, recall that both U and v are constant along the entire length of the bearing. Thus, they are simply defined as floating point miscellaneous system properties (MISCFL = 2).

As a numerical example consider the step bearing shown in Fig. 21.4.6, which has two constants gaps of different thicknesses but equal lengths. Select a mesh with three nodes (M = 3) and two elements (NE = 2). Let $l_1 = l_2 = 0.125$ ft, $U = 20$ ft/s, $v = 0.002$ lb s/ft^2, $h_1 = 0.025$ ft, and $h_2 = 0.036$ ft. The two boundary conditions are $P_1 = P_3 = 0$ and we desire to calculate the pressure, P_2, at the step. The calculated pressure is $P_2 =$ 5.299 psf, which is the exact value, and the total force on the bearing is $F_y = 0.66$ ppf. The accuracy is not surprising since the exact solution for this problem gives a linear pressure variation over each of the two segments of the bearing. The typical output data are shown in Fig. 21.2.7. Table 21.3 summarizes this element.

21.5 Ordinary Differential Equations *

21.5.1 Linear Example

There are several weighted residual methods available for formulating finite element solutions of differential equations. As an example, consider the ordinary differential equation $y' + ay = b$, which can be written symbolically in operator form as $L(y) = Q(x)$. Akin [1] and Lynn and Arya [4] have presented a least squares finite element solution of such problems. Let the value of y within an element by $y = \mathbf{H}^e(x)\,\mathbf{y}^e$, where $\mathbf{H}$ is the element interpolation functions matrix, and $\mathbf{y}^e$ is the array of nodal values for the element. These authors show that the corresponding element matrices are

$$\mathbf{S}^e = \int_{l^e} \mathbf{G}^e(x)^T\,\mathbf{G}^e(x)\,dx\,, \quad \mathbf{C}^e = \int_{l^e} \mathbf{G}^e(x)^T\,Q(x)\,dx\,, \tag{21.14}$$

where $\mathbf{G}^e$ is the matrix obtained from the differential operator acting on the element interpolation functions, that is, $\mathbf{G}^e = L(\mathbf{H}^e)$. For the equation given above $L = (d/dx) + a$ and $Q = b$. Since the operator is of first order the above integral definitions of the element matrices will only require C^0 continuity of the approximate finite element solution. Thus, one can utilize a linear element with a node at each end. Substituting yields

$$\mathbf{G}^e = [(a(x_j - x) - 1)\ \ (a(x - x_i) + 1)]/l$$

so that the resulting element matrices are

$$\mathbf{C}^e = \frac{b}{2}\begin{Bmatrix} al - 2 \\ al + 2 \end{Bmatrix}, \quad \mathbf{S}^e = \frac{1}{3l}\begin{bmatrix} (3 - 3al + a^2l^2) & (a^2l^2 - 6)/2 \\ sym. & (3 + 3al + a^2l^2) \end{bmatrix}. \tag{21.15}$$

Figures 21.5.1 and 21.5.2 show the problem-dependent subroutines for calculating the above element matrices. Note that since a and b are constant over the whole solution domain, they are treated as miscellaneous system properties (MISCFL = 2). As a specific example, consider the solution of the problem $y' + 2y = +10$, where $y(0) = 0$. Let the domain of interest, $0 \le x \le 0.5$, be divided into five elements of equal length (NE = 5, M = 6, NSPACE = 1, N = 2, NG = 1). The exact solution is $y(x) = 5(1 - e^{-2x})$. The program output is summarized in Fig. 21.5.3. Table 21.4 gives the problem summary.

21.5.2 Nonlinear Example

The above procedure can be extended to solve nonlinear operators as well. If one utilizes an iterative solution of a differential operator of the form $L(y) + N(y) = Q(x)$, where L and N denote linear and nonlinear differential operators, then the element matrices for iteration n become

$$\mathbf{S}_n^e = \int_{l^e} \left[\mathbf{E}^{eT}\,\mathbf{E} + \mathbf{E}^{eT}\,\mathbf{F}_{n-1}^e + \mathbf{F}_{n-1}^{eT}\,\mathbf{E}^e + \mathbf{F}_{n-1}^{eT}\,\mathbf{F}_{n-1}^e \right] dx\,,$$

$$\mathbf{C}_n^e = \int_{l} Q(x) \left[\mathbf{E}^{eT} + \mathbf{F}_{n-1}^{eT} \right] ds\,, \tag{21.16}$$

where $\mathbf{E}$ and $\mathbf{F}$ denote the matrices resulting from the element interpolation functions being acted upon by the operators L and N, respectively. As an example, Akin presents the solution of $2yy'' - (y')^2 + 4y^2 = 0$ subject to the two boundary conditions of $y(\pi/6) = 1/4$ and $y(\pi/2) = 1$. Since this operator is of the second order, the above least square integral definitions will require a finite element model with C^1 continuity; that is,

Table 21.3 Parameter definitions for a slider bearing (* Default option)

CONTROL:

NSPACE = 1	NNPFLO = 0*, or 1	NG = 1
N = 2	NLPFLO = 1*, or 0	MISCFL = 2

DEPENDENT VARIABLES:
  1 = pressure, p

PROPERTIES:
  Nodal point level:
    Floating point:
      1 = none,* or film thickness at the node, h
  Element level:
    Floating point:
      1 = film thickness of the element,* h, or none

MISCELLANEOUS:
  Floating point:
    1 = viscosity, v
    2 = velocity, U

CONSTRAINTS:
  Type 1: specified nodal pressures

POST-SOLUTION CALCULATIONS:
  1 = resultant force on the element

```
C       ................................................
C        ***  ELSQ PROBLEM DEPENDENT STATEMENTS FOLLOW ***
C       ................................................
C       APPLICATION:  LEAST SQUARES SOLUTION OF Y'+A*Y=B
C           NSPACE = 1, N = 2, NG = 1
C       DL = LENGTH OF ELEMENT
C       ARRAY FLTMIS CONTAINS PROBLEM COEFFICIENTS A AND B
        COMMON /ELARG/  A, B, DL
        DL = COORD(2,1) - COORD(1,1)
        A  = FLTMIS(1)
        B  = FLTMIS(2)
        S(1,1) = (3. - 3.*A*DL + A*A*DL*DL)/3./DL
        S(2,2) = (3. + 3.*A*DL + A*A*DL*DL)/3./DL
        S(1,2) = (A*A*DL*DL - 6.)/6./DL
        S(2,1) = S(1,2)
        RETURN
        END
```

Figure 21.5.1 Linear ODE square matrix

```
C       ...................................................
C       ***  ELCOL PROBLEM DEPENDENT STATEMENTS FOLLOW ***
C       ...................................................
C       APPLICATION:  LEAST SQUARES SOLUTION OF Y'+A*Y=B
C          NSPACE = 1, N = 2, NG = 1
        COMMON /ELARG/  A, B, DL
        C(1) = 0.5*B*(A*DL - 2.)
        C(2) = 0.5*B*(A*DL + 2.)
        RETURN
        END
```

Figure 21.5.2 ODE source vector

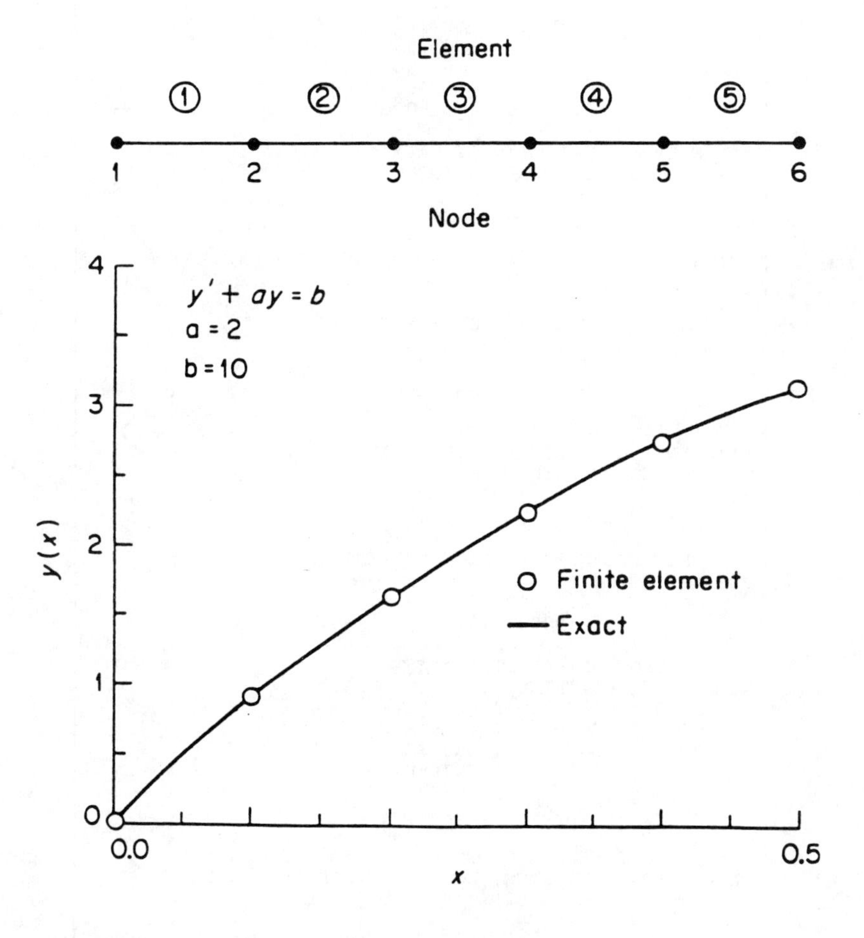

Figure 21.5.3 Mesh and results for weighted residual procedure

```
LEAST SQ.S SOLUTION OF Y'+2Y=10, Y(0) = 0

NUMBER OF NODAL POINTS IN SYSTEM =..........    6
NUMBER OF ELEMENTS IN SYSTEM =..............    5
NUMBER OF NODES PER ELEMENT =...............    1
NUMBER OF PARAMETERS PER NODE =.............    2
DIMENSION OF SPACE =........................    1
NUMBER OF ROWS IN B MATRIX =................    1
STIFFNESS STORAGE MODE: SKY, BAND =.........    1
NUMBER OF REAL MISCELLANEOUS  PROPERTIES =...   2

*** NODAL POINT DATA ***
NODE, CONSTRAINT FLAG, 1 COORDINATES
         1          1    0.0000
         2          0    0.1000
         3          0    0.2000
         4          0    0.3000
         5          0    0.4000
         6          0    0.5000

*** ELEMENT CONNECTIVITY DATA ***
ELEMENT NO., 2 NODAL INCIDENCES.
    1    1    2
    2    2    3
    3    3    4
    4    4    5
    5    5    6

***  NODAL PARAMETER CONSTRAINT LIST   ***
TYPE        EQUATIONS
   1           1
   2           0
   3           0

*** CONSTRAINT EQUATION DATA ***
CONSTRAINT TYPE ONE
EQ. NO.    NODE1    PAR1           A1
      1        1       1   0.00000E+00

***  MISCELLANEOUS SYSTEM PROPERTIES   ***
PROPERTY     VALUE
       1      2.00000E+00
       2      1.00000E+01

***  OUTPUT OF RESULTS   ***
NODE, 1 COORDINATES, 1 PARAMETERS.
    1  0.00000E+00  0.00000E+00
    2  1.00000E-01  9.07491E-01
    3  2.00000E-01  1.65018E+00
    4  3.00000E-01  2.25798E+00
    5  4.00000E-01  2.75537E+00
    6  5.00000E-01  3.16236E+00

*** EXTREME VALUES OF THE NODAL PARAMETERS ***
PARAMETER      MAXIMUM, NODE        MINIMUM, NODE
      1    3.1624E+00,    6    0.0000E+00,    1
```

Figure 21.5.4 Solution for ODE example

both y and y' must be continuous between elements. Thus, select the one-dimensional element which has two nodes (N = 2) and utilizes both y and y' as nodal parameters (NG = 2). Thus, the element interpolation functions for this cubic element are the first order Hermite polynomials, that is,

$$y(s) = \mathbf{H}(s)^e\, \mathbf{Y}^e\,, \quad \mathbf{Y}^{eT} = \begin{bmatrix} y_i & y_i' & y_j & y'_j \end{bmatrix} \tag{21.17}$$

and

$$\mathbf{H}(s)^e = \begin{bmatrix}(1-3r^2+2r^3) & l\,(r-2r^2+r^3) & (3r^2-2r^3) & l\,(r^3-r^2)\end{bmatrix}$$

and where $r = s/l$, $l = x_j - x_i$, and $s = x - x_i$. These are programmed in the shape function library included on the disk.

The above definitions of $\mathbf{S}^e$ and $\mathbf{C}^e$ are too complicated to be integrated in closed form so the calculations in subroutine ELSQ, shown in Fig. 21.5.5, are based on a four point Gaussian quadrature. The initial estimates of $y(x)$ and $y'(x)$ for the first iteration are calculated by the problem-dependent subroutine START, which is shown in Fig. 21.5.6. These starting values simply represent a linear interpolation between the two boundary conditions. Select a mesh with eleven equally spaced nodes (M = 11), and ten elements (NE = 10). In Fig. 21.5.7 the results of a three iteration solution (NITER = 3) are compared with an exact solution of $y(x) = \text{Sin}^2(x)$. Note that the iterative approach rapidly approaches the exact values when given a reasonably good starting estimate. The element summary is given in Table 21.5. The MODEL program uses routines CHANGE and CORECT, Fig. 21.4.1, to control the over-relaxation iteration procedure. Clearly, other procedures such as the Newton-Raphson method are better suited to the solution of nonlinear problems.

21.6 Plane Frame Structures

Weaver [9] has considered the detailed programming aspects of several one-dimensional structural elements. Among them is the plane frame element which has two displacements and one rotation per node. A typical element is shown in Fig. 21.6.1 where XM, YM are the member axes and x, y are the global axes. A frame element with an area of A, having a moment of inertia I, a length L, and made of a material having a modulus of E has an element square matrix (stiffness matrix) of

$$\mathbf{S}_m^e = \frac{E}{L}\begin{bmatrix} A & 0 & 0 & -A & 0 & 0 \\ & 12\beta/L & 6\beta & 0 & -12\beta/L & 6\beta \\ & & 4I & 0 & -6\beta & 2I \\ & & & A & 0 & 0 \\ & & & & 12\beta/L & -6\beta \\ sym. & & & & & 4I \end{bmatrix} \tag{21.18}$$

where $\beta = I/L$, and where the element displacements have been ordered as $\delta^{eT} = [\,u_1\; v_1\; \theta_1\; u_2\; v_2\; \theta_2\,]$. Weaver shows that to convert this expression to one valid in the global $(x,\,y)$ coordinates one must consider how the nodal degrees of freedom transform from the element axes to the global axes. This relation is shown to be $\mathbf{S}^e = \mathbf{T}^T\,\mathbf{S}_m^e\,\mathbf{T}$, where the transformation matrices are

Table 21.4 Parameter definitions for least squares solution of $y' + ay = b$

CONTROL:
NSPACE = 1 N = 2 NG = 1

DEPENDENT VARIABLES:
1 = y

PROPERTIES:
MISCELLANEOUS:
1 = coefficient a

CONSTRAINTS:
Type 1: specified y-value

```
C     .............................................................
C       ***  ELSQ PROBLEM DEPENDENT STATEMENTS FOLLOW ***
C     .............................................................
C     APPLICATION: LEAST SQ. SOL. OF 2YY''-Y'Y'+4YY=0
      COMMON /ELARG/  D2H(4), F(4), LSQ
      DATA  KALL / 1 /
C-->  EXTRACT QUADRATURE DATA ON THE FIRST CALL
      IF ( KALL .EQ. 1 )  THEN
        KALL = 0
        NIP = 4
        CALL  GAUSCO (NIP, PT, WT)
        LSQ = NELFRE*NELFRE
      ENDIF
      XI = COORD(1, 1)
      XJ = COORD(2, 1)
      DL = XJ - XI
C      NUMERICAL INTEGRATION LOOP
      CALL ZEROA (LSQ, S)
      DO 50  IP = 1, NIP
C-->     FIND LOCAL COORDINATES OF INTEGRATION POINT
        BETA = 0.5*(1. + PT(1,IP))
C-->     EVALUATE SHAPE FUNCTIONS AND DERIVATIVES
        CALL  SHPCU  (BETA, DL, H)
        CALL  DERCU  (BETA, DL, DLH)
        CALL  DER2CU (BETA, DL, D2H)
C-->     FORM NONLINEAR MATRIX F USING PREVIOUS D
        CALL  MMULT  (H, D, YOLD, 1, NELFRE, 1)
        CALL  MMULT  (DLH, D, DYOLD, 1, NELFRE, 1)
        DO 20  I = 1, NELFRE
   20   F(I) = 2.*YOLD*D2H(I) - DYOLD*DLH(1,I) + 4.*YOLD*H(I)
C        DX = DX/D_BETA*D_BETA/DU*DU = 2.*L*DU
        W = WT(IP)*2.*DL
        DO 40  I = 1, NELFRE
          DO 30  J = 1, NELFRE
   30     S(I,J) = S(I,J) + W*F(I)*F(J)
   40   CONTINUE
   50 CONTINUE
      RETURN
      END
```

Figure 21.5.5 Nonlinear ODE square matrix

```
      FUNCTION  START (IG, NSPACE, COORD)
C     * * * * * * * * * * * * * * * * * * * * * * * * * *
C     DEFINE STARTING VALUE OF PARAMETER IG IN TERMS OF
C     COORDINATES OF THE NODE  (FOR ITERATIVE SOLUTIONS)
C     * * * * * * * * * * * * * * * * * * * * * * * * * *
C        A PROBLEM DEPENDENT ROUTINE
CDP   IMPLICIT REAL*8(A-H,O-Z)
      DIMENSION  COORD(1,NSPACE)
C     NSPACE = DIMENSION OF SPACE
C     COORD  = SPATIAL COORDINATE ARRAY OF NODE
C     ...................................................
C     ** PROBLEM DEPENDENT START STATEMENTS FOLLOW **
C     ...................................................
C     APPLICATION: LEAST SQ. SOL. OF 2YY''-Y'Y'+4YY=0
C-->   STRAIGHT LINE FIT THROUGH TWO BOUNDARY VALUES
      X     = COORD(1,1)
      START = 0.7162D0*X - 0.125D0
      IF ( IG .EQ. 2 )  START = 0.7162D0
      RETURN
      END
```

Figure 21.5.6 Starting guess definition

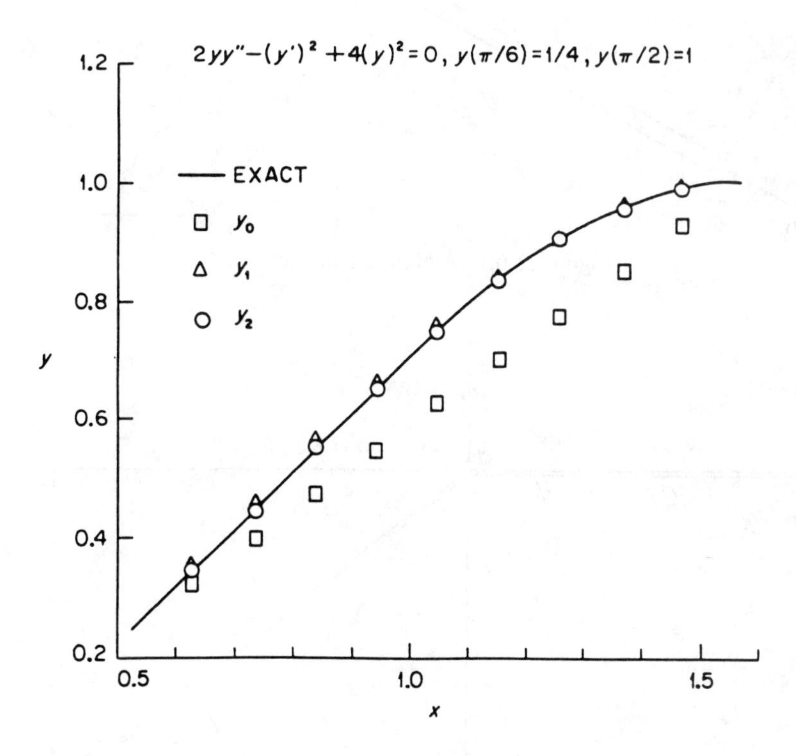

Figure 21.5.7 Iteration solution results

Table 21.5 Parameter definitions for least square solution of $L(y) + N(y) = Q(x)$

CONTROL:
NSPACE = 1 N = 2 NG = 2

DEPENDENT VARIABLES:
$1 = y$ $2 = y' = dy/dx$

CONSTRAINTS:
Type 1: specified values of y and/or y'

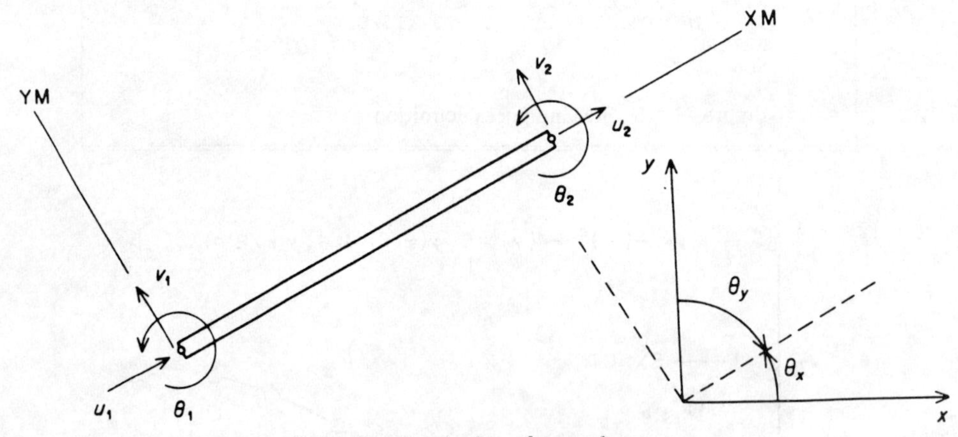

Figure 21.6.1 A plane frame element

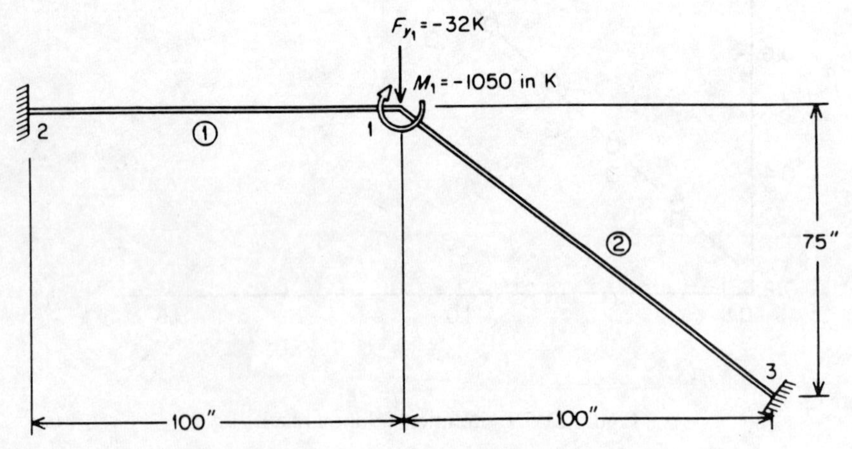

Figure 21.6.2 A plane frame structure

```
C     ......................................................
C      ***  ELSQ PROBLEM DEPENDENT STATEMENTS FOLLOW ***
C     ......................................................
C     APPLICATION:  PLANE FRAME ANALYSIS
C     ELPROP(1) = AREA, ELPROP(2) = MOMENT OF INERTIA
C     ELPROP(3) = MODULUS OF ELASTICITY
C     SM = LOCAL STIFFNESS, T = TRANSFORMATION MATRIX
      DIMENSION  SM(6,6), T(6,6), DUMMY(6,6)
      DATA  SM, T  / 36*0.0, 36*0.0 /
C-->   ELEMENT PROPERTIES,LENGTH, AND DIRECTION COSINES
      A  = ELPROP(1)
      ZI = ELPROP(2)
      E  = ELPROP(3)
      DX = COORD(2,1) - COORD(1,1)
      DY = COORD(2,2) - COORD(1,2)
      FL = SQRT (DX*DX + DY*DY)
      CX = DX/FL
      CY = DY/FL
C-->   DEFINE NON-ZERO TERMS IN STIFFNESS MATRIX
      EDL  = E/FL
      BETA = ZI/FL
      SM(1,1) = A*EDL
      SM(2,2) = 12.*EDL*BETA/FL
      SM(2,3) = 6.*EDL*BETA
      SM(3,2) = SM(2,3)
      SM(3,3) = 4.*EDL*ZI
      SM(1,4) = -A*EDL
      SM(4,1) = SM(1,4)
      SM(4,4) = A*EDL
      SM(2,5) =  -SM(2,2)
      SM(5,2) = SM(2,5)
      SM(3,5) = -SM(2,3)
      SM(5,3) = SM(3,5)
      SM(5,5) = SM(2,2)
      SM(2,6) = SM(2,3)
      SM(6,2) = SM(2,6)
      SM(3,6) = 2.*EDL*ZI
      SM(6,3) = SM(3,6)
      SM(5,6) = -SM(2,3)
      SM(6,5) = SM(5,6)
      SM(6,6) = 4.*EDL*ZI
C-->   TRANSFORM STIFFNESS MATRIX TO GLOBAL COORDINATES
      T(1,1) = CX
      T(2,2) = CX
      T(3,3) = 1.
      T(4,4) = CX
      T(5,5) = CX
      T(6,6) = 1.
      T(1,2) =  CY
      T(2,1) = -CY
      T(4,5) =  CY
      T(5,4) = -CY
      CALL  MMULT  (SM, T, DUMMY, NELFRE, NELFRE, NELFRE)
      CALL  MTMULT (T,DUMMY,S,NELFRE,NELFRE,NELFRE)
C      S = (T)T*SM*T, COULD CALL BTDB INSTEAD
      RETURN
      END
```

Figure 21.6.3 Plane frame stiffness

$$\mathbf{T} = \begin{bmatrix} R & 0 \\ 0 & R \end{bmatrix}, \quad \mathbf{R} = \begin{bmatrix} cx & cy & 0 \\ -cy & cx & 0 \\ 0 & 0 & 1 \end{bmatrix},$$

where $cx = \text{Cos}\,\theta_x$ and $cy = \text{Cos}\,\theta_y$ are the member direction cosines. If one is only interested in the deflections resulting from joint forces and moments these are the only equations that must be programmed. As an example application consider example structure three presented by Weaver. The structure and its loads are shown in Fig. 21.6.2. The homogeneous element properties are E = 10,000 ksi, A = 10 in^2, and I = 1,000 in^4. Nodes 2 and 3 are completely restrained from translating or rotating. The displacements and rotation at node 1 given by Weaver are -0.0202608 in, -0.0993600 in, and -0.00179756 radians, which are in good agreement with the MODEL results. The element matrices are presented in subroutine ELSQ in Fig. 21.6.3.

This problem involves three nodes (M = 3) with three degrees of freedom per node (NG = 3), two elements (NE = 2) with two nodes per element (N = 2), and three (A, I, E) homogeneous properties per element (NLPFLO = 3, LHOMO = 1). There are no element column matrix contributions (NULCOL = 1) and the applied loads at point 1 must be considered as initial contributions to the system matrix (ISOLVT = 1). The MODEL output is in Fig. 21.6.4. The element is summarized in Table 21.6.

Table 21.6 Parameter definitions for a plane frame

CONTROL:		
NSPACE = 2	NLPFLO = 3	NG = 3
N = 2	NULCOL = 1	INRHS = 1
DEPENDENT VARIABLES:		
1 = x-component of displacement, u		
2 = y-component of displacement, v		
3 = z-rotation, θ		
PROPERTIES:		
Element level:		
1 = cross-sectional area, A		
2 = moment of inertia, I		
3 = modulus of elasticity, E		
CONSTRAINTS:		
Type 1: specified nodal displacements and/or rotations		
Type 2: inclined rollers, or rigid members, etc.		
INITIAL VALUES OF COLUMN VECTOR:		
Specified external nodal forces or moments		

```
NUMBER OF NODAL POINTS IN SYSTEM =...........      3
NUMBER OF ELEMENTS IN SYSTEM =...............      2
NUMBER OF NODES PER ELEMENT =................      3
NUMBER OF PARAMETERS PER NODE =..............      2
DIMENSION OF SPACE =.........................      2
INITIAL FORCING VECTOR INPUT FLAG =..........      1
NUMBER OF ROWS IN B MATRIX =.................      1
NUMBER OF REAL PROPERTIES PER ELEMENT =......      3

*** NODAL POINT DATA ***
NODE, CONSTRAINT FLAG, 2 COORDINATES
         1         0  100.0000   75.0000
         2       111    0.0000   75.0000
         3       111  200.0000    0.0000

*** ELEMENT CONNECTIVITY DATA ***
ELEMENT NO., 2 NODAL INCIDENCES.
    1    2    1
    2    1    3

*** CONSTRAINT EQUATION DATA ***
CONSTRAINT TYPE ONE
EQ. NO.   NODE1   PAR1        A1
     1       2      1   0.00000E+00
     2       2      2   0.00000E+00
     3       2      3   0.00000E+00
     4       3      1   0.00000E+00
     5       3      2   0.00000E+00
     6       3      3   0.00000E+00

***  ELEMENT  PROPERTIES   ***
ELEMENT  PROPERTY       VALUE
    1         1     1.00000E+01
    1         2     1.00000E+03
    1         3     1.00000E+04

*** INITIAL FORCING VECTOR DATA ***
NODE    PARAMETER        VALUE        DOF
    1     2      -3.20000E+01       2
    1     3      -1.05000E+03       3
    3     3       0.00000E+00       9

***  OUTPUT OF RESULTS  ***
NODE, 2 COORDINATES, 3 PARAMETERS.
    1  1.00000E+02  7.50000E+01 -2.02608E-02 -9.93600E-02 -1.79756E-03
    2  0.00000E+00  7.50000E+01  0.00000E+00  0.00000E+00  0.00000E+00
    3  2.00000E+02  0.00000E+00  0.00000E+00  0.00000E+00  0.00000E+00

*** EXTREME VALUES OF THE NODAL PARAMETERS ***
PARAMETER     MAXIMUM, NODE        MINIMUM, NODE
      1    0.0000E+00,    3    -2.0261E-02,    1
      2    0.0000E+00,    3    -9.9360E-02,    1
      3    0.0000E+00,    3    -1.7976E-03,    1
```

Figure 21.6.4 Plane frame results

21.7 References

[1] Akin, J.E., "A Least Squares Finite Element Solution of Nonlinear Equations," pp. 153–162 in *The Mathematics of Finite Elements and Applications*, ed. J.R. Whiteman, London: Academic Press (1973).

[2] Allan, T., "The Application of Finite Element Analysis to Hydrodynamic and Externally Pressurized Pocket Bearings," *Wear*, **19**, pp. 169–206 (1972).

[3] Huebner, K.H. and Thornton, E.A., *Finite Element Method for Engineers*, New York: John Wiley (1982).

[4] Lynn, P.P. and Arya, S.K., "Finite Elements Formulation by the Weighted Discrete Least Squares Method," *Int. J. Num. Meth. Eng.*, **8**, pp. 71–90 (1974).

[5] Meek, J.L., *Matrix Structural Analysis*, New York: McGraw-Hill (1972).

[6] Myers, G.E., *Analytical Methods in Conduction Heat Transfer*, New York: McGraw-Hill (1971).

[7] Reddi, M.M., "Finite Element Solution of the Incompressible Lubrication Problem," *J. Lubrication Technology*, **53**(3), pp. 524–532 (July 1969).

[8] Wada, S. and Hayashi, H., "Application of Finite Element Method to Hydrodynamic Lubrication Problems," *Bulletin of Japanese Soc. Mech. Eng.*, **14**(77), pp. 1222–1244 (1971).

[9] Weaver, W.F., Jr. and Johnston, P.R., *Finite Elements for Structural Analysis*, Englewood Cliffs: Prentice-Hall (1984).

[10] Zienkiewicz, O.C., *The Finite Element Method,* 3rd edition, New York: McGraw-Hill (1979).

Chapter 22

MODEL APPLICATIONS IN 2- AND 3-D

22.1 Introduction

From the very beginning finite element methods have been applied to two-dimensional problems. The three node (linear) triangle has been utilized in the derivation of the element matrices for several engineering applications. Most of the current texts present the results for such applications as stress analysis, heat conduction and ground water seepage. The four node rectangular (bilinear) element has also been presented in some detail, but it is not often used due to the difficulty of fitting complex engineering shapes with rectangles. Meek [14] made a useful contribution in presenting the closed form element matrices for a six node (quadratic) triangle applied to the Poisson equation. The equations given by Meek are quite useful and should be of interest to a beginner. However, one soon learns that it is easier to program such elements when numerical integration is utilized. A significant feature of this section is the detailed formulation of isoparametric elements. This section, combined with Sec. 9.4, should supply the reader with the ability to master this very useful concept. The extensions of the isoparametric element to a three-dimensional application will be considered later to illustrate how easy it is to change from one element type to another.

This chapter will also begin to introduce the reader to some of the useful capabilities of the MODEL program that may not be immediately apparent. For example, it will be shown how one can reduce the data requirements for non-homogeneous problems by assigning to each element a material number and a corresponding array of floating point properties.

22.2 Plane Stress Analysis

The application of the finite element method to the plane stress analysis of solids has been considered in detail in several texts [4, 7, 14, 20]. The element matrices for the three node linear displacement (constant strain) triangle have been presented by Ural [17], Gallagher [8], and others. The nodal parameters are u and v, the nodal displacement components in the x– and y-directions, respectively. If one orders the element degrees of freedom as $\delta^{eT} = [u_1\ v_1\ u_2\ v_2\ u_3\ v_3]$ it has been shown that the element load vector due to distributed body forces per unit volume are

$$\mathbf{C}^e_{6\times 1} = t\Delta/3 \begin{Bmatrix} X \\ Y \\ X \\ Y \\ X \\ Y \end{Bmatrix} \tag{22.1}$$

where Δ is the area of the triangle and where X and Y denote the components of the distributed body force. The element square matrix (or *stiffness matrix*) for an element of constant thickness, t, is

$$\mathbf{S}^e_{6\times 6} = \mathbf{B}^{eT}\mathbf{D}^e\mathbf{B}^e t\Delta,$$

where

$$\mathbf{B}^e_{6\times 3} = \frac{1}{2\Delta}\begin{bmatrix} b_i & 0 & b_j & 0 & b_m & 0 \\ 0 & c_i & 0 & c_j & 0 & c_m \\ c_i & b_i & c_j & b_j & c_m & b_m \end{bmatrix} \tag{22.2}$$

denotes the *strain-displacement matrix* (i.e. $\varepsilon^e = \mathbf{B}^e\,\delta^e$), and where the material constitutive matrix for an isotropic material is

$$\mathbf{D}^e_{3\times 3} = \begin{bmatrix} c & \nu c & 0 \\ & c & 0 \\ sym. & & G \end{bmatrix} \tag{22.3}$$

with $c \equiv E/(1 - \nu^2)$ and $G = E/2(1 + \nu)$. In the latter equation, E equals the modulus of elasticity, ν is Poisson's ratio, and G is the shear modulus. Although G, in theory, is not an independent variable, one often encounters applications where it is desirable to treat G as a third material property. Thus, the MODEL code requires the user to define E, ν, and G in the input data.

The constants in the **B** matrix are defined in terms of the spatial coordinates of the element's nodes. For example, $b_i = y_j - y_m$ and $c_i = x_m - x_i$, and the other terms are obtained by a cyclic permutation of the subscripts. The strains and stresses under consideration are $\boldsymbol{\varepsilon}^{\mathrm{T}} = [\varepsilon_x\ \ \varepsilon_y\ \ \gamma_{xy}]$ and $\boldsymbol{\sigma}^{\mathrm{T}} = [\sigma_x\ \ \sigma_y\ \ \tau_{xy}]$ where $\boldsymbol{\sigma}^{\mathrm{e}} = \mathbf{D}^e\boldsymbol{\varepsilon}^{\mathrm{e}}$. If one has applied surface traction components on an element boundary segment then the contributions to the boundary column matrix are as shown in Fig. 20.2.14.

Thus far, the problem under consideration is two-dimensional (NSPACE = 2), involves triangular elements (N = 3), with six $[E, \nu, G, t, X, Y)$ floating point properties (NLPFLO = 6), and two nodes per typical boundary segments (LBN = 2). In addition, each node has two nodal parameters (NG = 2) so that the element has a total of six degrees of freedom. The above discussion provides enough data to calculate the nodal displacements of the mesh. Generally, the post-solution calculation of stresses and strains is important to the analyst. Since these quantities are constant in this type of element, they are usually listed as occurring at the centroidal coordinates of the element. These quantities are passed through the named common ELARG to subroutine ELPOST along with the global coordinates of the element centroid. The constant stress-displacement and strain-displacement matrices are generated by subroutine ELSQ. If NTAPE1 > 0, subroutine ELPOST simply writes these data on unit NTAPE1 for later use

```
C     .......................................................
C      ***  ELSQ PROBLEM DEPENDENT STATEMENTS FOLLOW ***
C     .......................................................
C     APPLICATION: PLANE STRESS, ISOTROPIC MATERIAL, CST
C     NSPACE = 2, N= 3, NG = 2, NELFRE = 6, NLPFLO = 4
C     ELPROP(1) = ELASTIC MODULUS ELPROP(2) = POISSON RATIO
C     ELPROP(3) = SHEAR MODULUS   ELPROP(4) = THICKNESS
C     ELPROP(5) = X-BODY FORCE    ELPROP(6) = Y-BODY FORCE
C     B       = ELEMENT STRAIN-DISPLACEMENT MATRIX
C     E       = MATERIAL CONSTITUTIVE MATRIX
C     EB      = ELEMENT STRESS-DISPLACEMENT MATRIX
      COMMON /ELARG/ XB, YB, EE, V, G, T, FX, FY, AREA
      DATA  KALL / 1 /
C-->  DEFINE NODAL COORDINATES
      XI = COORD(1,1)
      XJ = COORD(2,1)
      XK = COORD(3,1)
      YI = COORD(1,2)
      YJ = COORD(2,2)
      YK = COORD(3,2)
C-->  DEFINE PROPERTIES
      EE = ELPROP(1)
      V  = ELPROP(2)
      G  = ELPROP(3)
      T  = ELPROP(4)
      FX = ELPROP(5)
      FY = ELPROP(6)
C     DEFINE CENTROID COORDINATES
      XB = (XI + XJ + XK)/3.
      YB = (YI + YJ + YK)/3.
C-->  DEFINE GEOMETRIC PARAMETERS
      AI = XJ*YK-XK*YJ
      AJ = XK*YI-XI*YK
      AK = XI*YJ-XJ*YI
      BI = YJ-YK
      BJ = YK-YI
      BK = YI-YJ
      CI = XK-XJ
      CJ = XI-XK
      CK = XJ-XI
C     CALCULATE ELEMENT AREA
      TWOA = AI + AJ + AK
      AREA = TWOA*0.5
      TA   = T*AREA
C-->  ON THE FIRST CALL DEFINE ZERO TERMS IN E AND B
      IF ( KALL .EQ. 1 )  THEN
        KALL   = 0
        E(1,3) = 0.0
        E(3,1) = 0.0
        E(2,3) = 0.0
        E(3,2) = 0.0
        B(1,2) = 0.0
        B(1,4) = 0.0
        B(1,6) = 0.0
        B(2,1) = 0.0
        B(2,3) = 0.0
        B(2,5) = 0.0
      ENDIF
```

Figure 22.2.1a Plane stress stiffness for a T3 element

```
C-->   DEFINE VARIABLE TERMS IN E AND B
       E(1,1) = EE/(1.-V*V)
       E(2,2) = EE/(1.-V*V)
       E(3,3) = G
       E(1,2) = V*EE/(1.-V*V)
       E(2,1) = V*EE/(1.-V*V)
       B(1,1) = BI/TWOA
       B(1,3) = BJ/TWOA
       B(1,5) = BK/TWOA
       B(2,2) = CI/TWOA
       B(2,4) = CJ/TWOA
       B(2,6) = CK/TWOA
       B(3,1) = CI/TWOA
       B(3,3) = CJ/TWOA
       B(3,5) = CK/TWOA
       B(3,2) = BI/TWOA
       B(3,4) = BJ/TWOA
       B(3,6) = BK/TWOA
C      FORM EB-DISPLACEMENT MATRIX
       CALL  MMULT (E, B, EB, NRB, NRB, NELFRE)
C-->   COMPLETE STIFFNESS MATRIX, K= BT*E*B
       CALL  MTMULT (B, EB, S, NRB, NELFRE, NELFRE)
       CALL  MSMULT (TA, S, NELFRE, NELFRE)
       RETURN
       END
```

Figure 22.2.1b Plane stress stiffness for a T3 element

```
C      ....................................................
C       ***  ELCOL PROBLEM DEPENDENT STATEMENTS FOLLOW ***
C      ....................................................
C      APPLICATION: PLANE STRESS, ISOTROPIC MATERIAL, CST
C      NSPACE = 2, N= 3, NG = 2, NELFRE = 6, NLPFLO = 4
C      ELPROP(1) = ELASTIC MODULUS ELPROP(2) = POISSON RATIO
C      ELPROP(3) = SHEAR MODULUS   ELPROP(4) = THICKNESS
C      ELPROP(5) = X-BODY FORCE    ELPROP(6) = Y-BODY FORCE
       COMMON /ELARG/ XB, YB, EE, V, G, T, FX, FY, AREA
       TA = T*AREA
       CX = TA*FX/3.
       CY = TA*FY/3.
C---> BODY FORCE LOADS
       C(1) = CX
       C(2) = CY
       C(3) = CX
       C(4) = CY
       C(5) = CX
       C(6) = CY
       RETURN
       END
```

Figure 22.2.2 Plane stress body load for a T3 element

```
C      ......................................................
C       *** ELPOST PROBLEM DEPENDENT STATEMENTS FOLLOW ***
C      ......................................................
C      APPLICATION: PLANE STRESS, ISOTROPIC MATERIAL, CST
C-->   STORE DATA FOR STRESS AND STRAIN CALCULATIONS
C      B   = ELEMENT STRAIN-DISPLACEMENT MATRIX
C      EB  = ELEMENT STRESS-DISPLACEMENT MATRIX
       COMMON /ELARG/ XB, YB, EE, V, G, T, FX, FY, AREA
       WRITE (NTAPE1)  XB, YB, B, EB
       RETURN
       END
```

Figure 22.2.3 Storing T3 stress data

by subroutine POSTEL. Similarly, if NTAPE1 > 0, subroutine POSTEL is called after the nodal displacements have been calculated. It reads the above data, extracts the element's nodal displacements and multiplies them by the strain-displacement and stress-displacement matrices to obtain the element strains and stresses, respectively. Then the element's number, centroidal coordinates, stresses and strains are printed. After the post-solution calculations have been completed, one may elect to request the calculation of the displacement contours. This element is summarized in Table 22.1.

```
C      ......................................................
C       *** POSTEL PROBLEM DEPENDENT STATEMENTS FOLLOW ***
C      ......................................................
C      APPLICATION: PLANE STRESS, ISOTROPIC MATERIAL, CST
C      NSPACE = 2, N= 3, NG = 2, NELFRE = 6, NLPFLO = 4
C-->   CALCULATE STRESSES AND STRAINS AT ELEMENT CENTROID
       COMMON /ELARG/ XB, YB, EE, V, G, T, FX, FY, AREA
C      STRAIN = ELEMENT CENTROIDAL STRAINS
C      STRESS = ELEMENT CENTROIDAL STRESSES
       DATA  KALL / 1 /
C-->   PRINT HEADINGS ON FIRST CALL
       IF ( KALL .EQ. 1 )  THEN
         WRITE (6,5000)
 5000    FORMAT ('* * * ELEMENT STRESSES * * *',/,
     1   ' ELEMENT',5X,'X',10X,'Y',14X,'STRAIN',33X,'STRESS',/,
     2    26X,'XX',9X,'YY',9X,'XY',9X,'XX',9X,'YY',9X,'XY',//)
         KALL = 0
       ENDIF
       READ ( NTAPE1)  XB, YB, B, EB
C-->    CALCULATE STRAINS
       CALL  MMULT (B, D, STRAIN, NRB, NELFRE, 1)
C-->    CALCULATE STRESSES
       CALL  MMULT (EB, D, STRESS, NRB, NELFRE, 1)
C       PRINT RESULTS
       WRITE (6, 5010) IE, XB, YB, (STRAIN(K), K=1, NRB),
     1                             (STRESS(L), L=1, NRB)
 5010 FORMAT ( I5, 2F10.2, 6(1PE11.3) )
       RETURN
       END
```

Figure 22.2.4 Compute T3 element stresses and strains

Table 22.1 Parameter definitions for plane stress (* Default, ** Optional)

CONTROL:

NSPACE = 2	NLPFLO = 6	LBN = 0*, or 2** 0
N = 3	ISOLVT = 0*, or 1**	
NG = 2	NTAPE1 = 8	

DEPENDENT VARIABLES:
1 = x-component of displacement, u
2 = y-component of displacement, v

BOUNDARY FLUX COMPONENTS AT NODES** :
1 = x-component of surface traction, T_x
2 = y-component of surface traction, T_y

PROPERTIES: Element level:
Floating point:
1 = modulus of elasticity, E
2 = Poisson's ratio, ν
3 = shear modulus, G
4 = element thickness, t
5 = x-body force component, X
6 = y-body force component, Y

CONSTRAINTS:
Type 1: specified nodal displacements
Type 2: inclined roller supports

INITIAL VALUES OF COLUMN VECTOR** :
Specified external nodal forces

POST-SOLUTION CALCULATIONS:
1 = centroidal coordinates
2 = centroidal strains
3 = centroidal stresses

The problem-dependent portions of the element subroutines are shown in Figs. 22.2.1 to 22.2.4. As a numerical example consider the problem presented in detail by Ural [17]. As shown in Fig. 22.2.5, the problem has two elements (NE = 2) and four nodes (M = 4). The two nodes (1 and 2) on the left are completely fixed and at node 3 there is a 25 k = 25,000 lb load in the positive x-direction. No other loads are present (X = Y = 0) and the material has properties of E = 30,000 ksi and ν = 0.25 so that G = 12,000 ksi. The properties are homogeneous (LHOMO = 1) and each element has a thickness of t = 2 in. The output is presented in Fig. 22.2.6. The results agree with those

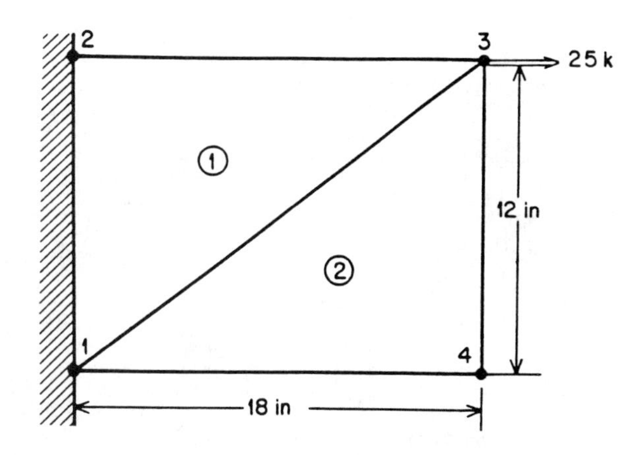

Figure 22.2.5 Plane stress problem with corner force

given by Ural, but the present solution is more accurate since only slide-rule accuracy was utilized in the former solution.

Many stress analysis problems involve non-homogeneous properties. It might prove inconvenient to specify all these properties for each element. Thus, several finite element codes allow a user to assign a single material number code to each element. This code is utilized along with a table of material properties to define the properties of each element. This same concept is easily implemented in the MODEL code. Subroutine MATPRT can be used for this purpose, and MATWRT will print the data for validation. The list of material properties for each material number if stored in order of material number in the miscellaneous floating point system array FLTMIS. That is, if there is a maximum of NMAX materials each having NLPFLT floating point properties then array FLTMIS contains NMAX sub-arrays each having NLPFLT floating point numbers (MISCFL = NMAX*NLPFLT). Given the material number, NUM, this subroutine extracts the appropriate properties list from FLTMIS and stores it in a material array called PRTMAT. Of course, the latter array must be dimensioned in one of the problem-dependent subroutines. The integer material number of each element is input by the user in array LPFIX (i.e., NLPFIX = 1) and it exists at the element level in LPROP(1). Recall that array LPFIX is initially zeroed; thus, one only has to input element material numbers that are greater than 1 if one defines LPROP(1) = 0 or = 1 to both imply a material number of unity in the problem-dependent subroutines that call MATPRT. The typical changes in the previous program to utilize this concept are shown in later applications.

To illustrate the effects of a surface traction load, consider the problem shown in Fig. 22.2.7. The problem-dependent subroutine BFLUX for this loading condition is shown in Fig. 22.2.8. This problem was worked out in great detail by Ural in Example 5.2 of his second text [17]. The problem involves five nodes (M = 5), and four elements (NE = 4, N = 3, NG = 2). The plate has a thickness of $t = 1.5$ in., a Young's modulus of $E = 29{,}000$ ksi, and a Poisson's ratio of $\nu = 0.3$. There are no body force loads ($X = Y = 0$) and no externally applied nodal loads (ISOLVT = 0). Nodes 1 and 2 are completely fixed and a surface traction is applied to edge 5-4 (NSEG = 1, LBN = 2). The traction consists of constant x- and y-components of 1/3 k/in ($T_{x4} = T_{x5} = 0.333$ k/in,

```
NUMBER OF NODAL POINTS IN SYSTEM =..........     4
NUMBER OF ELEMENTS IN SYSTEM =..............     2
NUMBER OF NODES PER ELEMENT =...............     2
NUMBER OF PARAMETERS PER NODE =.............     3
INITIAL FORCING VECTOR INPUT FLAG =.........     1
NUMBER OF ROWS IN B MATRIX =................     3
NUMBER OF REAL PROPERTIES PER ELEMENT =.....     6
OPTIONAL UNIT NUMBERS: NTAPE1 =   8

*** NODAL POINT DATA ***
NODE, CONSTRAINT FLAG, 2 COORDINATES
         1         11     0.0000      0.0000
         2         11     0.0000     12.0000
         3          0    18.0000     12.0000
         4          0    18.0000      0.0000

*** ELEMENT CONNECTIVITY DATA ***
ELEMENT NO.,   3 NODAL INCIDENCES.
   1     1    3   2
   2     1    4   3

*** CONSTRAINT EQUATION DATA ***
CONSTRAINT TYPE ONE
EQ. NO.    NODE1    PAR1          A1
      1        1       1   0.00000E+00
      2        1       2   0.00000E+00
      3        2       1   0.00000E+00
      4        2       2   0.00000E+00

***   ELEMENT   PROPERTIES    ***
ELEMENT   PROPERTY        VALUE
   1          1     3.00000E+04
   1          2     2.50000E-01
   1          3     1.20000E+04
   1          4     2.00000E+00
   1          5     0.00000E+00
   1          6     0.00000E+00

*** INITIAL FORCING VECTOR DATA ***
NODE    PARAMETER         VALUE         DOF
   3    1          2.50000E+01         5
   4    2          0.00000E+00         8

***   OUTPUT OF RESULTS   ***
NODE, 2 COORDINATES,   2 PARAMETERS.
   1  0.00000E+00  0.00000E+00  0.00000E+00  0.00000E+00
   2  0.00000E+00  1.20000E+01  0.00000E+00  0.00000E+00
   3  1.80000E+01  1.20000E+01  9.06186E-04 -4.72336E-04
   4  1.80000E+01  0.00000E+00  2.51913E-04 -5.09073E-04

* * * ELEMENT STRESSES * * *
ELEM        X            Y      STRAIN:  XX          YY           XY
ELEM                            STRESS:  XX          YY           XY
   1  6.000E+00   8.000E+00   5.034E-05   0.000E+00  -2.624E-05
   1                          1.611E+00   4.027E-01  -3.149E-01
   2  1.200E+01   4.000E+00   1.400E-05   3.061E-06   2.624E-05
   2                          4.723E-01   2.099E-01   3.149E-01
```

Figure 22.2.6 Plane stress T3 point load output

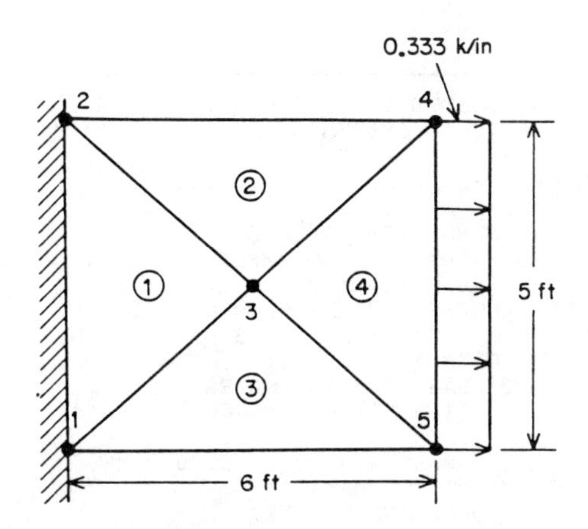

Figure 22.2.7 Plane stress T3 mesh with edge load

```
C       .................................................
C          ** BFLUX PROBLEM DEPENDENT STATEMENTS FOLLOW **
C       .................................................
C       LINEAR VARIATION OF 2 COMPONENTS ON STRAIGHT LINE
CDP     SQRT(Z) = DSQRT(Z)
        DX = COORD(2,1) - COORD(1,1)
        DY = COORD(2,2) - COORD(1,2)
        DL = SQRT (DX*DX + DY*DY)
        DLD6 = DL/6.
C       ODD = X FORCE, EVEN = Y FORCE
        C(1) = DLD6*( 2.*FLUX(1,1) + FLUX(2,1) )
        C(2) = DLD6*( 2.*FLUX(1,2) + FLUX(2,2) )
        C(3) = DLD6*( FLUX(1,1) + 2.*FLUX(2,1) )
        C(4) = DLD6*( FLUX(1,2) + 2.*FLUX(2,2) )
        RETURN
        END
```

Figure 22.2.8 Compute T3 edge traction resultants

$T_{y4} = T_{y5} = 0.0$). The calculated displacements and stresses agree with the results of Ural as expected. The output data for this example are in Fig. 22.2.9.

In general one would want to define the surface tractions in normal and tangential components. The additional computations in this approach involve finding the direction cosines of the boundary segment and using these data to calculate the corresponding resultant x- and y-components. Note that the stiffness matrix involved a product of the form $\mathbf{B}^T\mathbf{DB}$ where the matrix $\mathbf{D}$ is symmetric. This form will occur in most of the linear finite element problems. Thus, it is desirable to calculate this product in an efficient manner. Subroutine BTDB is designed for that purpose, since $\mathbf{D}$ is often a diagonal matrix, a special version, BTDIAB, is also given on the disk.

```
NUMBER OF NODAL POINTS IN SYSTEM =..........     5
NUMBER OF ELEMENTS IN SYSTEM =..............     4
NUMBER OF PARAMETERS PER NODE =.............     2
NUMBER OF NODES PER ELEMENT =...............     3
DIMENSION OF SPACE =........................     2
NUMBER OF BOUNDARIES WITH GIVEN FLUX =......     1
NUMBER OF NODES ON BOUNDARY SEGMENT =.......     2
NUMBER OF ROWS IN B MATRIX =................     3
NUMBER OF REAL PROPERTIES PER ELEMENT =.....     6
OPTIONAL UNIT NUMBERS: NTAPE1 =  8
*** NODAL POINT DATA ***
NODE, CONSTRAINT FLAG, 2 COORDINATES
         1         11    0.0000     0.0000
       ...        ...      ...        ...
         5          0   72.0000     0.0000
*** ELEMENT CONNECTIVITY DATA ***
ELEMENT NO.,  3 NODAL INCIDENCES.
    1    1    3    2
   ...  ...  ...  ...
    4    3    5    4
*** CONSTRAINT EQUATION DATA ***
CONSTRAINT TYPE ONE
EQ. NO.   NODE1   PAR1          A1
      1       1      1   0.00000E+00
      2       1      2   0.00000E+00
      3       2      1   0.00000E+00
      4       2      2   0.00000E+00
***  ELEMENT  PROPERTIES   ***
ELEMENT  PROPERTY      VALUE
    1          1    2.90000E+04
    1          2    3.00000E-01
    1          3    1.11540E+04
    1          4    1.50000E+00
*** ELEMENT BOUNDARY FLUXES ***
SEGMENT      2 NODES ON THE SEGMENT
SEGMENT      2 FLUX COMPONENTS PER NODE
   1    5    4
   1  3.333E-01  0.000E+00
   1  3.333E-01  0.000E+00
***  OUTPUT OF RESULTS  ***
NODE, 2 COORDINATES,  2 PARAMETERS.
    1  0.00000E+00  0.00000E+00  0.00000E+00  0.00000E+00
    2  0.00000E+00  6.00000E+01  0.00000E+00  0.00000E+00
    3  3.60000E+01  3.00000E+01  2.43878E-04  5.45697E-12
    4  7.20000E+01  6.00000E+01  5.34552E-04 -9.02307E-05
    5  7.20000E+01  0.00000E+00  5.34552E-04  9.02307E-05
* * * ELEMENT STRESSES * * *
ELEM        X          Y       STRAIN:    XX         YY         XY
ELEM                           STRESS:    XX         YY         XY
   1  1.200E+01  3.000E+01  6.774E-06  0.000E+00  1.516E-13
   1                        2.159E-01  6.477E-02  1.691E-09
   2  3.600E+01  5.000E+01  7.424E-06 -1.504E-06 -4.733E-07
   2                        2.222E-01  2.306E-02 -5.279E-03
   3  3.600E+01  1.000E+01  7.424E-06 -1.504E-06  4.733E-07
   3                        2.222E-01  2.306E-02  5.279E-03
   4  6.000E+01  3.000E+01  8.074E-06 -3.008E-06  7.958E-13
   4                        2.286E-01 -1.866E-02  1.211E-08
```

Figure 22.2.9 Plane stress T3 edge load output

22.3 Heat Conduction

Several authors, including Zienkiewicz [20], Desai and Abel [7], and Huebner [9], have considered two-dimensional heat conduction. The problem is governed by the two-dimensional Poisson equation

$$\frac{\partial}{\partial x}\left[K_x \frac{\partial T}{\partial x}\right] + \frac{\partial}{\partial y}\left[K_y \frac{\partial T}{\partial y}\right] + g = 0 \tag{22.4}$$

where $T(x, y)$ denotes the temperature, K_x and K_y are the material conductivities in the x– and y–directions, respectively, and $g(x, y)$ is the heat generation per unit area. Usually one encounters boundary conditions on T on some segment of the boundary and specifies heat fluxes, q_n, on the remaining segments. The latter quantity is positive in the direction of the outward normal vector. It has been shown [15] that the equivalent functional to be minimized is

$$I = \frac{1}{2}\int_A \left[K_x \left(\frac{\partial T}{\partial x}\right)^2 + K_y \left(\frac{\partial T}{\partial y}\right)^2\right] da - \int_A gT\,da - \int_{\Gamma_n} q_n T\,ds \tag{22.5}$$

where the last boundary integral is evaluated over the boundary segments on which q_n is specified. Other types of boundary conditions, such as convective heat losses, can be accounted for in the functional. The first term in the functional leads to the definition of the element square matrix, the second contributes to the element column matrix, and the third contributes to a boundary segment column matrix.

A four node bilinear isoparametric element is used and r and s are taken to be the local element coordinates. The element square matrix and column matrix are

$$\mathbf{S}^e_{4\times 4} = \int_{-1}^{1}\int_{-1}^{1} [K^e_x \mathbf{d}^{eT}_x \mathbf{d}^e_x + K^e_y \mathbf{d}^{eT}_y \mathbf{d}^e_y]\, |J^e|\, dr\, ds$$

and

$$\mathbf{C}^e_{4\times 1} = \int_{-1}^{1}\int_{-1}^{1} g^e \mathbf{H}^{eT}\, |J^e|\, dr\, ds \tag{22.6}$$

where $\mathbf{H}^e(r, s)$ is the element interpolation matrix, and $\mathbf{d}^e_x$ and $\mathbf{d}^e_y$ denote the global derivatives of $\mathbf{H}^e$; that is, $\mathbf{d}^e_x \equiv \partial/\partial x\, \mathbf{H}^e$, etc., as defined in Chap. 5. The column matrix associated with a specified flux on an exterior boundary segment of an element is

$$\mathbf{C}^b_{2\times 1} = \int_{-1}^{1} q^b_n \mathbf{H}^{bT}\, |J_b|\, d\Gamma\,. \tag{22.7}$$

So far, we have defined a two-dimensional element (NSPACE = 2), with four nodes (N = 4), one parameter, t, per node (NG = 1), and three floating point properties, g, K_x, and K_y per element (NLPFLO = 3). In addition, a typical element boundary segment has two nodes (LBN = 2). These three matrices are evaluated in subroutines ELSQ, and BFLUX, respectively. The ELSQ subroutine is shown in Fig. 22.3.1. The first two matrices are evaluated by Gaussian quadratures and the last is evaluated by using the closed form expression shown in Fig. 20.2.14. It would be more economical to form both $\mathbf{S}^e$ and $\mathbf{C}^e$ in the same integration loop ,but since this is not a production code the alternative approach has been selected for clarity. Since distorted (non-rectangular) elements may require higher order integration rules the user is required to input the desired integration

```
C     ..........................................................
C       ***  ELSQ PROBLEM DEPENDENT STATEMENTS FOLLOW ***
C     ..........................................................
C     SOLUTION OF POISSON EQUATION IN TWO DIMENSIONS
C     WITH Q4 ELEMENT, NSPACE = 2, N = 4, NG = 1, NELFRE = 4
C     ELPROP(1) = CONDUCTIVITY IN X-DIRECTION
C     ELPROP(2) = CONDUCTIVITY IN Y-DIRECTION
C     ELPROP(3) = HEAT GENERATION PER UNIT AREA
      COMMON /ELARG/ XK, YK, G, LSQ, TOTAL
      DATA  KALL, NGPOLD / 1, 0 /
C-->   CALCULATE QUADRATURE DATA ( IF REQUIRED )
      NGP = LPROP(1)
C     DEFAULT NGP = 2
      IF ( NGP .LT. 2 )  NGP = 2
      IF ( NGP. GT. 4 )  NGP = 4
      IF ( NGPOLD .NE. NGP )  THEN
        NGPOLD = NGP
        CALL GAUS2D (NGP, GPT, GWT, NIP, PT, WT)
        LSQ = NELFRE*NELFRE
      ENDIF
C-->   DEFINE PROPERTIES
      XK = ELPROP(1)
      YK = ELPROP(2)
      G  = ELPROP(3)
      LSQ = NELFRE*NELFRE
C-->   ZERO CALCULATED ARRAYS OF ELEMENT
      CALL ZEROA (LSQ, S)
      CALL ZEROA (NELFRE, C)
      CALL ZEROA (NELFRE, HINTG)
C-->   NUMERICAL INTEGRATION LOOP
      DO 60  IP = 1, NIP
        CALL  SHP4Q (PT(1,IP), PT(2,IP), H)
        CALL  DER4Q (PT(1,IP), PT(2,IP), DLH)
        CALL  JACOB (N, NSPACE, DLH, COORD, AJ)
        CALL  I2BY2 (AJ, AJINV, DET)
        DETWT = DET*WT(IP)
        CALL  GDERIV (NSPACE, N, AJINV, DLH, DGH)
        DO 50  J = 1, NELFRE
C          COLUMN MATRIX & AREA INTEGRAL FOR POST PROCESS
          C(J)     = C(J) + DETWT*G*H(J)
          HINTG(J) = HINTG(J) + DETWT*H(J)
C          SQUARE MATRIX
          DO 40  I = 1, NELFRE
   40     S(I,J) = S(I,J) + DETWT*( XK*DGH(1,I)*DGH(1,J)
     1                            + YK*DGH(2,I)*DGH(2,J))
   50   CONTINUE
   60 CONTINUE
C      SAVE HERE INSTEAD OF IN ELPOST
       IF ( NTAPE1 .GT. 0 ) WRITE(NTAPE1) HINTG
      RETURN
      END
```

Figure 22.3.1 Isoparametric matrices for a 2-D Poisson equation

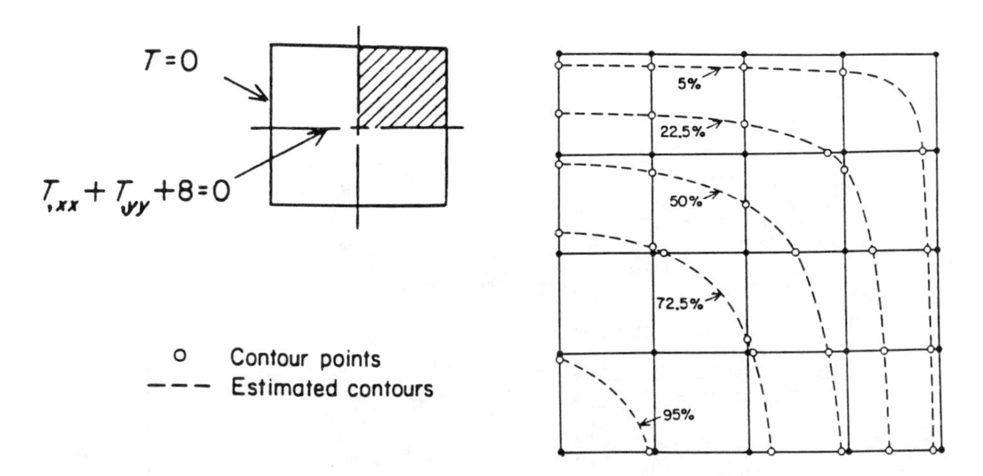

Figure 22.3.2 Solution domain and isothermal curves

order for each element, that is, there is one fixed point property per element (NLPFIX = 1). Subroutine ELSQ tests for this option. Note from Fig. 22.3.1 that subroutine ELSQ calls subroutines GAUS2D, SHP4Q, DER4Q, JACOB, I2BY2, and GDERIV. All of these subroutines have been presented in the previous chapters.

In this case no specific post-solution calculations are required (NTAPE1 = 0). However, if one desired to calculate the x- and y-thermal gradients at each integration point within the element it could be easily accomplished. Since the global derivative matrices $\mathbf{d}_x^e$ and $\mathbf{d}_y^e$ must be generated at each integration point they could be written on NTAPE1 from subroutine ELSQ for later use in subroutine POSTEL. Similarly, since $\mathbf{H}^e$ must be evaluated at each integration point in ELCOL, it is simple to find the global coordinates of the integration points, i.e., $\mathbf{x} : \mathbf{y} = \mathbf{H}\ \mathbf{COORD}$, and write them on NTAPE1 for later use by subroutine POSTEL. The non-zero value of NTAPE1 would cause subroutine ELPOST to be called, but it would consist of only a single RETURN statement. Once the nodal temperatures had been calculated, subroutine POSTEL would be called for each element and then three operations would be performed for each integration point in the element. First, $\mathbf{d}_x$ and $\mathbf{d}_y$ would be read from NTAPE1 and multiplied by the nodal temperatures to obtain the thermal gradients at the point, e.g., $\partial t/\partial x = \mathbf{d}_x^e \mathbf{t}^e$. Next the global coordinates of the point would be read from NTAPE1. Finally, the element number, integration point number, point coordinates, and thermal gradients would be printed by subroutine POSTEL.

It would be desirable to obtain contours of constant temperatures. Thus, we request five contours (NCURVE = 5) between the minimum and maximum temperatures calculated by the program. In closing, note that this formulation is summarized in Table 22.2. As the first example of two-dimensional heat conduction, consider the problem of uniform heat generation in a unit square plate. In non-dimensional form, let the governing equation be $T_{,xx} + T_{,yy} + 8 = 0$, i.e., $K_x = K_y = 1$ and $g = 8$, and let $T = 0$ around the perimeter of the square. Myers [15] presents the closed form solution (see problem 3.39) and shows that the steady state temperature at the center is 0.5894. The problem is shown in Fig. 22.3.2 along with the finite element mesh. Since the problem has symmetric geometry, properties, and boundary conditions only one quarter of the

```
NUMBER OF NODAL POINTS IN SYSTEM =..........   25
NUMBER OF ELEMENTS IN SYSTEM =..............   16
NUMBER OF PARAMETERS PER NODE =.............    1
NUMBER OF NODES PER ELEMENT =...............    4
DIMENSION OF SPACE =........................    2
NUMBER OF CONTOURS BETWEEN 5 & 95% =........    5
NUMBER OF ROWS IN B MATRIX =................    2
NUMBER OF QUADRATURE POINTS =...............    2
ELEMENT SHAPE: LINE, TRI, QUAD, HEX, TET =...   3
NUMBER OF INTEGER PROPERTIES PER ELEMENT =...   1
NUMBER OF REAL PROPERTIES PER ELEMENT =......   3

OPTIONAL UNIT NUMBERS: NTAPE1 =  8

*** NODAL POINT DATA ***
NODE, CONSTRAINT FLAG, 2 COORDINATES
        1         0     0.5000     0.5000
        2         0     0.6250     0.5000
       ...       ...     ...        ...
       24         1     0.8750     1.0000
       25         1     1.0000     1.0000

*** ELEMENT CONNECTIVITY DATA ***
ELEMENT NO.,   4 NODAL INCIDENCES.
     1   1     2    7    6
     2   2     3    8    7
     3   3     4    9    8
     4   4     5   10    9
     5   6     7   12   11
     6   7     8   13   12
     7   8     9   14   13
     8   9    10   15   14
     9  11    12   17   16
    10  12    13   18   17
    11  13    14   19   18
    12  14    15   20   19
    13  16    17   22   21
    14  17    18   23   22
    15  18    19   24   23
    16  19    20   25   24

*** CONSTRAINT EQUATION DATA ***
CONSTRAINT TYPE ONE
EQ. NO.   NODE1    PAR1          A1
      1       5      1    0.00000E+00
      2      10      1    0.00000E+00
      3      15      1    0.00000E+00
      4      20      1    0.00000E+00
      5      21      1    0.00000E+00
      6      22      1    0.00000E+00
      7      23      1    0.00000E+00
      8      24      1    0.00000E+00
      9      25      1    0.00000E+00
```

Figure 22.3.4a Output for uniform source example

```
***  ELEMENT  PROPERTIES   ***
ELEMENT  PROPERTY      VALUE
    1         1           3
END INTEGER PROPERTIES OF ELEMENTS
    1         1    1.00000E+00
    1         2    1.00000E+00
    1         3    8.00000E+00
END REAL PROPERTIES OF ELEMENTS

*** REACTION RECOVERY ***
NODE  DOF  REACTION   EQUATION
   5   1 -1.6807E-01      5
  10   1 -3.2143E-01     10
  15   1 -2.7731E-01     15
  20   1 -1.7857E-01     20
  21   1 -1.6807E-01     21
  22   1 -3.2143E-01     22
  23   1 -2.7731E-01     23
  24   1 -1.7857E-01     24
  25   1 -1.0924E-01     25
*RESULTANTS*
DOF        SUM
  1   -2.0000E+00

***  OUTPUT OF RESULTS  ***
NODE, 2 COORDINATES,  1 PARAMETERS.
    1  5.00000E-01  5.00000E-01  6.05042E-01
    2  6.25000E-01  5.00000E-01  5.71429E-01
    3  7.50000E-01  5.00000E-01  4.70588E-01
    4  8.75000E-01  5.00000E-01  2.85714E-01
    5  1.00000E+00  5.00000E-01  0.00000E+00
    6  5.00000E-01  6.25000E-01  5.71429E-01
    7  6.25000E-01  6.25000E-01  5.42542E-01
    8  7.50000E-01  6.25000E-01  4.46429E-01
    9  8.75000E-01  6.25000E-01  2.73634E-01
   10  1.00000E+00  6.25000E-01  0.00000E+00
   11  5.00000E-01  7.50000E-01  4.70588E-01
   12  6.25000E-01  7.50000E-01  4.46429E-01
   13  7.50000E-01  7.50000E-01  3.73950E-01
   14  8.75000E-01  7.50000E-01  2.32143E-01
   15  1.00000E+00  7.50000E-01  0.00000E+00
   16  5.00000E-01  8.75000E-01  2.85714E-01
   17  6.25000E-01  8.75000E-01  2.73634E-01
   18  7.50000E-01  8.75000E-01  2.32143E-01
   19  8.75000E-01  8.75000E-01  1.55987E-01
   20  1.00000E+00  8.75000E-01  0.00000E+00
   21  5.00000E-01  1.00000E+00  0.00000E+00
   22  6.25000E-01  1.00000E+00  0.00000E+00
   23  7.50000E-01  1.00000E+00  0.00000E+00
   24  8.75000E-01  1.00000E+00  0.00000E+00
   25  1.00000E+00  1.00000E+00  0.00000E+00

*** EXTREME VALUES OF THE NODAL PARAMETERS ***
PARAMETER      MAXIMUM, NODE        MINIMUM, NODE
     1   6.0504E-01,     1   0.0000E+00,    25
```

Figure 22.3.4b Output for uniform source example

Table 22.2 Parameter definitions for two-dimensional heat conduction
(* Default, ** Optional)

CONTROL:

NSPACE = 2	NLPFIX = 1	NG = 1
N = 4	NLPFLO = 3	LBN = 0*, or 2**

DEPENDENT VARIABLES:
1 = temperature, T

BOUNDARY FLUX COMPONENTS AT NODES**:
1 = specified normal outward heat flux per unit length, q_n

PROPERTIES:
Element level:
Fixed point:
1 = Gaussian quadrature order in each dimension, (2*)
Floating point:
1 = thermal conductivity in the x- direction, K_x
2 = thermal conductivity in the y-direction, K_y
3 = heat generation per unit area, g

CONSTRAINTS:
Type 1: specified nodal temperatures

region was utilized (although one-eighth could have been used). Along the two lines of symmetry one has a new boundary condition that the heat flux, q_n, must be zero. Since this leads to null contributions on those boundary lines, these conditions require no special consideration. The problem under consideration is two-dimensional (NSPACE = 2), involves 25 nodes (M = 25) with one temperature per node (NG = 1). There are 16 quadrilateral elements (NE = 16) with four nodes each (N = 4) and three floating point properties that are homogeneous throughout the mesh (NLPFLO = 3, LHOMO = 1). The internal heat generation will define a non-zero element column matrix (NULCOL = 0). There are no boundary segments with a specified non-zero normal heat flux (NSEG = 0). The output of the model code is shown in Fig. 22.3.4. Note that the center temperature at node 1 differs by only 1 1/4% from the exact value stated by Myers. The calculated contour data points from subroutine CONTUR are illustrated in Fig. 22.3.2. The contour curves exhibit the required symmetry along the diagonal line.

Before leaving this application it is desirable to consider the programming changes necessary to utilize a higher order element. For example, assume that one wishes to utilize an eight node quadratic element instead of the present linear element. Then it would be necessary to set N = 8, and replace the calls to shape function and local derivative subroutines, SHP4Q and DER4Q, with an equivalent set of subroutines, say SHP8Q and DER8Q, for the quadratic element. With a higher order element one would usually need to use a higher order quadrature rule in integrating the element matrices. The simplicity of introducing higher elements is a major practical advantage of programming with numerical integration. The changes necessary to convert to a three-dimensional analysis are almost as simple. If one is utilizing an eight node hexahedron (or *brick* element) one would again supply shape function and local derivative routines, say SHP8H and DER8H. Since NSPACE = 3, the Jacobian would be a 3 $\times$ 3 matrix and to obtain its inverse we would replace the call to I2BY2 with a similar subroutine, say I3BY3, to invert the matrix. Finally, the local coordinate definition of the quadrature points would be obtained by an alternative routine, say GAUS3D. The three-dimensional formulation will be illustrated in a later section.

22.4 Viscous Flow in Straight Ducts

Consider the flow of a viscous fluid in a straight duct with an arbitrary cross-section. The transverse velocity component, w, is of course zero on the perimeter of the cross-section. The distribution of the velocity $w(x, y)$ over the area has been investigated by Schechter [16]. He shows that the distribution of the transverse velocity is governed by the equation

$$\mu \frac{\partial^2 w}{\partial x^2} + \mu \frac{\partial^2 w}{\partial y^2} - \frac{\partial P}{\partial z} = 0 \tag{22.8}$$

where μ denotes the viscosity of the fluid, and $\partial P/\partial Z$ is the pressure gradient in the direction of flow (z-direction). This is another example of the application of the Poisson equation. The previous section presented the isoparametric element matrices for the general Poisson equation

$$\frac{\partial}{\partial x}\left[K_x \frac{\partial w}{\partial x}\right] + \frac{\partial}{\partial y}\left[K_y \frac{\partial w}{\partial y}\right] + G = 0. \tag{22.9}$$

Clearly, the flow problem is a special case of the above equation where $K_x = K_y = \mu$ = constant and $G = -\partial P/\partial z$. For the three node linear triangle, the element column matrix is

$$\mathbf{C}^e = -\frac{G\Delta}{3}\begin{Bmatrix} 1 \\ 1 \\ 1 \end{Bmatrix}, \tag{22.10}$$

where Δ is the area of the triangle. Similarly, the element square matrix is

$$\mathbf{S}^e = \frac{K_x}{4\Delta}\begin{bmatrix} b_i b_i & b_i b_j & b_i b_m \\ & b_j b_j & b_j b_m \\ sym. & & b_m b_m \end{bmatrix} + \frac{K_y}{4\Delta}\begin{bmatrix} c_i c_i & c_i c_j & c_i c_m \\ & c_j c_j & c_j c_m \\ sym. & & c_m c_m \end{bmatrix}, \tag{22.11}$$

where i, j, and m denote the three nodes of the triangle and where the constants b_i, and c_i are defined the geometric constants defined in Chap. 9. in terms of the coordinates of the nodes as

In this particular application one might be interested in a post-solution calculation of the flow rate, Q, which is defined as $Q = \int w\, da$. In this approximation, w is defined within a typical element by the interpolation functions, i.e., $w(x, y) = \mathbf{H}^e(x, y)\, \mathbf{W}^e$. Thus, the contribution of a typical flow rate is

$$Q^e = \int_{\Delta^e} w^e\, da = \int_{\Delta^e} \mathbf{H}^{eT}\, da\ \mathbf{W}^e \equiv \mathbf{q}^e\, \mathbf{W}^e \tag{22.12}$$

where

$$\mathbf{q}^e = \Delta\Big[\,(a_i + b_i\bar{x} + c_i\bar{y}) \quad (a_j + b_j\bar{x} + c_j\bar{y}) \quad (a_m + b_m\bar{x} + c_m\bar{y})\,\Big]. \qquad (22.13)$$

The equation for $\mathbf{q}^e$ could be calculated by subroutine ELPOST or ELSQ and stored for later use. After the nodal velocities have been calculated, subroutine POSTEL could read $\mathbf{q}^e$ from NTAPE1, calculate $Q^e = \mathbf{q}^e\mathbf{W}^e$, print the element number and Q^e, and end with a summary total of the total flow rate.

Rather than change element types at this point the previous isoparametric element will be used. Only the post-solution calculations need to be added, that is, we need to generate, store and recover, $\mathbf{Q}^e$. The calculation of $\mathbf{Q}^e$ has already been included in Fig. 22.3.1 as the standard argument, HINTG. Thus, it could be stored there or in ELPOST. After the nodal values have been calculated POSTEL recovers $\mathbf{Q}^e$, carries out the multiplication in Eqn. (22.10) and prints the results for each element. In this example the integral is the volume flow rate since it is the integral of the velocity over the element area. The additional POSTEL subroutine is illustrated in Fig. 22.4.1. The previous mesh was used to consider flow in a square duct. Figure 22.4.2 shows the comparison with the results of Schechter. The element area integrals are included in Fig. 22.4.3.

```
C      .........................................................
C        *** POSTEL PROBLEM DEPENDENT STATEMENTS FOLLOW ***
C      .........................................................
C-->    EVALUATE AREA INTEGRAL OF DEPENDENT VARIABLE
C       HINTG = QE = AREA INTEGRAL OF ELEMENT SHAPE FUNCTIONS
        COMMON /ELARG/ XK, YK, G, LSQ, TOTAL
        DATA  KALL / 1 /
        IF ( KALL .EQ. 1 )  THEN
C           PRINT TITLES ON THE FIRST CALL
          KALL  = 0
          TOTAL = 0.0
          WRITE (6,1000)
 1000     FORMAT ('AREA INTEGRAL OF DEPENDENT VARIABLE',//,
     1            ' ELEMENT        INTEGRAL')
   10 READ (NTAPE1)  HINTG
C-->     CALCULATE ELEMENT CONTRIBUTION
        CALL  MMULT (HINTG, D, VALUE, 1, NELFRE, 1)
        TOTAL = TOTAL + VALUE
        WRITE (6,1010)  IE, VALUE
 1010 FORMAT ( I8, 1PE18.6)
        IF ( IE .EQ. NE )  WRITE (6,1020)  TOTAL
 1020     FORMAT ('TOTAL AREA INTEGRAL = ', 1PE16.6)
        RETURN
        END
```

Figure 22.4.1 Postprocessing for area integral of solution

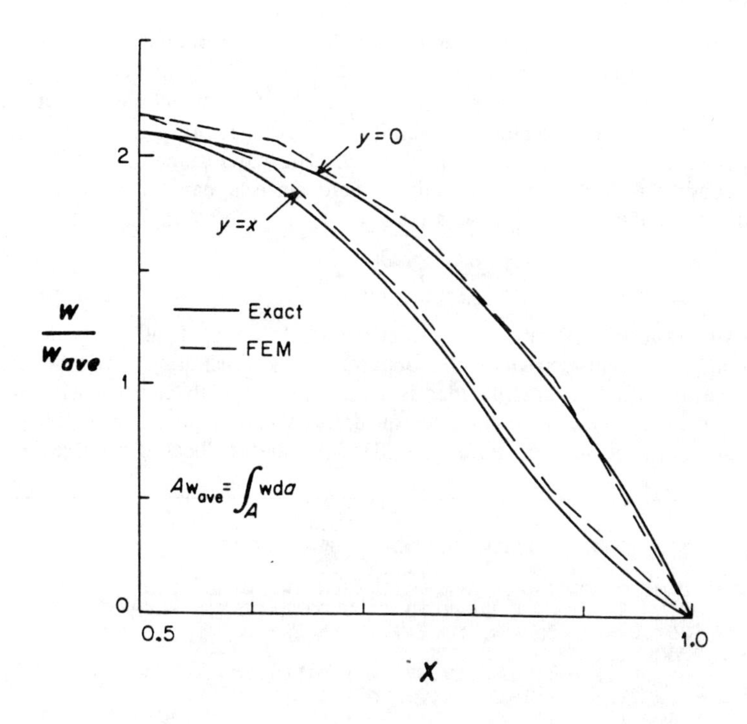

Figure 22.4.2 Velocity estimates in a square duct

```
** AREA INTEGRAL OF DEPENDENT VARIABLE **
ELEMENT        INTEGRAL
      1         8.947037E-03
      2         7.933546E-03
      3         5.767053E-03
      4         2.184956E-03
      5         7.933546E-03
      6         7.067770E-03
      7         5.180295E-03
      8         1.975693E-03
      9         5.767053E-03
     10         5.180296E-03
     11         3.883682E-03
     12         1.516134E-03
     13         2.184956E-03
     14         1.975693E-03
     15         1.516134E-03
     16         6.093258E-04
TOTAL AREA INTEGRAL =     6.962317E-02
```

Figure 22.4.3 Element volumetric flow rates

22.5 Potential Flow

Another common class of problem which can be formulated in terms of the Poisson equation is that of potential flow of ideal fluids. The diffusion coefficients, K_x and K_y, become unity and the source term, G, is usually zero so that the problem reduces to a solution of Laplace's equation. Potential flow can be formulated in terms of the velocity potential, ϕ, or the stream function, ψ. The latter sometimes yields simpler boundary conditions, but ϕ will be utilized here since it can be extended to three dimensions. Thus, the governing equation is

$$\frac{\partial^2\phi}{\partial x^2} + \frac{\partial^2\phi}{\partial y^2} = 0. \tag{22.14}$$

In this section it is desirable to consider potential flow applications, which involve curved boundaries. This provides the opportunity of illustrating how the previous program developed for the straight-sided isoparametric elements can be easily extended to curved elements. Select the eight-sided quadratic quadrilateral element. These shape functions are implemented in subroutine SHP8Q and the local derivatives of these

```
C      .........................................................
C       ***  ELSQ PROBLEM DEPENDENT STATEMENTS FOLLOW ***
C      .........................................................
C      SOLUTION OF LAPLACE EQUATION IN TWO DIMENSIONS
C      USING 8 NODE ISOPARAMETRIC QUADRILATERIAL ELEMENT
C      NSPACE = 2, N = 8, NG = 1, NELFRE = 8
       COMMON /ELARG/  XK, YK
C      ELPROP(1) = CONDUCTIVITY IN X-DIRECTION   = XK
C      ELPROP(2) = CONDUCTIVITY IN Y-DIRECTION   = YK
C      LPROP(1) = GAUSSIAN INTEGRATION ORDER IN EACH DIM
       DATA  KALL, NGPOLD, LSQ / 1, 0, 64 /
       IF ( KALL .EQ. 1 )  THEN
C-->        ON FIRST CALL EXTRACT PLOTTER DATA (IF ANY)
          KALL = 0
          IF ( NTAPE2 .GT. 0 )  CALL PLTSET (FLTMIS)
       ENDIF
C-->   DEFINE PROPERTIES
       XK = ELPROP(1)
       YK = ELPROP(2)
       NGP = LPROP(1)
C-->   CALCULATE QUADRATURE DATA, DEFAULT NGP = 2
       IF ( NGP .LT. 2 )  NGP = 2
       IF ( NGP .GT. 4 )  NGP = 4
       IF ( NGPOLD .NE. NGP )  THEN
         NGPOLD = NGP
         CALL  GAUS2D (NGP, GPT, GWT, NIP, PT, WT)
       ENDIF
C       STORE NUMBER OF POINTS FOR GRADIENT CALCULATIONS
       IF ( NTAPE1 .GT. 0 )  WRITE (NTAPE1)  NIP
C-->    NUMERICAL INTEGRATION LOOP
       CALL ISOPAR (N, NSPACE, NELFRE, NIP, S, COL, PT, WT, H,
     1      DLH, DGH, COORD, XYZ, AJ, AJINV, NTAPE1, NGRAND)
       RETURN
       END
```

Figure 22.5.1 Laplace equation square matrix for Q8 element

```
      SUBROUTINE  NGRAND (WT,DET,H,DGH,XPT,N,NSPACE,
     1                    NELFRE,COL,SQ,NTAPE1)
C     * * * * * * * * * * * * * * * * * * * * * * * * * *
C         PROBLEM DEPENDENT INTEGRAND EVALUATION IN
C                AN ISOPARAMETRIC ELEMENT
C     * * * * * * * * * * * * * * * * * * * * * * * * * *
      DIMENSION COL(NELFRE), SQ(NELFRE,NELFRE),
     1          H(N), DGH(NSPACE,N), XPT(NSPACE)
C     N      = NUMBER OF NODES PER ELEMENT
C     NSPACE = NUMBER OF SPATIAL DIMENSIONS
C     NELFRE = NO OF ELEMENT DEGREES OF FREEDOM
C     H      = ELEMENT INTERPOLATION FUNCTIONS
C     DGH    = GLOBAL DERIVATIVES OF H
C     XPT    = GLOBAL COORDS OF THE POINT
C     WT     = QUADRATURE WEIGHT AT POINT
C     DET    = JACOBIAN DETERMINATE AT POINT
C     COL    = PROB DEP COLUMN MATRIX INTEGRAND
C     SQ     = PROB DEP SQUARE MATRIX INTEGRAND
C     NTAPE1 = STORAGE UNIT FOR POST SOLUTION DATA
C     ....................................................
C     *** NGRAND PROBLEM DEPENDENT STATEMENTS FOLLOW ***
C     ....................................................
      COMMON /ELARG/  XK, YK
C     STORE COORDINATES AND DERIVATIVE MATRIX AT THE POINT
      IF ( NTAPE1 .GT. 0 )  WRITE (NTAPE1)  XPT, DGH
C      EVALUATE PRODUCTS
      DETWT = DET*WT
      DO 40  J = 1, N
        DO 30  I = 1, N
   30   SQ(I,J) = SQ(I,J) + DETWT*( XK*DGH(1,I)*DGH(1,J)
     1                          + YK*DGH(2,I)*DGH(2,J) )
   40 CONTINUE
      RETURN
      END
```

Figure 22.5.2 Laplace model kernel

relations are presented in subroutine DER8Q. To modify the previous set of isoparametric subroutines to utilize this element one simply replaces the original references to SHP4Q and DER4Q with calls to SHP8Q and DER8Q, respectively.

Recall that the potential flow problem is formulated in terms of either the stream function, ψ, or the *velocity potential,* ϕ. These quantities are usually of secondary interest and the analyst generally requires information on the velocity components. They are defined by the global derivatives of ψ and ϕ. For example, having formulated the problem in terms of ϕ one has

$$u \equiv \frac{\partial \phi}{\partial x}, \qquad v \equiv \frac{\partial \phi}{\partial y} \tag{22.15}$$

where u and v denote the x- and y-components of the velocity vector. Thus, although the program will yield the nodal values of ϕ one must also calculate the above global derivatives. This can be done economically since these derivative quantities must be generated at each quadrature point during construction of the element square matrix, **S**. Hence, one can simply store this derivative information, i.e., matrix **DGH**, and retrieve it later for calculating the global derivatives of ϕ or ψ. The new element square matrix subroutine and the new post-solution routine POSTEL are shown in Figs. 22.5.1 to

```
C     ....................................................
C      *** POSTEL PROBLEM DEPENDENT STATEMENTS FOLLOW ***
C     ....................................................
C-->          GRADIENT CALCULATION AND PLOTS FOR
C     SOLUTION OF LAPLACE EQUATION IN TWO DIMENSIONS
C     USING 8 NODE ISOPARAMETRIC QUADRILATERIAL ELEMENT
      COMMON /ELARG/  XK, YK
C      CONNECT TO PLOTTER COMMONS
      COMMON /PLTKAL/ XLEN, YLEN, FIRSTX, FIRSTY, DELTAX,
     1                DELTAY, XLAST, YLAST
C     NE     = TOTAL NUMBER OF ELEMENTS
C     STRESS = GRAD = GLOBAL COMPONENTS OF GRADIENT VECTOR
      DATA  KALL, SIZE, DERMAX / 1, 0.07, 0.0 /
      IF ( KALL .EQ. 1 )  THEN
C-->     PRINT TITLES ON THE FIRST CALL
        KALL = 0
        WRITE (6,5000)
 5000   FORMAT ('*** DGH DERIVATIVES AT INTEGRATION'
     1          ' POINTS ***',/,' POINT        X           Y',
     2          '          DP/DX          DP/DY',/)
      ENDIF
C-->   BEGIN ELEMENT POST SOLUTION ANALYSIS
      WRITE (6,*) 'ELEMENT NUMBER', IE
C      READ NUMBER OF POINTS TO BE CALCULATED
      READ (NTAPE1)  NIP
C      STORE FOR PLOT USE
      IF ( NTAPE2 .GT. 0 )  WRITE (NTAPE2)  NIP
      DO 20  J = 1, NIP
C-->     READ COORDS. AND DERIVATIVE MATRIX
        READ (NTAPE1)  XY, DGH
C        CALCULATE DERIVATIVES,  GRAD=DGH*D
        CALL  MMULT (DGH, D, GRAD, NRB, NELFRE, 1)
C-->      PRINT COORDINATES AND GRADIENT AT THE POINT
        WRITE (6,5020)  J, XY, GRAD
 5020   FORMAT ( I7, 2(1PE12.4), 2(1PE15.5) )
C-->      STORE RESULTS TO BE PLOTED LATER
        IF ( NTAPE2 .GT. 0 )  THEN
          WRITE (NTAPE2)  XY, GRAD
          IF ( ABS(GRAD(1)).GT.DERMAX ) DERMAX = ABS(GRAD(1))
          IF ( ABS(GRAD(2)).GT.DERMAX ) DERMAX = ABS(GRAD(2))
        ENDIF
   20 CONTINUE
C-->   ARE GRADIENT CALCULATIONS COMPLETE FOR ALL ELEMENTS
      IF ( IE .LT. NE )  RETURN
C-->   PRODUCE GRADIENT VECTOR PLOTS  (IF DESIRED)
      IF ( NTAPE2 .LT. 1 )  RETURN
        WRITE (6,*) '  BEGIN PLOT'
        CALL PAGNEW
        CALL VECT2D (NTAPE2,XLEN,YLEN,FIRSTX,FIRSTY,DELTAX,
     1               DELTAY,XLAST,YLAST,DERMAX,SIZE,NE)
        WRITE (6,*) '  END PLOT'
        RETURN
      END
```

Figure 22.5.3 Printing and plotting element gradient vectors

22.5.3. There are some new features in each of these routines. They include the option to store, within ELSQ, the global derivatives at each integration point as well as the global coordinates of the point. Thus, ELPOST is a dummy routine in this case. The square matrix calculations also utilize ISOPAR and NGRAND. Subroutine ELSQ also recovers installation-dependent plotting data through the miscellaneous properties.

Subroutine POSTEL also executes some site-dependent plots of the velocity vectors at the integration points. This is done only if the user has made unit NTAPE2 available for storing the data to be plotted. The plots are begun only after the gradients have been evaluated in all elements. Thus, it was necessary to utilize the value of the total number of elements, NE, to flag the call on which the plots are generated. Of course, the velocity vectors are also printed by POSTEL.

We wish to test this Laplace equation solver by running a patch test. Originally based on engineering judgement, the *patch test* has been proven to be a mathematically valid convergence test [10, 11]. Consider a patch (or sub-assembly) of finite elements containing at least one internal node. An internal node is one completely surrounded by elements. Let the problem be formulated by an integral statement containing derivatives of order n. Assume an arbitrary function, $P(x)$, whose n^{th} order derivatives are constant. Use this function to prescribe the dependent variable on all external nodes of the patch (i.e., $\theta_e = P(x_e)$). Solve for the internal nodal values of the dependent variable, θ_I, and

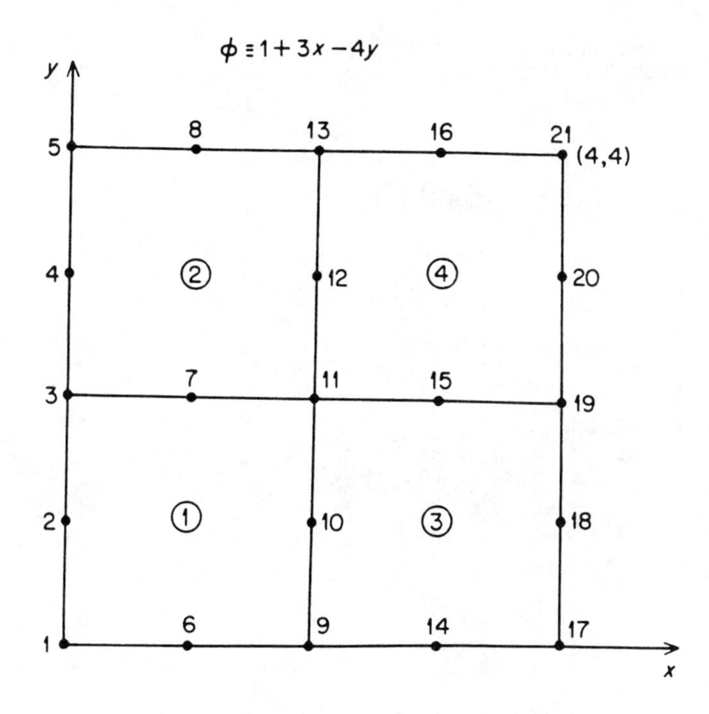

Figure 22.5.4 Patch test mesh

```
NUMBER OF NODAL POINTS IN SYSTEM =..........   21
NUMBER OF ELEMENTS IN SYSTEM =..............    4
NUMBER OF PARAMETERS PER NODE =.............    1
NUMBER OF NODES PER ELEMENT =...............    8
DIMENSION OF SPACE =........................    2
NUMBER OF ROWS IN B MATRIX =................    2
NUMBER OF QUADRATURE POINTS =...............    9
ELEMENT SHAPE: LINE, TRI, QUAD, HEX, TET =...   3
NUMBER OF INTEGER PROPERTIES PER ELEMENT =...   1
NUMBER OF REAL PROPERTIES PER ELEMENT =.....    3

OPTIONAL UNIT NUMBERS: NTAPE1 =  8

*** NODAL POINT DATA ***
NODE, CONSTRAINT FLAG, 2 COORDINATES
         1        1    0.0000     0.0000
        ...      ...    ...        ...
        21        1    4.0000     4.0000

*** ELEMENT CONNECTIVITY DATA ***
ELEMENT NO., 8 NODAL INCIDENCES.
    1    1    9   11    3    6   10    7    2
    2    3   11   13    5    7   12    8    4
    3    9   17   19   11   14   18   15   10
    4   11   19   21   13   15   20   16   12

*** CONSTRAINT EQUATION DATA ***
CONSTRAINT TYPE ONE
EQ. NO.   NODE1   PAR1         A1
      1       1      1   1.00000E+00
     ...     ...    ...    ....
     16      21      1  -3.00000E+00

***  ELEMENT  PROPERTIES    ***
ELEMENT  PROPERTY       VALUE
    1          1          3
    1          1    1.00000E+00
    1          2    1.00000E+00
    1          3    0.00000E+00

*** REACTION RECOVERY ***
NODE  DOF  REACTION   EQUATION
   1   1  3.3333E-01     1
   2   1 -4.0000E+00     2
   3   1 -2.0000E+00     3
   4   1 -4.0000E+00     4
   5   1 -2.3333E+00     5
   6   1  5.3333E+00     6
   8   1 -5.3333E+00     8
   9   1  2.6667E+00     9
  13   1 -2.6667E+00    13
  14   1  5.3333E+00    14
  16   1 -5.3333E+00    16
  17   1  2.3333E+00    17
  18   1  4.0000E+00    18
  19   1  2.0000E+00    19
  20   1  4.0000E+00    20
  21   1 -3.3333E-01    21
```

Figure 22.5.5a Output for Q8 patch test

```
*RESULTANTS*
DOF          SUM
  1    -2.1458E-06

*** EXTREME VALUES OF THE NODAL PARAMETERS ***
PARAMETER     MAXIMUM, NODE         MINIMUM, NODE
      1  1.3000E+01,    17  -1.5000E+01,      5

***  OUTPUT OF RESULTS  ***
NODE, 2 COORDINATES,   1 PARAMETERS.
    1  0.00000E+00  0.00000E+00  1.00000E+00
    2  0.00000E+00  1.00000E+00 -3.00000E+00
    3  0.00000E+00  2.00000E+00 -7.00000E+00
    4  0.00000E+00  3.00000E+00 -1.10000E+01
    5  0.00000E+00  4.00000E+00 -1.50000E+01
    6  1.00000E+00  0.00000E+00  4.00000E+00
    7  1.00000E+00  2.00000E+00 -4.00000E+00
    8  1.00000E+00  4.00000E+00 -1.20000E+01
    9  2.00000E+00  0.00000E+00  7.00000E+00
   10  2.00000E+00  1.00000E+00  3.00000E+00
   11  2.00000E+00  2.00000E+00 -1.00000E+00
   12  2.00000E+00  3.00000E+00 -5.00000E+00
   13  2.00000E+00  4.00000E+00 -9.00000E+00
   14  3.00000E+00  0.00000E+00  1.00000E+01
   15  3.00000E+00  2.00000E+00  2.00000E+00
   16  3.00000E+00  4.00000E+00 -6.00000E+00
   17  4.00000E+00  0.00000E+00  1.30000E+01
   18  4.00000E+00  1.00000E+00  9.00000E+00
   19  4.00000E+00  2.00000E+00  5.00000E+00
   20  4.00000E+00  3.00000E+00  1.00000E+00
   21  4.00000E+00  4.00000E+00 -3.00000E+00

    *** GLOBAL DERIVATIVES AT INTEGRATION POINTS ***
 POINT        X            Y            DP/DX          DP/DY

ELEMENT NUMBER   1

      1  1.7746E+00  1.7746E+00    3.00000E+00  -4.00000E+00
      2  1.0000E+00  1.7746E+00    3.00000E+00  -4.00000E+00
      3  2.2540E-01  1.7746E+00    3.00000E+00  -4.00000E+00
      4  1.7746E+00  1.0000E+00    3.00000E+00  -4.00000E+00
      5  1.0000E+00  1.0000E+00    3.00000E+00  -4.00000E+00
      6  2.2540E-01  1.0000E+00    3.00000E+00  -4.00000E+00
      7  1.7746E+00  2.2540E-01    3.00000E+00  -4.00000E+00
      8  1.0000E+00  2.2540E-01    3.00000E+00  -4.00000E+00
      9  2.2540E-01  2.2540E-01    3.00000E+00  -4.00000E+00

                      ...
ELEMENT NUMBER   4

      1  3.7746E+00  3.7746E+00    3.00000E+00  -4.00000E+00
      2  3.0000E+00  3.7746E+00    3.00000E+00  -4.00000E+00
      3  2.2254E+00  3.7746E+00    3.00000E+00  -4.00000E+00
      4  3.7746E+00  3.0000E+00    3.00000E+00  -4.00000E+00
      5  3.0000E+00  3.0000E+00    3.00000E+00  -4.00000E+00
      6  2.2254E+00  3.0000E+00    3.00000E+00  -4.00000E+00
      7  3.7746E+00  2.2254E+00    3.00000E+00  -4.00000E+00
      8  3.0000E+00  2.2254E+00    3.00000E+00  -4.00000E+00
      9  2.2254E+00  2.2254E+00    3.00000E+00  -4.00000E+00
```

Figure 22.5.5b Output for Q8 patch test

its nth order derivatives in each element. To be a convergent formulation:

1. The internal nodal values must agree with the assumed function evaluated at the internal points (i.e., $\phi_I = P(x_I)$?); and
2. The calculated n^{th} order derivatives *must* agree with the assumed constant values.

It has been found that some non-conforming elements will yield convergent solutions for only one particular mesh pattern. The patch mesh should be completely arbitrary for a valid numerical test. The patch test is very important from the engineering point of view since it can be executed numerically. Thus, one obtains a numerical check of the entire program used to formulate the patch test.

The patch of elements shown in Fig. 22.5.4 was utilized. It was assumed that

$$\phi(x, y) \equiv 1 + 3x - 4y \tag{22.16}$$

such that the derivatives $\phi_{,x} = 3, \phi_{,y} = -4$ are constant everywhere. All 16 points on the exterior boundary were assigned values by substituting their coordinates into Eqn. (22.16), that is, the boundary conditions that $\phi_1 \equiv \phi(x_1, y_1)$, etc., were applied on the exterior boundary. Then the problem was solved numerically to determine the value of ϕ at the interior points (7, 10, 11, 12, 15) and the values of its global derivatives at each integration point. The output results of this patch test are shown in Fig. 22.5.5. The output shows clearly that the global derivatives at all integration points have the assumed values. It is also easily verified that all the interior nodal values of ϕ are consistent with the assumed form. Thus, the patch test is satisfied and the subroutines pass a necessary numerical test.

Example — Flow Around a Cylinder

Martin and Carey [13] were among the first to publish a numerical example of a finite element potential flow analysis. This same example is also discussed by others [5, 7]. The problem considers the flow around a cylinder in a finite rectangular channel with a uniform inlet flow. The geometry is shown in Fig. 22.5.6. The present mesh is compared with that of Martin and Carey in Fig. 22.5.7. By using centerline symmetry and midstream antisymmetry it is possible to employ only one fourth of the flow field. The stream function boundary conditions are discussed by Martin and Carey [13] and Chung [5]. For the velocity potential one has four sets of Neumann (boundary flux) conditions and one set of Dirichlet (nodal parameter) conditions. The first involve zero normal flow, $q_n = \phi_{,n} \equiv 0$, along the centerline ab and the solid surfaces bc and de, and a uniform unit inflow, $q_n \equiv -1$, along ad. At the mid-section, ce, antisymmetry requires that $v = 0$. Thus, $\phi_{,y} = 0$ so that $\phi = \phi(x)$, but in this special case x is constant along that line so we can set ϕ to any desired constant, say zero, along ce. The output results are illustrated in Fig. 22.5.8. The velocity vector plots are shown in Fig. 22.5.9a. By changing only the specified inlet flux conditions, other problems can be considered. If one keeps the same total flow but introduces a parabolic boundary flow one obtains the vector plots shown in Fig. 22.5.9b.

As before, the presence of a boundary flux, q_n, makes it necessary to evaluate the flux column matrix

$$\mathbf{C}^b = \int_{l^b} \mathbf{H}^{bT} q_n \, ds. \tag{22.17}$$

The variation of q_n along the boundary segment is assumed to be defined by the nodal

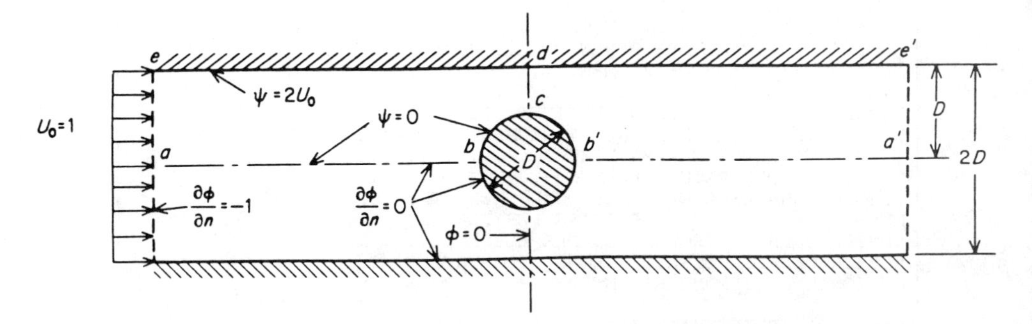

Figure 22.5.6 Cylinder between two parallel walls

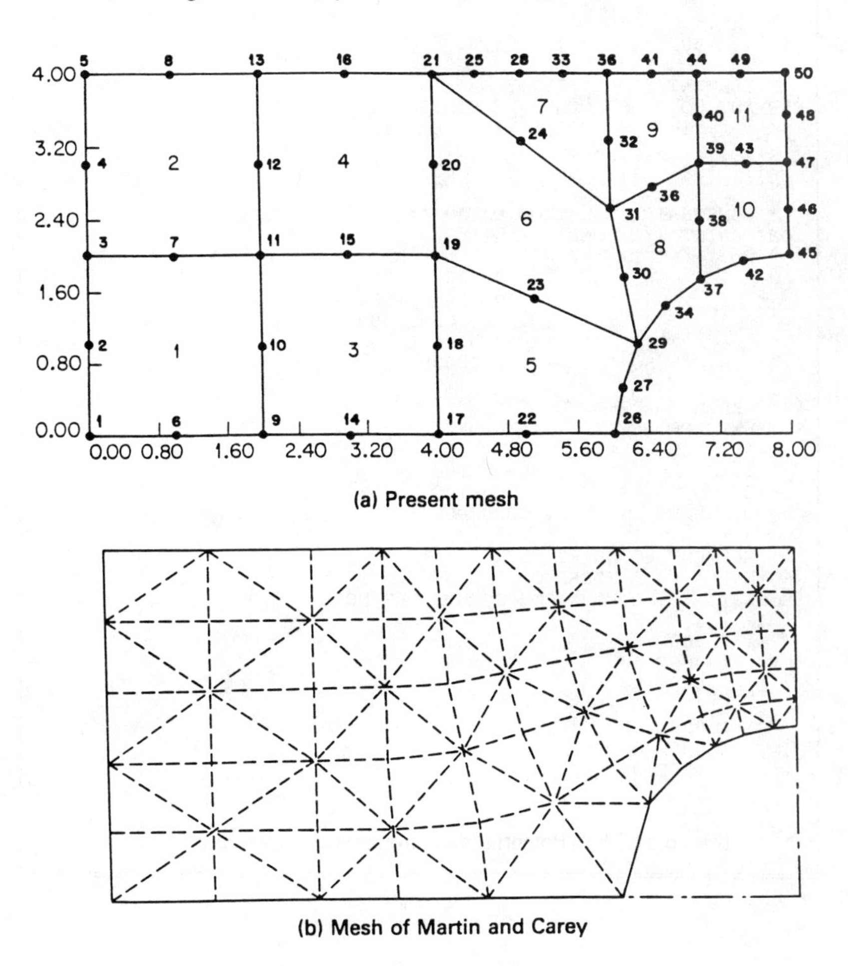

Figure 22.5.7 Sample meshes for cylinder flow problem

```
NUMBER OF NODAL POINTS IN SYSTEM =..........   50
NUMBER OF ELEMENTS IN SYSTEM =..............   11
NUMBER OF PARAMETERS PER NODE =.............    1
NUMBER OF NODES PER ELEMENT =...............    8
DIMENSION OF SPACE =........................    2
NUMBER OF BOUNDARIES WITH GIVEN FLUX =......    2
NUMBER OF NODES ON BOUNDARY SEGMENT =.......    3
NUMBER OF ROWS IN B MATRIX =................    2
NUMBER OF QUADRATURE POINTS =...............    9
ELEMENT SHAPE: LINE, TRI, QUAD, HEX, TET =...   3
NUMBER OF INTEGER PROPERTIES PER ELEMENT =...   1
NUMBER OF REAL PROPERTIES PER ELEMENT =......   3
NUMBER OF REAL MISCELLANEOUS  PROPERTIES =...   7
OPTIONAL UNIT NUMBERS: NTAPE1 =  8

*** NODAL POINT DATA ***
NODE, CONSTRAINT FLAG, 2 COORDINATES
        1         0    0.0000    0.0000
       ...       ...    ...        ...
       50         1    8.0000    4.0000

*** ELEMENT CONNECTIVITY DATA ***
ELEMENT NO.,  8 NODAL INCIDENCES.
    1    1    9   11    3    6   10    7    2
  ...  ...  ...  ...  ...  ...  ...  ...  ...
   11   39   47   50   44   43   48   49   40

*** CONSTRAINT EQUATION DATA ***
CONSTRAINT TYPE ONE
EQ. NO.   NODE1   PAR1         A1
      1     45       1   0.00000E+00
      2     46       1   0.00000E+00
      3     47       1   0.00000E+00
      4     48       1   0.00000E+00
      5     50       1   0.00000E+00

***  ELEMENT  PROPERTIES    ***
ELEMENT  PROPERTY        VALUE
    1          1            3
    1          1       1.00000E+00
    1          2       1.00000E+00
    1          3       0.00000E+00

*** ELEMENT BOUNDARY FLUXES ***
SEGMENT        3 NODES ON THE SEGMENT
SEGMENT        1 FLUX COMPONENTS PER NODE
   1    1    2    3
   1 -1.000E+00
   1 -1.000E+00
   1 -1.000E+00
   2    3    4    5
   2 -1.000E+00
   2 -1.000E+00
   2 -1.000E+00
```

Figure 22.5.8a Potential flow around a cylinder output

```
*** REACTION RECOVERY ***
NODE  DOF  REACTION    EQUATION
  45    1  4.2213E-01     45
  46    1  1.4443E+00     46
  47    1  6.2585E-01     47
  48    1  1.2124E+00     48
  50    1  2.9532E-01     50
*RESULTANTS*
DOF        SUM
  1     4.0000E+00

***  OUTPUT OF RESULTS  ***
NODE, 2 COORDINATES,  1 PARAMETERS.
    1  0.00000E+00  0.00000E+00 -9.97940E+00
  ...      ...          ...          ...
   50  8.00000E+00  4.00000E+00  0.00000E+00

*** GLOBAL DERIVATIVES AT INTEGRATION POINTS ***
 POINT          X            Y            DP/DX           DP/DY

ELEMENT NUMBER    1

       1  1.7746E+00  1.7746E+00    9.96609E-01     2.35997E-02
       2  1.0000E+00  1.7746E+00    9.97731E-01     1.53451E-02
       3  2.2540E-01  1.7746E+00    9.98855E-01     1.16549E-02
       4  1.7746E+00  1.0000E+00    9.86984E-01     1.55461E-02
       5  1.0000E+00  1.0000E+00    9.92672E-01     9.11808E-03
       6  2.2540E-01  1.0000E+00    9.98359E-01     7.25520E-03
       7  1.7746E+00  2.2540E-01    9.79186E-01     7.49326E-03
       8  1.0000E+00  2.2540E-01    9.89440E-01     2.89154E-03
       9  2.2540E-01  2.2540E-01    9.99694E-01     2.85721E-03

                   ...

ELEMENT NUMBER   10

       1  7.8866E+00  2.8869E+00    1.95174E+00     3.09016E-02
       2  7.4982E+00  2.8796E+00    1.87690E+00     1.32065E-01
       3  7.1120E+00  2.8635E+00    1.79784E+00     2.08992E-01
       4  7.8841E+00  2.4981E+00    2.13533E+00     7.80041E-02
       5  7.4921E+00  2.4660E+00    2.03036E+00     3.31006E-01
       6  7.1095E+00  2.3943E+00    1.88184E+00     5.23377E-01
       7  7.8817E+00  2.1093E+00    2.46961E+00     1.24155E-01
       8  7.4860E+00  2.0523E+00    2.30944E+00     5.28086E-01
       9  7.1071E+00  1.9251E+00    2.02313E+00     8.37462E-01

ELEMENT NUMBER   11

       1  7.8873E+00  3.8873E+00    1.78576E+00     4.83441E-03
       2  7.5000E+00  3.8873E+00    1.74344E+00     1.30610E-02
       3  7.1127E+00  3.8873E+00    1.70111E+00     8.29220E-03
       4  7.8873E+00  3.5000E+00    1.82113E+00     1.68485E-02
       5  7.5000E+00  3.5000E+00    1.76581E+00     6.63616E-02
       6  7.1127E+00  3.5000E+00    1.71049E+00     1.02882E-01
       7  7.8873E+00  3.1127E+00    1.89778E+00     2.88634E-02
       8  7.5000E+00  3.1127E+00    1.82947E+00     1.19662E-01
       9  7.1127E+00  3.1127E+00    1.76116E+00     1.97473E-01
```

Figure 22.5.8b Potential flow around a cylinder output

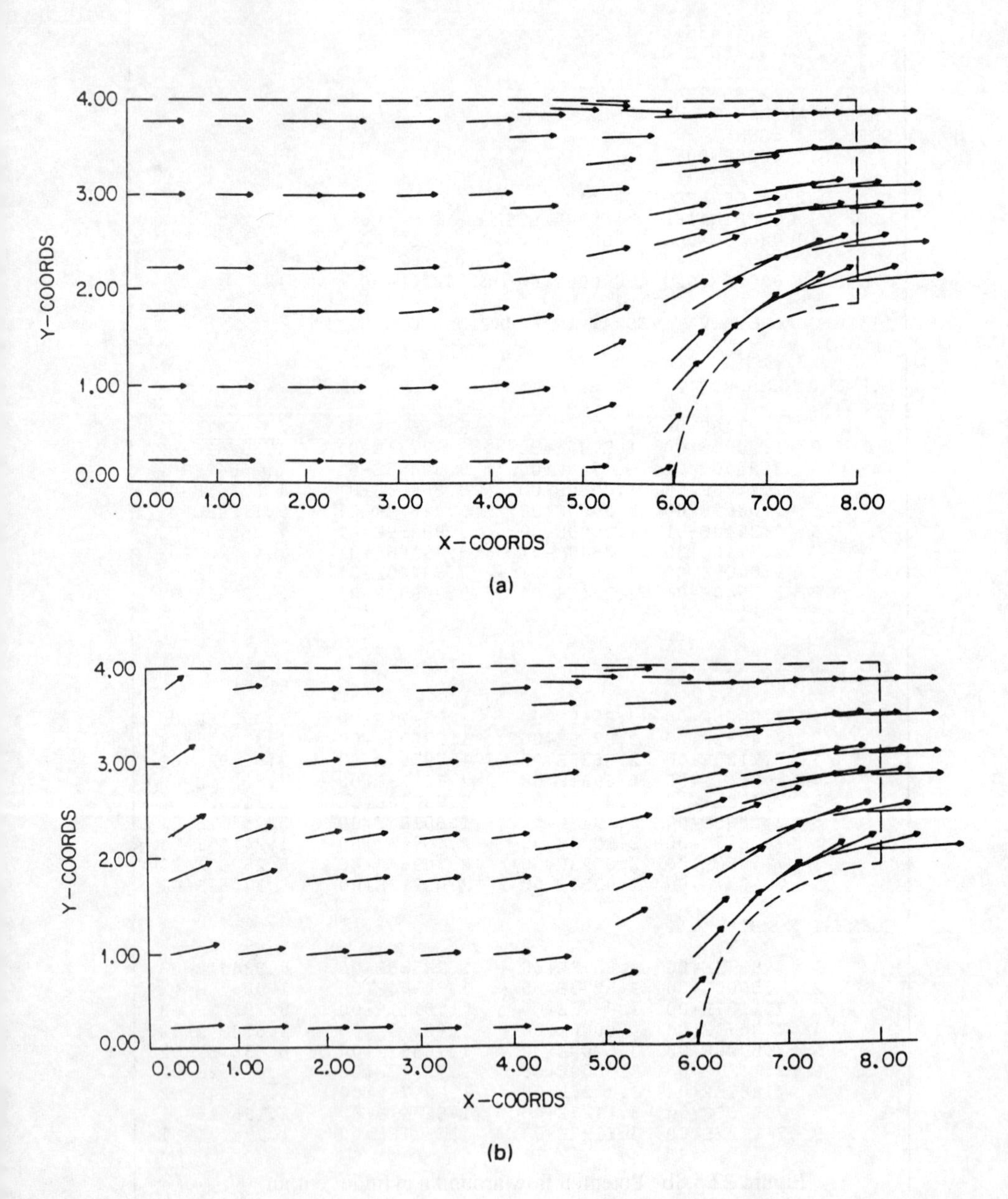

Figure 22.5.9 Velocity vector plots

```
C      ..................................................
C        ** BFLUX PROBLEM DEPENDENT STATEMENTS FOLLOW **
C      ..................................................
C      QUADRATIC NORMAL FLUX OVER A QUADRATIC BOUNDARY SEGMENT
C      LBN = 3, NSPACE = 2, NG = 1, NFLUX = 2
C      FLUX(K,1) = SPECIFIED OUTWARD NORMAL FLUX AT NODE K
C      TANGENTIAL FLUX NOT USED       1......2......3 ----> S
C      BL = BOUNDARY SEGMENT LENGTH   ]
C      ( MUST BE STRAIGHT )           N      SEGMENT SKETCH
       IOPT = 1
C       FIND DISTANCE BETWEEN FIRST AND THIRD POINTS
       DX = COORD(3,1) - COORD(1,1)
       DY = COORD(3,2) - COORD(1,2)
       BL = SQRT(DX*DX+DY*DY)
C       CALCULATE SEGMENT COLUMN MATRIX (UNIT THICKNESS)
       C(1) = BL*(FLUX(1,1)*4.+FLUX(2,1)*2.-FLUX(3,1))/30.
       C(2) = BL*(FLUX(1,1)*2.+FLUX(2,1)*16.+FLUX(3,1)*2.)/30.
       C(3) = BL*(-FLUX(1,1)+FLUX(2,1)*2.+FLUX(3,1)*4.)/30.
       RETURN
       END
```

Figure 22.5.10 Flux matrix for quadratic source on straight line

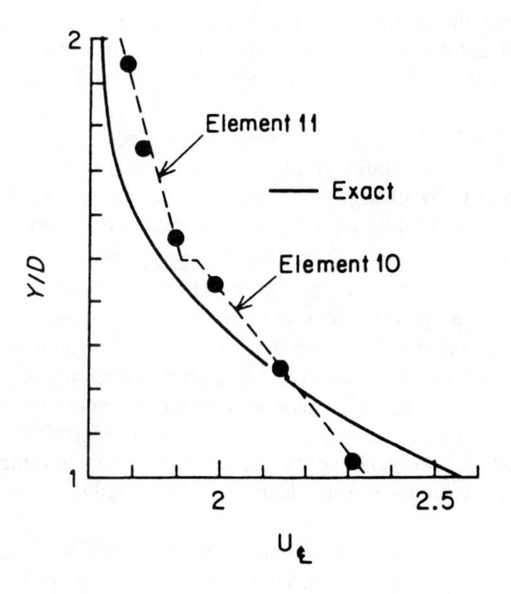

Figure 22.5.11 Outflow velocity estimates

(input) values and the segment interpolation equations, i.e.,

$$q_n(s) \equiv \mathbf{H}^b(s)\,\mathbf{q}_n^b.$$

Therefore, the segment column matrix (for a straight segment) becomes

$$\mathbf{C}^b = \int_{l^b} \mathbf{H}^{bT}\mathbf{H}^b \, ds \, \mathbf{q}_n^b = \frac{l^b}{30}\begin{bmatrix} 4 & 2 & -1 \\ 2 & 16 & 2 \\ -1 & 2 & 4 \end{bmatrix} \mathbf{q}_n^b . \tag{22.18}$$

These calculations are executed in subroutine BFLUX of Fig. 22.5.10.

Since the second set of post-solution calculations involved vector plots the routines for this purpose are given in PLTSET and VECT2D on the disk. Labeled element COMMONs beginning with PLT are used to transmit plot data. The parameter definitions used in the previous examples are given in Table 22.3. The values of the center velocity estimates for the flow around the cylinder are shown in Fig. 22.5.11. With two elements across the section there was about a 10% error in the maximum velocity. Since ϕ in each element is quadratic, the corresponding element velocity estimates are linear, as illustrated by the dashed lines. Clearly, one should have used several elements across the outflow section to get accurate results.

22.6 Electromagnetic Waveguides *

The calculation of electrical and magnetic fields for practical geometries often leads to problems which contain known boundary singularities of the type discussed in Sec. 6.8. For example, waveguides often contain re-entrant corners that introduce point singularities that affect the calculation of the cut-off frequency and mode shapes. Similarly, the transverse electromagnetic line (T.E.M.) analysis of microstrips or coaxial lines often have re-entrant corners. For a uniform mesh it has been shown analytically and experimentally that the error due to the singularity predominates the standard mesh errors.

The analysis of T.E.M. line again involves the use of Laplace's equation for the potential, ϕ. Thus, again one could employ the formulation presented in Sec. 11.5. However, it is necessary to change the program to correct for the known point singularity. If the solution domain contains a re-entrant corner with an internal angle $\pi < \alpha \leq 2\pi$, then near the corner ϕ is of order $0(\rho^{1-a})$, where $a = \pi/\alpha$ and ρ is the distance from the corner. Its derivatives are $0(\rho^{-a})$, and it is this loss of regularity of the derivatives that causes the singularity error to predominate.

The minor changes required in the programming are illustrated in Fig. 22.6.1. The modifications of the shape functions and their local derivatives are made by the routine SINGLR given in Fig. 16.3.2. It is also necessary to flag the presence of the singularity. One way of doing this is to let the order, a, be an element property. If it is zero, then the element is not adjacent to a singular point and no changes are required. Of course, it would also be necessary to input the incidences of the singularity elements so that the singular node is listed first.

As an example application consider a microstrip consisting of a center strip within a rectangular case. Let the boundary conditions be $\phi = 0$ on the outer case and $\phi = 1$ on the inner strip. At the edge of the inner strip one has $\phi = 0(\rho^{1/2})$ and $\partial\phi/\partial\rho = 0(\rho^{-1/2})$. Thus, there is a strong singularity at such a point. This problem has been analyzed by several authors including Daly [6] and Akin [1]. It was modeled using quarter symmetry, assuming an isotropic, homogeneous dielectric as shown in Fig. 22.6.2. In addition to ϕ, the capacitance per unit length, c, where

Table 22.3 Parameter definitions for quadratic quadrilateral potential flow
(* Default, ** Optional)

CONTROL:

NSPACE = 2	NLPFIX = 1	NSEG = 0*
NLPFLO = 2	LBN = 0*, or 3**	NG = 1
MISCFX = 0*, or 7**	NTAPE1 = 0*, or 8**	NTAPE2 = 0*, or 9**

DEPENDENT VARIABLES:
1 = velocity potential,

PROPERTIES:
Element level:
Fixed point:
1 = Gaussian quadrature order in each dimension, (2*)
Floating point:
1 = conductivity in x-direction, Kx
2 = conductivity in y-direction, Ky
Miscellaneous (NTAPE2 > 0, Plot Data)** :
Floating point:
1 = length of plot in x-direction
2 = length of plot in y-direction
3 = x-coordinate of plot origin
4 = y-coordinate of plot origin
5 = change in x-coordinate per inch of plot
6 = change in y-coordinate per inch of plot
7 = maximum length of plot

BOUNDARY FLUX DATA (when NSEG 0) :
Fixed point:
Three nodal points (domain on left)
Floating point:
Three nodal outward flux values

CONSTRAINTS:
Type 1: Specified potential values

POST-SOLUTION CALCULATIONS (when NTAPE1 0) :
1 Global coordinates and velocities at each quadrature point
2 (NTAPE2 0) Velocity vector plots**

$$c = \varepsilon_0 \kappa \iint_A | (\phi_{,x})^2 + (\phi_{,y})^2 | \, da, \tag{22.19}$$

was obtained from the post-solution calculations. Here ε_0 and κ denote the free-space and dielectric permittivity, respectively. The exact value of c is known and was compared with values calculated on a uniform grid. The accuracy of standard higher order elements and singularity elements are compared in Fig. 22.6.1.

To illustrate a milder $r^{-1/3}$ singularity, and compare local grid refinement with singularity modifications, consider an L-shaped region as did Whiteman and Akin [19] and Akin [3], as in Fig. 22.6.1 Both used bilinear quadrilaterals in a finite element solution of the Laplace equation. The standard finite element results were compared against singularity modifications, using Eqn. (5.28), and local grid refinement near the corner. The latter employed five node bilinear transition quadrilaterals developed by Whiteman [18]. Here, only the solutions near the singularity will be considered. They are compared in Fig. 22.6.3 with an *analytic* solution obtained by numerical conformed mapping. As expected, both singularity modifications and local grid refinement give significant improvements in the accuracy of the finite element solutions. In this case, the modification procedure is more cost-effective. In general it would probably be best to have limited local grid refinement, say two or three levels, combined with a singularity modification.

22.7 Axisymmetric Plasma Equilibria *

Nuclear fusion is being developed as a future source of energy. The heart of the fusion reactors will be a device for confining the reacting plasma and heating it to thermonuclear temperatures. This confinement problem can be solved through the use of magnetic fields of the proper geometry which generate a so-called "magnetic bottle". The tokamak containment concept employs three magnetic field components to confine the plasma. An externally applied toroidal magnetic field, B_T, is obtained from coils through which the torus passes. A second field component is the polodial magnetic field, B_P, which is produced by a large current flowing in the plasma itself. This current is induced in the plasma by transformer action and assists in heating the plasma. Finally, a vertical (axial) field, B_V, is also applied. These typical fields are illustrated in Fig. 22.7.1. For many purposes a very good picture of the plasma behavior can be obtained by treating it as an ideal magnetohydrodynamic (MHD) media. The equations governing the steady state flow of an ideal MHD plasma are

$$grad\, \mathbf{B} = 0, \quad \nabla P = \mathbf{J} \times \mathbf{B}, \quad curl\, \mathbf{B} = \mu \mathbf{J} \tag{22.20}$$

where P is the pressure, **B** the magnetic flux density vector, **J** the current density vector, and μ a constant that depends on the system of units being employed. Consider an axisymmetric equilibria defined in cylindrical coordinates (ρ, z, θ) so that $\partial/\partial\theta = 0$. Equation (22.20) implies the existence of a vector potential, **A**, such that curl **A** = **B**. Assuming that $\mathbf{A} = \mathbf{A}(\rho, z)$ and $A_\theta = \psi/\rho$, where ψ is a stream function, we obtain

$$B_\rho = -\psi_{,z}/\rho, \quad B_z = \psi_{,\rho}/\rho, \quad B_\theta = A_{\rho,z} - A_{z,\rho} = B_T \tag{22.21}$$

Therefore Eqn. (22.18) simplifies to

$$\frac{\partial^2\psi}{\partial\rho^2} - \frac{1}{\rho}\frac{\partial\psi}{\partial\rho} + \frac{\partial^2\psi}{\partial z^2} = -\mu\rho^2 P' - xx' = J_{\theta\rho}\,. \tag{22.22}$$

A typical stream function for a dee-shaped plasma torus is shown in Fig. 22.7.2. The above is the governing equation for the steady equilibrium flow of a plasma. For certain simple choices of P and B_θ, Eqn. (22.22) will be linear but in general it is nonlinear. The essential boundary condition on the limiting surface, Γ_1, is

$$\psi = K + 1/2 r^2 B_V \quad on\, \Gamma_1$$

```
C       SET SINGULARITY ORDER (NONE IF=0)
        ORDER = ELPROP(4)
        . . .
C-->     NUMERICAL INTEGRATION
        DO 60 IP = 1, NQP
C          FORM LOCAL DERIV, JACOBIAN, INVERSE
          CALL DER4Q (PT(1,IP), PT(2,IP), DLH)
          CALL JACOB (N, NSPACE, DLH, COORD, AJ)
          CALL I2BY2 (AJ, AJINV, DET)
          DETWT = DET*WT(IP)
C          *** MODIFY FOR POINT SINGULARITY ***
          IF ( ORDER .GT. 0.0 )  THEN
            CALL SHP4Q  (PT(1,IP), PT(2,IP), H)
            CALL SINGLR (ORDER, H, DLH, N, NSPACE)
          ENDIF
C          ***   MODIFICATION COMPLETE  ***
C          FORM GLOBAL DERIV,  ELEM MATRICES
          CALL GDERIV (NSPACE, N, AJINV, DLH, DGH)
          . . .
```

Figure 22.6.1 Typical modifications for point singularities

Percent error in capacitance

Corner nodes	Number of singularity elements	Types of element			
		LT	QT	CT	BQ
3×5	0	—	4.02	—	—
3×5	4	—	0.76	—	—
5×9	0	7.00	4.01	—	3.73
5×9	2	—	—	—	3.03
7×13	0	4.57	2.66	2.01	—
8×15	0	—	—	—	2.15
8×15	2	—	—	—	1.75
9×17	0	3.39	1.29	—	—

Uniform grid
$B = 2H = 2W$
LT = Linear Triangle
QT = Quadratic Triangle
CT = Cubic Triangle
BQ = Bilinear Quadrilateral

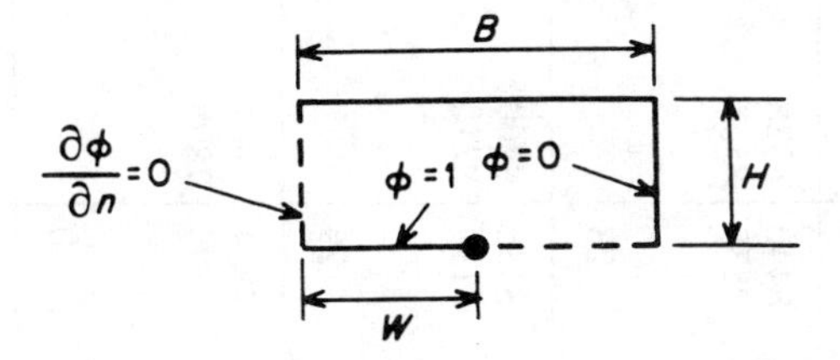

Figure 22.6.2 A comparison of singularity effects in TEM

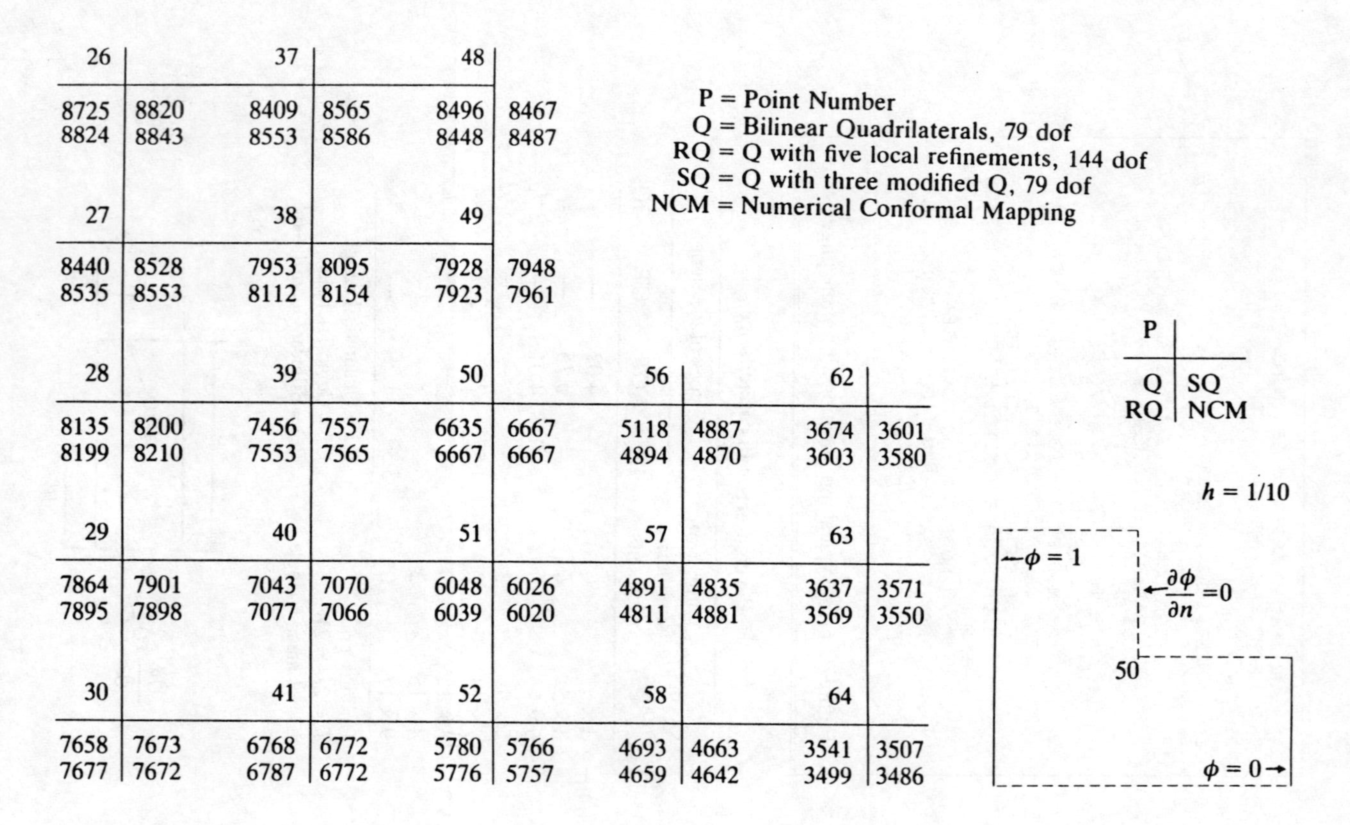

Figure 22.6.3 Comparison of standard elements, refinement, and singularity modifications near singularity of O($r^{-1/3}$)

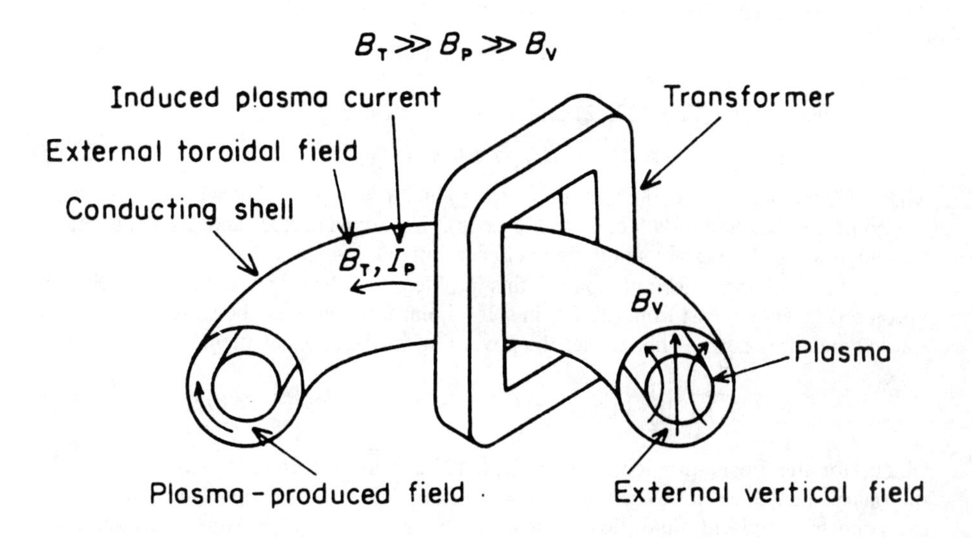

Figure 22.7.1 Schematic of tokamak fields and currents

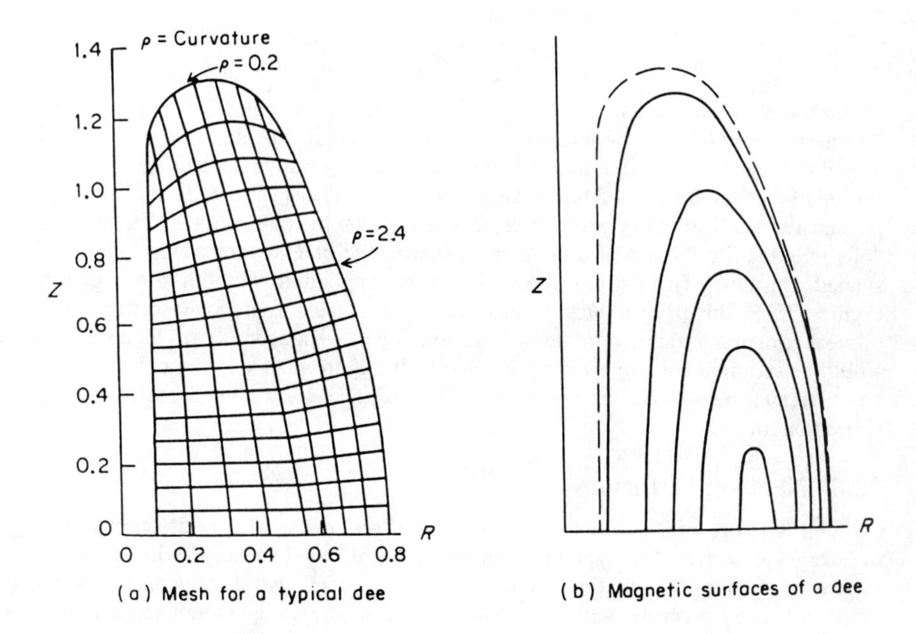

Figure 22.7.2 Stream function in a torus

where K is a constant and B_V is a superimposed direct current vertical (z) field. On planes of symmetry one also has vanishing normal gradients of ψ, i.e.,

$$\frac{\partial \psi}{\partial n} = 0 \; on \; \Gamma_2.$$

The right-hand side of Eqn. (22.22) can often be written as

$$H_{\theta\rho} = g\psi + h \tag{22.23}$$

where, for the above special cases, $g = g(\rho, z)$ and $h = h(\rho, z)$, but where in general h is a nonlinear function of ψ, i.e., $h = h(\rho, z, \psi)$. Equations (22.22) and (22.23) are those for which we wish to establish the finite element model.

A finite element formulation of this problem has been presented by Akin and Wooten [2]. They recast Eqn. (22.22) in self-adjoint form, applied the Galerkin criterion, and integrated by parts. This defines the governing variational statement

$$I = \iint_{\Omega} \left[\frac{1}{2} \{ (\psi_{,\rho})^2 + (\psi_{,z})^2 + g\psi^2 \} + h\psi \right] \frac{1}{\rho} \, d\rho \, dz \tag{22.24}$$

which, for the linear problem, yields Eqn. (22.22) as the Euler equation when I is stationary, i.e., $\delta I = 0$. When $g = h = 0$, Eqn. (22.24) also represents the case of axisymmetric inviscid fluid flow. Flow problems of this type were considered by Chung [5] using a similar procedure. For a typical element with N nodes the element contributions for Eqn. (22.23) are

$$\mathbf{S}^e = \int_{\Omega^e} [\mathbf{H}_{,\rho}^T \mathbf{H}_{,\rho} + \mathbf{H}_{,z}^T \mathbf{H}_{,z} + g(\rho, z) \mathbf{H}^T \mathbf{H}] \frac{1}{\rho} \, d\rho \, dz,$$

$$\mathbf{C}^e = \int_{\Omega} h(\rho, z) \mathbf{H}^T \frac{1}{\rho} \, d\rho \, dz, \tag{22.25}$$

where the N element interpolation functions, $\mathbf{H}$, define the value of ψ within an element by interpolating from its nodal values, ψ^e. Several applications of this model to plasma equilibria are given by Akin and Wooten [2]. The major advantage of the finite element formulation over other methods such as finite differences is that it allows the plasma physicist to study arbitrary geometries. Some feel that the fabrication of the toroidal field coils may require the use of a circular plasma, while others recommend the use of dee-shaped plasmas. The current model has been applied to both of these geometries. Figure 22.7.2 illustrates typical results for a dee-shape toruns cross-section where the bilinear isoparametric quadrilateral was employed. Biquadratic or bicubic elements would be required for some formulations which require post-solution calculations using the first and second derivatives of ψ. Figure 22.7.3 shows the integral calculations carried out in ELSQ.

22.8 Neutron Diffusion *

Fission reactor designs are usually based on the use of core assemblies arranged in hexagonal arrays. This type of geometry can be difficult to model with common finite difference techniques. In this section the steady state, two-dimensional, one-energy-group diffusion problem will be considered. The governing monoenergetic diffusion equation is

```
C     . . . . . . . . . . . . . . . . . . . . . . . . . . . . . . . . . . . . . . . .
C          ***  ELSQ PROBLEM DEPENDENT STATEMENTS FOLLOW ***
C     . . . . . . . . . . . . . . . . . . . . . . . . . . . . . . . . . . . . . . . .
C                  FOUR NODE ISOPARAMETRIC QUADRILATERAL
C               GALERKIN FORMULATION OF SELF-ADJOINT OPERATOR
C              L(D) = (D,R*1/R),R + D,ZZ*1/R - A*D/R - B/R = 0
C     FOR AXISYMMETRIC MHD PLASMA EQUILIBRIUM,     A,B .NE. 0  I.E.
C            A = -CONST*ALFA1*R**2 - 0.5*BETA1, = 0 IF IDEAL
C            B = -CONST*ALFA2*R**2 - 0.5*BETA2, = 0 IF IDEAL
C             N = 4, NG=1, NSPACE=2, NELFRE=4
C     CONST = FLTMIS(1), UNITS CONSTANT,ALFA1 = FLTMIS(2),
C     ALFA2 = FLTMIS(3), BETA1 = FLTMIS(4),BETA2 = FLTMIS(5)
      DATA  KALL / 1 /
      IF ( KALL .EQ. 1 )  THEN
C        SET SYSTEM & PLOT CONSTANTS ON THE FIRST CALL
        KALL  = 0
        CONST = FLTMIS(1)
        ALFA1 = FLTMIS(2)
        ALFA2 = FLTMIS(3)
        BETA1 = FLTMIS(4)
        BETA2 = FLTMIS(5)
      ENDIF
C      ZERO ELEMENT SQUARE AND COLUMN MATRIX
      CALL  ZEROA (LSQ, S)
      CALL  ZEROA (NELFRE, COL)
C-->    NUMERICAL INTEGRATION LOOP
      DO 60  IP = 1, NQP
C        EVALUATE INTERPOLATION FUNCTIONS
        CALL  SHP4Q ( PT(1,IP), PT(2,IP), H)
C        FIND GLOBAL COORDINATES OF THE POINT,  XYZ = H*COORD
        CALL  MMULT (H, COORD, XYZ, 1, N, NSPACE)
        R = XYZ(1)
        CALL  DER4Q (PT(1,IP), PT(2,IP), DLH)
C        FORM DETERMINATE AND INVERSE OF JACOBIAN
        CALL  JACOB (N, NSPACE, DLH, COORD, AJ)
        CALL  I2BY2 (AJ, AJINV, DET)
        WTDTDR = WT(IP)*DET/R
C        FORM GLOBAL DERIVATIVES,  DGH = AJINV*DLH
        CALL  GDERIV (NSPACE, N, AJINV, DLH, DGH)
C        EVALUATE CONTRIBUTIONS TO SQUARE AND COLUMN MATRICES
        A = -CONST*R*R*ALFA2 - 0.5*BETA2
        B = -CONST*R*R*ALFA1 - 0.5*BETA1
        DO 50  I = 1, N
          DO 40  J = 1, N
   40     S(I,J) = S(I,J) + (DGH(1,I)*DGH(1,J) + DGH(2,I)*DGH(2,J)
     1                      + A*H(I)*H(J) )*WTDTDR
        C(I) = C(I) - B*H(I)*WTDTDR
   50   CONTINUE
   60 CONTINUE
      RETURN
      END
```

Figure 22.7.3 Plasma element matrix evaluations

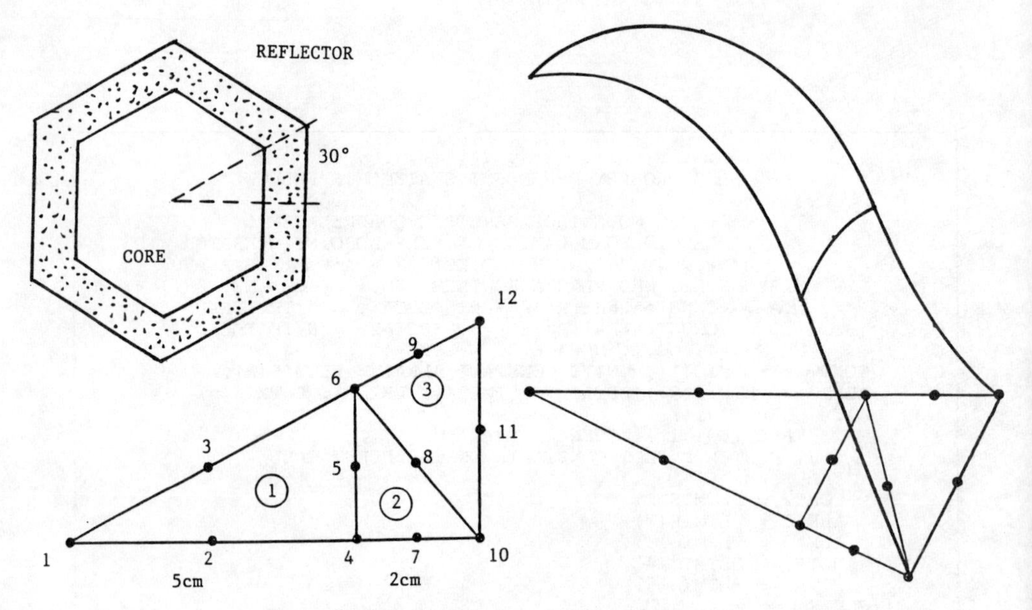

Figure 22.8.1 A two-material hexagonal reactor

NODE,	2 COORDINATES,	1 PARAMETERS	
1	0.00000E+00	0.00000E+00	2.30205E+00
2	2.50000E+00	0.00000E+00	2.64429E+00
3	2.50000E+00	1.25000E+00	2.61470E+00
4	5.00000E+00	0.00000E+00	1.70962E+00
5	5.00000E+00	1.25000E+00	1.63464E+00
6	5.00000E+00	2.50000E+00	1.38030E+00
7	6.00000E+00	0.00000E+00	7.80099E-01
8	6.00000E+00	1.25000E+00	7.47687E-01
9	6.00000E+00	3.00000E+00	4.97461E-01
10	7.00000E+00	0.00000E+00	0.00000E+00
11	7.00000E+00	1.75000E+00	0.00000E+00
12	7.00000E+00	3.50000E+00	0.00000E+00

Figure 22.8.2 Reactor output results

$$D(\phi_{,xx} + \phi_{,yy}) - \Sigma_A \, \phi + S_n = 0 \qquad (22.26)$$

where ϕ, D, Σ_A and S_n denote the neutron flux, coefficient of neutron diffusion, macroscopic absorption cross section and the neutron source per unit time, respectively. Applying the Galerkin method to the above equation yields the following definitions of the element matrices:

$$\mathbf{S}^e = \iint_{A^e} [\, D(\mathbf{H}_{,x}^{e^T}\mathbf{H}_{,x}^e + \mathbf{H}_{,y}^{e^T}\mathbf{H}_{,y}^e) + \Sigma_A \, \mathbf{H}^{e^T}\mathbf{H}^e]\, da\,, \quad \mathbf{C}_1^e = \iint_{A^e} S_n \mathbf{H}^{e^T} da. \qquad (22.27)$$

On a typical external boundary, where ϕ is not specified, one also has

$$\mathbf{C}_2^b = \int_{\Gamma_2^b} D \frac{\partial \phi}{\partial n} \mathbf{H}^{b^T} ds. \tag{22.28}$$

However, we will only consider problems where $\partial\phi/\partial n = 0$ on Γ_2.

This example will be implemented using a six node isoparametric triangle. The element integrals can be evaluated in closed form, but it is simpler to employ numerical integration. A typical application is the two region problem shown in Fig. 22.8.1. The quadratic element neutron flux results in Fig. 22.8.2 agree with a previous solution obtained using 49 linear elements. This represents 1/12-th of a hexagonal reactor with the following properties:

Material No.	Core 1	Reflector 2
D	0.3333 cm	0.3333 cm
Σ_A	0.375 cm^{-1}	0.0375 cm^{-1}
S_n	1.0 neut./sec.	0.0

The boundary conditions involve a vacuum, i.e., $\phi = 0$, at the outer edge of the reflector and symmetry, i.e., $\partial\phi/\partial n = 0$, on the other faces. The properties of the reflector and the inner core are also shown in the figure. The programming of this element is summarized in Figs. 22.8.3 where the problem-dependent parts of ELSQ are shown. There are some new examples here. Note that triangular quadrature rules are used. Also the interpolation functions and their local derivatives, are evaluated once at each point and stored for later use. If the storage of these data presents no problem this procedure can make the code much more efficient. Here the miscellaneous properties arrays were used to identify material properties by material number. Subroutines MATWRT and MATPRT were designed for this capability.

22.9 Axisymmetric Solenoid Magnet *

The finite element stress analysis of axisymmetric solids has been extended to consider the effects of spatially varying magnetic body forces. The procedure is applicable to all arbitrarily shaped axisymmetric magnets, but is probably most significant for high field solenoids. Example results for an E.P.R. polodial coil show that the axial stresses which are neglected in approximate solutions can represent a significant contribution to the maximum effective material stress. The effects of the structural supports on the stresses and natural frequency of the magnet can also be determined.

Magnet and coil engineering experiments for fusion reactors often involve the use of high field solenoid magnets. The stress analysis of such magnets is often based on simplified analytical models [12]. While such models are very useful in preliminary designs, they have some important disadvantages. The most significant shortcoming of these formulations is the fact that the axial loads on the conductors resulting from the radial component of the magnetic flux density, $\vec{B}$, are neglected. For some geometries the stresses associated with these loads can be significant. A second disadvantage is the difficulty of considering the effects of structural supports and orthotropic material properties.

The present example shows a more general analysis procedure based on a finite element formulation which utilizes isoparametric quadrilateral elements. Since a

```
C     .............................................................
C      ***  ELSQ PROBLEM DEPENDENT STATEMENTS FOLLOW ***
C     .............................................................
C     2-D NEUTRON DIFFUSION VIA 6 NODE TRIANGULAR ELEMENT
C     PRTMAT(1) = COEFFICIENT OF NEUTRON DIFFUSION
C     PRTMAT(2) = MACROSCOPIC ABSORPTION CROSS-SECTION
C     PRTMAT(3) = NEUTRON SOURCE PER UNIT TIME
C     LPROP(1)  = ELEMENT MATERIAL NUMBER
      PARAMETER ( NREAL = 3 )
      DATA  MATOLD, KALL, LFRESQ / 0, 1, 36 /
C-->  ON FIRST CALL SELECT RULE, PRINT PROPERTIES
      IF ( KALL .EQ. 1 )  THEN
        KALL = 0
        LFRESQ = NELFRE*NELFRE
        CALL  MATWRT (NREAL, MISCFL, FLTMIS, PRTMAT)
C        SELECT SYMMETRIC TRIANGULAR RULE
        CALL  SYMRUL (NQP, PT, WT)
      ENDIF
C--->  DEFINE PROPERTIES, DEFAULT MATNO = 1
      MATNO = LPROP(1)
      IF ( MATNO .LT. 1 )  MATNO = 1
C      EXTRACT MATERIAL PROPERTIES ( IF NEW )
      IF ( MATOLD .NE. MATNO )  THEN
        MATOLD = MATNO
        CALL  MATPRT (MATNO, NREAL, MISCFL, FLTMIS, PRTMAT)
        CND  = PRTMAT(1)
        SN   = PRTMAT(2)
        CSMA = PRTMAT(3)
      ENDIF
C     ZERO ELEMENT SQUARE MATRIX AND SOURCE VECTOR
      CALL ZEROA (LFRESQ, S)
      CALL ZEROA (NELFRE, C)
C      NUMERICAL INTEGRATION LOOP
      DO 110  IP = 1, NQP
        CALL  DER6T (PT(1,IP), PT(2,IP), DLH)
        CALL  JACOB (N, NSPACE, DLH, COORD, AJ)
        CALL  I2BY2 (AJ, AJINV, DET)
        CALL  GDERIV (NSPACE, N, AJINV, DLH, DGH)
        CALL  SHP6T (PT(1,IP), PT(2,IP), H)
        DETWT = DET*WT(IP)
        DO 90  I = 1, N
          DO 80  J = 1, N
   80     S(I,J) = S(I,J) + DETWT*(CND*DGH(1,I)*DGH(1,J)
     1                           + CND*DGH(2,I)*DGH(2,J)
     2                           + CSMA*H(I)*H(J)          )
          C(I) = C(I) + DETWT*SN*H(I)
   90   CONTINUE
  110 CONTINUE
      RETURN
      END
```

Figure 22.8.3 Neutron diffusion matrix evaluations

```
C     .....................................................
C      ***  ELSQ PROBLEM DEPENDENT STATEMENTS FOLLOW ***
C     .....................................................
C     AXISYMMETRIC STRESS ANALYSIS OF A SOLENOID MAGNET
C      USING AN ISOPARAMETRIC ELEMENT WITH 4 NODES
C     BODY        = WILL RECEIVE MAGNETIC FIELD VECTOR
C     PRTMAT(1) = MODULUS OF ELASTICITY
C     PRTMAT(2) = POISSON'S RATIO
C     PRTMAT(3) = SHEAR MODULUS
C     PRTMAT(4) = CONSTANT CURRENT IN THE THETA DIRECTION
C     LPROP(1)  = ELEMENT MATERIAL NUMBER
C     PRTLPT(J,1) = R-COMP MAGNETIC FLUX DENSITY AT NODE J
C     PRTLPT(J,2) = Z-COMP MAGNETIC FLUX DENSITY AT NODE J
C     STRESS AND STRAIN COMPONENT ORDER: RR, ZZ, XYZ, TT
      PARAMETER ( NREAL = 4 )
      COMMON /ELARG/ BEB(8,8), YM, PR, G, CUR
      DATA  MATOLD, KALL, LFRESQ /0, 0, 1, 64/
C-->  ON FIRST CALL SET ZEROS IN E AND B, PRINT PROPERTIES
      IF ( KALL .EQ. 1 )  THEN
        KALL = 0
        DO 10 I = 1, N
          B(1,2*I)   = 0.0
          B(2,2*I-1) = 0.0
          B(4,2*I)   = 0.0
          E(3,I) = 0.0
          E(I,3) = 0.0
   10   CONTINUE
        CALL  MATWRT (NREAL, MISCFL, FLTMIS, PRTMAT)
      ENDIF
      IF ( NTAPE1 .GT. 0 )  WRITE (NTAPE1)  NIP
C---> DEFINE PROPERTIES, DEFAULT MATNO = 1
      MATNO = LPROP(1)
      IF ( MATNO .LT. 1 )  MATNO = 1
C      EXTRACT MATERIAL PROPERTIES ( IF NEW )
      IF ( MATOLD .NE. MATNO )  THEN
        MATOLD = MATNO
        CALL  MATPRT (MATNO, NREAL, MISCFL, FLTMIS, PRTMAT)
        YM  = PRTMAT(1)
        PR  = PRTMAT(2)
        G   = PRTMAT(3)
        IF ( G .EQ. 0.0 ) G = 0.5*YM/(1. + PR)
        CUR = PRTMAT(4)
C-->      DEFINE NON-ZERO TERMS IN CONSTITUTIVE MATRIX
        CONST  = YM/(1.+PR)/(1.-PR-PR)
        E(1,1) = CONST*(1.-PR)
        E(2,2) = CONST*(1.-PR)
        E(4,4) = CONST*(1.-PR)
        E(2,1) = CONST*PR
        E(4,1) = CONST*PR
        E(1,2) = CONST*PR
        E(4,2) = CONST*PR
        E(1,4) = CONST*PR
        E(2,4) = CONST*PR
        E(3,3) = G
      ENDIF
C     ZERO ELEMENT SQUARE MATRIX AND BODY FORCE VECTOR
      CALL ZEROA (LFRESQ, S)
      CALL ZEROA (NELFRE, C)
```

Figure 22.9.1a Solenoid magnet element matrices

```
C       NUMERICAL INTEGRATION LOOP
      DO 110  IP = 1, NQP
        CALL  DER4Q (PT(1,IP), PT(2,IP), DLH)
        CALL  JACOB (N, NSPACE, DLH, COORD, AJ)
        CALL  I2BY2 (AJ, AJINV, DET)
        CALL  GDERIV (NSPACE, N, AJINV, DLH, DGH)
C        CALCULATE MAGNETIC BODY FORCE VECTOR AT THE POINT
        CALL  SHP4Q (PT(1,IP), PT(2,IP), H)
        CALL  CALPRT (N, 2, H, PRTLPT, BODY)
C        FIND COORDINATES OF INTEGRATION POINT
        CALL  MMULT (H, COORD, XYZ, 1, N, 2)
        DETWTR = DET*WT(IP)*XYZ(1)
C--->    FORM STRAIN DISPLACEMENT MATRIX,  B
        DO 70  L = 1, N
          J = L + L - 1
          B(3, 2*L) = DGH(1, L)
          B(1, J)   = DGH(1, L)
          B(2, 2*L) = DGH(2, L)
          B(3, J)   = DGH(2, L)
          B(4, J)   = H(L)/XYZ(1)
   70   CONTINUE
C        FORM STRESS-DISPLACEMENT MATRIX,  EB
        CALL  MMULT (E, B, EB, NRB, NRB, NELFRE)
C        STORE COORDINATES AND STRESS-DISPLACEMENT MATRIX
        IF ( NTAPE1 .GT. 0 )  WRITE (NTAPE1)  XYZ,  EB
C        EVALUATE INTEGRAN (COULD HAVE USED BTDB HERE)
        CALL MTMULT (B, EB, BEB, NRB, NELFRE, NELFRE)
        DO 90  I = 1, NELFRE
          DO 80  J = 1, NELFRE
   80   S(I,J) = S(I,J) + DETWTR*BEB(I,J)
   90   CONTINUE
C        FORCE PER VOLUME = J VECTOR CROSS B VECTOR
        FR =  DETWTR*CUR*BODY(2)
        FZ = -DETWTR*CUR*BODY(1)
        DO 100 I = 1, N
          C(2*I)   = C(2*I)    + FZ*H(I)
  100   C(2*I-1) = C(2*I-1) + FR*H(I)
  110 CONTINUE
      RETURN
      END
```

Figure 22.9.1b Solenoid magnet element matrices

standard finite element procedure automatically allows for arbitrarily shaped non-homogeneous materials, it is possible to consider the effects of the conductor, insulation, and reinforcement distributed throughout the magnet. In addition, the above materials can have orthotropic stress-strain relations. In the present analysis, the coil is divided into a number of axisymmetric finite elements. The magnetic flux density vector is assumed to be known at each node point. The current, $\mathbf{J}_\theta$, within a given conductor element is assumed constant, but each element can have a different value. To consider the full effects of the magnetic loading the body force per unit volume, $\mathbf{f} = \mathbf{J}_\theta \times \mathbf{B}$, is numerically integrated over the volume of each element. These magnetic body forces on the conductors are combined with any additional mechanical loads to obtain the total load matrix, $\mathbf{F}$. The governing matrix equations of equilibrium of the structure, $\mathbf{K}\,\mathbf{D} = \mathbf{F}$, are assembled and solved in the standard manner to yield the displacements, $\mathbf{D}$, of all nodes in the structure. From these data the stresses and strains within each element can be calculated.

According to the von Mises criterion, yielding of the material within a typical element begins when the effective stress, σ_E equals the yield point stress determined by a simple tension test. For a symmetrically-loaded solid of revolution

$$\sigma_E = 2^{-1/2}\left[(\sigma_R - \sigma_Z)^2 + (\sigma_Z - \sigma_\theta)^2 + (\sigma_\theta - \sigma_R)^2 + 6\tau_{RZ}^2\right]^{1/2}$$

where the subscripts R, Z, and θ denote stress components in the radial, axial and circumferential directions, respectively.

Since many magnet coils are subjected to pulsed currents (and, thus, loads), it is desirable to be able to estimate the natural frequency of the structure. To avoid large vibrations, it is important that the pulse period not be too close to the natural period of the structure. It has been shown that the static deflections can be utilized to obtain an upper bound estimate of the natural frequency. The Rayleigh quotient for the natural frequency, ω_n, is

$$\omega_n^2 = PE/KE$$

where the potential and kinetic energies are defined in terms of the nodal displacements, **D**, by which the potential and kinetic energies are defined in terms of the nodal displacements, **D**, by

$$PE = \mathbf{D}^T\mathbf{K}\mathbf{D}/2, \quad KE = \mathbf{D}^T\mathbf{M}\mathbf{D}/2$$

where **K** and **M** denote the structure stiffness matrix and mass matrix, respectively. Note that this allows the analyst to investigate the effects of displacement boundary conditions on the natural frequency of the magnet structure.

The contribution to the total matrix due to body forces acting on an element is

$$\mathbf{F}^e = \int_{V^e} \mathbf{N}^{e^T}\mathbf{f}\,dv$$

where $\mathbf{f}$ is the array containing the components of the body force per unit volume, $\mathbf{N}^e$ is the interpolation function matrix defining the displacements within the element in terms of its nodal displacements and V^e is the volume of the element. This integral is usually evaluated numerically by Gaussian quadratures. For an axisymmetric quadrilateral with Q quadrature points

$$\mathbf{F}^e = 2\pi \sum_{i=1}^{Q} W_i\,[\mathbf{N}_i^e]^T\,\mathbf{f}_i\,r_i\,|J|_i$$

where J is the determinant of the coordinate system Jacobian. For the present application one need only define the magnetic loads, $\mathbf{f}^T = [f_R \quad f_Z]$. Recalling that $\vec{f} = \vec{J}_\theta \times \vec{B}$ one must define the spatial variation, over the element of the current and the magnetic flux density, $\vec{B} = (\vec{B}_R + \vec{B}_Z)$. The current is assumed to be constant within a typical element. The most accurate evaluation of $\vec{B}$ would be obtained by coupling the stress analysis code with a magnetic field code for evaluating $\vec{B}$ at each quadrature point. However, this proved to be rather expensive. The computer time was reduced by a factor of three by inputing the magnetic flux density **B** at each node and then utilizing the element interpolation functions, $\mathbf{N}^e$, to obtain the components at each quadrature point. The results differed in only the sixth or seventh significant figure.

As a typical example consider a preliminary design of an experimental power reactor (EPR) pulsed polodial coil having the characteristics shown in Table 22.4. The stresses (MPa) and displacements (meters) were determined by three procedures for the

```
C       ...........................................................
C         *** POSTEL PROBLEM DEPENDENT STATEMENTS FOLLOW ***
C       ...........................................................
C       AXISYMMETRIC STRESS ANALYSIS OF A SOLENOID MAGNET
C       STRESS AND STRAIN COMPONENT ORDER: RR, ZZ, RZ, TT
        COMMON /ELARG/ BEB(8,8), YM, PR, G, CUR
        DATA  KALL / 1 /
C         PRINT TITLES ON FIRST CALL
        IF ( KALL .EQ. 1 )  THEN
          KALL = 0
          WRITE (6,5000)
 5000     FORMAT ('*** STRESSES AT INTEGRATION POINTS ***'
     1    ,'POINT        COORDINATES                    STRESSES'
     2    'NUMBER     R            Z              RR              ZZ',
     3    '         RZ            TT         EFFECTIVE',/)
        ENDIF
        WRITE (6,*) 'ELEMENT NUMBER', IE
C         READ NUMBER OF INTEGRATION POINTS
        READ (NTAPE1) NIP
        DO 20  J = 1, NIP
C           READ COORDINATES AND STRESS-DISPLACEMENT MATRIX
          READ (NTAPE1) XYZ, EB
C           CALCULATE STRESSES
          CALL  MMULT (EB, D, STRESS, NRB, NELFRE, 1)
C           CALCULATE EFFECTIVE STRESS
          STRESS(5)  =  0.7071068*SQRT( (STRESS(1)-STRESS(2))**2 +
     1      (STRESS(2)-STRESS(4))**2 + (STRESS(4)-STRESS(1))**2 +
     2      6.*STRESS(3)*STRESS(3) )
          WRITE (6,5020) J, XYZ(1), XYZ(2), (STRESS(I), I=1,5)
 5020     FORMAT  ( I7, 2(1PE12.4), 5(1PE14.5) )
   20   CONTINUE
        RETURN
        END
```

Figure 22.9.2 Recovering magnet stresses

Table 22.4 Pulsed EPR polodial coil

Inner Radius	1.5 m
Outer Radius	1.9 m
Half Height	6.0 m
Inner B_Z	7.1 T
Outer B_Z	–0.2 T
Current Density	1.46E7 A/m^2
Elastic Modulus	1.21E5 MPa
Poisson's Ratio	0.33
Mass Density	8.96E3 Kg/m^3

purpose of comparison and the results are shown in Table 22.5. The first method is based on the theory of Lontai and Marston [12]. It assumes that B_R is zero and that B_Z is a linear function of the radius. The next two solutions were obtained by utilizing the finite element method and considering the spatial variation of $\vec{B} = \vec{B}(R, Z)$. The second neglected the effects of the radial component, B_R, and the third included it. The finite element calculations were executed by the code MODEL. A uniform mesh with 305 nodes and 240 elements was utilized. The stress results shown in Table 22.5 show that the theory of Lontai and Marston is adequate for preliminary calculations. However, the axial loads resulting from B_R do become significant and must be considered in the final design. The deflection predictions were similar as shown in Fig. 22.9.3. The Lontai and Marston theory predicts no axial displacement and a uniform radial displacement of 1.19E-3 meters at the inner radius. The complete finite element analysis gave corresponding maximum values of 1.32E-3 and 2.55E-3 meters, respectively. It should be noted that the above analysis did not involve displacement boundary conditions, that is, the magnet was free to expand under the applied loads. The finite element analysis can automatically consider arbitrary displacement boundary conditions that can not be included in the elementary theory. The finite element method can also include thermal stresses or initial strain effects. The estimated natural frequency obtained from the finite element analysis was 1332 cycles per second which corresponds to a period of 7.5×10^{-4} seconds. The actual pulse period is expected to be 2.0 seconds. The estimated natural frequency would vary with the method of support. However, in this example, the pulse period is so large that the loads can be considered to be static.

22.10 Three-Dimensional Applications

Another advantage of the finite element procedure is the relative ease of extension to three-dimensional applications. Several three-dimensional solutions have been presented in the literature. The earliest applications involved the four node linear tetrahedral element. The simplicity of the interpolation functions for this element make it possible to derive the element matrices in closed form. For example, Zienkiewicz [20] presented the expressions for three-dimensional stress analysis using that element. The disadvantage of the tetrahedral elements is the great burden of data preparation that it places on the user. One can generally obtain a more accurate solution with less data preparation by utilizing a hexahedral element. The simplest element in this family is the eight node *linear* hexahedron. This element will be illustrated in the following section on heat conduction so as to reduce the number of new concepts to be considered.

The isoparametric formulation of the two-dimensional Poisson equation was presented in Sec. 11.3. The governing differential equation in three dimensions becomes

$$(K_x T_{,x})_{,x} + (K_y T_{,y})_{,y} + (K_z T_{,z})_{,z} + g = 0 \tag{22.29}$$

and the element matrices are

$$\mathbf{S}^e = \int_{V^e} [K_x \mathbf{H}_{,x}^T \mathbf{H}_{,x} + K_y \mathbf{H}_{,y}^T \mathbf{H}_{,y} + K_z \mathbf{H}_{,z}^T \mathbf{H}_{,z}]\, dv, \tag{22.30}$$

and

$$\mathbf{C}^e = \int_{V^e} g \mathbf{H}^T\, dv. \tag{22.31}$$

Of course, if there are convective or normal heat flux boundary conditions on the surface of the domain then the element matrices will contain additional terms. Comparing these equations with the corresponding two-dimensional equations one notes some important differences. First, the integrals are three-dimensional, which makes the relative

Table 22.5 Stress results

Method	R	Z	σ_E	σ_R/σ_E	σ_Z/σ_E	σ_θ/σ_E
Lontai and	1.50	0.	96.6	0.	0.	1.00
Marston	1.72	0.	80.8	−0.05	0.	0.98
$B_Z = B_Z(R)$	1.90	0.	70.0	0.	0.	1.00
Finite	1.55	0.05	91.8	−0.02	0.04	1.01
Elements	1.85	0.05	73.5	−0.02	0.	0.97
$B_R = 0$	1.55	5.95	52.1	−0.02	0.	0.99
Finite	1.55	0.05	115.2	−0.01	−0.31	0.81
Elements	1.85	0.05	99.4	−0.02	−0.04	0.72
$B_R = B_R(R, Z)$	1.55	5.95	54.4	−0.01	−0.04	0.97

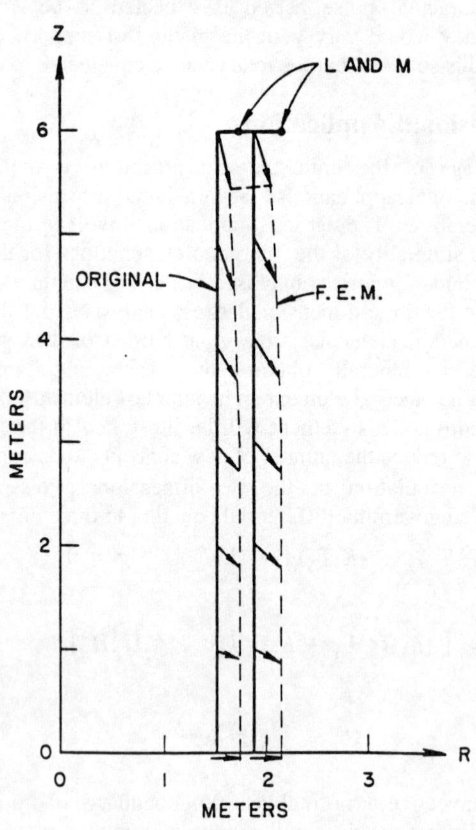

Figure 22.9.3 Magnet deformed shape

Table 22.6 Parameter definitions for three-dimensional heat conduction
(* Default, ** Optional)

CONTROL
NSPACE = 3 NLPFIX = 2 N = 8
MISFIX = 8 LBN - 0*, or 4** NCURVE = 3

DEPENDENT VARIABLES:
1 = temperature, T

BOUNDARY ELUX COMPONENTS AT NODES**:
1 = specified normal outward heat flux per unit area, qn

PROPERTIES:
Element level:
Fixed point:
1 = Gaussian quadrature order in each dimension, (2*)
2 = Element material number
Miscellaneous Level:
K + 1 = thermal conductivity in x-direction, kx
K + 2 = thermal conductivity in y-direction, ky
K + 3 = thermal conductivity in z-direction, kz
K + 4 = heat generation per unit volume
where for material number 1, K = 4*(1-1)

CONSTRAINTS:
Type 1: specified nodal temperatures

```
C      .....................................................
C       ***  NGRAND PROBLEM DEPENDENT STATEMENTS FOLLOW ***
C      .....................................................
       COMMON / ELARG / XK, YK, ZK, HG
C      XK = CONDUCTIVITY IN X-DIRECTION
C      YK = CONDUCTIVITY IN Y-DIRECTION
C      ZK = CONDUCTIVITY IN Z-DIRECTION
C      HG = HEAT GENERATION PER UNIT VOLUME
C       EVALUATE PRODUCTS
       DO 20  I = 1, N
         DO 10  J = 1, N
   10    S(I,J) = S(I,J) + DETWT*( XK*DGH(1,I)*DGH(1,J)
     1                           + YK*DGH(2,I)*DGH(2,J)
     2                           + ZK*DGH(3,I)*DGH(3,J))
   20 C(I) = C(I) + DETWT*HG*H(I)
       RETURN
       END
```

Figure 22.10.1 Poisson equation 3-D kernel

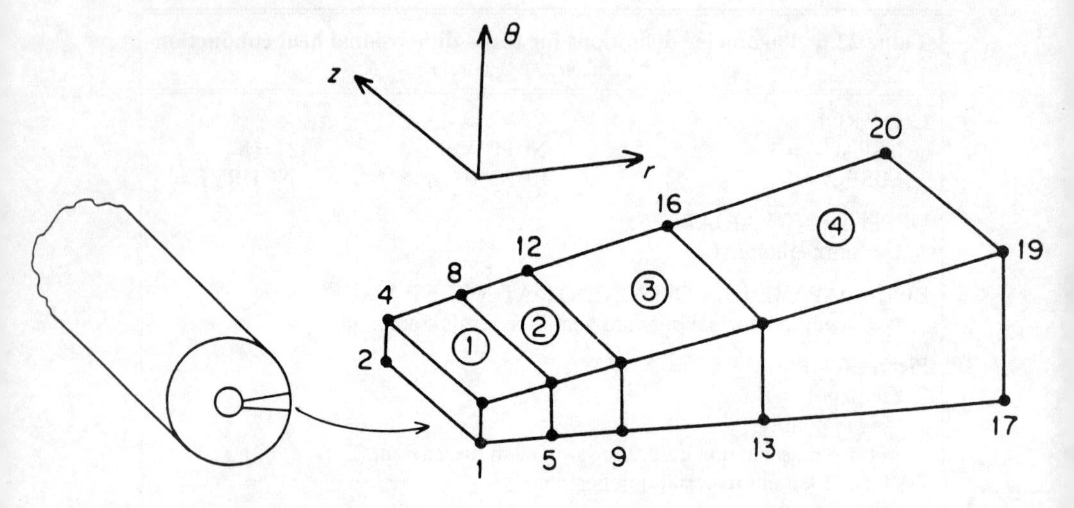

Figure 22.10.2 Wedge from a cylinder

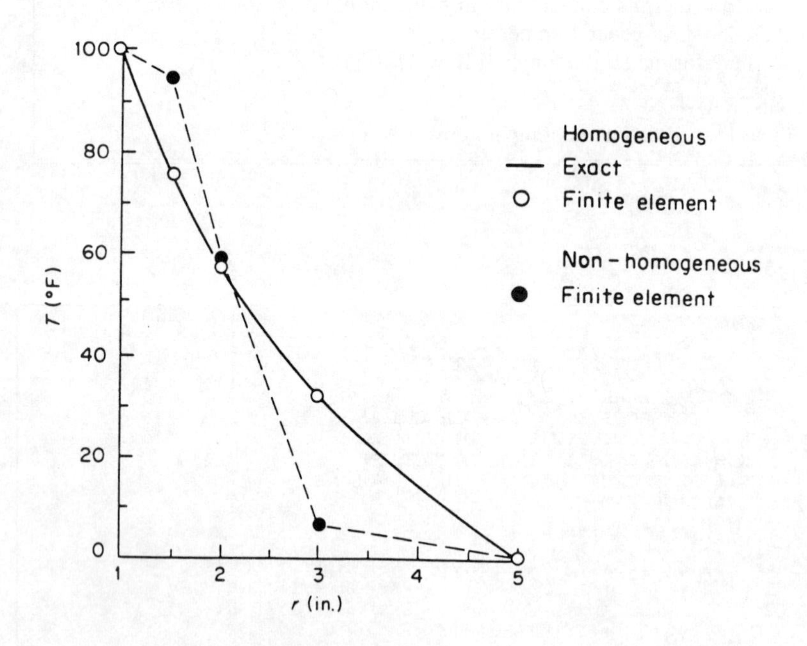

Figure 22.10.3 Comparison of exact and 3-D finite element solutions

```
*****  PROBLEM PARAMETERS  *****
NUMBER OF NODAL POINTS IN SYSTEM =..........    20
NUMBER OF ELEMENTS IN SYSTEM =..............     4
NUMBER OF PARAMETERS PER NODE =.............     1
NUMBER OF NODES PER ELEMENT =...............     8
DIMENSION OF SPACE =........................     3
NUMBER OF ROWS IN B MATRIX =................     3
NUMBER OF QUADRATURE POINTS =...............     8
SHAPE 1-LINE 2-TRI 3-QUAD 4-HEX 5-TET =.....     4
NUMBER OF INTEGER PROPERTIES PER ELEMENT =...    2
NUMBER OF REAL MISCELLANEOUS  PROPERTIES =...    8

*** NODAL POINT DATA ***
NODE, CONSTRAINT FLAG, 3 COORDINATES
         1        1    1.0000    0.0000    0.0000
        ...      ...    ...       ...       ...
        16        0    2.9995    0.0524    0.1000
        17        1    5.0000    0.0000    0.0000
        18        1    5.0000    0.0000    0.1000
        19        1    4.9992    0.0873    0.0000
        20        1    4.9992    0.0873    0.1000

*** ELEMENT CONNECTIVITY DATA ***
ELEMENT NO.,  8 NODAL INCIDENCES.
    1    1    5    7    3    2    6    8    4
    2    5    9   11    7    6   10   12    8
    3    9   13   15   11   10   14   16   12
    4   13   17   19   15   14   18   20   16

*** CONSTRAINT EQUATION DATA ***
EQ. NO.   NODE1   PAR1          A1
      1       1      1   1.00000E+02
      2       2      1   1.00000E+02
      3       3      1   1.00000E+02
      4       4      1   1.00000E+02
      5      17      1   0.00000E+00
      6      18      1   0.00000E+00
      7      19      1   0.00000E+00
      8      20      1   0.00000E+00

***  ELEMENT  PROPERTIES   ***
ELEMENT  PROPERTY       VALUE
    1         1            2
    2         1            2
    3         1            2
    4         1            2
    1         2            2
    2         2            1
    3         2            1
    4         2            2

***  MISCELLANEOUS SYSTEM PROPERTIES   ***
PROPERTY     VALUE
       1     1.20000E+02
       2     1.20000E+02
       3     1.20000E+02
       4     0.00000E+00
       5     1.20000E+03
       6     1.20000E+03
       7     1.20000E+03
       8     0.00000E+00
```

Figure 22.10.4a Selected solid results

```
*** INPUT SOURCE RESULTANTS ***
DOF        SUM
  1    0.0000E+00

*** REACTION RECOVERY ***
NODE  DOF   REACTION    EQUATION
   1    1  6.7368E+00      1
   2    1  6.7334E+00      2
   3    1  6.7588E+00      3
   4    1  6.7598E+00      4
  17    1 -6.7560E+00     17
  18    1 -6.7557E+00     18
  19    1 -6.7469E+00     19
  20    1 -6.7467E+00     20

***   OUTPUT OF RESULTS   ***
NODE, 3 COORDINATES,  1 PARAMETERS.
    1  1.00000E+00  0.00000E+00  0.00000E+00  1.00000E+02
    2  1.00000E+00  0.00000E+00  1.00000E-01  1.00000E+02
    3  9.99850E-01  1.74500E-02  0.00000E+00  1.00000E+02
    4  9.99850E-01  1.74500E-02  1.00000E-01  1.00000E+02
    5  1.50000E+00  0.00000E+00  0.00000E+00  9.48457E+01
    6  1.50000E+00  0.00000E+00  1.00000E-01  9.48456E+01
    7  1.49977E+00  2.61800E-02  0.00000E+00  9.48457E+01
    8  1.49977E+00  2.61800E-02  1.00000E-01  9.48457E+01
    9  2.00000E+00  0.00000E+00  0.00000E+00  5.80138E+01
   10  2.00000E+00  0.00000E+00  1.00000E-01  5.80138E+01
   11  1.99970E+00  3.49100E-02  0.00000E+00  5.80136E+01
   12  1.99970E+00  3.49100E-02  1.00000E-01  5.80136E+01
   13  3.00000E+00  0.00000E+00  0.00000E+00  6.44688E+00
   14  3.00000E+00  0.00000E+00  1.00000E-01  6.44688E+00
   15  2.99954E+00  5.23600E-02  0.00000E+00  6.44690E+00
   16  2.99954E+00  5.23600E-02  1.00000E-01  6.44690E+00
   17  5.00000E+00  0.00000E+00  0.00000E+00  0.00000E+00
   18  5.00000E+00  0.00000E+00  1.00000E-01  0.00000E+00
   19  4.99924E+00  8.72600E-02  0.00000E+00  0.00000E+00
   20  4.99924E+00  8.72600E-02  1.00000E-01  0.00000E+00
```

Figure 22.10.4b Selected solid results

numerical integration costs go from n^2 to n^3, where n is the number of quadrature points in each dimension. For this element one can utilize a value of $n = 2$ unless the element has a greatly distorted shape. The element square matrix contains an additional set of matrix products. Also, the size of that matrix has gone from 4×4 to 8×8, which also represents a significant increase in the number of terms to be calculated and assembled into the system equations. The problem-dependent subroutine for this element is shown in Fig. 22.10.1. This element is summarized in Table 22.4.

As a simple test problem consider the temperature distribution in an infinite cylinder. Figure 22.10.2 shows the mesh which is a one degree wedge segment divided into four elements. The inner radius has a specified temperature of 100° F and the outer surface has a temperature of 0° F. Due to symmetry, the normal heat flux on the other four surfaces is zero. This is a natural boundary condition that is automatically satisfied. For a homogeneous material the exact solution for this problem is shown by Myers [15] to be $t(r) = t_i - (t_i - t_o)\ ln(r/r_i)\ /ln(r_o/r_i)$ where where the subscripts i and o denote inner and outer surfaces, respectively. Figure 22.10.3 shows the comparison with this solution. The maximum error is 0.6%. The effect of non-homogeneous properties is also shown in that figure. The output data for the latter problem, where the inner and outer elements have conductivities that are ten times as large as the other two, are shown in Fig. 22.10.3.

Before leaving this section, it may be useful to comment on three-dimensional problems that involve surface integrals. Recall that if one has a specified normal heat flux on a surface Γ, then the element column matrix will contain a term

$$\mathbf{C}_q = \int_\Gamma q_n \mathbf{H}^T \, da \,. \tag{22.32}$$

Evaluation of da on the three-dimensional surface in terms of the two local coordinates, say ξ and η, requires the use of differential geometry. It can be shown that $da = \sqrt{A}\, d\xi\, d\eta$ where A is the determinant of the first fundamental magnitudes of the surface. The numerical integration of the above equation gives

$$\mathbf{C}_q = \sum_{i=1}^{NQP} W_i q_i \mathbf{H}(\xi_i,\, \eta_i)^T \sqrt{A_i} \,. \tag{22.33}$$

To evaluate the A_i term at an integration point one could utilize subroutine FMONE. In other applications one would also need the unit normal vector at the point to determine the global components of a given traction vector. The latter operation is performed by subroutine SNORMV. These two routines are included on the Disk.

22.11 References

[1] Akin, J.E., "Finite Element Analysis of Fields with Boundary Singularities," Int. Conf. Num. Meth. in Electric and Magnetic Field Problems (May 1976).

[2] Akin, J.E. and Wooten, J.W., "Tokamak Plasma Equilibria by Finite Elements," in *Finite Elements in Fluids III,* Chapter 21, ed. R.H. Gallagher, New York: John Wiley (1978).

[3] Akin, J.E., *Computer Assisted Mechanical Design*, Englewood Cliffs: Prentice-Hall (1990).

[4] Bathe, K.J., *Finite Element Procedures in Engineering Analysis*, Englewood Cliffs: Prentice-Hall (1982).

[5] Chung, T.J., *Finite Element Analysis in Fluid Dynamics*, New York: McGraw-Hill (1978).

[6] Daly, P., "Singularities in Transmission Lines," pp. 337–350 in *The Mathematics of Finite Elements and Applications*, ed. J.R. Whiteman, London: Academic Press (1973).

[7] Desai, C.S. and Abel, J.F., *Introduction to the Finite Element Method*, New York: Van Nostrand-Reinhold (1972).

[8] Gallagher, R.H., *Finite Element Analysis Fundamentals*, Englewood Cliffs: Prentice-Hall (1975).

[9] Huebner, K.H. and Thornton, E.A., *Finite Element Method for Engineers*, New York: John Wiley (1982).

[10] Irons, B.M. and Razzaque, A., "Experience with the Patch Test for Convergence of the Finite Element Method," pp. 557–587 in *Mathematical Foundation of the Finite Element Method*, ed. A.R. Aziz, New York: Academic Press (1972).

[11] Irons, B.M. and Ahmad, S., *Techniques of Finite Elements*, New York: John Wiley (1980).

[12] Lontai, L.M. and Marston, P.G., "A 100 Kilogauss Quasi-continuous Cryogenic Solenoid — Part I," Proc. Intern. Symp. Magnet Technology (1965).

[13] Martin, H.C. and Carey, G.F., *Introduction to Finite Element Analysis*, New York: McGraw-Hill (1974).

[14] Meek, J.L., *Matrix Structural Analysis*, New York: McGraw-Hill (1972).

[15] Myers, G.E., *Analytical Methods in Conduction Heat Transfer*, New York: McGraw-Hill (1971).

[16] Schechter, R.S., *The Variational Method in Engineering*, New York: McGraw-Hill (1967).

[17] Ural, O., *Finite Element Method: Basic Concepts and Application*, Scranton: Intext (1973).

[18] Whiteman, J.R., "Some Aspects of the Mathematics of Finite Elements," pp. 25–42 in *The Mathematics of Finite Elements and Applications, Vol. II*, ed. J.R. Whiteman, London: Academic Press (1976).

[19] Whiteman, J.R. and Akin, J.E., "Finite Elements, Singularities and Fracture," in *The Mathematics of Finite Elements and Applications, Vol. III*, ed. J.R. Whiteman, London: Academic Press (1978).

[20] Zienkiewicz, O.C., *The Finite Element Method,* 3rd edition, New York: McGraw-Hill (1979).

Appendix I

MATRIX NOTATION AND OPERATIONS

1. Introduction

Finite element analysis procedures are most commonly described using matrix notation. These procedures ultimately lead to the solution of a large set of simultaneous equations. Thus, for the sake of completeness and consistency with later sections, the basic forms of matrix notation and matrix operations will be described here.

2. Matrices

A *matrix* is defined as a rectangular array of quantities arranged in rows and columns. The array is enclosed in brackets, and thus if there are **m** rows and **n** columns, the matrix can be represented by

$$\mathbf{A} = \begin{bmatrix} a_{11} & a_{12} & a_{13} & \ldots & a_{1j} & \ldots & a_{1n} \\ a_{21} & a_{22} & a_{23} & \ldots & a_{2j} & \ldots & a_{2n} \\ a_{31} & a_{32} & a_{33} & \cdots & a_{3j} & \cdots & a_{3n} \\ \\ a_{i1} & a_{i2} & a_{i3} & \cdots & a_{ij} & \cdots & a_{in} \\ \\ a_{m1} & a_{m2} & a_{m3} & \cdots & a_{mj} & \cdots & a_{mn} \end{bmatrix} = [A] \tag{A.1.1}$$

where the typical element a_{ij} has two subscripts, of which the first denotes the row (i-th) and the second denotes the column (j-th) which the element occupies in the matrix. A matrix with m rows and n columns is defined as a matrix of order $m \times n$, or simply an $m \times n$ matrix. The number of rows is always specified first. In Eq. (A.1.1) the symbol **A** stands for the matrix of m rows and n columns, and it is usually printed in **boldface** type. If $m = n = 1$, then the matrix degenerates to a scalar. If only $m = 1$, the matrix **A** reduces to the single row

$$\mathbf{A} = [a_{11}\ a_{12}\ a_{13} \cdots a_{1j} \cdots a_{1n}] = (A) \tag{A.1.2}$$

which is called a *row matrix*. Similarly, if only $\mathbf{n} = 1$, the matrix **A** reduces to the column

$$\mathbf{A} = \begin{Bmatrix} a_{11} \\ a_{21} \\ \vdots \\ a_{m1} \end{Bmatrix} = \{A\} \tag{A.1.3}$$

which is called a *column matrix.* When all the elements of the matrix are equal to zero, the matrix is called *null* or *zero* and is indicated by 0. A null matrix serves the same function as zero does in ordinary algebra. If $\mathbf{m} = \mathbf{n}$, the matrix $\mathbf{A}$ is the square array

$$\mathbf{A} = \begin{bmatrix} a_{11} & a_{12} & \cdots & a_{1n} \\ \\ a_{n1} & n2 & \cdots & a_{nn} \end{bmatrix} \tag{A.1.4}$$

which is called a *square matrix.* A system of linear equations may be written as

$$\begin{aligned} a_{11}x_1 + a_{12}x_2 + a_{13}x_3 + \cdots + a_{1n}x_n &= c_1 \\ a_{21}x_1 + a_{22}x_2 + a_{23}x_3 + \cdots + a_{2n}x_n &= c_2 \\ a_{31}x_1 + a_{32}x_2 + a_{33}x_3 + \cdots + a_{3n}x_n &= c_3 \\ &\vdots \\ a_{n1}x_1 + a_{n2}x_2 + a_{n3}x_3 + \cdots + a_{nn}x_n &= c_n \end{aligned}$$

where the a_{ij}'s and c_i's represent known coefficients and the x_i's are unknowns. These equations may also be represented in the more compact form given by $\mathbf{AX} = \mathbf{C}$ where $\mathbf{A}$, $\mathbf{X}$, and $\mathbf{C}$ represent matrices. In expanded form this is

$$\begin{bmatrix} a_{11} & a_{12} & a_{13} & \ldots & a_{1n} \\ a_{21} & a_{22} & a_{23} & \ldots & a_{2n} \\ a_{31} & a_{32} & a_{33} & \cdots & a_{3n} \\ \vdots & & & & \vdots \\ a_{n1} & a_{n2} & a_{n3} & & a_{nn} \end{bmatrix} \begin{Bmatrix} x_1 \\ x_2 \\ x_3 \\ \vdots \\ x_n \end{Bmatrix} = \begin{Bmatrix} c_1 \\ c_2 \\ c_3 \\ \vdots \\ c_n \end{Bmatrix}. \tag{A.1.5}$$

Before considering some of the matrix algebra implied by the above equation, a few other matrix types will be defined for future reference.

A *diagonal matrix* is a square matrix which has zero elements outside the principal diagonal. It follows, therefore, that for a diagonal matrix $a_{ij} = 0$ when $i \neq j$, and not all a_{ii} are zero. A typical diagonal matrix may be represented by

$$\mathbf{A} = \begin{bmatrix} a_{11} & 0 & \cdots & 0 \\ 0 & a_{22} & \cdots & 0 \\ & & & \\ 0 & 0 & \cdots & a_{nn} \end{bmatrix} \tag{A.1.6}$$

or simply, in order to save space, as

$$\mathbf{A} = \left\lfloor a_{11} \quad a_{22} \cdots \quad a_{nn} \right\rfloor = \lceil A \rfloor .$$

A *unit matrix* is a square matrix whose elements are equal to 0 except those located on its main diagonal, which are equal to 1, that is: $a_{ij} = 1$ if $i = j$, and $a_{ij} = 0$ if $i \neq j$. The unit matrix will be given the symbol **I** throughout this book. An example of a 3×3 unit matrix is:

$$\mathbf{I} = \begin{bmatrix} 1 & 0 & 0 \\ 0 & 1 & 0 \\ 0 & 0 & 1 \end{bmatrix}.$$

A *symmetric matrix* is a square matrix whose elements $a_{ij} = a_{ji}$. For example:

$$\mathbf{S} = \begin{bmatrix} 12 & 2 & -1 \\ 2 & 33 & 0 \\ -1 & 0 & 15 \end{bmatrix}.$$

An *anti-symmetric matrix* (skew symmetric) is a square matrix whose elements $a_{ij} = -a_{ji}$ for $i \neq j$, and $a_{ii} = 0$. For example:

$$\mathbf{A} = \begin{bmatrix} 0 & 2 & -1 \\ -2 & 0 & 10 \\ 1 & -10 & 0 \end{bmatrix}.$$

The *transpose* $\mathbf{A}^T$, of a matrix **A** is obtained by interchanging the rows and columns. Thus, the transpose of an $\mathbf{m} \times \mathbf{n}$ matrix is an $\mathbf{n} \times \mathbf{m}$ matrix. For example:

$$\text{if } \mathbf{A} = \begin{bmatrix} 2 & 1 \\ 3 & 5 \\ 0 & 1 \end{bmatrix}, \quad \text{then } \mathbf{A}^T = \begin{bmatrix} 2 & 3 & 0 \\ & & \\ 1 & 5 & 1 \end{bmatrix}.$$

Column matrices are often written transposed so as to save space.

If all the elements on one side of the diagonal of a square matrix are zero, the matrix is called a *triangular matrix*. There are two types of triangular matrices; (1) an upper triangular **U** whose elements below the diagonal are all zero, and (2) a lower triangular **L**, whose elements above the diagonal are all zero. An example of a lower triangular matrix follows:

$$\mathbf{L} = \begin{bmatrix} 10 & 0 & 0 \\ 1 & 3 & 0 \\ 5 & 1 & 2 \end{bmatrix}.$$

A matrix may be divided into smaller arrays by horizontal and vertical lines. Such a matrix is then referred to as a *partitioned matrix*, and the smaller arrays are called *submatrices*. For example,

$$\mathbf{A} = \begin{bmatrix} a_{11} & a_{12} & \vdots & a_{13} \\ a_{21} & a_{22} & \vdots & a_{23} \\ \cdots & \cdots & \cdots & \cdots \\ a_{31} & a_{32} & \vdots & a_{33} \end{bmatrix} = \begin{bmatrix} \mathbf{A}_{11} & \mathbf{A}_{21} \\ \mathbf{A}_{21} & \mathbf{A}_{22} \end{bmatrix} = \begin{bmatrix} 2 & 1 & \vdots & 3 \\ 10 & 5 & \vdots & 0 \\ \cdots & \cdots & \cdots & \cdots \\ 4 & 6 & \vdots & 10 \end{bmatrix} \tag{A.1.7}$$

where

$$\mathbf{A}_{11} = \begin{bmatrix} a_{11} & a_{12} \\ a_{21} & a_{22} \end{bmatrix} = \begin{bmatrix} 2 & 1 \\ 10 & 5 \end{bmatrix}, \quad \mathbf{A}_{12}^T = [a_{13} \quad a_{23}] = [3 \quad 0]$$
$$\mathbf{A}_{21} = [a_{31} \quad a_{32}] = [4 \quad 6], \quad \mathbf{A}_{22} = [a_{33}] = [10]. \tag{A.1.8}$$

It should be noted that the elements of a partitioned matrix must be so ordered that they are compatible with the whole matrix **A** and with each other, that is, $\mathbf{A}_{11}$ and $\mathbf{A}_{12}$ must have an equal number of rows. Likewise, $\mathbf{A}_{21}$ and $\mathbf{A}_{22}$ must have an equal number of rows. Matrices $\mathbf{A}_{11}$ and $\mathbf{A}_{21}$ must have an equal number of columns. Likewise, for $\mathbf{A}_{12}$ and $\mathbf{A}_{22}$. It should be noted that $\mathbf{A}_{22}$ is a matrix even if it consists of only one element. Provided the general rules for matrix algebra are observed, the submatrices can be treated as if they were ordinary matrix elements.

3. Matrix Algebra

Addition of two matrices of the same order is accomplished by adding corresponding elements of each matrix. The matrix addition $\mathbf{C} = \mathbf{A} + \mathbf{B}$, where **A**, **B** and **C** are matrices of the same order $m \times n$ can be indicated by the equation

$$c_{ij} = a_{ij} + b_{ij} \tag{A.1.9}$$

where c_{ij}, a_{ij}, and b_{ij} are typical elements of the **C**, **A**, and **B** matrices, respectively. A typical matrix addition is:

$$\begin{bmatrix} 3 & 0 & 1 \\ 2 & -1 & 2 \\ 1 & 1 & 1 \end{bmatrix} + \begin{bmatrix} -1 & 1 & -1 \\ 2 & 5 & 6 \\ -3 & 4 & 9 \end{bmatrix} = \begin{bmatrix} 2 & 1 & 0 \\ 4 & 4 & 8 \\ -2 & 5 & 10 \end{bmatrix}.$$

The finite element procedures that will be developed will depend heavily on the use of the computer to manipulate matrices. Thus, now is a good time to begin thinking of equations like Eq. (A.1.9) from the point of view of a scientific language like FORTRAN. Therefore, we should note that any matrix would be a subscripted variable and should be listed in a DIMENSION statement. Similarly Eq. (A.1.9) implies a range over all values of the subscripts i and j. This would be accomplished with DO loops. To force one to be more conscious of this goal of utilizing the computer some standard shorthand mathematical notation will be used. For example, $\forall$ means "for every". Thus, the notation $\forall : 1 \le i \le m$ would be read as "for every i ranging from 1 to m". Therefore, Eq. (A.1.9) could be rewritten as:

$$c_{ij} = a_{ij} + b_{ij} \qquad \begin{matrix} \forall\, i: \ 1 \le i \le m \\ \forall\, j: \ 1 \le j \le n \end{matrix}.$$

This more clearly implies the required DO loops in a matrix addition subroutine.

One may want to note the economy considerations that columns add faster than rows and that subscripts should range on different line numbers to reduce termination checks. Matrix subtraction, $\mathbf{C} = \mathbf{A} - \mathbf{B}$, is performed in a similar manner:

$$c_{ij} = a_{ij} - b_{ij} \qquad \begin{matrix} \forall\, i: \ 1 \le i \le m \\ \forall\, j: \ 1 \le j \le n \end{matrix} .$$

Matrix addition and subtraction are associative and commutative: $\mathbf{A} + (\mathbf{B} + \mathbf{C}) = (\mathbf{A} + \mathbf{B}) + \mathbf{C}$, and $\mathbf{A} + \mathbf{B} + \mathbf{C} = \mathbf{C} + \mathbf{B} + \mathbf{A}$. This means that the order in which addition and subtraction is carried out has no effect on the result. Multiplication of the matrix **A** by a scalar c is defined as the multiplication of every element of the matrix by the scalar c. Thus, the elements of the product $\mathbf{B} = c\mathbf{A}$ are given by $b_{ij} = c\,a_{ij}$.

Two matrices **A** and **B** can be multiplied together in order **AB** only when the number of columns in **A** is equal to the number of rows in **B**. When this condition is fulfilled, the matrices **A** and **B** are said to be conformable for multiplication. Otherwise, matrix multiplication is not defined. The product of two conformable matrices **A** and **B** of order $m \times p$ and $p \times n$, respectively, is defined as a matrix $\mathbf{C} = \mathbf{AB}$ of order $m \times n$ calculated from

$$c_{ij} = \sum_{k=1}^{p} a_{ik} b_{kj} \qquad \begin{matrix} \forall\, i: \ 1 \le i \le m \\ \forall\, j: \ 1 \le j \le n \end{matrix} . \tag{A.1.10}$$

Notice that the appearance of the summation sign, Σ, implies the existence of another DO loop. When evaluating $\mathbf{AB} = \mathbf{C}$ by hand, it is helpful to arrange **A** and **B** matrices as shown in Eq. (A.1.10). The element c_{ik} results from multiplying elements of row i of the **A** matrix by elements of column k of the **B** matrix so the element c_{ik} is given in expanded form as:

$$c_{ik} = a_{i1} b_{1k} + a_{i2} b_{2k} + a_{i3} b_{3k} + \cdots + a_{ij} b_{jk} + \cdots + a_{in} b_{nk} .$$

Matrix multiplication is associative and distributive: $(\mathbf{AB})\mathbf{C} = \mathbf{A}(\mathbf{BC})$, and $\mathbf{A}(\mathbf{B} + \mathbf{C}) = \mathbf{AB} + \mathbf{AC}$. However, matrix multiplication is not commutative. In general, $\mathbf{AB} \ne \mathbf{BA}$, consequently, the order in which matrix multiplication is specified should not be changed. When two matrices **A** and **B** are multiplied, the product **AB** is referred to either as **B** premultiplied by **A** or as **A** postmultiplied by **B**. There may be cases when $\mathbf{AB} = \mathbf{BA}$, and the matrices **A** and **B** are then said to be *commutable*. For example, the unit matrix **I** commutes with any square matrix of the same order, that is, $\mathbf{AI} = \mathbf{IA} = \mathbf{A}$. The process of matrix multiplication can also be extended to partitioned matrices, provided the individual products of submatrices are conformable for multiplication. For example, the multiplication

$$\mathbf{AB} = \begin{bmatrix} \mathbf{A}_{11} & \mathbf{A}_{12} \\ \mathbf{A}_{21} & \mathbf{A}_{22} \end{bmatrix} \begin{bmatrix} \mathbf{B}_{11} & \mathbf{B}_{12} \\ \mathbf{B}_{21} & \mathbf{B}_{22} \end{bmatrix} = \begin{bmatrix} \mathbf{A}_{11}\mathbf{B}_{11} + \mathbf{A}_{12}\mathbf{B}_{21} & \mathbf{A}_{11}\mathbf{B}_{12} + \mathbf{A}_{12}\mathbf{B}_{22} \\ \mathbf{A}_{21}\mathbf{B}_{11} + \mathbf{A}_{22}\mathbf{B}_{21} & \mathbf{A}_{21}\mathbf{B}_{12} + \mathbf{A}_{22}\mathbf{B}_{22} \end{bmatrix}$$

is possible provided the products $\mathbf{A}_{11}\mathbf{B}_{11}$, $\mathbf{A}_{12}\mathbf{B}_{21}$, etc., are conformable. For this condition to be fulfilled, it is only necessary for the vertical partitions in **A** to include a number of columns equal to the number of rows in the corresponding horizontal partitions in **B**.

Often in finite element studies one encounters the transpose of a product of matrices (see MTMULT). The following relationship can be shown: $(\mathbf{AB})^T = \mathbf{B}^T\mathbf{A}^T$, or more generally $(\mathbf{ABC}\cdots\mathbf{YZ})^T = \mathbf{Z}^T\mathbf{Y}^T \cdots \mathbf{C}^T\mathbf{B}^T\mathbf{A}^T$. As an example of matrix multiplication, let $\mathbf{B}^T = [\,3 \quad 1 \quad 2\,]$, and let

$$\mathbf{A} = \begin{bmatrix} 2 & 1 & 0 \\ 1 & 0 & 1 \end{bmatrix},$$

then their product $\mathbf{C} = \mathbf{AB}$ is:

$$\begin{bmatrix} 2 & 1 & 0 \\ 1 & 0 & 1 \end{bmatrix} \begin{Bmatrix} 3 \\ 1 \\ 2 \end{Bmatrix} = \mathbf{C} = \begin{Bmatrix} 2*3+1*1+0*2 \\ 1*3+0*1+1*2 \end{Bmatrix}.$$

Every (non-singular) square matrix $\mathbf{A}$ has an *inverse* which will be indicated by $\mathbf{A}^{-1}$ such that by definition the product $\mathbf{AA}^{-1}$ is a unit matrix $\mathbf{AA}^{-1} = \mathbf{I}$. The reverse is also true $\mathbf{A}^{-1}\mathbf{A} = \mathbf{I}$. Inverse matrices are very useful in the solution of simultaneous equations $\mathbf{AX} = \mathbf{C}$ such as above where $\mathbf{A}$ and $\mathbf{C}$ are known and $\mathbf{X}$ is unknown. If the inverse of $\mathbf{A}$ is known, it is possible to solve for the unknowns of the $\mathbf{X}$ matrix by premultiplying both sides by the inverse $\mathbf{A}^{-1}\mathbf{AX} = \mathbf{A}^{-1}\mathbf{C}$ from which $\mathbf{X} = \mathbf{A}^{-1}\mathbf{C}$.

Various methods can be used to determine the inverse of a given matrix. For very large systems of equations it is probably more practical to avoid the calculation of the inverse and solve the equations by a procedure called *factorization*. Various procedures for computing an inverse matrix can be found in texts on numerical analysis. The inverse of 2×2 or 3×3 matrices can easily be written in closed form by using *Cramer's rule*. The result for a 3×3 matrix is illustrated in subroutine I3BY3. For future use a general matrix inversion program, INVERT, is included on the disk. Note that it requires two extra temporary work arrays, and it replaces the given matrix with its inverse. Our major interest is in systems of equations where the square coefficient matrix is symmetric. This special case allows some simplifications and symmetric matrices can be inverted by using a more efficient algorithm such as SYMINV.

4. Determinant of a Matrix

Every square matrix, say $\mathbf{A}$, has a single scalar quantity associated with it. That scalar is called the determinant, $|\mathbf{A}|$, of the matrix. The determinant is important in solving equations and inverting matrices since it can be shown that the inverse $\mathbf{A}^{-1}$ will not exist if $|\mathbf{A}| = 0$. In such a case, the matrix $\mathbf{A}$ or the set of equations is said to be *singular*. This implies that inversion algorithms, like I3BY3 or INVERT, must fail if $|\mathbf{A}| = 0$. Other algorithms are specialized for the sake of economy to work only when $|\mathbf{A}| > 0$. Fortunately, that special case occurs very often in finite element analysis. The beginner should be warned that some algorithms still return results when the system is singular. Usually these answers will be obvious garbage. This occurs, for example, when the problem is physically impossible because of incorrect boundary conditions. For future reference a few facts about determinant will be noted:

1. If two rows or columns are equal then the determinant is zero.
2. Interchanging two rows, or two columns, changes the sign of the determinant.
3. The determinant is unchanged if any row, or column, is modified by adding to it a linear combination of any of the other rows, or columns.
4. A singular square matrix may have nonsingular square partitions.

The last two items will become significant when we consider how to apply boundary conditions and how to solve a system of equations.

5. Matrix Calculus

At times we will find it necessary to differentiate or integrate matrices. These operations are simply carried out on each and every element of the matrix. Let the elements a_{ij} of $\mathbf{A}$ be a function of a parameter t. Then $\mathbf{B} = d\mathbf{A}/dt$ implies

$$b_{ij} = \frac{da_{ij}}{dt} \qquad \begin{matrix} \forall\, i:\ 1 \le i \le m \\ \forall\, j:\ 1 \le j \le n \end{matrix} \tag{A.1.11}$$

just as $\mathbf{C} = \int \mathbf{A}\, dt$ implies

$$c_{ij} = \int a_{ij}\, dt \qquad \begin{matrix} \forall\, i:\ 1 \le i \le m \\ \forall\, i:\ 1 \le j \le n \end{matrix} . \tag{A.1.12}$$

Clearly, if $\mathbf{A}$ is a matrix product the matrix operations must be completed before the calculus operations begin.

Often one encounters a scalar matrix, U, defined by a symmetric square $n \times n$ matrix, $\mathbf{A}$, a column vector $\mathbf{B}$, and a column vector of $\mathbf{n}$ parameters $\mathbf{X}$. The usual form is

$$U = ½\, \mathbf{X}^T \mathbf{A} \mathbf{X} + \mathbf{X}^T \mathbf{B} + \mathbf{C}. \tag{A.1.13}$$

When dealing with functional relations the concept of rate of change is often very important. If we have a function, f, of a single independent variable, say x, then we call the rate of change the derivative with respect to x. That quantity is written as df/dx. This concept is easily generalized. Suppose we have a function z of two variables x and y, $z = f(x, y)$. Here we may define two distinct rates of change. One is a rate of change with respect to y, with x being held constant. Thus, we can define two *partial derivatives*. When x is allowed to vary the derivative is called the *partial derivative* with respect to x. This is denoted by $\partial z/\partial x$. By analogy with the usual definition in calculus this can be written as

$$f_x = \frac{\partial f}{\partial x} = \underset{\Delta x \to 0}{\text{limit}} \frac{f(x + \Delta x, y) - f(x, y)}{\Delta x}.$$

A similar definition describes the partial derivative with respect to y, $\partial f/\partial y$. The total derivative is the sum of the partial derivatives, e.g.,

$$df = \frac{\partial f}{\partial x} dx + \frac{\partial f}{\partial y} dy.$$

The previous definition can be extended to include a function of any number of independent variables. If we take the derivative of the scalar U with respect to each $X_i\ \forall\, i:\ 1 \le i \le n$, the result is a column vector

$$\frac{\partial U}{\partial \mathbf{X}} = \mathbf{A}\mathbf{X} + \mathbf{B}. \tag{A.1.14}$$

This is verified by expanding Eq. (A.1.13), differentiating with respect to every X_i in $\mathbf{X}$ and re-writing the result as a matrix product. An outline of these steps follows:

$$U = \frac{1}{2} \sum_{i=1}^{n} \sum_{j=1}^{n} A_{ij} X_i X_j + \sum_{i=1}^{n} X_i B_i + C$$

or

$$U = \frac{1}{2}\left[A_{11}X_1^2 + A_{12}X_1X_2 + A_{21}X_2X_1 + \cdots + A_{13}X_1X_3 + A_{31}X_3X_1 + A_{33}X_3X^2 + \cdots\right]$$

$$+ \left[X_1B_1 + X_2B_2 + \cdots + X_nB_n\right] + C.$$

so that

$$\frac{\partial U}{\partial X_1} = \frac{1}{2}\left[2A_{11}X_1 + (A_{12}+A_{21})X_2 + \cdots + (A_{13}+A_{31})X_3 + \cdots\right] + B_1 + 0$$

which from symmetry of **A** reduces to

$$\frac{\partial U}{\partial X_1} = \frac{2}{2}\left[A_{11}X_1 + A_{12}X_2 + A_{13}X_3 + \cdots\right] + B_1.$$

Similarly for any typical X_i:

$$\frac{\partial U}{\partial X_i} = \sum_{j=1}^{n} A_{ij}X_j + B_i \qquad \forall\, i:\ -1 \le i \le n.$$

Combining all these rows into a set of equations gives the matrix form

$$\frac{\partial U}{\partial \mathbf{X}} = \left\{\frac{\partial U}{\partial X}\right\} = \mathbf{AX} + \mathbf{B}. \qquad \text{(A.1.14)}$$

6. Storage of Matrices in FORTRAN

When utilizing the computer to assist us with the manipulation of matrices (or other subscripted variables) it is sometimes necessary to known how an array is actually stored and retrieved. The procedure used to store the elements of an array is to sequentially store all the elements of the first column, then all of the second, etc. Another way of saying this is that the first (left most) subscript ranges over all its values before the second in incremented. After the second subscript has been incremented then the first again ranges over all its values. The main reason to note this concept is that it requires one to use care when extracting submatrices in partitioned arrays. However, the above knowledge can be used to execute some operations more efficiently. For example, the matrix addition procedure could be written as

$$c_k = a_k + b_k \qquad \forall\, k:\ 1 \le k \le m*n.$$

Such a subroutine would DIMENSION the arrays with a single subscript.

Appendix II

FUNCTIONS OF SUBROUTINES IN MODEL LIBRARY

PROGRAM	Description of its main function
ADFLUX	Assemble flux column and/or square matrices.
APLYBC	Apply nodal parameter (dof) boundary constraints.
ASYMBL	Assemble system eqs from the element matrices.
AT	Debug tool to print an integer.
BANCHK	Check banden system for invalid equations.
BANDBC	Apply essential boundary conditions to banded system.
BANMLT	Multiply a symmetric band matrix by a vector.
BANSUB	Convert sq matrix subscript to upper band subscript.
BARPRT	Print bar chart of dof vs node distance.
BELAST	Elasticity strain-displacement, B, matrix options.
BFLUX	Compute problem dependent boundary flux contributions.
BTDB	Matrix triple product, S=(B)T*D*B.
BTDIAB	Same as BTDB except D is diagonal matrix.
CALPRT	Find property at a pt by interpolation from nodal values.
CCOUNT	Count the number of dof constraints.
CEQBAN	Compute bandwidth of constraint equations.
CEQSKY	Compute skyline of constraint equations.
CHANGE	Change dof index to new order from REVISE.
CHKSHP	Check consistency of element shape type data.
CONDSE	Static condensation of the element matrices.
CONPLT	Plot contours on elements via linear interpolation.
CONTROL	Read the job control data for an application.
CONTUR	Print coords of contour pts on element boundaries.
CONVRT	Convert original dof numbers to reordered ones from REVISE.
CORECT	Correct dof estimates during an iterative solution.
CROUT	Crout factorization of a symmetric matrix.
CTOMQ	Corner to midpoint interpolation for a quadrilateral.
CTOMT	Corner to midpoint interpolation for a triangle.
DCHECK	Numerical check of isoparametric derivative functions.
DEGPAR	Compute dof number of given parameter at given node.
DELAST	Isotropic elasticity constitutive options.
DERCU	Derivatives of 1-D Hermite cubic in natural coord.
DERHQL	Derivatives of arbitrary hex, quad, or line.
DERIV	A general routine for computing local derivatives.

```
DERSHPH Derivatives & shape of arbitrary hex, quad, or line.
DER2CU  Second derivative of 1-D Hermite cubic in natural coord.
DER2L   Local derivatives of 2 node linear line element.
DER208  Derivatives of 8 to 20 node isoparametric brick.
DER3L   Derivatives of 3 node quadratic line element.
DER3T   Derivative of a three node triangle in local coords.
DER4P   Local derivatives of four node pyramid (tetrahedra).
DER4Q   Local derivatives of four node quadrilateral.
DER4T   Local derivatives of four node triangle.
DER412  Local derivatives of a 4 to 12 noded quadrilateral.
DER6T   Local derivatives of a six node triangle.
DER7T   Local derivatives of a seven node triangle.
DER8H   Local derivatives of an eight node hexahedra.
DER8Q   Local derivatives of an eight node quadrilateral.
DER9Q   Local derivatives of an nine node quadrilateral.
DER12Q  Local derivatives of an 12 node quadrilateral.
DER15T  Local derivatives of an 15 node triangle.
DER16Q  Local derivatives of an 16 node quadrilateral.
DER16R  Global derivatives of an 16 node C1 rectangle.
DER16QS Derivatives of 16 node serendipity quadrilateral.
DER16R  Derivatives of C1 rectangle.
DER17Q  Derivatives of 17 node serendipity quadrilateral.
DIAGSQ  Diagonalize and element square (mass) matrix.
DIRECT  Time integration of a dynamic system.
DOT     Dot product of two vectors.
DQRULE  Dunavant quadrature in area or unit coordinates.
DRIVER  Set problem memory limits, start analysis.
DSTART  Initalize dof for an iterative solution.
DYNDIM  Set up dynamic array storage, call MODEL.

EFLUX   Compute flux matrices on element edge.
ELBAND  Find half bandwidth of a given element.
ELCOL   Problem dependent element column matrix formation.
ELCORD  Find nodal coordinates for a given element.
ELFIND  Find elements connected to a given node.
ELFRE   Extract nodal parameters for a given element.
ELHIGH  Compute column heights due to element stiffness.
ELPOST  Generate prob dependent post solution data for element.
ELPRTY  Extract properties for a given element.
ELPSET  Set values of element properties.
ELSQ    Formulate problem dependent element square matrix.
ELTOEL  Find list of elements attached to a given element.
ELWAVE  Find wavefront size for an element.
ELWIDE  Width of element matrix row, for PVSOLVE.
ERROR   Get error estimates for asdaptive solutions.
EULER   Time history integration of eqs by Euler method.

FACTOR  Perform triangular factorization of system square matrix.
FFLUX   Compute flux matrices on element face.
FILL    Fill the shape and deriv arrays at quadrature points.
FMONE   Compute first fundamental form of a surface.
FORCER  Define a problem dependent time forcing function.
FRAME   Space frame element stiffness matrix.
FRBODY  Body force loads on a space frame element.
FULFAC  Factorization of full symmetric system.
FULSOL  Forward and backward substitution of a full system.

GAUSCO  Tabulate 1-D Gaussian quadrature data.
GAUS1D  Compute 1-D Gaussian quadrature using GAUSCO.
GAUS2D  Compute 2-D Gaussian quadrature using GAUSCO.
```

```
GAUS3D  Compute 3-D Gaussian quadrature using GAUSCO.
GDERIV  Compute global derivatives in an isoparametric element.
GENELM  Control generation of element matrices.
GETLT   Get data about a given element type.
GETQD   Get quadrature data for a given element type.
GUS3D   Gauss point upwind streamlining algorithm.
G2DER   Compute 2nd global derivatives from 2nd local derivatives.

HEAT    Element matrices for heat transfer.
HOOKE   Calculate stresses at a point, with initial strain.

ICOMB   Fast comb sort of integer list.
IDNODE  Find node associated with a given dof.
INCEQ   Input the nodal constraint data.
INDXEL  Find dof numbers associated with a given element.
INDXPT  Finf dof numbers at a given node point.
INFLUX  Input specified boundary flux data.
INLTYP  Read data for multiple element types.
INPROP  Input the problem dependent properties.
INPUT   Input nodal coordinates and element topology.
INSIDE  Determine if a point lies inside a 2-D polyogon.
INVDET  Compute inverse and determinant of 1, 2, or 3 rank matrix.
INVECT  Input initial values in system column matrix.
INVERT  Invert a square matrix.
IN3T    Find if a point is inside a T3 element.
IN4Q    Find if a point is in a Q4 quadrilateral.
IPRINT  Print an integer matrix.
IREMRK  Read and print a set of remarks cards.
ISOPAR  Control automatic generation of isoparametric matrices.
I2BY2   Invert 2X2 matrix and find its determinant.
I3BY3   Invert 3X3 matrix and find its determinant.

JACOB   Compute Jacobian matrix for isoparametric element.

LAME    Get Lame' differential geometry coefficients.
LASTL   Find elements where each node makes its last appearance.
LCONTR  Contour an isoparametric element.
LINE2D  Find tangent and normal vectors for a 2D line.
LISTI   List contents of integer via its pointer.
LISTR   List contents of real via its pointer.
LNODES  Extract nodal topology for a given element.
LOBATO  Lobatto quadrature on a natural coordinate line.
LPTPRT  Extract properties at nodes of a given element.
LRESEQ  Reorder data for wave or skyline reduction.

MADD    Matrix addition, C=A+B.
MAKDIM  Write most of DYNDIM and main.
MATPRT  Extract properties for a given material number.
MATWRT  Print properties for a given material number.
MAXMIN  Find maximun and minimum nodal parameter values.
MDMULT  Diagonal times a matrix.
MMDIFF  Matrix times the difference in two matrices.
MMULT   Matrix multiplication, C=A*B.
MMULTT  Matrix multiplication with transpose, C=A*(B)T
MODBAN  Compute bandwidth from constraint modifications.
MODSKY  Compute skyline from constraint modifications.
MODEL   Controlling program for common finite element problems.
MODFUL  Apply Type 1 constraint modification to full equations.
MODFY1  Apply Type 1 constraint to banded equations.
MODFY2  Apply Type 2 constraint to banded equations.
MODFY3  Apply Type 3 constraint to banded equations.
```

```
MODLFL  Modify full eqs for arbitrary linear constraint.
MSMULT  Multiply a matrix by a scalar, B=s*A.
MTMULT  Matrix transpose and multiplication, C=(A)t*B.
MUNIT   Create an idenity matrix.

NGRAND  Dummy EXTERNAL integran for isoparametric elem.
NTERPO  Linear interpolation from tabulated data.

PENLTY  Compute constraint elements for use in penalty method.
PENMOD  Modify eqs by adding constraint penalties.
PLTSET  Set plotter control from float misc properties.
POST    Control post-solution calculations.
POSTEL  Problem dependent element solutions using ELPOST data.
PRTYEL  Insert nodal properties into elem float properties.
PRTYMT  Insert mat number prop into misc floating properties.
PRTYPT  Update nodal property with float prop from a point.
PTCODE  Extract dof constraint flags for a node.
PTCORD  Extract coordinates of a node.
PTPRTY  Extract properties at a node point.
PTPSET  Set nodal properties via a material number.

QDGEOM  Calculate exact geometric properties of a quadrilateral.
QTOCQ   Extrapolate 4 gauss pts to corner of quad.
QTOCT   Extrapolate 3 gauss pts to corner of triangle.

RADAU   Tabulated data for quadrature on a triangle.
REACT   Compute reactions for full matrix essential bc.
REACTEL Compute reactions on an element.
REACTS  Compute reactions for banded matrix essential bc.
REVISE  Reorder dof numbers so constrained dof come last.
RPRINT  Print a real matrix.

SAVBAN  Save a row of banded matrix for reaction recovery.
SAVFUL  Save a row of full matrix for reaction recovery.
SCHECK  Numerical check of isoparametric shape functions.
SEC4Q   Second derivative of 1-D cubic Hermite polynomial.
SET     Set sizes of various arrays.
SHAPE   General element shape function calculation.
SHPCU   Shapes of 1-D Hermite cubic in natural coord.
SHPHQL  Shapes for an arbitrary hex, quad, or line.
SHPLEG  Shape functions & derivatives of Legrendre polynomials.
SHP2L   Local shapes of 2 node linear line element.
SHP208  Shapes of 8 to 20 node isoparametric brick.
SHP3L   Shapes of 3 node quadratic line element.
SHP3T   Shapes of a three node triangle in local coords.
SHP4P   Local shapes of four node pyramid (tetrahedra).
SHP4Q   Local shapes of four node quadrilateral.
SHP4R   Local shapes of four node C0 rectangle.
SHP4T   Local shapes of four node triangle.
SHP412  Local shapes of a 4 to 12 noded quadrilateral.
SHP6T   Local shapes of a six node triangle.
SHP7T   Local shapes of a seven node triangle.
SHP8H   Local shapes of an eight node hexahedra.
SHP8Q   Local shapes of an eight node quadrilateral.
SHP9Q   Local shapes of an nine node quadrilateral.
SHP12Q  Local shapes of an 12 node quadrilateral.
SHP12R  Local shapes of an 12 node C0 rectangle.
SHP15T  Local shapes of an 15 node triangle.
SHP16QS Local shapes of an 16 node quadrilateral.
SHP16R  Local shapes of an 16 node C1 rectangle.
SHP17Q  Local shapes of an 17 node quadrilateral.
```

```
SINGLR   Modify shape functions to include known singulatity.
SINGL2   Modify for singularity at any node of element.
SIZEI    Print size and name of integer array via pointer.
SIZER    Print size and name of real array via pointer.
SKYCHK   Check that sky diagonals are positive.
SKYDIA   Compute diagonal locations for skyline solver.
SKYEBC   Apply type 1 essential bc to skyline.
SKYFAC   Factorization routine for skyline solver.
SKYFUL   Copy skyline mode into full array and print.
SKYHI    Computes system skylines from ELHIGH.
SKYLCE   Modify skyline mode for linear constraint equation.
SKYPAD   Pad skyline for NASA PVSOLVE code.
SKYPVD   Diagonal location for PVSOLVE code.
SKYROW   Row width of skyline for PVSOLVE.
SKYSOL   Back substitutions for skyline solver.
SKYSOLVE Skyline Cholesky linear equation solver.
SKYSTR   System assembly control for skyline storage.
SKYSUB   Convert sq matrix subscripts to skyline pointer.
SKYTAL   Find top and bottom of a column in skyline.
SNORMV   Computer unit normal vector on a surface.
SOLVE    Solve eqs by forward & back substitution of factored system.
START    Start an iterative solution with an initial guess.
STORCL   Insert element column matrix into system column matrix.
STORPV   Insert element sq matrix into PVSOLVE sq matrix vector.
STORSQ   Insert element sq matrix into banded system sq matrix.
STRFUL   Insert element sq matrix into full system sq matrix.
SUMIN    Sum input values of system column vector.
SYMINV   Inversion of a symmetric matric.
SYMRUL   Symmetric quadrature data for triangles.
SYSBAN   Compute half bandwidth of system equations.

TANVEC   Get vector tangent to parametric curve.
TETRUL   Tetrahedra quadrature rule data in unit coord.
TRGEOM   Compute exact geometric properties of a triangle
TRINTG   Seven point symmetric quadrature rule for a triangle.

UNITCO   Convert natural coord quadrature to unit coord data.
USERQD   User supplied quadrature data rules.

VUNIT    Convert a vector to a unit vector.
VECT2D   Scale and draw an arrow in 2D.
VONMIS   Get Von Mises stress from global stress vector.

WRTELM   Print computed dof's element by element.
WRTPT    List coordinates and dof's at all nodes.

ZEROA    Zero a real vector.
ZEROI    Zero an integer vector.
```

Appendix III

Notation in MODEL

```
C     ...................... NOTATION ......................
C
C     AD    = VECTOR CONTAINING FLOATING POINT VARIABLES
C     AJ    = JACOBIAN MATRIX
C     AJINV = INVERSE JACOBIAN MATRIX
C
C     B      = STRAIN-DISPLACEMENT (GRADIENT) MATRIX
C     BODY   = BODY FORCE VECTOR
C
C     C      = ELEMENT COLUMN MATRIX
C     CB     = BOUNDARY SEGMENT COLUMN MATRIX
C     CC     = COLUMN MATRIX OF SYSTEM EQUATIONS
C     CEQ    = CONSTRAINT EQUATIONS COEFFICENTS ARRAY
C     COORD  = SPATIAL COORDINATES OF A SELECTED SET OF NODES
C     CP     = PENALTY CONSTRAINT COLUMN MATRIX
C     CUTOFF = NUMBER FOR CUTTING OFF ITERATIONS
C
C     D      = NODAL PARAMETERS ASSOCIATED WITH A GIVEN ELEMENT
C     DD     = SYSTEM LIST OF NODAL PARAMETERS
C     DGH    = GLOBAL DERIV.S OF INTERPOLATION FUNCTIONS H
C     DDOLD  = SYSTEM LIST OF NODAL DOF FROM LAST ITERATION
C     DLG    = LOCAL DERIVATIVES OF GEOMETRY FUNCTIONS G
C     DLH    = LOCAL DERIVATIVES OF INTERPOLATION FUNCTIONS H
C
C     E      = CONSTITUTIVE MATRIX
C     EB     = PRODUCT OF E AND B
C     ELPROP = ELEMENT ARRAY OF FLOATING POINT PROPERTIES
C
C     FLTNP  = REAL PROPERTIES OF SYSTEM NODES
C     FLTEL  = SYSTEM STORAGE OF FLOATING PT ELEMENT PROP
C     FLTMIS = SYSTEM STORAGE OF FLOATING PT MISC. PROP
C     FLTNP  = SYSTEM STORAGE OF FLOATING PT NODAL PROP
C     FLUX   = SPATIAL COMPONENTS OF SPECIFIED BOUNDARY FLUX
C
C     G      = INTERPOLATION FUNCTIONS FOR GEOMETRY
C     GLOBAL = GLOBAL DERIV.S OF INTERPOLATION FUNCTIONS H
C
C     H      = INTERPOLATION FUNCTIONS FOR AN ELEMENT SOLUTION
C     HINTG  = INTEGRAL OF INTERPOLATION FUNCTIONS
C
```

```
C      IBC    = NODAL POINT BOUNDARY RESTRAINT INDICATOR ARRAY
C      ID     = VECTOR CONTAINING FIXED POINT ARRAYS
C      IDIAG  = DIAGONAL LOCATION IN SKYLINE VECTOR
C      INDEX  = SYSTEM DEGREE OF FREEDOM NUMBERS ARRAY
C      INRHS  > 0, IF INITIAL VALUES OF CC ARE INPUT
C      IPTEST > 0, IF SOME PROPERTIES ARE DEFINED
C      ISAY   = NO. OF USER REMARKS TO BE I/O
C
C      KFIXED = ALLOCATED SIZE OF ARRAY ID
C      KFLOAT = ALLOCATED SIZE OF ARRAY AD
C      KODES  = LIST OF DOF RESTRAINT INDICATORS AT A NODE
C
C      LBN    = NUMBER OF NODES ON AN ELEMENT BOUNDARY SEGMENT
C      LEMWRT = 0, IF LIST NODAL PARAMETERS BY ELEMENTS
C      LHOMO  = 1, IF ELEMENT PROPERTIES ARE HOMOGENEOUS
C      LNODE  = THE N ELEMENT INCIDENCES OF THE ELEMENT
C      LPFIX  = SYSTEM STORAGE ARRAY FOR FIXED PT ELEMENT PROP
C      LPPROP = INTEGER PROPERTIES AT EACH ELEMENT NODE
C      LPROP  = ARRAY OF FIXED POINT ELEMENT PROPERTIES
C      LPTEST > 0, IF ELEMENT PROPERTIES ARE DEFINED
C
C      M      = NUMBER OF SYSTEM NODES
C      MAXACT = NO ACTIVE CONSTRAINT TYPES (<=MAXTYP)
C      MAXBAN = MAX. HALF BANDWIDTH OF SYSTEM EQUATIONS
C      MAXTIM > 0, CALCULATE CPU TIMES OF MAJOR SEGMENTS
C      MAXTYP = MAX NODAL CONSTRAINT TYPE (=3 NOW)
C      MISCFL = NO. MISC. FLOATING POINT SYSTEM PROPERTIES
C      MISCFX = NO. MISC. FIXED  POINT SYSTEM PROPERTIES
C      MISFIX = SYSTEM ARRAY OF MISC. FIXED POINT PROPERTIES
C      MODE   = MODE OF STORAGE, 0-SKYLINE, 1-BANDED
C      MTOTAL = REQUIRED SIZE OF ARRAY AD
C      M1 TO MNEXT = POINTERS FOR FLOATING POINT ARRAYS
C
C      N      = NUMBER OF NODES PER ELEMENT
C      NCURVE = NO. CONTOUR CURVES CALCULATED PER PARAMETER
C      NDFREE = TOTAL NUMBER OF SYSTEM DEGREES OF FREEDOM
C      NDXC   = CONSTRAINT EQS DOF NUMBERS ARRAY
C      NE     = NUMBER OF ELEMENTS IN SYSTEM
C      NELFRE = NUMBER OF DEGREES OF FREEDOM PER ELEMENT
C      NG     = NUMBER OF NODAL PARAMETERS (DOF) PER NODE
C      NGEOM  = NUMBER OF GEOMETRY NODES
C      NHOMO  > 0, IF NODAL SYSTEM PROPERTIES ARE HOMOGENEOUS
C      NITER  = NO. OF ITERATIONS TO BE RUN
C      NLPFIX = NO.   FIXED  POINT ELEMENT PROPERTIES
C      NLPFLO = NO. FLOATING POINT ELEMENT PROPERTIES
C      NMAT   = NUMBER OF MATERIAL TYPES
C      NNPFIX = NO. FIXED POINT NODAL PROPERTIES
C      NNPFLO = NO. FLOATING POINT NODAL PROPERTIES
C      NOCOEF = NO COEFF IN SYSTEM SQ MATRIX
C      NODES  = ELEMENT INCIDENCES OF ALL ELEMENTS
C      NOTHER = TOTAL NO. OF BOUNDARY RESTRAINTS .GT. TYPE1
C      NPARM  = DIMENSION OF PARAMWETRIC SPACE
C      NPFIX  = INTEGER PROPERTIES AT ALL NODES
C      NPROP  = NODAL ARRAY OF FIXED POINT PROPERTIES
C      NPTWRT = 0, LIST NODAL PARAMETERS BY NODES
C      NQP    = NUMBER OF QUADRATURE POINTS
C      NRANGE = ARRAY CONTAINING NODE NO.S OF EXTREME VALUES
C      NRB    = NUMBER OF ROWS IN B AND E MATRICES
```

```
C     NREQ   = NO. OF CONSTRAINT EQS. OF EACH TYPE
C     NRES   = NO. OF CONSTRAINT FLAGS OF EACH TYPE
C     NSEG   = NO OF ELEM BOUNDARY SEGMENTS WITH GIVEN FLUX
C     NSPACE = DIMENSION OF SPACE
C     NTAPE1 = UNIT FOR POST SOLUTION MATRICES STORAGE
C     NTAPE2,3,4 = OPTIONAL UNITS FOR USER (USED WHEN > 0)
C     NTOTAL = REQUIRED SIZE OF ARRAY ID
C     NULCOL > 0, IF ELEMENT COLUMN MATRIX IS ALWAYS ZERO
C     NUMCE  = NUMBER OF CONSTRAINT EQS
C     N1 TO NNEXT = POINTERS FOR FIXED POINT ARRAYS
C
C     PTPROP = NODAL ARRAY OF FLOATING PT PROPERTIES
C     PRTMAT = REAL ELEM PROPERTIES BASED ON MATERIAL NUMBER
C     PT     = QUADRATURE COORDINATES
C     PRTLPT = FLOATING PT PROP ARRAY OF ELEMENT'S NODES
C
C     RANGE  =  1-MAXIMUM VALUE, 2-MINIMUM VALUE OF DOF
C
C     S      = ELEMENT SQUARE MATRIX
C     SB     = BOUNDARY SEGMENT SQUARE MATRIX
C     SS     = 'SQUARE' MATRIX OF SYSTEM EQUATIONS
C     STRAIN = STRAIN OR GRADIENT VECTOR
C     STRAN0 = INITIAL STRAIN OR GRADIENT VECTOR
C     STRESS = STRESS VECTOR
C
C     TITLE  = PROBLEM TITLE
C
C     WT     = QUADRATURE WEIGHTS
C
C     X      = COORDINATES OF SYSTEM NODES
C     XYZ    = SPACE COORDINATES AT A POINT
C     ......................................................
```

Appendix IV

Programs Disk

```
==================================================================
                      Program Limitations

The source on this diskette is distributed with the book entitled
"Finite Elements for Analysis and Design" by J. E. Akin, Academic
Press, London.  Copyright 1993.  All rights reserved.

Although due care has been taken to present accurate programs
this software is provided "as is" WITHOUT WARRENTY OF ANY KIND,
EITHER EXPRESSED OR IMPLIED, AND EXPLICITLY EXCLUDING ANY IMPLIED
WARRANTIES OF MERCHANTABILITY OR FITNESS FOR A PARTICULAR USE.
The entire risk as to the quality and performance of the software
is with you the user.  The software is licensed for the personal
or academic use of the reader.  Academic institutions can utilize
the software in teaching and research.  The software may not be
transferred or sold to others.
====================================================================

                           model.hlp
                          revised 1993

A help file describing input data for the finite element code MODEL.
A typical problem will require: 1. a dummy main code to set the maximum
dimensions, 2. the problem-dependent element subroutines, 3. the problem
data.  This file contains information sections on: 1. Input data formats,
2. Main program, DRIVER, 3. Problem-dependent subroutines and functions.

++++ ++++ ++++  SECTION 1, INPUT DATA  ++++ ++++ ++++

==>> CARD TYPE 1    (Subroutine MODEL)
      TITLE, A general description of the problem.(20A4)

==>> CARD TYPE 2    (Subroutine MODEL) Control parameters. (16I5)

      M      = Number of nodal points in the system.
      NE     = Number of elements in the system.
      NG     = Number of parameters per node.
      N      = Number of nodes per element.

      NSPACE = Dimension of solution space.
```

```
NSEG   = No. of elem boundary segments with given flux.
LBN    = Number of nodes on an element boundary segment.
NITER  = Number of iterations to be run. (If any.)

NCURVE = No. of contours per parameter.
INRHS  = Initial system column matrix flag.
       = 0, Not to be input. (Default)
       > 0, Will be input by CARD TYPE 18.
ISAY   = Number of user remark cards to be read and
         printed. See CARD TYPE 4.
NRB    = Number of rows in element B matrix. (Default=1)

NQP    = Number of quadrature points. (Default=0)
LSHAPE = Shape 1-line 2-tri 3-quad 4-hex 5-tet
NLTYPE = Number of element types. (Default=1)
MODE   = Storage 1-banded 2-skyline    (Default=1)
```

Note, a negative value of NITER turns on debug printing. The absolute value of NITER is used.

```
==>> CARD TYPE 3   (Subroutine MODEL)       Control parameters. (16I5)

NNPFIX = Number of integer properties per node.
NNPFLO = Number of real properties per node.
NLPFIX = Number of integer properties per elem.
NLPFLO = Number of real properties per elem.

MISCFX = Number of miscellaneous integer properties.
MISCFL = Number of miscellaneous real properties
NHOMO  = Nodal properties flag.
       > 0, Properties are homogeneous. Only values for the first
         node will be input.
       = 0, Non-homogeneous properties. (Default)
LHOMO  = Element properties flag.
       > 0, Properties are homogeneous. Only values for the first
         element will be input.
       = 0, Non-homogeneous properties. (Default)

NPTWRT = Nodal parameter print flag.
       = 0, List calculated values by nodes. (Default) Otherwise
         omit nodal list.
LEMWRT = Nodal parameter print flag.
       = 0, List node and calculated nodal values for each element.
         Otherwise omit list.
NTAPE1 = I/O Unit for post solution calculation data.
       = 0, No post solution calculations. (Default)
       > 0, Unit to be utilized.
NTAPE2 = Extra scratch unit when > 0. (Note: Above units are for
       the analyst's problem-dependent data.)

NGF    = Number of generalized flux components (NSEG>0)
NULCOL = Element column matrix flag.
       = 0, Problem requires the matrix. (Default)
       > 0, The matrix is always a null vector.
NBSFIX = Number of integer properties per segment.
NBSFLO = Number of real properties per segment.

==>> CARD TYPE 4   (Subroutine IREMRK)      User comments.
```

Utilized only when ISAY > 0 on CARD TYPE 2. (20A4)

REMARK = Additional information to be printed with problem output.
Used to describe meaning of prameters, properties, etc.

NOTE *** There will be ISAY TYPE 4 cards.

==>> CARD TYPE 5 (Subroutine INPUT) Nodal point data. (2I10,(6F10.0))

J = System node number.
IBC(J) = Parameters constraint indicators at node J. (Right justified) Digit K equals constraint type of parameter K.
X(J,K) = Spatial coordinates (K=1,NSPACE) of node J

NOTE *** There will be one TYPE 5 card for each nodal point in the system. Example, IBC = 12213 and IBC = 10330 implies 3 Type 1, 1 Type 2, 1 Type 3. Mesh generation programs are available to supply these data.

==>> CARD TYPE 6 (Subroutine INPUT) Element Connectivity Data (I5,(15I5))

If NLTYPE = 1:
J = Element number.
NODES(J,K) = Nodal incidences (K=1,N) of elem. J

If NLTYPE > 1:
J = Element number.
LTYPE(J) = Element type number
NODES(J,K) = Nodal incidences (K=1,N) of elem. J

NOTE *** There will be one TYPE 6 card for each elem. Mesh generation programs are available to supply these data.

==>> CARD TYPE 7 (Subroutine INCEQ) Essential Boundary Conditions
Input data for a Type 1 nodal constraint, (2I10,F10.0)

NODE1 = system nodal point number.
IPAR1 = Constrainted parameter number, 1.le.IPAR1.le.NG
A1 = Value assigned to the parameter.

NOTES *** These constraints are of the form D(J) = A1, where D(J) is the system degree of freedom corresponding to parameter IPAR1 at node NODE1. That is, J = NG*(NODE1-1)+IPAR1. There will be one TYPE 7 card for each Type 1 constraint flagged in IBC(NODE1). See CARD TYPE 5.

==>> CARD TYPE 8 (Subroutine INCEQ) Type 2 linear constraint, if any.
(4I10,3F10.0)

NODE1 = First system node number.
IPAR1 = Constrained parameter number at NODE1.
NODE2 = System node number of second node.
IPAR2 = Constrained parameter number at NODE2.
A1 = Coefficient of first parameter.
A2 = Coefficient of second parameter.
A3 = Coefficient of right hand side.

NOTES *** These constraints are of the form A1*D(J1)+A2*D(J2)=A3. There will be one TYPE 8 card for each Type 2 constraint equation.

```
==>> CARD TYPE 9    (Subroutine INCEQ) Type 3 linear constraint, if any.
    (6I10, /,4 F10.0)

     NODE1 = First node point number.
     IPAR1 = First constrained parameter number.
     NODE2 = Second nodal point.
     IPAR2 = Second constrained parameter number.
     NODE3 = Third nodal number.
     IPAR3 = Third constrained parameter.
     A1    = Coefficient of first parameter.
     A2    = Coefficient of second parameter.
     A3    = Coefficient of third parameter.
     A4    = Constant on right hand side.

NOTES *** These constraints are of the form A1*D(J1)+A2*D(J2)+A3*D(J3)=A4.
There will be one TYPE 9 card for each Type 3 constraint equation.

==>> CARD TYPE 10   (Subroutine INPROP) Integer nodal properties.
Utilized only when NNPFIX > 0 on CARD TYPE 3. (I10,(7I10))

     J           = Nodal point number.
     NPFIX(J,K)  = Integer properties. K=1,NNPFIX

NOTE *** Reading terminated when J=M, unless NHOMO > 0 . Then only J=1
is read.  Omitted values default to zero when NHOMO=0.

==>> CARD TYPE 11   (Subroutine INPROP) Real nodal properties.
Utilized only when NNPFLO > 0 on CARD TYPE 3. (I10,(7F10.0))

     J           = Nodal point number.
     FLTNP(J,K)  = Real properties. K=1,NNPFLO

NOTE *** Reading terminated when J=M, unless NHOMO > 0. Then only J=1 is
read.  Omitted values default to zero when NHOMO=0.

==>> CARD TYPE 12   (Subroutine INPROP) Integer elem properties.
Utilizied only when NLPFIX > 0 on CARD TYPE 3. (I5,(7I10))

     J           = Element number.
     LPFIX(J,K)  = Integer properties. K=1,NLPFIX

NOTE *** Reading terminated when J=NE, unless LHOMO > 0. Then only J=1
is read.  Omitted values default to zero when LHOMO=0.

==>> CARD TYPE 13 (Subroutine INPROP) Real element properties.
Utilized only when NLPFLO > 0 on CARD TYPE 3. (I5,7(F10.0))

     J           = Element number.
     FLTEL(J,K)  = Real properties. K=1,NLPFLO

NOTE *** Reading terminated when J=NE, unless LHOMO > 0. Then only J=1
is read.  Omitted values default to zero when LHOMO=0.

==>> CARD TYPE 14   (Subroutine INPROP) Integer segment properties.
Utilizied only when NBSFIX > 0 on CARD TYPE 3. (I5,(7I10))

     J           = Segment number.
     NBSPFX(J,K) = Integer properties. K=1,NBSFIX
```

NOTE *** Reading terminated when J=NBS, unless LHOMO > 0. Then only J=1 is read. Omitted values default to zero when LHOMO=0.

==>> CARD TYPE 15 (Subroutine INPROP) Real segment properties.
Utilized only when NBSFLO > 0 on CARD TYPE 3. (I5,7(F10.0))

```
      J           = Segment number.
      FLTBS(J,K)  = Real properties. K=1,NBSFLO
```

NOTE *** Reading terminated when J=NBS, unless LHOMO > 0. Then only J=1 is read. Omitted values default to zero when LHOMO=0.

==>> CARD TYPE 16 (Subroutine INPROP) Integer miscellaneous properties.
Utilized only if MISCFX > 0 on CARD TYPE 3. (8I10.0)

```
      MISFIX(K) = Misc. fixed point properties. K=1,MISCFX
```

==>> CARD TYPE 17 (Subroutine INPROP) Real miscellaneous properties.
Utilized only when MISCFL > 0 on CARD TYPE 3. (8F10.0)

```
      FLTMIS(K) = Misc. floating point properties. K=1,MISCFL
```

==>> CARD TYPE 18 (Subroutine INVECT) Initial system column matrix.
Utilized only if INRHS > 0 on CARD TYPE 2. (2I10,F10.0)

```
      NODE  = System node number of the point.
      IPARM = Parameter number associated with the value.
      VALUE = Specified value to be added to column matrix.
```

NOTES *** Addition is of the form of C(J) = C(J) + VALUE where J=NG*(NODE-1)+IPARM. Reading is terminated when NODE=M and IPARM=NG. Omitted values default to zero.

==>> CARD TYPE 19 (Subroutine INFLUX) Specified boundary flux.
Utilized only when NSEG > 0 and LBN > 0 on CARD TYPE 2.

```
    TYPE 19A  (16I5)
      LNODE(K) = List of nodes on the segment. K=1,LBN.

    TYPE 19B  (8F10.4)
      (FLUX(K,L),L=1,NG) = The NG components of flux at each of the
                           above nodes. K=1,LBN.
```

NOTES *** These data are related to problem-dependent calculations. Thus the user assigns meanings to the nodal order and component order. There will be NSEG TYPE 19 CARD SETs.

===== End of Help =====
**** DUMMY MAIN PROGRAM ****

DRIVER, a dummy driver to set array sizes. The main program should have the following form. It sets the application array size limits and then calls DYNDIM to set the dynamic dimensions needed in MODEL. DYNDIM also sets various control parameters and calls MODEL. To change the maximum application size the user sets the values for the numbers maxr and maxi, as desired for the computer in use. If they are too small an error message is printed.

```
      PROGRAM  DRIVER
      CHARACTER*8 RN, IN
      PARAMETER (NUMR=       99, MAXR=       4600)
      PARAMETER (NUMI=       49, MAXI=       2300)
      DIMENSION  R(MAXR), RN(NUMR), I(MAXI), IN(NUMI)
      DIMENSION  J(NUMR), K(NUMI)
      COMMON / REAL /    R, RN, MMAXR, NNUMR, LLASTR, J
      COMMON / INTEGER / MMAXI, NNUMI, LLASTI, I, IN, K
C     NUMR   = ALLOWED NUMBER OF REAL ARRAYS
C     MAXR   = STORAGE ALLOWED FOR ALL REAL ARRAYS
C     LASTR  = NUMBER OF THE LAST ASSIGNED REAL ARRAY
C     R      = VECTOR HOLDING ALL REAL ARRAYS
C     RN     = NAMES OF EACH REAL ARRAY
C     J      = POINTER TO BEGINNING OF EACH REAL ARRAY
C     NUMI   = ALLOWED NUMBER OF INTEGER ARRAYS
C     MAXI   = STORAGE ALLOWED FOR ALL INTEGER ARRAYS
C     LASTI  = NUMBER OF THE LAST ASSIGNED INTEGER ARRAY
C     I      = VECTOR HOLDING ALL INTEGER ARRAYS
C     IN     = NAMES OF EACH INTEGER ARRAY
C     K      = POINTER TO BEGINNING OF EACH INTEGER ARRAY
      MMAXR  = MAXR
      MMAXI  = MAXI
      NNUMR  = NUMR
      NNUMI  = NUMI
      LLASTI = 0
      LLASTR = 0
      CALL  DYNDIM (NUMR, MAXR, NUMI, MAXI, R, RN, I, IN, J, K )
      STOP 'NORMAL END'
      END
```

The DRIVER should be followed at this point with any required problem dependent subroutines or functions. They include subroutine ELSQ and optionally ELCOL, ELPOST, POSTEL, and BFLUX. Standard versions of them are included with the extensions DD. Complete source, data, and output files for selected applications are given in the examples directories on the disk.

The commons REAL and INTEGER are never used but are available for access deeper into MODEL if needed. The DRIVER should be followed at this point with any required problem-dependent subroutines or functions. A "make" file is provided to compile and link all the needed files on a UNIX operating system. A typical version of the true main progam follows:

```
      SUBROUTINE  MODEL (MAXR, MAXI, NUMR, NUMI, LASTR, LASTI,
     1  NEXTR, NEXTI, RARRAY, RNAME, IARRAY, INAME, JRARAY, JIARAY,
     2  TITLE, NSEG, LBN, NITER, NCURVE, INRHS, ISAY, NNPFIX, NLPFIX,
     3  MISCFX, MISCFL, NHOMO, LHOMO, NPTWRT, LEMWRT, NTAPE1, NTAPE2,
     4  NTAPE3, NTAPE4, NTAPE5, NULCOL, MAXTYP, NLTYPE, NUMCE, IPTEST,
     5  LPTEST, MODE, M, MAXACT, MISCFL, MISCFX, N, NC, NCOEFF,
     6  NDFREE, NE, NELFRE, NF, NFLUX, NG, NGEOM, NLPFLO, NNPFLO,
     7  NOMAT, NPARM, NPLT, NQP, NRB, NSEG, NSPACE, NSYS, NTMP, NUMCE,
     8  X, AJ, AJINV, AVE, B, BODY, C, CC, CEQ, COORD, D, DDOLD,
     9  DGH, DLG, DLH, E, EB, ELPROP, FLTEL, FLTMIS, FLTNP, FLUX,
     1  FLUXBS, G, GPT, GWT, H, HINTG, PLTSET, PRTLPT, PRTMAT,
     2  PT, RANGE, S, SATPT, STRAIN, STRAN0, STRESS, SYSDAT,
     3  TMP, VALC, VALE, WT, XPT, XYZ, SS, IBC, KODES, NODES, NRES,
     4  IADD, IDIAG, INDEX, LFIRST, LLAST, LNODE, LPFIX, LPPROP,
     5  LPROP, LTDATA, LTYPE, MISFIX, NDXC, NODEF, NRANGE, NREQ,
```

```
     6  NPFIX, LSHAPE, IBUG, DD, LHIGH )
C
C      ***************************************************
C      *                  -M-O-D-E-L-                    *
C      *    MODULAR PROGRAMS FOR FINITE ELEMENT ANALYSES *
C      *          COPYRIGHT J.E. AKIN, 1992              *
C      ***************************************************
C              A SET OF BUILDING BLOCK PROGRAMS
C                             BY
C                     DR. J.E. AKIN, P.E.
C        DEPT. OF MECHANICAL ENGINEERING & MATERIALS SCIENCE
C                       RICE UNIVERSITY
C                   HOUSTON, TEXAS 77251-1892
C
      PARAMETER  ( NREACT = 20, CUTOFF = 1.E-6 )
      DIMENSION  TITLE(15)
      DIMENSION  RARRAY(MAXR), RNAME(NUMR),  IARRAY(MAXI),
     1           INAME(NUMI),  JRARAY(NUMR), JIARAY(NUMI)
      DIMENSION  X(M,NSPACE), AJ(NSPACE,NSPACE), AJINV(NSPACE,NSPACE),
     1  AVE(0:M,NSPACE+1), B(NRB,NELFRE), BODY(NSPACE), C(NELFRE),
     2  CC(NDFREE), CEQ(MAXACT,NUMCE), COORD(N,NSPACE), D(NELFRE),
     3  DD(NDFREE), DDOLD(NDFREE), DGH(NSPACE,N,0:NQP),
     4  DLG(NPARM,NGEOM,0:NQP), DLH(NSPACE,N,0:NQP), E(NRB,NRB),
     6  EB(NRB,NELFRE), ELPROP(0:NLPFLO), FLTEL(NE,0:NLPFLO),
     7  FLTMIS(0:MISCFL), FLTNP(M,0:NNPFLO), FLUX(0:NF),
     8  FLUXBS(0:NSEG,0:NFLUX), G(NGEOM,0:NQP), GPT(0:NQP), GWT(0:NQP),
     9  H(N,0:NQP), HINTG(N), PLTSET(0:NPLT), PRTLPT(N,0:NNPFLO)
      DIMENSION  PRTMAT(0:NLPFLO,0:NOMAT), PT(NPARM,0:NQP), RANGE(NG,2),
     1  S(NELFRE,NELFRE), SATPT(NRB+2,N), SS(1), STRAIN(NRB+2),
     2  STRAN0(NRB), STRESS(NRB+2), SYSDAT(0:NSYS), TMP(0:NTMP),
     3  VALC(NRB,0:NC), VALE(NRB,0:NC), WT(0:NQP), XPT(NSPACE,2),
     4  XYZ(NSPACE)
      DIMENSION  IBC(M), KODES(NG), NODES(NE,N), NRES(MAXTYP),
     1  IADD(0:M), IDIAG(NDFREE), INDEX(NELFRE), LFIRST(M), LLAST(M),
     2  LNODE(N), LPFIX(NE,0:NLPFIX), LPPROP(0:NNPFIX), LPROP(0:NLPFIX),
     3  LTDATA(6,NLTYPE), LTYPE(NE), MISFIX(0:MISCFX),
     4  NDXC(MAXACT,NUMCE), NODEF(0:NSEG,0:LBN), NRANGE(NG,2),
     5  NREQ(MAXACT), NPFIX(M,0:NNPFIX), LHIGH(NELFRE)
C     ..................... NOTATION .....................
C     SEE TEXT
C     ....................................................
      WRITE (6,5030) M, NE, NG, N, NSPACE, NSEG, LBN, NITER,
     1               NCURVE, INRHS, ISAY, NRB, NQP, LSHAPE,
     2               NLTYPE, MODE
 5030 FORMAT ( /, '*****  PROBLEM PARAMETERS  *****',/,
     1 'NUMBER OF NODAL POINTS IN SYSTEM =...........',I5,/,
     2 'NUMBER OF ELEMENTS IN SYSTEM =...............',I5,/,
     4 'NUMBER OF PARAMETERS PER NODE =..............',I5,/,
     3 'NUMBER OF NODES PER ELEMENT =................',I5,/,
     5 'DIMENSION OF SPACE =.........................',I5,/,
     6 'NUMBER OF BOUNDARIES WITH GIVEN FLUX =.......',I5,/,
     7 'NUMBER OF NODES ON BOUNDARY SEGMENT =........',I5,/,
     8 'NUMBER OF ITERATIONS TO BE RUN =.............',I5,/,
     9 'NUMBER OF CONTOURS BETWEEN 5 & 95% =.........',I5,/,
     + 'INITIAL FORCING VECTOR INPUT FLAG =..........',I5,/,
     1 'NUMBER OF USER REMARKS LINES =...............',I5,/,
     2 'NUMBER OF ROWS IN B MATRIX =.................',I5,/,
     3 'NUMBER OF QUADRATURE POINTS =................',I5,/,
```

```
     4 'SHAPE 1-LINE 2-TRI 3-QUAD 4-HEX 5-TET =......',I5,/,
     5 'NUMBER OF DIFFERENT ELEMENT TYPES =..........',I5,/,
     6 'STIFFNESS STORAGE MODE: SKY, BAND =..........',I5)
      IF ( LBN .GT. N )    WRITE (6,*)
     1     'INCONSISTANT VALUES OF LBN AND N.'
      WRITE (6,5080) NNPFIX, NNPFLO, NLPFIX,
     1               NLPFLO, MISCFX, MISCFL
 5080 FORMAT (
     1 'NUMBER OF INTEGER PROPERTIES PER NODE =......',I5,/,
     2 'NUMBER OF REAL PROPERTIES PER NODE =.........',I5,/,
     3 'NUMBER OF INTEGER PROPERTIES PER ELEMENT =...',I5,/,
     4 'NUMBER OF REAL PROPERTIES PER ELEMENT =......',I5,/,
     5 'NUMBER OF INTEGER MISCELLANEOUS  PROPERTIES =',I5,/,
     6 'NUMBER OF REAL MISCELLANEOUS  PROPERTIES =...',I5)
      WRITE (6,5081) NELFRE, NFLUX, NDFREE
 5081 FORMAT (
     1 'NUMBER OF D.O.F. FOR ELEMENT =...............',I5,/,
     2 'NUMBER OF D.O.F. ON FLUX SEGMENT =...........',I5,/,
     3 'NUMBER OF D.O.F. IN TOTAL SYSTEM =...........',I5)
      IF ( NHOMO .EQ. 1 )  WRITE (6,*)
     1   'NODAL POINT PROPERTIES ARE HOMOGENEOUS.'
      IF ( LHOMO .EQ. 1 )  WRITE (6,*)
     1    'ELEMENT PROPERTIES ARE HOMOGENEOUS.'
      NSUM = NTAPE1 + NTAPE2 + NTAPE3 + NTAPE4 + NTAPE5
      IF ( NSUM .GT. 0 )
     1    WRITE (6,5180) NTAPE1, NTAPE2, NTAPE3, NTAPE4, NTAPE5
 5180 FORMAT ( /, 'OPTIONAL UNIT NUMBERS (UTILIZED IF > 0)',/,
     1 'NTAPE1 = ',I2,' NTAPE2 = ',I2,' NTAPE3 = ',I2,
     2 'NTAPE4 = ',I2,' NTAPE5 = ',I2)
      IF ( NPTWRT .EQ. 0 )  WRITE (6,*)
     1     'NODAL PARAMETERS TO BE LISTED BY NODES'
      IF ( LEMWRT .EQ. 0 )  WRITE (6,*)
     1     'NODAL PARAMETERS TO BE LISTED BY ELEMENTS'
      IF ( NULCOL .NE. 0 )  WRITE (6,*)
     1    'ALL ELEMENT COLUMN MATRICES ARE ZERO.'
      IF ( IBUG .GT. 0 )  THEN
        LOC=0
        CALL SIZER(LOC,NEXTR,RNAME,JRARAY)
        CALL SIZEI(LOC,NEXTI,INAME,JIARAY)
      ENDIF
C      ZERO SYSTEM SOURCE VECTOR
      CALL  ZEROA (NDFREE, CC)
C-->     SET OR READ ELEMENT TYPE DATA
      CALL  INLTYP  (NLTYPE, LTDATA, N, NQP, NGEOM, NPARM, LSHAPE )
C-->     *** READ NODAL PARAMETER CONSTRAINT EQUATIONS ***
      CALL INCEQ (NG, MAXACT, NUMCE, NREQ, CEQ, NDXC, M)
      IF ( MODE .EQ. 1 )  THEN
C         DETERMINE SYSTEM HALF-BANDWIDTH
        CALL SYSBAN (NE, N, NG, IBW, NODES, LNODE, LID)
        JBW = 1
        IF ( MAXACT.GT.1 )
     1       CALL  CEQBAN (JBW, NREQ, MAXACT, NUMCE, NDXC, NDFREE)
        MAXBAN = MAX0 ( JBW,IBW )
        WRITE (6,5110)  IBW, LID, JBW, MAXBAN
 5110   FORMAT ( /,
     1  'EQUATION HALF BANDWIDTH =................',I5,/,
     2  'OCCURS IN ELEMENT NUMBER = ..............',I5,/,
     3  'CONSTRAINT HALF BANDWIDTH =..............',I5,/,
```

```
      4  'MAXIMUM HALF BANDWIDTH OF SYSTEM =.......',I5)
           NCOEFF = NDFREE*MAXBAN
         ELSE
C          SKYLINE
           CALL   SYSSKY (NDFREE, NE, N, NG, NELFRE, NODES,
      1                   LNODE, INDEX, LHIGH, IDIAG )
           IF ( MAXACT.GT.1 )
      1         CALL  CEQSKY (IDIAG, NREQ, MAXACT, NUMCE, NDXC, NDFREE)
           NCOEFF = IDIAG(NDFREE)
         ENDIF
           WRITE (6,5111)  NDFREE, NCOEFF
 5111      FORMAT (
      1  'TOTAL NUMBER OF SYSTEM EQUATIONS =.......',I5,/,
      2  'NUMBER OF STIFFNESS COEFFICIENTS =.......',I5)
C                         COMPLETE POINTERS
         LASTR = LASTR + NCOEFF
 10      WRITE (6,5120) LASTR, MAXR, LASTI, MAXI
 5120 FORMAT ( /, '*** ARRAY STORAGE ***',/,
      1 'TYPE    REQUIRED  AVAILABLE',/,
      2 'REAL    ', 2X, I5, 2X, I5,/,
      3 'INTEGER', 2X, I5, 2X, I5)
C
C            CHECK DATA AGAINST DIMENSION STATEMENTS
         IF ( LASTR .GT. MAXR .OR. LASTI .GT. MAXI )
      1      STOP 'STORAGE EXCEEDED, ABNORMAL PROGRAM END'
C            ZERO THE SYSTEM SQUARE MATRIX
         CALL ZEROA (NCOEFF, SS)
C-->                    *** READ PROPERTIES ***
         IF ( IPTEST .GT. 0 )  THEN
           CALL  INPROP (M, NE, NNPFIX, NNPFLO, NLPFIX, NLPFLO,
      1                  MISCFX, MISCFL, FLTNP, FLTEL, FLTMIS,
      2                  NPFIX, LPFIX, MISFIX, NHOMO, LHOMO)
         ELSE
           WRITE (6,*) 'WARNING, NO PROPERTY INPUT'
         ENDIF
C-->              ** INPUT INITIAL FORCING VECTOR **
         IF ( INRHS .GT. 0 ) CALL INVECT (NDFREE, NG, CC, M, D)
C-->     ** READ FLUX BOUNDARY COND. AND ADD TO SYSTEM EQS **
         IF ( NSEG .GT. 0 )
      1      CALL  INFLUX (NSEG, LBN, LNODE, FLUX, NG, COORD,
      2                    NSPACE, X, M, INDEX, C, CC, NDFREE,
      3                    S, SS, NCOEFF, NFLUX, MODE, N, IOPT,
      4                    NQP, NPARM, H, DGH, PT, WT, XYZ, DLH,
      5                    G, DLG, AJ, AJINV )
C           INITIALIZE SYSTEM DOF FOR ITERATIVE SOLUTION
         IF ( NITER .GT. 1 )
      1      CALL  DSTART (1, M, NG, NSPACE, NDFREE, INDEX, X,
      2                    COORD, DDOLD)
C                     *** BEGIN ITERATION LOOP ***
         RATIO = 1.0
         DO 30  IT = 1, NITER
C-->       *** CALCULATE AND ASSEMBLE ELEMENT MATRICES ***
C-->       *** GENERATE POST SOLUTION MATRICES & STORE ***
         CALL  ASYMBL ( NG, NCOEFF, MODE, IDIAG, NODES, SS,
      1     CC, M, NE, NDFREE, NITER, LPTEST, LHOMO,
      2     NHOMO, NULCOL, N, NSPACE, NELFRE, NRB, NQP, NGEOM,
      3     NPARM, NNPFIX, NNPFLO, MISCFX, MISCFL, NLPFIX, NLPFLO,
      4     LNODE, INDEX, X, DDOLD, COORD, S, C, H, DGH, B, E, EB,
```

```
     5    STRAIN, STRAN0, STRESS, BODY, PT, WT, XYZ, DLH, G, DLG,
     6    AJ, AJINV, HINTG, D, PRTLPT, FLTNP, FLTEL, FLTMIS,
     7    ELPROP, PRTLPT, PRTMAT, MISFIX,         NPFIX, LPFIX,
     8    LPROP, LPPROP, NTAPE1, NTAPE2, NTAPE3, NTAPE4, NTAPE5,
     9    LTYPE, NLTYPE, LTDATA, LSHAPE, GPT, GWT )
C          *** ASSEMBLY COMPLETED, CHECK SOURCES ***
      IF ( NULCOL .EQ. 0 )  CALL SUMIN  (NDFREE, M, NG, CC, D)
C           ** SAVE DATA FOR REACTION RECOVERY **
      IF ( IT .GT. 1 .AND. NREACT .GT. 0 )  REWIND (NREACT)
      IF ( MODE .EQ. 1 )  THEN
C        BANDED MODE
        CALL  SAVBAN (NREACT, M, NDFREE, NG, MAXBAN, IBC,
     1                INDEX, KODES, SS, CC)
      ELSE
C        SKYLINE MODE
        CALL  SAVSKY (NREACT, M, NDFREE, NG, NCOEFF, IBC,
     1                INDEX, KODES, SS, CC, IDIAG)
      ENDIF
C-->  ** APPLY BOUNDARY CONSTRAINTS TO NODAL PARAMETERS **
      CALL  APLYBC (MAXACT, NUMCE, NREQ, CEQ, NDXC,
     1              NDFREE, NCOEFF, SS, CC, IBW, IDIAG, MODE )
C              ** CHECK OR FIX SQUARE MATRIX **
      IF ( MODE .EQ. 1 )  THEN
C        BANDED MODE
        CALL  BANCHK (NDFREE, MAXBAN, M, NG, SS, CC)
      ELSE
C        SKYLINE MODE
        CALL  SKYCHK (NDFREE, NCOEFF, M, NG, SS, CC, IDIAG)
      ENDIF
C-->       *** SOLVE FOR UNKNOWN NODAL PARAMETERS ***
      IF ( MODE .EQ. 1 )  THEN
C        BANDED MODE
        MAXBAN = NCOEFF/NDFREE
        CALL  FACTOR (NDFREE, MAXBAN, SS)
        CALL  SOLVE  (NDFREE, MAXBAN, SS, CC, DD)
      ELSE
C       SKYLINE MODE
        FACT = .TRUE.
        BACK = .TRUE.
        CALL  SKYSOLVE (SS, CC, DD, IDIAG, NDFREE, FACT, BACK,
     1                  NCOEFF)
      ENDIF
C-->        *** SOLUTION COMPLETE, GET REACTIONS ***
      IF ( NREACT .GT. 0 )
     1     CALL  REACTS (NREACT, NDFREE, NG, DD, D)
C           *** PRINT RESULTS ***
      CALL  MAXMIN (M, NG, NDFREE, 1, RANGE, DD, INDEX, NRANGE)
      IF ( NPTWRT .EQ. 0 )
     1     CALL  WRTPT (M, NG, NDFREE, NSPACE, X, DD, INDEX)
      IF ( LEMWRT .EQ. 0 )
     1     CALL WRTELM (NE, N, NG, NDFREE, NELFRE, DD, INDEX,
     2                  NODES, LNODE)
C-->           *** POST SOLUTION CALCULATIONS ***
      IF ( NTAPE1 .GT. 0 )
     1  CALL  POST ( M, NE, NG, NDFREE, NODES, LNODE, INDEX,
     1    DD, N, NSPACE, NELFRE, NRB, NQP, NGEOM, NPARM, NNPFIX,
     2    NNPFLO, MISCFX, MISCFL, NLPFIX, NLPFLO, COORD, S, C, H,
     3    DGH, B, E, EB, STRAIN, STRAN0, STRESS, BODY, PT, WT,
```

```
     4    XYZ, DLH, G, DLG, AJ, AJINV, HINTG, D, PRTLPT, FLTMIS,
     5    ELPROP, PRTLPT, PRTMAT, MISFIX, LSHAPE, LPROP, LPPROP,
     6    NTAPE1, NTAPE2, NTAPE3, NTAPE4, NTAPE5, IT, NITER,
     7    LTYPE, NLTYPE, LTDATA, GPT, GWT )
      IF ( NITER .GT. 1 )  THEN
        IF ( IT .EQ. 1 )  RTEST = RATIO
C        *** UPDATE VALUES FOR NEXT ITERATION (IF ANY) ***
        WRITE (6,*)  ' '
        WRITE (6,*)  'ITERATION NUMBER = ', IT
        CALL  CHANGE (NDFREE, DD, DDOLD, TOTAL, DIFF, RATIO, 1)
        IF ( (RATIO/RTEST) .LT. CUTOFF ) GO TO 35
          CALL  CORECT (NDFREE, DD, DDOLD)
          CALL  ZEROA  (NCOEFF, SS)
          CALL  ZEROA  (NDFREE, CC)
      ENDIF
 30   CONTINUE
 35   IF ( NCURVE .GT. 0 )  THEN
C        ** CALCULATE CONTOUR CURVES FOR NODAL PARAMETERS **
        CALL  CONTUR (M, NE, N, NG, NSPACE, NDFREE, NELFRE, NCURVE,
     1                X, COORD, XYZ, DD, D, RANGE, NODES, LNODE, INDEX)
      ENDIF
      IF ( NSPACE .EQ. 1 .AND. NG .EQ. 1 )  THEN
        IBAR   = 1
        IPARM  = 1
        NODIST = 1
        CALL  BARPRT (M, NDFREE, NG, NSPACE, IBAR, IPARM,
     1                NODIST, X, DD, NODBAR)
      ENDIF
C                *** PROBLEM COMPLETED ***
      WRITE (6,*) 'NORMAL ENDING OF MODEL PROGRAM.'
      RETURN
      END
```

Closure

To provide the maximum potential utility the software has been placed on the diskette in source form along with typical input data and output results. Since executable files would be hardware dependent and quickly outdated none are provided. Therefore it will be necessary to run the source files through a Fortran compiler and linker to make an executable program. On a UNIX system that process is automated by using a "makefile" included with the source files. Two subroutines (dyndim and model) have more than twenty continuation lines and thus some compilers need an extra flag (-N140 on UNIX) set to compile them without false errors.

The numerous different types of application illustrated in the text each require a user written set Fortran subroutines. Dummy versions of these codes are provided to pass the arguments and set dimensions. The user simply must insert their source code before the return statement. They must be compiled by the user and combined with the main library of source programs on the disk.

A wider range of source files and example applications with data can be obtained by electronic mail as described in the Preface. Once an executable system has been established this text can be used to verify that the correct output results are being obtained. The estimated file sizes (in Kb) for the software on the disk by name, source size, and executable size, respectively are: Model, 350, 340; Mesh2d, 74, 320; View2d, 12, 240; View3d, 13, 256; and Reduce, 54, 224. The executable sizes would increase if the user increases the one dimension statement that controls the real and integer array sizes.

Index

WARNING

BY OPENING THIS SEALED PACKAGE YOU ARE AGREEING TO BECOME BOUND BY THE TERMS OF THIS LICENCE AGREEMENT which contains restrictions and limitations of liability.

IF YOU DO NOT AGREE with the terms set out here DO NOT OPEN THE PACKAGE OR BREAK THE SEAL. RETURN THE UNOPENED PACKAGE AND DISKS TO WHERE YOU BOUGHT THEM AND THE PRICE PAID WILL BE REFUNDED TO YOU.

THE MATERIAL CONTAINED IN THIS PACKAGE AND ON THE DISKS IS PROTECTED BY COPYRIGHT LAWS. HAVING PAID THE PURCHASE PRICE AND IF YOU AGREE TO THE TERMS OF THIS LICENCE BY BREAKING THE SEALS ON THE DISKS, YOU ARE HEREBY LICENSED BY ACADEMIC PRESS LIMITED TO USE THE MATERIAL UPON AND SUBJECT TO THE FOLLOWING TERMS:

1. The licence is a non-exclusive right to use the written instructions, manuals and the material recorded on the disc (all of which is called the "software" in this Agreement). This licence is granted personally to you for use on a single computer (that is, a single Central Processing Unit).

2. This licence shall continue in force for as long as you abide by its terms and conditions. Any breach of the terms and conditions by you automatically revokes the licence without any need for notification by the copyright owner or its authorised agents.

3. As a licensee of the software you do not own what has been written nor do you own the computer program stored on the disks and it is an express condition of this licence that no right in relation to the software other than the licence set out in this Agreement is given, transferred or assigned to you.

   Ownership of the material contained in any copy of the software made by you (with authorisation or without it) remains the property of the copyright owner and unauthorised copies may also be seized in accord with the provisions of the copyright laws.

4. Copying of the software is expressly forbidden under all circumstances except one. The one exception is that you are permitted to make ONE COPY ONLY of the software on the disks for security purposes and you are encouraged to do so. This copy is referred to as the back-up copy.

5. You may transfer the software which is on the disks between computers from time to time but you agree not to use it on more than one computer at any one time and you will not transmit the contents of the disks electronically to any other. You also agree that you will not, under any circumstances, modify, adapt, translate, reverse engineer, or decompile the software or attempt any of those things.

6. Neither the licence to use the Software nor any back-up copy may be transferred to any other person without the prior written consent of Academic Press Limited which consent shall not be withheld provided the recipient agrees to be bound by the terms of a licence identical to this licence. Likewise, you may not transfer assign, lend, rent, lease, sell or otherwise dispose of the software in any packaging or upon any carrying medium other than as supplied to you by Academic Press Limited. The back-up copy may not be transferred and must be destroyed if you part with possession of the software. Failure to comply with this clause will constitute a breach of this licence and may also be an infringement of copyright law.

7. It is an express condition of the licence granted to you that you disclaim any and all rights to actions or claims against Academic Press Limited and/or the copyright owner with regard to any loss, injury or damage (whether to person or property and whether consequential, economic or otherwise) where any such loss, injury or damage arises directly or indirectly as a consequence (whether in whole or part):

   a) of the failure of any software contained in these disks to be fit for the purpose it has been sold whether:
      i) as a result of a defect in the software, howsoever caused; or
      ii) any defect in the carrying medium howsoever caused; and/or

   b) any other act or omission of the copyright owner or its authorised agents whatsoever.

   and you agree to indemnify Academic Press Limited and the copyright owner against any and all claims, actions, suits or proceedings brought by any third party in relation to any of the above matters.

8. This agreement is to be read as subject to all local laws which are expressed to prevail over any agreement between parties to the contrary, but only to the extent that any provision in this Agreement is inconsistent with local law and no further.

9. This agreement shall be governed by the laws of the United Kingdom.